Benchmark Papers in Electrical Engineering and Computer Science

Series Editor: John B. Thomas
Princeton University

Published Volumes

SYSTEM SENSITIVITY ANALYSIS
J. B. Cruz, Jr.
RANDOM PROCESSES: Multiplicity Theory and Canonical Decompositions
Anthony Ephremides and John B. Thomas
ALGEBRAIC CODING THEORY: History and Development
Ian F. Blake
PATTERN RECOGNITION: Introduction and Foundations
Jack Sklansky
COMPUTER-AIDED CIRCUIT DESIGN: Simulation and Optimization
S. W. Director
ENVIRONMENTAL MODELING: Analysis and Management
Douglas Daetz and Richard H. Pantell
AUTOMATIC CONTROL: Classical Linear Theory
George J. Thaler
CIRCUIT THEORY: Foundations and Classical Contributions
M. E. Van Valkenburg
DATA COMMUNICATION: Fundamentals of Baseband Transmission
L. E. Franks
NONLINEAR SYSTEMS: Processing of Random Signals—Classical Analysis
A. H. Haddad

Additional Volumes in Preparation

Benchmark Papers
in Electrical Engineering
and Computer Science / 10

A *BENCHMARK*® Books Series

NONLINEAR SYSTEMS:
Processing of Random Signals—
Classical Analysis

Edited by
A. H. HADDAD
University of Illinois
at Urbana–Champaign

Published by
Dowden, Hutchinson & Ross, Inc.
523 Sarah Street, Stroudsburg, Pa. 18360

Distributed by ACADEMIC PRESS

Benchmark Papers in Electrical Engineering and Computer Science, Volume 10
Library of Congress Catalog Card Number: 75–1014
ISBN: 0–470–33848–2

77 76 75 1 2 3 4 5

Manufactured in the United States of America.

LIBRARY OF CONGRESS CATALOGING IN PUBLICATION DATA

Haddad, Abraham H comp.
Nonlinear systems.

(Benchmark papers in electrical engineering and computer science ; v. 10)
Bibliography: p.
Includes indexes.
1. Signal processing. 2. Nonlinear theories. 3. Stochastic processes. I. Title.
TK5101.H244 621.38'043 75-1014
ISBN 0-470-33848-2

Distributed by
ACADEMIC PRESS
A Subsidiary of Harcourt Brace Jovanovich, Publishers

Acknowledgments and Permissions

ACKNOWLEDGMENTS

MIT RESEARCH LABORATORY OF ELECTRONICS—*MIT Research Laboratory of Electronics Technical Report 216*

Crosscorrelation Functions of Amplitude-Distorted Gaussian Signals

PERMISSIONS

The following papers have been reprinted with the permission of the authors and copyright holders.

ACOUSTICAL SOCIETY OF AMERICA—*Journal of the Acoustical Society of America*

Response of a Linear Rectifier to Signal and Noise

AMERICAN INSTITUTE OF PHYSICS—*Journal of Applied Physics*

On the Distributions of Signals and Noise After Rectification and Filtering
On the Theory of Noise in Radio Receivers with Square Law Detectors
Signal-to-Noise Ratios in Band-Pass Limiters

AMERICAN MATHEMATICAL SOCIETY

Proceedings of the American Mathematical Society Symposium on Applied Mathematics

The Construction of a Class of Stationary Markoff Processes

Quarterly of Applied Mathematics

Some General Results in the Theory of Noise Through Non-linear Devices

AMERICAN TELEPHONE AND TELEGRAPH COMPANY—*Bell System Technical Journal*

Spectra of Quantized Signals

INSTITUTE OF ELECTRICAL AND ELECTRONICS ENGINEERS, INC.

IEEE Transactions on Information Theory

Comment on "A Useful Theorem for Nonlinear Devices Having Gaussian Inputs"
The Correlation Function of Gaussian Noise Passed Through Nonlinear Devices
Functional Calculus in the Theory of Nonlinear Systems with Stochastic Signals
General Results in the Mathematical Theory of Random Signals and Noise in Nonlinear Devices
Nonlinear Transformations of Random Processes
On the Expansion of a Bivariate Distribution and Its Relationship to the Output of a Nonlinearity
The Signal × Signal, Noise × Noise, and Signal × Noise Output of a Nonlinearity
The Uncorrelated Output Components of a Nonlinearity

IRE Transactions on Information Theory

The Effect of Instantaneous Nonlinear Devices on Cross-Correlation
An Expansion for Some Second-Order Probability Distributions and Its Application to Noise Problems
A Systematic Approach to a Class of Problems in the Theory of Noise and Other Random Phenomena: Part I
A Systematic Approach to a Class of Problems in the Theory of Noise and Other Random Phenomena: Part II. Examples

A Systematic Approach to a Class of Problems in the Theory of Noise and Other Random Phenomena: Part III. Examples
A Useful Theorem for Nonlinear Devices Having Gaussian Inputs
Proceedings of the IEEE
The Output Properties of Volterra Systems (Nonlinear Systems with Memory) Driven by Harmonic and Gaussian Inputs
The Spectrum of Clipped Noise

PRINCETON UNIVERSITY PRESS—*Annals of Mathematics*
The Orthogonal Development of Non-linear Functionals in Series of Fourier–Hermite Functionals

SOCIETY FOR INDUSTRIAL AND APPLIED MATHEMATICS—*Journal of the Society for Industrial and Applied Mathematics*
The Distribution of Quadratic Forms in Normal Variates: A Small Sample Theory with Applications to Spectral Analysis

TAYLOR & FRANCIS LTD.—*Journal of Electronics and Control*
The Use of Functionals in the Analysis of Non-linear Physical Systems

Series Editor's Preface

This Benchmark Series in Electrical Engineering and Computer Science is aimed at sifting, organizing, and making readily accessible to the reader the vast literature that has accumulated. Although the series is not intended as a complete substitute for a study of this literature, it will serve at least three major critical purposes. In the first place, it provides a practical point of entry into a given area of research. Each volume offers an expert's selection of the critical papers on a given topic as well as his views on its structure, development, and present status. In the second place, the series provides a convenient and time-saving means for study in areas related to but not contiguous with one's principal interests. Last, but by no means least, the series allows the collection, in a particularly compact and convenient form, of the major works on which present research activities and interests are based.

Each volume in the series has been collected, organized, and edited by an authority in the area to which it pertains. In order to present a unified view of the area, the volume editor has prepared an introduction to the subject, has included his comments on each article, and has provided a subject index to facilitate access to the papers.

We believe that this series will provide a manageable working library of the most important technical articles in electrical engineering and computer science. We hope that it will be equally valuable to students, teachers, and researchers.

This volume, *Nonlinear Systems: Processing of Random Signals—Classical Analysis,* has been edited by A. H. Haddad of the University of Illinois. It contains twenty-six papers on the classical analysis of nonlinear systems with random inputs. Included are the pioneering efforts of Bennett, Middleton, Barrett, Siegert, and others, as well as more recent contributions. Despite the complexity of the subject, Professor Haddad has succeeded admirably in organizing this volume and in providing a unifying set of editorial comments and explanations for this important area of communications and signal processing.

John B. Thomas

Contents

Acknowledgments and Permissions **v**
Series Editor's Preface **vii**
Contents by Author **xiii**

Introduction **1**

I. REPRESENTATIONS OF RANDOM PROCESSES APPLIED TO NONLINEAR SYSTEMS

Editor's Comments on Papers 1 Through 5 **6**

1 **BUSSGANG, J. J.:** Crosscorrelation Functions of Amplitude-Distorted Gaussian Signals **8**
MIT Res. Lab. Electronics Tech. Rept. 216, 1–14 (Mar. 26, 1952)

2 **BARRETT, J. F., and D. G. LAMPARD:** An Expansion for Some Second-Order Probability Distributions and Its Application to Noise Problems **23**
IRE Trans. Information Theory, **IT-1**(1), 10–15 (1955)

3 **LEIPNIK, R.:** The Effect of Instantaneous Nonlinear Devices on Cross-Correlation **29**
IRE Trans. Information Theory, **IT-4**(2), 73–76 (1958)

4 **WONG, E.:** The Construction of a Class of Stationary Markoff Processes **33**
Proc. Amer. Math. Soc. Symp. Appl. Math., **16,** 264–276 (1963)

5 **CAMBANIS, S., and B. LIU:** On the Expansion of a Bivariate Distribution and Its Relationship to the Output of a Nonlinearity **46**
IEEE Trans. Information Theory, **IT-17**(1), 17–25 (1971)

II. OUTPUT PROPERTIES OF ZNL SYSTEMS

Editor's Comments on Papers 6 Through 12 **56**

6A **PRICE, R.:** A Useful Theorem for Nonlinear Devices Having Gaussian Inputs **59**
IRE Trans. Information Theory, **IT-4**(2), 69–72 (1958)

6B **PRICE, R.:** Comment on "A Useful Theorem for Nonlinear Devices Having Gaussian Inputs" **63**
IEEE Trans. Information Theory, **IT-10**(2), 171 (1964)

7 **GORMAN, C. D., and J. ZABORSZKY:** Functional Calculus in the Theory of Nonlinear Systems with Stochastic Signals **64**
IEEE Trans. Information Theory, **IT-14**(4), 528–531 (1968)

8 **SHUTTERLY, H. B.:** General Results in the Mathematical Theory of Random Signals and Noise in Nonlinear Devices **68**
IEEE Trans. Information Theory, **IT-9**(2), 74–84 (1963)

9 **BAUM, R. F.:** The Correlation Function of Gaussian Noise Passed Through Nonlinear Devices **79**
IEEE Trans. Information Theory, **IT-15**(4), 448–456 (1969)

10 **ABRAMSON, N.:** Nonlinear Transformations of Random Processes **88**
IEEE Trans. Information Theory, **IT-13**(3), 502–505 (1967)

11 **BLACHMAN, N. M.:** The Signal × Signal, Noise × Noise, and Signal × Noise Output of a Nonlinearity **92**
IEEE Trans. Information Theory, **IT-14**(1), 21–27 (1968)

12 **BLACHMAN, N. M.:** The Uncorrelated Output Components of a Nonlinearity **99**
IEEE Trans. Information Theory, **IT-14**(2), 250–255 (1968)

III. FUNCTIONALS WITH RANDOM INPUTS

Editor's Comments on Papers 13, 14, and 15 **106**

13 **CAMERON, R. H., and W. T. MARTIN:** The Orthogonal Development of Non-linear Functionals in Series of Fourier–Hermite Functionals **108**
Ann. Math., **48**(2), 385–392 (1947)

14 **BARRETT, J. F.:** The Use of Functionals in the Analysis of Non-linear Physical Systems **116**
J. Electronics and Control, **15**(6), 567–615 (1963)

15 **BEDROSIAN, E., and S. O. RICE:** The Output Properties of Volterra Systems (Nonlinear Systems with Memory) Driven by Harmonic and Gaussian Inputs **167**
Proc. IEEE, **59**(12), 1688–1707 (1971)

IV. BASIC PAPERS ON OUTPUT SPECTRA OF SPECIAL NONLINEARITIES

Editor's Comments on Papers 16, 17, and 18 188

16 **BENNETT, W. R.:** Response of a Linear Rectifier to Signal and Noise 190
J. Acoust. Soc. Amer., **15**(3), 164–172 (1944)

17 **VAN VLECK, J. H., and D. MIDDLETON:** The Spectrum of Clipped Noise 199
Proc. IEEE, **54**(1), 2–19 (1966)

18 **BENNETT, W. R.:** Spectra of Quantized Signals 217
Bell Syst. Tech. J., **27,** 446–472 (July 1948)

V. BANDPASS NONLINEAR SYSTEMS

Editor's Comments on Papers 19 and 20 246

19 **MIDDLETON, D.:** Some General Results in the Theory of Noise Through Non-linear Devices 248
Quart. Appl. Math., **5**(4), 445–498 (1948)

20 **DAVENPORT, W. B., Jr.:** Signal-to-Noise Ratios in Band-Pass Limiters 302
J. Appl. Phys., **24**(6), 720–727 (1953)

VI. OUTPUT DISTRIBUTIONS OF ZADEH'S CLASS η_1 FUNCTIONALS

Editor's Comments on Papers 21 Through 26 312

21 **KAC, M., and A. J. F. SIEGERT:** On the Theory of Noise in Radio Receivers with Square Law Detectors 315
J. Appl. Phys., **18**(4), 383–397 (1947)

22 **MEYER, M. A., and D. MIDDLETON:** On the Distributions of Signals and Noise After Rectification and Filtering 330
J. Appl. Phys., **25**(8), 1037–1052 (1954)

23 **DARLING, D. A., and A. J. F. SIEGERT:** A Systematic Approach to a Class of Problems in the Theory of Noise and Other Random Phenomena: Part I 346
IRE Trans. Information Theory, **IT-3**(1), 32–37 (1957)

24 **SIEGERT, A. J. F.:** A Systematic Approach to a Class of Problems in the Theory of Noise and Other Random Phenomena: Part II. Examples 352
IRE Trans. Information Theory, **IT-3**(1), 38–43 (1957)

25 **SIEGERT, A. J. F.:** A Systematic Approach to a Class of Problems in the Theory of Noise and Other Random Phenomena: Part III. Examples **358**
IRE Trans. Information Theory, **IT-4**(1), 4–14 (1958)

26 **GRENANDER, U., H. O. POLLAK, and D. SLEPIAN:** The Distribution of Quadratic Forms in Normal Variates: A Small Sample Theory with Applications to Spectral Analysis **369**
J. Soc. Ind. Appl. Math., **7**(4), 374–401 (1959)

Bibliography **397**
Author Citation Index **403**
Subject Index **407**

Contents by Author

Introduction

The topic nonlinear processing of random signals is, by its definition, of too wide a scope to fit into one volume. Any paper concerned with some aspects of nonlinear systems with random inputs naturally belongs to this topic. Therefore, some form of classification is required to reduce the subject to manageable proportions. Here again the problem is complicated by the fact that unlike the linear case there is no unifying theme, as the only common property is that the systems considered are *not* linear. There is, however, a natural division into two types of problems:

1. Analysis: Nonlinearities are assumed to be an inherent part of a given system, and it is desired to derive the properties of the outputs of such systems subject to random inputs.
2. Synthesis and optimization: Nonlinearities are introduced in a given system so as to optimize certain performance criteria, such as optimum signal extraction or detection in the presence of noise.

The borderline between these two types of problems is rather hazy, as in many problems of design a given ad hoc nonlinear scheme is proposed so that the problem reduces to one of analysis. However, for lack of a more systematic way of division, the above classification is adopted in restricting this volume to the analysis problem. For both problems, in most of the early investigations, the nonlinear systems were characterized by the input–output functional relations. A particular emphasis has been placed on systems composed of two basic elementary subsystems: linear systems with memory and zero-memory nonlinear (ZNL) systems. In recent years a greater interest has been focused on the state-variable or dynamic model of nonlinear systems. The basis for the dynamic approach dates back to the 1930s [see, for example, Chandrasekhar (1943) and Wang and Uhlenbeck (1945) and their references to earlier work] when the Fokker–Planck method had been used for the analysis of Brownian motion and Markov processes. However, the renewed interest in the dynamic approach, and the overwhelming number of papers on its application to the analysis and synthesis of dynamic nonlinear filters, make it more appropriate for a separate treatment. Consequently, this volume is restricted to the classical approaches to the analysis of nonlinear systems with random inputs. It should be noted, though, that some duplication or overlap with other related topics is unavoidable for obvious reasons.

The problem in general may be stated as follows. Given an input vector random process $\mathbf{x}(t)$ with known statistical properties, and given that the output vector process $\mathbf{y}(t)$ satisfies a known nonlinear functional relation to the input, namely,

$$\mathbf{y}(t) = F\{\mathbf{x}(t)\}$$

it is desired to derive some statistical properties of $\mathbf{y}(t)$ such as distributions, covariance functions, moments, power spectral densities, etc. The problem is rather simple in theory, but difficult in practice. Since the general problem and its solution does not yield meaningful results, the problem is reformulated in terms of subproblems or special cases of general enough interest and applications. The reduction to special cases is obtained either by restricting the class of input random processes (such as Gaussian, sinusoidal, Markov, etc.) or by restricting the class of nonlinear functionals (such as ZNL systems, ZNL systems followed by linear systems, polynomial functionals, etc.). Consequently, the selection and classification of papers cannot be done chronologically, since contributions to any aspect or special case of the problem are widespread in time from the early 1930s—Bennett and Rice (1934)—to the present. Two of the most important classical contributions to this subject have not been included in this volume: Wiener's work in the 1950s on nonlinear functionals with random inputs and Rice's (1944) basic paper, which includes a general approach to nonlinear systems with random noise and is the most widely used reference on this subject. These two pioneering works are not included, since they would have occupied almost the entire volume, and they are readily accessible in book form—Wiener's work as a (1958) monograph and the Rice paper as a central part of Wax's (1954) volume.

The reasons that make the classification of the nonlinear problem difficult also cause difficulties in dividing the papers in this volume into groups. Chronological grouping is not appropriate, while topical grouping may be accomplished in several satisfactory ways. Thus a natural grouping into four subjects is possible, two of which are further subdivided into two groups each, to form the six parts of this volume:

1. Papers dealing with various characterizations of processes having the cross-correlation property is the subject of Part I. Since the general nonlinear processing problem is rather complex for arbitrary input processes, classes of random processes have been considered, for which general results for the nonlinear problem have been derived. The class that received the greatest attention is the "separable class," which satisfies the cross-correlation property first investigated by Bussgang (Paper 1). Generalizations and applications of this class or of the cross-correlation property extends from the early 1950s to the present [Masry (1973)]. Part I presents a chronological development of this subject.

2. The output second-order properties, such as autocorrelation function, power spectrum, bandwidth, and signal-to-noise ratio of ZNL systems occupy a great deal of the literature on the nonlinear problem. The results range from the very general for a given class of processes or a given class of ZNL systems, to the very special cases of a certain nonlinearity (square-law device, hard limiter) with Gaussian or other given process. The papers in this group are divided into two subgroups: the first considers typical general results on ZNL systems with certain classes of inputs and is the subject of Part II. The second is concerned with special nonlinearities of particular significance, especially to the development of the subject: quantizers, limiters, and rectifiers. Since the second group is also closely related to the topic of Part V, both topically and historically, it is deemed more appropriate as Part IV.

3. Part III is concerned with papers on general nonlinear functionals with random inputs. The most general treatment of nonlinear systems by the input–output method is the functional approach—Wiener (1958). The major contributions to the general results in this area were mostly limited to the 1950s. The generality of the system representation limits the application mostly to Gaussian or sinusoidal signals.

4. Since the results for general nonlinear functionals are rather limited, many papers are concerned with special cases of functionals, namely those that are a generalization of the system shown in Figure 1, which is a cascade of ZNL system followed by a linear system with memory. One generalization, called class η_1 by Zadeh (1953), has the functional form

$$y(t) = \int K\left[x(t-\tau), \tau\right]d\tau$$

A large number of papers investigated the output properties of subclasses of such systems for a variety of inputs (single, multiple, Gaussian, sinusoidal, etc.) with special emphasis on applications to receivers. The results may be divided naturally into two groups: The first is the subject of Part V, which considers the second-order properties of the output, such as the power spectral density and the signal-to-noise ratio. It primarily applies techniques used also in the papers of Parts II and IV, and its major reference is the Rice (1944) paper. The second is the subject of Part VI, which considers the output distributions of such functionals. Contributions to this group range from 1947 to the present, so Part VI provides a chronological development of the subject.

It should be emphasized that the parts are not mutually exclusive, and papers may belong to more than one subject, depending on the generality of the topic. The ordering of the parts may also be achieved in more than one way. The selection of papers for this volume was not an easy task, as there is a large number of excellent

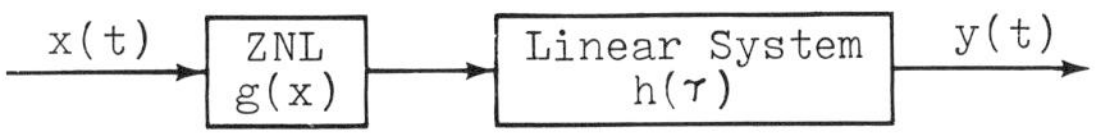

Figure 1

papers to choose from. In selecting most of the papers, therefore, one or more of the following qualities were considered:

1. Basic papers that set the trend or give an impetus for additional contributions in the area.
2. Significant techniques or results that are widely applied in future papers.
3. Important generalizations of earlier results, so that together they form a cohesive part of the development of the subject.

In some cases, survey papers coherently summarizing earlier results are included in place of or in addition to the early papers. Since most problems in nonlinear analysis cannot afford a general solution, there are many papers with special results or applications of a given method. Typical papers that illustrate the type of special results obtainable in the nonlinear processing problem are also included, especially in Part II.

Among the major topics not included in this volume are the applications of nonlinear processing to angle-modulated signals and envelope and phase of random processes [see, for example, Rice (1948), Middleton (1949b, 1950)]. Such topics fit more naturally with other angle demodulators, such as the phase-lock loop (PLL), which mostly utilize the dynamic approach for analysis [Viterbi (1963)]. Finally, the important subject of zero crossings or level crossings of stochastic processes, although intimately related to nonlinear processing [see survey by Blake and Lindsey (1973)], belongs naturally to the general subject of stochastic processes and their properties, and is therefore not included here.

In addition to the references mentioned in the comments, a bibliography is included which contains both these references and other papers related to the topic of this volume, classical nonlinear analysis. Notable among these are the books by Stratonovich (1963) and Kuznetsov et al. (1965), which contain collections of most of the results in the Russian literature and hence are not included here. Furthermore, each bibliographical item is classified, where appropriate, into one of the six parts of the volume. Earlier major summaries of the subject deserve to be mentioned specifically, the book by Deutsch (1962) and a more comprehensive treatment of the subject with an excellent bibliography in Chapter 6 by Thomas (1969).

Helpful suggestions from contributors to this volume are acknowledged. I would also like to thank J. B. Thomas for his conscientious and helpful advice and his interest in this work.

I

Representations of Random Processes Applied to Nonlinear Systems

Editor's Comments on Papers 1 Through 5

1 **Bussgang:** *Crosscorrelation Functions of Amplitude-Distorted Gaussian Signals*

2 **Barrett and Lampard:** *An Expansion for Some Second-Order Probability Distributions and Its Application to Noise Problems*

3 **Leipnik:** *The Effect of Instantaneous Nonlinear Devices on Cross-Correlation*

4 **Wong:** *The Construction of a Class of Stationary Markoff Processes*

5 **Cambanis and Liu:** *On the Expansion of a Bivariate Distribution and Its Relationship to the Output of a Nonlinearity*

Many important contributions to the analysis of nonlinearly processed random signals are concerned with the input–output second-order properties. The input–output cross-correlation is a very useful parameter in nonlinear systems applications, such as mean-squared-error analysis and system optimization or identification. Therefore, there has been wide interest in classes of processes for which the cross-correlation (or other second-order properties) satisfies simple rules, allowing the derivation of general results for such processes. One such case is the "cross-correlation property" first applied by Bussgang (Paper 1) for Gaussian processes. Bussgang's unpublished report has been the basis for later generalizations in two directions: generalization of the property itself and of the classes of random processes satisfying the property. The cross-correlation property simply states that the cross-correlation of the outputs of two ZNL systems, one of which is linear, is proportional to the inputs cross-correlation. The first extension of this property to a class of random processes was given by Barrett and Lampard (Paper 2). They derived a sufficient condition on the second-order probability density function of the input process for the cross-correlation property to hold. The second-order density function of the resulting class of processes has a diagonal expansion in orthogonal polynomial series. These processes include the Gaussian as a special case, and several other cases had already been studied in the framework of random walk and Brownian motion [see, for example, Chandrasekhar (1943) and Uhlenbeck and Ornstein (1930)]. Brown (1957) obtained a necessary and sufficient condition on the series expansion of the second-order density for the cross-correlation property to hold. Leipnik (Paper 3) provides the first generalization of the property, resulting in his (m, n)-cross-correlation property, where the ZNL systems are polynomials of degrees m and n, respectively. He also derives conditions for the property to hold for finite nonunity m, n. The resulting processes are interesting, but very few meaningful examples exist that satisfy the (m, n) property. Nuttall (1958) studied extensively a class of processes which he terms the "separable class," in the sense that its conditional expectation separates into a product of two functions:

$$E\{x(t_2) \mid x(t_1) = x\} = \rho(t_2, t_1)g(x)$$

The separability property is an alternative characterization of the cross-correlation property. Nuttall's main contributions are: first, the study of conditions for which the separability property is preserved under addition, multiplication, or special nonlinear operations; second, the optimization and identification of systems of class η_1 with inputs of the separable class.

An exhaustive study of the class of second-order density with diagonal orthogonal polynomial expansion was accomplished by Wong and Thomas (1962). The Fokker–Planck equation is used to characterize the second-order density, and thus provides the relationship to Markov processes and dynamical systems. Wong (Paper 4) further completes the study by the construction of Markov processes with densities having the diagonal expansion. In addition to its completeness, the paper also provides a physical interpretation of these processes and considers applications to the output distributions of nonlinear functionals based on Kac and Siegert (Paper 21, Part VI). The latest contribution to the study of the class of processes satisfying the cross-correlation property is given by Cambanis and Liu (Paper 5). The paper's central theme relates to the conditions for the existence of the biorthogonal series expansion of second-order density functions. These expansions are then related to the analysis of second-order output properties of ZNL systems. An important contribution is the necessary and sufficient condition for the cross-correlation property to hold, regardless of the existence of a series expansion for the second-order density function. Hence the results of Barrett and Lampard, Leipnik, and Brown (1957) can be derived without the restrictive assumption of the existence of the orthogonal expansion of the second-order densities. In this sense the class of processes that satisfies the cross-correlation property is extended by Cambanis.

The class of processes that satisfies the cross-correlation property has been applied to problems of analysis [Blachman (Papers 11 and 12)] and optimization [see Zadeh (1957)]. Many results dealing with the analysis and optimization of nonlinear systems are based on the cross-correlation property. One of the latest contributions to the application of the cross-correlation property is Masry's (1973) result on the equivalence of nonlinearly distorted Gaussian processes. Another important application is to Booton's (1954) linearization method, whereby a ZNL system with feedback is replaced by an equivalent linear system.

This section illustrates a basic approach to the ZNL analysis problem—the derivation of special classes of random processes, which allows for general results to be obtained. Since in many of the papers, especially those by Wong and by Cambanis and Liu, there are additional applications of these processes to nonlinear analysis problems, these papers may also belong to other parts of this collection.

1

Reprinted from *Mass. Inst. Technol. Res. Lab. Electronics Tech. Rept. 216,* 1–14 (Mar. 26, 1952)

CROSSCORRELATION FUNCTIONS OF AMPLITUDE-DISTORTED GAUSSIAN SIGNALS

Julian J. Bussgang

Abstract

This report treats the crosscorrelation functions of gaussian signals. It is shown that the crosscorrelation function of two such signals taken after one of them has undergone a nonlinear amplitude distortion is identical except for a factor of proportionality to the crosscorrelation function taken before the distortion. Possible practical applications of this property are indicated. A number of correlation functions associated with some of the most common types of distortion are computed.

Julian J. Bussgang

CROSSCORRELATION FUNCTIONS OF AMPLITUDE-DISTORTED GAUSSIAN SIGNALS

I. Introduction

Considerable attention has recently been given to the application of statistical methods to communication problems (1). Some of these problems involve stationary random noise signals possessing gaussian distributions of amplitudes. It is sometimes desirable to determine the autocorrelation and crosscorrelation functions of such signals after they undergo nonlinear amplitude distortion in terms of the original correlation functions. Analysis based on the ergodic hypothesis and on some well-known properties of the gaussian variable can be applied (2, 3, 4).

It is the purpose of this report to present some of the results of this analysis which do not seem to have been noted yet in the literature and to indicate where they may be found useful. We begin by outlining the technique of correlation analysis of distorted gaussian signals.

II. Correlation Function Analysis of Distorted Gaussian Signals

Suppose that we are dealing with a signal f(t) which is generated by a stationary random process. If we consider an ensemble of such time functions, generated by the same process, the amplitude at an arbitrary time t_1 can be treated as a random variable x. Associated with this random variable x are a certain probability density p(x), a mean value $\bar{x}$, and a mean-square value σ_x^2. Let us assume that the distribution of amplitudes is normal with a zero mean value and a unity mean-square value. We then have

$$p(x) = \frac{1}{\sqrt{2\pi}} e^{-x^2/2} \tag{1}$$

$$\sigma_x^2 = \int_{-\infty}^{\infty} x^2 \, p(x) \, dx = 1 \tag{2}$$

$$\bar{x} = \int_{-\infty}^{\infty} x \, p(x) \, dx = 0 \tag{3}$$

where the bar over a variable is used to denote its statistical average. We shall call such a signal a gaussian signal.

Let g(t) be another gaussian signal. Then this will also be the case with the delayed signal $g(t+\tau)$. If y is the random variable associated with $g(t+\tau)$, then the probability density of y will be p(y).

Now, f(t) and g(t) will be, in general, coherent functions of time, and therefore there will be some dependence between the variables x and y. This means that the mean of the product (or covariance) of the two variables, which we shall call r, is nonzero. Writing $p(x, y; \tau)$ for the density of the joint distribution of x and y, where τ is inserted to indicate clearly the dependence, we have

$$r = \int_{-\infty}^{\infty}\int_{-\infty}^{\infty} xy\, p(x, y; \tau)\, dx\, dy. \tag{4}$$

Since we are dealing with stationary processes, the ensemble average and the time average can be equated, that is

$$\int_{-\infty}^{\infty}\int_{-\infty}^{\infty} xy\, p(x, y; \tau)\, dx\, dy = \lim_{T\to\infty} \frac{1}{2T}\int_{-T}^{T} f(t)\, g(t+\tau)\, dt. \tag{5}$$

Now, the right-half member of Eq. 5 is by definition (ref. 1, p. 10, Eq. 45) the cross-correlation function $\phi_{fg}(\tau)$ of the two signals f(t) and g(t). With this in mind Eq. 5 can be rewritten as

$$\phi_{fg}(\tau) = \int_{-\infty}^{\infty}\int_{-\infty}^{\infty} xy\, p(x, y; \tau)\, dx\, dy \tag{6}$$

that is

$$\phi_{fg}(\tau) = r. \tag{7}$$

It is a property of the two-dimensional normal distribution that the joint probability density of x and y depends only on their covariance and is given explicitly by the expression (ref. 5, p. 395, Eq. 29.6)

$$p(x, y; \tau) = \frac{1}{2\pi\sqrt{1 - r^2}}\, e^{-(x^2+y^2-2rxy)/\left[2(1-r^2)\right]}. \tag{8}$$

In Eq. 8, r is written without an indication of the time functions with which the variables x and y are associated in order to simplify the notation.

Let the signal f(t) be passed through a device whose output, at any time t_1, is some single-valued function U of the amplitude of f(t) at that instant, defined for all values of f(t). We shall term such a device an amplitude-distorting device and U will be referred to as the distortion function. If F(t) is the resulting time function at the output, we have

$$F(t_1) = U\left[f(t_1)\right]. \tag{9}$$

If g(t) is passed through a similar device with a characteristic V, instead of U, the resulting output G(t) is related to g(t), at any instant t_1 by

$$G(t_1) = V\left[g(t_1)\right]. \tag{10}$$

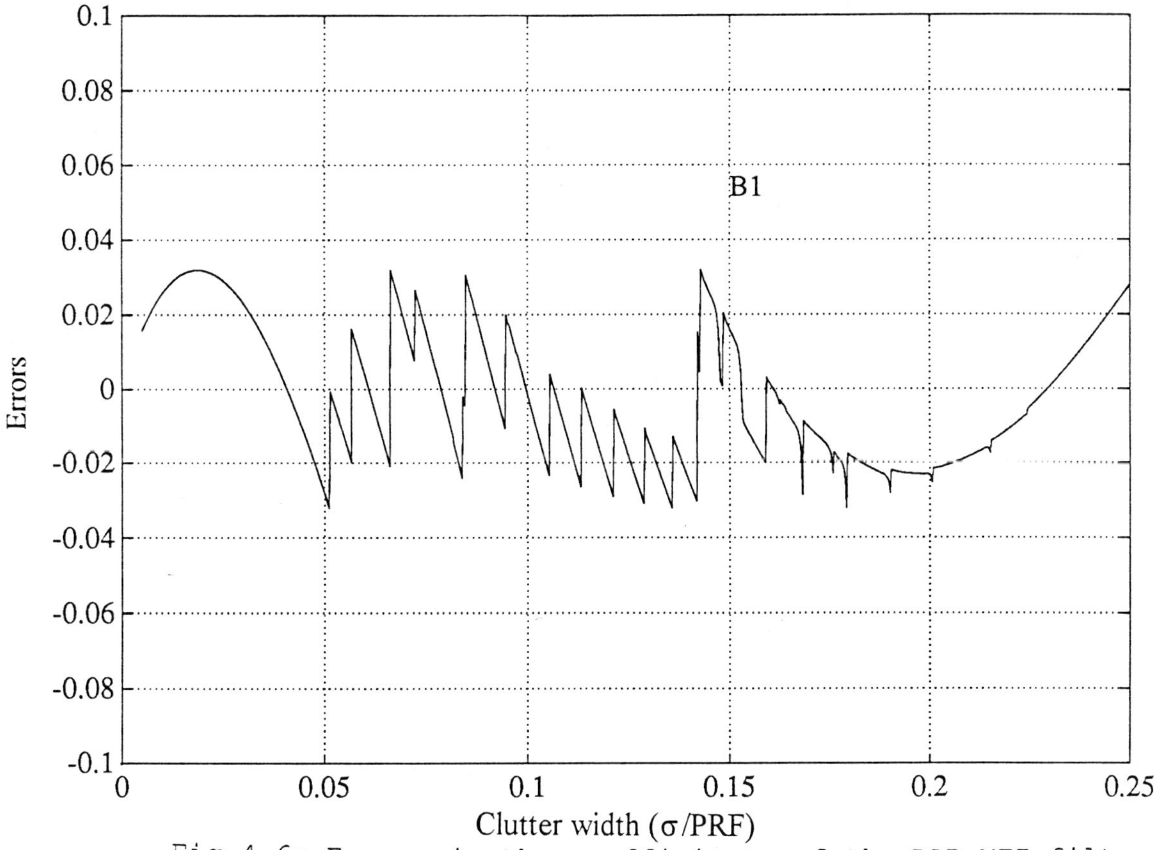

Fig 4.6c Errors in the coefficients of the DAR-MTI filter for 7 samples ~~No 7~~

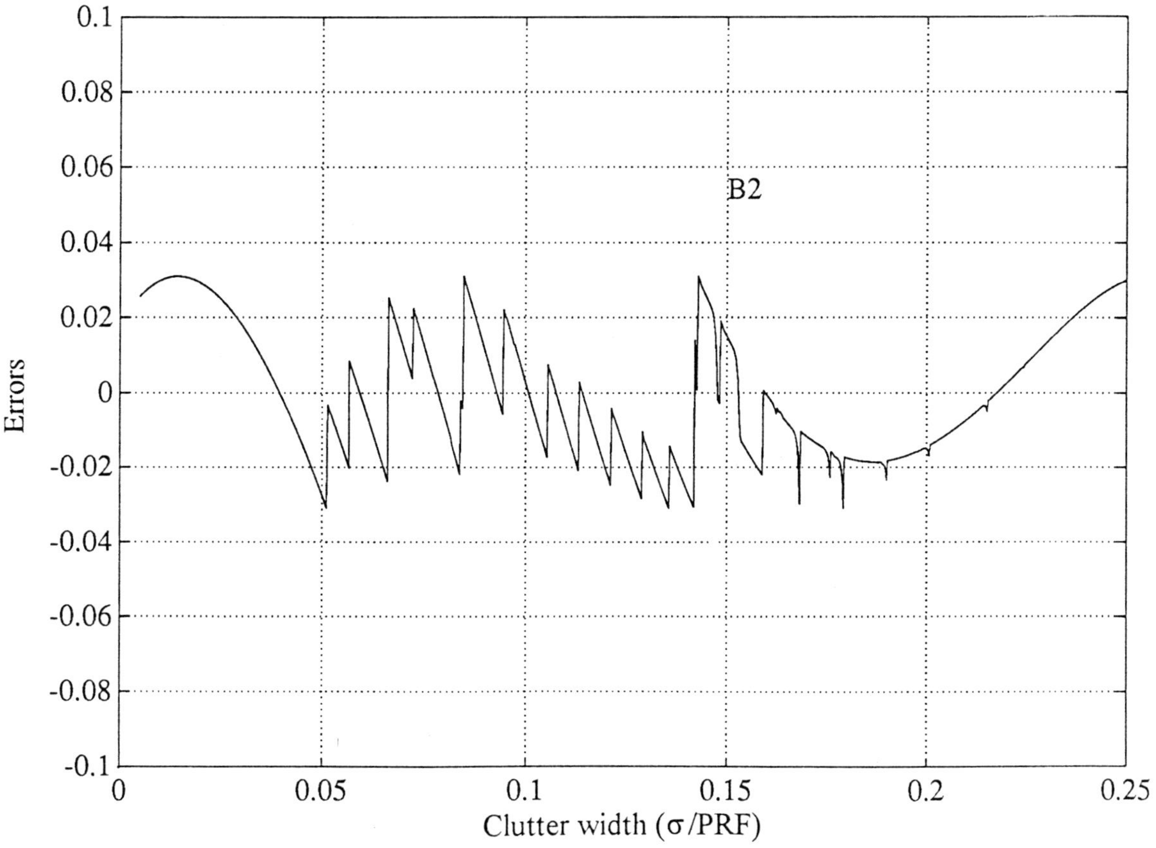

Fig 4.6d Errors in the coefficients of the DAR-MTI filter for 7 samples ~~No 7~~

$x(0) = 1$

$\frac{Y(z)}{X(z)} =$

$y(n) = .362\,x(n) + .8746\,x(n-1)$

$y(0) = ~~.362~~\ 0$

$y(1) = .8746$

$y(2) =$

Y

It is not difficult to see that

$$G(t_1 + \tau) = V\left[g(t_1 + \tau)\right]. \tag{11}$$

The amplitudes of F(t) and G(t+τ), at an arbitrary time t_1, can then be thought of as random variables U(x) and V(y), respectively, with x and y as defined above. Figure 1 shows a schematic illustrating the notation.

The crosscorrelation function of F(t) and G(t) can be written, by definition, as

$$\phi_{FG}(\tau) = \lim_{T\to\infty} \frac{1}{2T} \int_{-T}^{T} F(t)\, G(t+\tau)\, dt. \tag{12}$$

Despite distortion, the processes are still stationary, so that the time and the ensemble averages are equal, that is

$$\phi_{FG}(\tau) = \overline{U(x)\, V(y)}. \tag{13}$$

But

$$\overline{U(x)\, V(y)} = \int_{-\infty}^{\infty}\int_{-\infty}^{\infty} U(x)\, V(y)\, p(x, y; \tau)\, dx\, dy. \tag{14}$$

Substituting Eq. 8 for p(x, y; τ), we finally obtain

$$\phi_{FG}(\tau) = \frac{1}{2\pi\sqrt{1 - r^2}} \int_{-\infty}^{\infty}\int_{-\infty}^{\infty} U(x)\, V(y)\, e^{-(x^2+y^2-2rxy)/\left[2(1-r^2)\right]}\, dx\, dy. \tag{15}$$

Equation 15 permits us to express the crosscorrelation function of two distorted gaussian signals in terms of their crosscorrelation function before distortion, provided the distortion characteristics are specified. In the case of U = V and f = g, that is, when we are dealing with the input f(t) and the output F(t) of a single device, one can compute the autocorrelation function $\phi_{FF}(\tau)$ after distortion in terms of the original autocorrelation function $\phi_{ff}(\tau)$.

The normalizations of the mean and of the mean-square values of the distributions are not really essential; they do, however, considerably simplify computations. A nonzero mean value will introduce additive constants and error-function type integrals in the formulas, while a different mean-square value condition will call for the multiplication of results by a constant factor.

The same method can thus be extended to any gaussian distribution. In addition, it will also cover correlation functions of higher orders, provided all the covariances are given.

Note, that if V and U comprise a periodic time dependence one can compute the

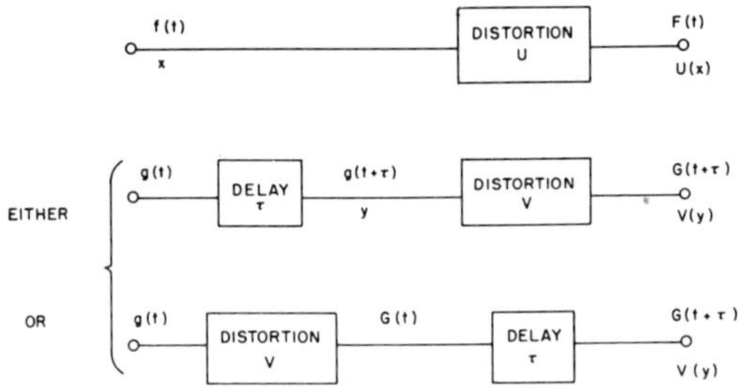

Fig. 1

Schematic explaining the notation.

integral in Eq. 12 and take the time average of the result to obtain the correlation function (6).

In general, it appears desirable to leave the results in terms of correlation functions, since the transformation to power spectra will involve an additional step and the explicit knowledge of the original correlation function.

III. A Theorem Regarding Crosscorrelations

> For two gaussian signals, the crosscorrelation function taken after one of them has undergone nonlinear amplitude distortion is identical, except for a factor of proportionality, to the crosscorrelation function taken before the distortion.

This means, for example, that the limiting of one signal in amplifiers, no matter how severe, has no effect on the shape of the crosscorrelation curve, provided the other signal is undistorted.

We next proceed to prove this, by considering simply a special case of Eq. 15, in which F = f. It follows that U(x) = x. For this case Eq. 15 becomes

$$\phi_{fG}(\tau) = \frac{1}{2\pi\sqrt{1-r^2}} \int_{-\infty}^{\infty}\int_{-\infty}^{\infty} x\, V(y)\, e^{-(x^2+y^2-2rxy)/\left[2(1-r^2)\right]}\, dx\, dy. \qquad (16)$$

Since

$$\int_{-\infty}^{\infty} x\, e^{-x^2/\left[2(1-r^2)\right] + \left[ry/(1-r^2)\right] x}\, dx = r\sqrt{2\pi(1-r^2)}\; y\, e^{r^2y^2/\left[2(1-r^2)\right]} \qquad (17)$$

Eq. 16 may be simplified to read

$$\phi_{fG}(\tau) = \frac{r}{\sqrt{2\pi}} \int_{-\infty}^{\infty} y\, V(y)\, e^{-y^2/2}\, dy. \qquad (18)$$

The integral above is seen to depend only on the distortion characteristic V and is independent of r. Remembering that $r = \phi_{fg}(\tau)$, we observe that for a given distortion characteristic V the ratio $\phi_{fG}(\tau)/\phi_{fg}(\tau)$ is constant for all values of τ. Let us denote this constant by K_V. We then have

$$K_V = \frac{1}{\sqrt{2\pi}} \int_{-\infty}^{\infty} y\, V(y)\, e^{-y^2/2}\, dy. \tag{19}$$

Equation 18 can now be rewritten in the form

$$\phi_{fG}(\tau) = K_V \phi_{fg}(\tau). \tag{20}$$

Equation 20 is the mathematical expression of the stated theorem. We see that it applies to nonlinear distortions of a very general character and that no restrictions have been imposed on the frequency range of the signals considered.

In particular, if the input and output of a distorting device are crosscorrelated, the result will be proportional to the autocorrelation of the input signal, that is

$$\phi_{gG}(\tau) = K_V \phi_{gg}(\tau). \tag{21}$$

It also follows directly from Eq. 21, that for the systems discussed above, the order of correlation between input and output is of no importance, that is

$$\phi_{gG}(\tau) = \phi_{Gg}(\tau) \quad . \tag{22}$$

In comparing the input autocorrelation with the input-output crosscorrelation, it is convenient to consider normalized quantities. We shall thus define a normalized coefficient k_V, characteristic of any system V, as

$$k_V = K_V \left[\frac{\overline{y^2}}{\overline{V^2(y)}} \right]^{1/2} . \tag{23}$$

It will be seen that k_V is the correlation coefficient (ref. 5, p. 265, Eq. 21.2.8) of the two variables y and V(y).

If the time notation is used, we have

$$k_V = \frac{\phi_{fG}(\tau_1)}{\phi_{fg}(\tau_1)} \left[\frac{\phi_{gg}(0)}{\phi_{GG}(0)} \right]^{1/2} \tag{24}$$

where the first ratio can be taken at any time τ_1. The values of k_V for several specific distorting devices have been computed.

IV. Power Series Devices

It is useful to consider a family of distortion functions defined as follows:

$$V_1(x) = \begin{cases} a_i x^i & x > 0 \\ b_i x^i & x < 0 \end{cases} \qquad i = 0, 1, 2, \ldots . \tag{25}$$

The first member of this family, V_o, represents a perfect peak clipper; the second, V_1, the linear rectifier; the third, V_2 (if $b_2 = 0$), the square law rectifier; and so on.

Given a certain range $x_1 < x < x_2$, we can approximate the distortion characteristic V by a series of V_i, so that by suitably choosing the coefficients a_i, b_i

$$V(x) = \sum_i V_i(x) \qquad x_1 < x < x_2. \tag{26}$$

Applying the result of Eq. 20 to the distortion function V of Eq. 26, we find

$$\phi_{fG}(\tau) = \phi_{fg}(\tau) \sum_{i=0,1,2\ldots} K_{v_i} \left[a_i - (-1)^i b_i\right] \tag{27}$$

where

$$K_{v_i} = \begin{cases} \dfrac{2^{i/2}}{\sqrt{2\pi}}\left(\dfrac{i}{2}\right)! & \text{for i even} \\ \dfrac{i!}{\left(\frac{i-1}{2}\right)!\; 2^{(i+1)/2}} & \text{for i odd.} \end{cases} \tag{28}$$

Values of the factor K_{v_i} for i from 0 to 8 are given in Table 1.

Table 1

Values of K_{v_i} for i = 0, 1, 2, . . . , 8.

i	K_{v_i}
0	$1/\sqrt{2\pi} = 0.399$
1	$1/2 = 0.500$
2	$\sqrt{2/\pi} = 0.798$
3	$3/2 = 1.50$
4	$8/\sqrt{2\pi} = 3.20$
5	$15/2 = 7.50$
6	$24/\sqrt{2\pi} = 9.60$
7	$35/2 = 17.50$
8	$384/\sqrt{2\pi} = 153.5$

V. Linear Rectification

The characteristic of a linear rectifier is shown in Fig. 2. The distortion function $V_1(x)$ is defined as follows

$$V_1(x) = \begin{cases} a_1 x & x \geqslant 0 \\ b_1 x & x \leqslant 0 \end{cases} \tag{29}$$

where a_1 and b_1 can be positive, negative or zero.

The crosscorrelation function of two rectified signals can be computed in terms of the original crosscorrelation function $\phi_{fg}(\tau)$. We have

$$\phi_{FG}(\tau) = \int_{-\infty}^{\infty}\int_{-\infty}^{\infty} V_1(x)\, V_1(y)\, p(x, y; \tau)\, dx\, dy. \tag{30}$$

Substituting for the expressions in Eq. 30 from Eqs. 29 and 8, we arrive at

$$\phi_{FG}(\tau) = \frac{1}{2\pi\sqrt{1 - r^2}} \left\{ (a_1^2 + b_1^2) \int_0^{\infty}\int_0^{\infty} xy\, e^{-(x^2+y^2-2rxy)/\left[2(1-r^2)\right]}\, dx\, dy \right.$$

$$\left. - 2a_1 b_1 \int_0^{\infty}\int_0^{\infty} xy\, e^{-(x^2+y^2+2rxy)/\left[2(1-r^2)\right]}\, dx\, dy \right. . \tag{31}$$

Let us denote the first of these integrals by $H(r)$, that is

$$H(r) = \frac{1}{2\pi\sqrt{1 - r^2}} \int_0^{\infty}\int_0^{\infty} xy\, e^{-(x^2+y^2-2rxy)/\left[2(1-r^2)\right]}. \tag{32}$$

On carrying out the integration, we find

$$H(r) = \frac{1}{2\pi}\left[\sqrt{1 - r^2} + \frac{r\pi}{2} + r \tan^{-1} \frac{r}{\sqrt{1 - r^2}}\right]. \tag{33}$$

Thus remembering Eq. 7, we obtain

$$\phi_{FG}(\tau) = (a_1^2 + b_1^2)\, H\left[\phi_{fg}(\tau)\right] - 2a_1 b_1\, H\left[-\phi_{fg}(\tau)\right]. \tag{34}$$

It can be verified that for $a_1 = b_1 = 1$, $\phi_{FG}(\tau)$ reduces to $\phi_{fg}(\tau)$.

The crosscorrelation function of a rectified signal with a nonrectified signal follows immediately from Eq. 27. We obtain

$$\phi_{fG} = \frac{1}{2}(a_1 + b_1)\, \phi_{fg}. \tag{35}$$

If we normalize the rms value of the output power to unity, so that

$$a_1^2 + b_1^2 = 1 \tag{36}$$

and in addition let

$$\frac{b_1}{a_1} = \tan\theta \tag{37}$$

Eq. 35 can be rewritten in the form

$$\phi_{fG}(\tau) = \phi_{fg}(\tau)\cos\left(\theta - \frac{\pi}{4}\right). \tag{38}$$

This means that

$$k_v = \cos\left(\theta - \frac{\pi}{4}\right). \tag{38a}$$

Figure 3 illustrates Eq. 38.

There are several types of distortion not of the power series type which are of practical importance. Two of these, symmetrical peak clipping and symmetrical center clipping, are considered below.

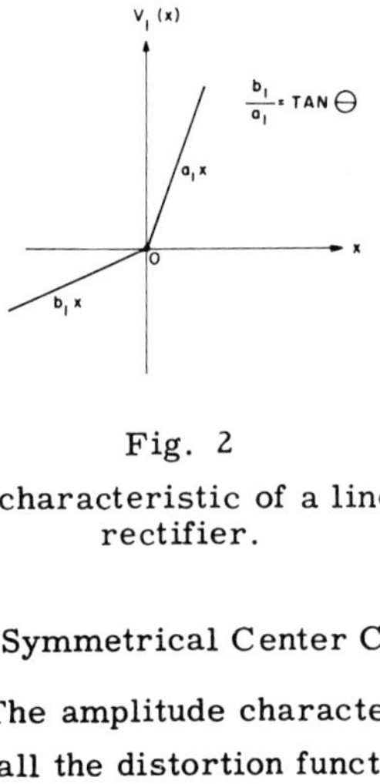

Fig. 2

The characteristic of a linear rectifier.

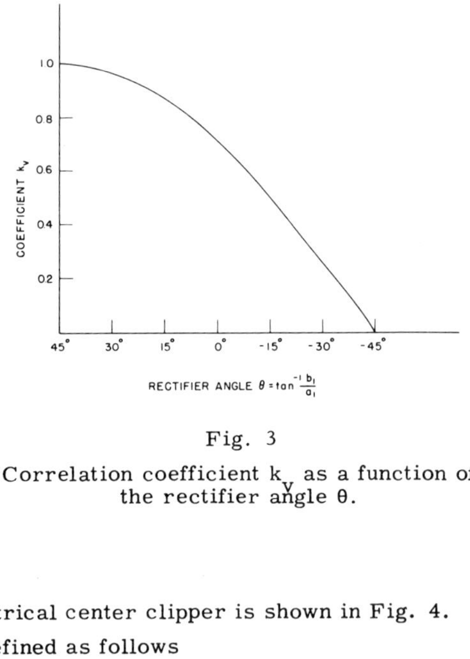

Fig. 3

Correlation coefficient k_v as a function of the rectifier angle θ.

VI. Symmetrical Center Clipping

The amplitude characteristic of a symmetrical center clipper is shown in Fig. 4. If we call the distortion function V, it can be defined as follows

$$V(x) = \begin{cases} \beta(x-h) & x > h \\ 0 & |x| < h \\ \beta(x+h) & x < -h. \end{cases} \tag{39}$$

If the output rms value is normalized to unity, β and h are related by the condition

$$\frac{\beta^2}{\sqrt{2\pi}} \int_h^{\infty} (x-h)^2 e^{-x^2/2}\, dx = \frac{1}{2}. \tag{40}$$

On integration one obtains

$$2\beta^2 \left[\frac{1}{2}(1 - a_h)(1 + h^2) - \frac{he^{-h^2/2}}{\sqrt{2\pi}} \right] = 1 \tag{41}$$

where a_h is the well-tabulated probability integral given by the expression

$$a_h = \frac{1}{\sqrt{2\pi}} \int_{-h}^{h} e^{-x^2/2} \, dx. \tag{42}$$

The crosscorrelation of an undistorted signal with a center-clipped signal can be computed from Eq. 18

$$\phi_{fG}(\tau) = \frac{\phi_{fg}}{\sqrt{2\pi}} \left[\int_{h}^{\infty} y\, \beta(y-h)\, e^{-y^2/2}\, dy + \int_{-\infty}^{-h} y\, \beta(y+h)\, e^{-y^2/2}\, dy \right]. \tag{43}$$

This turns out to be

$$\phi_{fG}(\tau) = \beta(1 - a_h)\, \phi_{fg}(\tau). \tag{44}$$

The correlation coefficient k_v computed on the basis of Eqs. 41 and 44 is plotted in Fig. 5, as a function of h.

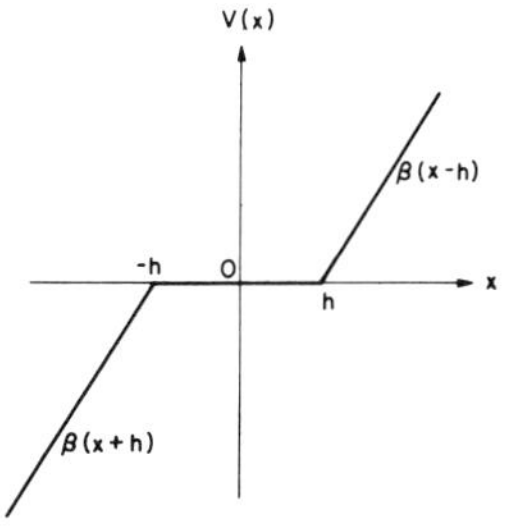

Fig. 4
The characteristic of a linear center clipper.

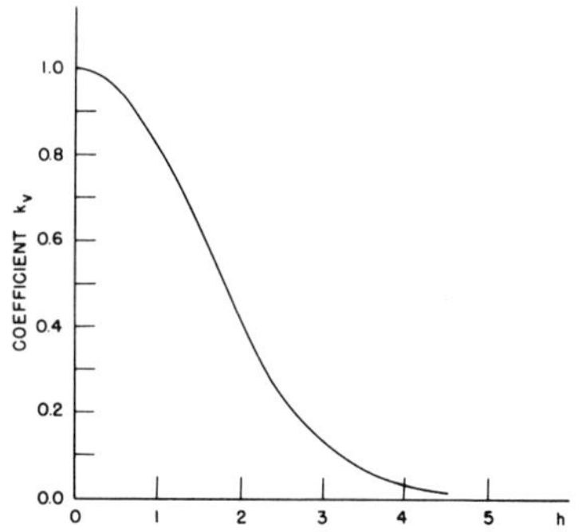

Fig. 5
Correlation coefficient k_v as a function of the clipping level for a symmetrical center clipper.

VII. Symmetrical Peak Clipping

We shall consider the abrupt peak clipping whose characteristic is shown in Fig. 6 and the gradual clipping with the characteristic shown in Fig. 7. The distortion function for the abrupt clipping is taken to be

$$V(x) = \begin{cases} c & x > d \\ \frac{c}{d}x & x < d \\ -c & x < -d. \end{cases} \tag{45}$$

Van Vleck (ref. 2, Eq. 41, p. 35) gives the autocorrelation function of the output of this clipper for the case c = d. One can immediately extend this result to the crosscorrelation of two identically clipped signals. Using the notation introduced above, we have

$$\phi_{FG}(\tau) = \frac{c^2}{d^2}\left\{\phi_{fg}\,\alpha_{d/\sqrt{2}} + \sum_{n=3,5\ldots}^{\infty} \frac{\phi_{fg}^n}{n!}\left[H_{n-2}(d)\,e^{-d^2/2}\right]^2\right\} \tag{46}$$

where $H_n(y)$ is the Hermitian polynomial of order n. That is

$$H_n(y) = \left[(-1)^n \frac{d^n}{dy^n} e^{-y^2/2}\right] e^{y^2/2}. \tag{47}$$

In the limiting case of "extreme" clipping when the output signal of each clipper is two-valued, following Van Vleck, we obtain the compact relation (ref. 2, Eq. 42, p. 35)

$$\phi_{FG}(\tau) = \frac{2}{\pi}c^2 \sin^{-1}\phi_{fg}(\tau). \tag{48}$$

Figure 8 illustrates Eq. 48 for the case in which the input crosscorrelation is given by $\phi_{fg} = e^{-a|\tau|}$. It is interesting to notice that the slope of ϕ_{FG} is infinite at the origin.

The crosscorrelation of a distorted signal with an undistorted signal is obtained by substituting our particular V in Eq. 19. Since the distortion is symmetrical, we have

$$\frac{1}{2}K_v = \frac{c}{d}\int_0^d y\,y\,e^{-y^2/2}\,dy + c\int_d^{\infty} y\,e^{-y^2/2}\,dy. \tag{49}$$

This leads to

$$\phi_{fG}(\tau) = \frac{c}{d}\,\alpha_d\,\phi_{fg}(\tau). \tag{50}$$

If it is desired to normalize the crosscorrelation of clipped signals to be unity at $\tau = 0$, c and d must be related by

$$\frac{1}{\sqrt{2\pi}}\int_0^d \frac{c^2}{d^2}x^2 e^{-x^2/2}\,dx + \frac{c^2}{\sqrt{2\pi}}\int_d^{\infty} e^{-x^2/2}\,dx = \frac{1}{2} \tag{51}$$

that is,

$$2c^2\left[-\frac{1}{d}\frac{e^{-d^2/2}}{\sqrt{2\pi}} + \frac{1}{d^2}\frac{1}{2}\alpha_d + \frac{1}{2}(1 - \alpha_d)\right] = 1 \tag{52}$$

where α_d is defined by Eq. 42.

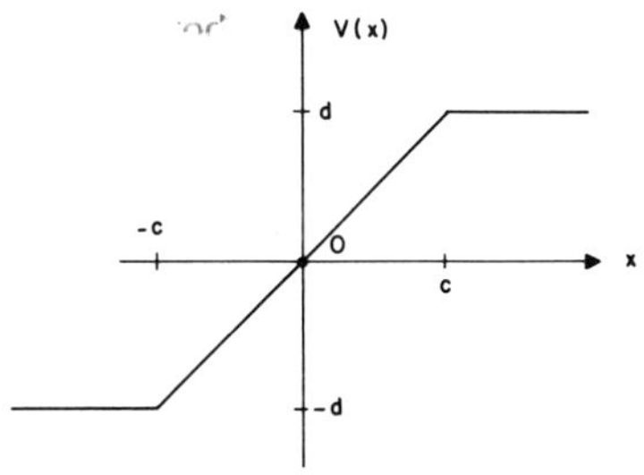

Fig. 6

The characteristic of an abrupt peak clipper.

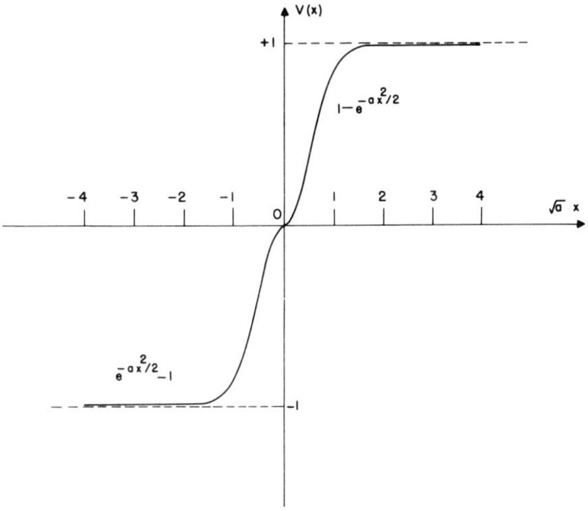

Fig. 7

The characteristic of a gradual peak clipper.

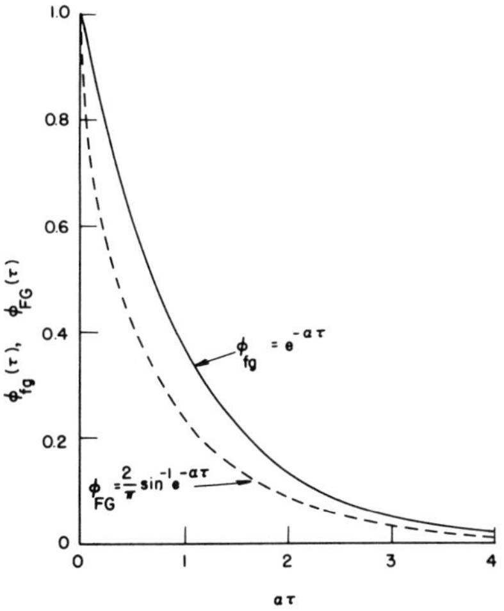

Fig. 8

Correlation functions before and after extreme clipping.

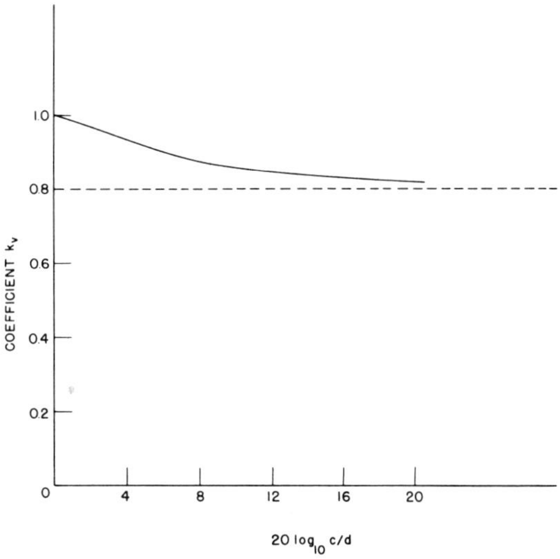

Fig. 9

Correlation coefficient k_v as a function of the peak clipper slope.

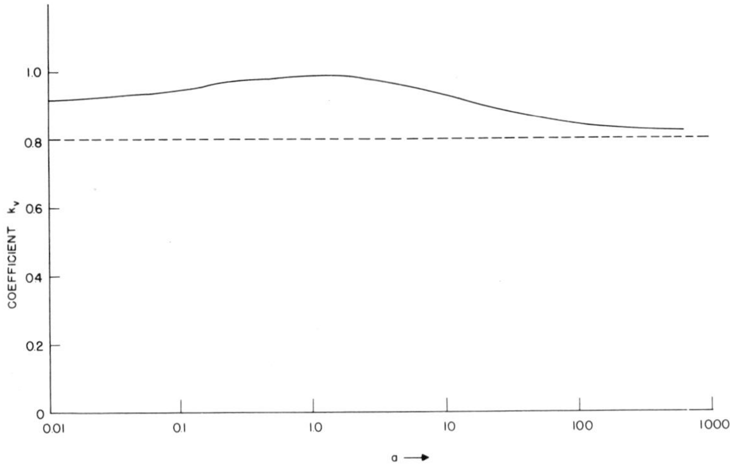

Fig. 10

Correlation coefficient k_v of a gradual clipper as a function of the exponent a.

On the basis of Eqs. 50 and 52 the normalized coefficient k_V is plotted in Fig. 9.

In some cases it is of interest to consider gradual rather than abrupt clipping. Van Vleck computes the autocorrelation for a distortion function V(x) = c tanh (x/c). His result is (ref. 2, Eq. 12, p. 20)

$$\phi_{FF} = \phi_{ff} + \frac{1}{c^2}(-2\phi_{ff}) + \frac{1}{c^4}\left(5\phi_{ff} + \frac{2}{3}\phi_{ff}^3\right) + \dots . \tag{53}$$

Equations 46 and 53 are most useful when we are dealing with amplifiers. A distortion characteristic which may be suitable for analyzing actual clippers is shown in Fig. 7. It is defined as follows

$$V(x) = \begin{cases} 1 - e^{-(a/2)x^2} & x > 0 \\ e^{-(a/2)x^2} - 1 & x < 0 \end{cases} \qquad a > 0. \tag{54}$$

This definition is convenient because the autocorrelation function of the output can be found in a closed form. The flat region near the origin must be remembered. It extends approximately from $x = -(1/\sqrt{a})$ to $x = (1/\sqrt{a})$ and may be made as small or as large as we please.

The expression for the autocorrelation of the output is

$$\phi_{FF}(\tau) = 2\int_0^\infty\int_0^\infty \left(1 - e^{-ax^2/2}\right)\left(1 - e^{-ay^2/2}\right)\left[p(x, y; \tau) - p(x, -y; \tau)\right] dx\, dy. \tag{55}$$

If we write r for $\phi_{ff}(\tau)$ for the sake of simplicity, this leads to

$$\phi_{FF}(\tau) = \frac{2}{\pi}\left[\sin^{-1} r - \frac{2}{\sqrt{1+a}}\sin^{-1}\frac{r}{\sqrt{1+a(1-r^2)}} + \frac{1}{\sqrt{1+2a+a^2(1-r^2)}}\sin^{-1}\frac{r}{1+a(1-r^2)}\right]. \tag{56}$$

We notice that the first term is the "extreme" clipper term of Eq. 48, and that the two other terms are correction terms which decrease as a increases.

K_v is easily found from Eq. 19 and is given by

$$K_v = \frac{2}{\sqrt{2\pi}}\int_0^{\infty} y\left(1 - e^{-ay^2/2}\right) e^{-y^2/2}\, dy$$

$$= \sqrt{\frac{2}{\pi}}\,\frac{1}{1+\frac{1}{a}}. \tag{57}$$

It follows that

$$k_v = \sqrt{\frac{2}{\pi}}\,\frac{1}{1+\frac{1}{a}}\,\frac{1}{\left[1+\frac{1}{\sqrt{1+2a}}-\frac{2}{\sqrt{1+a}}\right]^{1/2}}. \tag{58}$$

The curve of k_v is plotted in Fig. 10.

VIII. Gaussian Correlator

On the basis of the relation expressed by Eq. 20, a simplified correlator can be designed capable of handling gaussian signals. One of the signals can be deliberately peak clipped. It should be, of course, the one which is not corrupted by noise. One can then exchange the problem of continuously multiplying two signals, which often presents difficulties, for the problem of gating one signal with another. In addition, it might also be advantageous if only two fixed amplitudes of the signal need be stored in the delay circuits and if the limitations of amplitude range in the unclipped signal channel can be overcome by boosting the clipped signal level with the resulting high sensitivity.

Acknowledgment

The author wishes to express his indebtedness to Professor J. B. Wiesner under whose supervision he worked at the Research Laboratory of Electronics and to Professor Y. W. Lee who gave so generously of his time in guiding the preparation of this report. He also wishes to thank his colleagues Messrs. C. Desoer, B. Howland and R. F. Schreitmueller for their valuable comments and suggestions.

References

1. Y. W. Lee: Application of Statistical Methods to Communication Problems, Technical Report No. 181, Research Laboratory of Electronics, M. I. T. Sept. 1950
2. J. H. Van Vleck: The Spectrum of Clipped Noise, Radio Research Laboratory Report 51, Harvard, July 21, 1943
3. S. O. Rice: Mathematical Analysis of Random Noise, Bell Telephone System Monograph B-1589
4. D. Middleton: Noise and Nonlinear Communication Problems, Woods Hole Symposium on Application of Autocorrelation Analysis to Physical Problems, June 13-14, 1949, NAVEXOS-P-735
5. H. Cramer: Mathematical Methods of Statistics, Princeton University Press, 1946
6. J. J. Bussgang: Notes on Frequency Modulation (to be published)

Copyright © 1955 by the Institute of Electrical and Electronics Engineers, Inc.

Reprinted from IRE Trans. Information Theory, IT-1(1), 10–15 (1955)

An Expansion for Some Second-Order Probability Distributions and its Application to Noise Problems*

J. F. BARRETT† AND D. G. LAMPARD‡

Summary—In this paper it is shown that, in general, second-order probability distributions may be expanded in a certain double series involving orthogonal polynomials associated with the corresponding first-order probability distributions. Attention is restricted to those second-order probability distributions which lead to a "diagonal" form for this expansion.

When such distributions are joint probability distributions for samples taken from a pair of time series, some interesting results can be demonstrated. For example, it is shown that if one of the time series undergoes an amplitude distortion in a time-varying "instantaneous" nonlinear device, the covariance function after distortion is simply proportional to that before distortion.

Some simple results concerning conditional expectations are given and an extension of a theorem, due to Doob, on stationary Markov processes is presented.

The relation between the "diagonal" expansion used in this paper and the Mercer expansion of the kernel of a certain linear homogeneous integral equation, is pointed out and in conclusion explicit expansions are given for three specific examples.

INTRODUCTION

THE SOLUTION of most problems involving noise in nonlinear devices is difficult. However, a few years ago Bussgang[1] demonstrated an interesting and useful result. He showed that, if one of a pair of stationary time series with Gaussian probability distributions was amplitude distorted in a fixed "instantaneous" nonlinear device, then the cross-correlation function after the distortion is proportional to the cross-correlation function before the distortion.

An attempt to extend the scope of Bussgang's results to distributions other than Gaussian, led to a class of distributions which is discussed in this paper. Such distributions exhibit a number of interesting properties some of which are studied here. It is believed that many results will be useful in dealing with certain noise problems.

ANALYSIS

We suppose that $p(x_1 ; x_2)$ is a second-order probability distribution. The corresponding first-order probability distributions are given by

$$\left. \begin{array}{l} p_1(x_1) = \int_{\text{w.r.}} p(x_1 ; x_2) \, dx_2 \\ p_2(x_2) = \int_{\text{w.r.}} p(x_1 ; x_2) \, dx_1 \end{array} \right\} . \quad (1)$$

---

* Some of the material of this paper is taken from a Dissertation submitted by D. G. Lampard for the Ph.D. degree at the University of Cambridge.
† The Engineering Lab., Cambridge, Eng.
‡ Dept. of Electrical Engineering, Columbia Univ., New York, N. Y. Formerly at the Engineering Lab., Cambridge. At present on leave from CSIRO Div. Electrotechnology, Sydney, Australia.
[1] J. J. Bussgang, "Cross correlation Functions of Amplitude Distorted Gaussian Signals," Tech. Rep. No. 216, Res. Lab. Elec., MIT; March, 1952.

We construct[2] two sets of normalized orthogonal polynomials $\theta_n^{(1)}(x_1)$ and $\theta_n^{(2)}(x_2)$ such that

$$\left. \begin{array}{l} \int_{\text{w.r.}} p_1(x_1) \theta_m^{(1)}(x_1) \theta_n^{(1)}(x_1) \, dx_1 = \delta_{mn} \\ \int_{\text{w.r.}} p_2(x_2) \theta_m^{(2)}(x_2) \theta_n^{(2)}(x_2) \, dx_2 = \delta_{mn} \end{array} \right\} . \quad (2)$$

We now expand the second-order probability distribution in a double "Fourier" series involving these polynomials. Thus

$$p(x_1 ; x_2) = p_1(x_1) p_2(x_2) \sum_{m=0}^{\infty} \sum_{n=0}^{\infty} a_{mn} \theta_m^{(1)}(x_1) \theta_n^{(2)}(x_2). \quad (3)$$

The coefficients a_{mn} may be determined in the usual way by multiplying both sides by $\theta_k^{(1)}(x_1) \theta_l^{(2)}(x_2)$ and integrating using the orthogonal property (2). We find

$$a_{mn} = \iint_{\text{w.r.}} p(x_1 ; x_2) \theta_m^{(1)}(x_1) \theta_n^{(2)}(x_2) \, dx_1 \, dx_2 . \quad (4)$$

Thus we see from (3) that the second-order distribution is now determined completely by the two first-order distributions and the coefficient matrix $[a_{mn}]$.

In this paper we restrict our attention[3] to the class, Λ say, of distributions $p(x_1 ; x_2)$ which have the property that the matrix $[a_{mn}]$ is diagonal. As we shall see from examples, some important distributions that occur in physical problems are in this class.

Thus for all $p(x_1 ; x_2)$ in Λ, (3) may be written[4]

$$p(x_1 ; x_2) = p_1(x_1) p_2(x_2) \sum_{n=0}^{\infty} a_n \theta_n^{(1)}(x_1) \theta_n^{(2)}(x_2), \quad (5)\Lambda$$

where the coefficient a_n is now given by

$$a_n = \iint_{\text{w.r.}} p(x_1 ; x_2) \theta_n^{(1)}(x_1) \theta_n^{(2)}(x_2) \, dx_1 \, dx_2 . \quad (6)\Lambda$$

We now determine the form of the polynomials of degree 0 and 1. As $p_1(x_1)$ and $p_2(x_2)$ are probability distributions we must have

$$\int_{\text{w.r.}} p_1(x_1) \cdot 1 \cdot 1 \cdot dx_1 = 1 \quad (7)$$

---

[2] By applying the Gram-Schmidt procedure to the sequence $1, x, x^2 \cdots$. For details see Courant and Hilbert, "Methods of Mathematical Physics," Interscience Publishers, Inc., New York, N. Y., vol. I, p. 50; 1953.
[3] The authors have not been able so far to find what general restrictions must be placed on $p(x_1 ; x_2)$ in order that it may belong to Λ.
[4] Throughout this paper, equations which are only true when $p(x_1 ; x_2)$ belongs to Λ, will be denoted by having Λ added after the equation number.

$$\int_{w.r.} p_1(x_1)\cdot(x_1 - \mu_1)\cdot 1\cdot dx_1 = 0 \quad (8)$$

$$\int_{w.r.} p_1(x_1)(x_1 - \mu_1)(x_1 - \mu_1)\, dx_1 = \sigma_1^2 , \quad (9)$$

with corresponding results for $p_2(x_2)$. Here we have used the symbols μ, and σ_1 for the mean and standard deviation respectively. It follows that

$$\left.\begin{aligned} \theta_0^{(1)}(x_1) &= 1; & \theta_0^{(2)}(x_2) &= 1 \\ \theta_1^{(1)}(x_1) &= \frac{x_1 - \mu_1}{\sigma_1}; & \theta_1^{(2)}(x_2) &= \frac{x_2 - \mu_2}{\sigma_2} \end{aligned}\right\}. \quad (10)$$

On using (10) and (6), it is easy to see that the coefficients a_0 and a_1 are just

$$a_0 = 1 \quad (11)\Lambda$$

$$a_1 = \frac{\langle (x_1 - \mu_1)(x_2 - \mu_2)\rangle}{\sigma_1\sigma_2} = \rho. \quad (12)\Lambda$$

That is a_1 is just the normalized covariance[5] *or* correlation coefficient.

We now prove an important inequality[6] for the coefficients a_n. Let λ be a real variable and consider the expectation

$$\langle[\theta_n^{(1)}(x_1) + \lambda\theta_n^{(2)}(x_2)]^2\rangle$$

$$= \iint_{w.r.} p(x_1\, ;\, x_2)[\theta_n^{(1)}(x_1) + \lambda\theta_n^{(2)}(x_2)]^2\, dx_1\, dx_2 \quad (13)$$

$$= \iint_{w.r.} p(x_1\, ;\, x_2)[\theta_n^{(1)}(x_1)]^2\, dx_1\, dx_2$$

$$+ 2\lambda \iint_{w.r.} p(x_1\, ;\, x_2)\theta_n^{(1)}(x_1)\theta_n^{(2)}(x_2)\, dx_1\, dx_2$$

$$+ \lambda^2 \iint_{w.r.} p(x_1\, ;\, x_2)[\theta_n^{(2)}(x_2)]^2\, dx_1\, dx_2 . \quad (14)$$

On introducing the expansion (5) for $p(x_1\, ; x_2)$ and carrying out the integration, making use of the orthogonal properties (2) we obtain

$$\langle[\theta_n^{(1)}(x_1) + \lambda\theta_n^{(2)}(x_2)]^2\rangle = 1 + 2\lambda a_n + \lambda^2. \quad (15)\Lambda$$

As the expression on the left-hand side must be positive for all real λ we must have

$$a_n^2 \leq 1, \qquad \text{for all } n\cdot \quad (16)\Lambda$$

which is the required result.[7]

Application to Time Series

In the remainder of this paper we shall think of x_1 and x_2 as being sample values from a pair of time series $x_1(t)$, $x_2(t)$ at times t_1 and t_2 respectively, as it is this interpretation which has most physical significance. Thus we write

$$\left.\begin{aligned} x_1 &\equiv x_1(t_1) \\ x_2 &\equiv x_2(t_2) \end{aligned}\right\}. \quad (17)$$

We then consider an ergodic ensemble of such pairs of time series and take $p(x_1\, ;\, x_2)$ to be the second-order probability distribution of this ensemble. In general this probability distribution will be a function of t_1 and t_2. It then follows that the coefficients defined by (6) will be functions of t_1 and t_2 so we may write

$$a_n \equiv a_n(t_1\, ;\, t_2). \quad (18)\Lambda$$

In particular if the time series is stationary, only time differences are significant so that we have

$$a_n \equiv a_n(t_2 - t_1). \quad (19)\Lambda$$

The first-order probability distributions $p_1(x_1)$ and $p_2(x_2)$, [and the corresponding sets of polynomials $\theta_n^{(1)}(x_1)$ and $\theta_n^{(2)}(x_2)$], will also, in general, be functions of t_1 and t_2, but in the stationary case will be independent of time.

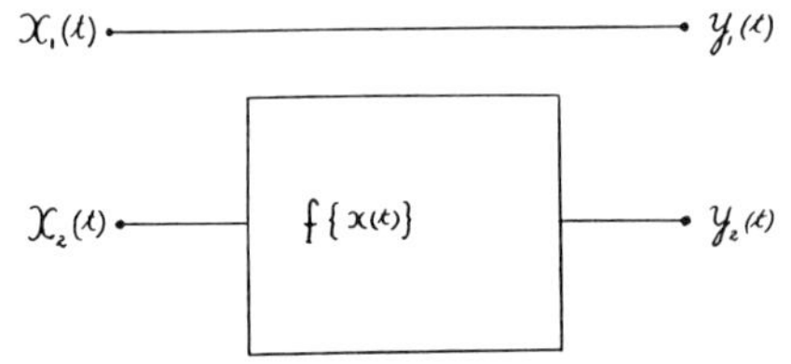

Fig. 1

Noise in "Instantaneous" Nonlinear Devices

Let us consider the system shown in Fig. 1.

The input-output relations of this system are given by

$$\left.\begin{aligned} y_1(t) &= x_1(t) \\ y_2(t) &= f\{x_2(t)\} \end{aligned}\right\}, \quad (20)$$

where the function f is assumed not to involve differential, integral, or delay operators. In general f may be an explicit function of time, assuming that such temporal dependence is statistically independent of the input.

A device characterized by such a function f will be called[8] an "instantaneous" time-varying nonlinear device.

Let us now find the (un-normalized) covariance function for the outputs. We have

$$\Psi_{12}(t_1\, ;\, t_2)$$

$$= \langle[y_1(t_1) - \langle y_1(t_1)\rangle][y_2(t_2) - \langle y_2(t_2)\rangle]\rangle \quad (21)$$

$$= \langle[x_1 - \mu_1][f(x_2) - \langle f(x_2)\rangle]\rangle \quad (22)$$

[5] Cramér, "Mathematical Methods of Statistics," Princeton University Press, Princeton, N. J., p. 265; 1946.

[6] This is just the Schwarz inequality adapted to this particular problem. See for example, Lovitt, "Linear Integral Equations," Dover Publications, New York, N. Y., p. 125; 1950.

[7] The authors are grateful to S. O. Rice for pointing out these inequalities in a private communication.

[8] The term "zero-memory" is also commonly used in place of "instantaneous."

$$= \iint_{\text{w.r.}} (x_1 - \mu_1)\{f(x_2) - \langle f(x_2)\rangle\}p(x_1 ; x_2)\, dx_1\, dx_2 \quad (23)$$

$$= \sigma_1 \iint_{\text{w.r.}} \theta_1^{(1)}(x_1)\{f(x_2) - \langle f(x_2)\rangle\}p(x_1 ; x_2)\, dx_1\, dx_2 . \quad (24)$$

To make further progress, we now expand $f(x_2)$ in a series of the polynomials $\theta_n^{(2)}(x_2)$. (We assume that the expansion can be justified.) Thus

$$f(x_2) = \sum_{m=0}^{\infty} c_m \theta_m^{(2)}(x_2) \quad (25)$$

where the coefficient $c_m \equiv c_m(t_2)$ is given by

$$c_m = \int_{\text{w.r.}} f(x_2)p_2(x_2)\theta_m^{(2)}(x_2)\, dx_2 . \quad (26)$$

In particular, remembering from (10) that $\theta_0^{(2)}(x_2) = 1$, we see that

$$c_0 = \int_{\text{w.r.}} f(x_2)p_2(x_2)\, dx_2 = \langle f(x_2)\rangle. \quad (27)$$

Thus

$$f(x_2) - \langle f(x_2)\rangle = \sum_{m=1}^{\infty} c_m \theta_m^{(2)}(x_2). \quad (28)$$

Thus using (28) in (24) we have

$$\Psi_{12}(t_1 ; t_2) = \sigma_1 \iint_{\text{w.r.}} \theta_1^{(1)}(x_1)\left\{\sum_{m=1}^{\infty} c_m \theta_m^{(2)}(x_2)\right\}p(x_1 ; x_2)\, dx_1\, dx_2 . \quad (29)$$

We now use expansion (5) for all $p(x_1 ; x_2)$ in Λ and find

$$\Psi_{12}(t_1 ; t_2) = \sigma_1 \iint_{\text{w.r.}} \theta_1^{(1)}(x_1)p_1(x_1)p_2(x_2) \cdot \sum_{m=1}^{\infty}\sum_{n=0}^{\infty} c_m a_n \theta_n^{(1)}(x_1)\theta_m^{(2)}(x_2)\theta_n^{(2)}(x_2)\, dx_1\, dx_2 \quad (30)\Lambda$$

$$= \sigma_1 a_1 c_1 , \quad (31)\Lambda$$

where, in carrying out the integrations, use has been made of the orthogonal properties (2).

We note from (12) that a_1 is just the normalized covariance of the inputs, that is

$$a_1 = \frac{\langle (x_1 - \mu_1)(x_2 - \mu_2)\rangle}{\sigma_1\sigma_2} = \frac{\psi_{12}(t_1 ; t_2)}{\sigma_1\sigma_2}. \quad (32)\Lambda$$

Thus (31) may be written

$$\Psi_{12}(t_1 ; t_2) = C \cdot \psi_{12}(t_1 ; t_2), \quad (33)\Lambda$$

when

$$C \equiv C(t_2) = \int_{\text{w.r.}} f(x_2)\frac{(x_2 - \mu_2)}{\sigma_2^2}\, dx_2 . \quad (34)\Lambda$$

For stationary time series and fixed "instantaneous" nonlinear devices (33) may be written as

$$\Psi_{12}(\tau) = C \cdot \psi_{12}(\tau), \quad (35)\Lambda$$

where the constant C is independent of time and when we have written $\tau = t_2 - t_1$. The ψ and Ψ are just the ordinary cross-correlation functions of the fluctuating parts of the input and output respectively, of the system of Fig. 1. In the still more special case in which the input time series are not only stationary but identical, that is $x_1(t) = x_2(t)$, the result (35) becomes identical to that obtained by Bussgang[9] for Gaussian noise.

Luce[10] has demonstrated that Bussgang's result applies to a certain class of distributions, of which the Gaussian distribution is a special case. However Luce's distributions are of a fairly simple form[11] and it is easy to show, by means of specific examples, that there are distributions in Λ which are not of the type considered by Luce.

Some Results on Conditional Expectations

For a distribution in Λ it is easy to see from (5) that the conditional probability density for x_2 given x_1 is

$$p(x_2 \mid x_1) = \frac{p(x_1 ; x_2)}{p_1(x_1)} = p_2(x_2)\sum_{n=0}^{\infty} a_n \theta_n^{(1)}(x_1)\theta_n^{(2)}(x_2). \quad (36)\Lambda$$

Then the conditional expectation for $\theta_m^{(2)}(x_2)$ given x_1 is

$$\langle \theta_m^{(2)}(x_2) \mid x_1\rangle = \int_{\text{w.r.}} \theta_m^{(2)}(x_2)p_2(x_2)\sum_{n=0}^{\infty} a_n \theta_n^{(1)}(x_1)\theta_n^{(2)}(x_2)\, dx_2 \quad (37)\Lambda$$

$$= a_m \theta_m^{(1)}(x_1). \quad (38)\Lambda$$

If in particular we take $m = 1$, we have, using (10) and (12)

$$\left\langle \frac{x_2 - \mu_2}{\sigma_2} \,\middle|\, x_1 \right\rangle = \rho(t_1 ; t_2)\frac{x_1 - \mu_1}{\sigma_1}. \quad (39)\Lambda$$

When $x_1(t) = x_2(t)$ is a stationary time series with normalized auto-correlation function $\rho(\tau)$, (39) becomes

$$\langle x_2 - \mu_2 \mid x_1\rangle = \rho(\tau)\cdot(x_1 - \mu_1), \quad (40)\Lambda$$

where, as usual, we have written $\tau = t_2 - t_1$. This result has an obvious interpretation.

Stationary Markov Processes

It has been shown by Doob[12] that a stationary one-dimensional Markov process with a Gaussian probability distribution must have a correlation function which is an exponential function of time. In this section we show that this result is true for any distribution in Λ.

Let $x_1 , x_2 \cdots x_n$ be samples from a stationary time series at times $t_1 , t_2 , \cdots t_n$.[13] If this time series is a Markov process then

$$p(x_n \mid x_1 , x_2 \cdots x_{n-1}) = p(x_n \mid x_{n-1}), \quad (41)$$

by definition of a Markov process. Thus

$$\frac{p(x_1 , x_2 \cdots x_n)}{p(x_1 , x_2 \cdots x_{n-1})} = \frac{p(x_{n-1} , x_n)}{p(x_{n-1})} \quad (42)$$

[9] Bussgang, *loc. cit.*

[10] R. D. Luce, "Amplitude Distorted Signals," Res. Lab. Elec. Quarterly Progress Report, p. 37; April 15, 1953.

[11] A simple change of variables always reducing the second-order probability distribution to the product of independent first-order probability distributions.

[12] J. L. Doob, "Brownian motion and stochastic equations," *Ann. Math.*, vol. 43, p. 351; 1942.

[13] It is assumed that $t_n \geq t_{n-1} \geq \cdots \geq t_1$.

$$p(x_1, x_2 \cdots x_n) = p(x_1, x_2 \cdots x_{n-1}) \frac{p(x_{n-1}, x_n)}{p(x_{n-1})} \quad (43)$$

$$= \frac{p(x_1, x_2)p(x_2, x_3) \cdots p(x_{n-1}, x_n)}{p(x_2) \cdots p(x_{n-1})} \quad (44)$$

by induction.

In particular we find

$$p(x_1, x_2, x_3) = \frac{p(x_1, x_2)p(x_2 ; x_3)}{p(x_2)}. \quad (45)$$

If the two-dimensional distributions $p(x_1 ; x_2)$ is in Λ, we can use the expansion (5) in (45). We find[14]

$$p(x_1, x_2, x_3)$$

$$= p(x_1)p(x_2)p(x_3) \sum_{m=0}^{\infty} \sum_{n=0}^{\infty} a_m(t_2 - t_1)a_n(t_3 - t_2) \cdot \theta_m(x_1)\theta_m(x_2)\theta_n(x_2)\theta_n(x_3). \quad (46)\Lambda$$

Let us now find the two-dimensional probability density, $p(x_1, x_3)$ by integrating out x_2 from both sides of (46).

$$p(x_1, x_3) = \int_{w.r.} p(x_1, x_2, x_3)\, dx_2 \quad (47)$$

$$= p(x_1)p(x_3) \sum_{n=0}^{\infty} a_n(t_3 - t_2)a_n(t_2 - t_1)\theta_n(x_1)\theta_n(x_3). \quad (48)\Lambda$$

But as $p(x_1, x_3)$ is an Λ we may write

$$p(x_1, x_3) = p(x_1)p(x_3) \sum_{n=0}^{\infty} a_n(t_3 - t_1)\theta_n(x_1)\theta_n(x_3). \quad (49)\Lambda$$

For (48) and (49) to be consistent we must have

$$a_n(t_3 - t_1) = a_n(t_3 - t_2)a_n(t_2 - t_1). \quad (50)\Lambda$$

It is clear that the only continuous solution of this equation is of the form

$$a_n(\tau) = e^{-\beta_n \tau}, \quad (51)\Lambda$$

where β_n is a constant depending on n.

In particular when we remember that $a_1(\tau) = \rho(\tau)$, the normalized auto correlation function, we see that

$$\rho(\tau) = e^{-\beta\tau}, \qquad \tau \geq 0, \quad (52)\Lambda$$

thus establishing the required result.

An Integral Equation Related to the Expansion (5)

The obvious similarity in form between the expansion (8) and the Mercer expansion[15] for the kernel of a linear homogeneous integral equation suggests that there may be an approach to our problem which is based on the theory of integral equations.

To demonstrate this connection we shall assume $p(x_1, x_2)$ to be a *symmetric*[16] two-dimensional probability distribution. That is, we assume

$$p(x_1, x_2) = p(x_2, x_1) \quad (53)$$

Then the expansion (5) may be written

$$p(x_1, x_2) = p(x_1)p(x_2) \sum_{n=0}^{\infty} a_n\theta_n(x_1)\theta_n(x_2). \quad (54)\Lambda$$

Multiplying both sides by $\theta_m(x_2)$ and integrating with respect to x_2 we have, using the orthogonal property (2)

$$\int_{w.r.} p(x_1, x_2)\theta_m(x_2)\, dx_2 = a_m p(x_1)\theta_m(x_1) \quad (55)\Lambda$$

which may be written

$$\int_{w.r.} \frac{p(x_1, x_2)}{\sqrt{p(x_1)p(x_2)}} \{\sqrt{p(x_2)}\, \theta_m(x_2)\}\, dx_2 = a_m\{\sqrt{p(x_1)}\, \theta_m(x_1)\} \quad (56)\Lambda$$

which is now in the sta'ndard form for a linear homogeneous integral equation

$$\int_{w.r.} K(x_1, x_2)\phi_m(x_2)\, dx_2 = \lambda_m\phi_m(x_1), \quad (57)$$

with

$$K(x_1, x_2) = \frac{p(x_1, x_2)}{\sqrt{p(x_1)p(x_2)}} \quad (58)\Lambda$$

$$\phi_m(x) = \sqrt{p(x)}\, \theta_m(x) \quad (59)\Lambda$$

$$\lambda_m = a_m. \quad (60)\Lambda$$

It follows from the definition (2) for $\theta_n(x)$ that

$$\int_{w.r.} \phi_m(x)\phi_n(x)\, dx = \int_{w.r.} p(x)\theta_m(x)\theta_n(x)\, dx = \delta_{mn} \quad (61)\Lambda$$

so that the ϕ_m defined by (59) are normalized orthogonal eigenfunctions of the linear homogeneous integral (57) with corresponding eigenvalues λ_m given by (60).

Then, providing it can be shown that the set of eigenfunctions defined by (59) is complete, the equivalence of the expansion (54) and the Mercer expansion

$$K(x_1, x_2) = \sum_{n=0}^{\infty} \lambda_n\phi_n(x_1)\phi_n(x_2) \quad (62)$$

is demonstrated.

It must be pointed out, however, that the integral equation approach does not seem to be a very suitable starting point because of the difficulties of justifying[17] the Mercer expansion and of showing when the eigenfunctions are of the form (59), which is essential for the applications discussed in this paper.

Some Examples of Distributions in Λ

In this final section we give three examples of second-order probability distributions which are in the class Λ. We shall choose distributions which are "symmetric" in the sense of (53) in order to keep the details of the analysis as simple as possible.

[14] We have written $a_n(t_2 - t_1)$, etc., in place of our more usual a_n to avoid confusion.

[15] Courant and Hilbert, *loc. cit.*, p. 138.

[16] This symmetry restriction can be removed by using theory of adjoint orthogonal functions. Courant and Hilbert, *loc. cit.*, p. 159.

[17] Sufficient, but not necessary conditions are the continuity and definiteness of the kernel $K(x_1, x_2)$. Courant and Hilbert, *loc. cit.*, p. 138.

Example 1

The second order Gaussian probability distribution has the well known form[18]

$$p(x_1, x_2) = \frac{1}{2\pi\sigma^2}(1-\rho^2)^{-1/2} \exp -\frac{1}{2}\left\{\frac{x_1^2 + x_2^2 - 2\rho x_1 x_2}{\sigma^2(1-\rho^2)}\right\} \tag{63}$$

where we have written σ, ρ for the standard deviation and normalized correlation function respectively.

The corresponding first-order probability distribution is

$$p(x) = \frac{1}{\sigma\sqrt{2\pi}} \exp\left\{-\frac{1}{2}\frac{x^2}{\sigma^2}\right\}. \tag{64}$$

We now make use of Mehler's expansion,[19] namely

$$(1-p^2)^{-1/2} \exp -\frac{1}{2}\left\{\frac{p^2(x_1^2+x_2^2) - 2px_1x_2}{(1-p^2)}\right\} = \sum_{n=0}^{\infty} p^n \frac{H_n(x_1)H_n(x_2)}{n!}, \tag{65}$$

where $H_n(x)$ is the Hermite Polynomial of degree n defined[20] by

$$\frac{d^n}{dx^n}\{e^{-x^2/2}\} = (-1)^n H_n(x) e^{-x^2/2}. \tag{66}$$

Then using (65) it follows that (63) may be written

$$p(x_1, x_2) = \frac{1}{2\pi\sigma^2} \exp -\frac{1}{2}\left\{\frac{x_1^2 + x_2^2}{\sigma^2}\right\} \cdot \sum_{n=0}^{\infty} \rho^n \frac{H_n(x_1/\sigma)H_n(x_2/\sigma)}{n!}. \tag{67}$$

The Hermite Polynomials have the orthogonal property[21]

$$\frac{1}{\sqrt{2\pi}}\int_{-\infty}^{+\infty} e^{-x^2/2} H_m(x)H_n(x)\,dx = \delta_{mn} n! \tag{68}$$

so it is seen that (67) can finally be written in the standard form[22]

$$p(x_1, x_2) = p(x_1)p(x_2)\sum_{n=0}^{\infty} a_n \theta_n(x_1)\theta_n(x_2) \tag{69}\Lambda$$

with

$$\left.\begin{aligned}\theta_n(x) &= (n!)^{-1/2}H_n(x)\\ a_n &= p^n\end{aligned}\right\} \tag{70}$$

showing that $p(x_1, x_2)$ is in Λ.

Example 2

When Gaussian noise is applied to a narrow bandpass filter, the output of the filter behaves as a sine wave at the midband frequency with random, relatively slowly varying amplitude and phase. Rice has shown[23] that the second-order probability distribution for the envelope R of the filter output is

$$p(R_1, R_2) = \frac{R_1R_2}{\sigma^4}(1-\mu^2)^{-1} I_0\left\{\frac{R_1R_2}{\sigma^2}\frac{\mu}{1-\mu^2}\right\} \cdot \exp -\frac{1}{2}\left\{\frac{R_1^2 + R_2^2}{\sigma^2(1-\mu^2)}\right\}. \tag{71}$$

Here $I_0(x)$ is the modified Bessel Function of zero order and

$$\mu^2 \equiv \mu^2(\tau) = \frac{\int_0^\infty\int_0^\infty w(f_1)w(f_2)\cos 2\pi(f_1 - f_2)\tau\, df_1\, df_2}{\left\{\int_0^\infty w(f)\,df\right\}^2} \tag{72}$$

$$\sigma^2 = \int_0^\infty w(f)\,df, \tag{73}$$

when $w(f)$ is the power spectrum of the noise at the output of the filter.

The corresponding first order distribution has the very simple form (Rayleigh distribution)

$$p(R) = \frac{R}{\sigma^2}\exp\left\{-\frac{1}{2}\frac{R^2}{\sigma^2}\right\}. \tag{74}$$

We now suppose that an "instantaneous" square law envelope detector is connected to the output of our narrow band filter. Such a detector would have a time constant, long compared to the reciprocal of the filter midband frequency, but short compared to the reciprocal of the filter bandwidth. These are conflicting requirements, but are usually closely approximated in practice. Then denoting the outputs at times t_1, t_2 by x_1 and x_2 respectively, we may write

$$\left.\begin{aligned}x_1 &= R_1^2\\ x_2 &= R_2^2\end{aligned}\right\}. \tag{75}$$

From (71), we find immediately

$$p(x_1, x_2) = \frac{(1-\mu^2)^{-1}}{4\sigma^4} I_0\left\{\frac{\sqrt{x_1x_2}}{\sigma^2}\frac{\mu}{1-\mu^2}\right\}\exp -\frac{1}{2}\left\{\frac{x_1 + x_2}{\sigma^2(1-\mu^2)}\right\} \tag{76}$$

and from (74), we have for the first-order distribution

$$p(x) = \frac{1}{2\sigma^2}\exp\left\{-\frac{1}{2}\frac{x}{\sigma^2}\right\}. \tag{77}$$

We now make use of the identity[24]

$$(1-t)^{-1}\exp\left\{-(x+y)\frac{t}{1-t}\right\} I_0\left\{\frac{2\sqrt{xyt}}{1-t}\right\} = \sum_{n=0}^{\infty} t^n L_n(x)L_n(y), \tag{78}$$

when $L_n(x)$ is the Laguerre Polynomial defined[25] by

$$\frac{d^n}{dx^n}\{x^n e^{-x}\} = n! e^{-x} L_n(x). \tag{79}$$

[18] S. O. Rice, *Bell Sys. Tech. Jour.*, vol. 23, p. 50; 1945. We have assumed zero means in this example for simplicity.

[19] G. N. Watson, "Generating functions for polynomials II," *Jour. London Math. Soc.*, vol. 8, p. 194; 1933.

[20] Cramér, *loc. cit.*, p. 133.

[21] Cramér, *loc. cit.*, p. 133.

[22] This expansion is a fairly well-known one (e.g. Cramér, p. 133) and has been used by Siegert in noise problems; see "On the evaluation of noise samples," *Jour. Appl. Phys.*, vol. 23, no. 7; 1952.

[23] S. O. Rice, *Bell Sys. Tech. Jour.*, vol. 23, p. 78; 1945.

[24] G. N. Watson, "Generating functions of polynomials I," *Jour. London Math. Soc.*, vol. 8, p. 189; 1933. This is a special case of the more general result given by Watson.

[25] G. Szego, "Orthogonal polynomials," *Amer. Math. Soc. Colloquium Pub.*, vol. 23, p. 97; 1939.

Using (78) we may write (76) in the form

$$p(x_1, x_2) = \frac{1}{4\sigma^4} \exp -\frac{1}{2}\left\{\frac{x_1 + x_2}{\sigma^2}\right\} \sum_{n=0}^{\infty} (\mu^2)^n L_n\left(\frac{x_1}{2\sigma^2}\right) L_n\left(\frac{x_2}{2\sigma^2}\right) \tag{80}$$

$$= p(x_1)p(x_2) \sum_{n=0}^{\infty} (\mu^2)^n L_n\left(\frac{x_1}{2\sigma^2}\right) L_n\left(\frac{x_2}{2\sigma^2}\right) \tag{81}$$

where use has been made of (77). When it is recalled that the orthogonal property for the Laguerre Polynomials is[26]

$$\int_0^{\infty} e^{-x} L_m(x) L_n(x)\, dx = \delta_{mn} \tag{82}$$

it is seen that (81) is already in standard form with

$$\left.\begin{aligned} \theta_n(x) &= L_n\left(\frac{x}{2\sigma^2}\right) \\ a_n &= (\mu^2)^n \end{aligned}\right\} \tag{83}$$

showing that $p(x_1, x_2)$ defined by (75) is in Λ.

Example 3

As a final example, we show that the second-order probability distribution for a sine wave of constant amplitude P is in Λ. It will be most convenient in this example to start from the characteristic function of the distribution. If we denote our sine wave by

$$x(t) = p \cos (wt + \phi), \tag{84}$$

the second-order characteristic function defined as

$$g(u, v) = \langle e^{iux_1 + ivx_2} \rangle \tag{85}$$

has been shown by Rice[27] to be

$$g(u, v) = J_0\{p\sqrt{u^2 + v^2 + 2uv \cos w\tau}\}. \tag{86}$$

It follows that the second order probability distribution $p(x_1, x_2)$ is given by the Fourier transform,

$$p(x_1, x_2) = \frac{1}{4\pi^2} \int_{-\infty}^{+\infty} \int_{-\infty}^{+\infty} J_0\{p\sqrt{u^2 + v^2 + 2uv \cos w\tau}\} \cdot e^{-iux_1 - iux_2}\, du\, dv. \tag{87}$$

To make further progress, we use Neumann's addition theorem,[28] namely

$$J_0\{\sqrt{u^2 + v^2 - 2uv \cos \phi}\} = \sum_{n=0}^{\infty} \epsilon_n J_n(u) J_n(v) \cos n\phi, \tag{88}$$

where

$$\left.\begin{aligned} \epsilon_n &= 1 \quad & n &= 0 \\ &= 2 \quad & n &= 1, 2, \cdots \end{aligned}\right\}. \tag{89}$$

Using (88) in (87) and reversing the order of summation and integration, we find

$$p(x_1, x_2) = \sum_{n=0}^{\infty} (-1)^n \epsilon_n \cos nw\tau \cdot \left\{\frac{1}{2\pi} \int_{-\infty}^{+\infty} J_n(pu) e^{-iux_1}\, du\right\}\left\{\frac{1}{2\pi} \int_{-\infty}^{+\infty} J_n(pv) e^{-ivx_2}\, dv\right\}. \tag{90}$$

We now use the result[29]

$$\left.\begin{aligned} \frac{1}{2\pi} \int_{-\infty}^{+\infty} J_n(pu) e^{-iux}\, du &= \frac{(-i)^n T_n(x/p)}{\pi(p^2 - x^2)^{1/2}}, \quad & \frac{x^2}{p^2} &< 1 \\ &= 0, & \frac{x^2}{p^2} &> 1 \end{aligned}\right\}. \tag{91}$$

Here $T_n(x)$ is the Tchebycheff Polynomial of the first kind defined by

$$T_n(x) = \cos \{n \cos^{-1} x\}. \tag{92}$$

Using (91), we see that (90) may be written

$$p(x_1, x_2) = \frac{1}{\pi^2} (p^2 - x_1^2)^{-1/2} (p^2 - x_2^2)^{-1/2} \cdot \sum_{n=0}^{\infty} \epsilon_n T_n\left(\frac{x_1}{p}\right) T_n\left(\frac{x_2}{p}\right) \cos nw\tau, \tag{93}$$

when

$$\frac{x_1^2}{p^2} < 0, \quad \frac{x_2^2}{p^2} < 0$$

otherwise

$$p(x_1, x_2) = 0.$$

The first-order probability distribution for the sine wave has the well-known form[30]

$$p(x) = \frac{1}{\pi} (p^2 - x^2)^{-1/2} \tag{94}$$

and the Tchebycheff Polynomial has the orthogonal property

$$\frac{1}{\pi} \int_{-1}^{+1} \epsilon_n T_m(x) T_n(x) (1 - x^2)^{-1/2}\, dx = \delta_{mn} \tag{95}$$

so it follows that (93) may be written in the standard form

$$p(x_1, x_2) = p(x_1)p(x_2) \sum_{n=0}^{\infty} a_n \theta_n(x_1) \theta_n(x_2), \tag{96}$$

where

$$\left.\begin{aligned} \theta_n(x) &= \sqrt{\epsilon_n}\, T_n\left(\frac{x}{p}\right) \\ a_n &= \cos nw\tau \end{aligned}\right\}. \tag{97}$$

Note that, from (92), we may write $a_n = T_n(a_1)$.

ACKNOWLEDGMENT

The authors have benefited from discussions with many friends during the course of this work and to them we offer our sincere thanks. In particular, we appreciate very much many helpful suggestions from D. W. Moore of Jesus College, Cambridge, S. O. Rice of the Bell Telephone Laboratories. Finally, we wish to thank Prof. L. A. Zadeh of Columbia University for reading the manuscript and suggesting several improvements and such reference works as that of R. D. Luce.

[26] *Ibid.*
[27] S. O. Rice, "Mathematical analysis of random noise," *Bell Sys. Tech. Jour.*, vol. 24, p. 138; 1945.
[28] G. N. Watson, "Theory of Bessel Functions," Cambridge University Press, Cambridge, Eng., p. 358; 1922.
[29] P. R. Clement, "The Chebyshev Approximation Method," *Quart. Appl. Math.*, vol. 11; July, 1953.
[30] Rice, "Mathematical analysis of random noise," *op. cit.*, p. 48.

3

Reprinted from *IRE Trans. Information Theory,* **IT-4**(2), 73–76 (1958)

The Effect of Instantaneous Nonlinear Devices on Cross-Correlation*

ROY LEIPNIK†

Summary—If $X_1(t)$, $X_2(t)$ are two noises (stochastic processes), f and g are functions describing the action of two instantaneous nonlinear devices, we say that the (m, n) cross-correlation property holds in case the cross-correlation of $f(X_1(t_1))$ with $g(X_2(t_2))$ is proportional to the cross-correlation of $X_1(t_2)$ with $X_2(t_2)$, whenever f and g are polynomials of degrees not exceeding m and n, respectively. We take $m = \infty$ or $n = \infty$ to mean that f or g is any continuous function.

The Barrett-Lampard expansion[2] of the second-order joint density of $X_1(t_1)$ and $X_2(t_2)$ is used to derive an expression for the cross-correlation of $f(X_1(t_1))$ and $g(X_2(t_2))$. This expression yields necessary and sufficient conditions for the validity of the cross-correlation property in three cases: $X_1(t)$ and $X_2(t)$ stationary, m, n unrestricted; $X_1(t)$ stationary, m, n unrestricted; $X_1(t)$ stationary, $n = 1$.

Examples are constructed with the help of special orthonormal polynomials illustrating the necessity and sufficiency of the conditions.

Introduction

Bussgang[1] showed that if one of a pair of stationary Gaussian processes is amplitude-distorted in an instantaneous nonlinear device, that the cross-correlations before and after distortion are proportional. In our terminology, this is the $(\infty, 1)$ cross-correlation property.

Barrett and Lampard[2] introduced a bi-orthonormal expansion method in a successful attempt to extend Bussgang's result to nonstationary processes with joint distribution whose bi-orthonormal coefficient matrix is diagonal. The latter is sufficient, but not necessary for the $(\infty, 1)$ cross-correlation property. Brown[3] found a necessary and sufficient condition for the $(\infty, 1)$ property when $X_1(t)$ and $X_2(t)$ are both stationary, and for a partially time-dependent $(\infty, 1)$ property when $X_1(t)$ is stationary, $X_2(t)$ nonstationary.

It is natural to ask whether a like property holds when both processes are distorted by instantaneous nonlinear devices and whether conditions can be relaxed if one or both devices are of restricted complexity (such as quadratic detectors).

We find that the results of Brown can be fully generalized in both of these directions (Theorems 1 and 2). We further discover (Theorem 3) that the partial time dependence allowed by Brown in the proportionality constant is absent when a stationary process is distorted and a nonstationary process is undistorted.

Analysis[4]

Let $p_{t_1,t_2}(x_1, x_2)$ be the joint density of two stochastic processes $X_1(t)$ and $X_2(t_2)$. If both are stationary, the time dependence involves $t_2 - t_1$ only. With Barrett and Lampard[2] we introduce the marginal densities $p_1(x_1)$, $p_2(x_2)$, sequences $\{\theta^{(1)}_{k,t_1}(x_1)\}$, $\{\theta^{(2)}_{k,t_2}(x_2)\}$ of polynomials orthonormal with respect to $p_1(x_1)$, $p_2(x_2)$, and the matrices $A_{t_1,t_2} = [a_{m,n:t_1,t_2}]$ such that

$$\frac{p_{t_1,t_2}(x_1, x_2)}{p_{t_1}(x_1)p_{t_2}(x_2)}$$

has the formal bi-orthonormal expansion

$$\sum_{m,n} a_{m,n,t_1,t_2}\, \theta^{(1)}_{m,t_1}(x_1)\, \theta^{(2)}_{n,t_2}(x_2)$$

We say the (m, n) cross-correlation property holds if for each polynomial f of degree $\leq m$, each polynomial g of degree $\leq n$, there is a constant $k(f, g)$ such that $\rho_{t_1,t_2}(f, g) = k(f, g)\, \rho_{t_1,t_2}$, where

$$\rho_{t_1,t_2}(f, g) = \frac{\operatorname{cov}\,(f(X_1(t_1)),\, g(X_2(t_2)))}{\sqrt{\operatorname{var}\,(f(X_1(t_1)))\cdot \operatorname{Var}\,(g(X_2(t_2)))}}$$

$$\rho_{t_1,t_2} = \frac{\operatorname{cov}\,(X_1(t_1),\, X_2(t_2))}{\sqrt{(\operatorname{var} X_1(t_1))(\operatorname{var} X_2(t_2))}}. \tag{1}$$

The conditions derived below depend on a lemma stated here and proved in the Appendix.

Lemma: If

$$c_{u,t_1} = \int \theta^{(1)}_{u,t_1}(x_1) f(x_1) p_{1,t_1}(x_1)\, dx_1$$

$$d_{v,t_2} = \int \theta^{(2)}_{v,t_2}(x_2) g(x_2) p_{2,t_2}(x_2)\, dx_2$$

$$\text{for} \quad u, v = 0, 1, 2, \cdots,$$

where f and g are polynomials of degrees $\leq m$ and $\leq n$ respectively, then

$$\rho_{t_1,t_2}(f, g) = \frac{\sum_{u=1}^{M} \sum_{v=1}^{n} c_{u,t_1} d_{v,t_2} a_{u,v,t_1,t_2}}{\sqrt{\sum_{u=1}^{m} c^2_{u,t_1} \sum_{v=1}^{n} d^2_{v,t_2}}}. \tag{2}$$

* Manuscript received by the PGIT, September 16, 1957; revised manuscript received, February 25, 1958.

† Dept. of Mathematics, University of Washington, Seattle, Wash.

[1] J. J. Bussgang, "Cross Correlation Functions of Amplitude Distorted Gaussan Signals," Res. Lab. of Electronics, M.I.T., Cambridge, Mass., Tech. Rep. No. 216; March 26, 1952.

[2] J. F. Barrett and D. G. Lampard, "An expansion for some second order probability distributions and its application to noise problems," IRE Trans. on Information Theory, vol. IT-1, pp. 10–15; March, 1955.

[3] J. L. Brown, Jr., "On a cross-correlation property for stationary random processes," IRE Trans. on Information Theory, vol. IT-3, pp. 28–31; March, 1957.

[4] Questions of convergence are avoided in the formal calculations here. A rigorous treatment of the Barrett-Lampard expansion may be found in a forthcoming paper by the author in the *J. SIAM*.

This simple bilinear expression immediately yields

Theorem 1

If $X_1(t)$, $X_2(t)$ are stationary, then the (m, n) cross-correlation property holds if and only, if there is a time independent $m \times n$ matrix $Q = [q_{u,v}]_{m,n}$ with $q_{1,1} = 1$ such that the upper left $m \times n$ submatrix $B_{t_1,t_2} = [A_{t_1,t_2}]_{m,n}$ of A_{t_1,t_2} satisfies the equation $B_{t_1,t_2} = a_{1,1,t_1,t_2}Q$.

Proof: Since $X_1(t)$ and $X_2(t)$ are stationary, the Fourier coefficients of f and g are time-independent: $c_{u,t_1} = c_u$, $d_{v,t_2} = d_v$ for $u = 1, \cdots m$, $v = 1, \cdots n$. Hence

$$\rho_{t_1,t_2}(f, g) = \frac{\sum_{u=1}^{m}\sum_{v=1}^{n} c_u\, d_v a_{u,v,t_1,t_2}}{\sqrt{\sum_{u=1}^{m} c_u^2 \sum_{v=1}^{n} d_v^2}}.$$

Note that $\rho_{t_1,t_2} = a_{1,1,t_1,t_2}$. If the matrix equation holds, then

$$\rho_{t_1,t_2}(f, g) = \frac{\sum_{u=1}^{m}\sum_{v=1}^{n} c_u\, d_v q_{uv}}{\sqrt{\sum_{u=1}^{m} c_u^2 \sum_{v=1}^{n} d_v^2}}\,\rho_{t_1,t_2}$$

$$= K(f, g)\rho_{t_1,t_2}.$$

(Actually, the correlations depend on $t_2 - t_1$ only).

Conversely, if $\rho_{t_1,t_2}(f, g) = K(f, g)\ \rho_{t_1,t_2}$, for polynomials f, g of degrees $\leqq m, n$, pick $f_i = \theta_i^{(1)}$, $g_k = \theta_k^{(2)}$ so that $c_u = \delta_{u,i}$, $d_v = \delta_{v,k}$, $\rho_{t_1,t_2}(f_i, g_k) = a_{i,k,t_1,t_2} = K(f_i, g_k)\ a_{1,1,t_1,t_2}$. Thus we have $q_{u,v} = K(f_u, g_v)$ for $u = 1, \cdots m; v = 1, \cdots n$ and $B_{t_1,t_2} = a_{1,1,t_1,t_2}Q$.

If only one of the processes is stationary, results are still obtainable.

Theorem 2

If $X_1(t)$ is stationary, then $\rho_{t_1,t_2}(f, g) = K_{t_2}(f, g)\ \rho_{t_1,t_2}$ for all polynomials f, g of degrees $\leqq m, n$ if and only if there exists an $m \times n$ matrix Q_{t_2} such that $B_{t_1,t_2} = a_{1,1,t_1,t_2}Q_{t_2}$.

Proof: We have $c_{u,t_1} = c_u$ for $u = 1, \cdots m$, so that

$$\rho_{t_1,t_2}(f, g) = \frac{\sum_{u=1}^{m}\sum_{v=1}^{n} c_u\, d_{v,t_2}a_{u,v,t_1,t_2}}{\sqrt{\sum_{u=1}^{m} c_u^2 \sum_{v=1}^{n} d_{v,t_2}^2}}.$$

If the matrix equation holds, then

$$\rho_{t_1,t_2}(f, g) \doteq \frac{\sum_{u=1}^{m}\sum_{v=1}^{n} c_u\, d_{v,t_2}q_{t_2,u,v}}{\sqrt{\sum_{u=1}^{m} c_u^2 \sum_{v=1}^{n} d_{v,t_2}^2}}\,\rho_{t_1,t_2}$$

$$= K_{t_2}(f, g)\rho_{t_1,t_2}.$$

The converse is proved as above.

The time dependence of the proportionality constant in Theorem 2 above drops out, yielding the cross-correlation property of primary interest, in case it is only the stationary process X_1 which is nonlinearly distorted.

Theorem 3

If $X_1(t)$ is stationary, then the $(m, 1)$ cross-correlation property holds if and only if there is a m-vector $q = [q_v]m$ with $q_1 = 1$ such that the $m \times 1$ upper left vector b_{t_1,t_2} of A_{t_1,t_2} satisfies $b_{t_1,t_2} = a_{1,1,t_1,t_2}q$.

Proof: Since $X_1(t)$ is stationary, $c_{u,t_1} = c_u$ for $u = 1, \cdots m$. Since $X_2(t)$ is passed through at worst a linear device, $d_{v,t_2} = \delta_{v,1}$. Thus

$$\rho_{t_1,t_2}(f, g) = \frac{\sum_{u=1}^{m} c_u a_{u,1,t_1,t_2}}{\sqrt{\sum_{u=1}^{m} c_u^2}}.$$

If the vector equation holds, then

$$\rho_{t_1,t_2}(f, g) = \frac{\sum_{u=1}^{m} c_u q_u}{\sqrt{\sum_{u=1}^{m} c_u^2}}\,\rho_{t_1,t_2}.$$

The converse follows as in Theorem 1.

The results of Brown follow from Theorems 1 and 2 on setting $m = \infty$, $n = 1$.

Examples

1) The first example considered is where $X_1(t_1)$ and $X_2(t_2)$ have a joint Gaussian density

$$p_{t_1,t_2}(x_1, x_2) = \frac{1}{2\pi\sigma_{t_1}^{(1)}\sigma_{t_2}^{(2)}\sqrt{1-\rho_{t_1,t_2}^2}} \cdot \exp\left\{-\frac{1}{2(1-\rho_{t_1,t_2}^2)}\left[\frac{x_1^2}{\sigma_{t_1}^{(1)2}} - 2\rho_{t_1,t_2}\frac{x_1}{\sigma_{t_1}^{(1)}}\frac{x_2}{\sigma_{t_2}^{(2)}} + \frac{x_2^2}{\sigma_{t_2}^{(2)2}}\right]\right\}$$

with marginal densities

$$p_{t_1}(x_1) = \left(\frac{1}{2\pi\sigma_{t_1}^{(1)2}}\right)^{1/2}\exp\left\{-\frac{x_1^2}{2\sigma_{t_1}^{(1)2}}\right\}$$

$$p_{t_2}(x_2) = \left(\frac{1}{2\pi\sigma_{t_2}^{(2)2}}\right)^{1/2}\exp\left\{-\frac{x_2^2}{2\sigma_{t_2}^{(2)2}}\right\}.$$

The orthonormal polynomials are Hermite polynomials:

$$\theta_k^{(1)}(x_1) = \frac{1}{(2^k\cdot k!)^{1/2}}H_k\left(\frac{x_1}{\sigma_{t_1}^{(1)}\sqrt{2}}\right)$$

$$\theta_k^{(2)}(x_2) = \frac{1}{(2^k\cdot k!)^{1/2}}H_k\left(\frac{x_2}{\sigma_{t_2}^{(2)}\sqrt{2}}\right).$$

We then have

$$\frac{p_{t_1,t_2}(x_1, x_2)}{p_{t_1}(x_1)p_{t_2}(x_2)} = \frac{1}{\sqrt{1-\rho_{t_1,t_2}^2}} \cdot\exp\left\{-\frac{1}{1-\rho_{t_1,t_2}^2}\left[\rho_{t_1,t_2}^2\cdot\left(\frac{x_1^2}{2\sigma_{t_1}^{(1)2}} + \frac{x_2^2}{2\sigma_{t_2}^{(2)2}}\right) - 2\rho_{t_1,t_2}\frac{x_1x_2}{2\sigma_{t_1}^{(1)}\sigma_{t_2}^{(2)}}\right]\right\} = \sum_{n=0}^{\infty}\rho_{t_1,t_2}^n\theta_n^{(1)}(x_1)\theta_n^{(2)}(x_2).$$

Hence $a_{t_1,t_2,u,v} = \rho^{u}_{t_1,t_2}\delta_{u,v}$. It follows that if $X_1(t_1)$ and $X_2(t_2)$ are uncorrelated ($\rho_{t_1,t_2} = 0$), then so are $f(X_1(t_1))$ and $g(X_2(t_2))$ for all choices of f and g.

Choose $f(x) = g(x) = x^2$, so that

$$c_{0,t_1} = d_{0,t_2} = 1, \quad c_{1,t_1} = d_{0,t_2} = 0, \quad c_{2,t_1} = \sqrt{2}\sigma^{(1)^2}_{t_1},$$

$$d_{2,t_2} = \sqrt{2}\sigma^{(2)^2}_{t_2}, \quad c_{k,t_1} = d_{k,t_2} = 0 \quad \text{for} \quad k = 3, 4, \cdots .$$

We thus have

$$\rho_{t_1,t_2}(f, g) = 2\sigma^{(1)^2}_{t_1}\sigma^{(2)^2}_{t_2}\rho^2_{t_1,t_2}.$$

If $X_1(t_1)$ is stationary, then

$$\rho_{t_1,t_2}(f, g) = 2\sigma^{(1)^2}\sigma^{(2)^2}_{t_2}\rho^2_{t_1,t_2}.$$

If also $X_2(t_2)$ is stationary, then

$$\rho_{t_1,t_2}(f, g) = 2\sigma^{(1)^2}\sigma^{(2)^2}\rho^2_{t_1,t_2}.$$

The 2 × 2 matrix of a's is

$$\begin{bmatrix} \rho_{t_1,t_2} & 0 \\ 0 & \rho^2_{t_1,t_2} \end{bmatrix}.$$

Clearly the matrix equations and (2,2) cross-correlation conditions both reduce to $\rho_{t_1,t_2} = k$ and $\rho_{t_1,t_2} = k_{t_2}$ in the cases taken up in Theorems 1 and 2. It follows that if the cross-correlation is nonconstant in the doubly stationary case, the (m, n) property fails if $m \geqq 2$ and $n \geqq 2$.

2) The second example, or family of examples, is constructed with the help of Legendre polynomials. We find distributions satisfying: a) the (m, n) property but not the $(m, n + 1)$ or $(m + 1, n)$ properties, and b) the (∞, ∞) property.

It must be admitted that these examples are rather artificial.

Let $\{s_k(x)\}$ be the polynomials orthonormal on [0, 1] with respect to the rectangular distribution (unit weight function). Thus[5]

$$s_k(x) = \sqrt{2k+1}\,\frac{(2k)!}{2^k(k!)^2}P_k(2x - 1),$$

where the $P_k(x)$ are Legendre polynomials.

Let $T_k(y) = \int_0^y s_k(x)dx$. Since $s_0(x) = 1$, we have $T_0(y) = y$.

Clearly $T_k(0) = 0$ for all k. Moreover, $T_k(1) = \int_0^1 s_k(x)dx = \int_0^1 s_k(x)s_0(x)dx = \delta_{k0}$, so that $T_k(1) = 0$ for $k \geqq 1$. Since $s_k(x)$ is a polynomial, $T_k(y)$ is a polynomial, and we can write $T_k(y) = y(1 - y)\,U_k(y)$, where $U_k(y)$ is a polynomial of degree $k - 1$ for $k \geqq 1$.

Now set

$$F(x_1, x_2) = \sum_{i,j=0}^{\infty} b_{i,j}T_i(x_1)T_j(x_2),$$

$$\text{where} \quad b_{0,0} = 1, \qquad b_{i,0} = b_{0,i} = 0, \qquad i \geqq 1,$$

and (x_1, x_2) is in the unit square.

[5] Follows from the normalization found in A. Erdelyi, "Higher Transcendental Functions," McGraw-Hill Book Co., Inc., New York, N. Y., vol. 2, p. 179; 1953.

We wish to find conditions for $F(x_1, x_2)$ to be a joint cumulative distribution on the unit square. Note that $F(x_1, 0) = F(0, x_2) = 0$, and that

$$F(x_1, 1) = \sum_{i=0}^{\infty} b_{i0}T_i(x_1) = T_0(x_1) = x_1, \qquad F(1, x_2) = x_2.$$

Thus the marginal cumulative distributions $F_1(x_1)$ and $F_2(x_2)$ are rectangular, with densities $f_1(x_1) = f_2(x_2) = 1$. It remains only to show that $F(x_1, x_2)$ is jointly monotone in both variables on the unit square. If the joint density $f(x_1, x_2) = \partial^2F/\partial_{x_1}\partial_{x_2}$ exists in the interior of the unit square, the monotone character of $F(x_1, x_2)$ is insured by the nonnegativity of $f(x_1, x_2)$. Observe that

$$\frac{f(x_1, x_2)}{f_1(x_1)f_2(x_2)} = f(x_1, x_2) = 1 + \sum_{i,j=1}^{\infty} b_{ij}s_i(x_1)s_j(x_2).$$

Because of the alternating character of Legendre polynomials, not every choice of (b_{ij}) will make $f(x_1, x_2) \geqq 0$, so that a certain discretion must be exercised. However, a wide latitude is available. The classical inequality[6] $|P_k(x)| \leqq 1$ for $-1 \leqq x \leqq 1$ shows that $|s_k(x)| \leqq \sqrt{2k+1}\,(2k)!/2^k\cdot(k!)^2$ for $k \geqq 0$, $0 \leqq x \leqq 1$. Hence it is sufficient to so choose the b_{ij} that

$$1 + \sum_{i,j=1}^{\infty} \pm\, b_{ij}\sqrt{(2i+1)(2j+1)}\cdot\frac{(2i)!(2j)!}{2^{i+j}(i!)^2(j!)^2} \geqq 0$$

for all choices of sign. This crude condition suffices for the construction of the first example.

a) Let

$$b_{1,1} = \rho_{t_1,t_2} = \rho_{t_2-t_1}$$

and let

$$b_{m+1,n+1} = \alpha_{t_1,t_2} = \alpha_{t_2-t_1}$$

where $\rho_{t_2-t_1}$ and $\alpha_{t_2-t_1}$ are functions such that

$$\frac{\rho_{t_2-t_1}}{\alpha_{t_2-t_1}}$$

is nonconstant and

$$1 \pm 3\rho_{t_2-t_1} \pm \alpha_{t_2-t_1}\sqrt{(2m+3)(2n+3)} \cdot\frac{(2m+2)!(2n+2)!}{((m+1)!(n+1)!)^2 2^{m+n+2}}$$

is nonnegative for all four choices of sign. We set all the other $b_{ij} = 0$ for $i, j \geqq 1$. It follows from Theorem 1 that the (m, n) cross-correlation property holds, while the $(m + 1, n)$ and $(m, n + 1)$ properties fail.

The other type of example, in the opposite direction, is more elegant.

b) We begin with the little-known identity[7]

$$\sum_{k=1}^{\infty}\frac{2k+1}{k(k+1)}P_k(x)P_k(y) = -1 + 2\log 2 - \log\left[(1 - \min(x, y))(1 + \max(x, y))\right]$$

[6] *Ibid.*, p. 205.
[7] *Ibid*, p. 183.

valid for (x, y) in $[-1, 1] \times [-1, 1]$. Shifted to the unit square, this becomes, after some manipulation,

$$\sum_{k=1}^{\infty} \frac{2^{2k}(k!)^4}{k(k+1)[(2k)!]^2} s_k(x_1)s_k(x_2) = -1 - \log G(x_1, x_2),$$

where G is the Green's kernel

$$G(x_1, x_2) = \begin{cases} x_1(1 - x_2), & x_1 \geqq x_2 \\ x_2(1 - x_1), & x_1 \leqq x_2 \end{cases}$$

Since $0 \leqq G(x_1, x_2) \leqq 1$ in the unit square, $-\log G(x_1, x_2) \geqq 0$.

Now choose as coefficients

$$b_{i,j} = \rho_{t_2-t_1} \frac{2^{2i+1}(i!)^4}{i(i+1)((2i)!)^2} \delta_{i,j} \quad \text{for} \quad i, j \geqq 1.$$

Thus we have

$$\frac{f(x_1, x_2)}{f_1(x_1)f_2(x_2)} = f(x_1, x_2) = 1 + \sum_{i,j=1}^{\infty} b_{ij}s_i(x_1)s_j(x_2)$$

$$= 1 - 2\rho_{t_2-t_1} + 2\rho_{t_2-t_1}[-\log G(x_1, x_2)].$$

If the cross-correlation $\rho_{t_2-t_1}$ satisfies the inequalities $0 \leqq \rho_{t_2-t_1} \leqq \frac{1}{2}$ for all $t_2 - t_1$, then $f(x_1, x_2) \geqq 0$, and we have a legitimate distribution. (If $\rho_{t_2-t_1} = 0$, the example reduces to independent rectangular distributions). Since the conditions of Theorem 1 are satisfied for all m and all n, the above joint distribution has the (∞, ∞) cross-correlation property. In other words, each of two processes with the above joint distribution can be distorted through arbitrary instantaneous nonlinear devices without altering the time dependence of the cross correlation.

Appendix

We now derive the Lemma. Let

$$c_{u,t_1} = \int \theta_{u,t_1}^{(1)}(x_1)f(x_1)p_{1,t_1}(x_1)\,dx_1,\ d_{v,t_2}$$

$$= \int \theta_{v,t_2}^{(2)}(x_2)g(x_2)p_{2,t_2}(x_2)\,dx_2,$$

where f, g are polynomials of degrees m, n. Since $\theta_{0,t}^{(1)}(x) = \theta_{0,t}^{(2)}(x) = 1$, we have

$$c_{0,t_1} = \int f(x_1)p_{1,t_1}(x_1)\,dx_1 = E[f(X_1(t_1))],\ d_{0,t_2}$$

$$= \int g(x_2)p_{2,t_2}(x_2)\,dx_2 = E[g(X_2(t_2))].$$

Hence

$$f(x_1) - E[f(X_1(t_1))]$$

$$= \sum_{u=1}^{m} c_{u,t_1}\theta_{u,t_1}^{(1)}(x_1),\ g(x_2) - E[g(X_2(t_2))]$$

$$= \sum_{v=1}^{n} d_{v,t_2}\theta_{v,t_2}^{(2)}(x_2),$$

and

$$E[(f(X_1(t_1)) - E[f(X_1(t_1))])(g(X_2(t_2)) - E[g(X_2(t_2))])]$$

$$= \iint p_{t_1,t_2}(x_1, x_2) \sum_{u=1}^{m}\sum_{v=1}^{n} c_{u,t_1}^{(1)}(x_1)\, d_{v,t_2}\theta_{v,t_2}^{(2)}(x_2)\,dx_1\,dx_2$$

$$= \sum_{u=1}^{m}\sum_{v=1}^{n} c_{u,t_1} d_{v,t_2} \iint \sum_{i,j} a_{i,j,t_1,t_2}p_{t_1}(x_1)p_{t_2}(x_2)$$

$$\cdot\,\theta_{i,t_1}^{(1)}(x_1)\theta_{j,t_2}^{(2)}(x_2)\theta_{u,t_1}^{(1)}(x_1)\theta_{v,t_2}^{(2)}(x_2)\,dx_1\,dx_2$$

$$= \sum_{u=1}^{m}\sum_{v=1}^{n}\sum_{i,j} c_{u,t_1} d_{v,t_2}a_{i,j,t_1,t_2}\,\delta_{iu}\,\delta_{jv}$$

$$= \sum_{u=1}^{m}\sum_{v=1}^{n} c_{u,t_1} d_{v,t_2}a_{u,v,t_1,t_2}$$

on using the biorthonormal expansion for $p_{t_1,t_2}(x_1, x_2)$ and taking account of the orthonormality of $\{\theta_{m,t_1}^{(1)}(x_1)\}$ and $\{\theta_{m,t_2}^{(2)}(x_2)\}$, Similarly

$$E[(f(X_1(t_1)) - E[f(X_1(t_1))])^2]$$

$$= \int p_{t_1}(x_1)\left(\sum_{u=1}^{m} c_{u,t_1}\theta_{u,t_1}^{(1)}(x_1)\right)^2 dx_1$$

$$= \sum_{u=1}^{m}\sum_{v=1}^{n} c_{u,t_1}c_{v,t_1} \int p_{t_1}(x_1)\theta_{u,t_1}^{(1)}(x_1)\theta_{v,t_1}^{(1)}(x_1)\,dx_1$$

$$= \sum_{u=1}^{m} c_{u,t_1}^2$$

and

$$E[g(X_2(t_2)) - E[g(X_2(t_2))])^2] = \sum_{v=1}^{n} d_{v,t_2}^2$$

from which the desired result follows.

4

Reprinted with permission of the publisher from *Proc. Amer. Math. Soc. Symp. Appl. Math.*, **16**, 264–276 (1963)

THE CONSTRUCTION OF A CLASS OF STATIONARY MARKOFF PROCESSES

BY

EUGENE WONG

Introduction. We take a forward, or Fokker–Planck, equation of diffusion theory

$$\frac{\partial^2}{\partial x^2}[B(x)p] - \frac{\partial}{\partial x}[A(x)p] = \frac{\partial p}{\partial t}, \qquad p = p(x|x_0, t), \qquad 0 < t < \infty, \tag{1}$$

on an interval I with end points x_1 and x_2, and consider the solution $p(x|x_0, t)$ corresponding to an initial value $\delta(x - x_0)$, i.e., the principal solution. The function $B(x)$ can be interpreted as a variance **[15]**, and is therefore non-negative on (x_1, x_2). In general, we take, as boundary conditions, those corresponding to reflecting barriers **[9]**,

$$\frac{\partial}{\partial x}[B(x)p] - A(x)p = 0, \qquad x = x_1, x_2. \tag{2}$$

The unique solution $p(x|x_0, t)$ is the density function for the transitional probability of a stationary Markoff process $X(t)$ in one dimension,

$$\Pr\{X(t + \tau) \in E | X(\tau) = x_0\} = \int_E p(x|x_0, t)\, dx, \qquad x_0 \in I, \qquad E \subseteq I. \tag{3}$$

Next, we consider the class of first-order probability density functions $W(x)$ which satisfy the Pearson equation **[4]**

$$\frac{dW(x)}{dx} = \frac{ax + b}{cx^2 + dx + e} W(x). \tag{4}$$

In §1 we shall show that by suitably identifying the functions $A(x)$ and $B(x)$ in (1) with the polynomials $ax + b$ and $cx^2 + dx + e$ in (4), a class of stationary Markoff processes is constructed for which

$$\lim_{t\to\infty} p(x|x_0, t) = \int_{x_1}^{x_2} W(x_0)p(x|x_0, t)\, dx_0 = W(x), \tag{5}$$

where $W(x)$ satisfies (4). The identification scheme is such that the Pearson equation (4) uniquely specifies the Fokker–Planck equation (1). In a sense, the Pearson equation serves a rather natural role in bridging the gap between the first-order statistical description characterized by $W(x)$ and the transitional properties represented by $p(x|x_0, t)$.

In §2 we shall exhibit some specific processes of this class. These examples include several well-known ones, as well as some that appear not to have been investigated previously.

In §3 we shall examine some properties of processes of this class and give some physical interpretation to processes in §2. Finally, the distribution of some functionals involving processes of this class are discussed in §4.

1. **Construction.** We note that a straightforward application of separations of variables to (1) yields an equation of the Sturm–Liouville type

$$\frac{d}{dx}\left[B(x)\rho(x)\frac{d\varphi(x)}{dx}\right] + \lambda\rho(x)\varphi(x) = 0. \tag{6}$$

The boundary conditions (2) imply that

$$B(x)\rho(x)\frac{d\varphi(x)}{dx} = 0, \qquad x = x_1, x_2. \tag{7}$$

The function $\rho(x)$ is taken to be non-negative on (x_1, x_2) and satisfies

$$\frac{d}{dx}[B(x)\rho(x)] - A(x)\rho(x) = 0. \tag{8}$$

A comparison of (8) and (4) shows that they become identical if we make the following identifications:

$$\rho(x) = W(x), \tag{9}$$

$$B(x) = \beta(cx^2 + dx + e) \tag{10}$$

and

$$A(x) = \frac{dB(x)}{dx} + \beta(ax + b), \tag{11}$$

subject to the condition that $B(x)$ be non-negative on (x_1, x_2). It is clear that through (10) and (11), $A(x)$ and $B(x)$ are uniquely specified by the Pearson equation (4) up to a common positive multiplicative constant β, which represents a scaling factor in the variable t.

With the substitution of $W(x)$ for $\rho(x)$, (6) and (7) become

$$\frac{d}{dx}\left[B(x)W(x)\frac{d\varphi(x)}{dx}\right] + \lambda W(x)\varphi(x) = 0, \tag{12}$$

and

$$B(x)W(x)\frac{d\varphi(x)}{dx} = 0, \qquad x = x_1, x_2. \tag{13}$$

From the classical Sturm–Liouville theory it is known that the spectrum of (12) is discrete if x_1 and x_2 are finite. If one or both of the boundaries are infinite, then a continuous range of eigenvalues may be present. In terms of the solutions of (12) the principal solution $p(x|x_0, t)$ of (1) can be written as

$$p(x|x_0, t) = W(x)\left\{\sum_n e^{-\lambda_n t}\varphi_n(x_0)\varphi_n(x) + \int e^{-\lambda t}\varphi(\lambda, x_0)\varphi(\lambda, x)\, d\lambda\right\}, \tag{14}$$

where the summation is taken over all discrete eigenvalues and the integral is taken over the continuous range of eigenvalues. The eigenfunctions $\varphi(x)$ in (14) are assumed to be normalized so that corresponding to discrete eigenvalues we have

$$\int_{x_1}^{x_2} W(x)\varphi_m(x)\varphi_n(x) = \delta_{mn}, \tag{15}$$

and corresponding to continuous eigenvalues

$$\int_{x_1}^{x_2} W(x)\varphi(\lambda, x)\varphi(\lambda', x)\, dx = \delta(\lambda - \lambda'). \tag{16}$$

It can be verified directly that corresponding to boundary conditions (13), the Sturm–Liouville equation (12) has at least one discrete eigenvalue, namely, $\lambda = 0$ with corresponding eigenfunction $\varphi(x) = 1$. In addition, if we assume that the spectrum is discrete, then

$$\begin{aligned} \lambda_n &= -\int_{x_1}^{x_2} \varphi_n(x)\frac{d}{dx}\left[B(x)W(x)\frac{d\varphi_n(x)}{dx}\right] dx \\ &= \int_{x_1}^{x_2} B(x)W(x)\left[\frac{d\varphi_n(x)}{dx}\right]^2 dx, \end{aligned} \tag{17}$$

where we have made use of the boundary conditions (13). Since $B(x)W(x)[d\varphi(x)/dx]^2$ is non-negative on (x_1, x_2), it follows from (17) that

$$\lambda_n \geqq 0. \tag{18}$$

Furthermore, the case where a continuous range of eigenvalues is present can be arrived at by taking cases with finite boundaries, thus with discrete spectra, and considering the limiting situation when one or both of the boundaries becomes infinite. It is clear that, in general, we have

$$\lambda \geqq 0. \tag{19}$$

Further, even in the limiting cases $\lambda = 0$ must remain a discrete eigenvalue, since the corresponding eigenfunction $\varphi = 1$ is always square-integrable with respect to the density function $W(x)$. It follows, from the fact that $\lambda = 0$, $\varphi = 1$ is always a discrete solution, that

$$\int_{x_1}^{x_2} p(x|x_0, t)W(x_0)\, dx_0 = W(x). \tag{20}$$

Further, from the fact that $\lambda = 0$ is also the minimum eigenvalue, it follows that

$$\lim_{t\to\infty} p(x|x_0, t) = W(x). \tag{21}$$

The construction procedure is seen to consist of two parts. First, starting with a first-order probability density function $W(x)$ and a corresponding Pearson equation, we identify the functions $A(x)$ and $B(x)$ in (1) with the polynomials in the Pearson equation according to (10) and (11). Secondly, the density function $p(x|x_0, t)$ of the transitional probability is obtained as the principal solution to (1). (For a brief discussion along similar lines, see Kolmogorov [**13**].)

2. **Some specific processes.** If we consider the roots of the equation $B(x) = 0$, five distinct situations are possible, namely: (a) no root, in which case $B(x)$ is a constant, (b) a single real root, (c) two unequal real roots, (d) two equal real roots, and (e) a pair of complex-conjugate roots. In this section we shall consider six specific processes which, while by no means exhausting all the possibilities, include at least one example of each of the five situations corresponding to the roots of $B(x) = 0$. In every case except one, the choice of boundaries will be a natural one in that each boundary is either at a real root of "$B(x) = 0$" or is infinite. Without loss of generality we shall set $\beta = 1$, thus normalizing the time scale.

A. Consider the first-order density function

$$W(x) = e^{-x}, \qquad 0 \leqq x < \infty, \tag{22}$$

with a corresponding Pearson equation

$$\frac{dW(x)}{dx} = -\frac{1}{1}W(x).$$

The resulting Fokker–Planck equation is

$$\frac{\partial^2 p}{\partial x^2} + \frac{\partial p}{\partial x} = \frac{\partial p}{\partial t}$$

with boundary condition

$$\frac{d}{dx}p(x|x_0, t) = 0, \qquad x = 0.$$

The principal solution $p(x|x_0, t)$ is given by

$$p(x|x_0, t) = e^{-x}\left\{1 + e^{-t/4}\int_0^\infty e^{-\mu^2 t}\psi(\mu, x_0)\psi(\mu, x)\,d\mu\right\}, \tag{23}$$

where

$$\psi(\mu, x) = \sqrt{\frac{2}{\pi}}\frac{2\mu}{\sqrt{1 + 4\mu^2}}\left[\cos \mu x - \frac{1}{2\mu}\sin \mu x\right]e^{x/2}.$$

The function $p(x|x_0, t)$ can also be written as

$$\begin{aligned} p(x|x_0, t) = \frac{1}{2\sqrt{\pi t}}\exp\left[-\tfrac{1}{2}(x - x_0)\right]\exp\left[-\tfrac{1}{4}t\right]\left\{\exp\left[-\frac{(x - x_0)^2}{4t}\right]\right. \\ \left. + \exp\left[-\frac{(x + x_0)^2}{4t}\right]\right\} + \frac{1}{\sqrt{\pi}}e^{-x}\int_{(x+x_0-t)/2\sqrt{t}}^\infty e^{-z^2}\,dz. \end{aligned} \tag{24}$$

B. Let the first-order density function be given by

$$W(x) = \frac{1}{\sqrt{2\pi}} \exp\left[-\tfrac{1}{2}x^2\right], \qquad -\infty < x < \infty, \tag{25}$$

with an associated Pearson equation

$$\frac{dW(x)}{dx} = -\frac{x}{1} W(x).$$

The resulting Fokker–Planck equation can be written as

$$\frac{\partial^2 p}{\partial x^2} + \frac{\partial}{\partial x}(xp) = \frac{\partial p}{\partial t}.$$

The density function $p(x|x_0, t)$ is given by

$$p(x|x_0, t) = \frac{1}{\sqrt{2\pi}} e^{-x^2/2} \sum_{n=0}^{\infty} \frac{e^{-nt}}{n!} H_n(x_0)H_n(x), \tag{26}$$

where $H_n(x)$ are the Hermite polynomials **[14]**

$$H_n(x) = (-1)^n e^{-x^2/2} \frac{d^n}{dx^n}(e^{-x^2/2}).$$

By the use of Mehler's formula we can write $p(x|x_0, t)$ as

$$p(x|x_0, t) = \frac{1}{\sqrt{2\pi(1 - e^{-2t})}} \exp\left[-\frac{1}{2(1 - e^{-2t})}(x - x_0 e^{-t})^2\right]. \tag{27}$$

C. Consider the first-order density function

$$W(x) = \frac{x^\alpha}{\Gamma(\alpha + 1)} e^{-x}, \qquad \alpha > -1, 0 \leqq x < \infty \tag{28}$$

with corresponding Pearson equation

$$\frac{dW(x)}{dx} = \left(\frac{\alpha - x}{x}\right) W(x).$$

The associated Fokker–Planck equation is

$$\frac{\partial^2}{\partial x^2}(xp) - \frac{\partial}{\partial x}[(\alpha + 1 - x)p] = \frac{\partial p}{\partial t}.$$

The principal solution $p(x|x_0, t)$ is given by

$$p(x|x_0, t) = x^\alpha e^{-x} \sum_{n=0}^{\infty} \frac{n!}{\Gamma(n + \alpha + 1)} e^{-n\beta t} L_n^\alpha(x_0)L_n^\alpha(x), \tag{29}$$

where

$$L_n^\alpha(x) = \frac{1}{n!} x^{-\alpha} e^{x} \frac{d^n}{dx^n}(x^\alpha e^{-x}),$$

are the Laguerre polynomials **[14]**. The function $p(x|x_0, t)$ can also be written as

$$p(x|x_0, t) = \frac{1}{1 - e^{-t}} \left(\frac{x}{x_0 \exp[-t]}\right)^{\alpha/2} \cdot \exp\left[-\frac{1}{(1 - e^{-t})}(x + x_0 e^{-t})\right] I_\alpha\left(\frac{2\, e^{-t/2}\sqrt{x_0 x}}{1 - \exp[-t]}\right), \tag{30}$$

where $I_\alpha(z)$ is the modified Bessel function.

D. Corresponding to the situation when "$B(x) = 0$" has two unequal real roots, we take $W(x)$ to be

$$W(x) = \frac{\Gamma(\alpha + \gamma + 2)}{\Gamma(\alpha + 1)\Gamma(\gamma + 1)} \frac{(1 + x)^\alpha (1 - x)^\gamma}{2^{\alpha+\gamma+1}}, \quad \alpha, \gamma \geqq -1, \ -1 \leqq x \leqq +1, \tag{31}$$

with associated Pearson equation

$$\frac{dW(x)}{dx} = \frac{(\alpha - \gamma) - (\alpha + \gamma)x}{(1 - x^2)} W(x).$$

The resulting Fokker–Planck equation becomes

$$\frac{\partial^2}{\partial x^2}[(1 - x^2)p] + (\alpha + \gamma + 2)\frac{\partial}{\partial x}[xp] - (\alpha - \gamma)\frac{\partial p}{\partial x} = \frac{\partial p}{\partial t}.$$

The solution $p(x|x_0, t)$ is given by

$$p(x|x_0, t) = \frac{(1 + x)^\alpha (1 - x)^\gamma}{2^{\alpha+\gamma+1}} \sum_{n=0}^{\infty} e^{-n(n+\alpha+\gamma+1)t} A_n P_n^{\alpha,\gamma}(x_0) P_n^{\alpha,\gamma}(x), \tag{32}$$

where $P_n^{\alpha,\gamma}(x)$ are the Jacobi polynomials[1] **[14]**

$$P_n^{\alpha,\gamma}(x) = \frac{(-1)^n}{2^n}(1 + x)^\alpha (1 - x)^\gamma \frac{d^n}{dx^n}[(1 + x)^{\alpha+n}(1 - x)^{\gamma+n}],$$

and

$$A_n = \frac{(2n + \alpha + \gamma + 1)\Gamma(n + \alpha + \gamma + 1)}{\Gamma(n + \alpha + 1)\Gamma(n + \gamma + 1)n!}.$$

E. Corresponding to a pair of complex conjugate-roots of "$B(x) = 0$," we take

$$W(x) = \frac{\Gamma(\alpha + \frac{1}{2})}{\Gamma(\frac{1}{2})\Gamma(\alpha)}(1 + x^2)^{-(\alpha+1/2)}, \qquad \alpha > 0, -\infty < x < \infty, \tag{33}$$

and associate with it a Pearson equation

$$\frac{dW(x)}{dx} = -\frac{(2\alpha + 1)x}{(1 + x^2)} W(x).$$

[1] The normalization for $P_n^{\alpha,\gamma}$ here is not the conventional one.

The Fokker–Planck equation for this case becomes

$$\frac{d^2}{dx^2}[(1 + x^2)p] + (2\alpha - 1)\frac{d}{dx}(xp) = \frac{\partial p}{\partial t}.$$

The Sturm–Liouville equation (12) in this case has $N + 1$ discrete eigenvalues ($\alpha - 1 \leqq N < \alpha$), and a continuous range of eigenvalues. More precisely, we have

$$\lambda_n = n(2\alpha - n), \qquad n = 0, 1, 2, \cdots, N,$$

and

$$\lambda = \alpha^2 + \mu^2, \qquad \mu \geqq 0.$$

The solution $p(x|x_0, t)$ can be written as

$$\begin{aligned} p(x|x_0, t) = (1 + x^2)^{-(\alpha+1/2)} \Bigg\{ &\frac{1}{\pi} \sum_{n=0}^{N} \frac{(\alpha - n)}{n!\Gamma(2\alpha + 1 - n)} e^{-n(2\alpha-n)t} \theta_n(x_0)\theta_n(x) \\ &+ \frac{1}{2\pi}\int_0^\infty e^{-(\alpha^2+\mu^2)t}[\psi(\mu, x_0)\psi(-\mu, x) + \psi(-\mu, x_0)\psi(\mu, x)]\, d\mu \Bigg\}, \end{aligned} \tag{34}$$

where

$$\theta_n(x) = 2^{\alpha-n}\Gamma(\alpha - n + \tfrac{1}{2})(-1)^n(1 + x^2)^{\alpha+1/2}\frac{d^n}{dx^n}[(1 + x^2)^{n-\alpha-1/2}], \tag{35}$$

are polynomials of degree n, and $\psi(\mu, x)$ is given by

$$\psi(\mu, x) = (x + \sqrt{1 + x^2})^{i\mu}(1 + x^2)^{1/2} {}_2F_1\left(-\alpha, \alpha + 1; 1 + i\mu; \frac{1}{2} + \frac{1}{2}\frac{x}{\sqrt{1 + x^2}}\right),$$

${}_2F_1$ being the Gauss hypergeometric series [7].

For $\alpha = K$, a positive integer, $p(x|x_0, t)$ can be written somewhat more explicitly as

$$\begin{aligned} p(x|x_0, t) = \frac{1}{(1 + x^2)^{K+1/2}} \Bigg\{ &[(1 + x_0^2)(1 + x^2)]^{K/2} \frac{1}{2\sqrt{\pi t}} e^{-K^2 t} e^{-u^2} \\ &+ \frac{1}{\pi}\sum_{n=0}^{K-1} \frac{(K - n)}{n!\Gamma(2K + 1 - n)} e^{-n(2K-n)t}\theta_n(x_0)\theta_n(x) f_n(x_0, x, t) \Bigg\}, \end{aligned} \tag{36}$$

where

$$u = u(x_0, x, t) = \frac{\text{arc sinh } x - \text{arc sinh } x_0}{2\sqrt{t}},$$

$$f_n(x_0, x, t) = \frac{1}{\sqrt{\pi}} \int_{u-(K-n)\sqrt{t}}^{u+(K-n)\sqrt{t}} e^{-z^2}\, dz,$$

and θ_n are polynomials defined by (35).

F. Corresponding to the situation where "$B(x) = 0$" has a double real root, we consider the first-order density function

$$W(x) = \frac{1}{\Gamma(2\alpha)} x^{-(2\alpha+1)} e^{-1/x}, \qquad \alpha > 0,\ 0 \leqq x < \infty, \tag{37}$$

with corresponding Pearson equation

$$\frac{dW(x)}{dx} = \frac{1 - (2\alpha + 1)x}{x^2} W(x).$$

The resulting Fokker–Planck equation is given by

$$\frac{\partial^2}{\partial x^2}[x^2 p] - \frac{\partial}{\partial x}\{[1 - (2\alpha - 1)x]p\} = \frac{\partial p}{\partial t}.$$

The solution of the Sturm–Liouville equation for this case consists of $N + 1$ discrete eigenvalues ($\alpha - 1 \leqq N < \alpha$), and a continuous range of eigenvalues:

$$\lambda_n = n(2\alpha - n), \qquad n = 0, 1, \cdots, N,$$
$$\lambda = \alpha^2 + \mu^2, \qquad \mu \geqq 0.$$

The transitional probability density function $p(x|x_0, t)$ is found to be

$$\begin{aligned} p(x|x_0, t) = x^{-(2\alpha+1)} e^{-1/x} \Bigg\{ &\sum_{n=0}^{N} \frac{2(\alpha - n)}{\Gamma(2\alpha + 1 - n)} \frac{1}{n!} e^{-n(2\alpha-n)t} \theta_n(x_0)\theta_n(x) \\ &+ \frac{1}{2\pi} \int_0^\infty e^{-(\alpha^2+\mu^2)t} A(\mu)\psi(\mu, x_0)\psi(\mu, x)\, d\mu \Bigg\}. \end{aligned} \tag{38}$$

Here, $\theta_n(x)$ are orthogonal polynomials of degree n,

$$\theta_n(x) = (-1)^n x^{2\alpha+1} e^{1/x} \frac{d^n}{dx^n}(x^{2n-2\alpha-1} e^{-1/x}),$$

and

$$\psi(\mu, x) = {}_2F_0(-\alpha - i\mu, -\alpha + i\mu, -x),$$

${}_2F_0$ being the generalized hypergeometric series [7]. Alternative representations of $\psi(\mu, x)$ in terms of confluent hypergeometric functions follow from properties of ${}_2F_0$, e.g.,

$${}_2F_0(-\alpha - i\mu, -\alpha + i\mu, -x) = x^{(\alpha+i\mu)}\Psi\left(-\frac{\alpha}{2} - i\mu, 1 - 2i\mu, \frac{1}{x}\right),$$

where Ψ is the hypergeometric Ψ-function [7]. The quantity $A(\mu)$ is a normalization factor given by

$$A(\mu) = \frac{\Gamma(-\alpha + i\mu)\Gamma(-\alpha - i\mu)}{\Gamma(i\mu)\Gamma(-i\mu)}.$$

3. **Physical interpretation and properties.** The processes outlined in (A) through (D) of §2 are familiar processes with some well-known interpretations. The process of (2 A) represents Brownian motion in a constant force-field (e.g., a gravitational field) with a reflecting barrier [**3**; **11**]. The process of (2 B) is the Ornstein–Uhlenbeck process.

The process of (2 C) has been studied in connection with population growth [**8**]. For $\alpha = N/2 - 1$, $N = 1, 2, \cdots$, it can also be considered as the sum of the squares of N independent and identical Ornstein–Uhlenbeck processes, or alternatively as the square of the radial component of a Brownian motion in N dimensions with radial restoring force proportional to the distance from the origin [**12**].

The process of (2 D), whose transitional density function involves Jacobi polynomials, has been studied by Bochner, and Karlin and McGregor [**2**; **12**]. For special values of the parameters α and γ, $\alpha = \gamma = (N - 3)/2$, $N = 2, 3, \cdots$, a geometric interpretation of the Jacobi diffusion process in terms of Brownian motion on a unit sphere in N dimensions has been given by Karlin and McGregor. It has also been used in connection with biological applications (see Karlin and McGregor).

The process of (2 E) and (2 F) appear to be new. It is of some interest to note that the density function $W(x)$ in (2 E) and (2 F), for $\alpha = \frac{1}{2}$ and $\alpha = \frac{1}{4}$ respectively, become $1/\pi(1 + x^2)$ and $(1/\sqrt{\pi})x^{-3/2}\, e^{-1/x}$. Both of these are known to be density functions for stable distributions [**10**].

The process $X(t)$ of (2 E) can be interpreted in terms of a Langevin type stochastic differential equation. Specifically, if we make the transformation $Y(t) = \sinh X(t)$, then $Y(t)$ is the solution of a differential equation [**6**]

$$dY(t) + 2\alpha \tanh Y(t)\, dt = dU(t),$$

where $U(t)$ is a Brownian motion process. The term $2\alpha \tanh Y(t)$ is interesting from the point of view of application, since it approximates a saturating linear element (e.g., a saturating linear amplifier).

For $\alpha = K$ an integer, the form of $p(x|x_0, t)$ given by (36) suggests that a simple interpretation of the process (2 E) in terms of Brownian motion should be possible. However, no such interpretation has yet been found.

The process $X(t)$ of (2 F) can also be interpreted in terms of a differential equation

$$dY + [2\alpha - \exp(-Y)]\, dt = dU(t),$$

where $Y(t) = \ln X(t)$ and $U(t)$ is a Brownian motion process.

A feature of processes of this class is that the transitional probability density function can always be represented in terms of relatively simple classical orthogonal functions. This is due to the fact that $B(x)$ and $A(x)$ are chosen to be polynomials of second and first degrees, thus simplifying the resulting Sturm–Liouville equation. In particular, we note the presence of classical orthogonal polynomials. The transitional probability density function $p(x|x_0, t)$ given by (26), (29) and (32) have the following common form of representation:

$$p(x|x_0, t) = W(x)\sum_{n=0}^{\infty} e^{-\lambda_n t}\theta_n(x_0)\theta_n(x), \tag{39}$$

where $\theta_n(x)$ are normalized orthogonal polynomials. Representations of second-order probability density functions as a single sum of orthogonal polynomials were studied by Barrett and Lampard [**1**], where they discussed the properties and application of such density functions. It has been shown [**16**] that if $p(x|x_0, t)$ satisfies a Fokker–Planck equation of the form given by (1), and if

$$\lim_{t\to\infty} p(x|x_0, t) = W(x),$$

then $p(x|x_0, t)$ has the representation given by (39) if and only if

(1) $B(x) = ax^2 + bx + c$, $A(x) = dx + e$,
(2) $B(x_1)W(x_1) = B(x_2)W(x_2) = 0$,
(3) $\int_{x_1}^{x_2} W(x)x^n\,dx < \infty$, $x = 0, 1, 2, \cdots$.

In effect, (2 B), (2 C) and (2 D) exhaust all such possibilities.

A property, satisfied by most of the processes discussed in §2, is that the normalized covariance function is an exponential function of t, i.e.,

$$E\left(\frac{X(t+\tau)-m}{\sigma}\right)\left(\frac{X(\tau)-m}{\sigma}\right) = e^{-\beta t}; \qquad \beta > 0,\ t \geqq 0, \tag{40}$$

where m and σ^2 are the mean and variance respectively. This property has long been known for Ornstein–Uhlenbeck process, and is closely connected with the question of representing $p(x|x_0, t)$ in terms of orthogonal polynomials. It can be shown that if the transitional probability density function $p(x|x_0, t)$ of a Markoff process $X(t)$ satisfies (1) and if $\lim_{t\to\infty} p(x|x_0, t) = W(x)$, then (40) is satisfied if and only if

(1) $B(x) = ax^2 + bx + c$, $A(x) = dx + e$,
(2) $B(x_1)W(x_1) = B(x_2)W(x_2) = 0$,
(3) $\int_{x_1}^{x_2} W(x)x^n\,dx < \infty$, $n = 0, 1, 2$.

In addition to (2 B), (2 C) and (2 D), the process of (2 E) for $\alpha > 1$, and the process of (2 F) for $\alpha > 1$, also satisfy (40). In each of these two cases the representation of $p(x|x_0, t)$ is in part in terms of a sum of orthogonal polynomials.

4. **The distribution of functionals of Markoff processes.** We consider a functional of the form

$$Y(t) = \int_0^t f[X(\tau)]\,d\tau, \tag{41}$$

where $X(\tau)$ is a stationary Markoff process of the class being considered in this paper. Darling and Siegert [**5**] have defined a function

$$r(x|x_0, t, \eta) \equiv E\{e^{-\eta Y(t)}|X(0) = x_0, X(t) = x\}\cdot p(x|x_0, t),$$

which in our case is the principal solution of

$$\frac{\partial^2}{\partial x^2}[B(x)r] - \frac{\partial}{\partial x}[A(x)r] - \eta f(x)r = \frac{\partial r}{\partial t}. \tag{42}$$

The corresponding Sturm–Liouville equation becomes

$$\frac{d}{dx}\left[B(x)W(x)\frac{d\varphi(x)}{dx}\right] + W(x)[\lambda - \eta f(x)]\varphi(x) = 0. \tag{43}$$

A comparison of (43) and (12) shows that the only difference is the presence of a term $\eta f(x)$ in (43). If the addition of the term $\eta f(x)$ does not significantly complicate the Sturm–Liouville equation, then it is to be expected that the distribution of the corresponding functional is not too difficult to find.

With a standard transformation, (43) can be rewritten in the form

$$\frac{d^2\psi}{dz^2} + [\lambda - V(z) - \eta f]\psi = 0, \tag{44}$$

with

$$z = \int \frac{1}{\sqrt{B(x)}}\,dx,$$

$$q = [B(x)W^2(x)]^{1/4},$$

$$\psi = q\varphi,$$

$$V(z) = q^{-1}(z)\frac{d^2}{dz^2}q(z)$$

and

$$f = f(x(z)).$$

If the kernel f of the functional is such that

$$V(z) + \eta f(x(z)) = a^2V(az + b) + c, \tag{45}$$

where a, b, c are constants (i.e., independent of z), then (44) can be rewritten as

$$\frac{d^2\psi}{d\zeta^2} + \left[\left(\frac{\lambda - c}{a^2}\right) - V(\zeta)\right]\psi = 0, \tag{46}$$

with

$$\zeta = az + b.$$

In that case the solutions of (43) follow immediately from the solution of (12).

As an example, consider the functional

$$Y(t) = \int_0^t X(\tau)\,d\tau, \tag{47}$$

where $X(\tau)$ is the process of (2 C) with first-order probability density function

$$W(x) = \frac{1}{\Gamma(\alpha + 1)}x^\alpha e^{-x}.$$

For this case (43) becomes

$$\frac{d}{dx}\left[x^{\alpha+1} e^{-x} \frac{d\varphi}{dx}\right] + x^{\alpha} e^{-x}[\lambda - \eta x]\varphi(x) = 0. \tag{48}$$

It can be verified that $f(x) = x$ for this case satisfies (45) with

$$\begin{aligned} a &= (1 + 4\eta)^{1/4}, \\ b &= c = 0. \end{aligned} \tag{49}$$

The solutions of (48) are

$$\lambda_n = a^2(n + \alpha + 1) - (\alpha + 1) \tag{50}$$

and

$$\varphi_n(x) = e^{(1-a^2)x/2} L_n^{\alpha}(a^2 x), \tag{51}$$

with

$$L_n^{\alpha}(x) = \frac{1}{n!} x^{-\alpha} e^{x} \frac{d^n}{dx^n} (x^{n+\alpha} e^{-x}).$$

The function $r(x|x_0, t; \eta)$ can be written as

$$r(x|x_0, t; \eta) = a^2(a^2 x)^{\alpha} e^{-x} \sum_{n=0}^{\infty} \frac{n!}{\Gamma(n + \alpha + 1)} \varphi_n(x_0)\varphi_n(x). \tag{52}$$

From (52) we find

$$\begin{aligned} F(\eta, t) = E(e^{-\eta Y(t)}) &= \int_0^{\infty} \int_0^{\infty} W(x_0) r(x|x_0, t, \eta)\, dx_0\, dx \\ &= \left[\frac{4a^2 e^{-(a^2-1)t/2}}{(a^2 + 1)^2 - (a^2 - 1) e^{-a^2 t}}\right]^{\alpha+1}. \end{aligned} \tag{53}$$

Finally, the probability density function for $Y(t)$ can be found from (53) using the inversion integral of Laplace transform. Specifically, let $W(y, t)$ be the density function. Then

$$W(y, t) = \frac{1}{2\pi i} \int_{c-i\infty}^{c+i\infty} e^{\eta y} F(\eta, t)\, d\eta, \qquad c > -\tfrac{1}{4}. \tag{54}$$

References

1. J. F. Barrett and D. G. Lampard, *An expansion for some second-order probability distributions and its application to noise problems*, IRE Trans. **IT-1** (1955), 10–15.

2. S. Bochner, *Sturm-Liouville and heat equations whose eigenfunctions and ultraspherical polynomials or associated Bessel functions*, Proc. Conf. Differential Equations, pp. 23–48, University of Maryland Book Store, College Park, Md., 1955.

3. S. Chandrasekhar, *Stochastic problems in physics and astronomy*, Rev. Modern Phys. **15** (1943), 1–89.

4. H. Cramer, *Mathematical methods of statistics*, Princeton Univ. Press, Princeton, N. J., 1946.

5. D. A. Darling and A. J. F. Siegert, *A systematic approach to a class of problems in the theory of noise and other random phenomena*. I, IRE Trans. **IT-3** (1957), 32–37.

6. J. L. Doob, *The Brownian movement and stochastic equations*, Ann. of Math. (2) **43** (1942), 351–369.

7. A. Erdelyi, ed., *Higher transcendental functions*, Vol. I, McGraw-Hill, New York, 1953.

8. W. Feller, *Diffusion processes in genetics*, Proc. 2nd Berkeley Sympos. Math. Statist. and Prob., pp. 227–246, Univ. of California Press, Berkeley, Calif., 1951.

9. ———, *Diffusion processes in one dimension*, Trans. Amer. Math. Soc. **77** (1954), 1–31.

10. B. V. Gnedenko and A. N. Kolmogorov, *Limit distributions for sums of independent random variables*, Addison-Wesley, Cambridge, 1954.

11. M. Kac, *Random walk and the theory of Brownian motion*, Amer. Math. Monthly **54** (1947), 369–391.

12. S. Karlin and J. McGregor, *Classical diffusion processes and total positivity*, J. Math. Anal. Appl. **1** (1960), 163–183.

13. A. N. Kolmogorov, *Über die analytische Methoden in der Wahrscheinlichkeitsrechrung*, Math. Ann. **104** (1931), 415–458.

14. G. Szegö, *Orthogonal polynomials*, Amer. Math. Soc. Colloq. Publ. Vol. 23, Amer. Math. Soc., Providence, R. I., 1939.

15. M. C. Wang and G. E. Uhlenbeck, *On the theory of the Brownian motion*. II, Rev. Modern Phys. **17** (1945), 323–342.

16. E. Wong and J. B. Thomas, *On polynomial expansions of second-order distributions*, J. Soc. Indust. Appl. Math. **10** (1962), 507–516.

5

Reprinted from *IEEE Trans. Information Theory,* **IT-17**(1), 17–25 (1971)

On the Expansion of a Bivariate Distribution and its Relationship to the Output of a Nonlinearity

STAMATIS CAMBANIS, MEMBER, IEEE, AND BEDE LIU, MEMBER, IEEE

***Abstract*—The series expansion of a bivariate distribution and the series expansion of the output of a nonlinearity are considered, as well as the relationship between these two problems. Three distinct expansions of bivariate distributions are presented along with a constructive procedure to obtain them explicitly. The cross-covariance property and certain results on the expansion of the output of a nonlinearity are extended to a larger class of random processes.**

I. Introduction

HISTORICALLY, the study of the series expansion of the output of a nonlinearity has been linked with the series expansion of bivariate distributions. Under the assumption of existence of series expansions of bivariate distributions, several results related to nonlinearities have been obtained. However, without taking into consideration the mode of convergence in the expansions of bivariate distributions, the precise conditions under which the results are derived tend to be overlooked. More important, the question of how restrictive is the assumption of existence of these expansions has not been given enough attention in the literature. This paper treats the general problem of series expansion of bivariate distributions and the problem of series expansion of the output of a nonlinearity. It shows that there is no intrinsic connection between these two problems by generalizing certain results on nonlinearities to the case where no assumption is made on the bivariate distributions. Also the question of the restrictiveness of the assumptions mentioned above is clarified by providing expansions of bivariate distributions under different sets of conditions.

Section II considers the problem of series expansion of bivariate distributions in terms of their marginal distributions. Lancaster gave sufficient conditions for the existence of a diagonal series expansion with respect to properly chosen complete sets of orthonormal functions [9], [10]. Examples of bivariate density functions that admit diagonal expansions in terms of orthonormal polynomials are given in [2], [13], [14]. Brown [6] obtained necessary and sufficient conditions for an expansion of a bivariate density in terms of orthonormal polynomials to be diagonal. We prove a number of propositions on the expansion of bivariate distributions and related topics. Proposition 1 contains a general result, which is applied in Section III to the series expansion of the autocorrelation function of the output of a nonlinearity. A number of expansions of bivariate distributions in terms of their marginals are then obtained under various assumptions in Propositions 2, 3, and 4. Proposition 2 requires the minimal assumptions and the result is therefore most general, while Proposition 3 gives the simplest result, but requires additional assumptions. Proposition 4 represents an improvement over Proposition 1 under assumptions that are met in many practical cases, leading to an expansion very similar to Proposition 3. A constructive procedure to obtain these expansions explicitly is then presented.

Barrett and Lampard [2] considered a diagonal expansion of bivariate densities in terms of orthonormal polynomials and applied it to obtain an interesting cross-covariance property for a nonlinearity with a class of input processes. Brown [5] derived a necessary and sufficient condition for the cross-covariance property to hold by using a double series expansion of the bivariate density in terms of orthonormal polynomials. In a recent paper, Blachman [4] considered a diagonal expansion of bivariate density functions with equal marginal densities and applied it to obtain useful expansions of the output of a nonlinearity and of its autocorrelation function when the input is second-order stationary. These results on the output of a nonlinearity were obtained under the assumption that the bivariate distribution of the input processes admits series expansions of particular forms. Section III extends the class of random processes for which the cross-covariance property holds [5] and the class of processes for which some results on the expansion of the output of a nonlinearity have been obtained [4]. No assumption on the stationarity of the process and no assumption on the existence of a series expansion, diagonal or double, of the bivariate distribution of the input processes are necessary.

II. Expansions of Bivariate Distribution

We first make clear our notation. Let $(\Omega, \mathscr{M}, P)$ be the probability space, R^1 the real line, $\mathscr{B}^1$ the σ-algebra of Borel sets on the real line, $R^2 = R^1 \times R^1$, and $\mathscr{B}^2 = \mathscr{B}^1 \times \mathscr{B}^1$.

Manuscript received February 26, 1969; revised February 13, 1970. This research was supported by the National Science Foundation, Grant GK-1439, and the A.F. Office of Scientific Research, Grant AFOSR-1333-67 with Princeton University.

S. Cambanis was with the Department of Electrical Engineering, Princeton University, Princeton, N.J. He is now with the Department of Statistics, University of North Carolina, Chapel Hill, N.C. 27514.

B. Liu is with the Department of Electrical Engineering, Princeton University, Princeton, N.J. 08540.

Let $x = x(\omega)$, $\omega \in \Omega$, be a real random variable, μ the probability measure induced on $(R^1, \mathscr{B}^1)$ by x: $\mu(S) = P[x^{-1}(S)]$ for all $S \in \mathscr{B}^1$, and $G(\cdot)$ its distribution function defined by $G(u) = \mu\{(-\infty, u]\}$. The measure μ will be denoted sometimes by dG. Let $(R^1, \mathscr{B}^1, v)$ be the probability measure space induced by the random variable $y = y(\omega)$, and $H(\cdot)$ its distribution function.

Let λ be the probability measure induced on $(R^2, \mathscr{B}^2)$ by the pair of random variables $x(\omega)$ and $y(\omega)$, $\lambda(S_1 \times S_2) = P[x^{-1}(S_1) \cap y^{-1}(S_2)]$, and $F(u, v) = \lambda\{(-\infty, u] \times (-\infty, v]\}$ their bivariate distribution function. $G(u)$ and $H(v)$ are the marginal distribution functions of $F(u, v)$ and μ and v are the marginal probability measures of λ, i.e., $\mu(S) = \lambda(S \times R^1)$, $v(S) = \lambda(R^1 \times S)$ for all $S \in \mathscr{B}^1$.

Let us note that the relation

$$\mathscr{E}[A(x, y)] = \iint_{-\infty}^{\infty} A(u, v)\, dF(u, v)$$

where $\mathscr{E}$ denotes the expectation, implies that A has finite mean-square value if and only if it belongs to $L^2(R^2, \mathscr{B}^2, dF) = L^2(dF)$. Similarly $A(x)$ has finite mean-square value if and only if it belongs to $L^2(R^1, \mathscr{B}^1, dG) = L^2(dG)$.

Our first general result is presented as Proposition 1.

Proposition 1: Let $F(u, v)$ be a bivariate distribution function with marginal distributions $G(u)$ and $H(v)$. Let $\{g_i(u)\}_{i=0}^{\infty}$ and $\{h_j(v)\}_{j=0}^{\infty}$ be orthonormal and complete sets of functions in the real Hilbert spaces $L^2(dG)$ and $L^2(dH)$, respectively. Then for all $g \in L^2(dG)$ and $h \in L^2(dH)$

$$\iint_{-\infty}^{\infty} g(u)h(v)\, dF(u, v) = \sum_{i,j=0}^{\infty} a_i b_j \rho_{ij}$$

$$= \lim_{N,M} \iint_{-\infty}^{\infty} g(u)h(v)\, d_{N,M}(u, v)\, dG(u)\, dH(v) \quad (1)$$

where

$$a_i = \int_{-\infty}^{\infty} g(u) g_i(u)\, dG(u) \quad (2a)$$

$$b_j = \int_{-\infty}^{\infty} h(v) h_j(v)\, dH(v) \quad (2b)$$

$$\rho_{ij} = \iint_{-\infty}^{\infty} g_i(u) h_j(v)\, dF(u, v) \quad (3)$$

$$d_{N,M}(u, v) = \sum_{i=0}^{N} \sum_{j=0}^{M} \rho_{ij} g_i(u) h_j(v) \text{ a.e. } [dG \times dH]. \quad (4)$$

Proof: If $g(u) \in L^2(dG)$ and $h(v) \in L^2(dH)$, then $g(u)I_{R^1}(v)$ and $I_{R^1}(u)h(v)$ are in $L^2(dF)$, where I is the indicator function and $g(u)h(v) \in L^1(dF)$. In particular, $g_i(u)h_j(v) \in L^1(dF)$ and by (3)

$$|\rho_{ij}| \leq \|g_i h_j\|_{L^1(dF)} \leq \|g_i\|_{L^2(dG)} \|h_j\|_{L^2(dH)} = 1. \quad (5)$$

From the relations

$$g(u) = \lim_{N} \sum_{i=0}^{N} a_i g_i(u) = \lim_{N} g^{(N)}(u) \quad \text{in } L^2(dG)$$

$$h(v) = \lim_{M} \sum_{j=0}^{M} b_j h_j(v) = \lim_{M} h^{(M)}(v) \quad \text{in } L^2(dH)$$

where the a_i and b_j are defined in (2a) and (2b), it can be shown that

$$g(u)h(v) = \lim_{N,M} g^{(N)}(u) h^{(M)}(v) \quad \text{in } L^1(dF)$$

and that

$$g^{(N)}(u) h^{(M)}(v) = \sum_{i=0}^{N} \sum_{j=0}^{M} a_i b_j g_i(u) h_j(v) \quad \text{in } L^1(dF).$$

Hence

$$\iint_{-\infty}^{\infty} g(u)h(v)\, dF(u, v) = \lim_{N,M} \iint_{-\infty}^{\infty} g^{(N)}(u) h^{(M)}(v)\, dF(u, v)$$

$$= \lim_{N,M} \sum_{i=0}^{N} \sum_{j=0}^{M} a_i b_j \rho_{ij}$$

and by using (2a), (2b), and (4) the proof of (1) is completed.

The convergence in (1) is unconditional and hence absolute, since it does not depend on the enumerations of the bases $\{g_i\}$ and $\{h_j\}$. Q.E.D.

Series expansions of bivariate distributions have often been used in obtaining series expansions for the mean and correlation functions of stochastic processes. As will become clear in Section III, series expansions for correlation functions can be obtained by using Proposition 1 with no assumption on the bivariate distribution of the process.

By taking in particular $g(u) = I_{S_1}(u)$, $h(v) = I_{S_2}(v)$, $S_1, S_2 \in \mathscr{B}^1$, (1) can be written as

$$\lambda(S_1 \times S_2) = \lim_{N,M} \int_{S_1 \times S_2} d_{N,M}(u, v)(\mu \times v)(du, dv)$$

$$= \sum_{i,j=0}^{\infty} \rho_{ij} \int_{S_1} g_i(u)\, dG(u) \int_{S_2} h_j(v)\, dH(v). \quad (6)$$

Since a measure on $(R^2, \mathscr{B}^2)$ is uniquely determined by its values on the set of all measurable rectangles $S_1 \times S_2$ of $\mathscr{B}^2$, (6) leads to the following.

Corollary to Proposition 1: A bivariate distribution is completely characterized by its marginal distributions and the correlation ρ_{ij} of any pair of orthonormal and complete sets of functions on the marginal distributions.

This has been proven by Lancaster [10] by a different approach.

Consider now the Lebesgue decomposition of λ with respect to $\mu \times v$:

$$\lambda(S) = \int_S d(u, v)(\mu \times v)(du, dv) + \sigma(S) \quad (7)$$

for all $S \in \mathscr{B}^2$, where σ is the singular part and $d(u, v)$ the Radon–Nikodym derivative of the absolutely continuous part of λ.

It should be noted that, without additional assumptions, neither (6) holds for all $S \in \mathscr{B}^2$, nor (1) for all $A(u, v) \in L^1(dF)$, which would be useful from the applications viewpoint, since a necessary condition for both is the absolute continuity of λ with respect to $\mu \times \nu$.

For the sake of illustration we give an example of a bivariate distribution that is singular with respect to the product of its marginals and we verify (1) and (6) independent of the proof given in Proposition 1. This example will be referred to in Section III.

Example: Let the random variables x and y be distributed along the diagonal of the plane according to the law determined by any probability density function. Then $\mu = \nu$, $G = H$, $\lambda(S_1 \times S_2) = \lambda(S \times S) = \mu(S)$ where $S = S_1 \cap S_2$ and $S_1, S_2 \in \mathscr{B}^1$, G has density function, λ is singular with respect to $\mu \times \mu$, and

$$\iint_{-\infty}^{\infty} A(u, v)\, dF(u, v) = \int_{-\infty}^{\infty} A(u, u)\, dG(u).$$

The validity of (1) and (6) follows from the relations

$$\begin{aligned}\iint_{-\infty}^{\infty} g(u)h(v)\, dF(u, v) &= \int_{-\infty}^{\infty} g(u)h(u)\, dG(u) = (g, h)_{L^2(dG)}\\ &= \sum_{i=0}^{\infty} \int_{-\infty}^{\infty} g(u)g_i(u)\, dG(u) \cdot \int_{-\infty}^{\infty} h(v)g_i(v)\, dG(v)\end{aligned}$$

$$\rho_{ij} = \int_{-\infty}^{\infty} g_i(u)g_j(u)\, dG(u) = \delta_{ij}$$

$$\begin{aligned}\lim_N \int_{S_1 \times S_2} d_{N,N}(u, v)(\mu \times \nu)(du, dv) &= \sum_{i=0}^{\infty} \int_{S_1} g_i(u)\mu(du) \int_{S_2} g_i(v)\mu(dv)\\ &= (I_{S_1}, I_{S_2})_{L^2(\mu)} = \mu(S_1 \cap S_2) = \lambda(S_1 \times S_2).\end{aligned}$$

Let us note that the same results, including $\rho_{ij} = \delta_{ij}$, hold in the more general case where x and y are distributed along any one-to-one curve in the plane.

We turn now to the problem of series expansion of a bivariate distribution with respect to the product of its marginals. It is clear from (7) that a minimal assumption is the absolute continuity of λ with respect to $\mu \times \nu$. It should be noted that this assumption is always satisfied in the important case where F has a density function f. In this case, G and H have the density functions

$$g(u) = \int_{-\infty}^{\infty} f(u, v)m(dv) \text{ a.e. } [m]$$

and

$$h(v) = \int_{-\infty}^{\infty} f(u, v)m(du) \text{ a.e. } [m],$$

where m is the Lebesgue measure, and

$$d(u, v) = \left[\frac{dF(u, v)}{dG(u)\, dH(v)}\right] = \frac{f(u, v)}{g(u)h(v)} \text{ a.e. } [gdm \times hdm]. \quad (8)$$

A useful consequence of the absolute continuity assumption is given in the following.

Lemma: Let F, G, H, $\{g_i\}_{i=0}^{\infty}$ and $\{h_j\}_{j=0}^{\infty}$ be as in Proposition 1. If dF is absolutely continuous with respect to $dG \times dH$ with Radon–Nikodym derivative $d(u, v)$, then the set

$$\left\{ f_{ij}(u, v) = \frac{g_i(u)h_j(v)}{\sqrt{d(u, v)}} \quad i, j = 0, 1, 2, \cdots \right\} \quad (9)$$

is orthonormal and complete in $L^2(dF)$.

Proof: It is clear from (7) that $d(u, v) \geq 0$ a.e. $[dG \times dH]$. The orthogonality of the set $\{f_{ij}\}$ in $L^2(dF)$ follows from

$$\begin{aligned}\iint_{-\infty}^{\infty} f_{ij}(u, v)f_{kl}(u, v)\, dF(u, v) &= \iint_{-\infty}^{\infty} g_i(u)h_j(v)g_k(u)h_l(v)\, dG(u)\, dH(v)\\ &= \delta_{ik}\, \delta_{jl}.\end{aligned}$$

Its completeness follows from the facts that the set $\{g_i(u)h_j(v), i, j = 0, 1, \cdots\}$ is orthonormal and complete in $L^2(dG \times dH)$, that $A\sqrt{d} \in L^2(dG \times dH)$ for all $A \in L^2(dF)$, and from

$$\begin{aligned}\iint_{-\infty}^{\infty} \left| A(u, v) - \sum_{i=0}^{N} \sum_{j=0}^{M} a_{ij}f_{ij}(u, v)\right|^2 dF(u, v) &= \iint_{-\infty}^{\infty} \left| A(u, v)\sqrt{d(u, v)} - \sum_{i=0}^{N} \sum_{j=0}^{M} a_{ij}g_i(u)h_j(v)\right|^2\\ &\quad \cdot dG(u)\, dH(v)\end{aligned} \quad (10)$$

where

$$a_{ij} = (A, f_{ij})_{L^2(dF)} = (A\sqrt{d}, g_ih_j)_{L^2(dG \times dH)} \quad (11)$$

Q.E.D.

The following Proposition gives the most general series expansion of $d(u, v)$ and is a straightforward consequence of the foregoing Lemma and of the fact that $d \in L^1(dG \times dH)$ or $\sqrt{d} \in L^2(dG \times dH)$.

Proposition 2: Let F, G, H, $\{g_i\}$, $\{h_j\}$, d, and $\{f_{ij}\}$ be as in the Lemma. Then

$$\sqrt{d(u, v)} = \sum_{i,j=0}^{\infty} d_{ij}g_i(u)h_j(v) \quad \text{in } L^2(dG \times dH) \quad (12a)$$

$$d(u, v) = \sum_{i,j,k,l=0}^{\infty} d_{ij}d_{kl}g_i(u)g_k(u)h_j(v)h_l(v) \quad \text{in } L^1(dG \times dH) \quad (12b)$$

where

$$\begin{aligned}d_{ij} &= \iint_{-\infty}^{\infty} \sqrt{d(u, v)}g_i(u)h_j(v)\, dG(u)\, dH(v)\\ &= \iint_{-\infty}^{\infty} f_{ij}(u, v)\, dF(u, v)\end{aligned} \quad (13)$$

and

$$\sum_{i,j=0}^{\infty} d_{ij}^2 = 1. \quad (14)$$

It follows from (8) and Proposition 2 that every bivariate probability density $f(u, v)$, with marginal densities $g(u)$ and $h(v)$, admits the series expansions

$$\sqrt{f(u, v)} = \sum_{i,j=0}^{\infty} d_{ij} g_i(u) h_j(v) \sqrt{g(u)} \sqrt{h(v)} \quad \text{in } L^2(m^2) \quad (15a)$$

$$f(u, v) = \sum_{i,j,k,l=0}^{\infty} d_{ij} d_{kl} g_i(u) g_k(u) h_j(v) h_l(v) g(u) h(v) \quad \text{in } L^1(m^2). \quad (15b)$$

Proposition 2 can be used to obtain a series expansion for the mean of a function of x and y with finite mean-square value:

$$\mathscr{E}[A(x, y)] = \iint_{-\infty}^{\infty} A(u, v)\, dF(u, v) = \sum_{i,j=0}^{\infty} a_{ij} d_{ij} \quad (16)$$

where the a_{ij} and d_{ij} are given by (11) and (13), respectively.

A simpler expansion of d is given in Proposition 3 under the additional assumption that $d \in L^2(dG \times dH)$.

Proposition 3: Let F, G, H, $\{g_i\}$, $\{h_j\}$, and d be as in Proposition 2. If $d(u, v) \in L^2(dG \times dH)$ then

$$d(u, v) = \sum_{i,j=0}^{\infty} \rho_{ij} g_i(u) h_j(v) \quad \text{in } L^2(dG \times dH) \quad (17)$$

where the ρ_{ij} are given by (3).

Proof: Equation (17) follows from the fact that the set $\{g_i(u)h_j(v)\}_{i,j=0}^{\infty}$ is orthonormal and complete in $L^2(dG \times dH)$ and from the relation

$$\iint_{-\infty}^{\infty} d(u, v) g_i(u) h_j(v)\, dG(u)\, dH(v) = \iint_{-\infty}^{\infty} g_i(u) h_j(v)\, dF(u, v) = \rho_{ij}.$$

Q.E.D.

Lancaster [10] proved that, under the assumptions of Proposition 3, appropriate sets of orthonormal and complete functions can be found on the marginal distributions so that $\rho_{ij} = 0$ for all $i \neq j$. In this case (17) and (1) take a diagonal form.

In the case where F has a density function f and g and h are the density functions of G and H, respectively, it follows from Proposition 3 that if $[f^2(u, v)/g(u)h(v)] \in L^1(m^2)$ then

$$f(u, v) = \sum_{i,j=0}^{\infty} \rho_{ij} g_i(u) h_j(v) g(u) h(v) \quad (18)$$

where the convergence can be thought of in the following two ways. First, $\sum_{i,j=0}^{\infty} \rho_{ij} g_i(u) h_j(v)$ converges in $L^2(gdm \times hdm)$ to $d(u, v)$ and $f(u, v) = d(u, v) g(u) h(v)$ a.e. $[m^2]$. Second, $\sum_{i,j=0}^{\infty} \rho_{ij} g_i(u) h_j(v) g(u) h(v)$ converges to $f(u, v)$ in $L^1(m^2)$.

In the literature, (17) has been employed frequently to derive (1). However, it is clear from Proposition 1 that the expansion (1) can be obtained without the restrictive assumptions of Proposition 3, nor those implicit in assuming the existence of an expansion of the form (17).

It is interesting to note that a condition weaker than (17), namely, the convergence in $L^1(dG \times dH)$ of the sequence $d_{N,M}(u, v)$ defined by (4), implies that dF is absolutely continuous with respect to $dG \times dH$ with Radon–Nikodym derivative $d(u, v)$:

$$d(u, v) = \sum_{i,j=0}^{\infty} \rho_{ij} g_i(u) h_j(v) \quad \text{in } L^1(dG \times dH). \quad (19)$$

This is seen as follows. The convergence of $d_{N,M}$ to d in $L^1(dG \times dH)$ and (6) imply that

$$\lambda(S_1 \times S_2) = \int_{S_1 \times S_2} d(u, v)(\mu \times \nu)(du, dv) \quad (20)$$

for all $S_1, S_2 \in \mathscr{B}^1$. It now follows that (20) holds for all $S \in \mathscr{B}^2$ ([15], p. 239) and hence d is the Radon–Nikodym derivative of λ with respect to $\mu \times \nu$.

Since convergence in $L^2(dG \times dH)$ implies convergence in $L^1(dG \times dH)$ and since $d_{N,M}(u, v)$ converges in $L^2(dG \times dH)$ if and only if $\sum_{i,j=0}^{\infty} \rho_{ij}^2 < \infty$, we obtain the following.

Corollary to Proposition 3: Let F, G, H, $\{g_i\}$, $\{h_j\}$, and $\{\rho_{ij}\}$ be as in Proposition 1. Then dF is absolutely continuous with respect to $dG \times dH$ with Radon–Nikodym derivative $d(u, v)$ in $L^2(dG \times dH)$, given by (17), if and only if

$$\sum_{i,j=0}^{\infty} \rho_{ij}^2 < \infty. \quad (21)$$

An expansion very similar to (17) is obtained in the following Proposition under assumptions that are met in many practical cases.

Proposition 4: Let F, G, H, and d be as in Proposition 2. If there exist orthonormal and complete sets of functions $\{g_i(u)\}_{i=0}^{\infty}$ and $\{h_j(v)\}_{j=0}^{\infty}$ in $L^2(dG)$ and $L^2(dH)$, respectively, which are uniformly bounded, i.e.,

$$|g_i(u)| \leq C_1 \qquad |h_j(v)| \leq C_2 \quad (22)$$

for all $u, v \in R^1$ and $i, j = 0, 1, 2, \cdots$, where C_1, C_2 are positive constants, then

$$d(u, v) = \lim_{N,M} D_{N,M}(u, v) \quad \text{in } L^1(dG \times dH) \quad (23)$$

where

$$D_{N,M}(u, v) = \sum_{i,j=0}^{\infty} \rho_{ij}(N, M) g_i(u) h_j(v) \quad \text{in } L^2(dG \times dH) \quad (24)$$

$$\rho_{ij}(N, M) = \sum_{n=0}^{N} \sum_{m=0}^{M} d_{nm} c(i, j; n, m) \quad (25)$$

$$c(i, j; n, m) = \iint_{-\infty}^{\infty} \sqrt{d(u, v)} g_i(u) g_n(u) h_j(v) h_m(v) \cdot dG(u)\, dH(v) \quad (26)$$

$$\lim_{N,M} \rho_{ij}(N, M) = \rho_{ij} \tag{27}$$

for all $i, j = 0, 1, 2, \cdots$, uniformly in i, j, and the d_{nm} and the ρ_{ij} are given by (13) and (3), respectively.

Proof: It follows from (12) that

$$d(u, v) = \sum_{n,m=0}^{\infty} d_{nm} g_n(u) h_m(v) \sqrt{d(u, v)} \quad \text{in } L^1(dG \times dH).$$

Conditions (22) imply $g_n(u)h_m(v)\sqrt{d(u, v)} \in L^2(dG \times dH)$. Hence

$$g_n(u)h_m(v)\sqrt{d(u, v)} = \sum_{i,j=0}^{\infty} c(i, j; n, m) g_i(u) h_j(v) \quad \text{in } L^2(dG \times dH) \tag{28}$$

where the $c(i, j; n, m)$ are given by (26). If $D_{N,M}(u, v)$ is defined by

$$D_{N,M}(u, v) = \sum_{n=0}^{N} \sum_{m=0}^{M} d_{nm} g_n(u) h_m(v) \sqrt{d(u, v)} = \sum_{i,j=0}^{\infty} \left[\sum_{n=0}^{N} \sum_{m=0}^{M} d_{nm} c(i, j; n, m) \right] g_i(u) h_j(v)$$

in $L^2(dG \times dH)$, and $\rho_{ij}(N, M)$ by (25), then (23) and (24) are satisfied. It is easily seen from (25) and (28) that $\rho_{ij}(N, M) = (g_i h_j \sqrt{d}, \Delta_{N,M})L^2(dG \times dH)$, where

$$\Delta_{N,M}(u, v) = \sum_{i=0}^{N} \sum_{j=0}^{M} d_{ij} g_i(u) h_j(v),$$

and from (3) that $\rho_{ij} = (g_i h_j \sqrt{d}, \sqrt{d})_{L^2(dG \times dH)}$. Hence

$$|\rho_{ij}(N, M) - \rho_{ij}| \leq C_1 C_2 \|\Delta_{N,M} - \sqrt{d}\|_{L^2(dG \times dH)}.$$

Since, by Proposition 2, $\Delta_{N,M}$ converges to $\sqrt{d}$ in $L^2(dG \times dH)$, it follows that $\lim_{N,M} \rho_{ij}(N, M) = \rho_{ij}$ uniformly in i, j. Q.E.D.

As will be remarked at the end of Section II, conditions (22) are satisfied if both marginal distributions G and H have at most a finite number of jumps, i.e., of discrete masses. Hence, conditions (22) are met in many cases of practical interest, including the important case where both G and H have density functions.

It is interesting to compare (17) and (23), taking into account (24) and (27). The classes of bivariate distributions satisfying the conditions of Propositions 3 and 4 intersect, their intersection being properly contained in both classes. It is easily seen that, under the additional assumption $d \in L^2(dG \times dH)$ or $\sum_{i,j=0}^{\infty} \rho_{ij}^2 < \infty$, (23) reduces to (17).

Conditional Expectations

Let F, G, H, $\{g_i\}$, and $\{h_j\}$ be as in Proposition 1. It is seen that $A(u, v) \in L^2(dF)$ implies $\mathscr{E}[A(x, y) \| y = v] \in L^2(dH)$. Hence

$$\mathscr{E}[A(x, y) \| y = v] = \sum_{j=0}^{\infty} c_j h_j(v) \quad \text{in } L^2(dH) \tag{29a}$$

where

$$c_j = \int_{-\infty}^{\infty} \mathscr{E}[A(x, y) \| y = v] h_j(v)\, dH(u) = \mathscr{E}[A(x, y) h_j(y)]. \tag{30}$$

In particular,

$$\mathscr{E}[g(x) \| y = v] = \sum_{j=0}^{\infty} \left[\sum_{i=0}^{\infty} a_i \rho_{ij} \right] h_j(v) \quad \text{in } L^2(dH) \tag{29b}$$

for all $g \in L^2(dG)$ where the a_i are given by (4), and

$$\mathscr{E}[g_k(x) \| y = v] = \sum_{j=0}^{\infty} \rho_{kj} h_j(v) \quad \text{in } L^2(dH). \tag{29c}$$

Similar expressions hold for conditional expectations given $x = u$.

Let $P(\cdot | v)$ be the conditional probability of x given $y = v$. Then ([11], p. 366) or ([7], p. 158),

$$\lambda(S_1 \times S_2) = \int_{S_2} P(S_1 | v) \nu(dv) \tag{31}$$

for all $S_1, S_2 \in \mathscr{B}^1$. If dF is absolutely continuous with respect to $dG \times dH$, it follows from (20) and (31) that

$$\int_{S_2} [P(S_1 | v) - \int_{S_1} d(u, v) \mu(du)] \nu(dv) = 0$$

for all $S_1, S_2 \in \mathscr{B}^1$. Hence,

$$P(S_1 | v) = \int_{S_1} d(u, v) \mu(du) \quad \text{a.e. } [v]$$

for all $S_1 \in \mathscr{B}^1$ and

$$\mathscr{E}[A(x, y) \| y = v] = \int_{-\infty}^{\infty} A(u, v)\, d(u, v)\, dG(u) \quad \text{a.e. } [dH]. \tag{32}$$

Similarly for conditional expectations given $x = u$.

Orthonormal and Complete Sets in $L^2(dG)$

A general procedure to construct an orthonormal basis in $L^2(dG)$ is discussed here.[1] Once orthonormal bases in $L^2(dG)$ and $L^2(dH)$ have been constructed, the expansions given in Propositions 2, 3, and 4 can be obtained explicitly.

Let us first remark that we can always choose $g_0(u) = 1$ for all $u \in R^1$. Thus

$$\mathscr{E}[g_i(x)] = \int_{-\infty}^{\infty} g_0(u) g_i(u)\, dG(u) = 0 \qquad i \neq 0. \tag{33}$$

Similarly, by choosing $h_0(v) = 1$, $\mathscr{E}[h_j(y)] = 0$ for all $j \neq 0$. Furthermore,

$$\begin{aligned} \rho_{00} &= \mathscr{E}[g_0(x) h_0(y)] = 1 \\ \rho_{0j} &= \mathscr{E}[g_0(x) h_j(y)] = 0 \qquad \text{all } j \neq 0 \\ \rho_{i0} &= \mathscr{E}[g_i(x) h_0(y)] = 0 \qquad \text{all } i \neq 0. \end{aligned} \tag{34}$$

Consider now the decomposition of G: $G = G_1 + G_2$ into a step function G_1 including all jumps of G and a

[1] For certain particular cases, orthonormal and complete sets in $L^2(dG)$ are well known ([1], paragraphs 12 and 13).

continuous function G_2. The corresponding decomposition of μ is $\mu = \mu_1 + \mu_2$, where μ_1 is supported by the at most countable set of discontinuity points of G: $E_1 = \{\alpha_k\}_{k=0}^{\infty}$ and μ_2 assigns zero measure to all points of R^1. Let $E_2 \subseteq R^1$ be the support of μ_2, i.e., the Borel set on which G_2 is strictly increasing. Denote by $c = \mu_2(R^1)$ the total measure of μ_2. Then $0 \leq c < 1$ and the total measure of μ_1 is given by $\mu_1(R^1) = \mu(R^1) - \mu_2(R^1) = 1 - c$.

An orthonormal and complete set of functions in $L^2(E_2, \mathscr{B}^1/E_2, dG_2)$ has been given ([12], Theorem 2). For the complex Hilbert space, it is

$$\{g_k^{(2)}(u) = (1/\sqrt{c}) \exp [ik2\pi G_2(u)/c] \qquad k = 0 \pm 1, \pm 2, \cdots\}$$

and for the real Hilbert space considered here:

$$\begin{aligned} g_0^{(2)}(u) &= 1/\sqrt{c} \\ g_{2k-1}^{(2)}(u) &= \sqrt{2/c} \cos [k2\pi(G_2(u)/c)] \\ g_{2k}^{(2)}(u) &= \sqrt{2/c} \sin [k2\pi(G_2(u)/c)] \end{aligned} \tag{35}$$

for $k = 1, 2, 3, \cdots$. An obvious orthonormal and complete set in $L^2(E_1, \mathscr{B}^1/E_1, dG_1)$ is the following.

$$g_k^{(1)}(u) = [1/\sqrt{\mu_1(\{\alpha_k\})}] I_{\{\alpha_k\}}(u) \qquad k = 0, 1, 2, \cdots. \tag{36}$$

This can be easily transformed to give an orthonormal and complete set with $g_0^{(1)}(u) = (1/\sqrt{1-c}) I_{E_1}(u)$. For instance, $g_k^{(1)}(u)$, $k = 1, 2, \cdots$ may be chosen so as to be zero on α_n, $n \geq k + 2$. Such will be the $\{g_k^{(1)}\}$ set used in the following. A complete set of orthonormal functions in $L^2(R^1, \mathscr{B}^1, dG)$ is given by

$$\begin{aligned} g_0(u) &= 1 \\ g_1(u) &= \begin{cases} \sqrt{(1-c)/c} & \text{on } E_2 \sim E_1 \\ -\sqrt{c/(1-c)} & \text{on } E_1 \\ 0 & \text{otherwise} \end{cases} \\ g_{2k}(u) &= g_k^{(1)}(u) \quad \text{on } E_1 \\ &= 0 \quad \text{otherwise} \\ g_{2k+1}(u) &= g_k^{(2)}(u) \quad \text{on } E_2 \sim E_1 \\ &= 0 \quad \text{otherwise} \end{aligned} \tag{37}$$

$k = 1, 2, 3, \cdots$.

It follows from (35)–(37) that the orthonormal and complete functions $\{g_i(u)\}_{i=0}^{\infty}$ are uniformly bounded on R^1 if G has an at most finite number of jumps, i.e., if E_1 is a finite set of points. This includes the case where dG has a density function.

III. Random Processes and Nonlinearities

Let $\{x(t, \omega), t \in R^1, \omega \in \Omega\}$ be a real random process of second order, $G_t(\cdot)$ the distribution function of $x(t, \omega)$, $\{g_i(u; t)\}_{i=0}^{\infty}$ an orthonormal and complete set in $L^2(dG_t)$, and $F_{t,t'}(\cdot, \cdot)$ the bivariate distribution function of $x(t, \omega)$ and $x(t', \omega)$. Equation (3) can now be written as

$$\begin{aligned} \rho_{ij}(t, t') &= \mathscr{E}[g_i(x; t) g_j(x'; t')] \\ &= \iint_{-\infty}^{\infty} g_i(u; t) g_j(u'; t') \, dF_{t,t'}(u, u'). \end{aligned} \tag{38}$$

It is clear that in the general case the marginal distribution $G_t(\cdot)$ is a function of t and so is the corresponding orthonormal and complete set $\{g_i(\cdot; t)\}_{i=0}^{\infty}$, while the correlation coefficients $\rho_{ij}(t, t')$ depend on both t and t'. When $x(t, \omega)$ is stationary up to second order, the marginal distribution and the corresponding basis are independent of t, and the correlation coefficients $\rho_{ij}(t, t')$ depend only on $t - t'$.

Properties of the Correlation Coefficients $\rho_{ij}(t, t')$

Let us first introduce the following notation:

$$m_k(t) = \mathscr{E}[g_k(x; t)] \tag{39}$$

$$M(t) = \{m_k(t)\} \qquad \text{infinite column vector} \tag{40}$$

$$R(t, t') = \{\rho_{ij}(t, t')\} \qquad \text{infinite matrix}. \tag{41}$$

Then we have the following properties:

1) $|\rho_{ij}(t, t')| \leq 1$ for all t, t', i, j;

2) $M^T(t) M(t) = 1$; hence $\{m_k(t)\} \in l^2$ for all t;

3) $M(t) = R(t, t') M(t')$ for all t, t', $R(t, t) = I$ for all t;

4) the matrix R represents a bounded linear operator on l^2.

Proof: 1) It follows from (5).

2) Since $I_{R^1}(\cdot)$ belongs to $L^2(dG_t)$ for all t and has norm 1 and coefficients with respect to $\{g_k(\cdot; t)\}$,

$$a_k(t) = \mathscr{E}[I_{R^1}(x) g_k(x; t)] = m_k(t).$$

It follows that

$$\sum_{k=0}^{\infty} m_k^2(t) = 1 \qquad \text{all } t.$$

3) By taking $g(x) = g_k(x; t)$ and $h(x') = I_{R^1}(x')$ we obtain from (1)

$$m_k(t) = \mathscr{E}[g_k(x; t)] = \sum_{j=0}^{\infty} m_j(t') \rho_{kj}(t, t').$$

Hence, $M(t) = R(t, t') M(t')$. $R(t, t) = I$ follows from the orthonormality of the set $\{g_k(\cdot; t)\}$.

4) For any numbers $\{\alpha_i\}_1^p$, $\{\beta_j\}_1^q$, we have

$$\begin{aligned} &\left| \sum_{i=0}^{p} \sum_{j=0}^{q} \alpha_i \beta_j^* \rho_{ij}(t, t') \right| \\ &= \left| \mathscr{E}\left[\sum_{i=0}^{p} \alpha_i g_i(x; t) \sum_{j=0}^{q} \beta_j^* g_j(x'; t') \right] \right| \\ &\leq \sqrt{\mathscr{E}\left| \sum_{i=0}^{p} \alpha_i g_i(x; t) \right|^2} \cdot \sqrt{\mathscr{E}\left| \sum_{j=0}^{q} \beta_j g_j(x'; t') \right|^2} \\ &= \sqrt{\sum_{i=0}^{p} |\alpha_i|^2} \cdot \sqrt{\sum_{j=0}^{q} |\beta_j|^2}. \end{aligned}$$

The assertion follows from a theorem in ([1], p. 53). Incidentally, Landau's theorem ([1], pp. 39, 53) implies that

$$\sum_{i=0}^{\infty} \rho_{ij}^2(t, t') < \infty \qquad \text{all } j, t, t'$$

$$\sum_{j=0}^{\infty} \rho_{ij}^2(t, t') < \infty \qquad \text{all } i, t, t'.$$

Q.E.D.

Single Input to a Nonlinearity

Any real function $A(x)$ with finite mean-square value, i.e., $A(\cdot) \in L^2(dG_t)$ for all t, can be expanded in the form

$$A(x) = \sum_{i=0}^{\infty} a_i(t) g_i(x; t) \qquad \text{in } L^2(dG_t) \tag{42}$$

where

$$a_i(t) = \mathscr{E}[A(x) g_i(x; t)] = \int_{-\infty}^{\infty} A(u) g_i(u; t)\, dG_t(u). \tag{43}$$

It follows from (1) that its autocorrelation function has the corresponding expansion

$$R_{AA}(t, t') = \mathscr{E}[A[x(t)]A[x(t')]]$$

$$= \sum_{i,j=0}^{\infty} a_i(t) a_j(t') \rho_{ij}(t, t'). \tag{44}$$

It should be emphasized that these expansions for A and R_{AA} are always possible, the only assumption being that of the mean-square finiteness of A.

For fixed t the random variables $g_i(x; t)$ in (42) are orthogonal, i.e., $\rho_{ij}(t, t) = \delta_{ij}$ for all t, i, j. Since ρ_{ij} is the cross correlation function of g_i and g_j, the terms in the expansion (42) for A are uncorrelated if and only if $\rho_{ij}(t, t') = 0$ for all t, t' and $i \neq j$. In this case (44) takes a diagonal form, and so does the expansion (17), if it exists, i.e., if

$$\left[\frac{dF_{t,t'}}{dG_t \times dG_{t'}}\right] \in L^2(dG_t \times dG_{t'}).$$

This necessary and sufficient condition generalizes a result in [4] and is independent of the existence of a series expansion of the bivariate distribution of $x(t, \omega)$. This can be illustrated by examples where $\rho_{ij}(t, t') = 0$ for all t, t' and $i \neq j$ and $dF_{t,t'}$ is not absolutely continuous with respect to $dG_t \times dG_{t'}$. (Note that this absolute continuity is clearly a necessary condition for the existence of a series expansion of a bivariate distribution in terms of its marginals.) Consider, for instance, a random process of the form $x(t, \omega) = \varphi(t)\xi(\omega)$, where φ is a deterministic function and the random variable ξ is distributed according to any probability density law (x may be a sine wave with random amplitude). Whenever $\varphi(t) \neq 0$ it follows from $x(t', \omega) = [\varphi(t')/\varphi(t)]x(t, \omega)$ and the example given in Section II that $\rho_{ij}(t, t') = 0$ for all t, t' and $i \neq j$ and that $dF_{t,t'}$ is singular with respect to $dG_t \times dG_{t'}$; hence a series expansion of $F_{t,t'}$ in terms of its marginals cannot exist.

If the $g_i(u; t)$ are chosen so that $g_0(u; t) = 1$ for all t, then (42) and (33) imply that

$$\mathscr{E}[A(x)] = a_0(t). \tag{45}$$

The conditional expectation of $A[x(t)]$ given $x(t') = u'$ is obtained by using (29b):

$$\mathscr{E}[A[x(t)] \| x(t') = u'] = \sum_{j=0}^{\infty} \left[\sum_{i=0}^{\infty} a_i(t) \rho_{ij}(t, t')\right] g_j(u'; t') \tag{46}$$

in $L^2(dG_{t'})$. In particular, if $x(t, \omega)$ is stationary up to second order, we have

$$\mathscr{E}[x(t + \eta) \| x(t) = u] = \sum_{j=0}^{\infty} r_j(\eta) g_j(u) \qquad \text{in } L^2(dG) \tag{47}$$

where

$$r_j(\eta) = \sum_{i=0}^{\infty} \rho_{ij}(\eta) \mathscr{E}[x g_i(x)]. \tag{48}$$

If N is the upper limit of the summation in (47), then (47) is precisely the definition of the generalized separable class S_N of second-order stationary random processes introduced in [8]. Hence every second-order stationary process belongs to the generalized separable class S_∞. The results of [8] on the derivation of the optimum nonlinear predictor or interpolator with respect to a given class of nonlinear filters for random processes in S_N, N finite, can then be extended to all second-order stationary processes.

On a Cross-Covariance Property

Let $x(t, \omega)$ and $y(t, \omega)$ be two real random processes of second order with distribution functions $G_t(\cdot)$ and $H_t(\cdot)$, respectively, and let $\{g_i(\cdot; t)\}_{i=0}^{\infty}$ and $\{h_j(\cdot; t)\}_{j=0}^{\infty}$ be orthonormal and complete sets of functions in $L^2(dG_t)$ and $L^2(dH_t)$. We say that x and y have the cross-covariance property [2], [5] if and only if

$$C_{x,h(y)}(t, \tau) = K(h, \tau) C_{x,y}(t, \tau) \tag{49a}$$

for all instantaneous nonlinearities with finite mean-square value, i.e. for all $h(\cdot) \in L^2(dH_t)$ for all t, where $C_{x,h(y)}$ is the cross covariance of x and $h(y)$ defined by

$$C_{x,h(y)}(t, \tau) = \mathscr{E}[\{x(t) - \mathscr{E}[x(t)]\}\{h[y(\tau)] - \mathscr{E}(h[y(\tau)])\}] \tag{50}$$

and K is a real function of h and τ. The importance of (49a) lies in the fact that K depends only on the nonlinearity h and τ and not on t. In case x and y are jointly stationary, (49a) is replaced by

$$C_{x,h(y)}(t) = K(h) C_{x,y}(t) \tag{49b}$$

and the cross covariance given by (50) is a function of time differences.

Barrett and Lampard [2] proved that if the joint probability density of x and y has a diagonal expansion in terms of orthonormal polynomials then x and y have the cross-covariance property. Brown [5] assumed that the joint probability density of x and y has a double series expansion in terms of orthonormal polynomials and derived necessary and sufficient conditions for x and y to have the cross-covariance property. The approach of [2] and [5] is based on the restrictive assumptions that the joint distribution of x and y has a series expansion and that the expansion is in terms of orthonormal polynomials, which are obtained by orthonormalizing the set $\{1, u, u^2, \cdots\}$ in $L^2(dG)$. However, it is known that, even if $u^n \in L^2(dG)$ for all n, this set of powers is not necessarily complete in $L^2(dG)$ ([1], paragraph 12). Thus there are distributions for which the results of [2] and [5] can not be applied. It will be shown presently that none of these assumptions is essential and that the necessary and sufficient conditions given in [5] can be obtained in the general case with no assumptions whatever.

In Section II a constructive procedure to obtain explicitly the bases $\{g_i(\cdot\,; t)\}$ was presented. These bases can be transformed so as to have

$$g_0(x; t) = 1$$
$$g_1(x; t) = \frac{x(t) - \mathscr{E}[x(t)]}{\text{var}\,[x(t)]}. \tag{51}$$

Similarly for the bases $\{h_j(\cdot\,; t)\}$. It is clear that (51) can always be satisfied, provided x is of second order, and that no other g_i is necessarily polynomial in x. Noting that (51) implies

$$x(t) = \mathscr{E}[x(t)]g_0(x; t) + \text{var}\,[x(t)]g_1(x; t) \tag{52}$$

we obtain from (50), (1), (52), (45), and (33):

$$C_{x,h(y)}(t, \tau) = \mathscr{E}\{x(t)h[y(\tau)]\} - \mathscr{E}\{x(t)\}\mathscr{E}\{h[y(\tau)]\}$$
$$= \text{var}\,[x(t)] \sum_{j=1}^{\infty} b_j(\tau)\gamma_{1j}(t, \tau) \tag{53}$$

where

$$b_j(\tau) = \mathscr{E}[h[y(\tau)]h_j(y; \tau)] \tag{54}$$
$$\gamma_{ij}(t, \tau) = \mathscr{E}[g_i(x; t)h_j(y; \tau)]. \tag{55a}$$

From (55a) and (51)

$$\gamma_{11}(t, \tau) = \frac{C_{x,y}(t, \tau)}{\text{var}\,[x(t)]\,\text{var}\,[y(\tau)]}. \tag{55b}$$

Using (53) and (55b), it can be shown as in [5] that 1) x and y have the cross-covariance property (49a) if and only if there exist real-valued functions $k_j(\cdot)$, $j \geq 2$, on the real line such that

$$\gamma_{1j}(t, \tau) = K_j(\tau)\gamma_{11}(t, \tau) \qquad \text{all } t, \tau, j \geq 2; \tag{56a}$$

2) if x and y are stationary and jointly stationary, then they have the cross-covariance property (49b) if and only if there exists real constants k_j such that

$$\gamma_{1j}(t) = K_j\gamma_{11}(t) \qquad \text{all } t, j \geq 2. \tag{56b}$$

Booton's equivalent gain [5] $K_b(h, \tau)$ of an instantaneous nonlinearity $h[y(t)]$ is defined so as to minimize

$$\mathscr{E}[h[y(\tau)] - K_b(h, \tau)\{y(\tau) - \mathscr{E}[y(\tau)]\}]^2$$

and is given by

$$K_b(h, \tau) = \frac{\mathscr{E}[h[y(\tau)]\{y(\tau) - \mathscr{E}[y(\tau)]\}]}{\text{var}^2\,[y(\tau)]}$$
$$= \frac{b_1(\tau)}{\text{var}\,[y(\tau)]}. \tag{57}$$

If y is stationary, then K_b is a function of h only and not of τ.

It is easily shown from (53), (55b), and (57), as in [5], that 3) x and y have the cross-covariance property (49a) with $K(h, \tau) = K_b(h, \tau)$ if and only if

$$\gamma_{1j}(t, \tau) = 0 \qquad \text{all } t\ \tau, j \geq 2; \tag{58a}$$

4) if x and y are jointly stationary, then they have the cross-covariance property (49b) with $K(h) = K_b(h)$ if and only if

$$\gamma_{1j}(t) = 0 \qquad \text{all } t, j \geq 2. \tag{58b}$$

It is of interest to note that (6) shows how the conditions (56a), (56b), (58a), and (58b) characterize the joint distribution of the processes x and y.

One can find examples of joint distributions satisfying the stronger conditions (58a) or (58b) and yet not admitting series expansions. This is illustrated by an example in the very important case where $y = x$. If x has the form $x(t, \omega) = \varphi(t)\xi(\omega)$, it follows from the remark made after (44) that the joint distribution of x and y is singular with respect to the product of the distributions of x and of y and as such it does not admit a series expansion. Yet, $\gamma_{ij}(t, t') = 0$ for all t, t' and $i \neq j$ and therefore (49a) or (49b), the latter if $\varphi(t) = e^{i\alpha t}$, hold with $K = K_b$. By mixing two "sufficiently disjoint" joint probability distributions, of which one admits a diagonal expansion (see Examples 2 and 3 in [2]) and the other is singular with respect to the product of its marginals, one can generate joint probability distributions having both absolutely continuous and singular parts with respect to the product of their marginals and yet satisfying $\gamma_{ij}(t, t') = \delta_{ij}$ for all t, t'; exhibiting thus the cross-covariance property with coefficient Booton's equivalent gain.

Nonlinearities With Signal and Noise Input

Let $\{s(t, \omega),\ t \in R^1,\ \omega \in \Omega\}$ and $\{n(t, \omega),\ t \in R^1,\ \omega \in \Omega\}$ be two independent real random processes of second order, called signal and noise, respectively, and let $A(s, n)$ be a function with finite mean-square value, which can be thought of as the output of a memoryless nonlinearity with inputs s and n.

Let $G_t(\cdot)$ and $H_t(\cdot)$ be the distribution functions of s and n, and $\{g_i(u; t)\}_{i=0}^{\infty}$ and $\{h_j(v; t)\}_{j=0}^{\infty}$ be complete sets of orthonormal functions in $L^2(dG_t)$ and $L^2(dH_t)$, respectively. Then $\{g_i(u; t)h_j(v; t)\}_{i,j=0}^{\infty}$ is a complete set of orthonormal functions in $L^2(dG_t \times dH_t)$.

Since s and n are independent and since $A(s, n)$ has finite mean-square value for all t, it belongs to $L^2(dG_t \times dH_t)$.

Therefore

$$A(s, n) = \sum_{i,j=0}^{\infty} a_{ij}(t, t)g_i(s; t)h_j(n; t) \quad \text{in } L^2(dG_t \times dH_t) \quad (59)$$

where

$$a_{ij}(t, t) = \mathscr{E}\{A[s(t), n(t)]g_i(s; t)h_j(n; t)\}. \quad (60)$$

In a similar way as in Proposition 1 we obtain from (59) the following expression for the autocorrelation function of A:

$$R_{AA}(t, t') = \mathscr{E}\{A[s(t), n(t)]A[s(t'), n(t')]\}$$

$$= \sum_{i,j,k,l=0}^{\infty} a_{ij}(t, t)a_{kl}(t', t')\rho_{ik}(t, t')r_{jl}(t, t') \quad (61)$$

where

$$\rho_{ik}(t, t') = \mathscr{E}[g_i(s; t)g_k(s'; t')] \quad (62a)$$

$$r_{jl}(t, t') = \mathscr{E}[h_j(n; t)h_l(n'; t')]. \quad (62b)$$

Both expansions (59) and (61) are consequences only of the mean-square finiteness of A. For fixed t the random variables in (59) are orthogonal. The random terms in (59) are uncorrelated if and only if $\rho_{ij}(t, t') = 0$ and $r_{ij}(t, t') = 0$ for all t, t' and $i \neq j$.

If we take $g_0(u; t) = 1$ and $h_0(v; t) = 1$ for all t, and notice that the independence of $s(t)$ and $n(t)$ implies

$$\mathscr{E}[A(s, n)|s(t) = u] = \int_{-\infty}^{\infty} A(u, v)\, dH_t(v) \quad \text{a.e. } [dG_t]$$

$$\mathscr{E}[A(s, n)|n(t) = v] = \int_{-\infty}^{\infty} A(u, v)\, dG_t(u) \quad \text{a.e. } [dH_t],$$

it is seen that $A(s, n)$ can be separated into four uncorrelated components, namely:

$$\begin{aligned} &a_{00}(t, t) \\ &\sum_{i=1}^{\infty} a_{i0}(t, t)g_i(s; t) \\ &\sum_{j=1}^{\infty} a_{0j}(t, t)h_j(n; t) \\ &\sum_{i,j=1}^{\infty} a_{ij}(t, t)g_i(s; t)h_j(n; t), \end{aligned} \quad (63)$$

which are precisely the four uncorrelated components defined by Blachman [3] as the D.C.[2] component (D.C. $= \mathscr{E}[A(s, n)]$), the signal × signal component (S.S. $= \mathscr{E}[A(s, n)|s] -$ D.C.), the noise × noise component (N.N. $= \mathscr{E}[A(s, n)|n] -$ D.C.), and the signal × noise component (S.N. $= A(s, n) -$ D.C. $-$ S.S. $-$ N.N.) of the output, respectively.

[2] By D.C. output component, it is meant the mean value of the output A. If A is not stationary, its mean value may not be a constant.

As is clear from (63), the four uncorrelated components, into which the output of every nonlinearity can be resolved, can always be written as the sum of terms, which for fixed t, are orthogonal. These terms are uncorrelated if and only if $\rho_{ij}(t, t') = 0$ and $r_{ij}(t, t') = 0$ for all t, t' and $i \neq j$. In this case, (61) reduces to a double series. This necessary and sufficient condition generalizes a result stated in [4].

Comments

All the properties and expansions that have been developed in this section for instantaneous nonlinearities $A[x(t)]$, $h[x(t)]$, and $A[s(t), n(t)]$ clearly hold for time-dependent nonlinearities $A[x(t), t]$, $h[x(t), t]$, and $A[s(t), n(t), t]$ as well. Also the extension to complex-valued nonlinearities is clear.

ACKNOWLEDGMENT

The authors wish to thank Prof. E. Masry of the University of California, San Diego, for his constructive criticism and an anonymous reviewer for his helpful comments in indicating a loophole in a proof of the original manuscript and for pointing out [15].

REFERENCES

[1] N. I. Akhiezer and I. M. Glazman, *Theory of Linear Operators in Hilbert Space*, vol. 1. New York: Ungar, 1961.
[2] J. F. Barrett and D. G. Lampard, "An expansion for some second-order probability distributions and its application to noise problems," *IRE Trans. Inform. Theory*, vol. IT-1, pp. 10–15, March 1955.
[3] N. M. Blachman, "The signal × signal, noise × noise, and signal × noise output of a nonlinearity," *IEEE Trans. Inform. Theory*, vol. IT-14, pp. 21–27, January 1968.
[4] ——, "The uncorrelated output components of a nonlinearity," *IEEE Trans. Inform. Theory*, vol. IT-14, pp. 250–255, March 1968.
[5] J. L. Brown, "On a cross-correlation property for stationary random processes," *IRE Trans. Inform. Theory*, vol. IT-3, pp. 28–31, March 1957.
[6] ——, "A criterion for the diagonal expansion of a second-order probability distribution in orthogonal polynomials," *IEEE Trans. Inform. Theory*, vol. IT-4, p. 172, December 1958.
[7] W. Feller, *An Introduction to Probability Theory and Its Applications*, vol. 2. New York: Wiley, 1966.
[8] A. H. Haddad, "Nonlinear prediction of a class of random processes," *IEEE Trans. Inform. Theory*, vol. IT-14, pp. 664–668, September 1968.
[9] H. O. Lancaster, "The structure of bivariate distributions," *Ann. Math. Stat.*, vol. 29, pp. 719–736, 1958.
[10] ——, "Correlations and canonical forms of bivariate distributions," *Ann. Math. Stat.*, vol. 34, pp. 532–538, 1963.
[11] M. Loève, *Probability Theory*. Princeton, N.J.: Van Nostrand, 1963.
[12] E. Masry, B. Liu, and K. Steiglitz, "Series expansion of wide-sense stationary random processes," *IEEE Trans. Inform. Theory*, vol. IT-14, pp. 792–796, November 1968.
[13] D. K. McGraw and J. F. Wagner, "Elliptically symmetric distributions," *IEEE Trans. Inform. Theory*, vol. IT-14, pp. 110–120, January 1968.
[14] E. Wong and J. B. Thomas, "On polynomial expansions of second-order distributions," *SIAM J. Appl. Math.*, vol. 10, pp. 507–516, 1962.
[15] A. C. Zaanen, *Integration*. Amsterdam: North-Holland, 1967.

II
Output Properties of ZNL Systems

Editor's Comments on Papers 6 Through 12

6A Price: *A Useful Theorem for Nonlinear Devices Having Gaussian Inputs*

6B Price: *Comment on "A Useful Theorem for Nonlinear Devices Having Gaussian Inputs"*

7 Gorman and Zaborszky: *Functional Calculus in the Theory of Nonlinear Systems with Stochastic Signals*

8 Shutterly: *General Results in the Mathematical Theory of Random Signals and Noise in Nonlinear Devices*

9 Baum: *The Correlation Function of Gaussian Noise Passed Through Nonlinear Devices*

10 Abramson: *Nonlinear Transformations of Random Processes*

11 Blachman: *The Signal × Signal, Noise × Noise, and Signal × Noise Output of a Nonlinearity*

12 Blachman: *The Uncorrelated Output Components of a Nonlinearity*

The ZNL system has been one of the most widely analyzed nonlinear systems with random inputs. The analysis of the ZNL has been either as a separate system or in conjunction with other linear functional blocks. It has been used as the basic building block in all the classical analysis and synthesis of nonlinear systems. Consequently, most of the papers in other sections of this volume are also concerned with the analysis of ZNL systems with random inputs. This section is restricted to papers that treat several aspects of the analysis of the output properties of ZNL systems. Therefore, the papers in this section have no central theme (other than the ZNL system), and as a result the section is the least cohesive. All the papers consider some general output property of a class of nonlinearities with special classes of inputs. They are typical of the many papers treating this subject which is mostly concerned with output autocorrelation or power spectrum of ZNL systems with one or more inputs. The arrangement of the papers is not chronological but rather topical.

The first topic is Price's theorem, first established by Price (Papers 6a and 6b). It is a property of the autocorrelation of the outputs of ZNL systems with Gaussian inputs that simplifies the derivation of the desired autocorrelation function. The theorem relates the output covariance of the derivatives of the ZNL systems to the derivative of the covariance of the outputs with respect to the input covariance. Like many of the papers in nonlinear analysis, it is based on Rice's (1944) method. Price's theorem may be used to derive Bussgang's (Paper 1) result; however, it cannot be generalized to processes other than the Gaussian. The theorem has been generalized in several directions to extend its application, with the three most important extensions published in 1967 and 1968. Brown (1967) extends the theorem to the expectation of a single-multivariable ZNL system instead of the product of single-variable ZNL systems in the original version. Pawula's (1967) generalization allows the application of the theorem to the multivariable case with greater ease. Pawula's extension is significant in that his derivation is based on the partial differential equation satisfied by the multivariable Gaussian density function. While Price's and Brown's results involve the derivative of the desired expectation with respect to the covariance of any two input variables, Pawula's version involves the derivative with respect to a parameter that when set to zero results in the independence of the input variables. The latest and most general extension of Price's theorem is given by Gorman and Zaborszky (Paper 7), who applied the theorem to an arbitrary nonlinear functional with Gaussian inputs. Therefore, while the subject of the paper is nonlinear functionals and not ZNL systems, it is intimately related to Price's theorem or Bussgang's (Paper 1) results, and hence fits better in this section.

The second topic is concerned with the use of power-series expansions of ZNL systems in order to obtain the desired moments or autocorrelation function of the output. The method is rather brute force and receives widespread use in the modern approaches to nonlinear dynamic analysis and filtering. A typical illustration of the use of the series-expansion method is given by Shutterly (Paper 8). The results obtained are the same as the ones derived by other methods, such as Rice (1944) or Middleton (Paper 19). Other series-approximation methods have also been used, among them the orthogonal polynomial expansions for the class discussed by the papers in Part I. Thomson (1955) used the Hermite polynomial expansion for the evaluation of output properties. Booton's (1954) linearization techniques may also be considered as a form of series expansion where the contributions of the higher-order terms is minimized. Booton's technique has been used extensively for the analysis of feedback with a single ZNL system. The linearization is the stochastic equivalent of the describing function method for the deterministic case [see, for example, Barrett and Coales (1956), Leland (1960)]. While the orthogonal expansion may be applied to a single type of input, such as Gaussian, the power-series expansion may also be applied to mixed inputs such as a sinusoidal signal and Gaussian noise.

Baum (Paper 9) also considers the autocorrelation of the output of ZNL systems with Gaussian inputs. The paper illustrates the general results that can be obtained without restricting the discussion to a special case. Specifically, the paper studies the properties of the transformation relating the output autocorrelation to the input autocorrelation. The applications of the general results are to special cases that have been considered in earlier papers.

The remaining three papers are applicable to special classes of non-Gaussian inputs. The first, by Abramson (Paper 10), investigates the mean-squared bandwidth of outputs of general ZNL systems with a special class of inputs. It is an excellent illustration of the type of general results for special output properties. There is no restriction on the ZNL system, but there is a restriction on the input—that the signal and its derivative at the same instant be independent. Such inputs have also been considered in earlier papers.

Finally, Blachman's papers (11 and 12) treat the properties of the outputs of ZNL systems for inputs with diagonal expansion for their second-order density (see Part I). The first paper considers a systematic method, other than power-series expansion, for the decomposition of the output of ZNL systems with signal-and-noise inputs into three uncorrelated components. The result, of course, is readily applicable to earlier work on determining the signal and noise components of the output of ZNL systems. The second paper carries the decomposition one step further, in that the output of ZNL systems with processes satisfying the cross-correlation property is decomposed into a sum of uncorrelated components. The method used is the expansion of the ZNL output in the orthogonal polynomial series, which also form the basis for the diagonal expansion of the second-order density of the input. Additional work in this area considers either extending the class of processes such as in Paper 5 or the application of the properties to special nonlinearities, such as limiters, rectifiers, quantizers, and power-law devices. Three special nonlinearities are singled out for inclusion in this collection in Part IV, since they form a basis for many later results. The classic paper by Rice (1944) also belongs technically in this section, as it presents the transform method (characteristic function method) for the derivation of the autocorrelation of the outputs of ZNL systems with Gaussian inputs.

6A

Reprinted from *IRE Trans. Information Theory*, **IT-4**(2), 69–72 (1958)

A Useful Theorem for Nonlinear Devices Having Gaussian Inputs*

ROBERT PRICE†

Summary—**If and only if the inputs to a set of nonlinear, zero-memory devices are variates drawn from a Gaussian random process, a useful general relationship may be found between certain input and output statistics of the set. This relationship equates partial derivatives of the (high-order) output correlation coefficient taken with respect to the input correlation coefficients, to the output correlation coefficient of a new set of nonlinear devices bearing a simple derivative relation to the original set. Application is made to the interesting special cases of conventional cross-correlation and autocorrelation functions, and Bussgang's theorem is easily proved. As examples, the output autocorrelation functions are simply obtained for a hard limiter, linear detector, clipper, and smooth limiter.**

IN THE COURSE of investigating the asymptotic frequency behavior of power spectra resulting from the passage of noise through zero-memory nonlinear devices, an interesting, unique property of Gaussian processes has been encountered, which does not appear to have been previously reported.

Statement of the Theorem

Assume $x_1, x_2, \cdots, x_n$ to be random variables from a Gaussian process whose nth order joint probability density is given by:[1]

$$p(x_1, x_2, \cdots, x_n) = (2\pi)^{-n/2} \mid M_n \mid^{-1/2} \cdot \exp\left\{-\frac{1}{2}\sum_{r=1}^{n}\sum_{s=1}^{n}\frac{M_{rs}}{\mid M_n \mid}(x_r - \overline{x_r})(x_s - \overline{x_s})\right\} \tag{1}$$

where $\mid M_n \mid$ is the determinant of $M_n = [\rho_{rs}]$ and $\rho_{rs} = \overline{x_r x_s} - \overline{x_r}\ \overline{x_s} = \rho_{sr}$ is the correlation coefficient of x_r and x_s. The means of x_r and x_s are $\overline{x_r}$ and $\overline{x_s}$, respectively. M_{rs} is the cofactor of ρ_{sr} in M_n.

Let there be n zero-memory nonlinear devices (linearity of course being included as a special case) specified by the input-output relationship $f_i(x)$, $i = 1, 2, \cdots, n$. Let each x_i be the single input to a corresponding $f_i(x)$, and designate the nth-order correlation coefficient of the outputs as:

$$R = \overline{\prod_{i=1}^{n} f_i(x_i)} \tag{2}$$

where the bar denotes the average taken over all x_i. Then, with weak restrictions on the $f_i(x)$, we have the following theorem for the partial derivatives of R with respect to the input correlation coefficients:

$$\frac{\partial^k R}{\prod_{m=1}^{N}(\partial\rho_{r_m s_m})^{k_m}} = \left(\frac{1}{2}\right)^{\sum_{m=1}^{N} k_m \delta_{r_m s_m}} \left[\overline{\prod_{i=1}^{n} f_i^{\left(\sum_{m=1}^{N}\epsilon_{im}k_m\right)}(x_i)}\right] \tag{3}$$

where r_m and s_m, $m = 1, 2, \cdots, N$, are integers lying between 1 and n, inclusive, and are not necessarily distinct. The k_m are positive integers, with $k = \sum_{m=1}^{N} k_m$. ϵ_{im} is the number of times i appears in (r_m, s_m). $\delta_{r_m s_m}$ is the Kronecker δ function, $\delta_{r_m s_m} = 1$ for $r_m = s_m$, 0 for $r_m \neq s_m$. The symbol $f_i^{(q)}(x_i)$ denotes the qth derivative of $f_i(x)$, taken at x_i.

Furthermore, not only is the above theorem true for inputs having an nth-order joint Gaussian distribution, but it holds true *only* for such inputs if the $f_i(x)$ are allowed to be of general form.

Proof

We now prove that in order for (3) to be satisfied it is both sufficient and necessary that the x_i have the joint probability density given by (1). Assume that each $f_i(x)$ can be represented by the sum of two Laplace transforms,[2]

$$f_i(x) = \frac{1}{2\pi j}\int_{C_{i+}} h_{i+}(u)e^{jux}\,du + \frac{1}{2\pi j}\int_{C_{i-}} h_{i-}(u)e^{jux}\,du \tag{4}$$

where

$$\left.\begin{aligned} h_{i+}(u) &= \int_0^\infty f_i(x)e^{-jux}\,dx \\ h_{i-}(u) &= \int_{-\infty}^0 f_i(x)e^{-jux}\,dx \end{aligned}\right\} \tag{5}$$

and the C_{i+} and C_{i-} are appropriate contours. Without assuming any particular form for $p(x_1, x_2, \cdots, x_n)$ for the present,

$$R = \int_{-\infty}^{+\infty}\int_{-\infty}^{+\infty}\cdots\int_{-\infty}^{+\infty} \cdot \prod_{i=1}^{n} f_i(x_i)p(x_1, x_2, \cdots, x_n)\,dx_1\,dx_2\cdots dx_n. \tag{6}$$

Substituting (4) in (6) and inverting the order of integration, following Rice's characteristic function method,[3]

* Manuscript received by the PGIT, January 3, 1958. The research in this paper was supported jointly by the Army, Navy, and Air Force under contract with Mass. Inst. Tech.

† Lincoln Lab., M.I.T., Lexington, Mass.

[1] H. Cramér, "Mathematical Methods of Statistics," Princeton University Press, Princeton, N. J., sec. 24.2; 1946.

[2] D. V. Widder, "The Laplace Transform," Princeton University Press, Princeton, N. J., ch. 6; 1946.

[3] S. O. Rice, "Mathematical analysis of random noise," *Bell Sys. Tech. J.*, vol. 23, pp. 282–332, July, 1944; and vol. 24, pp. 46–156; January, 1945. See sec. 4.8.

$$R = \frac{1}{(2\pi j)^n} \sum{}' \int_{C_{1*}} \int_{C_{2*}} \cdots \int_{C_{n*}} \cdot \prod_{i=1}^{n} h_{i*}(u_i)\theta(u_1, u_2, \cdots, u_n)\, du_1\, du_2 \cdots du_n \quad (7)$$

where $\sum'$ denotes a summation over all possible $\pm$ combinations and $\theta(u_1, u_2, \cdots, u_n)$ is the nth-order characteristic function:

$$\theta(u_1, u_2, \cdots, u_n) = \int_{-\infty}^{+\infty} \int_{-\infty}^{+\infty} \cdots \int_{-\infty}^{+\infty} p(x_1, x_2, \cdots, x_n) \cdot \exp\left(j \sum_{i=1}^{n} u_i x_i\right) dx_1\, dx_2 \cdots dx_n \quad (8)$$

with $j = \sqrt{-1}$.

We find a necessary condition for (3) to be satisfied by setting $N = 1 = k = k_1$. The partial derivative of the left-hand side of (3) is taken on θ in the integrand of (7), and the derivatives of the right-hand side are taken using (4). Thus the necessary condition:

$$\sum{}' \int_{C_{1*}} \int_{C_{2*}} \cdots \int_{C_{n*}} \prod_{i=1}^{n} h_{i*}(u_i) \left\{ \frac{\partial \theta(u_1, u_2, \cdots, u_n)}{\partial \rho_{r_1 s_1}} + \left(\frac{1}{2}\right)^{\delta_{r_1 s_1}} u_{r_1} u_{s_1} \theta(u_1, u_2, \cdots, u_n) \right\} du_1\, du_2 \cdots du_n = 0 \quad (9)$$

is obtained. The term in braces must be zero in order to satisfy (9) for arbitrary $f_i(x)$ and hence $h_{i*}(u)$. Integrating the resulting equation for all (r_1, s_1) (but taking into account that $\rho_{rs} = \rho_{sr}$),

$$\log \theta(u_1, u_2, \cdots, u_n) = -\frac{1}{2} \sum_{r=1}^{n} \sum_{s=1}^{n} \rho_{rs} u_r u_s + g(u_1, u_2, \cdots, u_n) \quad (10)$$

where g is some function which must now be found.

Let $\rho_{rs} = 1$ for all (r, s). Then all the x_i are completely correlated, and $p(x_1, x_2, \cdots, x_n)$ can be written:

$$p(x_1, x_2, \cdots, x_n) = p(x_1) \prod_{i=2}^{n} \delta(x_i - x_1 + \overline{x_1} - \overline{x_i}) \quad (11)$$

where $\delta(x)$ is the Dirac δ function. Substituting (11) in (8), θ is of the form:

$$\theta(u_1, u_2, \cdots, u_n) = \exp\left(j \sum_{i=1}^{n} u_i \overline{x_i}\right) g_1\left(\sum_{i=1}^{n} u_i\right), \quad \text{for all } \rho_{rs} = 1 \quad (12)$$

where

$$g_1(u) = \int_{-\infty}^{+\infty} p_1(x_1 - \overline{x_1}) e^{ju(x_1 - \overline{x_1})}\, d(x_1 - \overline{x_1}). \quad (13)$$

Similarly, when $\rho_{11} = 1$ r, $\rho_{1r} = \rho_{r1} = -1$ for all $r \neq 1$, and $\rho_{rs} = 1$ for all r or $s \neq 1$, then $x_2, x_3, \cdots, x_n$ are completely correlated with $(-x_1)$ and we obtain:

$$\theta(u_1, u_2, \cdots, u_n) = \exp\left(j \sum_{i=1}^{n} u_i \overline{x_i}\right) g_1\left(u_1 - \sum_{i=2}^{n} u_i\right),$$

$$\text{for } \rho_{11} = 1,\ \rho_{1r} = \rho_{r1} = -1 \text{ for all } r \neq 1, \text{ and } \rho_{rs} = 1 \text{ for all } r, \quad s \neq 1. \quad (14)$$

Substituting (12) in (10), we find:

$$g(u_1, u_2, \cdots, u_n) = j \sum_{i=1}^{n} u_i \overline{x_i} + g_2\left(\sum_{i=1}^{n} u_i\right) \quad (15)$$

where $g_2(u) = \log g_1(u) + u^2/2$. On the other hand, substituting (14) in (10) yields

$$g(u_1, u_2, \cdots, u_n) = j \sum_{i=1}^{n} u_i \overline{x_i} + g_2\left(2u_1 - \sum_{i=1}^{n} u_i\right). \quad (16)$$

Since u_1 and $\sum_{i=1}^{n} u_i$ may be considered as independent variables, the only solution which renders (15) and (16) compatible is $g_2(u) = K$, a constant. Thus, finally, we have from (10) and (15) the necessary condition:

$$\theta(u_1, u_2, \cdots, u_n) = \exp\left[-\frac{1}{2} \sum_{r=1}^{n} \sum_{s=1}^{n} \rho_{rs} u_r u_s + j \sum_{i=1}^{n} u_i \overline{x_i} + K \right]. \quad (17)$$

This is recognized to be the characteristic function of the n-dimensional Gaussian distribution[4] of (1) ($K = 0$ for proper normalization).

It is now a simple matter to prove the sufficiency of (17), and hence (1), for satisfying (3). Using (17) in (7), and remembering that $\rho_{rs} = \rho_{sr}$,

$$\frac{(-1)^k\, \partial k_R}{\prod_{m=1}^{N} (\partial \rho_{r_m s_m})^{k_m}} = \left(\frac{1}{2}\right)^{\sum_{m=1}^{N} k_m \delta_{r_m s_m}} \sum{}' \int_{C_{1*}} \int_{C_{2*}} \cdots \int_{C_{n*}} \cdot \prod_{i=1}^{n} u_1^{\sum_{m=1}^{N} \epsilon_{im} k_m} h_{i*}(u_i)\theta(u_1, u_2, u_n)\, du_1\, du_2 \cdots du_n. \quad (18)$$

By analogy to (6) and (7), and differentiating (4) with respect to x, the right side of (18) is seen to be equal to

$$(-1)^k \left(\frac{1}{2}\right)^{\sum_{m=1}^{N} k_m \delta_{r_m s_m}} \int_{-\infty}^{+\infty} \int_{-\infty}^{+\infty} \cdots \int_{-\infty}^{+\infty} \cdot \prod_{i=1}^{n} f_i^{\left(\sum_{m=1}^{N} \epsilon_{im} k_m\right)}(x_i) p(x_1, x_2, \cdots, x_n)\, dx_1\, dx_2 \cdots dx_n \quad (19)$$

thus yielding (3).

A Special Case and Its Applications

Consider the familiar situation where $n = 2$, and let ρ denote the crosscorrelation coefficient of x_1 and x_2. Then (3) yields

$$\frac{\partial^k R}{\partial \rho^k} = \overline{f_1^{(k)}(x_1) f_2^{(k)}(x_2)}. \quad (20)$$

Suppose that x_1 and x_2 are values of a stationary Gaussian time series $x(t)$ whose autocorrelation function is $\rho(\tau)$. x_1 is taken at time t and x_2 at time $(t + \tau)$. $R(\tau)$ will denote the crosscorrelation function between the outputs

[4] Cramér, *op. cit.*, sec. 24.1.

of two zero-memory nonlinear devices whose inputs are $x(t)$ and $x(t+\tau)$, respectively.

Taking the particular case where $\overline{x(t)} = 0$, $\overline{x^2(t)} = 1$, and using (1),[5]

$$\frac{\partial^k R(\tau)}{\partial \rho(\tau)^k} = \overline{f_1^{(k)}[x(t)]f_2^{(k)}[x(t+\tau)]} = \int_{-\infty}^{+\infty}\int_{-\infty}^{+\infty} \frac{f_1^{(k)}(x_1)f_2^{(k)}(x_2)\exp\{-[x_1^2+x_2^2-2\rho(\tau)x_1x_2]/2[1-\rho^2(\tau)]\}}{2\pi\sqrt{1-\rho^2(\tau)}}\,dx_1\,dx_2. \tag{21}$$

Eq. (21) is particularly simple when the $f_i(x)$ are piecewise-polynomial functions and k is sufficiently high. Then the $f_i^{(k)}(x)$ consist entirely of δ functions of various orders and the integral can be easily evaluated.

It is often of interest to obtain the derivatives of a crosscorrelation function with respect to τ. It is convenient to break down such τ derivatives into a series of products of derivatives of $R(\tau)$ with respect to $\rho(\tau)$, and $\rho(\tau)$ with respect to τ, using

$$\frac{dR(\tau)}{d\tau} = \frac{\partial R(\tau)}{\partial \rho(\tau)}\cdot\frac{d\rho(\tau)}{d\tau}. \tag{22}$$

This enables the nonlinear devices to be treated independently of the shape of the input correlation function $\rho(\tau)$, using (21). Similarly, the derivatives of $\rho(\tau)$ with respect to τ do not involve the $f_i(x)$.

As an example, Cohen[6] shows that in general, for autocorrelation functions $R(\tau)$, the limiting behavior of the corresponding power spectrum $\Phi(\omega)$ is given by:

$$\lim_{\omega\to\infty} \omega^2\Phi(\omega) = -\frac{1}{\pi}\frac{dR(\tau)}{d\tau}\bigg|_{\tau=0+}$$
$$\lim_{\omega\to\infty} \omega^2\left[\omega^2\Phi(\omega) + \frac{1}{\pi}\frac{dR(\tau)}{d\tau}\bigg|_{\tau=0+}\right] = -\frac{1}{\pi}\frac{d^3R(\tau)}{d\tau^3}\bigg|_{\tau=0+} \tag{23}$$

and so on, where the derivatives are with respect to τ.

Another application of (20) is in deriving Bussgang's interesting result[7] that the crosscorrelation function between the input and the output of a nonlinear device driven by Gaussian noise has the same shape as the input autocorrelation function. In this case $f_1(x) = x$ and $f_2(x)$ is arbitrary. Then $f_1'(x)$ is unity, and all higher derivatives of $f_1(x)$ are zero. Putting this into (21) and evaluating the integral,

$$\frac{\partial^k R(\tau)}{\partial\rho(\tau)^k} = \begin{cases} \displaystyle\int_{-\infty}^{+\infty} \frac{f_2'(x)\exp(-x^2/2)}{\sqrt{2\pi}}\,dx & k = 1 \\ 0 & k > 1. \end{cases} \tag{24}$$

We find easily

$$R(\tau) = \rho(\tau)\int_{-\infty}^{+\infty} \frac{xf_2(x)e^{-x^2/2}}{\sqrt{2\pi}}\,dx \tag{25}$$

thus yielding Bussgang's result. Unlike Bussgang's theorem, (20) cannot be generalized to hold for probability distributions other than Gaussian.[8-10]

Some Simple Autocorrelation Examples [for $\overline{x(t)} = 0$, $\overline{x^2(t)} = 1$]

Hard Limiter

Van Vleck's well-known result on the autocorrelation function of the output of a hard limiter[11] can be derived very simply, using (21). If

$$f_1(x) = f_2(x) = \begin{cases} 1; & x \ge 0 \\ -1; & x < 0 \end{cases} \tag{26}$$

then $f_1^{(1)}(x)$ and $f_2^{(1)}(x)$ are first-order δ functions of area 2, at $x = 0$.

Substituting in (21) and integrating,

$$\frac{\partial R(\tau)}{\partial\rho(\tau)} = \frac{2}{\pi\sqrt{1-\rho^2(\tau)}}. \tag{27}$$

When $\rho(\tau) = 0$, $R(\tau) = 0$. Thus

$$R(\tau) = \frac{2}{\pi}\int_0^{\rho(\tau)} \frac{d\rho(\tau)}{\sqrt{1-\rho^2(\tau)}} = \frac{2}{\pi}\sin^{-1}[\rho(\tau)] \tag{28}$$

which is Van Vleck's result.

Linear Detector

Similarly, the autocorrelation function of the output of a linear detector can be easily found. If

$$f_1(x) = f_2(x) = \begin{cases} x; & x \ge 0 \\ 0; & x < 0 \end{cases} \tag{29}$$

then $f_1^{(2)}(x)$ and $f_2^{(2)}(x)$ are first-order δ functions of area unity at $x = 0$. Substituting in (21) and integrating:

$$\frac{\partial^2 R(\tau)}{\partial\rho(\tau)^2} = \frac{1}{2\pi\sqrt{1-\rho^2(\tau)}}. \tag{30}$$

Doubly-integrating (30) with the boundary conditions:

[5] I. N. Amiantov and V. I. Tikhonov, "The effect of normal fluctuations on typical nonlinear elements," *Bull. Acad. Sci. USSR*, pp. 33–42; April, 1956. Here, an autocorrelation case $[f_1(x) = f_2(x)]$ is studied by expanding the second-order joint Gaussian probability density of $x(t)$ and $x(t+\tau)$ in powers of $\rho(\tau)$, using Mehler's formula; see Bateman Manuscript Project, "Higher Transcendental Functions," McGraw-Hill Book Co., Inc., New York, N. Y., vol. 2, p. 194, (22); 1953. They then integrate by parts to obtain the output correlation function in a similar series but do not recognize the simple form (21) for this series. Using this method it is not required that $f_1(x)$ be Laplace transformable, rather than our proof.

[6] R. Cohen, "Some Analytical and Practical Aspects of Wiener's Theory of Prediction," Res. Lab. of Electronics, M.I.T., Cambridge, Mass., Tech. Rep. No. 69, ch. 4, sec. 2; June 2, 1948.

[7] J. J. Bussgang, "Crosscorrelation Functions of Amplitude-Distorted Gaussian Signals," Res. Lab. of Electronics, M.I.T., Cambridge, Mass., Tech. Rep. 216, sec. 3; March 26, 1952.

[8] J. F. Barrett and D. G. Lampard, "An expansion for some second-order probability distributions and its application to noise problems," IRE Trans. on Information Theory, vol. IT-1, pp. 10–15; March, 1955.

[9] J. L. Brown, Jr., "On a cross-correlation property for stationary random processes," IRE Trans. on Information Theory, vol. IT-3, pp. 28–31; March, 1957.

[10] A. H. Nuttall, "Invariance of Correlation Functions under Nonlinear Transformations," Res. Lab. of Electronics, M.I.T., Cambridge, Mass., Quart. Progress Rep., p. 63; October 15, 1957.

[11] J. L. Lawson and G. E. Uhlenbeck, "Threshold Signals," McGraw-Hill Book Co., Inc., New York, N. Y., p. 58; 1950.

$$\left.\begin{aligned}\frac{\partial R(\tau)}{\partial\rho(\tau)} &= \left[\int_0^\infty f_1^{(1)}(x)\frac{e^{-x^2/2}}{\sqrt{2\pi}}dx\right]^2 = \frac{1}{4}\\ R(\tau) &= \left[\int_0^\infty \frac{xe^{-x^2/2}}{\sqrt{2\pi}}dx\right]^2 = \frac{1}{2\pi}\end{aligned}\right\} \text{ for } \rho(\tau) = 0 \quad (31)$$

we obtain:

$$R(\tau) = \int_0^{\rho(\tau)}\left[\frac{1}{4} + \int_0^y \frac{dt}{2\pi\sqrt{1-t^2}}\right]dy + \frac{1}{2\pi}$$

$$= \frac{1}{2\pi}\left\{\rho(\tau)\cos^{-1}[\rho(\tau)] + \sqrt{1-\rho^2(\tau)}\right\} \quad (32)$$

which is in agreement with Rice's result.[12]

Clipper

The relations derived independently by Robin[13] and Laning and Battin[14] for the autocorrelation function of the output of a clipper may also be found by this method. With a clipper characteristic:

$$f_1(x) = f_2(x) = \begin{cases} l; & l \le x \\ x; & -l \le x \le l \\ -l; & x \le l \end{cases} \quad (33)$$

and $f_1^{(2)}(x)$ and $f_2^{(2)}(x)$ each are a pair of first-order δ functions at $x = -l$ and $x = l$, with areas 1 and -1, respectively. Substituting in (21) and integrating,

$$\frac{\partial^2 R(\tau)}{\partial\rho(\tau)^2} = \frac{\exp\left[-\frac{l^2}{1+\rho(\tau)}\right] - \exp\left[-\frac{l^2}{1-\rho(\tau)}\right]}{\pi\sqrt{1-\rho^2(\tau)}} \quad (34)$$

which is Robin's result, for input noise of unit variance.

Smooth Limiter

Finally, Baum's recent interesting result[15] for the autocorrelation function of the output of a device having an error-function characteristic will be derived. With

$$f_1(x) = f_2(x) = \frac{1}{\sqrt{2\pi}}\int_0^x e^{-t^2/2l^2}\,dt \quad (35)$$

we have

$$f_1^{(1)}(x) = f_2^{(1)}(x) = \frac{1}{\sqrt{2\pi}}e^{-x^2/2l^2}. \quad (36)$$

Substituting in (21):

$$\frac{\partial R(\tau)}{\partial\rho(\tau)} = \frac{1}{2\pi}\sqrt{\frac{\rho_1^2-\rho_2^2}{1-\rho^2(\tau)}}$$

$$\cdot\left\{\int_{-\infty}^{+\infty}\int_{-\infty}^{+\infty}\frac{\exp\left[-\frac{\rho_1x_1^2+\rho_1x_2^2-2\rho_2x_1x_2}{2(\rho_1^2-\rho_2^2)}\right]}{2\pi\sqrt{\rho_1^2-\rho_2^2}}dx_1\,dx_2\right\} \quad (37)$$

where

$$\left.\begin{aligned}\rho_1 &= \frac{\{l^{-2}[1-\rho^2(\tau)]+1\}[1-\rho^2(\tau)]}{\{l^{-2}[1-\rho^2(\tau)]+1\}^2-\rho^2(\tau)}\\ \rho_2 &= \frac{\rho(\tau)[1-\rho^2(\tau)]}{\{l^{-2}[1-\rho^2(\tau)]+1\}^2-\rho^2(\tau)}\end{aligned}\right\} \quad (38)$$

The term in braces in (37) must equal unity, since it is the integral of a second-order Gaussian probability density. Thus, from (38),

$$\frac{\partial R(\tau)}{\partial\rho(\tau)} = \frac{1}{2\pi}\sqrt{\frac{\rho_1^2-\rho_2^2}{1-\rho^2(\tau)}}$$

$$= \frac{1}{2\pi\sqrt{(1+l^{-2})^2 - l^{-4}\rho^2(\tau)}}. \quad (39)$$

Integrating and using the condition that when $\rho(\tau) = 0$, $R(\tau) = 0$,

$$R(\tau) = \frac{l^2}{2\pi}\int_0^{\rho(\tau)}\frac{d\rho(\tau)}{\sqrt{(l^2+1)^2-\rho^2(\tau)}}$$

$$= \frac{l^2}{2\pi}\sin^{-1}\left[\frac{\rho(\tau)}{1+l^2}\right] \quad (40)$$

which is in agreement with Baum's result.

[12] Rice, *op. cit.*, eq. (4.7-5).

[13] L. Robin, "The autocorrelation function and power spectrum of clipped thermal noise. Filtering of simple periodic signals in this noise," *Ann. Telecomm.*, vol. 7, pp. 375–387; September, 1952.

[14] J. H. Laning, Jr. and R. H. Battin, "Random Processes in Automatic Control," McGraw-Hill Book Co., Inc., New York, N. Y., p. 362, eq. (B-8); 1956.

[15] R. F. Baum, "The correlation function of smoothly limited Gaussian noise," IRE TRANS. ON INFORMATION THEORY, vol. IT-3, pp. 193–197; September, 1957.

Errata

Page 69, third line below "Statement of the Theorem": should read "is given by,[1] for unit variance,"

Page 70, line following equation (13): should read "Similarly, when $\rho_{11} = 1, \rho_{1_r} = \rho_{r1} = -1$ for all"

Page 70, equation (18): should read

$$\prod_{i=1}^{n} u_1^{\sum_{m=1}^{N}\epsilon_{im}k_m} h_{i\pm}(u_i)\,\theta\,(u_1, u_2, \ldots, u_n)\,du_1\,du_2\ldots du_n.$$

Page 71, last line of footnote 5: should read "Laplace transformable, unlike our proof."

Page 71, equation (26): should read

$$f_1(x) = f_2(x) = \begin{cases} 1; & x \geq 0 \\ -1; & x < 0 \end{cases}$$

6B

Reprinted from *IEEE Trans. Information Theory,* **IT-10,** 171 (Apr. 1964)

Comment on: "A Useful Theorem for Nonlinear Devices Having Gaussian Inputs"

The theorem in question[1] deals with zero-memory nonlinear devices driven by Gaussian processes, and it relates derivatives of output correlation functions with respect to those of the inputs to derivatives of the nonlinear characteristics. Although Gaussianness is necessary for the theorem as stated to hold true, the author would like to point out that the utility of the theorem is much enhanced if it is modified as follows.

Let the input variates be drawn from a sum of two or more mutually independent random processes, the first process having the set of correlation coefficients $\{\rho_{rs}\}$ and unit variance as originally. Then the result (3) and its special case (20) are true so long as the first process is Gaussian, despite the statistics of the other processes. One now terminates the proof of necessity at (10), since letting $\rho_{rs} = 1$ or -1 no longer implies complete linear dependence among the inputs (provided that at least one of the other processes is of nonzero variance). Thus nonlinear problems involving input processes of a much broader class than the Gaussian, such as those containing sine waves imbedded in Gaussian noise, can be treated by the correlation-derivative method.

The author would like also to take this opportunity to cite prior related work[2,3] that was kindly brought to his attention some time ago by Prof. J. A. McFadden of Purdue University, and to correct several minor errors.

In connection with (1) and all that follows, the Gaussian processes should have specifically been stated (without loss of generality) to all be of unit variance. Just after (13) there is a superfluous r; in (18) the argument of θ should read $u_1, u_2, \cdots, u_n$; and in (26) $f_x(x)$ should read $f_2(x)$.

R. Price
Lincoln Lab.
Mass. Inst. Tech.
Lexington, Mass.

Manuscript received November 19, 1963.

[1] R. Price, "A useful theorem for nonlinear devices having Gaussian inputs," IRE Trans. on Information Theory, vol. IT-4, pp. 69–72; June, 1958.

[2] R. L. Plackett, "A reduction formula for normal multivariate integrals," *Biometrika*, vol. 41, pp. 351–360, December, 1954. [See Eq. (3).]

[3] H. R. Van der Vaart, "The content of some classes of non-Euclidean polyhedra for any number of dimensions, with several applications, I and II," *Akademie van Wetenschappen Proc.*, Series A, vol. 58, pp. 199–221, 564; 1955. (See Lemma 2 and the Corollary of Theorem 4.)

7

Reprinted from *IEEE Trans. Information Theory,* **IT-14,** 528–531 (July 1968)

Functional Calculus in the Theory of Nonlinear Systems with Stochastic Signals

C. DAVID GORMAN AND JOHN ZABORSZKY, FELLOW, IEEE

***Abstract*—Connections between the theorems of Bussgang, Price, McMahon, Novikov, and Wiener are established, and new theorems are introduced for nonlinear functionals of Gaussian random processes by the use of functional calculus. In particular, Novikov's formula is extended to obtain an expression for the Wiener- Hermite kernels of a functional in terms of its functional derivatives.**

Introduction

Statistical properties of Gaussian random processes are completely determined by their mean and covariance functions. Therefore, averages of differentiable nonlinear functionals of such Gaussian processes may be expressed as functionals of the mean and covariance. This realization is the principal basis of the results reported in this paper, where, by the introduction of the formalism of functional calculus, a number of isolated results concerning the averages of nonlinear functionals may be generalized and placed in the proper setting.

In particular, formulas proposed by Bussgang,[1] Price,[2] and McMahon[3] relate input and output correlation functions for nonlinear memoryless devices. The input-output relations of such devices define the output as a single-valued function of the input. On the other hand, input-output relations of nonlinearities with memory can be given as functional operators on a suitable linear topological function space. We have used functional calculus to establish equivalents of the Price and McMahon theorems for general functionals. Furthermore, we have shown that a formula proposed earlier by Novikov[4] represents a generalization of the Bussgang theorem. A general version of the Novikov formula is introduced and related to Wiener's work on the orthogonal expansions of nonlinear functionals.[5] Proofs are presented only for the class of functional polynomials. The results, however, are valid for general differentiable functionals.

A Functional Analog of the Price-McMahon Formula

Let $\{x(t),\ 0 \leq t \leq T\}$ be a Gaussian random process with mean $m(t) = Ex(t)$ and correlation function $r(t, t') = Ex(t)x(t')$, where E denotes expectation, m is continuous in t, and r is continuous in t and t'. Let $F[x]$ be an arbitrary twice differentiable functional. The following formulas then apply:

Manuscript received April 18, 1967; revised December 18, 1967. This study was supported by the Applied Mathematics Branch of the USAF Office of Scientific Research, under Grant AF-AFOSR-482-66.

The authors are with the School of Engineering and Applied Science, Washington University, St. Louis, Mo. 63130

$$\frac{\delta}{\delta m(t)} EF[x] = E\,\frac{\delta F[x]}{\delta x(t)}, \tag{1a}$$

$$\frac{\delta}{\delta r(t, t')} EF[x] = E\,\frac{\delta^2 F[x]}{\delta x(t)\,\delta x(t')}, \tag{1b}$$

where $\delta/\delta r(t, t')$ and $\delta/\delta x(t)$ denote functional derivatives or gradients. The functional derivative of $F[x]$ with respect to $x(t)$ can be defined as the following heuristic generalization of the concept of the gradient:

$$\frac{\delta F[x]}{\delta x(t)} = \lim_{dx,dt\to 0} \frac{F\left[x(\cdot) + dx\,\mathrm{rect}\left(\frac{\cdot - t}{dt}\right)\right] - F[x]}{dx\,|dt|}$$

where rect (τ) is one for $-\frac{1}{2} < \tau \leq \frac{1}{2}$ and zero otherwise. For example, if

$$F[x] = \int_0^T dt_1 \int_0^T dt_2\, K(t_1, t_2)x(t_1)x(t_2)$$

where K is symmetric, then

$$\frac{F\left[x(\cdot) + dx\,\mathrm{rect}\left(\frac{\cdot - t}{dt}\right)\right] - F[x]}{dx\,|dt|}$$

$$= \int_0^T dt_1 \int_0^T dt_2\, K(t_1, t_2)x(t_1)\,\frac{\mathrm{rect}\left(\frac{t_2 - t}{dt}\right)}{|dt|}$$

$$+ \int_0^T dt_1 \int_0^T dt_2\, K(t_1, t_2)x(t_2)\,\frac{\mathrm{rect}\left(\frac{t_1 - t}{dt}\right)}{|dt|} + O(dx).$$

Thus,

$$\frac{\delta F[x]}{\delta x(t)} = 2\int_0^T dt_1\, K(t, t_1)x(t_1).$$

More detailed definitions and examples may be found in the literature.[6]–[8]

As an example, note that a functional operator such as $F[x]$ may describe the input-output relation of a nonlinear dynamic system over $0 \leq t \leq T$, or it may define the square of the output or the square of the input integrated over $0 \leq t \leq T$.

For example,

$$F[x] = \int_0^T dt_1 \left[\int_0^T dt_2\, g(t_1, t_2)x(t_2)\right]^2$$

$$= \int_0^T dt_1 \int_0^T dt_2 \int_0^T dt_3\, g(t_1, t_2)g(t_1, t_3)x(t_2)x(t_3) \tag{2}$$

r

$$F[x](t) = \int_0^t dt_1 \int_0^T dt_2 \int_0^T dt_3\, g(t_1, t_2)g(t_1, t_3)x(t_2)x(t_3), \qquad 0 \le t \le T \tag{3}$$

where $g(t_1, t_2)$ is the impulse response of some linear system, $x(t_2)$ is its input, and $F[x]$ is then its integrated square output as a nonlinear functional. In (3) note the variable upper limit of integration; this indicates that $F[x](t)$ is a nonlinear functional operator or mapping, whereas $F[x]$ in (2) is a functional associating a number with each x. Further,

$$E\frac{\delta F[x]}{\delta x(t')} = 0, \qquad E\frac{\delta^n F[x]}{\delta x(t_1)\cdots\delta x(t_n)} = 0, \quad n > 2,$$

$$E\frac{\delta^2 F[x]}{\delta x(t')\,\delta x(t'')} = \int_0^T dt_1\, g(t_1, t')g(t_1, t'') = \frac{\delta EF[x]}{\delta r(t', t'')}.$$

The last expression may represent the sensitivity of the integral square output to the covariance function (that is, to the specific character of the Gaussian random input). More complex examples of functional derivatives may be found in the literature.[6]-[8],[11]

The formula (1b) is analogous for nonlinear *dynamic systems* to certain relations proposed earlier[2],[3] for nonlinear *memoryless systems*. Specifically, Price[2] has given a formula relating to the outputs of memoryless nonlinear devices with Gaussian inputs. The simplest version of this formula can be stated as follows.

Let x_1, x_2 be random variables with Gaussian joint distribution and zero mean, and let $f_1(x)$, $f_2(x)$ be continuously differentiable real-valued functions. Then

$$\frac{\partial}{\partial r_{12}} Ef_1(x_1)f_2(x_2) = E\frac{df(x_1)}{dx_1}\frac{df_2(x_2)}{dx_2}, \tag{4}$$

where $r_{12} = Ex_1x_2$.

McMahon[3] has generalized Price's formula to the form

$$\frac{\partial}{\partial r_{12}} Ef(x_1, x_2) = E\frac{\partial^2 f}{\partial x_1\,\partial x_2}, \tag{5}$$

where $f(x_1, x_2)$ is a twice continuously differentiable function of two variables.

The proof of formula (1b) will now be given. Although it holds for the general nonzero mean case, for the sake of the exposition we restrict ourselves here to proving the $m(t) = 0$ case.

A functional polynomial is an expression of the form

$$P[x] = K_0 + \sum_{n=1}^{N} \int_0^T dt_1 \cdots \int_0^T dt_n\, K_n(t_1, \cdots, t_n) \prod_{i=1}^{n} x(t_i). \tag{6}$$

Since the formula (1b) is linear, the proof for a general functional polynomial follows from the proof of the case of a homogeneous functional of degree n; that is, the nth term on the right side of (6).

$$F(x) = \int_0^T dt_1 \cdots \int_0^T dt_n\, K_n(t_1, \cdots, t_n) \prod_{i=1}^{n} x(t_i). \tag{7}$$

Clearly, one need only consider the case when n is even, since it holds trivially in the case of odd n. Let $n = 2m$. Note that, by a well-known[1] property of Gaussian processes,

$$E\prod_{i=1}^{2m} x(t_i) = \sum r(t_{i_1}, t_{i_1'}) \cdots r(t_{i_m}, t_{i_m'}) \tag{8}$$

where the sum is overall $(2m-1)(2m-3)\cdots 3\cdot 1$ ways of grouping the indices $1, \cdots, 2m$ into m pairs. Since the kernel K_{2m} of F is symmetric[2] in all its variables, we may apply the expectation operator to both sides of (7) and take into consideration[9] that, by the continuity of r, integration can be exchanged with the expectation operator E to obtain

$$EF[x] = \frac{(2m)!}{2^m m!}\int dt_1 \int dt_1' \cdots \int dt_m \int dt_m'\, K_{2m}(t_1, t_1', \cdots, t_m, t_m') \prod_{i=1}^{m} r(t_i, t_i'). \tag{9}$$

On the other hand,

$$\frac{\delta F[x]}{\delta x(t)\,\delta x(t')} = 2m(2m-1)\int dt_1 \cdots \int dt_{2m-2}\, K_{2m}(t, t', t_1, t_1', \cdots, t_{2m-2}') \prod_{i=1}^{2m-2} x(t_i). \tag{10}$$

Taking the expection of (10) yields

$$E\frac{\delta F[x]}{\delta x(t)\,\delta x(t')} = \frac{(2m)!}{2^{m-1}(m-1)!}\int dt_1 \int dt_1' \cdots \int dt_{m-1}\int dt_{m-1}'\, K_{2m}(t, t', t_1, t_1', \cdots, t_{m-1}, t_{m-1}') \prod_{i=1}^{m-1} r(t_i, t_i'). \tag{11}$$

Using the definition of functional derivative and the symmetry of r, we see that (11) is the functional derivative of (9) with respect to $r(t, t')$. This completes the proof of (1b) for functional polynomials with $m(t) = 0$.

The validity of the formula (1a) may be similarly tested by trying it for the homogeneous functional (7). We compute the expectation of F and find

$$EF[x] = \int dt_1 \cdots \int dt_n\, K_n \cdot E\{(m(t_1) + \xi(t_1)) \cdots (m(t_n) + \xi(t_n))\}$$

where $\xi(t)$ is a mean zero Gaussian random process. Thus, since ξ is independent of $m(t)$,

[1] This may be derived from the moment generating function by differentiation. The moment generating function of a multivariable Gaussian density is given, for example, in K. S. Miller, *Multidimensional Gaussian Distributions*. New York: Wiley, 1964, p. 71.

[2] K_{2m} may be assumed symmetric; otherwise, it may be replaced by an equivalent symmetric kernel.

$$\frac{\delta}{\delta m(t)} EF[x] = n \int dt_2 \cdots \int dt_n K_n(t, t_2, \cdots, t_n) \cdot E\{(m(t_2) + \xi(t_2)) \cdots (m(t_n) + \xi(t_n))\} \tag{12}$$

where, as before, K_n has been assumed symmetric.

On the other hand,

$$\frac{\delta}{\delta x(t)} F[x] = n \int dt_2 \cdots \int dt_n K_n(t, t_2, \cdots, t_n) x(t_2) \cdots x(t_n). \tag{13}$$

The expectation of (13) is identical to (12). This establishes (1a) for homogeneous functionals and, by linearity, for the class of all functional polynomials.

Novikov's Formula and its Connection with Bussgang's Theorem

Novikov's formula can be summarized as follows: Let $\{x(t), 0 \leq t \leq T\}$ be a Gaussian random process with mean zero and covariance function $r(t, t')$ continuous in t and t'. Let $F[x]$ be a functional which is once differentiable. Then

$$Ex(t)F[x] = \int_0^T r(t, t') E \frac{\delta F[x]}{\delta x(t')} dt'. \tag{14}$$

This formula was first reported by Novikov[4] in a slightly different form. The reader may refer to Novikov's paper for its proof.

Bussgang's theorem[1] states the following: Let $f(x)$ be an arbitrary, real-valued continuously differentiable function of x for all real x. Then, if $x(t)$ is a sample function of a Gaussian process as described above,

$$Ex(t)f(x(t_1)) = r(t, t_1)f'(x(t_1)), \tag{15}$$

where $f'(x) = df/dx$. Note that a function $f(x)$ of this type may represent a memoryless nonlinearity.

This theorem is clearly a special case of the Price-McMahon theorem,[2],[3] as may be seen by using the substitution $f_1(x) = x$, $f_2(x) = f(x)$ in (4).

Now it is easy to show that (15) is a consequence of the more general (14) for the case in which $f(x)$ is differentiable. Let $F[x] = f(x(t_1))$. The left-hand sides of (14) and (15) are then the same, while

$$\frac{\delta}{\delta x(t')} f(x(t)) = f'(x(t_1)) \frac{\delta x(t_1)}{\delta x(t')} = f'(x(t_1))\, \delta(t' - t_1) \tag{16}$$

where $\delta(\cdot)$ is the Dirac delta function.

When (16) is substituted into (14) and the integration over t' is carried out (by using the conventional definitions associated with the delta function), the right-hand side of (15) is obtained.

Hence, it is seen that Novikov's theorem[4] is a generalization to functionals of Bussgang's theorem,[1] although the former was introduced directly and not based on its antecedent.

Generalization of Novikov's Formula

Let the *Hermite-Wiener* functionals $H_n[Q_n; x]$, $n = 0, 1, \cdots$ corresponding to an arbitrary sequence Q_0, $Q_1(t_1), \cdots, Q_n(t_1, \cdots, t_n), \cdots$ of square-integrable symmetric kernels be defined as

$$H_n[Q_n; x] = \int_0^T dt_1 \cdots \int_0^T dt_n\, Q_n(t_1, \cdots, t_n) \cdot H^{(n)}(x(t_1) \cdots x(t_n)),$$

where the $H^{(n)}$ are the generalization of the n-variant Hermite polynomials[10]-[12] $H^{(0)} = 1$, $H^{(1)}(x_i) = x_i$, $H^{(2)}(x_i, x_j) = x_i x_j - \delta_{ij}$, etc. and are obtained from them by replacing the Kronecker delta function δ_{ij} by the covariance function $r(t_i, t_j)$.

Now the *generalized Novikov formula* can be stated as follows: Let $\{x(t), 0 \leq t \leq T\}$ be a Gaussian random process with mean zero and covariance function $r(t, t')$ which is continuous in t and t'. Let $F[x]$ be a functional which is differentiable n times. Then

$$EH_n[Q_n, x]F[x] = \int_0^T dt_1 \int_0^T dt_1' \cdots \int_0^T dt_n \int_0^T dt_n'\, Q_n(t_1, \cdots, t_n) \cdot E \frac{\delta^n F[x]}{\delta x(t_1') \cdots \delta x(t_n')} \prod_{i=1}^{n} r(t_i, t_i'). \tag{17}$$

Note that Novikov's formula is the $n = 1$ case.

This may be proved using the orthogonality of the Hermite-Wiener functionals and the expansion of functionals using such an orthogonal series; however, here we give a simpler proof which depends on (14) and the following recurrence relation, which holds for Hermite-Wiener functionals:

$$H_n[Q_n; x] = -\int_0^T dt \int_0^T dt'\, r(t, t') \frac{\delta}{\delta x(t')} H_{n-1}[Q_n(t, \cdots); x] + \int_0^T dt\, x(t) H_{n-1}[Q_n(t, \cdots); x], \tag{18}$$

where $H_{n-1}[Q_n(t, \cdots); x]$ means that $Q_n(t, t_1, \cdots, t_{n-1})$ for t fixed is to be used as the kernel of the functional H_{n-1}.

This recurrence relation (18) follows from the corresponding one for n-variate Hermite polynomials proved by Grad[10] and the definition of the functional Hermite-Wiener polynomials. It will be shown that (17) follows by induction on n. The case $n = 0$ is trivial. Assume (17) holds for Hermite-Wiener functional polynomials up to and including those of degree $n - 1$. To prove that it holds for those of degree n, proceed as follows using (18):

$$EH_n[Q_n; x]F[x]$$

$$= -\int_0^T dt \int_0^T dt'\, r(t, t')EF[x] \frac{\delta}{\delta x(t')} H_{n-1}[Q_n(t, \cdots); x]$$

$$+ \int_0^T dt\, Ex(t)H_{n-1}[Q_n(t, \cdots); x]F[x]$$

$$= -\int_0^T dt \int_0^T dt'\, r(t, t')E \frac{\delta}{\delta x(t')} F[x]H_{n-1}[Q_n(t, \cdots); x]$$

$$+ \int_0^T dt \int_0^T dt'\, r(t, t')EH_{n-1}[Q_n(t', \cdots); x] \frac{\delta F[x]}{\delta x(t')}$$

$$+ \int_0^T dt\, Ex(t)H_{n-1}[Q_n(t, \cdots); x]F[x], \qquad (19)$$

or, applying (11) and the inductive hypothesis,

$$EH_n[Q_n; x]F[x] = \int_0^T dt \int_0^T dt'\, r(t, t') \int_0^T dt_1 \int_0^T dt_1' \cdots$$

$$\cdot \int_0^T dt_{n-1} \int_0^T dt_{n-1}' Q_n(t, t_1 \cdots t_{n-1})$$

$$\cdot E \frac{\delta F[x]}{\delta x(t')\, \delta x(t_1') \cdots \delta x(t_{n-1}')} \prod_{i=1}^{n-1} r(t_i, t_i')$$

$$= \int_0^T dt_1 \int_0^T dt_1' \cdots \int_0^T dt_n \int_0^T dt_n'\, Q_n(t_1, \cdots, t_n)$$

$$\cdot E \frac{\delta F[x]}{\delta x(t_1') \cdots \delta x(t_n')} \prod_{i=1}^{n} r(t_i, t_i'). \qquad (20)$$

This completes the proof of (17).

Note, as a simple application of formula (17), that the orthogonality[5] $EH_n[Q_n; x]H_m[Q_m; x] = 0 \; n \neq m$ of the Hermite-Wiener functionals follows trivially.

The formula (17) may be used directly to compute, for example, the kernels in the Hermite-Wiener expansion of a functional. If it is desired to expand the functional operator $F[x](t)$ in a series of orthogonal functionals of the form

$$F[x](t) = \sum_{n=0}^{\infty} \int_{-\infty}^{\infty} dt_1 \cdots \int_{-\infty}^{\infty} dt_n\, K_n(t_1, \cdots, t_n)$$

$$\cdot H^{(n)}(x(t - t_1) \cdots x(t - t_n)), \qquad (21)$$

then the kernels K_n are given by[11]

$$K_n(\tau_1, \cdots, \tau_n)$$

$$= \frac{1}{n!} EF[x](t)H^{(n)}(x(t - \tau_1), \cdots, x(t - \tau_n)). \qquad (22)$$

Here $x(t)$ is a stationary Gaussian random process with mean zero. The expectation of the product on the right-hand side of (22) is identical to the left-hand side of (17) if the substitution

$$Q_n(t_1, \cdots, t_n) = \prod_{i=1}^{n} \delta(t_i - (t - \tau_i))$$

is made in (17), where $\delta(\cdot)$ is the usual Dirac delta function. If expressions (22) and (17) are thus combined, we find the following formula for the kernels K_n:

$$K_n(\tau_1, \cdots, \tau_n) = \frac{1}{n!} \int_{-\infty}^{\infty} dt_1' \cdots$$

$$\cdot \int_{-\infty}^{\infty} dt_n'\, E \frac{\delta^n F[x](t)}{\delta x(t_1') \cdots x(t_n')} \prod_{i=1}^{n} r(t - \tau_i - t_i'). \qquad (23)$$

For example, if $F[x](t) = x^2(t + t_0)$ where t_0 is a fixed positive constant, then, according to (23), $K_n \equiv 0, n > 2$, and

$$K_2(\tau_1, \tau_2) = r(\tau_1 - t_0)r(t_2 - t_0)$$

$$K_1(\tau_1, \tau_2) \equiv 0$$

$$K_0 = r(0)\left[\int_{-\infty}^{+\infty} dt\, r(t)\right]^2.$$

Conclusions

The principal results of the paper fall into two categories. Connections are pointed out for the first time between developments which took place in quite widely divergent fields. The theorems of Bussgang, Price, and McMahon resulted from information theory approaches to memoryless nonlinearities. Novikov pursued functional analysis studies applied to turbulence, and Wiener was concerned with nonlinear dynamic systems by functional expansion.

In addition to establishing connections between these areas, the paper also establishes two original relationships. One is a functional calculus analog of the Price-McMahon theorem. The other is a general relation involving Wiener's homogeneous functionals. This latter includes Novikov's theorem as a special case.

References

[1] J. J. Bussgang, "Crosscorrelation functions of amplitude distorted Gaussian signals," Res. Lab. Elec., M.I.T. Tech. Rept. 216, sec. 3, March 1952.

[2] R. Price, "A useful theorem for nonlinear devices having Gaussian inputs," *IRE Trans. Information Theory*, vol. IT-4, pp. 69–72, June 1958.

[3] E. L. McMahon, "An extension of Price's theorem," *IEEE Trans. Information Theory (Correspondence)*, vol. IT-10, p. 168, April 1964.

[4] E. A. Novikov, "Functionals and the random-force method in turbulence theory," *Soviet Physics—JETP*, vol. 20, p. 1290, 1965.

[5] N. Wiener, *Nonlinear Problems in Random Theory*. New York: Wiley, 1958, p. 28.

[6] D. Gorman and J. Zaborszky, "Functional representation of nonlinear systems, interpolation, and Lagrange expansion for functionals," *ASME Trans. Basic Engrg.*, pp. 429–436, June 1966.

[7] J. Zaborszky and D. Gorman, "Control by functional Lagrange expansion," *Proc. 1966 IFAC Cong.*, vol. 1. London: Institution of Mechanical Engineers, book 2, paper 18D.

[8] D. Gorman and J. Zaborszky, "Functional Lagrange expansion in state space and the *s* domain," *IEEE Trans. Automatic Control*, vol. AC-11, pp. 498–505, July 1966.

[9] M. Loeve, *Probability Theory*. Princeton, N. J.: Van Nostrand, 1960, p. 472.

[10] H. Grad, "Note on *N*-dimensional Hermite polynomials," *Commun. Pure Appl. Math.*, vol. 2, p. 325, 1949.

[11] L. A. Zadeh, "On the representation of nonlinear operators," *1957 IRE WESCON Conv. Rec.*, pt. 2, pp. 105–113.

[12] D. A. Chester, Res. Lab. Elec., M.I.T., Tech. Rept. 366, 1960.

8

Reprinted from *IEEE Trans. Information Theory*, **IT-9,** 74–84 (Apr. 1963)

General Results in the Mathematical Theory of Random Signals and Noise in Nonlinear Devices*

H. B. SHUTTERLY†, MEMBER, IEEE

Summary—**An analysis is made of the output resulting from passing signals and noise through general zero memory nonlinear devices. New expressions are derived for the output time function and autocorrelation function in terms of weighted averages of the nonlinear characteristic and its derivatives. These expressions are not restricted to Gaussian noise and apply to any nonlinearity having no more than a finite number of discontinuities. The method of analysis used is heuristic.**

I. Introduction

Rice,[1] Bennett,[2] Middleton,[3,4] Campbell[5] and Price[6] have determined the output autocorrelation function and power spectrum for sinusoidal signals and Gaussian noise through a number of specific nonlinear characteristics, principally the half-wave ν-th law rectifier. Most of this work has been done using the transform method described by Rice. In this paper a real-plane method of analysis is used and expressions for the output time and autocorrelation functions are obtained which are applicable to general nonlinear devices with general inputs.

The organization of the paper is as follows: Section II states the basic results and presents one illustrative example.

In Section III the various results of the paper are derived. These consist of series expressions for the time and autocorrelation functions corresponding to general nonlinear transformations of random signal and noise processes.

Section IV discusses the relation between the expressions obtained in this paper and the expressions previously obtained by the transform method. It is shown that one of the expressions obtained in this paper for an input of Gaussian noise and a single sinusoidal signal is easily obtainable from the general transform solution. Readers familiar with the transform method may wish to read this section before reading the derivation in Section III, since a general solution for a Gaussian noise input is obtained here very simply and quickly.

* Received August 10, 1961; revised manuscript received, May 21, 1962.

† Westinghouse Research and Development Center, Pittsburgh, Pa. Formerly with Sperry Gyroscope Company, Great Neck, L. I., N. Y.

[1] S. O. Rice, "Mathematical analysis of random noise," *Bell Syst. Tech. J.*, vol. 24, pp. 46–156; January, 1946.

[2] W. R. Bennett, "Response of a linear rectifier to signal and noise," *J. Acoust. Soc. Am.*, vol. 15, pp. 164–172; January, 1944.

[3] D. Middleton, "Rectification of a sinusoidally modulated carrier in the presence of noise," Proc. IRE, vol. 36, pp. 1467–1477; December, 1948.

[4] D. Middleton, "Some general results in the theory of noise through nonlinear devices," *Quart. Appl. Math.*, pp. 445–498; January, 1948.

[5] L. Lorne Campbell, "Rectification of two signals in random noise," IRE Trans. on Information Theory, vol. IT-2, pp. 119–124; December, 1956.

[6] Robert Price, "A useful theorem for nonlinear devices having Gaussian inputs," IRE Trans. on Information Theory, vol. IT-4, pp. 69–72; June, 1958.

II. Fundamental Results

A. General Series Expansion for the Output Time Function

It is assumed that the transfer characteristic of the nonlinear device has no more than a finite number of discontinuities and can be represented by $y(t) = g[z(t)]$. The input to the device is assumed to be the sum of a signal and a noise, $z(t) = s(t) + x(t)$ where $s(t)$ and $x(t)$ are sample functions from random processes. The signal process is assumed statistically independent of the noise process but is otherwise unrestricted. The noise process is assumed to be ergodic and to have no dc or periodic components. Under these conditions, the output time function can be expressed as

$$y(t) = \sum_{p=0}^{\infty} \sum_{k=0}^{\infty} \frac{C_{pk} s(t)^p x(t)^k}{p!\, k!} \tag{1}$$

where

$$C_{pk} = \sum_{v=0}^{\infty} \alpha_v \overline{g^{(p+k+v)}(x)}, \tag{2}$$

$g^{(p+k+v)}(x)$ denotes the $(p + k + v)$th derivative of $g(x)$.[7] The bar over it indicates the statistical average with respect to x, *i.e.*, over the noise ensemble.

The α_v are constants determined by the moments of the noise process, which are independent of the nonlinear characteristic

$$\alpha_0 = 1, \quad \alpha_1 = -\bar{x} = 0, \quad \alpha_2 = -\frac{\overline{x^2}}{2!}, \quad \alpha_3 = -\frac{\overline{x^3}}{3!}$$

$$\alpha_4 = -\frac{\overline{x^4}}{4!} + \frac{\overline{x^2}\cdot\overline{x^2}}{2!\,2!}, \qquad \alpha_5 = -\frac{\overline{x^5}}{5!} + \frac{2\overline{x^3}\cdot\overline{x^2}}{3!\,2!}$$

and, in general, for $v \geq 4$,

$$\alpha_v = -\frac{\overline{x^v}}{v!} + \sum_{r_1=1}^{v-1} \sum_{r_2=1}^{v-r_1} \frac{\overline{x^{r_1}}\cdot\overline{x^{r_2}}}{r_1!\,r_2!} - \sum_{r_1=1}^{v-2} \sum_{r_2=1}^{v-r_1-1} \sum_{r_3=1}^{v-r_1-r_2} \frac{\overline{x^{r_1}}\cdot\overline{x^{r_2}}\cdot\overline{x^{r_3}}}{r_1!\,r_2!\,r_3!} + \cdots (-1)^v \bar{x}^v. \tag{3}$$

It should be understood that, in general, the usefulness of (1) is based on the assumption that one is interested in certain selected terms of the output, *e.g.*, $s(t)$ or the principle terms contributing energy to a certain frequency band, etc. It may also be noted that, in general, the number of significant terms in the series expression for C_{pk} is expected to decrease as the input noise power increases, although the coefficients α_v may behave oppositely. Alternate expressions for $y(t)$ are given in Section III for the special case of $s(t) = A \cos(wt + \phi)$, with ϕ uniformly distributed, 0 to 2π. Relative advantages of the various expressions given depends on the input snr.

[7] When the derivatives of $g(x)$ are discontinuous, $\overline{g^{(p+k+v)}(x)}$ can be obtained as a limit for a continuous function or, more simply, by the introduction of suitable impulse functions. An example of the latter method follows in Section II-C; one limitation is that the interval over which $g^{(p+k+v)}(x)$ is averaged must include all discontinuities which are encountered by the sum of signal plus noise (see Section III-G).

B. Expressions for the Output Autocorrelation Function

Under the same conditions as given in Section II-A, the output autocorrelation function is given by

$$R_y(t_1, t_2) = \overline{y(t_1)y(t_2)}$$

$$= \sum_{p,k,q,l=0}^{\infty} \frac{C_{pk}C_{ql}\, \overline{s(t_1)^p s(t_2)^q}\ \overline{x(t_1)^k x(t_2)^l}}{p!\, k!\, q!\, l!} \tag{4}$$

where the coefficients C_{pk} and C_{ql} are defined by (2), and the notation under the summation sign means that each of the four variables ranges independently over the positive integers.

In the important case of Gaussian input noise, the autocorrelation can be expressed much more simply as follows:

$$R_y(t, t + \tau) = \sum_{p,q,k=0}^{\infty} \frac{\overline{g^{(k+p)}(x)}\ \overline{g^{(k+q)}(x)}\ \overline{s(t)^p s(t+\tau)^q}\, R_x(\tau)^k}{p!q!k!} \tag{5}$$

where t and $t + \tau$ have been substituted for t_1 and t_2, and $R_x(\tau)$ is the autocorrelation function of the Gaussian input noise.

Additional expressions for the autocorrelation function are given in Section III for $s(t) = A \cos(wt + \phi)$.

C. Example—Half-Wave Square-Law Device

The purpose of this example is to illustrate the evaluation of $\overline{g^{(k)}(x)}$ for a characteristic with discontinuous derivatives.

A half-wave square-law device can be represented by

$$g(z) = Kz^2 \quad z \geq 0; \qquad = 0 \quad z < 0. \tag{6}$$

It follows that the second derivative is discontinuous at the origin, and is given by

$$g^{(2)}(z) = 2K \quad z \geq 0; \qquad = 0 \quad z < 0. \tag{7}$$

The third derivative can therefore be represented by a unit impulse of magnitude $2K$

$$g^{(3)}(z) = 2K\delta(z). \tag{8}$$

Then making the assumption that the input noise is Gaussian, the statistical averages of the derivatives are given by

$$\overline{g(x)} = \int_0^\infty (Kx^2) \frac{1}{\sqrt{2\pi}\sigma} e^{-x^2/2\sigma^2}\, dx = \frac{K\sigma^2}{2} \tag{9}$$

$$\overline{g^{(1)}(x)} = \int_0^\infty (2Kx) \cdot \frac{1}{\sqrt{2\pi}\sigma} e^{-x^2/2\sigma^2}\, dx = \frac{2K\sigma}{\sqrt{2\pi}} \tag{10}$$

$$\overline{g^{(2)}(x)} = K \tag{11}$$

and

$$\overline{g^{(k)}(x)} = 2K\overline{\delta^{(k-3)}(x)} \quad \text{for } k \geq 3. \tag{12}$$

The ensemble averages of derivatives of an impulse function can be found from the following expression:[8]

$$\int_{-\infty}^{\infty} f(x)\delta^{(n)}(x - x_0)\,dx = (-1)^n f^{(n)}(x_0). \tag{13}$$

Thus, letting

$$f(x) = \frac{1}{\sqrt{2\pi}\sigma} e^{-x^2/2\sigma^2}$$

and $x_0 = 0$, we find

$$\overline{\delta^{(n)}(x)} = (-1)^n \frac{d^n}{dx^n}\left(\frac{1}{\sqrt{2\pi}\sigma} e^{-x^2/2\sigma^2}\right)\Bigg|_{x=0}. \tag{14}$$

This is most simply evaluated by noting that

$$e^{-x^2/2\sigma^2} = \sum_{k=0}^{\infty} \frac{(-1)^k}{k!}\left(\frac{x^2}{2\sigma^2}\right)^k$$

which gives

$$\overline{\delta^{(n)}(x)} = \frac{(-1)^{n/2} n!}{\sqrt{2\pi} 2^{n/2}\sigma^{n+1}\left(\frac{n}{2}\right)!} \qquad (n \text{ even}) \tag{15}$$

$$= 0 \qquad (n \text{ odd})$$

Thus,

$$\overline{g^{(k)}(x)} = 2K\overline{\delta^{(k-3)}(x)} = \frac{2\mathbf{K}(-1)^{(k-3)/2}(k-3)!}{\sqrt{2\pi}2^{(k-3)/2}\sigma^{k-2}\left(\frac{k-3}{2}\right)!} \tag{16}$$

$$\text{for } k \geq 3, k-3 \text{ even}$$

$$= 0 \qquad \text{for } k \geq 3, k-3 \text{ odd}.$$

Substituting the values of $\overline{g^{(k)}(x)}$ into (5) then determines the output autocorrelation function. For example, if we consider a noise-only input, *i.e.*, let $s(t) = 0$ in (5), then the output autocorrelation function is given by

$$R_y(\tau) = \sum_{k=0}^{\infty} \frac{[\overline{g^{(k)}_{(x)}}]^2}{k!} R_x(\tau)^k \tag{17}$$

$$= \frac{K^2\sigma^4}{4} + \frac{2K^2\sigma^2}{\pi} R_x(\tau) + \frac{K^2}{2} R_x(\tau)^2$$

$$+ \frac{K^2}{3\pi\sigma^2} R_x(\tau)^3 + \frac{K^2}{60\pi\sigma^6} R_x(\tau)^5$$

$$+ \frac{K^2}{280\pi\sigma^{10}} R_x(\tau)^7 + \cdots .$$

III. Derivation of Series Expansions for the Output Time and Autocorrelation Function

A. *Specification of the Input Signal and Noise Processes*

In obtaining the initial and most general results, the signal process $\{s(t)\}$ will be assumed to be independent of the noise process but will be otherwise unrestricted. For more specialized results, the signal process will be assumed to have sample functions of the form,

$$s(t) = A\cos(wt + \phi) \tag{18}$$

where the random variable ϕ is assumed independent of the noise process and uniformly distributed in the interval 0 to 2π.

The input noise process $\{x(t)\}$ is assumed to be ergodic and such that

$$\int_{-\infty}^{\infty} | R_x(\tau) | \, d\tau < \infty \tag{19}$$

where $R_x(\tau)$ is the autocorrelation function of the input noise process.

It will also be necessary in the analysis to express the sample noise function $x(t)$ in terms of a Fourier series. For convenience in doing this, we will work with a new function $x_T(t)$ which is defined to be periodic of period T and identical to $x(t)$ in the interval $-T/2$ to $T/2$. Then,

$$x_T(t) = \sum_{n=-\infty}^{\infty} x_n e^{jnw_0 t} \tag{20}$$

where

$$x_n = \frac{1}{T}\int_{-T/2}^{T/2} x(t)e^{-jnw_0 t}\,dt \tag{21}$$

and

$$w_0 = \frac{2\pi}{T}.$$

Results obtained with this series can then be made to apply to the original function $x(t)$ by letting the interval T approach infinity.

The coefficients x_n in the above expansion are random variables, since they may differ in value for each sample function of the noise process. Two important statistical properties of these coefficients that follow from (19) are[9]

$$\lim_{T\to\infty} \overline{x_n} = 0 \tag{22}$$

$$\lim_{T\to\infty} \overline{x_n^2} = 0. \tag{23}$$

B. *Expansion of the Output Function in Terms of a Taylor Series*

We will first assume that the nonlinear transfer characteristic $y = g(z)$ possesses continuous derivatives of all orders for every z, which permits the expansion of $g(z)$ in a Taylor series. (It will be shown in Section III-G that the results obtained are valid for nonlinear characteristics possessing finite discontinuities.) Working with $x_T(t)$ we have

$$g[x_T(t) + s(t)] = \sum_{p=0}^{\infty} \frac{s(t)^p}{p!} g^{(p)}[x_T(t)] \tag{24}$$

where $g^{(p)}[x]$ denotes the pth derivative of $g[x]$.

[8] See for example, Davenport and Root, "Random Signals and Noise," McGraw Hill Book Company, Inc., New York, N. Y., Appendix 1, pp. 369–370; 1958.

[9] Davenport and Root, *op. cit.*, pp. 93–96.

The purpose in making the above expansion is to separate the signal and noise terms and to show explicitly the various powers of $s(t)$. Next, we will expand $g^{(p)}[x_T(t)]$ in terms of the Fourier components of $x_T(t)$ to make explicit the various powers of the noise frequency components. To simplify the expressions to follow, we first introduce the following notation:

$$x_T(t) = \sum_{-\infty}^{\infty} x_n e^{jnw_0 t} \equiv \sum_a + \sum_\Delta \qquad (25)$$

where

$$\sum_\Delta \equiv \sum_\Delta x_n e^{jnw_0 t}$$

is a finite number of terms selected from the infinite series for $x_T(t)$, and

$$\sum_a \equiv \sum_a x_n e^{jnw_0 t}$$

$$= \sum_{-\infty}^{\infty} x_n e^{jnw_0 t} - \sum_\Delta x_n e^{jnw_0 t}.$$

Then,

$$g^{(p)}[x_T(t)] = \sum_{k=0}^{\infty} \frac{g^{(p+k)}(\sum_a)}{k!} (\sum_\Delta)^k. \qquad (26)$$

Thus, from (24) and (26)

$$g[x_T(t) + s(t)] = \sum_{P=0}^{\infty} \frac{s(t)^p}{P!} \sum_{k=0}^{\infty} \frac{g^{(p+k)}(\sum_a)}{k!} (\sum_\Delta)^k. \qquad (27)$$

C. Determination of the Output Magnitude of Products of Fourier Noise Components.

The output noise factor, $g^{(p)}[x_T(t)]$, consists of sums of products of the input Fourier components, as can be seen from (26) where $(\sum_\Delta)^k$ represents a sum of Fourier components raised to the kth power. The purpose of this section is to determine the magnitude, or coefficient, of each particular product of Fourier noise components contained in $g^{(p)}[x_T(t)]$.

We first note that all of the Fourier component products contained in $(\sum_\Delta)^n$ are of the nth order; these are the products for which the output magnitude will be determined. Referring to (26), it is apparent that $g^{(k+p)}(\sum_a)$ cannot contain any of the products in $(\sum_\Delta)^n$, since by definition $\sum_a$ does not contain the required Fourier components. (It should be understood that $g^{(k+p)}(\sum_a)$ may contain other products with resultant frequencies equal to those of the products in $(\sum_\Delta)^n$, but we are interested only in products of the particular Fourier components contained in $\sum_\Delta$.) It is also apparent that all terms $(\sum_\Delta)^k$ for which $k > n$ do contain all of the products of $(\sum_\Delta)^n$. However, from (22) and (23) it follows that

$$\lim_{T\to\infty} \sum_\Delta = 0.$$

Consequently,

$$\lim_{T\to\infty} \frac{(\sum_\Delta)^{n+q}}{(\sum_\Delta)^n} = 0 \quad \text{for} \quad q \geq 1$$

and, therefore, for T sufficiently large, the only significant term containing the desired products is $(\sum_\Delta)^n$.

From (26) the term multiplying $(\sum_\Delta)^n$ is $g^{(p+n)}(\sum_a)/n!$. This term, of course, consists of time-varying periodic functions. If $g^{(p+n)}(\sum_a)/n!$ is expanded in a Fourier series, only the constant or dc term will not change the frequencies of the products contained in $(\sum_\Delta)^n$. Consequently, for T large, the non-time-varying coefficient of each product term contained in $(\sum_\Delta)^n$ is $1/n! \langle g^{(p+n)}[\sum_a]\rangle$, where $\langle\ \rangle$ denotes the time average or dc value.

Since

$$\lim_{T\to\infty} \sum_\Delta = 0,$$

$$\lim_{T\to\infty} \langle g^{(p+n)}(\sum_a)\rangle = \lim_{T\to\infty} \langle g^{(p+n)}[x_T(t) - \sum_\Delta]\rangle$$

$$= \lim_{T\to\infty} \langle g^{(p+n)}[x_T(t)]\rangle.$$

Therefore, for T large, the dc coefficient of $(\sum_\Delta)^n$ is independent of the particular Fourier components contained in $\sum_\Delta$. Consequently, we may now state that, for T sufficiently large, the dc coefficient of every product of n Fourier components contained in $g^{(p)}[x_T(t)]$ is $1/n! \langle g^{(p+n)}[x_T(t)]\rangle$.

It may also be noted that $\langle g^{(p+n)}[x_T(t)]\rangle$ is itself made up of sums of products of Fourier components which happen to have a resultant frequency of zero.

D. Synthesis of the Output Noise Factor in Terms of Fourier Components

As is made evident by the Taylor series expansion, the noise portion of the output of the nonlinear device is made up of various weighted powers of the input Fourier components. Since we have determined the coefficient, or weighting factor, for any possible product, we can now specify the complete noise output as a sum of products. The advantage of such an expression is that the nonlinear characteristic enters only in terms of the average values of its derivatives.

In developing the expression for the total output, it is essential to avoid systematically adding the same noise component products to the expression more than once.

For example, $[x_T(t)]^k \equiv [\sum_{-\infty}^{\infty} x_n e^{in\omega_0 t}]^k$ contains all possible products of k Fourier components, and $[x_T(t)]^{k-2}$ contains all possible products of order $k-2$. For T large, these products appear in the output multiplied by $1/k!\langle g^{(k+p)}[x_T(t)]\rangle$ and $1/(k-2)!\langle g^{(k-2+p)}[x_T(t)]\rangle$, respectively, as we have shown. However, note that $[x_T(t)]^k$ includes the following terms, which we will denote by E:

$$E \equiv \frac{k!}{2!\,(k-2)!}\left[\sum_{-\infty}^{\infty} x_n x_{-n}\right][x_T(t)]^{k-2}.$$

If both

$$\frac{\langle g^{(k-2+p)}[x_T(t)]\rangle}{(k-2)!}[x_T(t)]^{k-2}$$

and

$$\frac{\langle g^{(k+p)}[x_T(t)]\rangle}{k!}[x_T(t)]^{k}$$

are included in the expression for $g^{(p)}[x_T(t)]$, the terms denoted by E above will be included twice. This would result in a finite error, since

$$\lim_{T\to\infty}\sum_{-\infty}^{\infty} x_n x_{-n} = \lim_{T\to\infty}\langle x_T(t)^2\rangle \neq 0.$$

This problem of double counting is eliminated if we use explicitly only those products of the input Fourier components which cannot be partitioned into a dc and an ac component. Thus, a term such as $x_3x_{-3}x_9e^{i9\omega_0 t}$ should not be explicitly included as a third-order product since it will be counted as the first-order product $x_9e^{i9\omega_0 t}$. Thus, we count as a kth order product all terms of the form $x_n x_m \cdots x_q e^{i(n+m+\cdots q)\omega_0 t}$, where there are k Fourier components in the product (not necessarily different; *e.g.*, it could be one component to the kth power) and no two or more terms in the frequency determining coefficient $(n+m+\cdots+q)$ sum to zero. It should be understood that products of components which do not meet the above requirements are not discarded, they are simply counted as lower-order products. Thus, if two terms in the frequency coefficient $(n+m+\cdots+q)$ sum to zero and no other partial summation equals zero, then this product is counted as a $(k-2)$th order product.

We will denote the sum of all possible kth order products, as defined above, by $[x_T(t)]^k_{AC}$. It is important to recognize that $[x_T(t)]^k_{AC}$ is *not* just $[x_T(t)]^k$ with its average value subtracted, as the notation might ordinarily suggest. Thus,

$$[x_T(t)]^k_{AC} = \underbrace{\sum_n \cdots \sum_q}_{k\text{-fold}} x_n x_m \cdots x_q e^{i(n+m+\cdots+q)\omega_0 t} \tag{28}$$

where the summations are extended over all possible values from $-\infty$ to ∞ such that no two or more terms in the frequency determining coefficient $(n+m+\cdots+q)$, add to zero.

The output noise term for large T is then given by the sum of all possible output Fourier component products,

$$g^{(p)}[x_T(t)] = \sum_{k=0}^{\infty}\frac{\langle g^{(k+p)}[x_T(t)]\rangle}{k!}[x_T(t)]^k_{AC}. \tag{29}$$

Substituting (29) in (24), we obtain

$$g[x_T(t)+s(t)] = \sum_{p=0}^{\infty}\frac{s(t)^p}{p!} \cdot \sum_{k=0}^{\infty}\frac{\langle g^{(k+p)}[x_T(t)]\rangle}{k!}[x_T(t)]^k_{AC}. \tag{30}$$

Thus, we have

$$y(t) = \lim_{T\to\infty} g[x_T(t)+s(t)] = \sum_{p=0}^{\infty}\frac{s(t)^p}{p!}\sum_{k=0}^{\infty}\frac{\langle g^{(k+p)}[x(t)]\rangle}{k!}[x(t)]^k_{AC} \tag{31}$$

where

$$[x(t)]^k_{AC} \equiv \lim_{T\to\infty}[x_T(t)]^k_{AC}.$$

In order to use (31) quantitatively we will require a direct expression for $[x(t)]^k_{AC}$ in terms of $x(t)$. It is shown in Appendix A that

$$[x(t)]^k_{AC} = x(t)^k - \sum_{r=1}^{k}\frac{k!\,\overline{x^r}}{(k-r)!\,r!}[x(t)]^{k-r}_{AC} \tag{32}$$

or

$$\begin{aligned}[x(t)]^k_{AC} &= x(t)^k - \sum_{r=1}^{k}\frac{k!\,\overline{x^r}}{(k-r)!\,r!}x(t)^{k-r} \\ &+ \sum_{r_1=1}^{k-1}\sum_{r_2=1}^{k-r_1}\frac{k!\,\overline{x^{r_1}}\;\overline{x^{r_2}}}{r_1!\,r_2!\,(k-r_1-r_2)!}x(t)^{k-r_1-r_2} \\ &- \sum_{r_1=1}^{k-2}\sum_{r_2=1}^{k-r_1-1}\sum_{r_3=1}^{k-r_1-r_2} \cdot\frac{k!\,\overline{x^{r_1}}\;\overline{x^{r_2}}\;\overline{x^{r_3}}}{r_1!\,r_2!\,r_3!\,(k-r_1-r_2-r_3)!}x(t)^{k-r_1-r_2-r_3} \\ &+ \cdots\end{aligned} \tag{33}$$

where the last term is given by $(-1)^k\,k!\,\bar{x}^k$ and $\overline{x^r}$ denotes the statistical or ensemble average of x^r.

It can be seen from (33) that $x(t)^n$ (weighted by a constant) is a component of every $[x(t)]^k_{AC}$ for which $k \geq n$. Substituting (33) in (31) and collecting terms for each power of $x(t)$ we obtain the following basic result:

$$y(t) = \sum_{p=0}^{\infty}\sum_{k=0}^{\infty}\frac{c_{pk}s(t)^p x(t)^k}{p!\,k!} \tag{34}$$

where

$$c_{pk} = \sum_{v=0}^{\infty}\alpha_v\langle g^{(p+k+v)}[x(t)]\rangle \tag{35}$$

and

$$\begin{aligned}\alpha_0 &= 1 \\ \alpha_v &= -\frac{\overline{x^v}}{v!} + \sum_{r_1=1}^{v-1}\sum_{r_2=1}^{v-r_1}\frac{\overline{x^{r_1}}\;\overline{x^{r_2}}}{r_1!\,r_2!} \\ &- \sum_{r_1=1}^{v-2}\sum_{r_2=1}^{v-r_1-1}\sum_{r_3=1}^{v-r_1-r_2}\frac{\overline{x^{r_1}}\;\overline{x^{r_2}}\;\overline{x^{r_3}}}{r_1!\,r_2!\,r_3!} \\ &+ \cdots \qquad (v \geq 1).\end{aligned} \tag{36}$$

It should be noted that the constants α_v are independent of the nonlinear characteristic and can therefore be tabulated independently for any desired noise probability density function. Also, since $\bar{x} = 0$ (by assumption and without loss of generality) many terms of (36) vanish. As an illustration, the expressions for the first six values of α_v are as follows:

$$\alpha_0 = 1$$

$$\alpha_1 = 0$$

$$\alpha_2 = -\frac{\overline{x^2}}{2!}$$

$$\alpha_3 = -\frac{\overline{x^3}}{3!}$$

$$\alpha_4 = -\frac{\overline{x^4}}{4!} + \frac{\overline{x^2}\cdot\overline{x^2}}{2!\,2!}$$

$$\alpha_5 = -\frac{\overline{x^5}}{5!} + \frac{2\overline{x^3}\cdot\overline{x^2}}{3!\,2!}.$$

E. Calculation of the Output Autocorrelation Function

The output autocorrelation function $R_y(\tau)$ is given by

$$R_y(t_1, t_2) = \overline{y(t_1)y(t_2)}.$$

Therefore, using (34) and noting from the ergodic theorem that

$$\overline{\langle g^{(r)}[x(t_1)]\rangle \langle g^{(s)}[x(t_2)]\rangle} = \langle g^{(r)}[x(t)]\rangle \langle g^{(s)}[x(t)]\rangle = \overline{g^{(r)}(x)}\cdot\overline{g^{(s)}(x)}, \tag{37}$$

we obtain

$$R_y(t_1, t_2) = \sum_{p,k,q,l=0}^{\infty} \frac{C_{pk}C_{ql}\overline{s(t_1)^p s(t_2)^q}\ \overline{x(t_1)^k x(t_2)^l}}{p!\,k!\,q!\,l!} \tag{38}$$

where the variables p, k, q and l range independently over the positive integers, and

$$C_{pk} = \sum_{v=0}^{\infty} \alpha_v \overline{g^{(p+k+v)}(x)} \tag{39}$$

$$C_{ql} = \sum_{v=0}^{\infty} \alpha_v \overline{g^{(q+l+v)}(x)}. \tag{40}$$

The nonlinear characteristic enters only into the determination of the weighting factors C_{pk} and C_{ql}, which, as shown by (39) and (40), are determined by various derivatives of the nonlinear characteristic averaged over the noise ensemble. In a later result it will be shown that $R_y(t_1, t_2)$ may also be expressed such that the derivatives of the nonlinear characteristic are averaged with respect to the signal process.

To proceed further we specialize to a particular signal process. The sample functions will be assumed to be given by (18), *i.e.*, $s(t) = A\cos(wt + \phi)$. Rather than making the substitution directly into (38), it is convenient to return to (34) and to expand the factor $[\cos(wt + \phi)]^p$ of $s(t)^p$ into its Fourier components. Noting that

$$[\cos(wt + \phi)]^p = \sum_{\substack{m=0\\(m+p\ \text{even})}}^{p} \frac{\epsilon_m p!\cos m(wt + \phi)}{\left(\frac{p+m}{2}\right)!\left(\frac{p-m}{2}\right)!\,2^p} \tag{41}$$

where ϵ_m is the Neuman factor,

$$\epsilon_0 = 1$$

$$\epsilon_m = 2 \qquad (m = 1, 2, 3, \cdots),$$

we obtain

$$y(t) = \sum_{\substack{m=0\\(m+p\ \text{even})}}^{\infty}\sum_{k=0}^{\infty}\sum_{p=m}^{\infty} \frac{\epsilon_m A^p C_{pk} x(t)^k \cos m(wt + \phi)}{\left(\frac{p+m}{2}\right)!\left(\frac{p-m}{2}\right)!\,2^p k!}. \tag{42}$$

Then making the substitution $r = (p - m)/2$, we obtain

$$y(t) = \sum_{m=0}^{\infty}\sum_{k=0}^{\infty}\sum_{r=0}^{\infty} \frac{\epsilon_m A^{m+2r} C_{m+2r,k} x(t)^k \cos m(wt + \phi)}{(m + r)!\,r!\,2^{m+2r}k!}. \tag{43}$$

Using (43) to calculate $R_y(t_1, t_2)$ and noting that

$$\overline{\cos m(wt + \phi)\cos n[w(t + \tau) + \phi]} = 0 \qquad (n \neq m)$$

$$= \frac{\cos mw\tau}{\epsilon_m} \qquad (n = m) \tag{44}$$

yields the following result:

$$R_y(t, t + \tau) = \sum_{m,k,l=0}^{\infty} \frac{\epsilon_m f_{mk} f_{ml} \cos mw\tau\, \overline{x(t)^k x(t + \tau)^l}}{k!\,l!} \tag{45}$$

where

$$f_{mk} = \sum_{r=0}^{\infty} \frac{C_{m+2r,k}A^{m+2r}}{(m + r)!\,r!\,2^{m+2r}} \tag{46}$$

$$f_{ml} = \sum_{s=0}^{\infty} \frac{C_{m+2s,l}A^{m+2s}}{(m + s)!\,s!\,2^{m+2s}}. \tag{47}$$

In general, the number of significant terms in the series expressions for $C_{m+2r,k}$ and $C_{m+2s,l}$ are expected to decrease with increasing input noise power. Also it is apparent that the number of significant terms in (46) and (47) is reduced as the signal amplitude A is reduced. Eq. (45) is therefore particularly useful for small input signal-to-noise ratios; this is also true for the initial expression given by (38).

To obtain a more useful result for large input signal-to-noise ratios we must return to (31). First, note that

$$g^{(k)}[A\cos(wt + \phi) + x] = \sum_{p=0}^{\infty} g^{(k+p)}[x]\frac{[A\cos(wt + \phi)]^p}{p!}. \tag{48}$$

It is apparent, therefore, from the ergodic theorem that

$$\int_{-\infty}^{\infty} f(x)g^{(k)}[A\cos(wt + \phi) + x]\,dx = \sum_{p=0}^{\infty}\frac{[A\cos(wt + \phi)]^p}{p!}\langle g^{(k+p)}[x(t)]\rangle \tag{49}$$

where $f(x)$ is the noise probability density function. Substituting (49) in (31) we obtain

$$y(t) = \sum_{k=0}^{\infty} \frac{[x(t)]_{AC}^k}{k!} \cdot \int_{-\infty}^{\infty} f(x) g^{(k)}[A \cos(wt + \phi) + x] \, dx. \quad (50)$$

Next, the following Fourier expansion can be made:

$$g^{(k)}[A \cos(wt + \phi) + x] = \sum_{m=0}^{\infty} \epsilon_m A_{mk}(x) \cos m(wt + \phi) \quad (51)$$

where

$$A_{mk}(x) = \frac{1}{\pi} \int_0^{\pi} g^{(k)}[A \cos \theta + x] \cos m\theta \, d\theta. \quad (52)$$

Substituting (51) in (50) we obtain

$$y(t) = \sum_{k=0}^{\infty} \sum_{m=0}^{\infty} \frac{\epsilon_m \overline{A_{mk}(x)[x(t)]_{AC}^k} \cos m(wt + \phi)}{k!}. \quad (53)$$

Then, substituting (33) into (53) and collecting terms for each power of $x(t)$ we obtain

$$y(t) = \sum_{k=0}^{\infty} \sum_{m=0}^{\infty} \sum_{v=0}^{\infty} \frac{\epsilon_m \alpha_v \overline{A_{m,k+v}(x)} x(t)^k \cos m(wt + \phi)}{k!}. \quad (54)$$

Comparison of (54) with (43) shows that $R_y(t, t + \tau)$ is again given by (45), where now

$$f_{mk} = \sum_{v=0}^{\infty} \alpha_v \overline{A_{m,k+v}(x)} \quad (55)$$

$$f_{ml} = \sum_{v=0}^{\infty} \alpha_v \overline{A_{m,l+v}(x)}, \quad (56)$$

and

$$\overline{A_{m,k+v}(x)} = \frac{1}{\pi} \int_{-\infty}^{\infty} f(x) \, dx \int_0^{\pi} g^{(k+v)}[A \cos \theta + x] \cos m\theta \, d\theta \quad (57)$$

$$\overline{A_{m,l+v}(x)} = \frac{1}{\pi} \int_{-\infty}^{\infty} f(x) \, dx \int_0^{\pi} g^{(l+v)}[A \cos \theta + x] \cos m\theta \, d\theta. \quad (58)$$

Eqs. (57) and (58) can be simplified for purposes of calculation by making the substitution

$$g^{(k)}[A \cos \theta + x] = \sum_{s=0}^{\infty} \frac{x^s}{s!} g^{(k+s)}[A \cos \theta], \quad (59)$$

which yields

$$\overline{A_{m,k+v}(x)} = \sum_{s=0}^{\infty} \frac{\overline{x^s}}{s!} \frac{1}{\pi} \int_0^{\pi} g^{(k+v+s)}[A \cos \theta] \cos m\theta \, d\theta \quad (60)$$

$$\overline{A_{m,l+v}(x)} = \sum_{s=0}^{\infty} \frac{\overline{x^s}}{s!} \frac{1}{\pi} \int_0^{\pi} g^{(l+v+s)}[A \cos \theta] \cos m\theta \, d\theta. \quad (61)$$

Eqs. (55), (56), (60) and (61) are expected to be useful when the input signal-to-noise ratio is large because the number of significant terms then tends to be small.

F. The Output Autocorrelation Function When the Input Noise is Gaussian

If we calculate the output autocorrelation function directly from (31) we obtain

$$R_y(t_1, t_2) = \sum_{p,q,k,l=0}^{\infty} \frac{\overline{g^{(k+p)}(x)} \, \overline{g^{(l+q)}(x)} \, s(t_1)^p s(t_2)^q \, \overline{[x(t_1)]_{AC}^k [x(t_2)]_{AC}^l}}{p!q!k!l!}. \quad (62)$$

In Appendix B it is shown that when the input noise is Gaussian,

$$\overline{[x(t)]_{AC}^k [x(t + \tau)]_{AC}^l} = k! R_x(\tau)^k \quad \text{for } l = k$$
$$= 0 \quad \text{for } l \neq k \quad (63)$$

where $R_x(\tau)$ is the autocorrelation function of the input noise.

Consequently, for the special case of Gaussian noise we have

$$R_y(t, t + \tau) = \sum_{p,q,k=0}^{\infty} \frac{\overline{g^{(k+p)}(x)} \, \overline{g^{(k+q)}(x)} \, s(t)^p s(t + \tau)^q R_x(\tau)^k}{p!q!k!}. \quad (64)$$

The relative simplicity of this result may be seen by comparing (64) with the corresponding general result given by (38).

Special expressions for $s(t) = A \cos(wt + \phi)$ are obtained as follows. Returning to (31) and substituting (41) results in

$$y(t) = \sum_{m=0}^{\infty} \sum_{r=0}^{\infty} \frac{\epsilon_m A^{m+2r} \cos m(wt + \phi)}{(m + r)! \, r! \, 2^{m+2r}} \cdot \sum_{k=0}^{\infty} \frac{\langle g^{(k+m+2r)}[x(t)] \rangle}{k!} [x(t)]_{AC}^k. \quad (65)$$

Then, using (37), (44) and (63) we obtain for a sinusoidal input with additive Gaussian noise

$$R_y(\tau) = \sum_{m=0}^{\infty} \sum_{k=0}^{\infty} \frac{\epsilon_m h_{mk}^2 R_x(\tau)^k \cos mw\tau}{k!} \quad (66)$$

where

$$h_{mk} = \sum_{r=0}^{\infty} \frac{\overline{g^{(k+m+2r)}(x)} A^{m+2r}}{(m + r)! \, r! \, 2^{m+2r}}. \quad (67)$$

These expressions correspond to (45)–(47) of the general noise analysis.

Eq. (67) is convenient when the input signal-to-noise ratio is small. For large input signal-to-noise ratios, a generally more useful expression for h_{mk} is obtained by using the expression for $y(t)$ given by (53). This yields the same expression for $R_y(\tau)$ but h_{mk} is then given by

$$h_{mk} = \frac{1}{\pi} \int_{-\infty}^{\infty} f(x) \, dx \int_0^{\pi} g^{(k)}[A \cos \theta + x] \cos m\theta \, d\theta \quad (68)$$

or equivalently,

$$h_{mk} = \sum_{s=0}^{\infty} \frac{\overline{x^s}}{s!} \frac{1}{\pi} \int_0^{\pi} g^{(k+s)}[A \cos \theta] \cos m\theta \, d\theta. \quad (69)$$

G. Extension of Results to Nonlinear Devices Having Discontinuities

The mathematical results obtained have been based on Taylor series expansions which are valid only if the nonlinear transfer characteristic possesses continuous derivatives of all orders at each point. The purpose of this section is to show that the results also apply when the nonlinear characteristic and its derivatives are only sectionally continuous.

Let the kth derivative, $k = 0, 1, 2, \cdots$ of $g(x)$ possess a discontinuity at $x = x_0$. Now define,

$$g_R(x) = g(x) \quad \text{for} \quad x > x_0 \tag{70}$$

$$g_L(x) = g(x) \quad \text{for} \quad x < x_0 \tag{71}$$

where $g_R(x)$ and $g_L(x)$ are bounded continuous functions with continuous derivatives of all orders for every x. Then let,

$$g_\sigma(x) = g_R(x) \int_{-\infty}^{x} \frac{1}{\sqrt{2\pi}\sigma} e^{-(u-x_0)^2/2\sigma^2}\, du + g_L(x) \int_{x}^{\infty} \frac{1}{\sqrt{2\pi}\sigma} e^{-(u-x_0)^2/2\sigma^2}\, du \tag{72}$$

Since $g_\sigma(x)$ consists of a sum of products of functions which are continuous with continuous derivatives for every x, it is itself continuous with continuous derivatives for every x. Thus, $g_\sigma(x)$ satisfies the assumptions made in deriving the output autocorrelation function. It is also evident from (70)–(72) that

$$\lim_{\sigma \to 0} [g_\sigma(x) - g(x)] = 0 \quad \text{for all} \quad x \tag{73}$$

where it is assumed that by definition

$$g(x_0) = \tfrac{1}{2}[g(x_0^-) + g(x_0^+)]. \tag{74}$$

Therefore,

$$R_y(\tau) = \overline{g[x(t)]g[x(t+\tau)]} = \lim_{\sigma \to 0} \overline{g_\sigma[x(t)]g_\sigma[x(t+\tau)]} \tag{75}$$

provided the indicated limit exists.

Since the nonlinear characteristic enters only in terms of the statistical average of its derivatives ((38)–(40)), the limit in (75) exists when the following limit, denoted by L_n, exists for all n:

$$L_n = \lim_{\sigma \to 0} \overline{g_\sigma^{(n)}(x)} = \lim_{\sigma \to 0} \int_{-\infty}^{\infty} f(x) g_\sigma^{(n)}(x)\, dx \tag{76}$$

where $f(x)$ = noise probability density. That the limit does exist for all n for a sectionally smooth nonlinear characteristic can be seen as follows:

Let $a > 0$ be a number such that

$$\text{Prob}\, [x > x_0 + a] > 0. \tag{77}$$

Then we note first that

$$\lim_{\sigma \to 0} \left[\int_{-\infty}^{x_0 - a} f(x) g_\sigma^{(k)}(x)\, dx + \int_{x_0 + a}^{\infty} f(x) g_\sigma^{(k)}(x)\, dx \right] \tag{78}$$

$$= \int_{-\infty}^{x_0 - a} f(x) g^{(k)}(x)\, dx + \int_{x_0 + a}^{\infty} f(x) g^{(k)}(x)\, dx < \infty \quad \text{for all} \quad k.$$

Thus, we need only show that

$$\lim_{\sigma \to 0} \int_{x_0 - a}^{x_0 + a} f(x) g_\sigma^{(k)}(x)\, dx = A_k < \infty \quad \text{for all} \quad k. \tag{79}$$

Proof

Assume that the converse is true, *i.e.*,

$$\lim_{\sigma \to 0} \int_{x_0 - a}^{x_0 + a} f(x) g_\sigma^{(k)}(x)\, dx = \infty. \tag{80}$$

Since

$$\int_{x_0 - a}^{x_0 + a} f(x)\, dx < 1, \tag{81}$$

this can be true only if

$$\lim_{\sigma \to 0} \int_{x_0 - a}^{x_0 + a} g_\sigma^{(k)}(x)\, dx = \infty. \tag{82}$$

Then, obviously, we have

$$\lim_{\sigma \to 0} g_\sigma^{(k-1)}(x) = \infty \quad \text{for} \quad x \geq x_0 + a \tag{83}$$

and, thus,

$$\lim_{\sigma \to 0} \int_{-\infty}^{x} f(x) g_\sigma^{(k-1)}(x)\, dx = \infty \quad \text{for} \quad x > x_0 + a \tag{84}$$

and, therefore,

$$\lim_{\sigma \to 0} g_\sigma^{(k-2)}(x) = \infty \quad \text{for} \quad x > x_0 + a. \tag{85}$$

By continuing this process, we see that a necessary consequence of our assumption is that

$$\lim_{\sigma \to 0} g_\sigma(x) = \infty \quad \text{for} \quad x > x_0 + a. \tag{86}$$

But this cannot be true since

$$\lim_{\sigma \to 0} g_\sigma(x) = g(x) \tag{87}$$

and $g(x)$, by definition, is bounded. Therefore, for all k,

$$\lim_{\sigma \to 0} \int_{x_0 - a}^{x_0 + a} f(x) g_\sigma^{(k)}(x)\, dx = A_k < \infty \qquad \text{Q.E.D.}$$

We have considered nonlinear characteristics with a single discontinuity only. However, it is apparent that multiple discontinuities can be handled in the same way. Thus, by forming $g_\sigma(x)$ we can remove one discontinuity, then by operating in the same way on the new function we could remove a second discontinuity, etc. The existence of a limit would then follow by applying the above argument to each point of discontinuity.

We have also not considered the equations in which derivatives of the nonlinear characteristic are averaged with respect to the signal process instead of the noise process. However, the only necessary requirement on the averaging function in the foregoing arguments is that it be finite at the points of discontinuity. Consequently, the same argument applies.

It is important to note that in practical calculation, rather than constructing $g_\sigma(x)$ and evaluating (75), (39) and (40) or (60) and (61) can be applied directly, using the proper impulse functions and their derivatives to determine the contributions due to the discontinuities. This is true because as $\sigma \to 0$ the difference between $g(x)$ and $g_\sigma(x)$ away from the points of discontinuity become negligible in comparison to the differences in the immediate vicinities of the discontinuities. This direct evaluation is valid, however, only when the range of values over which $g_\sigma(x)$ and its derivatives are averaged includes all of the points of discontinuity. Thus, the expressions given in (60) and (61) can be evaluated directly without forming $g_\sigma(x)$ only if all discontinuities lie inside the interval $-A < x < A$. For the usual case of Gaussian noise, using (64) or (66) and (67), the averaging interval is $-\infty$ to $+\infty$ so that the direct evaluation may always be used.

IV. Comparison of Results with Those Obtained by the Transform Method

Eq. (66), the output autocorrelation function for a single signal and Gaussian noise input, is identical to the general equation associated with the transform method of analysis.[10] The expression for h_{mk}, however, is quite different as can be seen by comparing (67) with the following contour integral, which is the expression obtained by the transform method of analysis:

$$h_{mk} = \frac{1}{2\pi j} \int_C G(w) w^k I_m(wA) e^{\sigma^2 w^2/2} \, dw \tag{88}$$

where $G(w)$ is the Fourier or bilaterial Laplace transform of the nonlinear characteristic, *i.e.*,

$$G(w) = \int_{-\infty}^{\infty} g(x) e^{-wx} \, dx, \qquad w = u + jv$$

and $I_m(wA)$ is a modified Bessel function of the first kind.

Eq. (88) has been evaluated for a number of particular nonlinear characteristics, as was discussed in Section I. Most of these results involve the confluent hypergeometric function, which is defined by a special infinite series. The interesting point is that the series of terms obtained in the transform analysis is identical to the series obtained from (67). The reason for this is made clear in the following, where it is shown that a general transformation of the contour integral exists which is identical to (67).

[10] See, for example, Davenport and Root, *op. cit.*, p. 290.

We begin by noting that

$$I_m(wA) = \sum_{r=0}^{\infty} \frac{(wA)^{m+2r}}{2^{m+2r} r! \, \Gamma(m + r + 1)}. \tag{89}$$

Substituting (89) in (88) and interchanging the order of summation and integration, we obtain

$$h_{mk} = \sum_{r=0}^{\infty} \frac{A^{m+2r}}{2^{m+2r} r! \Gamma(m + r + 1)} \frac{1}{2\pi j} \cdot \int_C G(w) w^{k+m+2r} e^{\sigma^2 w^2/2} \, dw. \tag{90}$$

Next, multiplying inside the integral by e^{wx} and e^{-wx} we have,

$$\frac{1}{2\pi j} \int_C G(w) w^{k+m+2r} e^{\sigma^2 w^2/2} e^{wx} e^{-wx} \, dw = F^{-1}[G(w) w^{k+m+2r} e^{\sigma^2 w^2/2} e^{-wx}] \tag{91}$$

where F^{-1} means bilateral inverse transform. Next, using the standard transform properties, we note that

$$F^{-1}[G(w) w^{k+m+2r}] = g^{(k+m+2r)}(x), \tag{92}$$

$$F^{-1}[e^{\sigma^2 w^2/2}] = \frac{1}{\sqrt{2\pi}\sigma} e^{-x^2/2\sigma^2} = f(x). \tag{93}$$

Therefore,

$$F^{-1}[G(w) w^{k+m+2r} e^{\sigma^2 w^2/2}] = \int_{-\infty}^{\infty} g^{(k+m+2r)}(x - t) f(t) \, dt. \tag{94}$$

And finally, by the shifting theorem

$$F^{-1}\{[G(w) w^{k+m+2r} e^{\sigma^2 w^2/2}] e^{-wx}\} = \int_{-\infty}^{\infty} g^{(k+m+2r)}[x - t] f(t - x) \, dt = \int_{-\infty}^{\infty} f(x) g^{(k+m+2r)}(x) \, dx. \tag{95}$$

Substituting (95) in (90) and replacing $\Gamma_{(m+r+1)}$ by $(m + r)!$, we obtain the desired result,

$$h_{mk} = \sum_{r=0}^{\infty} \frac{\overline{g^{(k+m+2r)}[x]} (A/2)^{m+2r}}{r! \, (m + r)!}. \tag{96}$$

V. Appendix

A. Evaluation of $[x(t)]_{AC}^k$

The definition of $[x_T(t)]_{AC}^k$, as given in (28) is

$$[x_T(t)]_{AC}^k = \underbrace{\sum_n \cdots \sum_q}_{k\text{-fold}} x_n x_m \cdots x_q e^{j(n+m+\cdots+q)w_0 t}$$

where the summations are extended over all possible values from $-\infty$ to $+\infty$ such that no two or more terms in the frequency determining coefficient $(n + m + \cdots + q)$ add to zero.

Using this definition, consider first

$$[x_T(t)]_{AC}^2 = \sum_{-\infty}^{\infty} \sum_{\substack{-\infty \\ (n+m\neq 0)}}^{\infty} x_n x_m e^{j(n+m)w_0 t}.$$

It is apparent that $[x_T(t)]^2$ is periodic of period T and that the dc component is given by

$$\sum_{-\infty}^{\infty} x_n x_{-n} = \frac{1}{T} \int_{-T/2}^{T/2} x_T(t)^2 \, dt = \langle [x_T(t)]^2 \rangle.$$

Therefore,

$$[x_T(t)]^2_{AC} = [x_T(t)]^2 - \langle [x_T(t)]^2 \rangle.$$

Thus,

$$[x(t)]^2_{AC} = \lim_{T \to \infty} [x_T(t)]^2_{AC} = x(t)^2 - \langle x(t)^2 \rangle$$
$$= x(t)^2 - \overline{x^2}.$$

Next, consider

$$[x_T(t)]^3_{AC} = \sum_{-\infty}^{\infty} \sum_{-\infty}^{\infty} \sum_{-\infty}^{\infty} x_n x_m x_p e^{i(n+m+p)\omega_0 t}$$
$$\substack{n+m \neq 0 \\ n+p \neq 0 \\ p+m \neq 0 \\ n+p+m \neq 0}.$$

Here, the terms to be removed from $[x_T(t)]^3$ are of the form,

$$\sum_{-\infty}^{\infty} \sum_{-\infty}^{\infty} \sum_{-\infty}^{\infty} x_n x_{-n} x_p e^{ip\omega_0 t} = \langle [x_T(t)]^2 \rangle \sum_{-\infty}^{\infty} x_p e^{ip\omega_0 t}$$

of which there are three such terms, corresponding to $n + m = 0$, $n + p = 0$ and $p + m = 0$, and

$$\sum_{-\infty}^{\infty} \sum_{-\infty}^{\infty} \sum_{-\infty}^{\infty} x_n x_m x_p = \langle [x_T(t)]^3 \rangle.$$
$$\substack{(n+m+p=0)}$$

Thus,

$$[x(t)]^3_{AC} = \lim_{T \to \infty} [x_T(t)]^3_{AC}$$
$$= x(t)^3 - 3\langle x(t)^2 \rangle x(t) - \langle x(t)^3 \rangle$$
$$= x(t)^3 - 3\overline{x^2} x(t) - \overline{x^3}.$$

If we continue in this way we find that $[x(t)]^k_{AC}$ can be defined in terms of the following recurrence formula:

$$[x(t)]^k_{AC} = x(t)^k - \sum_{r=1}^{k} \frac{k! \, \overline{x^r}}{(k-r)! \, r!} [x(t)]^{k-r}_{AC}. \quad (97)$$

Expansion of (97) results in

$$[x(t)]^k_{AC} = x(t)^k - \sum_{r=1}^{k} \frac{k! \, \overline{x^r}}{(k-r)! \, r!} x(t)^{k-r} \quad (98)$$
$$+ \sum_{r_1=1}^{k-1} \sum_{r_2=1}^{k-r_1} \frac{k! \, \overline{x^{r_1}} \, \overline{x^{r_2}}}{r_1! \, r_2! \, (k - r_1 - r_2)!} x(t)^{k-r_1-r_2}$$
$$- \sum_{r_1=1}^{k-2} \sum_{r_2=1}^{k-r_1-1} \sum_{r_3=1}^{k-r_1-r_2} \frac{k! \, \overline{x^{r_1}} \, \overline{x^{r_2}} \, \overline{x^{r_3}}}{r_1! \, r_2! \, r_3! \, (k - r_1 - r_2 - r_3)!}$$
$$\cdot x(t)^{k-r_1-r_2-r_3} + \cdots$$

where the final term is $(-1)^k k! \, \bar{x}^k$.

B. Evaluation of $\overline{[x(t)]^k_{AC}[x(t + \tau)]^l_{AC}}$ for Gaussian Noise

The noise term in the expression for the output autocorrelation function given by (62) is

$$\overline{[x(t)]^k_{AC}[x(t + \tau)]^l_{AC}}.$$

The expression for $[x(t)]^k_{AC}$ is given in (97) and (98). Evaluating $[x(t)]^k_{AC}$ for the first few values of k with $\bar{x}$ set equal to zero, we find

$$[x(t)]^1_{AC} = x(t)$$
$$[x(t)]^2_{AC} = x(t)^2 - \overline{x^2}$$
$$[x(t)]^3_{AC} = x(t)^3 - 3\overline{x^2} x(t) - \overline{x^3}$$
$$[x(t)]^4_{AC} = x(t)^4 - 6\overline{x^2} x(t)^2$$
$$- 4\overline{x^3} x(t) - \overline{x^4} + 6\overline{x^2} \cdot \overline{x^2}.$$

To proceed further, we need the moments and joint moments of the Gaussian process. By differentiating the joint characteristic function of a Gaussian process, it can be shown that

$$\overline{[x(t)]^{k+2m}[x(t+\tau)]^k}$$
$$= \sum_{r=0,1,2,\ldots}^{k/2} \frac{(k+2m)! \, k! \, (\sigma^2/2)^{2r+m} R_x(\tau)^{k-2r}}{r! \, (r+m)! \, (k-2r)!} \quad (99)$$

$$\overline{x(t)^{2m}} = \frac{(2m)!}{m!} \left(\frac{\sigma^2}{2}\right)^m \quad m = 1, 2, 3, \cdots \quad (100)$$

$$\overline{x(t)^{2m+1}} = 0 \quad m = 1, 2, 3, \cdots. \quad (101)$$

Using (99)–(101), we have for the Gaussian process

$$[x(t)]_{AC} = x(t)$$
$$[x(t)]^2_{AC} = x(t)^2 - \sigma^2$$
$$[x(t)]^3_{AC} = x(t)^3 - 3\sigma^2 x(t)$$
$$\overline{[x(t)]^4_{AC}} = x(t)^4 - 6\sigma^2 x(t)^2 + 3\sigma^4.$$

Using these expressions

$$\overline{[x(t)]_{AC}[x(t + \tau)]_{AC}} = R_x(\tau)$$
$$\overline{[x(t)]^2_{AC}[x(t + \tau)]^2_{AC}} = 2! \, R_x(\tau)^2$$
$$\overline{[x(t)]^3_{AC}[x(t + \tau)]^3_{AC}} = 3! \, R_x(\tau)^3$$
$$\overline{[x(t)]^4_{AC}[x(t + \tau)]^4_{AC}} = 4! \, R_x(\tau)^4$$

and

$$\overline{[x(t)]^n_{AC}[x(t + \tau)]^m_{AC}} = 0 \quad n \neq m \quad (1 \leq n \leq 4)$$
$$(1 \leq m \leq 4).$$

These results suggest, of course, that the general result for a Gaussian process is

$$\overline{[x(t)]^k_{AC}[x(t + \tau)]^l_{AC}} = k! \, R_x(\tau)^k \quad \text{for} \quad l = k \quad (102)$$
$$= 0 \quad l \neq k.$$

That this is true may be shown most easily by inverting the problem, *i.e.*, we will assume (102) to be true and show that this implies that the joint moments of the process are as given in (99) for the Gaussian process.

From (97) we have

$$x(t)^k = \sum_{r=0}^{k} \frac{k!\overline{x^r}}{(k-r)!\,r!} [x(t)]_{AC}^{k-r} \qquad (103)$$

thus,

$$\overline{[x(t)]^{k+2m}[x(t+\tau)]^k}$$

$$= \sum_{p=0}^{k+2m} \sum_{s=0}^{k} \frac{(k+2m)!\,k!\,\overline{x^p}\,\overline{x^s}}{(k+2m-p)!\,(k-s)!\,p!\,s!} \qquad (104)$$

$$\cdot \overline{[x(t)]_{AC}^{k+2m-p}[x(t+\tau)]_{AC}^{k-s}}.$$

Then, using (102)

$$\overline{[x(t)]^{k+2m}[x(t+\tau)]^k} = \sum_{s=0}^{k} \frac{(k+2m)!\,k!\,\overline{x^{s+2m}}\,\overline{x^s}}{(k-s)!\,(s+2m)!\,s!} R_x(\tau)^{k-s}.$$

Letting $s = 2r$ and substituting

$$\overline{x^{2r+2m}} = \frac{(2r+2m)!}{(r+m)!}\left(\frac{\sigma^2}{2}\right)^{r+m}$$

$$\overline{x^{2r}} = \frac{(2r)!}{r!}\left(\frac{\sigma^2}{2}\right)^{r}$$

we obtain the desired result,

$$\overline{[x(t)]^{k+2m}[x(t+\tau)]^k}$$

$$= \sum_{r=0}^{k/2} \frac{(k+2m)!\,k!\,(\sigma^2/2)^{2r+m}}{(r+m)!\,r!\,(k-2r)!} R_x(\tau)^{k-2r}.$$

VI. Acknowledgment

The author wishes to thank the IRE reviewers for their very specific and helpful suggestions for revising earlier drafts of this paper, and the Sperry Gyroscope Company for permission to publish this work.

9

Reprinted from *IEEE Trans. Information Theory*, **IT-15**(4), 448–456 (1969)

The Correlation Function of Gaussian Noise Passed Through Nonlinear Devices

RICHARD F. BAUM, SENIOR MEMBER, IEEE

***Abstract*—This paper is concerned with the output autocorrelation function R^y of Gaussian noise passed through a nonlinear device. An attempt is made to investigate in a systematic way the changes in R^y when certain mathematical manipulations are performed on some given device whose correlation function is known. These manipulations are the "elementary combinations and transformations" used in the theory of Fourier integrals, such as addition, differentiation, integration, shifting, etc. To each of these, the corresponding law governing R^y is established. The same laws are shown to hold for the envelope of signal plus noise for narrow-band noise with spectrum symmetric about signal frequency.**

Throughout the text and in the Appendix it is shown how the results can be used to establish unknown correlation function quickly with main emphasis on power-law devices $y = x^m$, with m either an integer or half integer. Some interesting recurrence formulas are given. A second-order differential equation is derived which serves as an alternative means for calculating R^y.

I. Introduction

THE THEORY of noise passed through a nonlinear memoryless device is based on formulas given by Rice [1] and Middleton [2], [3]. These formulas use a double-ended Laplace transform in order to derive the output autocorrelation function R^y of noise, or signal plus noise. Not all nonlinear devices, though, necessitate the application of the double-ended Laplace transform and in most cases we may replace it by the conventional Fourier transform [10].

The purpose of this investigation is to establish certain rules for the calculation of autocorrelation functions. These rules are obtained by a systematic application of basic rules governing Fourier transforms as given, for instance, in Campbell and Foster's tables [9], where they are called "Elementary Combinations and Transformations." We give our results in a corresponding table, using the same title. We also show that the envelope of the autocorrelation function (for signals plus noise) in the narrow-band case obeys the same rules.

The results show, indeed, that quite a number of autocorrelation functions can be written down by inspection. Many results can be used as cross checks for the validity of previously published results or for their extension.

One of the interesting parts of the investigation is the presentation of a differential equation for some R^y, which may be used instead of the customary evaluation of a rather complicated integral [3].

In Section II the theoretical background for noise alone will be reviewed. Section IV gives an extension to the case of signal plus noise. In Section III the elementary rules are derived, collected in a table, and commented upon. Appendix I contains a number of applications of the rules with the intention to display their practical usefulness.

Manuscript received November 9, 1967; revised December 2, 1968.

The author is with the Electronic Systems Division, TRW Systems, Redondo Beach, Calif. 90278.

II. Theoretical Background—Noise Alone

Let $x(t)$ be a stationary random process with zero mean, variance σ, and with Gaussian first and second distribution functions

$$f(x) = \frac{1}{\sqrt{2\pi R_0}} \exp\left[-\frac{x^2}{2R_0}\right] \tag{1}$$

$$f(x_1, x_2, \tau) = \frac{1}{2\pi R_0\sqrt{1-r^2}} \exp\left[-\frac{x_1^2 + x_2^2 - 2rx_1x_2}{2R_0(1-r^2)}\right] \tag{2}$$

and with autocorrelation $R(\tau)$, where τ is the time difference between two samples $x_1(t_1)$ and $x_2(t_1 + \tau)$. The correlation coefficient r is

$$r = \frac{R(\tau)}{R(0)} = \frac{R_\tau}{R_0} = \frac{R_\tau}{\sigma^2}. \tag{3}$$

The output power spectrum is the Fourier transform of $R^y(\tau)$. Let this process be transformed by a memoryless nonlinear device into $y(t)$ as indicated in Fig. 1(a). The device is characterized by a transfer characteristic $y = y[x(t)]$.

The average (or dc component) of the output is given by

$$\overline{y(t)} = \int_{-\infty}^{\infty} y(x)f(x)\, dx \tag{4}$$

and the autocorrelation of the output is, by definition,

$$R^y(\tau) = \overline{y(t_1)y(t_1 + \tau)}$$

$$= \iint_{-\infty}^{\infty} y(x_1)y(x_2)f(x_1, x_2, \tau)\, dx_1\, dx_2 \tag{5}$$

$$= \frac{1}{2\pi R_0\sqrt{1-r^2}} \int_{-\infty}^{\infty}\int_{-\infty}^{+\infty} y(x_1)y(x_2) \cdot \exp\left[-\frac{x_1^2 + x_2^2 - 2rx_1x_2}{2R_0(1-r^2)}\right] dx_1\, dx_2. \tag{6}$$

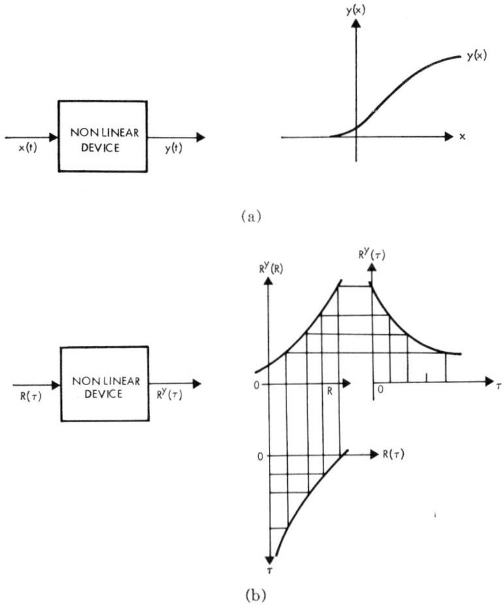

Fig. 1. (a) $y(t)$ is the output of the nonlinear device with input $x(t)$. (b) $R^y(\tau)$ is the output of a nonlinear device with input $R(\tau)$.

This equation can be transformed [10] into the equivalent expression

$$R^y(\tau) = \frac{1}{4\pi^2}\iint_{-\infty}^{\infty} F(ju)F(jv) \cdot \exp\left[-\frac{R_0}{2}(u^2+v^2) - R_0 r u v\right] du\, dv. \quad (7)$$

In this expression, $F(ju)$ is defined as the Fourier transform of the nonlinear characteristic $y(x)$:

$$F(ju) = \int_{-\infty}^{\infty} y(x) \exp(-jux)\, dx, \quad (8)$$

from which

$$y(x) = \frac{1}{2\pi}\int_{-\infty}^{\infty} F(ju) \exp(jux)\, du. \quad (9)$$

The exponential function inside the integral of (7) represents the bivariate characteristic function of the random variable and is defined as the two-dimensional Fourier transform of $f(x_1, x_2, \tau)$.

In the absence of signal, if we let $\tau \to \infty$, then $R_\tau \to 0$ and we obtain from (5) and (7), respectively,

$$R^y(\infty) = \left\{\frac{1}{2\pi R_0}\int_{-\infty}^{\infty} y(x) \exp\left(-\frac{x^2}{2R_0}\right) dx\right\}^2 \quad (10)$$

$$= \left\{\frac{1}{2\pi}\int_{-\infty}^{\infty} F(ju) \exp\left(-\frac{R_0}{2}u^2\right) du\right\}^2. \quad (11)$$

By comparison with (4) we recognize that the right sides of (10) and (11) are expressions for the dc output power $\{\overline{y(x)}\}^2$ [1, p. 275].

We observe that, although $R^y(\tau)$ is a function of the time difference τ, it would be more appropriate to write $R^y(R_\tau)$, since R_τ is the parameter that enters the integral (7). It makes no difference whatsoever *how* R_τ varies with τ. Hence, when the input to the nonlinear device is pure noise, the integral (7) may be considered as defining another nonlinear device $R^y(R_\tau)$. $R^y(R_\tau)$ can be obtained as shown by mirroring R_τ on $R^y(R_\tau)$ as indicated in Fig. 1(b).

III. Elementary Rules

Since the integrand in (7) depends upon the Fourier transform $F(ju)$, we may expect that some of the elementary rules that govern Fourier transforms will be reflected in some fundamental rules governing the calculation of the output correlation function R^y. These rules will enable us to find R^y from a given collection of "basic" R^y, if the nonlinear device $y(x)$ can be reduced to one or more "basic" nonlinear devices by the process of multiplication, addition, differentiation, shifting, etc. We thus hope to eliminate the necessity of reevaluating the integral (7) in every particular case. The results of the investigation are collected in Table I, to which we give the following comments.

A. Multiplication by a Constant

If $y(x)$ is multiplied by a constant λ, then R^y is multiplied by λ^2.

B. Change of Scale

If x is replaced by x/a $(a > 0)$, then the function $R^y(R_\tau)$ becomes $R^y(R_\tau/a^2)$.

C. Addition of Constant

The addition of a constant K to $y(x)$ (or a shift of the characteristic in the vertical direction) adds a constant K to the output $y(x)$. If the original autocorrelation was $R^y(\tau)$, then the autocorrelation after shift must have the dc power

$$R^{y\prime}(\infty) = (\sqrt{R^y(\infty)} + K)^2 = (\bar{y} + K)^2$$

where $\bar{y}$ is the mean of $y(x)$. Hence,

$$R^{y\prime}(\tau) = R^y(\tau) + 2K\bar{y} + K^2. \quad (12)$$

D. Addition Theorem

Assume that a given nonlinear characteristic $y(x)$ can be represented by the sum of two characteristics:

$$y(x) = y_1(x) + y_2(x)$$

with Fourier transforms $F_1(u)$ and $F_2(u)$, and with autocorrelation functions $R_1(\tau)$ and $R_2(\tau)$. The autocorrelation of $y(x)$ is

This result can be modified for the special case of an mth power device. In this case it will be shown that R^y has the form (41)

$$R^y = R_0^m \phi(r) \tag{37}$$

where $\phi(r)$ is not a function of R_0. Hence,

$$\frac{\partial R^y}{\partial R_0} = \frac{m}{R_0} R^y. \tag{38}$$

Therefore, for an mth power device $y' = xy^m$, we obtain

$$R^{y'} = R_0\left[(1 - r^2)\frac{\partial R^y}{\partial r} + (2m + 1)\, rR^y\right]. \tag{39}$$

Numerous applications of this formula are given in Appendix I.

I. Derivative $\partial R^y/\partial R_0$

The following expression is derived in Appendix II:

$$\frac{\partial R^y}{\partial R_0} = \frac{1}{4\pi^2 R_0}\iint F^*(ju)F(jv) \cdot \exp\left[-\frac{R_0}{2}(u^2 + v^2) - R_0 ruv\right] du\, dv \tag{40}$$

where $F^*(ju)$ is the Fourier transform of the product $x[dy(x)/dx]$.

If we apply this result to the mth power-law device we have

$$y = x^m \qquad \text{for } x > 0$$

and

$$x\frac{dy}{dx} = mx^m = my(x);$$

therefore,

$$F^*(ju) = mF(ju)$$

and

$$\frac{\partial R^y}{\partial R_0} = \frac{m}{R_0} R^y. \tag{41}$$

Any equation of the form

$$R^y = R_0^m \phi(r),$$

where $\phi(r)$ is *not* a function of R_0, will satisfy (41). Indeed, we encounter it in Tables II–IV which refer to power-law devices.

IV. Theoretical Background—Signal Plus Noise

It has been shown [1] that in the presence of a signal

$$e(t) = P \cos wt,$$

the output autocorrelation function becomes

$$R^y = \frac{1}{4\pi^2}\iint F(ju)F(jv) \cdot \exp\left[-\frac{R_0}{2}(u^2 + v^2) - R_0 ruv\right] \cdot J_0[P\sqrt{u^2 + v^2 + 2uv \cos wt}]\, du\, dv. \tag{42}$$

We shall assume that we deal with narrow-band noise with a symmetric spectrum centered about the signal frequency w. In this case the input noise autocorrelation function can be written as

$$R(\tau) = R_0 r(\tau) \cos w\tau; \tag{43}$$

$r(\tau)$ now plays the role of an envelope function for the periodic cos term. Our further derivations will strictly apply only for the limiting case of an extremely narrow bandwidth B where we assume that $r(\tau)$ is varying very slowly so that the assumption

$$\left|\frac{dr(\tau)}{d\tau}\right| \ll w\, |r| \tag{44}$$

is justified. In this case, points on the envelope and points where $R(\tau)$ has its maxima (or minima) can be assumed to coincide. At these points

$$\frac{dR(\tau)}{d\tau} = R_0\left[\frac{dr(\tau)}{d\tau}\cos w\tau - r(\tau)w \sin w\tau\right] = 0 \tag{45}$$

or, in view of (44),

$$\tau_n \cong \frac{n\pi}{w} \qquad n = 0, 1, 2 \cdots \tag{46}$$

and

$$R_0 r\left(\frac{n\pi}{w}\right) \cong R\left(\tau = \frac{n\pi}{w}\right). \tag{47}$$

The envelope, therefore, can be obtained by replacing the discrete time instants τ_n by a continuous time function τ. Similar reasoning will now be applied to the output correlation function $R^y(\tau)$. Our intention is to show that the basic rules, as derived in the preceding section and compiled in Table I, hold for the *envelope* $R_E^y(\tau)$ of the output $R^y(\tau)$.

Let us write (42) for short as

$$R^y = \iint g(u, v, R(\tau))J_0(z)\, du\, dv, \tag{48}$$

where

$$z = P\sqrt{u^2 + v^2 + 2uv \cos w\tau}$$

and J_0 is the Bessel function of the first kind, zero order. The time derivative of R^y becomes

$$\frac{dR^y}{d\tau} = \iint g(u, v, R(\tau))\left[\left(\frac{dr}{d\tau}\cos w\tau - rw \sin w\tau\right)J_0(z) - \frac{dJ_0(z)}{dz}\frac{P^2uvw \sin w\tau}{z}\right] du\, dv. \tag{49}$$

This shows that the integrand is zero whenever $\sin w\tau = 0$, since the cosine is multiplied by $dr/d\tau$, which tends to zero. The extremes of R^y, therefore, are obtained by setting $\cos w\tau = +1$ or -1 and we obtain the envelope of R^y:

$$R_E^y = \frac{1}{4\pi^2} \iint F(ju)F(jv) \cdot \exp\left[-\frac{R_0}{2}(u^2 + v^2) \pm R_0 r(\tau) uv \right] \cdot J_0[P(u \pm v)]\, du\, dv. \tag{50}$$

The signs correspond to the maxima ($\cos w\tau_n = +1$) and minima ($\cos w\tau_n = -1$), respectively, since R^y must have a maximum at $\tau = 0$. We replaced the discrete values of τ_n with a continuous time function τ, R_E^y may again be thought of as defining the envelope of R^y.

The factor $J_0\,[P(u + v)]$ in the integrand of (50) in no way alters the validity of the rules of the "Elementray Combinations and Transformations" given in Table I, but it alters the shape of R_E^y, which now is also a function of P.

Appendix I

In this section we outline the application of the rules by way of some examples.

A. *Exponential Devices*

Assume a nonlinear device of the form [8]:

$$y = \exp(x).$$

The derivative dy/dx equals y. The differentiation law (5) tells us that the same property must hold for the autocorrelation; hence,

$$R^y(R_0 r) = C \exp(R_0 r). \tag{51}$$

The proportionality constant C is obtained by letting $r \to 0$; it must equal the dc power $\exp(R_0)$. Hence,

$$R^y(R_0 r) = \exp(R_0 + R_0 r). \tag{52}$$

The same result is, of course, obtained for $y = \exp(-x)$.

By means of (16) and (17), we show that to $y = \cosh x$ and $y = \sinh x$ correspond, respectively,

$$R^y(R_0 r) = \exp(R_0) \cosh(R_0 r)$$

and

$$R^y(R_0 r) = \exp(R_0) \sinh(R_0 r). \tag{53}$$

Vice versa, (52) could be obtained from (53) by means of (18).

B. *Power-Law Devices*

Table II lists the output correlation R^y for the "half-wave rectifier" devices of the type

$$y^{(m)}(x) = x^m \quad \text{for } m = 0, 1, 2, \cdots \qquad x > 0. \tag{54}$$

Therefore, $y^{(m)}$ can be obtained from $y^{(m-1)}$ by integration and multiplication by m. According to the integration law (Table I, no. 6) and the multiplication law (Table I, no. 1) the corresponding $R^{y(m)}$ is found from $R^{y(m-1)}$ by multiplying by R_0, integrating with respect to r, and multiplying by m^2. The integration constant is found by letting $\tau \to \infty$ or $r \to 0$. The remaining expression is the dc power P_{dc}, which is known [3], [10]:

$$P_{dc} = \frac{1}{2\pi}\left(\frac{2^K K!}{m!}\right)^2 R_0^m \quad \text{for odd } m = 2K + 1 \tag{55}$$

TABLE II

Half-Wave Rectification: $y = x^m$, $x > 0$

No.	$y(x)$	R^y	Total Power $r \to 1$	DC Power $r \to 0$
A	$\delta(x)$ unit impulse	$\frac{1}{2\pi R_0}(1 - r^2)^{-1/2}$		$\frac{1}{2\pi R_0}$
0	$u(x)$ unit step	$\frac{1}{2\pi}\cos^{-1}(-r) = \frac{1}{2\pi}\left[\sin^{-1} r + \frac{\pi}{2}\right]$	$\frac{1}{2}$	$\frac{1}{4}$
1	x unit ramp	$\frac{R_0}{2\pi}[r\cos^{-1}(-r) + \sqrt{1 - r^2}]$	$\frac{R_0}{2}$	$\frac{R_0}{2\pi}$
2	x^2	$\frac{R_0^2}{2\pi}[(2r^2 + 1)\cos^{-1}(-r) + 3r\sqrt{1 - r^2}]$	$\frac{3}{2}R_0^2$	$\frac{R_0^2}{4}$
3	x^3	$\frac{R_0^3}{2\pi}[(6r^3 + 9r)\cos^{-1}(-r) + (11r^2 + 4)\sqrt{1 - r^2}]$	$\frac{15}{2}R_0^3$	$4\frac{R_0^3}{2\pi}$
4	x^4	$\frac{R_0^4}{2\pi}[(24r^4 + 72r^2 + 9)\cos^{-1}(-r) + (50r^3 + 55r)\sqrt{1 - r^2}]$	$\frac{105}{2}R_0^4$	$\frac{9}{4}R_0^4$
5	x^5	$\frac{R_0^5}{2\pi}[5(24r^5 + 120r^3 + 45r)\cos^{-1}(-r) + (274r^4 + 607r^2 + 64)\sqrt{1 - r^2}]$	$\frac{945}{2}R_0^5$	$64\frac{R_0^5}{2\pi}$
m	x^m, $x > 0$ 0, $x < 0$	$\frac{R_0^m}{2\pi}[P_1^{(m)}(r)\cos^{-1}(-r) + P_2^{(m)}(r)\sqrt{1 - r^2}]$		

TABLE III

$y = x^m, \quad x \gtrless 0$

No.	$y(x)$	R^y	Total Power	DC Power
0	1	1	1	1
1	x	$R_0 r$	R_0	0
2	x^2	$R_0^2\,(2r^2+1)$	$3R_0^2$	R_0^2
3	x^3	$R_0^3\,(6r^3+9r)$	$15R_0^3$	0
4	x^4	$R_0^4\,(24r^4+72r^2+9)$	$105R_0^4$	$9R_0^4$
5	x^5	$R_0^5\,5(24r^5+120r^3+45r)$	$945R_0^5$	0

TABLE IV

$y = x^m, \quad x > 0; y = -x^m, \quad x < 0$

No.	$\lvert y(x)\rvert$	R^y	Total Power	DC Power
0	1	$\frac{2}{\pi}\sin^{-1} r$	1	0
1	$\lvert x\rvert$	$\frac{2}{\pi} R_0\,[r\sin^{-1} r + \sqrt{1-r^2}]$	R_0	$\frac{2}{\pi} R_0$
2	$\lvert x^2\rvert$	$\frac{2}{\pi} R_0^2\,[(2r^2+1)\sin^{-1} r + 3r\sqrt{1-r^2}]$	$3R_0^2$	0
3	$\lvert x^3\rvert$	$\frac{2}{\pi} R_0^3\,[(6r^3+9r)\sin^{-1} r + (11r^2+4)\sqrt{1-r^2}]$	$15R_0^3$	$\frac{8}{\pi} R_0^3$
4	$\lvert x^4\rvert$	$\frac{2}{\pi} R_0^4\,[(24r^4+72r^2+9)\sin^{-1} r + (50r^3+55r)\sqrt{1-r^2}]$	$105R_0^4$	0
5	$\lvert x^5\rvert$	$\frac{2}{\pi} R_0^5\,[5(24r^5+120r^3+45r)\sin^{-1} r + (274r^4+607r^2+64)\sqrt{1-r^2}]$	$945R_0^5$	$\frac{128}{\pi} R_0^5$

and

$$P_{dc} = \left(\frac{1}{2^{K+1}K!}\right)^2 R_0^m \qquad \text{for even } m = 2K. \tag{56}$$

As a check on the correctness of R^y, we also may let $r = 0$ or $r = 1$ and obtain the known total output power P;

$$P = \frac{(2m)!}{2^{m+1}(m!)} R_0^m. \tag{57}$$

Table III contains the set $y = x^m$ for all x. The devices with even m may be classified as full-wave rectifiers.

If we now reverse the sign of $y(x)$ for $x < 0$, we obtain Table IV. The full-wave rectifier curves now are those with m odd.

The same rules regarding integration and multiplication by x apply. In addition we can derive the functions R^y by applying the rules of addition, (16) or (17), to the results of the preceding tables.

C. The "Ideal Limiter"

The "ideal limiter" is defined by

$$\begin{aligned} y &= 1 && \text{for} \quad x > 1 \\ y &= x && \text{for} \quad -1 < x < 1 \\ y &= -1 && \text{for} \quad x < -1. \end{aligned} \tag{58}$$

By differentiating twice we obtain a negative impulse at $x = +1$, and a positive one at $x = -1$. The output correlation corresponding to $\delta(x-1)$ is, using the shifting factor (28),

$$R^\delta = \frac{1}{2\pi}\exp\left(\frac{-1}{R_0(1+r)}\right)\frac{1}{R_0\sqrt{1-r^2}}. \tag{59}$$

Using the symmetry relation (17), we obtain immediately, for the desired output correlation R^y,

$$\frac{d^2R^y}{R_0^2\,dr^2} = \frac{1}{\pi R_0\sqrt{1-r^2}} \cdot\left[\exp\left(\frac{-1}{R_0(1+r)}\right) - \exp\left(\frac{-1}{R_0(1-r)}\right)\right]. \tag{60}$$

The integration of (60) by machine computation is discussed in [7]. Many other nonlinear devices can be reduced to a sum of shifted impulses by repeated differentiation. Unfortunately, an explicit integration formula for the shifted pulse (59) is not available.

D. Recurrence Formulas for Half-Wave Rectifiers

The general expression for R^y in Table II is given by

$$R^{y,m} = [P_1^m(r)\cos^{-1}(-r) + P_2^m(r)\sqrt{1-r^2}]R_0^m \tag{61}$$

where $P_1^m(r)$ and $P_2^m(r)$ are polynomials in r. The index

m refers to the power of x in $y = x^m$. It is either even or uneven. P_1^m is of highest power m and P_2^m is of highest power $(m - 1)$. By applying (39) we obtain immediately the recurrence formulas

$$P_1^{m+1}(r) = (1 - r^2)\frac{dP_1^m(r)}{dr} + (2m + 1)rP_1^m(r) \quad (62)$$

$$P_2^{m+1}(r) = P_1^m(r) + 2mrP_2^m(r) + (1 - r^2)\frac{dP_2^m(r)}{dr}. \quad (63)$$

On the other hand, by simple differentiation of (61), one obtains the additional relationships

$$P_1^{m-1}(r) = \frac{1}{m^2}\frac{dP_1^m(r)}{dr} \quad (64)$$

and

$$P_2^{m-1} = \left[\frac{P_1^m(r) - rP_2^m(r)}{1 - r^2} + \frac{dP_2^m(r)}{dr}\right]\frac{1}{m^2}. \quad (65)$$

E. Differential Equation for Power-Law Devices

Equation (39), together with the rule for differentiation, can be used to obtain a differential equation for $R^{\nu'}$ as follows. Since

$$y' = xy^m = y^{m+1},$$

we have

$$\frac{dy'}{dx} = (m + 1)y^m(x); \quad (66)$$

therefore, by no. 5 and no. 1 (Table I),

$$\frac{1}{R_0}\frac{\partial R^{\nu'}}{\partial r} = (m + 1)^2 R^{\nu'}. \quad (67)$$

Thus, by differentiating (39) we have

$$(m + 1)^2 R^{\nu'} = (1 - r^2)\frac{\partial^2 R^{\nu'}}{\partial r^2} + (2m - 1)r\frac{\partial R^{\nu'}}{\partial r} + (2m + 1)R^{\nu'}$$

or

$$(1 - r^2)\frac{\partial^2 R^{\nu'}}{\partial r^2} + (2m - 1)r\frac{\partial R^{\nu'}}{\partial r} - m^2 R^{\nu'} = 0. \quad (68)$$

The correlation function of the output of any mth power device must fulfill this equation. By using (46) twice, we may also write this result as a recursion formula

$$\underline{(m - 1)^2(1 - r^2)R^{\nu'(m-2)} + (2m - 1)rR^{\nu'(m-1)} - R^{\nu' m} = 0}. \quad (69)$$

F. Solutions Obtained From Differential Equations

In order to find a solution to (68) substitute either

$$s = \frac{1 + r}{2} \quad (70)$$

or

$$s = \frac{1 - r}{2} \quad (71)$$

and obtain an equation of the hypergeometric type:

$$s(1 - s)\frac{d^2 R^{\nu'}}{ds^2} + \left[\frac{1 - 2m}{2} - (1 - 2m)s\right]\frac{dR^{\nu'}}{ds} - m^2 R = 0. \quad (72)$$

It has the general form

$$s(1 - s)\frac{d^2 F}{ds^2} + [c - (a + b + 1)s]\frac{dF}{ds} - abF = 0 \quad (73)$$

and one solution is the Gaussian hypergeometric series [12]

$$_2F_1(a, b, c; s) = 1 + \frac{ab}{c}\frac{s}{1!} + \frac{a(a + 1)b(b + 1)}{c(c + 1)}\frac{s^2}{2!} + \cdots. \quad (74)$$

Comparing (71) with (72) we find

$$a = b = -m \quad (75)$$
$$c = -m + \tfrac{1}{2}$$

whence the solution of (72) becomes

$$R^{\nu'} = C\ {}_2F_1(-m, -m, -m + \tfrac{1}{2}; s). \quad (76)$$

Using the Quadratic Transformation Formula [12, 15.3.30]:

$$_2F_1\left(a, b, \frac{a + b}{2} + \frac{1}{2}; s\right) = {}_2F_1\left(\frac{a}{2}, \frac{b}{2}, \frac{a + b}{2} + \frac{1}{2}; 4s - 4s^2\right), \quad (77)$$

since

$$4s - 4s^2 = 1 - r^2 \quad (78)$$

we obtain

$$R^{\nu'} = C\ {}_2F_1\left(-\frac{m}{2}, -\frac{m}{2}, -m + \frac{1}{2}; 1 - r^2\right). \quad (79)$$

This again can be written as

$$R^{\nu'} = A\ {}_2F_1\left(-\frac{m}{2}, -\frac{m}{2}, \frac{1}{2}; r^2\right) + \sqrt{r^2}\, B\ {}_2F_1\left(\frac{1 - m}{2}, \frac{1 - m}{2}, \frac{3}{2}; r^2\right) \quad (80)$$

where we have used the Linear Transformation Formula [12, 15.3.6]:

$$_2F_1(a, b, c; z) = A_1\ {}_2F_1(a, b, a + b - c + 1; 1 - z) + (1 - z)^{c-a-b} \cdot A_2\ {}_2F_1(c - a, c - b, c - a - b + 1; 1 - z)$$
$$A_{1,2} = \text{const.} \quad (81)$$

Equations (70) and (71) show that the root can be either taken as $(+r)$ or $(-r)$ and since we are dealing with a linear differential equation we conclude that a general solution of (68) is of the form

$$R^{\nu'} = C_1\ {}_2F_1\left(-\frac{m}{2}, -\frac{m}{2}, \frac{1}{2}; r^2\right) + C_2 r\ {}_2F_1\left(\frac{1 - m}{2}, \frac{1 - m}{2}, \frac{3}{2}; r^2\right). \quad (82)$$

The constants C_1 and C_2 are to be chosen so as to obtain particular known values of $R^{\nu'}$ at $r = 0$ and $r = 1$. For one-way rectifiers one obtains [3]

$$C_1 = R_0^m \frac{2^m}{4\pi} \Gamma^2\left(\frac{m+1}{2}\right) \tag{83}$$

$$C_2 = R_0^m \frac{2^{m+1}}{4\pi} \Gamma^2\left(\frac{m}{2}+1\right). \tag{84}$$

This solution is valid for any m. For half-integer values the solution can be written explicitly by means of one or two complete elliptic integrals. Assume, for instance, that

$$m = \tfrac{1}{2}; \tag{85}$$

then (72) becomes

$$s(1-s)\frac{d^2R^{\nu'}}{ds^2} - \frac{R^{\nu'}}{4} = 0 \tag{86}$$

which is [11, p. 80] the differential equation for $sB(s)$, with s being the square of the modulus k of the complete elliptic integral $B(k^2)$. Therefore, we obtain the solution

$$R^{\nu'} = C_1 \frac{1+r}{2} B\left(\frac{1+r}{2}\right) + C_2 \frac{1-r}{2} B\left(\frac{1-r}{2}\right). \tag{87}$$

The solutions for other half-integer values of m can be obtained by repeated integration or differentiation of $R^{\nu'}$ with respect to r (or, after substitution, with respect to s). These transformations lead again to complete elliptic integrals [11, pp. 76 and 77].

Appendix II

Calculation of the Derivative ($\partial R^\nu / \partial R_0$)

Introduce into (7) the new variables $u' = \sigma u$ and $v' = \sigma v$:

$$R^\nu = \frac{1}{4\pi^2} \iint \frac{F(ju'/\sigma)}{\sigma} \frac{F(jv'/\sigma)}{\sigma} \cdot \exp\left[- \frac{u'^2 + v'^2}{2} - ru'v'\right] du'\, dv' \tag{88}$$

$$\begin{aligned}\frac{\partial R^\nu}{\partial R_0} &= \frac{1}{2\sigma}\frac{\partial R^\nu}{\partial \sigma} \\ &= \frac{1}{2\sigma}\frac{1}{4\pi^2} \iint \frac{\partial}{\partial \sigma}\left[\frac{F(ju'/\sigma)}{\sigma}\frac{F(jv'/\sigma)}{\sigma}\right] \cdot \exp\left[-\frac{u'^2+v'^2}{2} - ru'v'\right] du'\, dv'.\end{aligned} \tag{89}$$

Since

$$\frac{\partial}{\partial \sigma}\left[\frac{F(ju'/\sigma)}{\sigma}\right] = -\frac{1}{\sigma^2}\left[F\frac{ju'}{\sigma} + \frac{ju'}{\sigma}\frac{dF(ju'/\sigma)}{d(ju'/\sigma)}\right]$$

one obtains

$$\frac{\partial R^\nu}{\partial R_0} = \frac{-1}{\sigma}\frac{1}{4\pi^2} \iint \frac{F(jv'/\sigma)}{\sigma^3}\left[F\frac{ju'}{\sigma} + \frac{ju'}{\sigma}\frac{dF(ju'/\sigma)}{d(ju'/\sigma)}\right] \cdot \exp\left[-\frac{u'^2+v'^2}{2} - ru'v'\right] du'\, dv'. \tag{90}$$

Now, reintroduce u and v and $R_0 = \sigma^2$:

$$\frac{\partial R^\nu}{\partial R_0} = \frac{1}{4\pi^2 R_0} \iint -F(jv)\left[F(ju) + ju\frac{dF(ju)}{d(ju)}\right] \cdot \exp\left[-\frac{R_0}{2}(u^2+v^2) - R_0ruv\right] du\, dv. \tag{91}$$

Introduce the function $F^*(ju)$:

$$F^*(ju) = -\frac{d}{d(ju)}[juF(ju)]; \tag{92}$$

it is the Fourier transform of the produce $x[dy(x)/dx]$. Then,

$$\frac{\partial R^\nu}{\partial R_0} = \frac{1}{4\pi^2 R_0} \iint F^*(ju)F(jv) \cdot \exp\left[-\frac{R_0}{2}(u^2+v^2) - R_0ruv\right] du\, dv. \tag{93}$$

References

[1] S. O. Rice, "Mathematical analysis of random noise," *Bell Sys. Tech. J.*, vols. 23 and 24, ch. 4.6–4.9.

[2] D. Middleton, *Statistical Communication Theory*. New York: McGraw-Hill, 1960, ch. 5, p. 239.

[3] ——, "Some general results in the theory of noise through non-linear devices," *Quart. Appl. Math.*, vol. 5, p. 453, January 1948.

[4] R. Price, "A useful theorem for nonlinear devices having Gaussian inputs," *IRE Trans. Information Theory*, vol. IT-4, pp. 69–72, June 1958.

[5] Clavier, Panter, and Dite, "Reduction by limiters of amplitude modulation in an amplitude- and frequency-modulated wave," *Elec. Commun.*, p. 291, 1947.

[6] R. F. Baum, "The correlation function of smoothly limited Gaussian noise," *IRE Trans. Information Theory*, vol. IT-3, pp. 193–197, September 1957.

[7] Laning and Battin, *Random Processes in Automatic Control*. New York: McGraw-Hill, 1956, pp. 164, 362.

[8] A. Papoulis, *Probability, Random Variables and Stochastic Processes*. New York: McGraw-Hill, 1965, p. 476.

[9] Campbell and Foster, *Fourier Integrals for Practical Applications*. Princeton, N. J.: Van Nostrand, 1951, p. 43.

[10] H. Schlitt, *Systemtheorie fur regellose Vorgänge*. Berlin: Springer, 1960, p. 268.

[11] Jahnke and Emde, *Tables of Functions*. New York: Dover, 1945, pp. 73–80.

[12] M. Abramowitz and I. A. Stegun, Eds., *Handbook of Mathematical Functions*, Appl. Math. Ser. 55. Washington, D. C.: NBS, 1964, pp. 556–562.

10

Reprinted from *IEEE Trans. Information Theory*, **IT-13,** 502–505 (July 1967)

Nonlinear Transformations of Random Processes

NORMAN ABRAMSON, MEMBER, IEEE

***Abstract*—This paper provides a general method of calculating the mean-square bandwidth (and other spectral moments) of an arbitrary zero-memory nonlinear transformation of a stationary random process. The method is valid when the original process is an arbitrary combination of other random processes. It can be used to determine the mean-square bandwidth (or the spectral moments) of the transformed process either before or after that process is passed through a bandpass filter. Five examples of the application of this method are provided simplifying and generalizing known results, as well as providing new results.**

I. Introduction

THE CALCULATION of the power spectral density of the result of a nonlinear transformation of a stationary random process presents considerable analytic difficultes. Results of this sort are available in closed form for a small number of special cases.[1]–[4] If we are interested only in the bandwidth of the transformed random process however somewhat more is known, although results are available only when the original random process is Gaussian and the method of deriving these results can be somewhat involved.[5],[6] In this paper we derive a simple expression for the mean-square bandwidth of an arbitrary zero-memory nonlinear transformation of a stationary random process. Our method is valid for non-Gaussian as well as Gaussian random processes. It can be used to determine the mean-square bandwidth and other spectral moments of the transformed process either before or after that process is passed through a bandpass filter. Examples of the use of this expression to derive new results as well as to simplify and to generalize known results are given.

II. Mean-Square Bandwidth

Consider an arbitrary zero-memory nonlinear transformation of the stationary random process $x(t)$. By zero-memory we mean that the result of the transformation $y(t)$ is some function, $g(\cdot)$, of the instantaneous value of $x(t)$ (see Fig. 1).

Fig. 1. Nonlinear system.

The random process $x(t)$ is not necessarily Gaussian and, in fact, it may be an arbitrary combination of two other stationary random processes, $s(t)$ and $n(t)$. We are interested in calculating B_y^2, the mean-square bandwidth of $y(t)$. Let $R_y(\tau)$ and $S_y(f)$ be the autocorrelation and power spectral density of $y(t)$, respectively. Then

$$B_y^2 = \frac{\int f^2 S_y(f)\, df}{\int S_y(f)\, df} \tag{1}$$

or[6]

$$B_y^2 = -\frac{\ddot{R}_y(0)}{(2\pi)^2 E[y^2]} \tag{2}$$

where the integrals are taken over the interval $(-\infty, \infty)$ and $\ddot{R}_y(0)$ is the second derivative of $R_y(\tau)$ evaluated at $\tau = 0$. The mean-square bandwidth B_y^2 is the second moment of the power spectral density of $y(t)$ about zero frequency. For certain applications[6] we wish to define a mean-square bandwidth as the second moment about a carrier frequency f_0. If we define f_0 as the first absolute spectral moment (e.g., (49) in the Appendix) then we may define B_{y_0}, the second moment bandwidth of $y(t)$ about frequency f_0, by the equation

$$B_y^2 = B_{y_0}^2 + f_0^2. \tag{3}$$

III. Results

From (2) we see that the calculation of the mean-square bandwidth hinges upon the calculation of the second derivative of the autocorrelation $R_y(\tau)$. If $\ddot{R}_y(0)$ exists we can express[7] $\ddot{R}_y(\tau)$ in terms of $R_{\dot{y}}(\tau)$, the autocorrelation of the derivative of $y(t)$

$$\ddot{R}_y(\tau) = -R_{\dot{y}}(\tau) \tag{4}$$

so that (2) and (4) may be combined to obtain B_y^2 in terms of the mean-square values of the $y(t)$ and $\dot{y}(t)$ random processes.

$$B_y^2 = \frac{E[\dot{y}^2]}{(2\pi)^2 E[y^2]}. \tag{5}$$

Since

$$\begin{aligned} y &= g(x), \\ \dot{y} &= g'(x)\dot{x} \end{aligned} \tag{6}$$

where

$$g'(x) = \frac{dg}{dx} \tag{7}$$

and (5) may also be written

$$B_y^2 = \frac{E[(g'\dot{x})^2]}{(2\pi)^2 E[g^2]}. \tag{8}$$

Manuscript received March 14, 1966; revised March 23, 1966. This work was supported by Joint Services Electronics Program, Harvard University, Cambridge, Mass., Contract Nonr-1866(16); and by Air Force Office of Scientific Research, University of Hawaii, Grant AF-AFOSR-1251-67.
The author is with the Department of Electrical Engineering, University of Hawaii, Honolulu, Hawaii 96822

If the random process $x(t)$ is some function of two other random processes, $s(t)$ and $n(t)$, an obvious generalization of (8) yields

$$B_\nu^2 = \frac{E[(g_s\dot{s} + g_n\dot{n})^2]}{(2\pi)^2E[g^2]} \tag{9}$$

where g_s and g_n are the partial derivatives of $g[x(s, n)]$ with respect to $s(t)$ and $n(t)$. Note that $s(t)$ and $n(t)$ are not necessarily independent.

Equations (9), (8), or (5) may be used directly to compute B_ν^2. Under an additional condition however a further simplification is possible.

For many random processes $x(t)$, the random *variables* $x(t)$ and $\dot{x}(t)$ are statistically independent. (Note we are *not* assuming the random *processes* are independent.) We call such random processes, derivative independent (DI). Some examples of DI processes are[1]:

1) arbitrary stationary Gaussian random process (SGRP),
2) envelope of a SGRP,
3) derivative[8] of the envelope of a SGRP,
4) phase of a SGRP,
5) sum of two independent DI processes, and
6) sum of a DI process and a known function.

For a DI process the expectation of the product in the numerator of (8) factors into the product of expectations, and we have

$$B_\nu^2 = \frac{E[(g')^2]E[\dot{x}^2]}{(2\pi)^2E[g^2]}. \tag{10}$$

We may use (5) on the random process $x(t)$ to express $E[\dot{x}^2]$ in (10) in terms of B_x^2.

$$B_\nu^2 = \frac{E[(g')^2]E[x^2]}{E[g^2]} B_x^2. \tag{11}$$

We note that the same method used to derive (11), (9), (8), and (5) may be applied to find higher moments of the spectral density.

In the remainder of this paper we provide a sequence of examples illustrating the application of (11), (9), (8), and (5).

IV. Examples

A. *The Envelope of a SGRP (Bello)*

Let $x(t)$ be a zero mean, SGRP. Let $V(t)$ be the envelope of $x(t)$ as defined in the Appendix. In this example we calculate the mean-square bandwidth of $V(t)$. Without loss of generality we may normalize $x(t)$ so that $E[x^2] = 1$. A direct application of (5) yields

$$B_V^2 = \frac{E[\dot{V}^2]}{(2\pi)^2E[V^2]}. \tag{12}$$

[1] See the Appendix.

These expectations may be evaluated immediately using the Appendix.

$$B_V^2 = \frac{1}{2}\frac{\beta^2 - \gamma^2}{(2\pi)^2} - \frac{B_{x_0}^2}{2} \tag{13}$$

where $B_{x_0}^2$ is the mean-square bandwidth of $x(t)$ about the frequency f_0.

B. *νth Law Envelope Detectors With Gaussian Inputs (Bello)*

Let $x(t)$ and $V(t)$ be defined as in Example A. Let $y(t) = [V(t)]^\nu$. Then, on identifying V with x of (11), we have

$$B_\nu^2 = \frac{\nu^2E[V^{2\nu-2}]E[V^2]}{E[V^{2\nu}]} B_V^2. \tag{14}$$

The envelope V is Rayleigh with density given in the Appendix, so

$$E[V^n] = 2^{n/2}\Gamma\left(\frac{n}{2} + 1\right) \tag{15}$$

and

$$B_\nu^2 = \nu B_V^2 = \frac{\nu}{2} B_{x_0}^2. \tag{16}$$

This result due to Bello[5] was originally obtained using a more complicated procedure.

C. *Angle Modulated Sinusoids (Kahn and Thomas)*

Define

$$y(t) = \cos[\omega_1 t + x(t) + \theta] \tag{17}$$

where $\omega_1 = 2\pi f_1$, $x(t)$ is a stationary random process, and θ is a random variable, independent of $x(t)$, and uniformly distributed on $[0, 2\pi]$. If we let $x_1(t) = \omega_1 t + x(t) + \theta$, where $x_1(t)$ is taken modulo 2π, we see that $x_1(t)$ is DI. Note we do *not* have to assume that $x(t)$ is DI. Now we may apply (10) to obtain

$$B_\nu^2 = \frac{E[\sin^2(\cdot)]E[\dot{x}_1^2]}{(2\pi)^2E[\cos^2(\cdot)]} = f_1^2 + \frac{E[\dot{x}^2]}{(2\pi)^2} \tag{18}$$

and, using (5) once more

$$B_\nu^2 = f_1^2 + E[x^2]B_x^2 \tag{19}$$

or, in terms of the mean-square bandwidth of $y(t)$ about the carrier frequency f_1

$$B_{\nu_1}^2 = E[x^2]B_x^2. \tag{20}$$

Equations (19) and (20) have previously been obtained for the Gaussian case in another paper[6] and for deterministic as well as random processes by Kahn and Thomas.[9]

D. νth Law Devices With Gaussian Inputs

Let $x(t)$ be a SGRP with zero mean. Let $y(t) = g[x(t)]$, where

$$g(x) = \begin{cases} x^\nu & x \geq 0 \\ 0 & x \leq 0 \end{cases} \tag{21}$$

(half-wave νth law device), or

$$g(x) = |x|^\nu \tag{22}$$

(even νth law device), or

$$g(x) = \frac{x}{|x|} |x|^\nu \tag{23}$$

(odd νth law device). For any one of these three characteristics a direct application (11) yields

$$B_y^2 = \frac{\nu^2}{2\nu - 1} B_x^2 \tag{24}$$

for $\nu > \frac{1}{2}$.

E. Filtered Output of νth Law Devices With Gaussian Inputs

In this example we obtain the mean-square bandwidth of the random process $z(t)$ of Fig. 2. Let $x(t)$ be a stationary random process and let us write $x(t) = V(t) \cos \Phi(t)$, where the envelope $V(t)$ and phase $\Phi(t)$ are defined in the Appendix. (No narrowband assumption is necessary at this point.) We describe the nonlinearity by the function $g(x)$.

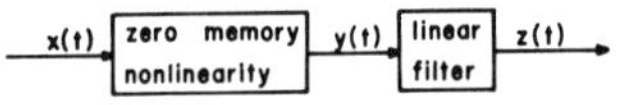

Fig. 2. Nonlinearity and filter.

$$\begin{aligned} y &= g(x) \\ &= g(V \cos \Phi) \end{aligned} \tag{25}$$

and since y is periodic in Φ we can expand y in a Fourier Series.[10]

$$y = \sum_{m=0}^{\infty} g_m(V) \cos m\Phi \tag{26}$$

where

$$g_m(V) = \frac{\epsilon_m}{2\pi} \int_{-\pi}^{\pi} g(V \cos \Phi) \cos m\Phi \, d\Phi \tag{27}$$

$\epsilon_m = 2$, $m \neq 0$ and $\epsilon_0 = 1$. For the half-wave νth law device, (27) becomes

$$g_m(V) = V^\nu c(\nu, m) \tag{28}$$

where

$$c(\nu, m) = \frac{\epsilon_m}{2\pi} \int_{-\pi/2}^{\pi/2} (\cos \Phi)^\nu \cos m\Phi \, d\Phi. \tag{29}$$

For the even νth law device

$$g_m(V) = \begin{cases} 2V^\nu c(\nu, m) & m \text{ even} \\ 0 & m \text{ odd} \end{cases} \tag{30}$$

while for the odd νth law device

$$g_m(V) = \begin{cases} 2V^\nu c(\nu, m) & m \text{ odd} \\ 0 & m \text{ even.} \end{cases} \tag{31}$$

In all three of these cases, (26) is of the form

$$y = \sum_{m=0}^{\infty} d(\nu, m) V^\nu \cos m\Phi \tag{32}$$

where the $d(\nu, m)$ are constants, depending only upon ν and m. At this point we make the assumption that $x(t)$ is narrowband. Then, each of the terms in (32) is concentrated in a narrow band of frequencies centered around mf_0. If the filter in Fig. 2 passes only those frequencies close to mf_0 its output is

$$z(t) = V^\nu \cos m\Phi. \tag{33}$$

We have arbitrarily set the constant term which should appear in (33) equal to one since a constant gain will be irrelevant when we calculate B_z^2. In this example, $z(t)$, the random process of interest, is the product of two different functions of the random processes $V(t)$ and $\Phi(t)$. We can use (9), (identifying $V(t)$ with $s(t)$ and $\Phi(t)$ with $n(t)$, to obtain

$$B_z^2 = \frac{E\{[\nu \dot{V} V^{\nu-1} \cos m\Phi - m\dot{\Phi} V^\nu \sin m\Phi]^2\}}{(2\pi)^2 E[V^{2\nu} \cos^2 m\Phi]}. \tag{34}$$

Now we assume $x(t)$ is a SGRP with zero mean so that the density of the four random variables Φ, $\dot{\Phi}$, V, $\dot{V}$ may be obtained from the Appendix. In this case Φ is independent of the other three random variables. Since

$$E[\cos^2 m\Phi] = E[\sin^2 m\Phi] = \tfrac{1}{2} \tag{35}$$

and

$$E[(\sin m\Phi)(\cos m\Phi)] = 0 \tag{36}$$

(34) becomes

$$B_z^2 = \frac{\nu^2 E[\dot{V}^2 V^{2\nu-2}] + m^2 E[\dot{\Phi}^2 V^{2\nu}]}{(2\pi)^2 E[V^{2\nu}]}. \tag{37}$$

The first term in the numerator of (37) can be evaluated immediately since V and $\dot{V}$ are independent random variables. The random variables V and $\dot{\Phi}$ are not independent. To evaluate $E[\dot{\Phi}^2 v^{2\nu}]$ however, we merely fix V and take the expectation first with respect to the density $p(\dot{\Phi} \mid V)$. From the Appendix we see this is Guassian so that the conditional expectation of $\dot{\Phi}^2 V^{2\nu}$ is just

$$\left[\gamma^2 + \frac{\beta^2 - \gamma^2}{V^2}\right] V^{2\nu}. \tag{38}$$

The expected value of (38) with respect to the density of V now merely involves moments of a Rayleigh distribution and (37) simplifies to

$$B_z^2 = (mf_0)^2 + \frac{\nu^2 + m^2}{2\nu} B_{x_0}^2. \tag{39}$$

if we let $B_{z_o}^2$ be the mean-square bandwidth of $z(t)$ about its carrier frequency mf_0

$$B_{z_o}^2 = \frac{\nu^2 + m^2}{2\nu} B_{x_o}^2. \tag{40}$$

Appendix

Let $x(t)$ be a stationary Gaussian random process with zero mean and autocorrelation $R(\tau)$. For simplicity we normalize so that $R(0) = 1$. Let $\tilde{x}(t)$ be the Hilbert Transform of $x(t)$. Then we define

$$z(t) = x(t) + i\tilde{x}(t) \tag{41}$$

and the envelope $V(t)$ and phase $\Phi(t)$ of $x(t)$ are given by

$$z(t) = V(t)e^{i\Phi(t)}. \tag{42}$$

Although no narrowband assumption is made the envelope and phase as defined previously correspond to the usual definition in the narrowband case. Kelly and Reed[11] derive the joint density of the four random variables V, $\dot{V}$, Φ, $\dot{\Phi}$. However the form in which their results are presented is somewhat inconvenient for our purposes. We shall state without proof a more useful form of the density $p(V, \dot{V}, \Phi, \dot{\Phi})$. The proof follows directly from some algebraic manipulations on the results of Kelly and Reed. We use the notation $N(m, \sigma^2)$ to indicate a normal density with mean m and variance σ^2.

$$p(V, \dot{V}, \Phi, \dot{\Phi}) = p(\Phi)p(\dot{V})p(\dot{\Phi} \mid V)p(V) \tag{43}$$

where

$$p(\Phi) = \begin{cases} \dfrac{1}{2\pi} & \Phi \in [0, 2\pi] \\ 0 & \text{elsewhere} \end{cases} \tag{44}$$

$$p(\dot{V}) = N(0, \beta^2 - \gamma^2) \tag{45}$$

$$p(\dot{\Phi} \mid V) = N\left(\gamma, \frac{\beta^2 - \gamma^2}{V^2}\right) \tag{46}$$

$$p(V) = \begin{cases} Ve^{-V^2/2} & V \geq 0 \\ 0 & V < 0. \end{cases} \tag{47}$$

Here we have defined

$$\beta^2 = -\ddot{R}(0) = (2\pi)^2 \int f^2 S(f)\, df = (2\pi B_x)^2 \tag{48}$$

$$\gamma = \dot{\tilde{R}}(0) = (2\pi) \int |f|\, S(f)\, df = 2\pi f_0. \tag{49}$$

$\tilde{R}(\tau)$ is the Hilbert Transform of $R(\tau)$, and $\dot{\tilde{R}}(0)$ is the derivative of $\tilde{R}(\tau)$ evaluated at $\tau = 0$. Thus $\beta^2/(2\pi)^2$ is the mean-square bandwidth of $x(t)$, $\gamma/2\pi$ is the center frequency of $x(t)$ and $(\beta^2 - \gamma)^2/(2\pi)^2$ is the mean-square bandwidth about the center frequency.

$$B_{x_o}^2 = \frac{\beta^2 - \gamma^2}{(2\pi)^2}. \tag{50}$$

References

[1] S. Rice, "Mathematical analysis of random noise," *Bell Sys. Tech. J.*, vol. 23, pp. 282–332, July 1944; also, vol. 24, pp. 46–156, January 1945.

[2] W. B. Davenport, Jr. and W. L. Root, *Random Signals and Noise.* New York: McGraw-Hill, 1958.

[3] D. Middleton, *An Introduction to Statistical Communication Theory.* New York: McGraw-Hill, 1960.

[4] R. Deutsch, *Nonlinear Transformations of Random Processes.* Englewood Cliffs, N. J.: Prentice-Hall, 1962.

[5] P. A. Bello, "On the rms bandwidth of nonlinearly envelope detected narrow-band gaussian noise," *IEEE Trans. Information Theory*, vol. IT-11, pp. 236–239, April 1965.

[6] N. Abramson, "Bandwidth and spectra of phase-and-frequency-modulated waves," *IEEE Trans. Communication Systems*, vol. CS-11, pp. 407-414, December 1963.

[7] J. L. Doob, *Stochastic Processes.* New York: Wiley, 1953, p. 535.

[8] B. N. Zvyaghintsev, "Some statistical properties of the second derivative of the envelope of a normal random process," *Radio Engrg. electronic Phys.*, vol. 10, pp. 635–637, April 1965.

[9] R. E. Kahn and J. B. Thomas, "Some bandwidth properties of simultaneous amplitude and angle modulation," *IEEE Trans. Information Theory*, vol. IT-11, pp. 516–520, October 1965.,

[10] N. M. Blachman, "Band-pass nonlinearities," *IEEE Trans Information Theory*, vol. IT-10, pp. 162–164, April 1964.

[11] E. J. Kelly and I. S. Reed, "Some properties of stationary Gaussian processes," M.I.T. Lincoln Lab., Lexington, Mass., Tech. Rept. 157, June 1957.

11

Reprinted from *IEEE Trans. Information Theory,* **IT-14**(1), 21–27 (1968)

The Signal X Signal, Noise X Noise, and Signal X Noise Output of a Nonlinearity

NELSON M. BLACHMAN, FELLOW, IEEE

Abstract—The signal × signal, noise × noise, and signal × noise output components of a nonlinearity are defined in a manner agreeing with intuition. The definitions are shown to accord with previous usage, for which they now provide a clear meaning. The case of a limiter is used as an illustrative example. The approach is applicable not only to fixed memoryless transformations but also to nonlinearities with memory that depend explicitly on time, such as frequency or phase modulation, demodulation, and bandpass nonlinearities.

Introduction

In the past, mathematical analyses of the effect of a nonlinearity upon a noise-corrupted signal have usually resolved the output autocorrelation function and spectrum into "signal × signal," "noise × noise," and "signal × noise" components simply by noticing whether each of its terms depends on the statistics of the input signal or the input noise or both.[1]-[3] Although Rice[4] showed that the part of the autocorrelation function later called "signal × signal" is in fact the autocorrelation function of the periodic output component when the input signal is sinusoidal, this procedure has never been fully justified, particularly with regard to the "noise × noise" and "signal × noise" components.

We shall first look at two examples of this procedure in order to formulate a better description of it, and then we shall see how the output of the nonlinearity itself can be resolved into suitable components. We shall show that these components are entirely uncorrelated and that their autocorrelation functions are the quantities obtained by the foregoing procedure. Finally, for the sake of illustration, we shall look at these output components in some particular cases.

Manuscript received October 27, 1966; revised July 11, 1967.
The author is with Sylvania Electronic Systems, Western Division, Mountain View, Calif. 94042

Resolution of Output Autocorrelation Function into Components

Square-Law Device with Gaussian Input Signal and Noise

As our first example we consider the case of a nonlinearity with input $u(t) = s(t) + n(t)$ and output $V(t) = u^2(t)$, where $s(t)$ and $n(t)$ are independent, zero-mean, stationary, Gaussian random processes with autocorrelation functions $R_s(\tau)$ and $R_n(\tau)$, respectively. Here the output autocorrelation function is

$$R_V(\tau) = R_s^2(0) + R_n^2(0) + 2R_s(0)R_n(0) + 2R_s^2(\tau) + 2R_n^2(\tau) + 4R_s(\tau)R_n(\tau).$$

Since the first three terms on the right-hand side are constant, we associate them with the dc output component $E\{V(t)\} = R_s(0) + R_n(0)$. The last three terms depend, respectively, on $R_s(\tau)$, on $R_n(\tau)$, and on both. Hence, we associate them with the output "signal $\times$ signal," "noise $\times$ noise," and "signal $\times$ noise," although these terms have not yet been defined. In the present case these three parts of $R_V(\tau)$ can be seen to come, respectively, from the three terms of $V = s^2 + n^2 + 2sn$, but, in general, their relation to V will not be as simple as this.

Limiter with Gaussian Input Signal and Noise

If we turn to the case of an ideal limiter whose output is $V(t) = \text{sgn } u(t)$, i.e., $V(t) = \pm 1$ according to the sign of the zero-mean Guassian input $u(t) = s(t) + n(t)$ with autocorrelation function $R_u(\tau) = R_s(\tau) + R_n(\tau)$, we find that output autocorrelation function[5] is

$$R_V(\tau) = \frac{2}{\pi} \arcsin \frac{R_s(\tau) + R_n(\tau)}{R_s(0) + R_n(0)}.$$

Here we seem to have only a single term, depending on both $R_s(\tau)$ and $R_n(\tau)$, and hence only signal $\times$ noise. However, $R_V(\tau)$ will appear as the sum of many terms if we express the arcsin as a power series. Expanding the result in powers of $R_s(\tau) + R_n(\tau)$ by means of the binomial theorem, we then have a double power series, as in the square-law case, whose terms can be separated into four types according to their dependence on $R_s(\tau)$ and $R_n(\tau)$.

Since arcsin 0 = 0, there is no constant term and hence no dc component. Combining all of the terms that involve only $R_s(\tau)$, we see that the signal $\times$ signal part of $R_V(\tau)$ is now, say,

$$R_S(\tau) = \frac{2}{\pi} \arcsin \frac{R_s(\tau)}{R_s(0) + R_n(0)}.$$

Similarly, the noise $\times$ noise part is, say,

$$R_N(\tau) = \frac{2}{\pi} \arcsin \frac{R_n(\tau)}{R_s(0) + R_n(0)}.$$

The remaining terms, which must add up to, say,

$$R_I(\tau) = \frac{2}{\pi} \arcsin \frac{R_s(\tau) + R_n(\tau)}{R_s(0) + R_n(0)} - \frac{2}{\pi} \arcsin \frac{R_s(\tau)}{R_s(0) + R_n(0)} - \frac{2}{\pi} \arcsin \frac{R_n(\tau)}{R_s(0) + R_n(0)},$$

will therefore represent signal $\times$ noise intermodulation.

More General Decomposition of Output Autocorrelation Function

When $s(t)$ and $n(t)$ are not Gaussian, their statistics will not be entirely determined by $R_s(\tau)$ and $R_n(\tau)$, and $R_V(\tau)$ cannot generally be expressed as a double power series in these quantities. Hence, the process by which we have separated $R_V(\tau)$ into four components appears to be limited in its applicability. To give it greater generality we shall now describe the same procedure somewhat differently.

We notice first that the dc component, say R_C, could have been obtained from $R_V(\tau)$ by setting both $R_s(\tau)$ and $R_n(\tau)$ equal to zero and that $R_S(\tau) + R_C$ and $R_N(\tau) + R_C$ could have been obtained by setting $R_n(\tau)$ or $R_s(\tau)$ equal to zero, respectively, since in this way we get the terms that do not depend on the zero quantity. Next we recall that, for any zero-mean ergodic random processes $s(t)$ and $n(t)$, we have $R_s(\infty) = 0$ and $R_n(\infty) = 0$ if these limits exist, i.e., setting $R_s(\tau)$ or $R_n(\tau)$ equal to zero is equivalent to passing to the limit $\tau \to \infty$ in the argument of the respective input autocorrelation function.

Thus, we can describe our procedure as follows. Instead of $R_V(\tau)$, we determine $R_V(\tau_s, \tau_n)$, the expectation of the product of the output when the inputs are $s(t)$ and $n(t)$ times the output when the inputs are $s(t + \tau_s)$ and $n(t + \tau_n)$, respectively. Then

$$R_C = R_V(\infty, \infty) = R_V(\infty),$$

$$R_S(\tau) = R_V(\tau, \infty) - R_C,$$

$$R_N(\tau) = R_V(\infty, \tau) - R_C,$$

and

$$R_I(\tau) = R_V(\tau) - R_S(\tau) - R_N(\tau) - R_C.$$

To ensure their existence whenever $s(t)$ and $n(t)$ are jointly ergodic, we may define $R_V(\infty)$ as

$$\lim_{T\to\infty} T^{-1} \int_0^T R_V(\tau)\, d\tau,$$

$R_V(\tau, \infty)$ as

$$\lim_{T\to\infty} T^{-1} \int_0^T R_V(\tau, \tau_n)\, d\tau_n,$$

and $R_V(\infty, \tau)$ as

$$\lim_{T\to\infty} T^{-1} \int_0^T R_V(\tau_s, \tau)\, d\tau_s.$$

Described in this way, our procedure for decomposing $R_V(\tau)$ is perfectly general and is applicable whether $s(t)$ and $n(t)$ are combined additively or otherwise to form the input to the nonlinearity. However, no clear rationale has yet been presented for such a procedure, nor has it yet been shown that the $R_S(\tau)$, $R_N(\tau)$, and $R_I(\tau)$ that it yields are necessarily autocorrelation functions themselves. We therefore turn our attention at this point from the output autocorrelation function $R_V(\tau)$ to the output $V(t)$ itself, and we consider the separation of $V(t)$ into suitable components.

Output Components

Direct Current Component

If $V(t)$ is ergodic, its time average, which is the dc output component, can also be expressed as

$$C = E\{V(t)\},$$

where $E\{\cdot\}$ denotes the mathematical expectation, i.e., the ensemble average for any given value of t. We shall take this equation as our definition of the "dc output component" even when $V(t)$ is not statistically stationary and C may therefore vary with t; it will in any case be completely predictable from the statistics of $V(t)$.

For the square-law device with zero-mean inputs having mean-squared values $E\{s^2(t)\} = R_s(0) = \sigma_s^2$ and $E\{n^2(t)\} = R_n(0) = \sigma_n^2$, as we have already noted, the dc output component is $C = \sigma_s^2 + \sigma_n^2$. For the limiter $C = 0$.

Signal × Signal

If $s(t)$ is periodic, we may associate with it all Fourier components of the output $V(t)$ having the same period T (or submultiples of it), and we may call their sum, exclusive of the dc component, the "signal × signal output" $S(t)$. To find their sum $S(t) + C$ experimentally (including the dc component now), we would feed $V(t)$ to a comb filter passing all of the harmonics of the fundamental frequency $1/T$ of $s(t)$. Such a filter uses a delay line and feedback to produce an output at any time t that is the average of $V(t)$, $V(t - T)$, $V(t - 2T)$, $\cdots$, i.e., the average of $V(t)$ over all of the times at which $s(t)$ had the same value as at that moment.

If $s(t)$ and $V(t)$ are jointly ergodic, this time average will equal the conditional ensemble average of $V(t)$ for a given value of $s(t)$. Thus, $S(t) + C = E\{V(t) \mid s(t)\}$, and we can express the signal × signal output as

$$S(t) = E\{V \mid s(t)\} - C.$$

We shall continue to use this expression to define the signal × signal output even when $s(t)$ is not periodic. A similar definition has been used previously[6] but it has not been related to the decomposition of the output autocorrelation function. Note that $E\{S(t)\} = C - C = 0$.

The conditional expectation $E\{V \mid s(t)\}$ is an average over the statistics of the input noise and hence does not depend on $n(t)$. Thus, $S(t)$ is a function of $s(t)$ alone and is the output that would be obtained from an appropriate nonlinearity with only $s(t)$ as its input. (Its form will in general depend on the univariate input-noise statistics.) For our square-law device we find $S(t) = s^2(t) - \sigma_s^2$; for the limiter with Gaussian inputs we get

$$S(t) = \operatorname{erf} \frac{s(t)}{\sigma_n},$$

where

$$\operatorname{erf} x = \sqrt{\frac{2}{\pi}} \int_0^x e^{-y^2/2}\, dy.$$

This is the output of a smooth limiter with input $s(t)$.

Total Output Noise

The remainder of the output,

$$T(t) = V(t) - S(t) - C,$$

is noise; it represents the fluctuations of $V(t)$ about its conditional mean $E\{V \mid s(t)\} = S(t) + C$. By noticing that the average of this output noise $T(t)$ over the statistics of $n(t)$ for any given $s(t)$ is

$$E\{T \mid s(t)\} = E\{V \mid s(t)\} - S(t) - C$$
$$= 0,$$

we see that the average value of the product of $T(t)$ times any function of the input signal, such as $s(t + \tau)$ or $S(t + \tau)$, vanishes for every τ, i.e., the crosscorrelation function of $T(t)$ and $s(t)$ or of $T(t)$ and $S(t)$ is identically zero, and $T(t)$ is entirely uncorrelated with either $s(t)$ or $S(t)$. It is likewise uncorrelated with C.

Noise × Noise

The total noise output $T(t)$ can be expressed as the sum of two components, signal × noise and noise × noise, the latter being defined by analogy with the signal × signal output as

$$N(t) = E\{V \mid n(t)\} - C.$$

Like $T(t)$, the noise × noise output $N(t)$ is entirely uncorrelated with C, $s(t)$, and $S(t)$, since it does not depend on $s(t)$ and so

$$E\{N \mid s(t)\} = E\{N(t)\}$$
$$= C - C$$
$$= 0.$$

The form of $N(t)$ as a function of $n(t)$ will, in general, depend on the univariate input-signal statistics, just as $S(t)$ depends on the input-noise statistics. For our square-law device we have $N(t) = n^2(t) - \sigma_n^2$; for the limiter with Gaussian inputs

$$N(t) = \operatorname{erf} \frac{n(t)}{\sigma_s}.$$

Signal × Noise

If the input signal and noise are periodic with incommensurable periods, the remainder

$$I(t) = T(t) - N(t)$$
$$= V(t) - S(t) - N(t) - C$$

will represent the output components that have neither the period of the input signal nor that of the input noise. Because the frequencies of its Fourier components will be sums and differences of harmonics of the input signal and noise frequencies, it may appropriately be called the "signal × noise output" or the "signal–noise intermodulation"

When averaged over the statistics of $n(t)$ for a given $s(t)$ (or vice versa), $I(t)$ becomes

$$E\{I \mid s(t)\} = [S(t) + C] - S(t) - 0 - C = 0.$$

Hence, $I(t)$ is completely uncorrelated with $s(t)$, $S(t)$, $n(t)$, $N(t)$, or C. Thus, except for those involving $s(t)$ and $S(t)$ or $n(t)$ and $N(t)$, the crosscorrelation function of any pair chosen from among the four output components and two input processes vanishes identically.

For our square-law device we find $I(t) = 2s(t)n(t)$; for the limiter with Gaussian inputs

$$I(t) = \text{sgn}\,[s(t) + n(t)] - \text{erf}\,\frac{s(t)}{\sigma_n} - \text{erf}\,\frac{n(t)}{\sigma_s}.$$

Autocorrelation Functions and Spectra

From the absence of crosscorrelation it follows that the autocorrelation function of $V(t)$ is the sum of the autocorrelation functions of $S(t)$, $N(t)$, $I(t)$, and C, that its power is the sum of their powers, and that its power spectral density is the sum of their power spectral densities. Thus, for example, the spectrum of the total output noise $T(t)$ can be found by subtracting the spectra of $S(t)$ and C from that of $V(t)$, and its power, which is the value taken by its autocorrelation function when $\tau = 0$, can be found similarly.

Comparison with Procedure Used Previously

Recalling our description of the procedure that has been used for separating output autocorrelation functions into four components, we can now observe that the resulting components are precisely the autocorrelation functions of the four output components we have just defined.

We consider first the constant component $R_C = R_V(\infty)$, which is the expectation of the product of the output when the input signal and noise are $s(t)$ and $n(t)$, respectively, times the output when they are $s(t + \infty)$ and $n(t + \infty)$. The infinite time difference, which may be regarded as uniformly distributed[1] over an interval of length $T \to \infty$, effectively makes $s(t + \infty)$ and $n(t + \infty)$ statistically independent of $s(t)$ and $n(t)$. Hence, the expectation of the product is the product of the expectations, each of which is $E\{V\} = C$, the dc output component. The constant component of the output autocorrelation function is therefore $R_C = C^2$, which is seen to be the autocorrelation function of the dc output component C.

Similarly, the signal $\times$ signal component of the output autocorrelation function was described as $R_S(\tau) = R_V(\tau, \infty) - R_C$, where $R_V(\tau, \infty)$ denotes the expectation of the product of the output when the input signal and noise are $s(t)$ and $n(t)$, respectively, times the output when they are $s(t + \tau)$ and $n(t + \infty)$. Averaging first over the distribution of $n(t)$ and $n(t + \infty)$, which are statistically independent of each other, we find that $R_S(\tau) + R_C$ is the expectation of the product of the average of the first factor, $E\{V \mid s(t)\} = S(t) + C$, times the average of the second factor, $E\{V \mid s(t + \tau)\} = S(t + \tau) + C$. This expectation is seen to be the autocorrelation function of $S(t) + C$, which is the sum of the autocorrelation functions of $S(t)$ and C. Hence, $R_S(\tau)$ is the autocorrelation function of $S(t)$.

It can be shown similarly that $R_N(\tau) = R_V(\infty, \tau) - R_C$ is the autocorrelation function of $N(t)$. Since the autocorrelation function of $I(t)$ equals that of $V(t)$ minus those of $S(t)$, $N(t)$, and C, it follows that the remaining component of the output autocorrelation function, $R_I(\tau)$, must be the autocorrelation function of the signal $\times$ noise output component, $I(t)$.

Versatility of the New Approach

While analysis of the output autocorrelation function will yield the power spectra of the output components, our definitions of these components yield their actual waveforms and in many cases lead to simpler calculations. Although, in the examples we have considered, the input to the nonlinearity has been the sum of a signal plus noise, the foregoing analysis is applicable whenever the output $V(t)$ is *any* function $V(s, n)$ of the input signal $s(t)$ and the input noise $n(t)$ at the same instant.

Furthermore V, s, and n can be vector-valued random processes, each with *any* number of components. Thus, n may represent several different independent or dependent noises or interfering signals that affect the desired signal in similar or in different ways. These quantities can also be two-dimensional vectors or phasors representing the instantaneous amplitude and phase of narrowband waveforms. In the case of an FM demodulator, V may have only a single component representing the low-frequency output, while s and n have four components representing the inputs' instantaneous amplitudes and phases and their time derivatives. For an AM demodulator, two components can suffice for s and for n. (Some examples are given in Blachman,[6] chs. 5 and 6.)

Extending the notion of vector-valued inputs $s(t)$ and $n(t)$, we see that these may represent the entire past behavior of the respective inputs and that the nonlinearity may have memory without altering the foregoing conclusions in any way. Moreover, the inputs need not be statistically stationary for the four output components to be uncorrelated, and the nonlinearity can depend explicitly on time, since time dependence does not differ from dependence on an additional deterministic input-noise component. Our resolution of the output into four uncorrelated components is thus very general.

Frequency modulation of a sinusoidal carrier by a signal and noise is an example of a time-dependent nonlinear transformation with memory to which the foregoing results apply. Here, as well as in the case of phase modulation, which is a memoryless, time-dependent, nonlinear transformation, the "dc" output component is the residual

[1] Recall that $R_V(\infty)$ was defined as
$\lim_{T\to\infty} T^{-1} \int_0^T R_V(\tau)\, d\tau = \lim_{T\to\infty} T^{-1} \int_0^T E\{V(t)V(t + \tau)\}\, d\tau$.

carrier-frequency component. Although not constant, this "dc" component is completely deterministic. Alternatively, an ergodic carrier process can be regarded as the input noise, and the residual carrier component after angle modulation will then be the noise × noise output.

In applying the foregoing analysis there is considerable freedom as to what should be called $s(t)$ and $n(t)$, and the choice will vary with the application. In the case of a demodulator, for example, $s(t)$ should denote the modulating wave, which is usually the desired output.

Analysis of the Signal × Signal Output

If $s(t)$ denotes not only one of the two quantities that together determine the output $V(t)$ but also, to within a constant multiplicative factor, say g, denotes the desired output, we can separate $S(t)$ into two components—one of them the "undistorted output signal"

$$U(t) = gs(t),$$

and the other the "distortion"

$$D(t) = S(t) - gs(t).$$

To determine the gain factor g, we suppose that $D(t)$ is uncorrelated with $U(t)$, i.e., that $E\{U(t)D(t)\} = 0$. It follows that

$$g = \frac{E\{sS\}}{E\{s^2\}}$$

and that the power $E\{S^2\}$ of $S(t)$ is the sum of the powers of its two components $U(t)$ and $D(t)$. This value of g, which is Booton's "equivalent gain,"[7] minimizes the power of $D(t)$.

Despite our assumption $E\{U(t)D(t)\} = 0$, we, in general, have $E\{U(t+\tau)D(t)\} \neq 0$, i.e., $U(t)$ and $D(t)$ are not completely uncorrelated, and the spectrum of $S(t)$ is not the sum of their spectra. However, this will be the case whenever $s(t)$ is a "separable random process" in Nuttall's sense,[8] i.e., whenever the conditional mean $m = E\{s(t+\tau) \mid s(t)\}$ is a linear function $A(\tau)s(t) + B(\tau)$ of $s(t)$ for every τ, as it is, for example, in the case of a Gaussian random process.

To prove this result, which is known as Bussgang's theorem,[9] we notice first that for any separable $s(t)$ whose mean, as we shall suppose, is zero we can express $m = E\{s(t+\tau) \mid s(t)\} = A(\tau)s(t) + B(\tau)$ as $m = E\{s(t) \cdot s(t+\tau)\}s(t)/E\{s^2(t)\}$ by equating $E\{m\}$, averaged over $s(t)$, to $E\{s(t+\tau)\} = 0$ and by setting $E\{ms(t)\}$ equal to $E\{s(t)s(t+\tau)\}$ to find $A(\tau)$ and $B(\tau)$. Then the cross-correlation function of $U(t)$ and $D(t)$ is

$$E\{gs(t+\tau)[S(t) - gs(t)]\}$$
$$= gE\{s(t+\tau)S(t)\} - g^2E\{s(t+\tau)s(t)\}.$$

Averaging the first term on the right-hand side over the distribution of $s(t+\tau)$ for a given $s(t)$ and, therefore, a given $S(t)$, we see that it becomes

$$g\frac{E\{s(t+\tau)s(t)\}}{E\{s^2(t)\}}E\{s(t)S(t)\} = g^2E\{s(t+\tau)s(t)\},$$

and so the crosscorrelation function vanishes identically.

Because $N(t)$ and $I(t)$ are completely uncorrelated with $s(t)$ and, therefore, with $U(t)$, it follows by subtraction that they are also completely uncorrelated with $D(t)$. Thus, whenever $s(t)$ is separable, the output $V(t)$ can be resolved into five completely uncorrelated components, $U(t)$, $D(t)$, $N(t)$, $I(t)$, and C.

For the same reason, g can be expressed not only as $E\{sS\}/E\{s^2\}$ but also as

$$g = \frac{E\{sV\}}{E\{s^2\}}.$$

In addition we may note that, if and only if the univariate distribution of s is normal with zero mean, g can also be expressed as

$$g = E\left\{\frac{dS}{ds}\right\},$$

the expectation of the slope of the signal × signal output as a function of the input signal.

To show this we note that only a probability density function $p(s)$ of the form $(2\pi\sigma^2)^{-1/2}\exp(-\frac{1}{2}s^2/\sigma^2)$ satisfies the differential equation

$$\sigma^2\frac{dp}{ds} = -sp(s),$$

and for it $\sigma^2 = E\{s^2\}$. When s has such a probability density function, we find by integration by parts that

$$g = \frac{E\{sS\}}{E\{s^2\}}$$
$$= \frac{1}{E\{s^2\}}\int_{-\infty}^{\infty} sS(s)p(s)\,ds$$
$$= -S(s)p(s)\Big|_{-\infty}^{\infty} + \int_{-\infty}^{\infty}\frac{dS}{ds}p(s)\,ds$$
$$= E\left\{\frac{dS}{ds}\right\}.$$

By expressing S here as $E\{V \mid s\} = \int_{-\infty}^{\infty} V(s, n)q(n)\,dn$, where $q(n)$ is the probability density function of the input noise, we see that g can also be expressed as

$$g = E\left\{\frac{\partial V}{\partial s}\right\}.$$

When V is a function of $u(t) = s(t) + n(t)$, this becomes the average value of dV/du.

For our square-law device with Gaussian inputs, we find that $g = 0$ and hence $S(t)$ is entirely distortion; for the limiter

$$g = \sqrt{\frac{2/\pi}{\sigma_s^2 + \sigma_n^2}}.$$

Notice that this is the square root of the coefficient of $R_s(\tau)$ in the double-power-series expansion of $R_V(\tau)$ or the power-series expansion of $R_S(\tau)$.

Autocorrelation Function of Undistorted Output Signal

When the input signal $s(t)$ is a separable random process in Nuttall's sense and the five output components are therefore completely uncorrelated, $R_V(\tau)$ will evidently contain a term $g^2R_s(\tau)$ representing the autocorrelation function of the undistorted output signal. A term of this form appearing in an output autocorrelation function has sometimes been regarded as obviously representing the undistorted output signal, i.e., the converse of the preceding statement has sometimes been assumed true.

That it is not generally true, however, can be seen by considering a simple counterexample. We suppose that the input signal $s(t)$ is a step function which takes a different, constant, statistically independent, zero-mean random value throughout each of an ergodic succession of intervals. Such a random process is separable in Nuttall's sense. We shall suppose that its values have a normal distribution with unit variance. If we pass this random process (without noise) through a nonlinearity whose output is $V = (s^2 - 1)/\sqrt{2}$, we find that the autocorrelation function of the output is identical with that of the input. Since $E\{s(s^2 - 1)\} = 0$, however, $g = 0$ and there is no undistorted output signal at all; the entire output represents distortion.

On the other hand, when the input signal is a Gaussian random process and the autocorrelation function of the signal $\times$ signal output is expanded as a power series in the autocorrelation function $R_s(\tau)$ of the input signal, it can be seen by the use of Price's theorem[10] that the coefficient of $R_s(\tau)$ is g^2 and that the linear term $g^2R_s(\tau)$ therefore represents just the undistorted output signal.

Additional Applications to Limiters

Limiter with Two Sinusoidal Inputs

The ideal limiter with $n(t)$ an interfering sinusoidal input of constant amplitude $b > 0$ provides an interesting application of our definition of signal $\times$ signal output. Since the output is $V(t) = \pm 1$ accordingly as the input $u(t) = s(t) + n(t)$ is positive or negative, we have

$$\begin{aligned} S(t) &= E\{V \mid s(t)\} \\ &= \Pr\{n > -s(t)\} - \Pr\{n < -s(t)\}. \end{aligned}$$

With θ uniformly distributed over $(-\pi, \pi)$, the probability that $b \cos \theta \gtrless -s$ is evidently $\frac{1}{2} \pm (1/\pi) \arcsin s/b$ whenever $|s| \leq 1$; hence,

$$S(t) = \begin{cases} 1 & \text{for} \quad s \geq b, \\ \dfrac{2}{\pi} \arcsin \dfrac{s}{b} & \text{for} \quad -b \leq s \leq b, \\ -1 & \text{for} \quad s \leq -b. \end{cases}$$

If $s(t)$ is a sinusoid of amplitude $a > 0$ whose frequency is incommensurable with that of $n(t)$, and if $a \ll b$, it can be seen that $S(t) \approx (2/\pi b)s(t)$, i.e., that the signal is simply attenuated by the factor $2/\pi b$ and is not distorted. With $a \ll b$, the noise output is evidently a square wave of unit amplitude, and the amplitude of its fundamental component is $4/\pi$, i.e., $4/\pi b$ times the amplitude of the $n(t)$ input. Thus, we see that the weaker input sinusoid is attenuated by 6.02 dB more than the stronger one.

When $a = b$ and both inputs have the same amplitude, it can be seen that $S(t)$ is a symmetric sawtooth wave of unit amplitude. Comparing the amplitude $8/\pi^2$ of the fundamental component of such a waveform with that which would have been obtained in the absence of the interference $n(t)$, viz., $4/\pi$, we see that the undistorted signal output is attenuated by the factor $2/\pi$ or 3.92 dB on account of the interference.

For these results to be valid it evidently suffices for the phases of the two inputs to vary sufficiently slowly; they need not be constant, but the bandwidths of $s(t)$ and $n(t)$ should be small compared to their carrier frequencies. The general case, where $a \neq b$, leads to more complicated results involving the complete elliptic integrals, in agreement with the work of Granlund[11] and Baghdady.[12]

Limiter with Symmetrically Distributed Inputs

The sawtooth signal $\times$ signal output we have just found for $a = b$ is a special case of an interesting general phenomenon that occurs whenever the univariate probability density functions of the two additive limiter inputs, $p(s)$ and $q(n)$, exist (i.e., have continuous integrals) and are mirror images, i.e., $p(s) = q(-s)$. In this case the total input $u = s + n$ has an even probability density function, and so the average (dc) output is $C = 0$. For the signal $\times$ signal output we have

$$\begin{aligned} S &= \Pr\{n > -s\} - \Pr\{n < -s\} \\ &= 1 - 2 \Pr\{n < -s\} \\ &= 1 - 2 \int_{-\infty}^{-s} q(n)\, dn \\ &= 2 \int_{-\infty}^{s} p(s)\, ds - 1, \end{aligned}$$

i.e., the signal $\times$ signal output is 1 less than twice the cumulative distribution function of the input signal s. Since a cumulative distribution function is always uniformly distributed between 0 and 1, it follows that S is uniformly distributed between -1 and 1. We find similarly that the noise $\times$ noise output is $N = 1 - 2 \int_{-\infty}^{-n} p(s)\, ds$ and that it, too, is uniformly distributed between -1 and 1, independently of S.

To determine the distribution of the signal $\times$ noise output $I = V - S - N$ we observe that $V = \pm 1$ accordingly as $u = s + n \gtrless 0$, i.e., accordingly as $S + N = 2 \int_{-n}^{s} p(s)\, ds \gtrless 0$. For any given s and, therefore, any given S, as we increase N uniformly from its least possible value, -1, up to $-S$, we thus see that I decreases uniformly from $-S$ to -1, and as we further increase N uniformly from $-S$ up to its maximum possible value, 1, we observe that I decreases from 1 to $-S$. Hence I

is uniformly distributed between -1 and 1 regardless of the value of S, and it is therefore statistically independent of S.

By symmetry we see that I is also statistically independent of N, though it is not statistically independent of the pair S and N, which together determine I. On account of the assumed independence of s and n, S and N are statistically independent of each other. Thus, $S(t)$, $N(t)$, and $I(t)$ are all uniformly distributed between -1 and 1, and every pair of them is statistically independent for any single, given t.

Substituting $\tau = 0$ and $R_s(0) = R_n(0)$ into the $R_S(\tau)$, $R_N(\tau)$, and $R_I(\tau)$ for our limiter with Gaussian inputs, we can partially verify the foregoing by noticing that each of these three output components[13] has mean-squared value $\frac{1}{3}$.

Conclusion

The topic of this paper might be described as a folk theorem—the widely believed but hitherto unproven notion that the autocorrelation function of the output of a nonlinearity can generally be meaningfully resolved into signal × signal, noise × noise, and signal × noise components. We have seen that this theorem is indeed true, that the components of the output autocorrelation function are in fact the autocorrelation functions of well-defined output components, that these output components may be of interest in their own right, that their definitions are applicable to a wide variety of cases, including time-dependent nonlinearities with memory having nonstationary inputs, and that they can sometimes lead to simplified analyses of the effect of a nonlinearity on a signal and noise.

References

[1] W. B. Davenport, Jr., and W. L. Root, *Random Signals and Noise.* New York: McGraw-Hill, 1958, pp. 258 and 291–292.

[2] D. Middleton, *Statistical Communication Theory.* New York: McGraw-Hill, 1960, p. 249.

[3] W. W. Harman, *Statistical Theory of Communication.* New York: McGraw-Hill, 1963, pp. 198 and 211–212.

[4] S. O. Rice, "Mathematical analysis of random noise," *Bell Sys. Tech. J.*, vol. 24, pp. 46–156, sec. 4.9, January 1945.

[5] J. H. Van Vleck and David Middleton, "The spectrum of clipped noise," *Proc. IEEE*, vol. 54, pp. 2–19, January 1966.

[6] N. M. Blachman, *Noise and Its Effect on Communication.* New York: McGraw-Hill, 1966, pp. 78–80.

[7] R. C. Booton, Jr., "Analysis of nonlinear control systems with random inputs," *IRE Trans. Circuit Theory*, vol. CT-1, pp. 9–18, March 1954.

[8] A. H. Nuttall, "Theory and application of the separable class of random processes," Research Lab. of Electronics, M.I.T., Cambridge, Mass., Tech. Rept. 343, May 1958.

[9] J. J. Bussgang, "Crosscorrelation functions of amplitude-distorted Gaussian signals," Research Lab. of Electronics, M.I.T., Cambridge, Mass., Tech. Rept. 216, March 1952.

[10] R. Price, "A useful theorem for nonlinear devices having Gaussian inputs," *IRE Trans. Information Theory*, vol. IT-4, pp. 69–72, June 1958.

[11] J. Granlund, "Interference in frequency-modulation reception," Research Lab. of Electronics, M.I.T., Cambridge, Mass., Tech. Rept. 42, January 1949.

[12] E. J. Baghdady, "Interference rejection in FM receivers," Research Lab. of Electronics, M.I.T., Cambridge, Mass., Tech. Rept. 252, September 1956.

[13] R. F. Baum, "The correlation function of smoothly limited Gaussian noise," *IRE Trans. Information Theory*, vol. IT-3, pp. 193–197, September 1957, in effect noted the uniform distribution of $S(t)$ in this case.

12

Reprinted from *IEEE Trans. Information Theory*, **IT-14**(2), 250–255 (1968)

The Uncorrelated Output Components of a Nonlinearity

NELSON M. BLACHMAN, FELLOW, IEEE

***Abstract*—Using his characteristic-function approach, Rice (1945) obtained a double series for the autocorrelation function of a sinusoidal signal and Gaussian noise after passage through a memoryless nonlinearity. It is shown here that the output of the nonlinearity can be expressed as the sum of uncorrelated terms whose autocorrelation functions are the terms of Rice's double series. Such a decomposition of the output is shown to be generally possible if and only if the bivariate probability density functions of the input signal and the input noise can both be expressed in the diagonal form studied by Barrett and Lampard (1955), though not necessarily involving polynomials, as they can in the sinusoidal and Gaussian cases. In addition, a more direct and meaningful equation is found for the coefficients in Rice's double series.**

A RECENT PAPER[1] showed how the output of a nonlinearity can be resolved into four uncorrelated components—signal × signal, noise × noise, signal × noise, and dc—whose autocorrelation functions are the portions of the output autocorrelation function that had come to be called by these names. In the present paper we, in effect, consider the resolution of each of these output components (except the dc) into a sum of uncorrelated terms whose autocorrelation functions are the respective terms of the output autocorrelation function as found by Rice's characteristic-function method.[2] To do this, we begin with the case of a single input—signal *or* noise.

Manuscript received January 31, 1967; revised August 8, 1967.
The author is with Sylvania Electronic Systems, Mountain View, Calif. 94042

Diagonal Expansion

For several important types of stationary random processes $u(t)$, the bivariate probability density function of

$$u = u(t) \quad \text{and} \quad u' = u(t + \tau)$$

can be expressed in the diagonal form

$$p(u, u'; \tau) = p(u)p(u') \sum_{m=0}^{\infty} \frac{a_m(\tau)}{F_m} f_m(u) f_m(u'), \qquad (1)$$

where $p(u)$ is the univariate (mariginal) probability density function of $u(t)$ and the $\{f_m(u)\}$ are linearly independent functions with mean-squared values

$$F_m = E\{f_m^2(u)\} = \int_{-\infty}^{\infty} f_m^2(u)p(u)\, du. \qquad (2)$$

By "linearly independent" we mean that $\sum c_m f_m(u) = 0$ wherever $p(u) \neq 0$ only if every c_m is zero; this restriction insures that the summation in (1) cannot in effect contain cross products like $f_m(u)f_k(u')$ with $m \neq k$. The factor $1/F_m$ in (1) merely serves to normalize $f_m(u)$ and $f_m(u')$; it could have been included in them or in $a_m(\tau)$, but it will be convenient to keep it separate. Likewise, the factors $p(u)$ and $p(u')$ could have been included in $f_m(u)$ and $f_m(u')$, respectively, but it will be convenient to keep them separate also.

Examples

Barrett and Lampard[3] have discussed a number of random processes having diagonalizable bivariate probability density functions, among them the stationary, zero-mean Gaussian, and the sinusoidal processes. For the former we have

$$p(u) = \frac{1}{\sqrt{2\pi}\,\sigma} e^{-u^2/2\sigma^2}; \quad f_m(u) = \mathrm{He}_m\left(\frac{u}{\sigma}\right), \quad F_m = m!,$$

$$a_m(\tau) = \frac{\psi^m(\tau)}{\sigma^{2m}}, \tag{3}$$

where $\mathrm{He}_m(x) = e^{\frac{1}{2}x^2}(-d/dx)^m e^{-\frac{1}{2}x^2}$ is the Hermite polynomial of order m, $\psi(\tau) = E\{uu'\}$ is the autocorrelation function of $u(t)$, and $\sigma^2 = \psi(0)$ is its mean-squared value. Substituted into (1), (3) gives Mehler's (1866) expansion of the bivariate normal density function.

For the stationary sinusoidal process

$$u(t) = A \cos(2\pi F t + \phi)$$

with $A > 0$ and F fixed and with the constant ϕ distributed uniformly between 0 and 2π, we have for $-A < u < A$

$$p(u) = \frac{1}{\pi\sqrt{A^2 - u^2}}, \quad f_m(u) = T_m\left(\frac{u}{A}\right),$$

$$F_m = \frac{1}{\epsilon_m}, \quad a_m(\tau) = \cos 2m\pi F\tau,$$

$$p(u) = 0 \quad \text{for } |u| > A,$$

where $T_m(x) = \cos(m \arccos x)$ is the Chebyshev polynomial of order m and ϵ_m is the Neumann factor, being 1 for $m = 0$ and 2 otherwise. Leipnik[4] has pointed out that, on account of the one-dimensionality of the bivariate distribution in this case, the series (1) converges to $p(u, u'; \tau)$ only in the mean-square sense.

Zadeh,[5] Wong and Thomas,[6] McFadden,[7] and McGraw and Wagner[8] have discovered other examples, such as the chi-squared process, with diagonal expansions (1).[1] Unlike these authors,[3]–[8] we shall have no need to restrict ourselves to cases in which the $\{f_m(u)\}$ are polynomials. A stationary random process clearly has a diagonalizable bivariate probability density function of the form (1) if and only if the associated characteristic function [the double Fourier transform of $p(u, u')$] can be expressed in a similar diagonal form. It is also evident that, if the random process $u(t)$ is diagonal and if $v(u)$ is any strictly monotonic function of u, then the random process $v(u(t))$, too, has a diagonalizable bivariate distribution.

Completeness

If $v(u)$ is now *any* function of u with finite mean-squared value and if $p(u, u'; \tau)$ is given by (1), we can express the autocorrelation function of $v(u(t))$ as

$$\begin{aligned}\psi_v(\tau) &= \iint_{-\infty}^{\infty} v(u)v(u')p(u, u'; \tau)\, du\, du' \\ &= \sum_{m=0}^{\infty} \frac{h_m^2}{F_m} a_m(\tau),\end{aligned} \tag{4}$$

where

$$\begin{aligned}h_m &= E\{v(u)f_m(u)\} \\ &= \int_{-\infty}^{\infty} v(u)f_m(u)p(u)\, du.\end{aligned} \tag{5}$$

Setting $\tau = 0$ in (4), we get the mean-squared value of v,

$$\int_{-\infty}^{\infty} v^2(u)p(u)\, du = \sum_{m=0}^{\infty} \frac{a_m(0)}{F_m} h_m^2. \tag{6}$$

Substituting (5) into (6) and differentiating both sides with respect to the value taken by $v(u)$ in the ϵ neighborhood of any particular u, we find, on canceling the factor $2\,\epsilon\, p(u)$ on both sides, that

$$v(u) = \sum_{m=0}^{\infty} a_m(0) \frac{h_m}{F_m} f_m(u) \tag{7}$$

wherever $p(u) \neq 0$. Since we are able to express an arbitrary $v(u)$ in this form, it follows that the $\{f_m(u)\}$ constitute a complete set of functions.

Orthogonality

Substituting $v(u) = f_k(u)$ into (5) and (7), we find that with

$$h_m = \int_{-\infty}^{\infty} f_m(u)f_k(u)p(u)\, du, \tag{8}$$

we have

$$f_k(u) = \sum_{m=0}^{\infty} a_m(0) \frac{h_m}{F_m} f_m(u).$$

Since the $\{f_m(u)\}$ were assumed linearly independent, all of the coefficients on the right-hand side of this equation must vanish except $a_k(0)h_k/F_k$, which must be unity. From (8) and (2) we here have $h_k = F_k$, and hence,[2] $a_k(0) = 1$ for all k. Thus (7) can be simplified to

[1] For some distributions (see Wong[9]), the range of the indexing parameter m is at least partly continuous. Our results will remain valid if, for the continuum, summation over m is replaced by integration and Kronecker deltas are replaced by Dirac delta functions of the difference of the two indexes.

[2] Substituting $a_m(0) = 1$ in (6) for all m, we see that the assumed finiteness of $E\{v^2\}$ implies the finiteness of all h_m.

$$v(u) = \sum_{m=0}^{\infty} \frac{h_m}{F_m} f_m(u) \tag{9}$$

with h_m given by (5). As we have seen with $v(u) = f_k(u)$, (8) must vanish whenever $m \neq k$. We therefore have the orthogonality relation

$$\int_{-\infty}^{\infty} f_m(u) f_k(u) p(u)\, du = 0 \quad \text{for} \quad m \neq k, \tag{10}$$

i.e., every distinct $f_m(u)$ and $f_k(u)$ are orthogonal to each other relative to the weighting function $p(u)$.

Uncorrelatedness

Making use of (1), (2), and (10), we find that

$$\begin{aligned} E\{f_m(u) f_k(u')\} &= \iint_{-\infty}^{\infty} f_m(u) f_k(u') p(u, u'; \tau)\, du\, du' \\ &= F_m a_m(\tau)\, \delta_{mk}, \end{aligned} \tag{11}$$

where δ_{mk} is the Kronecker delta, being 1 for $m = k$ and 0 otherwise. Because this crosscorrelation function vanishes for all τ whenever $m \neq k$, we describe the set of $\{f_m(u)\}$ as uncorrelated.

From (11) we see that $F_m a_m(\tau)$ is the autocorrelation function of $f_m(u)$ and is therefore an even function of τ. Hence, by (2) it must lie between $\pm F_m$, i.e., as Barrett and Lampard[3] pointed out, $-1 \leq a_m(\tau) \leq 1$ for every τ; and $a_m(-\tau) = a_m(\tau)$.

Uniqueness

Using (1) and (11), we can observe that the conditional expectation of $f_m(u')$ for a given value of u is

$$\begin{aligned} E\{f_m(u') \mid u\} &= \int_{-\infty}^{\infty} f_m(u') p(u' \mid u, \tau)\, du' \\ &= \int_{-\infty}^{\infty} f_m(u') \frac{p(u, u'; \tau)}{p(u)}\, du' \\ &= a_m(\tau) f_m(u). \end{aligned} \tag{12}$$

Thus, as Barrett and Lampard[3] noted, the $\{a_m(\tau)\}$ are the eigenvalues of an integral equation, and the $\{f_m(u)\}$ are the corresponding eigenfunctions. Since any $p(u, u'; \tau)$ of the form (1) equals $p(u, u'; -\tau) = p(u', u; \tau)$ for every u, u', and τ, the integral equation (12) can be written with a symmetric kernel $p(u, u'; \tau)/\sqrt{p(u)p(u')}$, eigenvalue $a_m(\tau)$, and eigenfunction $\sqrt{p(u)}\, f_m(u)$. The kernel is also bounded, for the expected product of $f_m(u)$ times (12) is

$$a_m(\tau) \int_{-\infty}^{\infty} f_m^2(u) p(u)\, du = \iint_{-\infty}^{\infty} f_m(u) f_m(u') p(u, u'; \tau)\, du\, du',$$

whose square, by Schwarz's inequality, cannot exceed

$$\iint_{-\infty}^{\infty} f_m^2(u) p(u, u'; \tau)\, du\, du' \iint_{-\infty}^{\infty} f_m^2(u') p(u, u'; \tau)\, du\, du' = \left[\int_{-\infty}^{\infty} f_m^2(u) p(u)\, du \right]^2,$$

and all eigenvalues $a_m(\tau)$ therefore lie between ± 1.

Hence, the $\{f_m(u)\}$ are unique except for any which may be associated with identically equal eigenvalues. In general, the eigenfunctions of the integral equation (12) will depend on τ; only in the case where they form a countable, complete set and do not depend on τ can we express the bivariate probability density function by Mercer's theorem in the diagonal form (1). If

$$E\{p(u, u'; \tau)/p(u)p(u')\} = \sum a_m^2(\tau)$$

is finite, the eigenvalues will be discrete and therefore countable. Even if the eigenfunctions depend on τ, we can still use them in diagonal series like (1) and (4) to express $p(u, u'; \tau)$ and $\psi_v(\tau)$, but an extraneous τ will appear on the right-hand side of (9) which results in crosscorrelation.

Constant $f_0(u)$

Multiplying both sides of (12) by $p(u)$ and integrating over all u, we find that $E\{f_m(u)\} = a_m(\tau) E\{f_m(u)\}$. Hence, for any m, either $E\{f_m(u)\} = 0$ or $a_m(\tau) = 1$ for all τ. In the latter case, which we may designate as $m = 0$, we have from (11) with $m = k = 0$

$$E\{f_0^2(u)\} = E\{f_0(u) f_0(u')\} = E\{f_0^2(u')\}$$

and, therefore,

$$E\{[f_0(u) - f_0(u')]^2\} = 0$$

for all τ, from which it follows that $f_0(u)$ is a nonzero constant.

Such an $f_0(u)$ clearly satisfies the integral equation (12) with eigenvalue $a_0(\tau) \equiv 1$ uniquely. That such a function must be included among the $\{f_m(u)\}$ can be seen by expanding $v \equiv 1$ in the form (9). Then all coefficients $h_m = E\{f_m(u)\}$ vanish except h_0, and so $f_0(u)$ is required for the completeness of the $\{f_m(u)\}$. Thus, $a_0(\tau)$ is identically equal to unity, $f_0(u)$ is a nonzero constant, and $E\{f_m(u)\} = 0$ for all nonzero m.

Output Autocorrelation Function

If $v(u)$ represents the output of a memoryless nonlinearity whose bivariate input distribution is given by (1), its autocorrelation function $\psi_v(\tau)$ is given by (4). We can now see that each term of the series (4) is the autocorrelation function of the corresponding term of the series (9) of uncorrelated terms whose sum is the output $v(u(t))$.

We may note that, if the output $v(t)$ is passed through a linear filter[5] with impulse response $g(t)$, the terms of the resulting output

$$w(t) = v(u(t)) * g(t) = \sum_{m=0}^{\infty} \frac{h_m}{F_m} f_m(u(t)) * g(t)$$

will remain uncorrelated, and the autocorrelation function of $w(t)$ will be[10]

$$\psi_w(\tau) = \sum_{m=0}^{\infty} \frac{h_m}{F_m} a_m(\tau) * g(\tau) * g(-\tau).$$

Necessary and Sufficient Condition

Campbell[11] showed the terms of (9) to be uncorrelated in the case of a stationary Gaussian $u(t)$; here we have shown that they are uncorrelated whenever $p(u, u'; \tau)$ can be expressed in the diagonal form (1). This diagonalizability is not only a sufficient but also a necessary condition for an arbitrary $v(u)$ to be expressible as a linear combination of uncorrelated terms of fixed form $f_m(u)$, for, if the $\{f_m(u)\}$ are complete and uncorrelated, satisfying (11), we find, on expressing $p(u, u'; \tau)/p(u)p(u')$ as a double series $\sum \sum c_{mk}(\tau) f_m(u) f_k(u')$, that the coefficients $c_{mk}(\tau)$ with $m \neq k$ vanish.

Nonlinearity With Signal and Noise Input

We turn now to the case in which the output of the memoryless nonlinearity is a function $v(s, n)$ of two independent stationary random processes $s(t)$ and $n(t)$, which may represent a signal and noise, respectively. It may, for example, be a function $v(s + n)$ of their sum. In effect, the input $u(t)$ with which we have been dealing is now a two-component vector.

We shall suppose that the bivariate probability density functions of both the input signal and the input noise can be expressed diagonally as

$$p(s, s'; \tau) = p(s)p(s') \sum_{m=0}^{\infty} \frac{a_m(\tau)}{F_m} f_m(s) f_m(s') \tag{13}$$

and

$$q(n, n'; \tau) = q(n)q(n') \sum_{k=0}^{\infty} \frac{b_k(\tau)}{G_k} g_k(n) g_k(n'), \tag{14}$$

in which the various quantities are defined analogously to those in (1). Being complete sets of orthogonal functions for the reasons given previously, the $\{f_m(s)\}$ and $\{g_k(n)\}$ can be used to express $v(s, n)$ as

$$v(s, n) = \sum_{m=0}^{\infty} \sum_{k=0}^{\infty} \frac{h_{mk}}{F_m G_k} f_m(s) g_k(n), \tag{15}$$

where

$$\begin{aligned} h_{mk} &= E\{v(s, n) f_m(s) g_k(n)\} \\ &= \iint_{-\infty}^{\infty} v(s, n) f_m(s) g_k(n) p(s) q(n)\, ds\, dn. \end{aligned} \tag{16}$$

Our procedure in obtaining this double series may be described as first expressing $v(s, n)$ as $\sum (h_m(n)/F_m) f_m(s)$, like (9) for any given n, and then expanding each $h_m(n)$ as $\sum (h_{mk}/G_k) g_k(n)$.

Because the $\{f_m(s)\}$ are uncorrelated and the $\{g_k(n)\}$ are also, and because $s(t)$ and $n(t)$ are statistically independent, the crosscorrelation function of any pair of terms of the double series (15) vanishes identically; i.e., (15) expresses $v(s, n)$ as a sum of uncorrelated terms. Its autocorrelation function is therefore the sum of the autocorrelation functions of the terms of (15), viz.,

$$\psi_v(\tau) = \sum_{m=0}^{\infty} \sum_{k=0}^{\infty} \frac{h_{mk}^2}{F_m G_k} a_m(\tau) b_k(\tau). \tag{17}$$

Signal × Signal, Noise × Noise, and Signal × Noise

Since $f_0(s)$ and $g_0(n)$ are nonzero constants, while $E\{f_m(s)\} = 0 = E\{g_k(n)\}$ for $m \neq 0 \neq k$, we can easily separate the terms of (15) or (17) into four types according to the values of their subscripts (m, k). The (0, 0) term evidently represents the dc output component. The remaining $(m, 0)$ terms represent the signal × signal output,[1] since they do not depend on the instantaneous value of $n(t)$, and the remaining $(0, k)$ terms represent the noise × noise output, since they are not functions of $s(t)$. The terms for which neither m nor k is zero depend on both $s(t)$ and $n(t)$ and represent the signal × noise output.

Necessary and Sufficient Condition

We have seen that, for an arbitrary $v(s, n)$ to be expressible as a linear combination of a fixed set of uncorrelated functions, it suffices for $s(t)$ and $n(t)$ to have diagonal bivariate probability density functions of the form (13) and (14). We shall now show that this condition is necessary as well as sufficient. For this purpose we suppose that $\{f_m(s, n)\}$ is a complete set of uncorrelated functions, and more particularly we suppose that

$$\begin{aligned} E\{f_m(s, n) f_k(s', n')\} &= \iiiint_{-\infty}^{\infty} f_m(s, n) f_k(s', n') p(s, s'; \tau) \\ &\qquad \cdot q(n, n'; \tau)\, ds\, dn\, ds'\, dn' \\ &= a_m(\tau) F_m\, \delta_{mk}. \end{aligned} \tag{18}$$

Then, wherever $p(s)q(n) \neq 0$, we can express any $v(s)$ as

$$v(s) = \sum_{m=0}^{\infty} \frac{h_m}{F_m} f_m(s, n) \tag{19}$$

with

$$\begin{aligned} h_m &= E\{v(s) f_m(s, n)\} \\ &= E\{v(s) f_m(s)\}, \end{aligned} \tag{20}$$

where $f_m(s)$ is now defined as the conditional expectation

$$\begin{aligned} f_m(s) &= E\{f_m(s, n) \mid s\} \\ &= \int_{-\infty}^{\infty} f_m(s, n) q(n)\, dn. \end{aligned} \tag{21}$$

Thus, averaging both sides of (19) over the distribution of n, we find

$$v(s) = \sum_{m=0}^{\infty} \frac{h_m}{F_m} f_m(s), \tag{22}$$

and hence, the $\{f_m(s)\}$ defined in (21) are a complete set of functions.

From the completeness and uncorrelatedness of the $\{f_m(s, n)\}$ it follows as in the case of a single-component input that $p(s, n, s', n'; \tau)$ can be expressed as a diagonal series in these functions and, hence, that

$$E\{f_m(s', n' \mid s, n\} = a_m(\tau) f_m(s, n), \tag{23}$$

a relationship analogous to (12), $a_m(\tau)$ being the same as in (18). Averaging both sides of (23) over the distribution of n and making use of (21), we obtain

$$E\{f_m(s') \mid s\} = a_m(\tau)f_m(s), \tag{24}$$

which shows that $f_m(s)$ satisfies the same integral equation as $f_m(s, n)$ and with the same eigenvalue $a_m(\tau)$. Hence, each $f_m(s)$, if it is not identically zero, must be a linear combination of the $\{f_m(s, n)\}$ having the same eigenvalue $a_m(\tau)$ for all τ. Thus, like the $\{f_m(s, n)\}$, the $\{f_m(s)\}$ are uncorrelated except possibly those having identical eigenvalues. On the other hand, from a set of $\{f_m(s)\}$ having identically the same $a_m(\tau)$, we can obtain an equivalent set of uncorrelated functions by the usual orthogonalization procedure if they are not already uncorrelated.

In this way we obtain a complete uncorrelated set of $\{f_m(s)\}$, and these can be used to express $p(s, s'; \tau)$ in the diagonal form (13). In a similar way we can obtain from the $\{f_m(s, n)\}$ a complete uncorrelated set $\{g_k(n)\}$, which can be used to express $q(n, n'; \tau)$ in the diagonal form (14). This completes the proof of necessity.

Sinusoidal Signal and Gaussian Noise

Applying the foregoing to the case of a sinusoidal input signal of amplitude A and frequency F and zero-mean Gaussian input noise of rms value σ and autocorrelation function $\psi(\tau)$, we see that the output (15) of the nonlinearity becomes

$$v(s, n) = \sum_{m=0}^{\infty} \sum_{k=0}^{\infty} \frac{\epsilon_m h_{mk}}{k!} T_m\left(\frac{s}{A}\right) \mathrm{He}_k\left(\frac{n}{\sigma}\right) \tag{25}$$

with the coefficients

$$h_{mk} = E\left\{v(s, n)\, T_m\left(\frac{s}{A}\right) \mathrm{He}_k\left(\frac{n}{\sigma}\right)\right\}$$

$$= \frac{1}{\sigma} \int_{-\infty}^{\infty} \int_{-A}^{A} v(s, n) T_m\left(\frac{s}{A}\right) \mathrm{He}_k\left(\frac{n}{\sigma}\right) \frac{e^{-n^2/2\sigma^2}\, ds\, dn}{\sqrt{2\pi^3(A^2 - s^2)}} \tag{26}$$

and that its autocorrelation function is

$$\psi_v(\tau) = \sum_{m=0}^{\infty} \sum_{k=0}^{\infty} \frac{\epsilon_m h_{mk}^2}{\sigma^{2k} k!} \psi^k(\tau) \cos 2m\pi F\tau. \tag{27}$$

This result is in complete agreement with that of Rice[2],[12] for this case except for a difference in the appearance of the formula for h_{mk}. Rice treated only the case of additive noise, with the output $v(u)$ a function of the sum $u = s + n$, and he represented this nonlinearity as a Fourier integral taken over a suitable contour

$$v(u) = \int_C V(z)e^{iuz}\, dz. \tag{28}$$

Substituting this expression for $v(s + n)$ in (26) and carrying out the integrations with respect to s and n, we get

$$h_{mk} = i^{m+k} \int_C V(z)(\sigma z)^k J_m(Az)e^{-\frac{1}{2}\sigma^2 z^2}\, dz, \tag{29}$$

which is the form obtained by Rice through the use of the characteristic-function approach. Here we have generalized his result to the case of nonadditive noise and have discovered where the individual terms of (27) come from; they are the autocorrelation functions of the corresponding terms of (25).

Putting $s = A \cos \theta$ in (26), substituting the explicit expressions for the polynomials, and integrating by parts, we find that

$$h_{mk} = (2\pi)^{-3/2}\sigma^{k-1} \int_{-\infty}^{\infty} \int_0^{2\pi} v(A \cos \theta, n) \cos m\theta \cdot(- d/dn)^k e^{-n^2/2\sigma^2}\, d\theta\, dn$$

$$= (2\pi)^{-3/2}\sigma^{k-1} \int_{-\infty}^{\infty} \int_0^{2\pi} e^{-n^2/2\sigma^2} \cos m\theta \cdot(\partial/\partial n)^k v(A \cos \theta, n)\, d\theta\, dn$$

$$= \sigma^k E\{v_m^{(k)}(n)\} \quad \text{with} \quad v_m(n) = \frac{1}{2\pi} \int_0^{2\pi} v(A \cos \theta, n) \cos m\theta\, d\theta; \tag{30}$$

i.e., h_{mk} is the average over the distribution of n of the kth derivative with respect to n of the mth Fourier coefficient of $v(A \cos \theta, n)$ as a function of θ (or the average of the Fourier coefficient of the derivative or the Fourier coefficient of the average derivative). For $\sigma = 0$ or for[11] $A = 0$, this reduces to the known result.

Modulated Signal

If the input signal is modulated, it takes the form

$$s(t) = A(t) \cos [2\pi Ft + \phi(t)],$$

and the h_{mk} become functions of t throught heir dependence on $A(t)$. However, the terms of (25) with $h_{mk}(A)$ still given by (26), (29), or (30) remain uncorrelated, $T_m(s/A)$ being $\cos [2m\pi Ft + m\phi(t)]$ in (25).

The output autocorrelation function retains the form (27) except that $\epsilon_m h_{mk}^2 \cos 2m\pi F\tau$ must be replaced by the autocorrelation function of $\epsilon_m h_{mk}(A(t)) \cos [2m\pi Ft + m\phi(t)]$.

Similar results can be obtained for a nonlinearity with other diagonal input-signal and input-noise statistics—both by the foregoing approach and by Rice's characteristic-function technique. In every case a double series for the output autocorrelation function will be obtained whose terms are the autocorrelation functions of the corresponding terms of (15). For the coefficients in this double series the two methods will again yield the same results—expressed by the characteristic-function technique in a form like (29) and by the foregoing approach in a form like (16) and (26).

Time-Dependent Nonlinearities

More generally, (15) [or (9)] can be used to express any time-dependent nonlinearity $v(s, n, t)$ [or $v(u, t)$] as the sum of a series of uncorrelated terms with coefficients $h_{mk}(t)$ [or $h_m(t)$] given by (16) [or (5)]. Then h_{mk}^2 in (17) [or h_m^2 in (4)] is replaced by the autocorrelation function of $h_{mk}(t)$ [or $h_m(t)$].[3]

In the case of phase modulation, for example, we have

$$v(u, t) = A \cos [2\pi Ft + u(t) + \theta].$$

For a zero-mean Gaussian modulating process $u(t)$ with variance σ^2 and autocorrelation function $\psi(\tau)$, (5) gives

$$h_m(t) = A\sigma^m e^{-\frac{1}{2}\sigma^2} \cos (2\pi Ft + \theta + \tfrac{1}{2}m\pi),$$

and (9) thus becomes

$$v(u, t) = Ae^{-\frac{1}{2}\sigma^2} \sum_{m=0}^{\infty} \frac{\sigma^m}{m!} \mathrm{He}_m\left(\frac{u}{\sigma}\right) \cos (2\pi Ft + \theta + \tfrac{1}{2}m\pi)$$

Notice that the term $m = 0$ represents the residual carrier-frequency component, since it is the average of $v(u, t)$ over the distribution of u.

Because of the uncorrelatedness of the Hermite polynomials with Gaussian arguments, the terms of this series are uncorrelated with each other. The autocorrelation function of $h_m(t)$ is evidently $\frac{1}{2}A^2\sigma^{2m}e^{-\sigma^2} \cos 2\pi F\tau$ with θ assumed uniformly distributed between $-\pi$ and π, and hence, the autocorrelation function of the phase-modulated wave is

$$\psi_v(\tau) = \tfrac{1}{2}A^2e^{-\sigma^2} \cos 2\pi F\tau \sum_{m=0}^{\infty} \frac{\psi^m(\tau)}{m!}$$
$$= \tfrac{1}{2}A^2e^{\psi(\tau)-\sigma^2} \cos 2\pi F\tau,$$

[3] Treating the case of $v(s, n)$ in this manner, with s regarded as explicitly dependent on t, Campbell[13] obtained (25) and (27) for sinusoidal $s(t)$ and additive Gaussian noise $n(t)$.

in agreement with the well-known result.[14] Here we see the significance of the individual terms that are obtained[15] when this result is expressed as a power series in $\psi(\tau)$.

References

[1] N. M. Blachman, "The signal × signal, noise × noise, and signal × noise output of a nonlinearity," *IEEE Trans. Information Theory*, vol. IT-14, pp. 21–27, January 1968.

[2] S. O. Rice, "Mathematical analysis of random noise," *Bell Sys. Tech. J.*, vol. 24, pp. 46–157, sec. 4.9, January 1945.

[3] J. F. Barrett and D. G. Lampard, "An expansion of some second-order probability distributions and its application to noise problems," *IRE Trans. Information Theory*, vol. IT-1, pp. 10–15, March 1955.

[4] R. Leipnik, "Integral equations, biorthononormal expansions, and noise," *J. SIAM*, vol. 7, pp. 6–30, March 1959.

[5] L. A. Zadeh, "On the representation of nonlinear operators," *1957 IRE WESCON Conf. Rec.*, vol. 1, pt. 2, pp. 105–113.

[6] E. Wong and J. B. Thomas, "On polynomial expansions of second-order distributions," *J. SIAM*, vol. 10, pp. 506–516, September 1962.

[7] J. A. McFadden, "A diagonal expansion in Gegenbauer polynomials for a class of second-order probability densities," *J. SIAM*, vol. 14, pp. 1433–1436, November 1966.

[8] D. K. McGraw and J. F. Wagner, "Elliptically symmetric distributions," *IEEE Trans. Information Theory*, vol. IT-14, pp. 110–121, January 1968.

[9] E. Wong, "The construction of a class of stationary Markoff processes," in *Stochastic Processes in Mathematical Physics and Engineering*, R. Bellman, Ed. Providence, R. I.: American Mathematical Society, 1964, pp. 264–276.

[10] N. M. Blachman, *Noise and Its Effect on Communication*. New York: McGraw-Hill, 1966, p. 27.

[11] L. L. Campbell, "A general analysis of post-detection correlation," *IEEE Trans. Information Theory*, vol. IT-11, pp. 409–417, July 1965.

[12] See Blachman,[10] p. 97.

[13] L. L. Campbell, "On the use of Hermite expansions in noise problems," *J. SIAM*, vol. 5, pp. 244–249, December 1957.

[14] See Blachman,[10] p. 44.

[15] N. M. Abramson, "Bandwidth and spectra of phase- and frequency-modulated waves," *IEEE Trans. Communication Systems*, vol. CS-11, pp. 407–414, December 1963.

III
Functionals with Random Inputs

Editor's Comments on Papers 13, 14, and 15

13 Cameron and Martin: *The Orthogonal Development of Non-linear Functionals in Series of Fourier–Hermite Functionals*

14 Barrett: *The Use of Functionals in the Analysis of Non-linear Physical Systems*

15 Bedrosian and Rice: *The Output Properties of Volterra Systems (Nonlinear Systems with Memory) Driven by Harmonic and Gaussian Inputs*

Functional power-series expansions of arbitrary nonlinear systems have been developed by Volterra (1930) for the general representation and identification of deterministic nonlinear systems or circuits. The development of Wiener's method for the expansion of functionals in orthogonal polynomial functional series and its application to the identification of nonlinear system by white Gaussian noise input gave an impetus to many contributions in the theory of functionals with random inputs, especially by the MIT group [see, for example, Bose (1956), George (1959), Chesler (1960), and Schetzen (1962)]. Wiener's basic results have been published in his famous monograph—Wiener (1958)—and hence are not included here. The mathematical basis of orthogonal series expansion of functionals is found in one of Wiener's references, the paper (13) by Cameron and Martin. The paper establishes the completeness of the orthogonal set of Fourier–Hermite functional polynomials, and provides the basis for the expansion of nonlinear functionals in polynomial functional series.

Most of the work following Wiener's on nonlinear functionals with random inputs has dealt with one of the three aspects of the problem:

1. The analysis of output statistical properties of functional polynomials both of the Volterra type and Fourier–Hermite type.

2. The classification of functionals, or the expansion in special series to match input statistical properties.

3. The identification of systems in terms of their coefficients in a functional polynomial series expansion with the use of random inputs. The results, of course, were also applied in general to the filtering problem of finding the functional of a given class to optimally extract signals from noise.

The Barrett paper (14) is another basic article that considers most of the aspects mentioned above. Barrett's work appeared in a report form in 1955 and was finally published in 1963. Its results parallel and complement the contributions of the MIT group; however, the early version of the paper preceded many of the MIT publications and served as their reference. The first part of the paper is concerned with the deterministic analysis of polynomial functional series and with the representation of given nonlinear systems by such series. It also investigates the possibility of realizing a given nonlinear functional polynomial by a combination of ZNL systems and linear systems with memory. The second part of the paper is the relevant part to the subject matter of this volume as it explores the output moments and autocorrelation of functional polynomials with Gaussian inputs. Finally, it also provides a concise statement of the generation of orthogonal polynomial functionals for non-white Gaussian input. It is difficult to evaluate the impact of Barrett's work on the MIT group, which considered applications of functional polynomial series to problems in optimization and analysis. Chesler (1960) developed orthogonal polynomial functionals for the case of signal-and-noise inputs, and applied such functionals to the optimum filtering problem. Hause (1960) used functional methods for the analysis and optimization of frequency-modulation systems. Expansion in other series, including gate functions and kernel identification, was considered by Schetzen (1962). Van Trees (1964) applied functional methods to the analysis of the phase-lock loop.

An alternative approach to functional classification was proposed by Zadeh (1953). The classification is based not on the degree of the nonlinearities involved as in the functional polynomials, but rather on the order of input distribution required for mean-squared identification or optimization. Zadeh defines a functional of class η_n if it has the form

$$y(t) = \int \cdots \int K_n\left[x(t - \tau_1), \cdots, x(t - \tau_n); \tau_1, \cdots, \tau_n\right] d\tau_1 \cdots d\tau_n$$

where $x(t)$ and $y(t)$ are the input and output, respectively, and K_n is the characteristic kernel. The specific form of the kernel may result in subclasses of η_n. Zadeh (1953) also provides applications to the optimal filtering problem for a given class of functionals. Obviously, the lack of general $2n$th-order densities of arbitrary random processes restricted the usefulness of the general class η_n to the case $n = 1$, which in its simplest form is given as a parallel structure of ZNL systems followed by linear systems with memory. The usefulness of this class, and orthogonal series representations of functionals of class η_1, were considered by Zadeh (1957) for processes with second-order densities having diagonal orthogonal expansions. The significance of Zadeh's classification is illustrated by the fact that results obtained for a subclass such as a ZNL system followed by a linear system were later generalized to Zadeh's class η_1 [see Darling and Siegert (Paper 23)]. However, its principal application has been to the problem of classical nonlinear filtering, and hence has not been included in this collection.

In recent years the emphasis in both analysis and optimization has shifted from the functional representation to the dynamical system representation. Despite this fact, there are novel contributions to the use of functionals, as seen, for example, in the paper by French and Butz (1974). This part concludes with a relatively recent paper by Bedrosian and Rice (Paper 15). It is an excellent survey paper on the analysis of functional polynomial series with sinusoidal and Gaussian inputs. It provides a coherent review of earlier results and relates the various aspects of nonlinear functionals, ranging from Wiener's method to the derivation of output distributions by Kac and Siegert (Paper 21).

13

Reprinted from *Ann. Math.*, **48**(2), 385–392 (1947)

THE ORTHOGONAL DEVELOPMENT OF NON-LINEAR FUNCTIONALS IN SERIES OF FOURIER-HERMITE FUNCTIONALS[1]

BY R. H. CAMERON AND W. T. MARTIN

(Received April 9, 1946)

1. Introduction

Denote by C the space of real functions $x(t)$ which are continuous on the interval $0 \leqq t \leqq 1$ and which vanish at $t = 0$. Using the Wiener measure on C and completeness properties of the Hermite polynomials over $(-\infty, \infty)$ we shall introduce a complete orthonormal set of functionals on C so that every real or complex valued functional $F[x(\cdot)]$ which belongs to $L_2(C)$,

$$\int_C^w |F[x]|^2 \, d_w x < \infty,$$

has a Fourier development in terms of this set which converges in the $L_2(C)$ sense to F.

2. A complete orthonormal set over C

Denote by $H_n(u)$ the (partially normalized) Hermite polynomial

$$H_n(u) = (-1)^n 2^{-n/2}(n!)^{-\frac{1}{2}} e^{u^2} \frac{d^n}{du^n}(e^{-u^2}), \quad n = 0, 1, 2, \cdots. \tag{2.1}$$

Then, as is well known, the set

$$\{\pi^{-\frac{1}{4}} H_n(u) e^{-u^2/2}\} \tag{2.2}$$

is a C. O. N. set on $(-\infty, \infty)$:

$$\pi^{-\frac{1}{2}} \int_{-\infty}^{\infty} H_m(u) H_n(u) e^{-u^2} \, du = \delta_{m,n}, \quad m, n = 0, 1, 2, \cdots; \tag{2.3}$$

(see, for example, Kaczmarz and Steinhaus [I, pp. 143–144]).

Let

$$\{\alpha_p(t)\} \qquad p = 1, 2, 3, \cdots \tag{2.4}$$

be any orthonormal set of real functions, each belonging to $L_2(0, 1)$. Paley and Wiener [II] have shown for each index $p = 1, 2, \cdots$ that $\int_0^1 \alpha_p(t) \, dx(t)$ exist as a generalized Stieltjes integral for almost all functions $x(\cdot)$ of C and that the equality

$$\begin{aligned} &\int_C^w G\left[\int_0^1 \alpha_1(t) \, dx(t), \cdots, \int_0^1 \alpha_p(t) \, dx(t)\right] d_w x \\ &\qquad = \pi^{-p/2} \int_{-\infty}^{\infty} \cdots \int_{-\infty}^{\infty} G(u_1, \cdots, u_p) e^{-u_1^2 \cdots u_p^2} \, du_1 \cdots du_p \end{aligned} \tag{2.5}$$

[1] Presented to the American Mathematical Society, November 23, 1945.

holds for every function $G(u_1, \cdots, u_p)$ for which the integral on the righ exists as an absolutely convergent Lebesgue integral. Using this result and the orthogonality properties of the Hermite polynomials as expressed by relation (2.3) we define a C. O. N. set of functionals as follows:

DEFINITION 2.1. Let (2.4) be any C. O. N. set of real functions, each belonging to $L_2(0, 1)$, and define

$$\Phi_{m,p}(x) = H_m\left[\int_0^1 \alpha_p(t)\, dx(t)\right]; \qquad m = 0, 1, 2, \cdots; \qquad p = 1, 2, \cdots, \tag{2.6}$$

and

$$\begin{aligned}\Psi_{m_1 \cdots m_p}(x) &\equiv \Psi_{m_1 \cdots m_p 0 \cdots 0}(x) \\ &= \Phi_{m_1,1}(x) \cdots \Phi_{m_p,p}(x).\end{aligned} \tag{2.7}$$

The index p in (2.7) may be any positive integer. The subscripts $m_1, \cdots, m_p$ may be any non-negative integers; for any particular functional Ψ at most a finite number of subscripts may be different from zero. Since $\Phi_{0,p}(x) \equiv 1$ for every $p = 1, 2, \cdots$, no ambiguity arises in the definition of the Ψ's. Since the α's belong to $L_2(0, 1)$, the Φ's and Ψ's are defined for almost all x in C.

We shall term the set (2.7) the *Fourier-Hermite set.*

Our main result is the following theorem.

THEOREM 1. *The Fourier-Hermite series of any (real or complex) functional $F[x]$ of $L_2(C)$ converges in the $L_2(C)$ sense to $F[x]$. This means that if $F[x]$ is any functional for which*

$$\int_C^w |F[x]|^2\, d_w x < \infty \tag{2.8}$$

then

$$\int_C^w \left| F[x] - \sum_{m_1, \cdots, m_N = 0}^{N} A_{m_1, \cdots, m_N} \Psi_{m_1 \cdots m_N}(x) \right|^2 d_w x \to 0 \quad \text{as} \quad N \to \infty, \tag{2.9}$$

where $A_{m_1, \cdots, m_N}$ is the Fourier-Hermite coefficient.

$$A_{m_1, \cdots, m_N} = \int_C^w F[x] \Psi_{m_1, \cdots, m_N}(x)\, d_w x. \tag{2.10}$$

In the remainder of this section we shall prove that the set (2.7) is O. N., and we shall prove the Bessel inequality and the best approximation theorem for the set. In the next section we shall prove Theorem 1 for a special class of funcionals $F[x]$ and in Section 4 we shall prove the theorem in its general form.

In order to show that the set (2.7) is O. N. in $L_2(C)$ it is sufficient to show that

$$\int_C^w \Psi_{m_1, \cdots, m_n}(x) \Psi_{j_1, \cdots, j_n}(x)\, d_w x = \delta_{m_1, j_1} \cdots \delta_{m_n, j_n}. \tag{2.11}$$

In (2.11) the indices $m_1, \cdots, m_n, j_1, \cdots, j_n$ may be any non-negative integers. Since zero is an admissible value for any index, there clearly is no loss in taking the same number, n, of m's and j's.

Now by (2.7), (2.6), (2.5), and (2.3), the left-hand member of (2.11) is equal to

$$\pi^{-n/2} \int_{-\infty}^{\infty} \cdots \int_{-\infty}^{\infty} H_{m_1}(u_1) H_{j_1}(u_1) e^{-u_1^2} \cdots H_{m_n}(u_n) \cdot H_{j_n}(u_n) e^{-u_n^2} \, du_1 \cdots du_n = \delta_{m_1, j_1} \cdots \delta_{m_n, j_n} . \tag{2.12}$$

This yields the desired orthonormality of the set (2.7). From this we obtain the Bessel inequality and the best approximation theorem by the usual formal analysis:

$$\begin{aligned} \int_C^w \left| F[x] - \sum_{m_1, \cdots, m_n = 0}^{n} B_{m_1, \cdots, m_n} \Psi_{m_1, \cdots, m_n}[x] \right|^2 d_w x \\ = \int_C^w | F[x] |^2 \, d_w x - 2Re \Sigma A_{m_1, \cdots, m_n} \bar{B}_{m_1, \cdots m_n} + \Sigma \, | B_{m_1, \cdots, m_n} |^2 \\ = \int_C^w | F[x] |^2 \, d_w x - \Sigma \, | A_{m_1, \cdots, m_n} |^2 + \Sigma \, | A_{m_1, \cdots, m_n} - B_{m_1, \cdots, m_n} |^2 \end{aligned} \tag{2.13}$$

so that the first integral is a minimum whenever $B_{m_1, \cdots, m_n} = A_{m_1, \cdots, m_n}$ for all $m_1, \cdots, m_n$. Also since the left member is non-negative we obtain

$$\sum_{m_1, \cdots, m_n = 0}^{n} | A_{m_1, \cdots, m_n} |^2 \leqq \int_C^w | F[x] |^2 \, d_w x; \qquad n = 0, 1, 2, \cdots .$$

3. A special case of Theorem 1

We shall first derive a very simple formula giving the Fourier-Hermite coefficients for a special class of functionals $F[x]$. Let $\Psi_{m_1, \cdots, m_p}(x)$ be defined as in Definition 2.1, with $\{\alpha_j(t)\}, j = 1, 2, \cdots$, a C. O. N. set of real functions each belonging to $L_2(0, 1)$. Let n be a fixed positive integer and let $f(u_1, \cdots, u_n)$ be any function such that

$$f(u_1, \cdots, u_n) e^{-(u_1^2 + \cdots + u_n^2)/2} \epsilon L_2(-\infty, \infty). \tag{3.1}$$

Finally consider the special "n dimensional" functional

$$F[x] \equiv f\left[\int_0^1 \alpha_1(t) \, dx(t), \cdots, \int_0^1 \alpha_n(t) \, dx(t) \right]. \tag{3.2}$$

We shall prove the following result.

Lemma 3.1. *For any non-negative indices $m_1, \cdots, m_p$ we have*

$$\int_C^w F[x] \Psi_{m_1, \cdots, m_p}[x] \, d_w x = \begin{cases} 0 \text{ if } n < p \text{ and } m_p \neq 0, \\ f_{m_1, \cdots, m_n} \text{ if } p = n \end{cases} \tag{3.3}$$

where $f_{m_1, \cdots, m_n}$ is the ordinary n-dimensional Hermite coefficient of $f(u_1, \cdots u_n)$ $\exp\{-(u_1^2 + \cdots + u_n^2)/2\}$, namely

$$f_{m_1, \cdots, m_n} = \pi^{-n/2} \int_{-\infty}^{\infty} \cdots \int_{-\infty}^{\infty} f(u_1, \cdots, u_n) e^{-(u_1^2 + \cdots + u_n^2)/2} \prod_{j=1}^{n} [H_{m_j}(u_j) e^{-u_j^2/2} du_j]. \tag{3.4}$$

Remark. Since zero is an admissible value for each m_j we can alway arrange in (3.3) to have at least n subscripts. Thus the two cases $p > n$ and $p = n$ cover all cases.

Proof of Lemma 3.1. By (2.5) the left member of (3.3) is equal to

$$\pi^{-p/2}\int_{-\infty}^{\infty}\cdots\int_{-\infty}^{\infty} f(u_1, \cdots, u_n)\left[\prod_{j=1}^{p} H_{m_j}(u_j)e^{-u_j^2}\, du_j\right]. \tag{3.5}$$

If $p > n$ and $m_p > 0$ the factor

$$\int_{-\infty}^{\infty} H_{m_p}(u_p)e^{-u_p^2}\, du_p$$

comes out and is zero because of (2.3) with $H_0(u_p) \equiv 1$ as the second factor of the integrand. If, on the other hand, $p = n$ then (3.5) agrees with the right member of (3.4). This yields Lemma 3.1.

We shall now prove the following lemma.

Lemma 3.2. *Let $F[x]$ be defined as in (3.2) and let*

$$A_{m_1,\cdots,m_p} = \int_C^w F[x]\Psi_{m_1,\cdots,m_p}(x)\, d_w x. \tag{3.6}$$

Then

$$\int_C^w \left| F[x] - \sum_{m_1,\cdots,m_n=0}^{N} A_{m_1,\cdots,m_n}\Psi_{m_1,\cdots,m_n}(x)\right|^2 d_w x \tag{3.7}$$

$$\to 0 \text{ as } N \to \infty.$$

Proof. The left member of (3.7) is equal to

$$\begin{aligned} &\pi^{-n/2}\int_{-\infty}^{\infty}\cdots\int_{-\infty}^{\infty}\left| f(u_1, \cdots, u_n) \right.\\ &\quad \left. - \sum_{m_1,\cdots,m_n=0}^{N} A_{m_1,\cdots,m_n}H_{m_1}(u_1)\cdots H_{m_n}(u_n)\right|^2 e^{-(u_1^2+\cdots+u_n^2)}\, du_1, \cdots, du_n \\ &= \pi^{-n/2}\int_{-\infty}^{\infty}\cdots\int_{-\infty}^{\infty}\left| f(u_1, \cdots, u_n)e^{-(u_1^2+\cdots+u_n^2)/2} - \sum_{m_1,\cdots,m_n=0}^{N} \right. \\ &\quad \left. f_{m_1,\cdots,m_n}H_{m_1}(u_1)\cdots H_{m_n}(u_n)e^{-(u_1^2+\cdots+u_n^2)/2}\right|^2 du_1 \cdots du_n. \end{aligned} \tag{3.8}$$

But this latter multiple integral approaches zero as $N \to \infty$ due to the completeness of the set $\{H_j(u_j)e^{-u_j^2/2}\}$. This yields Lemma 3.2.

If we now note that $A_{m_1,\cdots,m_p}$ as defined in (3.6) (with F defined by (3.2)) is zero when $p > n$ and $m_p > 0$ (see (3.3)), then we see that the number of subscripts on Ψ and A in (3.7) may be increased beyond n without changing the sum. Hence we may take N subscripts in (3.7) instead of n, and we find that (2.9) holds for any F defined by (3.2). Thus we have:

Theorem 1a. *For any $F[x]$ defined by* (3.2) *with f satisfying* (3.1), *the Fourier-*

Hermite series converges in the $L_2(C)$ *sense to* $F[x]$. *In other words,* (2.9) *holds with the A's defined as in* (2.10).

4. Proof of Theorem 1

We prove Theorem 1 by extending the class of functionals to which Theorem 1a applies. We do this by first proving the following lemma.

LEMMA 4.1. *Any functional of* $L_2(C)$ *can be approached in the* $L_2(C)$ *sense by functionals satisfying the hypotheses of Theorem* 1a, *i.e., these functionals are everywhere dense in* $L_2(C)$.

To see this we note that (by the usual Lebesgue argument) any functional of $L_2(C)$ can be approached arbitrarily closely by step-functionals (functionals taking on only a finite set of values). But such a step-functional is a finite linear combination of characteristic functionals of W-measurable sets in C. Moreover each characteristic functional of a W-measurable set in C can be approximated arbitrarily closely in the $L_2(C)$ sense by a finite linear combination of characteristic functionals of quasi-intervals. It therefore only remains to show that the characteristic functional of a quasi-interval can be approached arbitrarily closely by functionals satisfying the hypotheses of Theorem 1a.

Now (cf. [III, p. 371]) let Q be the quasi-interval

$$Q\colon \xi_j' \leqq x(t_j) \leqq \xi_j'' \begin{Bmatrix} j = 1, \cdots, n \\ 0 < t_1 < \cdots < t_n \leqq 1 \end{Bmatrix}.$$

(It is permissible for any ξ_j' to be $-\infty$ and any ξ_j'' to be $+\infty$, in which cases we, of course, drop the equality signs since x is always finite.) Let $\epsilon > 0$ and let $\varphi_{j,\epsilon}(\eta)$ be a continuous "trapezoidal" function which is zero outside the interval $\xi_j' - \epsilon < \eta < \xi_j'' + \epsilon$, equals unity inside the interval $\xi_j' \leqq \eta \leqq \xi_j''$, and is linear on the remaining intervals. (If ξ_j' takes on the improper value $-\infty$, so does $\xi_j' - \epsilon$, etc.).

Let

$$\chi_\epsilon(x) = \prod_{j=1}^{n} \varphi_{j,\epsilon}[x(t_j)] \tag{4.1}$$

and let

$$\chi(x) = \begin{cases} 1 \text{ if } x \in Q, \\ 0 \text{ if } x \in Q; \end{cases} \tag{4.2}$$

so that $\chi(x)$ is the characteristic functional of Q.

Now

$$\lim_{\epsilon \to 0} \chi_\epsilon(x) = \chi(x) \qquad \text{for each } x \text{ in } C, \tag{4.3}$$

and by bounded convergence we conclude that

$$\underset{\epsilon \to 0}{\text{L.I.M.}}\ \chi_\epsilon(x) = \chi(x) \text{ on } C. \tag{4.4}$$

Next, define

$$h_j(t) = \begin{cases} 1 & 0 \leqq t \leqq t_j \\ 0 & t_j < t \leqq 1 \end{cases} \tag{4.5}$$

and let $a_{j,\mu}$ be the coefficients of the orthogonal developments of $h_j(t)$ in terms of $\{\alpha_\mu(t)\}$:

$$h_j(t) \sim \sum_{\mu=1}^{\infty} a_{j,\mu}\,\alpha_\mu(t). \tag{4.6}$$

Define

$$H_{j,m}(t) = \sum_{\mu=1}^{m} a_{j,\mu}\,\alpha_\mu(t) \tag{4.7}$$

so that

$$\underset{m\to\infty}{\text{l.i.m.}}\; H_{j,m}(t) = h_j(t) \qquad \text{on } 0 \leqq t \leqq 1. \tag{4.8}$$

Now by a formula of Paley and Wiener [II]

$$\begin{aligned}\int_C^w \left\{\int_0^1 [H_{j,m}(t) - h_j(t)]\,dx(t)\right\}^2 d_w x \\ = \pi^{-\frac{1}{2}} \int_{-\infty}^{\infty} u^2 e^{-u^2}\,du \cdot \int_0^1 [H_{j,m}(t) - h_j(t)]^2\,dt \\ = \tfrac{1}{2}\int_0^1 [H_{j,m}(t) - h_j(t)]^2\,dt \to 0 \text{ as } m \to \infty.\end{aligned} \tag{4.9}$$

Thus

$$\int_0^1 H_{j,m}(t)\,dx(t) \to \int_0^1 h_j(t)\,dx(t) \tag{4.10}$$

in the $L_2(C)$ sense on C as $m \to \infty$, and hence there exists (at least) a subsequence $m_1, m_2, \cdots$ such that

$$\lim_{p\to\infty} \int_0^1 [H_{j,m_p}(t) - h_j(t)]\,dx(t) = 0 \tag{4.11}$$

for almost all functions $x(\cdot)$ in C. By exercising proper care in defining the subsequence $m_1, m_2, \cdots$ we can use the same subsequence for each value of $j = 1, \cdots, n$.

We now define (for fixed $\epsilon > 0$)

$$\chi_{\epsilon,p}(x) = \prod_{j=1}^{n} \varphi_{j,\epsilon}\left[\int_0^1 H_{j,m_p}(t)\,dx(t)\right] \tag{4.12}$$

and note that for almost all x in C

$$\begin{aligned}\lim_{p\to\infty} \chi_{\epsilon,p}(x) &= \prod_{j=1}^{n} \varphi_{j,\epsilon}\left[\int_0^1 h_j(t)\, dx(t)\right] \\ &= \prod_{j=1}^{n} \varphi_{j,\epsilon}[x(t_j)] \qquad (4.13) \\ &= \chi_\epsilon(x).\end{aligned}$$

But since $|\varphi_{j,\epsilon}(u)| \leqq 1$ we have by bounded convergence

$$\text{(4.14)} \qquad \underset{p\to\infty}{\text{L.I.M.}}\ \chi_{\epsilon,p}(x) = \chi_\epsilon(x) \qquad \text{on } C.$$

We have now shown that $\chi(x)$, the characteristic functional of Q, can be approximated in the $L_2(C)$ sense by functionals of the type $\chi_{\epsilon,p}(x)$ (cf. (4.14) and (4.4)). But these functionals clearly satisfy the hypotheses of Theorem 1a with

$$f(u_1, \cdots, u_{m_p}) = \prod_{j=1}^{n} \varphi_{j,\epsilon}\left[\sum_{\mu=1}^{m_p} a_{j,\mu} u_\mu\right],$$

and hence the lemma is proved.

The proof of Theorem 1 follows easily from Lemma 4.1, and the best approximation theorem derived at the end of Section 2. For, let $F[x] \in L_2(C)$ but otherwise unrestricted, and denote by $A_{m_1,\cdots,m_n}$ its Fourier-Hermite coefficients. Then if $\epsilon > 0$ is given, there is an $F^*[x]$ satisfying the hypotheses of Theorem 1a and such that

$$\int_c^w |F[x] - F^*[x]|^2\, d_w x < \epsilon/4.$$

If $A^*_{m_1,\cdots,m_n}$ are the Fourier-Hermite coefficients of $F^*[x]$, we can choose N so great that

$$\int_c^w \left| F^*[x] - \sum_{m_1,\cdots,m_N=0}^{N} A^*_{m_1,\cdots,m_N} \Psi_{m_1,\cdots,m_N}(x) \right|^2 d_w x < \epsilon/4,$$

and by the Minkowski inequality we have

$$\text{(4.15)} \qquad \int_c^w \left| F[x] - \sum_{m_1,\cdots,m_N=0}^{N} A^*_{m_1,\cdots,m_N} \Psi_{m_1,\cdots,m_N}(x) \right|^2 d_w x < \epsilon.$$

Then by the best approximation theorem it follows that the inequality (4.15) will remain true if we replace A^* by A, and hence (2.9) holds. This yields Theorem 1.

5. Implications of Theorem 1

An immediate consequence of Theorem 1 and (2.13) with $B = A$ is the Parseval equation

$$\int_c^w |F[x]|^2\, d_w x = \lim_{N\to\infty} \sum_{m_1,\cdots,m_N=0}^{N} |A_{m_1,\cdots,m_N}|^2.$$

Moreover we can readily verify that the functionals in $L_2(C)$ with the inner product

$$(F, G) = \int_C^w F[x]\overline{G[x]}\, d_w x$$

satisfy all the postulates of Hilbert space. However, the conclusion that $L_2(C)$ is a Hilbert space can be obtained more directly and simply by noting that the Wiener transformation takes $L_2(0, 1)$ into $L_2(C)$ in such a way as to take inner products into inner products without altering their numerical values. Therefore any C. O. N. set in $L_2(0, 1)$ goes into a C. O. N. set in $L_2(C)$. However, consideration of with Wiener mapping itself shows that the transform in $L_2(C)$ of even a very simple C. O. N. set in $L_2(0, 1)$ is an extremely complicated set which would not seem worthwhile considering when compared with the simple Fourier-Hermite series developed here.

In defining the Fourier-Hermite functionals in (2.7) we have the C. O. N. set $\{\alpha_p(t)\}$ at our disposal. A very convenient set would be the Haar functions. Since the Haar functions are step functions, the integrals in the Fourier-Hermite functionals become simply discrete sums. Thus, after the Fourier-Hermite coefficients have been calculated, the numerical calculation for the development of $F(x)$ for any one particular x involves only addition and multiplication and the use of a table of Hermite polynomials. No integration is needed, and the calculation becomes very simple.

6. Conclusion

This paper opens up the whole set of techniques of Orthogonal Developments and of Hilbert space for use in the development and study of non-linear functionals and non-linear operators. (The operators can be treated as single parameter families of functionals.)

The University of Minnesota
and
Syracuse University

Bibliography

I S. Kaczmarz and H. Steinhaus, *Theorie der Orthogonalreihen.* Warsaw (1935).

II R.E.A.C. Paley and N. Wiener, *Fourier transforms in the complex domain.* Am. Math. Soc. Coll. Pub., Vol. XIX, New York, (1934).

III R. H. Cameron and W. T. Martin, *Transformations of Wiener integrals under translations.* Annals of Math., 45 (1944), pp. 386-396.

14

Reprinted from *J. Electronics and Control,* **15**(6), 567–615 (1963)

The Use of Functionals in the Analysis of Non-linear Physical Systems†

By J. F. Barrett
Department of Mechanical Engineering, University of Birmingham

[Received February 4, 1963]

Abstract

This report is an attempt to develop a method of analysis applicable equally to linear or non-linear systems. The main method discussed is the expansion of the input–output relation in a functional power series—an idea first due to Volterra for general functional relationships and to Wiener in its application to non-linear communication problems. The report attempts to present a systematic development of this idea. The last part of the report discusses analogous expansions in a series of terms orthogonal with respect to input statistics.

---

§ 1. Statistical Design

1.1. *The Black Box*

The widely used 'black box' method of description of engineering systems shows that, for many purposes, the actual nature of an engineering system is unimportant and only the way it responds to certain input signals is of interest. Denote input signal symbolically by s and response by r. Then mathematically, the point of interest is the functional relation $r = F(s)$ between r and s.

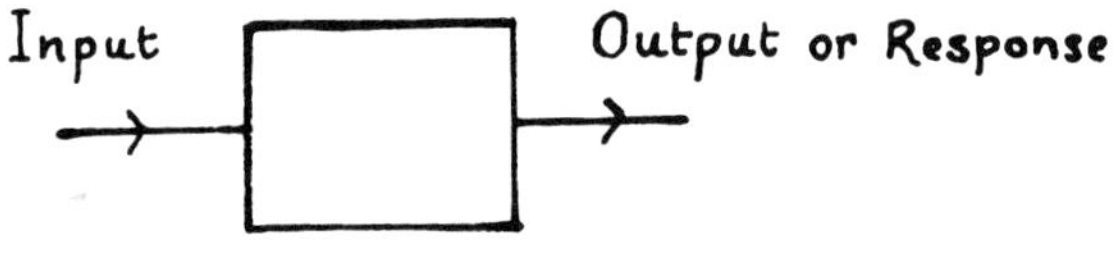

Example: If (i) the input and response can be represented as real or complex valued functions of time, e.g. they are fluctuating voltages, (ii) the system is linear and time-invariant (see below for the precise definition of these terms), then the functional relation F may be conveniently described by a transfer function, i.e. the ratio of the Laplace

---

† Communicated by A. J. Fuller. The present paper was first written in April 1955 as a Cambridge University Engineering Laboratory report and was duplicated and distributed in 1956 by the Ministry of Supply as S.A.U. report 1/57. The text is unchanged apart from the correction of errors, etc.

transforms of the response and input. At present no restriction is made to this class of systems and so the general F-notation must be retained.

1.2. *Performance Criteria*

In order to characterize the response of a system to some input s as a good or bad one, some method of comparison must be made between the actual response r and ideal response r_i. The nearer r is to r_i, in some sense, the better the response r. r_i depends on s, the input signal, and so it is possible to write $r_i = F_i(s)$, where F_i describes the ideal system. The measure of success of the system in dealing with signal s is then obtained by comparison of r with r_i ($F(s)$ with $F_i(s)$).

Examples:

1.2.1. *Duplicators*

In an important class of systems, sometimes called *duplicators* (Trimmer 1950), the system is required to reproduce the input signal exactly. Systems for transmitting speech and music come into this class as do also a large number of servo systems.

Here, with suitable choice of units, etc., the ideal response to a signal s is s itself: $r_i = s$. The success of the system in dealing with s is thus measured by the closeness in some sense of r to s. The method adopted to measure this closeness is largely arbitrary and can often be chosen for convenience. Thus, for a simple (angle) position control system, if $\theta_1(t)$ and $\theta_2(t)$ are input and response angles at time t, possible numerical measures of the amount of discrepancy between θ_2 and θ_1 in an interval of time, $0 \leqslant t \leqslant T$ are:

$$\max_{0\leqslant t\leqslant T} |\theta_2(t) - \theta_1(t)|, \quad \frac{1}{T}\int_0^T |\theta_2(t) - \theta_1(t)|\, dt$$

or

$$\frac{1}{T}\int_0^T (\theta_2(t) - \theta_1(t))^2\, dt.$$

All these are equivalent in the sense that if any one is small, so must the others be†.

1.2.2. *Filters*

In a system for filtering out noise, the input signal consists of a wanted signal and certain unwanted disturbances—'noise'. So, for example, if the input signal is in the form of a varying voltage,

$$V(t) = V_1(t) + V_2(t),$$

where $V(t)$, $V_1(t)$ and $V_2(t)$ are the input, wanted signal and noise voltages at time t. Here the input signal s is the record of $V(t)$ over the time interval of observation.

The ideal response r_i of the system to the signal $V(t)$ is a voltage $V_1(t)$ at time t or more generally $V(t - t_1)$ allowing for a delay of t_1. Let the actual response at time t be $V_3(t)$.

† θ_1 and θ_2 are assumed continuous functions of time.

As a measure of success of the system when operating on the particular signal $V(t)$ for the given time interval, any of the expressions of the preceding example may be used with $\theta_2(t)$ replaced by $V_3(t)$, $\theta_1(t)$ by $V_1(t)$ and the time interval $(0, T)$ by the interval of observation.

More complicated filtering problems arise, e.g. the received signal may be of the form $V(t) = a(t) \sin(\omega t + n_1(t)) + n_2(t)$, where $n_1(t)$ and $n_2(t)$ are now unwanted disturbances. Here the wanted signal $a(t)$ is modulated. The ideal system for simultaneous demodulation and filtering out noise would be one giving $a(t)$ from $V(t)$ at any time t (with possibly a delay).

1.2.3. *Predictors*

Here again, the received signal is a function of time. If the value of x at time t_1 ahead is required, the ideal response at time t is $x(t+t_1)$.

In the above examples, signals were in each case represented by time varying quantities and it was thus possible to give a numerical value to the discrepancy between two signals. This is the most common case. A signal may usually be adequately described by the time variation of some physical quantity. Thus a piece of music can be adequately described for present purposes by the voltage variations produced in a suitably adjusted microphone, or more basically by the pressure variations in the air near the microphone.

It will consequently be assumed that signals and responses are represented by the time variation of physical magnitudes and that a numerical measure of discrepancy between two responses is given. If the discrepancy between r and r_i is $\delta(r, r_i)$ the function will always be assumed to have the property :

$$\delta(r, r_i) \geqslant 0 \text{ for all } r \text{ and } \delta(r, r_i) = 0 \text{ if } r = r_i.$$

This condition is satisfied with the measures given above (§ 1.2.1). A function having this property has been called a *distance function* by Shannon and Weaver (1949).

The inaccuracy of the system in responding to a singal s is thus measured by $\delta(r, r_i) = \delta(F(s), F_i(s))$.

1.3. *Optical Systems*

In general, the problem is to make the system optimal with a given distance function. In other words an F must be found which minimizes the express $\delta(r, r_i) = \delta(F(s), F_i(s))$. A trivial solution is $F = F_i$ but in general it is not possible to achieve this because of the limitation of means in realizing F. Viewed mathematically, the problem is one of minimization under certain constraints which exclude the possibility of $F = F_i$.

This is merely the optimization of the response to a single input and will usually not be what is required. Any signal transmission system will be required to operate satisfactorily over a whole range of inputs. A message which is known before it is sent conveys no information and so a communication system designed to receive a single message is redundant. The signal which is actually the input on any one working run of the system must be one of a range of possible input signals. This range of possible input signals will be called the *signal ensemble.* When a signal transmission system is being designed, it must be designed to deal efficiently with every possible signal that might occur, i.e. with all signals of the signal ensemble. In this case it is necessary to make some compromise between the relative efficiencies for different signals (relative to some distance function). A system S_1 may have a better response than S_2 on some signals but a worse one on others. How is a choice to be made between the two systems?

The solution proposed by Wiener in 1942 (see Wiener 1949) is based on the relative frequencies of occurrence of the individual signals of the signal ensemble. In generalized form, the method is to choose F so as to minimize the average value $\overline{\delta(F(s), F_i(s))}$ of $\delta(F(s), F_i(s))$ the average value being over the relative frequencies of occurrence of the signals s†. A system designed on this principle will consequently have, on the average, a response at least as good, or better than, every other system of the range of systems considered.

This value of F will depend on δ, F_i, the input statistics and the range of variation allowed in F. The general mathematical problem in statistical design is to give an explicit expression for F in terms of these. The following theory is centred round the problem of finding this F when signals are functions of time (time-series) and δ is the mean square value.

The chief drawbacks of statistically designed systems should be mentioned. The two most important of these are probably:

(i) a system designed according to this method will only give a better performance than others on the average. On occasions its performance might be very poor;

(ii) a statistically designed system may be sensitive to changes in input statistics. If these change, the system will no longer be statistically optimal. Consequently it might appear that a system designed on the statistical principle is too highly specialized to justify the trouble of constructing it.

(i) seems quite valid in certain cases. The problem is often not to design a system which will be good on the average but one which will never be bad. If this is the case, the statistics of the signal ensemble do not have to be known and F will be chosen so as to minimize $\max \delta(f(s), F_i(s))$, the maximum being taken over all s of the signal ensemble. (ii) should be met by designing a system which adjusts itself, or can be adjusted, to changes of input statistics.

§ 2. Operators and Functionals

In the relation $r = F(s)$, if s and r are functions of time, as they will always be assumed to be from now on, F is a function with argument and value

† Convenient though not precise. $\overline{\delta(F(s),\ F_1(s))}$ depends on $\delta(F,\ F_1)$ and signal statistics, not s.

which are also functions. To avoid confusion of language, F will be termed an *operator*. An operator is defined by a rule for deriving one function from another, familiar examples being derivation and indefinite integration. Another way of expressing the relationship between s and r in the case where both are functions involves the idea of a *functional*. A functional is a function whose argument is a function and whose value is a number. Definite integration is an example: every definite integral will define a functional the argument being the integrand and the value being the value of the definite integral.

Given a general 'black box', the relation between the input signal s and the response r, both of which are time varying quantities x and y with values $x(t)$ and $y(t)$ at time t, may be described mathematically in two ways—both of which are really equivalent. In the first method, attention is given to the relation between the whole response as represented by a record or graph of y over the period of working of the system and the whole input signal as represented by a record or graph of x over the period of working. Here attention is directed to the functional relation between the two functions of time x and y. This is the *operational description*†. In the second method, the dependence of the response at a particular time t on the previous input is considered. This is a relation between a function and a number and is thus a *functional*. In this case the relation can be written $y(t)=f(t;s)$ or alternatively, $y(t)=f(t;\ x(t'), t' \leqslant t)$ since the only part of the signal s that y is dependent on at time t, is the part which occurs before time t. Generally speaking, the functional description is more suitable for concrete problems and more attention will be paid to it here. Before stating the most important results on functionals for the present theory, a few remarks are necessary on the representation of signals.

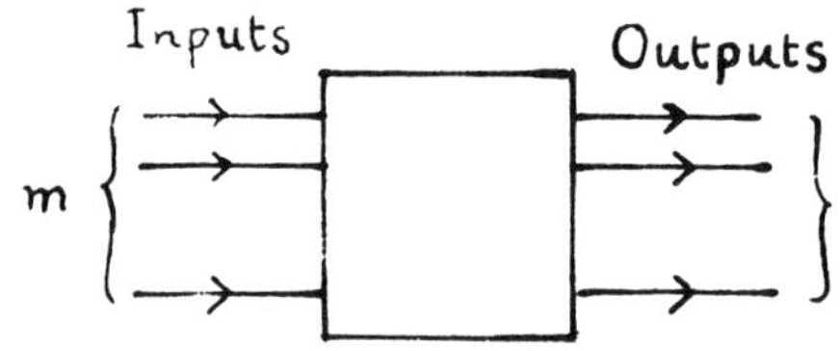

The input signal and response as represented by the time variation of physical quantities x and y will usually be real valued functions of time in the appropriate range of time. More generally they can be complex or vector quantities or both. The advantages of complex representation of electrical quantities is well known (see Gabor 1946 for the theory). Multi-channel transmission is a case where the signal is conveniently described by a vector: if there are n input signals to a system by the time variation of $x_1, x_2, \ldots, x_n$ the input may be described by the single vector quantity

† This usage of the term 'operational' should not be confused with that frequent in physics. 'Operational' here means 'described by a mathematical operator'.

$\mathbf{x}=(x_1, x_2, \ldots, x_n)$. The response may also be a vector. For instance, the response of the system may be recorded on n dials with readings $y_1, y_2, \ldots, y_n$. In what follows, functions for signals will usually be real valued functions of time, the extension to other cases being more or less straightforward. The range of time considered will be from a fixed time denoted by t_0, the time of switching on, to 'now', usually represented by the time t. t_0 may be at $-\infty$ in the case where only steady-state operation is considered.

With these preliminaries we return to the consideration of functionals. The most important result of this section is the power series expansion of a non-linear functional. This will be preceded by a characterization of two important classes of functionals, time-invariant and linear functionals respectively.

2.1. *Time-invariance*

In many systems the form of response is independent of the particular time at which the input signal is received. Electrical filters with lumped L, R, C elements with constant values come into this class. Such systems and the functionals (or operators) which describe them will be called *time-invariant*.

Let $f(t; s)$ be the response of the system at time t to a signal s. Let s' represent the same signal a time τ later (earlier if τ is negative). Then if f represents a time invariant functional the equation

$$f(t;\, s)=f(t+\tau;\, s')$$

must hold for all times t. Another way of expressing the same property is that in the notation $f(t;\, x(t'),\, t' \leqslant t)$ for the response at time t, the first t need not be written and the response is only dependent on the actual values of x before t and not on t itself. This may be deduced from the above equation as follows:

$$f(t;\, x(t'),\, t' \leqslant t)=f(t+\tau;\, x(t''),\, t'' \leqslant t+\tau)$$

from the above equation. Putting $t'''=t''-\tau$, the right-hand side of the equation becomes $f(t+\tau;\, x(t'''),\, t''' \leqslant t)$. This is the same as $f(t+\tau;\, x(t'),\, t' \leqslant t)$ since the primed t's are only dummy variables and can be replaced by any other variables without any change of meaning. Thus $f(t;\, x(t'),\, t' \leqslant t)=f(t+\tau;\, x(t'),\, t' \leqslant t)$ for any τ and so the first t is redundant: its value has no influence on the value of the functional f.

Time-invariant operators are defined in a similar way.

2.2. *Linearity*

Many systems, or more precisely the operators or functionals describing them, show the *additive property* or *principle of superposition* with respect to input signals, i.e. if input signals $x(t)$, $x'(t)$ give output signals $y(t)$, $y'(t)$, then an input signal $x(t)+x'(t)$ gives an output signal $y(t)+y'(t)$. (Here 'an input signal $x(t)$', etc., means a signal which is represented by a physical quantity denoted by x which has a value $x(t)$ at any time t, etc.)

It may be shown† that the additive property is equivalent to the *homogeneity property* with respect to input signals, i.e. that if a signal $x(t)$ gives output signal $y(t)$, then an input signal: constant $\times x(t)$, say $kx(t)$, gives an output signal $ky(t)$. This is equivalent to saying that, apart from a change of scale—in fact the same change of scale for both input and output—the performance of the system is independent of amplitude and there is no distortion due to changes of amplitude of the input‡.

A system (or the operator, functional describing it) will be called *linear* if it satisfies these conditions. If it is not linear in this sense, it will be called *non-linear*§.

2.3. *Examples of Functionals*

From now on the behaviour of systems will be discussed in terms of functionals rather than operators. (Any functional may be thought of as defining an operator.) The value of the functional at time t will be denoted by $y(t)$.

2.3.1. *The identity*

This is defined by the equation:

$$y(t) = f(t;\ x(t)',\ t' \leqslant t) = x(t).$$

It is obviously time-invariant and linear. The device described by this functional is the ideal duplicator: output always equals input.

2.3.2. *Time translation*

The defining equation is:

$$y(t) = f(t;\ x(t'),\ t' \leqslant t) = x(t+\tau).$$

Again this is obviously time-invariant and linear. The device described by this functional is the ideal predictor if $\tau > 0$ and a pure delay if $\tau < 0$. If $\tau = 0$ it reduces to the identity functional.

2.3.3. *Differentiation*

$$y(t) = f(t;\ x(t'),\ t' \leqslant t) = \dot{x}(t).$$

Once more, time-invariant and linear. In this example and in example 1, the value of the functional for time t depends only on the values of $x(t)$ in the immediate neighbourhood of the time t. A functional with this property is called *instantaneous*. So is the device it describes.

2.3.4. *Integral and convolution functionals*

f is a linear integral functional if it is of the form:

$$y(t) = \int_{t_0}^{t} g(t, t')x(t')\, dt.$$

$g(t, t')$ is called the *kernel function* of the integral functional, (kernel for short).

---

† Subject to certain mild continuity conditions.

‡ This behaviour is not solely characteristic of linear systems and is shared by those described by homogeneous operators (functionals) of any degree.

§ This definition of a non-linear system is by no means universal among engineers, e.g. see James *et al.* 1947, p. 29. A linear filter in the sense of this book is described here as a stable, time-invariant linear filter.

It is convenient to have the integral over the range $(-\infty, \infty)$, This is accomplished by defining $x(t)=0$ for $t<t_0$ if t_0 is finite and also defining $g(t,t)=0$ for $t<t'$. The integral is then:

$$\int_{-\infty}^{\infty} g(t,t')x(t)\,dt'.$$

Having g vanish for $t<t'$ ensures that only the past values of x at any time influence the value of the functional at that time. A kernel of this kind is said to be of *Volterra type.*

If $g(t,t')$ is a function only of the difference $t-t'$, the functional is time-invariant. This may be seen by a change of variable. Suppose $g(t,t')=h(t-t')$. Put $t-t'=t''$. Then

$$\begin{aligned}\int_{-\infty}^{\infty} g(t,t')x(t')\,dt' &= \int_{-\infty}^{\infty} h(t-t')x(t')\,dt' \\ &= \int_{-\infty}^{\infty} h(t'')x(t-t'')\,dt'' \\ &= \int_{-\infty}^{\infty} h(t'')x(t')\,dt''.\end{aligned}$$

Thus f depends only on the values of $x(t')$ for $t'\leqslant t$ (=values of $x(t-t'')$ for $t''\geqslant 0$). Such a functional will be called a *convolution integral* (convolution for short).

By introduction of the Dirac δ-function and its derivatives†, the functionals in examples 2.3.1, 2.3.2 and 2.3.3 may be written as convolutions. They are respectively:

$$\int_{-\infty}^{\infty} \delta(t-t')x(t')\,dt', \qquad \int_{-\infty}^{\infty} \delta(t+\tau_0-t')x(t')\,dt'$$

and

$$\int_{-\infty}^{\infty} \delta(t-t')x(t')\,dt.$$

2.3.5. *Modulation*

Defined by

$$f(t;\, x(t'), t'\leqslant t)=M(t)x(t).$$

It is linear, instantaneous, but not time-invariant if the modulating function $M(t)$ varies with time.

2.3.6. *Square law devices*

$$f(t;\, x(t'),\, t'\leqslant t)=kx^2(t),$$

where k is a constant. This is non-linear, instantaneous and time-invariant.

2.3.7. *Regular homogeneous functional of degree n*

$$f(t;\, x(t'),\, t'\leqslant t)=$$

$$\int_{t_0}^{t}\int_{t_0}^{t}\cdots\int_{t_0}^{t} g(t;\, t_1, t_2, \ldots, t_n)x(t_1)x(t_2)\ldots x(t_n)\,dt_1\,dt_2\ldots dt_n.$$

† The use of these may be justified by the theory of distributions of L. Schwartz (1950, 1951). See, e.g., Lafleur and Namias (1954).

g is again called the kernel function (kernel). It is of Volterra type, i.e. $g(t; t_1, t_2, \ldots, t_n) = 0$ if any $t_0 > t_i$, $i = 1, 2, 3, \ldots, n$. If the kernel is a function only of the time differences $t - t_i$, $i = 1, 2, 3, \ldots, n$, the functional is time-invariant. This follows from a change of variables as in the linear case.

The functional is said to be homogeneous of degree n because changing x to kx at all times multiplies the value of f by k^n. $n = 1$ gives the linear integral functional.

By using the Dirac δ-function again, the square law device may be described by a regular homogeneous functional of degree 2:

$$kx^2(t) = k\int \delta(t-t')x(t')\,dt' \quad \int \delta(t-t'')x(t'')\,dt''$$

$$= \int\int k\delta(t-t')\ \delta(t-t'')x(t')x(t'')\,dt'\,dt''.$$

Thus the kernel is $k\delta(t-t')\,\delta(t-t'')$. Note that the kernel depends only on the differences $t-t'$, $t-t''$, so verifying time-invariance.

In general the nth order device:

$$f(t;\ x(t'),\ t' \leqslant t) = kx^n(t)$$

may be expressed as a regular homogeneous functional with the kernel $k\delta(t-t_1)\ \delta(t-t_2)\ \delta(t-t_3)\ldots\delta(t-t_n)$.

2.3.8. *Functional power series*

These are defined by expressions of the form:

$$\Sigma\int\int\cdots\int g(t;\, t_1, t_2, \ldots, t_n)x(t_1)x(t_2)\ldots x(t_n)\,dt_1\,dt_2\ldots dt_n,$$

where the summation is over an infinity of terms. If the summation is over only a finite number of terms, the expression is called a *functional polynomial*. For the general expression to have a meaning, the series will have to be convergent.

The general instantaneous device with a characteristic given by a function $f(\ )$ may be written in this form, if f has a power series expansion:

$$f(x) = \sum_{n=0}^{\infty} k_n x^n.$$

Each term may now be expressed in integral form as noted in the last example.

Note: The same symbol g can be used for all the kernels. There is no risk of confusion because each is dependent on a different number of variables.

2.3.9. *Linear functionals*

Before going on to general functionals, the main results about the structure of linear functionals will be briefly summarized. It was seen from the preceding examples that many linear functionals may be expressed in integral form:

$$\int g(t, t')x(t')\,dt.$$

If the kernel is of Volterra type, viz. $g(t,t')=0$ if $t'>t$, the functional is dependent only on the past of x at any time. If $g(t,t')$ is a function of the time difference $t-t'$, the functional is time-invariant.

The main result is that these properties are quite general if delta functions and their derivatives are used in the kernel.

If delta functions are avoided, any linear functional is expressible in the form:

$$\int g(t, t')x(t')\,dt' + \sum_{m,n} a_{m,n}x^{(m)}(t-t_n),$$

where the summation is taken over a number, possibly infinite, of values of m and n.

If the functional acts only on past values of the input x at any time, g is of Volterra type and all the t_n are non-positive: $t_n \leqslant 0$ for all n.

If the functional is time-invariant, $g(t,t')$ depends only on the time difference $t-t'$.

The results may be expressed more concisely with δ-functions. The functional may be put in integral form with the kernel:

$$k(t,t') = g(t,t') + \sum_{m,n} a_{m,n}\delta^{(m)}(t'-t-t_n).$$

The customary derivation of the main result is as follows†: The function $x(t)$ may be thought of as the sum of an infinite number of impulses —one for each time t. The magnitude of the impulse corresponding to time t is proportional to $x(t)$.

This decomposition of x may be written:

$$x(t) = \sum x(t')\delta(t-t').$$

It is seen that this corresponds to the formula:

$$x(t) = \int \delta(t-t')x(t')\,dt'.$$

By an extension of the linearity property of f,

$$f(t;\, x(t'), t' \leqslant t) = \int x(t')g(t,t')\,dt',$$

where $g(t,t'')=f(t;\, \delta(t-t'')t'' \leqslant t)$, i.e. $g(t,t')$ is the effect at time t due to a unit impulsive input at time t'. This is the required integral form.

Since $f(t;\, x(t'), t' \leqslant t)$ is dependent only on the past, $g(t,t')=0$ if $t'>t$ by definition.

---

† A proof on these lines may be given using distribution theory (see footnote to page 574.) For continuous linear functionals, the result follows from an extension of the theorem of F. Riesz. See Bourbaki (1952), Book VI, p. 57 et seq.

If the system is time-invariant, a unit impulse applied at time $t'+\tau$ will give the same output at time $t+\tau$ as a unit impulse applied at time t' will give at time t: this follows immediately from the definition of time-invariance. In terms of the impulse response function, this statement is expressed by the equation: $g(t,+\tau, t'+\tau)=g(t,t')$ for all values of t, t' and τ. Putting $\tau=-t'$ gives $g(t,t')=g(t-t', 0)$, for all values of t and t', and so $g(t,t')$ depends only on the difference $t-t'$.

2.3.10. *Stability*

When dealing with input signals $x(t)$ which are non-zero over an infinite range of time, it is necessary to ensure that the infinite integral

$$\int_{-\infty}^{\infty} g(t,t')x(t')\,dt$$

exists at the lower limit. (The upper limit gives no trouble as the integral is only apparently an infinite one at $+\infty$ when the functional depends on the past values only.)

A sufficient condition for this is that

$$\int_{-\infty}^{\infty} |g(t,t')|\,dt' < \text{some constant},$$

say M, for all times t. For then

$$\left|\int_{\infty}^{\infty} g(t,t')x(t')\,dt'\right| \leqslant \int_{-\infty}^{\infty} |g(t,t')x(t')|\,dt'$$

$$\leqslant \max_{-\infty<t'<\infty}|x(t')| \int_{-\infty}^{\infty} |g(t,t')|\,d_it$$

$$< \max_{-\infty<t'<\infty}|x(t')|\,M.$$

Thus, if x is bounded, so is the infinite integral.

Alternatively we may state that every bounded ensemble of inputs will give a bounded ensemble of outputs. This is the definition of stability proposed by Zadeh (1952). For time-invariant linear systems, it may be shown equivalent to all the usual definitions (James *et al.* 1947). It has the advantage of applying equally well to non-linear systems whereas other definitions fail.

§ 3. General Functionals

We wish to find an explicit expression for a general functional which wil exhibit its structure in much the same way as the integral form does in the case of a linear functional. This is accomplished by the power series expansion which seems to be valid for a wide class of functionals†. The basic functionals turn out to be the regular homogeneous functionals already mentioned (p. 574). They share with the linear functionals the homogeneous (non-distortion) property, but not that of additivity.

† The power series expansion for a functional is due to Volterra. Accounts of the theory may be found in V. Volterra's book and in those of Lévy (1951) and Hille (1948) (abstract treatment).

The situation will be seen to be quite analogous to that which occurs with functions of many variables. If $f(x_1, x_2, \ldots, x_n)$ is a real valued function of n real variables $x_1, x_2, \ldots, x_n$, we may say that f is additive with respect to these variables if

$$f(x_1+x_1', x_2+x_2', \ldots, x_n+x_n')=f(x_1, x_2, \ldots, x_n)+f(x_1', x_2', \ldots, x_n')$$

and is homogeneous of degree 1 if

$$(kx_1, kx_2 \ldots, kx_n)=kf(x_1, x_2, \ldots, x_n) \quad (k \text{ any constant}).$$

If f is both additive and homogeneous of degree 1, it must be a linear sum of x's with constant coefficients.

For

$$f(x_1, x_2, \ldots, x_n)=f(x_1, 0, \ldots, 0)+f(0, x_2, \ldots, x_n).$$

by additivity,

$$= \sum_{i=1}^{n} f(0, 0, \ldots, x_i, \ldots, 0)$$

by repeated application of the additivity property:

$$= \sum_{i=1}^{n} x_i f(0, 0, \ldots, 1, \ldots, 0)$$

by the homogeneity property,

$$= \sum_{i=1}^{n} k_i x_i, \text{ where } k_i=f(0, 0, \ldots, 1, \ldots, 0)$$

which is the required result.

Subject to conditions on good behaviour† of the function f, it is easily shown that the properties of additivity and homogeneity of degree 1 are equivalent. Thus any sufficiently regular function satisfying either condition is a linear expression:

$$\sum_{i=1}^{n} k_i x_i.$$

This is the stage so far in the classification of functionals. The equivalent may be seen by regarding a functional as a function of a continuous infinity of variables $x_{t_1}, x_{t_2}, \ldots$, where x_t is $x(t)$, the value of x at time t. In this case, in the expression $\sum k_i x_i$, summation must be made over a continuous infinity of values of i, viz. all the values of t in the appropriate range. This implies an integration: $\int k(t)x(t)\,dt$, giving again the integral form for a linear functional.

The analogy between functionals and functions of many variables may be pushed further. It is known that if f is sufficiently regular (analytic) near the values $x_1=x_2=\ldots=x_n=0$ it has a power series expansion:

$$f(x_1, x_2, \ldots, x_n)=k_0+\sum_{i=1}^{n} k_i x_i+\sum_{i=1}^{n}\sum_{j=1}^{n} k_{ij} x_i x_j+\sum_{i=1}^{n}\sum_{j=1}^{n}\sum_{k=1}^{n} k_{ijk} x_i x_j x_k+\ldots,$$

which is convergent for sufficiently small values of the x's, i.e. there is some number $r>0$ with the property that the expansion is convergent if each of

† It is sufficient for f to be differentiable.

the x's is less than r in absolute value. The greatest such number r may be called the *modulus of convergence* of the series. r may, of course, be infinite.

To carry over this expansion to functionals, the previous procedure may be tentatively adopted, i.e. the method of replacing summation over a finite set of suffices by summation (integration) over a continuous infinity of values of the variable t:

$$\sum_{i=1}^{n} k_i x_i \quad \text{is replaced by} \quad \int k(t) x(t)\, dt,$$

$$\sum_{i=1}^{n} \sum_{j=1}^{n} k_{ij} x_i x_j \quad \text{is replaced by} \quad \int\int k(t_1, t_2) x(t_1) x(t_2)\, dt_1\, dt_2,$$

etc.

Thus the functional expansion might be expected to take the form:

$$f(x(t')\, t') = k_0 + \int k(t) x(t)\, dt + \int\int k(t_1, t_2) x(t_1) x(t_2)\, dt_1\, dt_2$$
$$+ \int\int\int k(t_1, t_2, t_3) x(t_1) x(t_2) x(t_3)\, dt_1\, dt_2\, dt_3$$
$$+ \ldots .$$

The modulus of convergence of the series is defined to be the greatest $r > 0$ with the property that the expansion is convergent if $|x(t)| < r$ (for the range of t considered). If it is possible to find some modulus of convergence of the above series, f will be called *analytic* at $x = 0$. It will be noticed that if f is a function analytic at $x = 0$, its power series expansion as a functional may be obtained from its power series expansion as a function (cf. example 2.3.8, page 575).

Not all functions of n variables can be expanded in a power series. Thus with $n = 1$, the function defined by $f(x) = 0$, $x < 0$ and $f(x) = mx$, $x > 0$ ($m \neq 0$), which has a simple discontinuity in the gradient at $x = 0$, has no power series expansion about the origin. It is, however, continuous, and so according to the theorem of Weierstrass can be approximated by polynomials to any required degree of accuracy in the neighbourhood of the origin. Weierstrass' theorem extends to continuous functionals. Any continuous functional may be approximated by polynomials, i.e. functional polynomials†. This approximation theorem should suffice in the majority of cases of engineering systems which do not respond critically to certain changes in input (e.g. as a flip-flop would). *From now on functionals will generally be assumed analytic.* It should be borne in mind that according to the above approximation theorem any continuous functional will be arbitrarily near to an analytic functional (in fact a polynomial).

---

† Volterra (1930), page 20. Continuity is defined as follows: $(f(x_1(t')) - f(x_2(t')))$ is to be small whenever $x_1(t') - x_2(t')$ is small (for all times t').

The power series expansion may be shown unique if all the kernels are completely symmetrical in the variables, i.e.

$$k(t_1, t_2, \ldots, t_n) = k(t_2, t_3, \ldots, t_n, t_1), \text{etc.},$$

for any rearrangement of the variables t. For example, for $n=3$ we would require that

$$k(t_1, t_2, t_3) = k(t_3, t_1, t_2) = k(t_2, t_3, t_1) = k(t_3, t_2, t_1)$$
$$= k(t_1, t_3, t_2) = k(t_2, t_1, t_3)$$

identically. Any unsymmetrical kernel may always be changed to a symmetrical kernel, i.e. if $k(t_1, t_2, \ldots, t_n)$ is not symmetrical, there is another kernel $k^*(t_1, t_2, \ldots, t_n)$ such that

$$\int\int \cdots \int k(t_1, t_2, \ldots, t_n) x(t_1) x(t_2) \ldots x(tn)\, dt_1\, dt_2 \ldots dt_n$$
$$= \int\int \cdots \int k^*(t_1, t_2, \ldots, t_n) x(t_1) x(t_2) \ldots x(t_n)\, dt_1\, dt_2 \ldots dt_n$$

identically. It is merely necessary to put

$$k^*(t_1, t_2, \ldots, t_n) = \frac{1}{n} \sum k(t_{i1}, t_{i2}, \ldots, t_{in}),$$

where the sum is over all permutations of the suffices. Thus, for $n=3$ $k^*(t_1, t_2, t_3) = \frac{1}{6}(k(t_1, t_2, t_3) + k(t_3, t_2, t_1) + \ldots)$. Clearly, if k is symmetrical, $k^* = k$.

Return now to the functional relation between the output signal $y(t)$ at any time say, and the input signal $x(t)$:

$$y(t) = f(t;\, x(t'), t' \leqslant t).$$

Expand f in a power series:

$$f(t;\, x(t'), t' \leqslant t) = \sum_n \int\int \cdots \int k(t_1, t_2, \ldots, t_n)\, x(t_1) x(t_2) \ldots x(t_n)\, dt_1\, dt_2 \ldots dt_n,$$

where the summation and integration are over the appropriate ranges.

The symmetric kernels in this expansion are uniquely determined by the functional f. Since the value of the left-hand side depends on t in the time-variant case, so do the kernels. This dependence should be explicit in the notation. Thus denoting $k(t_1, t_2, \ldots, t_n)$ corresponding to time t by $k(t;\, t_1, t_2, \ldots, t_n)$ we have:

$$f(t;\, x(t'), t' \leqslant t) = \sum \int\int \cdots \int k(t;\, t_1, t_2, \ldots, t_n) x(t_1) x(t_2) \ldots x(t_n)\, dt_1\, dt_2 \ldots dt_n.$$

Because f acts only on the past values of x, viz. the values $x(t')$, $t' < t$ at any time t, the kernels will have the property: $k(t:\, t_1, t_2, \ldots, t_n) = 0$ if any of the $t_1, t_2, \ldots, t_n$ is greater than t. The kernel with this property will be said to be of Volterra type.

If f is time-invariant, the kernels depend only on time differences $t-t_1$, $t-t_2$, etc. This may be shown as follows: if

$$y(t)=\sum\int\int\cdots\int k(t;\, t_1, t_2, \ldots, t_n)x(t_1)x(t_2)\ldots x(t_n)\, dt_1\, dt_2 \ldots dt_n, \tag{A}$$

and the functional on the right-hand side is time-invariant, then

$$y(t+\tau)=\sum\int\int\cdots\int k(t;\, t_1, t_2, \ldots, t_n)x\,(t_1+\tau)x(t_2+\tau)\ldots$$
$$x(t_1+\tau)\, dt_1\, dt_2 \ldots dt_n. \tag{B}$$

Since the right-hand side represents the output corresponding to an input $x(t'+\tau)$ at time t'; i.e. to $x(t')$ a time τ earlier. This by time-invariance is $y(t+\tau)$, i.e. $y(t)$ a time τ earlier.

Changing the variables in (A),

$$y(t+\tau)=\sum\int\int\cdots\int k(t+\tau;\, t_1+\tau, \ldots)x(t_1+\tau)\ldots dt_1 \ldots; \tag{C}$$

$\therefore$ by uniqueness of the expansion (symmetric kernels),

$$k(t;\, t_1, t_2, \ldots, t_n)=k(t+\tau;\, t_1+\tau, \ldots t_n+\tau) \text{ identically.}$$

Put $\tau=-t$:

$$k(t;\, t_1, t_2, \ldots, t_n)=k(0, t_1-t, t_2-t, \ldots t_n-t).$$

Hence putting

$$k(t;\, t_1, t_2, \ldots, t_n)=h(t-t_1, t-t_2, \ldots, t-t_n),$$

we have the general expression:

$$y(t)=\sum\int\int\cdots\int h(t-t_1, t-t_2, \ldots, t-t_n)x(t_1)x(t_2)\, x(t_n)\, dt_1\, dt_2 \ldots dt_n$$

in the time-invariant case which, of course, reduces to the ordinary convolution formula

$$y(t)=\int h(t-t_1)x(t_1)\, dt_1$$

for a linear time-invariant f.

Stability and convergence

Suppose that the functional power series

$$\sum\int\int\cdots\int f(t;\, t_1, t_2, \ldots, t_n)x(t_n)x(t_2)\ldots x(t_n)\, dt_1\, dt_2 \ldots dt_n$$

has modulus of convergence r and suppose that $\sum\limits_n a_n r^n$ is convergent, where

$$a_n=\int\int\cdots\int |f(t;\, t_1, t_2, \ldots, t_n)|\, dt_1\, dt_2 \ldots dt_n.$$

Put $\sum\limits_n a_n r^n = A$.

Then if

$$y(t)=\sum_n \underbrace{\int\int\cdots\int}_{n} f(t;\, t_1, t_2, \ldots, t_n)x(t_1)x(t_2)\ldots x(t_n)\, dt_1\, dt_2 \ldots dt_n$$

and if $|x(t)| < r$ for all t, then $|y(t)| < A$ for all t, i.e. the system described by the functional is stable (in the sense of Zadeh) for the regime $|x| < r$.

This connection between the stability of a system under continuous input and the modulus of convergence of the power series expansion of its functional makes the study of the last of some importance.

Analysis with functionals

It has been seen that in certain ways a functional is rather like an ordinary function of real variables. In fact a function of real variables can be considered as a special type of functional. As the literature shows, quite a large number of the properties of functions extend to functionals and in particular it is possible to define derivatives and show the validity of MacLaurin and Taylor expansions in some cases. Further discussion and references are to be found in the works of Volterra, Lévy and Hille.

Determination of functional expansions from differential equations

The relation between input and output of engineering systems is frequently given in the form of a differential equation or more generally, by a differential-difference equation. The connection between this form of relation and the relations of the type considered here will be briefly discussed.

A general differential equation relating input x and output y will be of the form:

$$f(x, \dot{x}, \ldots, x^{(m)}; y, \dot{y}, \ldots, y^{(n)}; t) = 0, \quad \quad (1)$$

where dots denote time-derivation†.

An example of a differential-difference equation would be

$$\dot{y}(t) + ay(t) = bx(t) + c\dot{x}(t)x(t-\tau),$$

where a, b, c and τ are constants.

If we only consider signals $x(t)$ and $y(t)$ which are zero up to a certain time, say t_0 (which will represent the switching-on time of the system), the equation will in general determine the value of y at any instant uniquely in terms of past values of x; i.e. $y(t)$ will be a functional depending only on past values of x. The explicit expression may be given if f is a linear sum of derivatives with constant coefficients: if

$$f = a_0 x + a_1 \dot{x} + \ldots + a_m x^{(m)} + b_0 y + b_1 \dot{y} + \ldots + b_n y^{(n)}$$

then the equation $f = 0$ may be solved by Laplace transformation to give:

$$y(t) = \int h(t - t_1) x(t_1)\, dt_1, \quad \quad (2)$$

where the Laplace transform of $h(t)$ is given by:

$$L_p(h) = \int_{-\infty}^{\infty} h(t) \exp(-pt)\, dt$$

$$= -\frac{a_0 + a_1 p + \ldots + a_m p^m}{b_0 + b_1 p + \ldots + b_n p^n}. \quad \quad (3)$$

† Millar gives some discussion on relations of this form in his paper. Existence theorems for equations of this sort are given in L. M. Graves *Theory of Functions of Real Variables* (McGraw-Hill), 1946, Chap. IX.

The method of Laplace transformation will extend to the treatment of linear differential-difference equations with constant coefficients (Bellman 1954) but not to any case of differential or differential-difference equation which involves t explicitly. However, it may easily be verified that the functional defined by such a differential equation must be linear and so by the general theory of linear functionals, the relation between x and y will be of the form:

$$y(t) = \int_{-\infty}^{\infty} k(t, t_1)x(t_1)\,dt_1,$$

where $k(t; t_1)$ is the output at time t due to a unit impulse input at time t_1. Thus in the linear case, the derivation of the functional expression from the differential or differential-difference equation is, at least in principle, determinable.

For a general function f, it is not at all clear yet when the functional relation between x and y is analytic. It seems that if $x = y = 0$ is a solution of eqn. (1) and that f is sufficiently regular for small values of x, y and their derivatives, then y is a continuous functional of x.

It can actually be shown in certain cases that if f is analytic in all the derivatives $x^{(r)}$, $y^{(s)}$, $r = 0, 1, \ldots, m$; $s = 1, 2, \ldots, n$, then $y(t)$ is an analytic functional of values of $x(t_1)$, $t_1 \leqslant t$ if $\max x(t)$ is sufficiently small and that the power series expansion of this functional may be obtained by:

(*a*) expanding f as a multiple power series in the derivatives $x^{(r)}$, $y^{(s)}$, $r = 0, 1, \ldots, m$; $s = 0, 1, \ldots, n$;

(*b*) writing $D^r x$ for $x^{(r)}$, $r = 0, 1, \ldots, m$; $D^s y$ for $y^{(s)}$, $s = 0, 1, \ldots, n$;

(*c*) formally solving the resulting double power series in x and y for y;

(*d*) interpreting the operators in the coefficients suitably.

Example:

$$\ddot{y} + a\dot{y} + by + ey^3 = x(t) \qquad (4)$$

(Duffing's equation with arbitrary forcing).

Step (*a*) is not necessary as the function f is a simple polynomial expression. Introduce the operator $D = d/dt$ and write the equation as:

$$Ly + ey^3 = x, \qquad (5)$$

where $L = D^2 + aD + b$. Now invert this relation in the neighbourhood of $x = y = 0$, treating L as an ordinary number to obtain a power series expansion:

$$y = c_1 x + c_2 x^2 + \ldots . \qquad (6)$$

The coefficients c may be obtained by substituting this power series for y in the left-hand side of eqn. (5) and equating coefficients of powers of x on left- and right-hand sides.

We get

$$L(c_1x+c_2x^2+\dots)+e(c_1x+c_2x^2+\dots)^3=x\ ;$$

i.e.

$$Lc_1x+Lc_2x^2+\dots+ec_1{}^3x^3+3ec_1{}^2c_2x^4+\dots=x.$$

$$\therefore\ Lc_1=1,$$

$$Lc_2=0,$$

$$Lc_3+ec_1{}^3=0.$$

$$\dots\dots\dots\dots$$

$$\therefore\ c_2=c_4=c_6=\dots=0 \quad\text{and}\quad c_1=L^{-1}, c_3=-L^{-1}eL^{-3},\dots .$$

The meaning of L^{-1} is fairly clear. For the equation $Ly=x$ has solution $y=L^{-1}x$.

Comparing this with the solution

$$y(t)=\int_{-\infty}^{\infty} h(t-t_1)x(t_1)\,dt_1$$

of $\ddot{y}+a\dot{y}+by=x$, is it seen that $L^{-1}x$ is to be interpreted as the linear functional operation defined by

$$\int_{-\infty}^{\infty} h(t-t_1)x(t_1)\,dt_1.$$

Here of course $h(t)$ is given by its Laplace transform :

$$L_p(h)=\frac{1}{p^2+ap+b}\,.$$

The first two non-zero terms of the solution are thus :

$$y(t)=\int_{-\infty}^{\infty} h(t-t_1)x(t_1)\,dt_1-e\int_{-\infty}^{\infty} h(t-t_1)\left(\int_{-\infty}^{\infty} h(t_1-t_2)x(t_2)\,dt_2\right)^3dt_1,$$

corresponding to $y=L^{-1}x-eL^{-1}(L^{-1}x)^3$. This may be re-written :

$$y(t)=\int_{-\infty}^{\infty} h(t-t_1)x({}_1t)\,dt_1-e\int_{-\infty}^{\infty} dt_1\iiint_{-\infty}^{\infty} h(t-t_1)h(t_1-t_2)$$

$$\times\, h(t_1-t_3)h(t_1-t_4)x(t_2)x(t_3)x(t_4)\,dt_2\,dt_3\,dt_4$$

$$=\int_{-\infty}^{\infty} h(t-t_1)x(t_1)\,dt_1-e\iiint_{-\infty}^{\infty} h(t-t_1,t-t_2,t-t_3)x(t_1)x(t_2)x(t_3)\,dt_1\,dt_2\,dt_3,$$

On writing

$$h(t-t_2,t-t_3,t-t_4)=\int_{-\infty}^{\infty} h(t-t_1)h(t-t_2)h(t_1-t_3)h(t_1-t_4)\,dt$$

and re-naming the variables t_2, t_3, t_4 as t_1, t_2, t_3.

This series expansion may be obtained rather more conveniently in this case by writing the equation :

$$Ly+ey^3=x \quad\text{as}\quad y+eL^{-1}y^3=L^{-1}x=\xi,$$

say, and then inverting the relation

$$y+eL^{-1}y^3=\xi,$$

for instance, by the method of successive approximations if e is small. In the case where $b>0$, $a^2>4b$ (linear part over-damped) the functional power series may quite easily be shown convergent (i) for all $x(t)$ when

$e \geqslant 0$, (ii) for all $x(t)$ satisfying

$$\max_{-\infty < t < \infty} |x(t)| < \frac{\sqrt{2b}}{3} \sqrt{\frac{b}{3|e|}}$$

when $e \leqslant 0$.

Examples of the Use of Functional Expansions

1. *The static characteristic of a device*

Suppose that a device is described by the analytical functional relation:

$$y(t) = \sum_{n=0}^{\infty} \int\int \ldots \int k_n(t;\, t_1, t_2, \ldots, t_n) x(t_1) x(t_2) \ldots x(t_n)\, dt_1\, dt_2 \ldots dt_n$$

between input ($x(t)$ at time t) and output ($y(t)$ at time t). The 'static characteristic' relating x and y is obtained by fixing the input $x(t)$ at a constant value say X, observing the resulting constant value of $y(t)$, say Y, and plotting the relation between X and Y. Put

$$\left.\begin{aligned} x(t) &= X \\ y(t) &= Y \end{aligned}\right\} \text{for all time } t.$$

Then

$$\begin{aligned} Y &= \sum_n \int\int \ldots \int k_n(t;\, t_1, t_2, \ldots, t_n) X . X \ldots X\, dt_1\, dt_2 \ldots dt_n \\ &= \sum_n X^n \int\int \ldots \int k_n(t;\, t_1, t_2, \ldots, t_n)\, dt_1\, dt_2 \ldots dt_n \\ &= \sum_n a_n X^n, \end{aligned}$$

where

$$a_n = \int\int \ldots \int k_n(t;\, t_1, t_2, \ldots, t_n)\, dt_1\, dt_2 \ldots dt_n.$$

Thus the characteristic is given in power series form. If the device is time-invariant, it is easily verified by the change of variables $t - t_i = \tau_i$; $i = 1, 2, \ldots, n$, that a_n is independent of time.

2. *Response to a sinusoidal input*

Put $x(t) = A \cos \omega t$.
Then

$$\begin{aligned} y(t) &= \sum_{n=0}^{\infty} \int\int \ldots \int k_n(t;\, t_1, t_2, \ldots, t_n) x(t_1) x(t_2) \ldots x(t_n)\, dt_1\, dt_2 \ldots dt_n \\ &= \sum \int\int \ldots \int k_n(t;\, t_1, t_2, \ldots, t_n) \cos \omega t_1 \cos \omega t_2 \ldots \cos \omega t_n\, dt_1\, dt_2 \ldots dt_n. \end{aligned}$$

The integral

$$\int\int \ldots \int k_n(t;\, t_1, t_2, \ldots, t_n) \cos \omega t_1 \cos \omega t_2 \ldots \cos \omega t_n\, dt_1\, dt_2 \ldots dt_n$$

may be evaluated by putting

$$\cos \omega t_r = \tfrac{1}{2}[\exp(i\omega t_r) + \exp(-i\omega t_r)],\ r = 1, 2, \ldots n,$$

and introducing the n-dimensional Fourier transform (or Laplace transform if k_n is of Volterra type).

Consider, e.g., the case $n=3$ and when the kernel is a function only of time differences.

Suppose

$$k_3(t\,;\,t_1,t_2,t_3)=h(t-t_1,t-t_2,t-t_3).$$

The 3-dimensional Fourier transform of h, $H(\omega_1,\omega_2,\omega_3)$ is defined by

$$H(\omega_1,\omega_2,\omega_3)=\iiint_{-\infty}^{\infty} h(\tau_1,\tau_2,\tau_3)\exp\left[-i(\omega_1\tau_1+\omega_2\tau_2+\omega_3\tau_3)\right]d\tau_1\,d\tau_2\,d\tau_3,$$

so that

$$\iiint k_3(t\,;\,t_1,t_2,t_3)\cos\omega t_1\cos\omega t_2\cos\omega t_3\,dt_1\,dt_2\,dt_3$$

$$=\frac{1}{2^3}\iiint h(t-t_1,t-t_2,t-t_3)[\exp(i\omega t_1)+\exp(-i\omega t_1)][\exp(i\omega t_2)$$

$$+\exp(-i\omega t_2)]\,[\exp(i\omega t_3)+\exp(-i\omega t_3)]\,dt_1\,dt_2\,dt_3$$

$$=\frac{1}{2^3}\iiint h(t-t_1,t-t_2,t-t_3)[\textstyle\sum\exp(\pm i\omega t_1\pm i\omega t_2\pm i\omega t_3)]\,dt_1\,dt_2\,dt_3,$$

where the summation is over all sign combinations:

$$=\frac{1}{2^3}\,[\exp(-3i\omega t)H(\omega,\omega,\omega)+\exp(-i\omega t)H(-\omega,\omega,\omega)$$

+ 6 more terms corresponding to all possible sign changes].

These eight terms may be grouped in four pairs corresponding to conjugate complex qualities and then can be replaced by sines and cosines of ωt and harmonics; e.g.

$$\tfrac{1}{2}[\exp(-i\omega t)H(-\omega,\omega,\omega)+\exp(i\omega t)H(\omega,-\omega,-\omega)$$

$$=\mathscr{R}H(\omega,-\omega,-\omega)\cos\omega t-\mathscr{I}H(\omega,-\omega,-\omega)\sin\omega t,$$

where $\mathscr{R}$ and $\mathscr{I}$ denote real and imaginary parts.

3. *Response to a random input*

Using an analytical function expansion it is possible to find the value of any output statistic in terms of the input statistics. This method has been used by Wiener (1930, 1942, 1949) and Ikehara (1951) to find the response of a non-linear device to noise.

For example, consider the auto-covariance of the output of a time-invariant non-linear device described by an odd analytical functional. Let

$$y(t)=\int k_1(t-t_1)x(t_1)\,dt_1+\iiint k_3(t-t_1,t-t_2,t-t_3)x(t_1)x(t_2)x(t_3)\,dt_1\,dt_2\,dt_3$$

$$+\dots\qquad\qquad (A)$$

Then, after a few changes of variable under the integral signs, we get:

$$y(t)y(t+\tau)=\iint k_1(t-t_1)k_1(t+\tau-t_2)x(t_1)x(t_2)\,dt_1\,dt_2$$

$$+\iiiint k_1(t-t_1)k_3(t+\tau-t_2,t+\tau-t_3,t+\tau-t_4)x(t_1)x(t_2)x(t_3)x(t_4)\,dt_1\,dt_2\,dt_3\,dt_4$$

$$+\iiiint k_1(t+\tau-t_1)k_3(t-t_2,t-t_3,t-t_4)x(t_1)x(t_2)x(t_3)x(t_4)\,dt_1\,dt_2\,dt_3\,dt_4$$

$$+\dots\qquad\qquad (B)$$

The required relation follows by taking statistical averages of both sides:

$$\overline{y(t)y(t+\tau)} = \int\int k_1(t-t_1)k_1(t+\tau-t_2)\overline{x(t_1)x(t_2)}\,dt_1\,dt_2$$
$$+\int\int\int\int k_1(t-t_1)k_3(t+\tau-t_2, t+\tau-t_3, t+\tau-t_4)\overline{x(t_1)x(t_2)x(t_3)x(t_4)}\,dt_1\,dt_2\,dt_3\,dt_4$$
$$+\int\int\int\int k_1(t+\tau-t_1)k_3(t-t_2, t-t_3, t-t_4),\overline{x(t_1)x(t_2)x(t_3)x(t_4)}\,dt_1\,dt_2\,dt_3\,dt_4$$
$$+\ldots \qquad \text{(C)}$$

If the input is a stationary stochastic process, moments such as $\overline{x(t_1)x(t_2)x(t_3)x(t_4)}$ will depend only on time differences t_1-t_2, t_1-t_3, etc. and the integrals may be re-written by using a Fourier transformation in a way which may be more useful for practical calculation, as follows:

By the n-dimensional form of Parseval's theorem (Bochner and Chandrasekharan 1949):

$$\int\int\ldots\int f_1(t_1, t_2, \ldots, t_n)f_2(t_1, t_2, \ldots, t_n)\,dt_1\,dt_2\ldots dt_n$$
$$= \frac{1}{(2\pi)^n}\int\int\ldots\int F_1(\omega_1, \omega_2, \ldots, \omega_n)F_2(\omega_1, \omega_2, \ldots, \omega_n)\,d\omega_1\,d\omega_2\ldots d\omega_n, \qquad \text{(D)}$$

where

$$\begin{cases} F_j(\omega_1, \omega_2, \ldots, \omega_n) = \int\int\ldots\int f_j(t_1, t_2, \ldots, t_n)\exp\left(-i\sum_{r=1}^{n}\omega_r t_r\right)dt_1\,dt_2\ldots dt_n, \\ f_j(t_1, t_2, \ldots, t_n) = \frac{1}{(2\pi)^n}\int\int\ldots\int F_j(\omega_1, \omega_2, \ldots, \omega_n)\exp\left(i\sum_{r=1}^{n}\omega_r t_r\right)d\omega_1\,d\omega_2\ldots d\omega_n, \end{cases}$$
$$j = 1,2$$

are inverse Fourier transforms.

In the first term on the right-hand side, of eqn. (C), $n=2$,

$\therefore$ put $f_1(t_1, t_2) = k_1(t-t_1)k_1(t+\tau-t_2)$, $f_2(t_1, t_2) = \overline{x(t_1)x(t_2)}$

in the Parseval formula:

$$F_1(\omega_1, \omega_2) = \int\int f_1(t_1, t_2)\exp\left[-i(\omega_1 t_1+\omega_2 t_2)\right]dt_1\,dt_2$$
$$= \int\int k_1(t-t_1)k_1(t+\tau-t_2)\exp\left[-i(\omega_1 t_1+\omega_2 t_2)\right]dt_1\,dt_2$$
$$= \left(\int k_1(t-t_1)\exp(-i\omega_1 t_1)\,dt_1\right)\left(\int k_2(t+\tau-t_2)\exp(-i\omega_2 t_2)\,dt_2\right)$$
$$= \left(K_1^*(\omega_1)\exp(-i\omega_1 t)\right)\left(K_1^*(\omega_2)\exp\left[-i\omega_2(t+\tau)\right)\right)$$
$$= \quad K_1^*(\omega_1)K_1^*(\omega_2)\exp\left[-i(\omega_1+\omega_2)t-i\omega_2\tau\right].$$

Here $K_1(\omega) = \int k_1(t)\exp(-i\omega t)\,dt$, the Fourier transform of $k_1(t)$ and the star denotes complex conjugate:

$$F_2(\omega_1, \omega_2) = \int\int \overline{x(t_1)x(t_2)} \exp[-i(\omega_1 t_1 + \omega_2 t_2)]\, dt_1\, dt_2$$
$$= \int\int \phi(\tau) \exp[-i(\omega_1 + \omega_2)t_2 - i\tau\omega_2)]\, d\tau\, dt_2,$$

where $\tau = t_1 - t_2$ and $\phi(\tau) = \overline{x(t)x(t+\tau)}$ is the auto-covariance function. It is dependent only on τ if $x(t)$ is a stationary process.

Now $\int_{-\infty}^{\infty} \exp(-i\omega t)\, dt = 2\pi\delta(\omega)$, where $\delta(t)$ is the Dirac δ-function.

$$\therefore\ F_2(\omega_1, \omega_2) = \left(\int \phi(\tau) \exp(-i\tau\omega_2)\, d\tau\right)\left(\int \exp[-i(\omega_1 + \omega_2)t_2)]\, dt_2\right)$$
$$= \Phi(\omega_2) 2\pi\delta(\omega_1 + \omega_2),$$

where $\Phi(\omega)$ is the spectral density function since $\Phi(\omega) = \int \phi(\tau) \exp(-i\tau\omega)\, d\tau$ by the Wiener Khinchin theorem.

$$\therefore \int\int f_1(t_1, t_2) f_2(t_1, t_2)\, dt_1\, dt_2 = \frac{1}{(2\pi)^2} \int\int K_1^*(\omega_1) K_1^*(\omega_2)$$
$$\times \exp[-i(\omega_1 + \omega_2)t - i\tau\omega_2] 2\pi\Phi(\omega_2)\, \delta(\omega_1 + \omega_2)\, d\omega_1\, d\omega_2$$
$$= \int K_1^*(-\omega_2) K_1^*(\omega_2) \exp(-i\tau\omega_2)\Phi(\omega_2)\, d\omega_2$$
$$= \int |K_1(\omega)|^2 \Phi(\omega) \exp(-i\tau\omega)\, d\omega \ldots \quad \text{(E)}$$

on changing from ω_2 to ω, because $K_1(-\omega) = K_1^*(\omega)$. This formula is well known from linear theory: it is the only term on the right-hand side of (C) if the functional is linear.

Consider the second term on the right-hand side of (C) which will only occur when the functional is non-linear.

In the Parseval formula, put $n = 4$ and

$$f_1(t_1, t_2, t_3, t_4) = k_1(t - t_1) k_3(t + \tau - t_2, t + \tau - t_3, t + \tau - t_4),$$
$$f_2(t_1, t_2, t_3, t_4) = \overline{x(t_1)x(t_2)x(t_3)x(t_4)},$$

$$F_1(\omega_1, \omega_2, \omega_3, \omega_4) = \int\int\int\int k_1(t - t_1) k_3(t + \tau - t_2, t + \tau - t_3, t + \tau - t_4)$$
$$\exp[-i(\omega_1 t_1 + \omega_2 t_2 + \omega_3 t_3 + \omega_4 t_4)]\, dt_1\, dt_2\, dt_3\, dt_4$$
$$= \left(\int k_1(t - t_1) \exp(-i\omega_1 t_1)\, dt_1\right)\left(\int\int\int k_3(t + \tau - t_2, t + \tau - t_3,\right.$$
$$\left. t + \tau - t_4) \exp[-i(\omega_2 t_2 + \omega_3 t_3 + \omega_4 t_4)]\, dt_2\, dt_3\, dt_4\right.$$
$$= (K_1^*(\omega_1) \exp(-i\omega_1 t))(K_3^*(\omega_2, \omega_3, \omega_4) \exp[-i(t + \tau)$$
$$\left.(\omega_2 + \omega_3 + \omega_4)\right)$$
$$= K_1^*(\omega_1) K_3^*(\omega_2, \omega_3, \omega_4) \exp[-it(\omega_1 + \omega_2 + \omega_3 + \omega_4)]$$
$$\exp[-i\tau(\omega_2 + \omega_3 + \omega_4)], \quad \text{(F)}$$

where

$$K_3(\omega_1, \omega_2, \omega_3) = \iiint k_3(t_1, t_2, t_3) \exp\left[-i(\omega_1 t_1 + \omega_2 t_2 + \omega_3 t_3)\right] dt_1\, dt_2\, dt_3$$

is the Fourier transform of $k_3(t_1, t_2, t_3)$:

$$F_2(\omega_1, \omega_2, \omega_3, \omega_4) = \iiiint \overline{x(t_1)x(t_2)x(t_3)x(t_4)} \exp\left[-i(\omega_1 t_1 + \ldots + \omega_4 t_4)\right] dt_1\, dt_2\, dt_3\, dt_4$$

$$= \iiiint \overline{x(t_1)x(t_1+\tau_2)x(t_1+\tau_3)x(t_1+\tau_4)} \exp\left[-i(\omega_1+\omega_2+\omega_3+\omega_4)t\right] \exp\left[-i(\omega_2\tau_2+\omega_3\tau_3+\omega_4\tau_4)\right] d\tau_1\, d\tau_2\, d\tau_3\, d\tau_4$$

(on putting $\tau_j = t_j - t_1,\ j = 2, 3, 4.$)

$$= \int \exp\left[-i(\omega_1+\omega_2+\omega_3+\omega_4)t_1\right] dt_1 \iiint \overline{x(t_1)x(t_1+\tau_1)x(t_1+\tau_2) x(t_1+\tau_3)} \exp\left[-i(\omega_2\tau_2+\omega_3\tau_3+\omega_4\tau_4)\right] d\tau_2\, d\tau_3\, d\tau_4$$

$$= 2\pi\delta(\omega_1+\omega_2+\omega_3+\omega_4)\Phi(\omega_2, \omega_3, \omega_4),$$

where

$$\Phi(\omega_2, \omega_3, \omega_4) = \iiint \overline{x(t_1)x(t_1+\tau_2)x(t_1+\tau_3)x(t_1+\tau_4)} \exp\left[-i(\omega_2\tau_2+\omega_3\tau_3+\omega_4\tau_4)\right] d\tau_2\, d\tau_3\, d\tau_4 \qquad . \; . \; . \quad \text{(G)}$$

is a generalized spectral density function.
∴ by Parseval's formula and eqns. (F) and (G):

$$\iiiint k_1(t-t_1)k_3(t+\tau-t_2, t+\tau-t_3, t+\tau-t_4)\, \overline{x(t_1)x(t_2)x(t_3)x(t_4)}\, dt_1\, dt_2\, dt_3\, dt_4$$

$$= \left(\frac{1}{(2\pi)^4}\right) \iiiint K_1{}^*(\omega_1)K_3{}^*(\omega_2, \omega_3, \omega_4) \exp\left[-it(\omega_1+\omega_2+\omega_3+\omega_4)\right] \times \exp\left[-i\tau(\omega_2+\omega_3+\omega_4)\right] . \, 2\pi\delta(\omega_1+\omega_2+\omega_3+\omega_4) \times \Phi(\omega_2, \omega_3, \omega_4)\, d\omega_1\, d\omega_2\, d\omega_3\, d\omega_4$$

$$= \left(\frac{1}{(2\pi)^3}\right) \iiint K_1{}^*(\omega_1)K_3{}^*(\omega_2, \omega_3, \omega_4)\Phi(\omega_2, \omega_3, \omega_4) \exp\left[i\tau(\omega_2+\omega_3+\omega_4)\right] d\omega_1\, d\omega_2\, d\omega_3\, d\omega_4 \quad \omega_1+\omega_2+\omega_3+\omega_4 = 0.$$

This is thus the second term of the expansion (C). Higher terms may be transformed in the same way. Note that the third term is obtained from the second by changing τ to $-\tau$.

3.1. *The Effect of a Non-linear Element on a Signal Disturbed by Noise*

As a further application of functional expansions to statistical calculations, the interaction terms between signal and noise due to passage through a non-linear device will be found.

If the input to a linear element consists of a signal perturbed by noise, it follows from the principle of superposition, that the output signal will be the sum of filtered signal and filtered noise. However, in the case where the element is not linear, this will not be the case and further terms will arise due to interaction between signal and noise.

Consider first the linear case. Let $x(t)$ be the input at time t which is

$$x(t)=s(t)+n(t), \qquad (1)$$

the sum of a signal component $s(t)$ and noise component, $n(t)$. Suppose that the input is related to the output $y(t)$ at time t by the equation

$$y(t)=\int k(t,t_1)x(t_1)\,dt_1. \qquad (2)$$

Then by substitution, it is seen that:

$$\left.\begin{aligned} y(t)&=s'(t)+n'(t),\\ s'(t)&=\int k(t,t_1)s(t_1)\,dt_1 \end{aligned}\right\} \qquad (3)$$

where

$$n'(t)=\int k(t,t_1)n(t_1)\,dt_1, \qquad (4)$$

which may be called filtered signal and filtered noise respectively.

The mean square deviation of the output from its true value due to the input disturbance $n(t)$ is:

$$\begin{aligned}\overline{[n'(t)]^2}&=\overline{\left[\int k(t,t_1)n(t_1)\,dt_1\right]^2}\\ &=\overline{\left[\int\int k(t,t_1)k(t_1,t_2)n(t_1)n(t_2)\,dt_1\,dt_2\right]}\\ &=\int\int k(t,t_1)k(t,t_2)\overline{n(t_1)n(t_2)}\,dt_1\,dt_2.\end{aligned}$$

It is seen to depend only on the value of the input disturbance. This will not hold in the non-linear case.

In the non-linear case the output will be assumed to be related to the input by a functional equation:

$$y(t)=f(t;x(t'),t'\leqslant t),$$

which, if analytic, may be expanded in a convergent power series. As the general case would lead to tedious computation, the special case will be considered where the functional is odd and almost linear so that it may be adequately represented by the linear and cubic terms of the expansion, viz.

$$y(t)=\int k_1(t,t_1)x(t_1)\,dt_1+\int\int\int k_3(t;t_1,t_2,t_3)x(t_1)x(t_2)x(t_3)\,dt_1\,dt_2\,dt_3.$$

Suppose, as before, that

$$x(t) = s(t) + n(t).$$

Then

$$y(t) = \int k_1(t, t_1)s(t_1)\,dt_1 + \int k_1(t, t_1)n(t_1)\,dt_1$$

linear part filtered signal + filtered noise

$$+ \iiint k_3(t; t_1, t_2, t_3)s(t_1)s(t_2)s(t_3)\,dt_1\,dt_2\,dt_3$$

non-linear part of filtered signal

$$+3\iiint k_3(t; t_1, t_2, t_3)s(t_1)s(t_2)n(t_3)\,dt_1\,dt_2\,dt_3$$

interaction terms

$$+3\iiint k_3(t; t_1, t_2, t_3)s(t_1)n(t_2)n(t_3)\,dt_1\,dt_2\,dt_3$$

$$+ \iiint k_3(t; t_1, t_2, t_3,)n(t_1)n(t_2)n(t_3)\,dt_1\,dt_2\,dt_3$$

non-linear part of filtered noise.

The output is seen to include a number of interaction terms due to the simultaneous presence of signal and noise. These remain, even if signal and noise are statistically independent.

The mean square deviation of the output from the value it would have in the absence of noise may be calculated as before. It is:

$$\overline{\left[y(t) - \int k_1(t, t_1)s(t_1)\,dt_1 - \iiint k_3(t; t_1, t_2, t_3)s(t_1)s(t_2)s(t_3)\,dt_1 dt_2\,dt_3\right]^2}$$

$$= \overline{\left[\int k_1(t, t_1)n(t_1)\,dt_3 + 3\iiint k_3(t; t_1, t_2, t_3)s(t_1)s(t_2)n(t_3)\,dt_1\,dt_2\,dt_3 + \ldots\right]^2},$$

and is seen to depend on the value of s as well as n. The right-hand side on expansion gives 10 terms involving multi-dimensional integrals of the higher cross-moments of s and n; e.g. one term is:

$$6\iiiint k_1(t, t_1)k_3(t; t_2, t_3, t_4)\overline{n(t_1)s(t_2)s(t_3)n(t_4)}\,dt_1\,dt_2\,dt_3\,dt_4.$$

To evaluate these integrals, the values of the kernels and the joint statistics of signal and noise need to be known.

3.2. *Chain Product of Functionals*

Since most signal transmission systems consist of sequences of devices in series, it is of interest to know the rule of combination of the corresponding functionals.

With the notation of the diagram, the basic relations are:

$$y(t) = f(t; x(t'), t' \leqslant t), \quad \ldots \ldots \quad \text{(A)}$$

$$z(t) = g(t; y(t'), t' \leqslant t). \quad \ldots \ldots \quad \text{(B)}$$

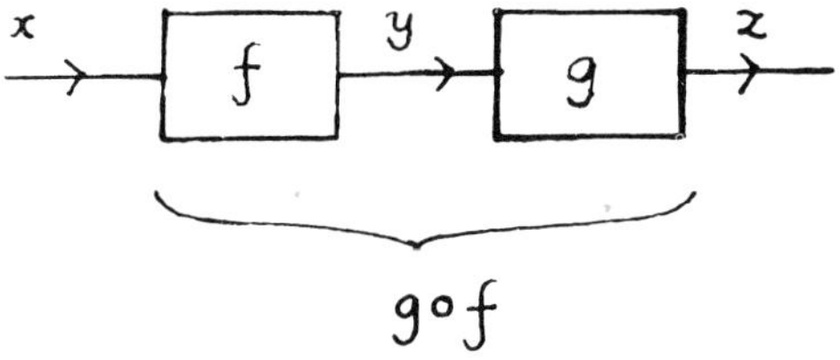

These two relations determine a relation of the same type between z and x, say:

$$z(t)=h(t\,;x(t'),\,t'\leqslant t)\,. \qquad \text{(C)}$$

The functional h will be called the chain product of f and g written $g{\circ}f$.

With three functionals f_1, f_2 and f_3, it is clear by definition that $f_3{\circ}(f_2{\circ}f_1)=(f_3{\circ}f_2){\circ}f_1$ and so these expressions may be written $f_3{\circ}f_2{\circ}f_1$ without ambiguity.

It is not true in general that $g{\circ}f=f{\circ}g$. If this is so, the functionals will be said to commute. The physical meaning of commuting functionals may be been in the case when the signals corresponding to the variables x and y are of the same nature, e.g. both voltages of roughly the same strength. If f and g commute then it does not matter whether the operation corresponding to f or that corresponding to g is applied first to the signal.

3.2.1. *f and g linear*

Denote the kernels of the functionals by $f(t,t_1)$ and $g(t,t_1)$ respectively. Then eqns. (A) and (B) are:

$$y(t)=\int f(t,t_1)x(t_1)\,dt_1, \qquad \text{(A')}$$

$$z(t)=\int g(t,t_1)y(t_1)\,dt_1. \qquad \text{(B')}$$

The result of eliminating y is a linear relation of the same form which may be obtained by straightforward substitution:

$$\begin{aligned} z(t) &= \int g(t,t_1)y(t_1)\,dt_1 \\ &= \int g(t,t_1)\left(\int f(t_1,t_2)x(t_2)\,dt_2\right)dt_1 \\ &= \int\int g(t,t_1)f(t_1,t_2)x(t_2)\,dt_1\,dt_2 \\ &= \int\left(\int g(t,t_1)f(t_1,t_2)\,dt_1\right)x(t_2)\,dt_2 \\ &= \int h(t,t_2)x(t_2)\,dt_2, \end{aligned}$$

where $h(t,t_2)=\int g(t,t_1)f(t_1,t_2)\,dt_1$. This is the kernel of $f{\circ}g$. The kernel of $f{\circ}g$ will be $\int f(t,t_1)g(t_1,t_2)\,dt_1$. These are not in general the same and so the functionals will not, in general, commute (Volterra 1930). However if f and g are time-invariant, the functionals will always commute. This is seen immediately, since in this case the Fourier transforms are related by a simple algebraic identity. Alternatively, it may be seen by a straightforward change of variable: f and g depend only on time differences in the time-invariant case. Write $f(t, t_1) = f_1(t - t_1)$. $g(t, t_1) = g_1(t - t_1)$. Then from above:

$$h(t, t_2) = \int g(t, t_1) f(t_1, t_2)\, dt_1$$

$$= \int g_1(t - t_1) f_1(t_1 - t_2)\, dt_1$$

$$= \int g_1(t_1' - t_2) f_1(t - t_1')\, dt_1'$$

(after the change of variable $t_1' = t - t_1 + t_2$)

$$= \int f(t, t_1') g(t_1,' t_2)\, dt_1'$$

as required.

3.2.2. *f and g analytic*

Let the expansions of f and g be:

$$y(t) = \sum_{n=0}^{\infty} \int\int \cdots \int f(t; t_1, t_2, \ldots, t_n) x(t_1) x(t_2) \ldots x(t_n)\, dt_1\, dt_2 \ldots dt_n \quad \text{(A}''\text{)}$$

$$z(t) = \sum_{n=0}^{\infty} \int\int \cdots \int g(t; t_1, t_2, \ldots, t_n) y(t_1) y(t_2) \ldots y(t_n)\, dt_1\, dt_2 \ldots dt_n, \quad \text{(B}''\text{)}$$

where the letters f and g have been used also for the kernels corresponding to the functionals f and g. The relation between z and x may be obtained by direct substitution of the values of y from the first equation in the second equation. As in the corresponding case with ordinary power-series (Goursat 1904) it is seen that if both series have non-zero moduli of convergence the resultant has a non-zero modulus of convergence. The result tends to be algebraically complicated. To reduce the length of the equations, an abbreviated notation will be introduced.

Equations (A″) and (B″) will be written, in this abbreviated notation:

$$\left.\begin{aligned} y_t &= \sum f_{t;\, t_1 t_2 \ldots t_n} x_{t_1} x_{t_2} \ldots x_{t_n}, \\ z_t &= \sum g_{t;\, t_1, t_2, \ldots, t_n} y_{t_1} y_{t_2} \ldots y_{t_n}. \end{aligned}\right\} \quad \ldots\ldots \quad \text{(C}''\text{)}$$

The convention is that functional dependence is indicated by the suffices and any suffix which occurs just twice in the same term is to be integrated over the appropriate range of time†. Note that any repeated suffix in the same term may be replaced by any other pair of repeated suffices which are not already present. For example,

$$g_{t;\, t_1, t_2,} x_{t_1} x_{t_2} = g_{t;\, \alpha,\, t_2} x_\alpha x_{t_2},$$

† This is a generalization of the summation convention of the tensor calculus (the quantities are actually tensors).

both being contracted forms of the integral:

$$\iint g(t;\, t_1, t_2)x(t_1)x(t_2)\, dt_1\, dt_2.$$

Changing the variable t_1 to α merely means that α is used as a variable of integration instead of t_1.

If attention is restricted to the case where $f_t = g_t = 0$ (this means that for both devices, zero input implies zero output), the first few terms of the expansion are given by the equations:

$$z_t = h_{t\alpha}x_\alpha + h_{t\alpha_1\alpha_2}x_{\alpha_1}x_{\alpha_2} + \ldots,$$

where

$$\left.\begin{aligned} h_{t\alpha} &= g_{t\beta}f_{\beta\alpha}, \\ h_{t\alpha_1\alpha_2} &= g_{t\beta_1\beta_2}f_{\beta_1\alpha_1}f_{\beta_2\alpha_2} + g_{t\beta}f_{\beta\alpha_1\alpha_2}, \\ h_{t\alpha_1\alpha_2\alpha_3} &= g_{t\beta_1\beta_2\beta_3}f_{\beta_2\alpha_1}f_{\beta_2\alpha_2}f_{\beta_3\alpha_3} + g_{t\beta_1\beta_2}f_{\beta_1\alpha_1}f_{\beta_2\alpha_2\alpha_3} + g_{t\beta_1\beta_2}f_{\beta_1\alpha_1\alpha_2}f_{\beta_2\alpha_3} + g_{t\beta}f_{\beta\alpha_1\alpha_2\alpha_3}. \end{aligned}\right\} \quad \text{(D)}$$

Greek letters are used here to avoid too many suffices on the t's. As remarked above, it does not matter what symbols are used. Thus the third kernel is, in full notation:

$$\begin{aligned} h(t;\, \alpha_1, \alpha_2, \alpha_3) = &\iiint g(t;\, \beta_1, \beta_2, \beta_3)f(\beta_1, \alpha_1)f(\beta_2, \alpha_2)f(\beta_3, \alpha_3)\, d\beta_1\, d\beta_2\, d\beta_3 \\ &+ \iint g(t;\, \beta_1, \beta_2)f(\beta_1, \alpha_1)f(\beta_2;\, \alpha_2, \alpha_3)\, d\beta_1\, d\beta_2 \\ &+ \iint g(t;\, \beta_1, \beta_2)f(\beta_1;\, \alpha_1, \alpha_2)f(\beta_2, \alpha_3)\, d\beta_1\, d\beta_2 \\ &+ \int g(t, \beta)f(\beta;\, \alpha_1, \alpha_2, \alpha_3)\, d\beta. \qquad \ldots\ldots \quad \text{(E)} \end{aligned}$$

Time-invariant case: Use of Fourier transform:

In the time-invariant case, all these integrals will take the form of generalized convolutions, the linear parts of the expressions naturally corresponding to the result found above for the linear case. By taking Fourier transforms of both sides of eqns. (D) it is possible to arrive at algebraic relations between the Fourier transforms of the kernels.

Thus suppose:

$$f(t, t_1) = f_1(t - t_1), \qquad g(t, t_1) = g_1(t - t_1), \qquad h(t, t_1) = h_1(t - t_1),$$

$$f(t;\, t_1, t_2) = f_1(t - t_1, t - t_2), \qquad \ldots, \qquad \ldots,$$

$$\ldots.$$

Without loss of generality, the kernels of f and g may be assumed symmetrical so that, e.g. $f(t;\, t_1, t_2) = f(t;\, t_2, t_1)$.

Corresponding to eqn. (E) is the equation:

$$
\begin{aligned}
h_1(t-\alpha_1, t-\alpha_2, t-\alpha_3) = & \iiint g_1(t-\beta_1, t-\beta_2, t-\beta_3) f_1(\beta_1-\alpha_1) f_1(\beta_2-\alpha_2) \\
& \quad \times f_1(\beta_3-\alpha_3)\, d\beta_1\, d\beta_2\, d\beta_3 \\
& + \iint g_1(t-\beta_1, t-\beta_2) f_1(\beta_1-\alpha_1) f_1(\beta_2-\alpha_2, \beta_2-\alpha_3) \\
& \quad d\beta_1\, d\beta_2 \\
& + \iint g_1(t-\beta_1, t-\beta_2) f_1(\beta_1-\alpha_1, \beta_1-\alpha_3) \\
& \quad f_1(\beta_2-\alpha_3)\, d\beta_1\, d\beta_2 \\
& + \int g_1(t-\beta) f_1(\beta-\alpha_1, \beta-\alpha_2, \beta-\alpha_3)\, d\beta. \qquad \text{(E}'\text{)}
\end{aligned}
$$

Taking the Fourier transform of both sides, we have:

$$
\begin{aligned}
H(\omega_1, \omega_2, \omega_3) = & \, G(\omega_1, \omega_2, \omega_3) F(\omega_1) F(\omega_2) F(\omega_3) \\
& + G(\omega_1, \omega_2+\omega_3) F(\omega_1) F(\omega_2, \omega_3) \\
& + G(\omega_1+\omega_2, \omega_3) F(\omega_1, \omega_2) F(\omega_3) \\
& + G(\omega_1+\omega_2+\omega_3) F(\omega_1, \omega_2, \omega_3), \qquad \text{(F)}
\end{aligned}
$$

where $$F(\omega) = \int f_1(\tau) \exp(-i\omega\tau)\, d\tau,$$

$$F(\omega_1, \omega_2) = \iint f_1(\tau_1, \tau_2) \exp[-i(\omega_1\tau_1 + \omega_2\tau_2)]\, d\tau_1\, d\tau_2,$$

etc. (with the same notation for g's and h's) are the Fourier transforms of the various kernels.

It is clear that, since the kernels corresponding to f and g are symmetrical in their variables, so are the corresponding Fourier transforms F and G. Since the kernels of h are not, in general, symmetrical, the Fourier transforms may not be expected to be symmetrical.

Thus the right-hand side of (F) appears unsymmetrical. By linearity, it is seen that the Fourier transform of the symmetrical kernel corresponding to $h_1(t-t_1, t-t_2, t-t_3)$ is:

$$
\begin{aligned}
H(\omega_1, \omega_2, \omega_3) = & \, G(\omega_1, \omega_2, \omega_3) F(\omega_1) F(\omega_2) F(\omega_3) \\
& + \tfrac{2}{3} \begin{bmatrix} G(\omega_1, \omega_2+\omega_3) F(\omega_1) F(\omega_2, \omega_3) \\ + G(\omega_2, \omega_3+\omega_1) F(\omega_2) F(\omega_3, \omega_1) \\ + G(\omega_3, \omega_1+\omega_2) F(\omega_3) F(\omega_1, \omega_2) \end{bmatrix} \\
& + G(\omega_1+\omega_2+\omega_3) F(\omega_1, \omega_2, \omega_3).
\end{aligned}
$$

§ 4.

4.1. *Inverse Functional*

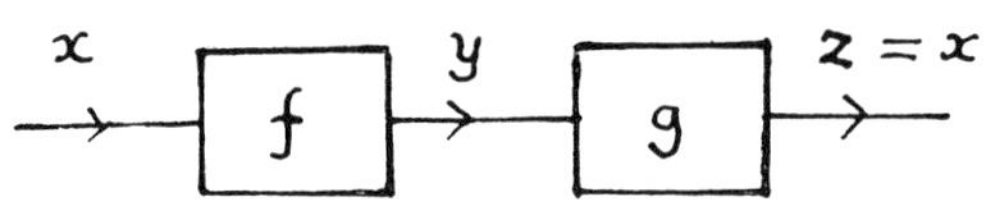

Two functionals, f and g, will be said to be inverse to one another if, $f \circ g$ and $g \circ f$ are both the identity. In this case, if a device representing

f is put in series with one represented by g, it will restore the input of this element from its output—and vice versa.

The functional relational between z and x is thus:

$$z(t) = x(t)$$
$$= \int \delta(t, \alpha) x(\alpha)\, d\alpha, \qquad \text{(A)}$$

where $\delta(t, \alpha) = \delta(t - \alpha)$ is the δ-function kernel.

4.1.1. *Linear case*

It is easily seen that if a linear functional has an inverse, this must be linear too. Suppose that the functionals f and g have kernels $f(t, t_1)$ and $g(t, t_1)$. Then if they are inverse to one another, it follows from § 3.2.1 that

$$\left.\begin{aligned} \int f(t, t_1) g(t_1, t_2)\, dt_2 &= \delta(t, t_2), \\ \int g(t, t_1) f(t_1, t_2)\, dt_1 &= \delta(t, t_2). \end{aligned}\right\} \qquad \text{(B)}$$

From this it is seen that f and g cannot both be proper functions since the integral of their product is a δ-function.

A case of practical importance is when

$$f(t, t_1) = \delta(t, t_1) + k(t, t_1), \qquad \text{(C)}$$

where $k(t, t_1)$ is a proper function. In this case the equation

$$y(t) = \int f(t, t_1) x(t_1)\, dt_1 \qquad \text{(D)}$$

is

$$y(t) = x(t) + \int k(t, t_1) x(t_1)\, dt_1; \qquad \text{(E)}$$

i.e. a linear integral equation which will be of the Volterra type when f acts only on the past. It is known (Lovitt 1950) that an equation of this type has a unique solution of the form:

$$x(t) = y(t) - \int k^{\dagger}(t, t_1) y(t_1)\, dt_1, \qquad \text{(F)}$$

where $k^{\dagger}(t, t_1)$ is related to $k(t, t_1)$ by an equation:

$$k^{\dagger}(t, t_1) = k(t, t_1) - \int k(t, t_2) k(t_2, t_1)\, dt_2$$
$$+ \int\int k(t, t_2) k(t_2, t_3) k(t_3, t_1)\, dt_2\, dt_3 - \dots$$

(Neumann–Liouville Series). (G)

Moreover the inversion of eqn. (F) gives eqn. (E).

Thus the inverse kernel corresponding to (C) is:

$$g(t, t_1) = \delta(t, t_1) - k^{\dagger}(t, t_1). \qquad \text{(H)}$$

Time-invariant case: In this case $f(t, t_1)$ and $g(t, t_1)$ are functions only of the time difference $t - t_1$. Suppose $f(t, t_1) = f_1(t - t_1)$, $g(t, t_1) = g_1(t - t_1)$, then eqns. (B) both give, on a Fourier transformation of both sides:

$$F(\omega)G(\omega)=1, \qquad \text{(B}')$$

where $$F(\omega)=\int f_1(\tau)\exp(-i\omega\tau)\,d\tau,\quad G(\omega)=\int g_1(\tau)\exp(-i\omega\tau)\,dt$$

are the Fourier transforms of $f_1(t)$ and $g_1(t)$.

Thus

$$G(\omega)=\frac{1}{F(\omega)} \qquad \text{(I)}$$

and so $g_1(t)$ may be found by an inverse Fourier transformation. It has yet to satisfy certain conditions for stability: $G(\omega)$ must have no poles in the region $\mathscr{I}(\omega)<0$ of the complex ω-plane. This means that $F(\omega)$ must have no zeros in this region. Also if $F(\omega)$ corresponds to a stable system, it too must have no poles in the region $\mathscr{I}(\omega)<0$. Therefore it must have no poles or zeros in this region. This condition is satisfied by minimum-phase networks (Bode 1945).

4.1.2. *Analytic case*

If f is an analytic functional with an invertible linear part then f has an analytic inverse. This follows from eqns. (D) of § 3.2.2 and relation (A) above. Thus, considering $g\circ f$ it follows that, in contracted notation:

$$\left.\begin{aligned}
\delta_{t,\alpha}&=g_{t\beta}f_{\beta\alpha},\\
0&=g_{t\beta_1\beta_2}f_{\beta_1\alpha_1}f_{\beta_2\alpha_2}+g_{t\beta}f_{\beta\alpha_1\alpha_2},\\
0&=g_{t\beta_1\beta_2\beta_3}f_{\beta_1\alpha_1}f_{\beta_2\alpha_2}f_{\beta_3\alpha_3}\\
&\quad+g_{t\beta_1\beta_2}f_{\beta_1\alpha_1}f_{\beta_2\alpha_2\alpha_3}\\
&\quad+g_{t\beta_1\beta_2}f_{\beta_1\alpha_1\alpha_2}f_{\beta_2\alpha_3}\\
&\quad+g_{t\beta}f_{\beta\alpha_1\alpha_2\alpha_3}.
\end{aligned}\right\} \qquad \text{(J)}$$

Suppose that the linear part of f, viz. the linear functional determined by the kernel $f(\beta,\alpha)$, is invertible. Then it is possible to find a kernel $g(t,\beta)$ satisfying the first of these equations and also

$$\delta_{t_1\alpha}=f_{t\beta}g_{\beta\alpha} \qquad \text{(K)}$$

and it is then possible to solve for all the higher kernels.

Thus from the second of the set of eqns. (J):

$$\begin{aligned}
0&=(g_{t\beta_1\beta_2}f_{\beta_1\alpha_1}f_{\beta_2\alpha_2}+g_{t\beta}f_{\beta\alpha_1\alpha_2})g_{\alpha_1r_1}g_{\alpha_2r_2}.\\
&=g_{t\beta_1\beta_2}f_{\beta_1\alpha_1}\,g_{\alpha_1r_1}f_{\beta_2\alpha_2}g_{\alpha_2r_2}+g_{t\beta}f_{\beta\alpha_1\alpha_2}g_{\alpha_1r_1}g_{\alpha_1r_2}\\
&=g_{t\beta_1\beta_2}\delta_{\beta_1r_1}\delta_{\beta_2r_2}+g_{t\beta}f_{\beta\alpha_1\alpha_2}g_{\alpha_1r_1}g_{\alpha_2r_2}\ \text{from eqn. (K)}\\
&=g_{tr_1r_2}+g_{t\beta}f_{\beta\alpha_1\alpha_2}g_{\alpha_1r_1}g_{\alpha_2r_2};\\
\therefore\ g_{tr_1r_2}&=-g_{t\beta}f_{\beta\alpha_1\alpha_2}g_{\alpha_1r_1}g_{\alpha_2r_2}.
\end{aligned}$$

This solves for $g_{tr_1r_2}$ in terms of kernels which are assumed known. The same procedure of using eqn. (J) to isolate the required kernel will also solve the equations for the higher kernels which may be found by this method one by one. The next equation gives:

$$\begin{aligned} g_{tr_1r_2r_3} = &+g_{t\delta}f_{\delta\epsilon_1\epsilon_2}g_{\epsilon_1r_1}g_{\epsilon_2\beta_2}f_{\beta_2\alpha_2\alpha_3}g_{\alpha_2r_2}g_{\alpha_3r_3} \\ &+g_{t\delta}f_{\delta\epsilon_1\epsilon_2}g_{\epsilon_1\beta_1}g_{\epsilon_2r_3}f_{\beta_1\alpha_2\alpha_3}g_{\alpha_2r_1}g_{\alpha_3r_2} \\ &-g_{t\beta}f_{\beta\alpha_1\alpha_2\alpha_3}g_{\alpha_1r_1}g_{\alpha_2r_2}g_{\alpha_3r_3}. \end{aligned}$$

Time-invariant case: Here it is possible, by taking Fourier transforms of the appropriate dimension, to convert the set of eqns. (J) into a set of algebraic equations between the Fourier transforms of kernels. To avoid expressions of great complexity this method will be illustrated in the case when the functional f is odd. The power series expansion of f has then only odd powers. Let it be:

$$f_{t\alpha}x_\alpha+f_{t\alpha_1\alpha_2\alpha_3}x_{\alpha_1}x_{\alpha_2}x_{\alpha_3}+f_{t\alpha_1\alpha_2\alpha_3\alpha_4\alpha_5}x_{\alpha_1}x_{\alpha_2}\dots x_{\alpha_5}+\dots.$$

Equations (J) for this case are:

$$\begin{aligned} \delta_{t\alpha} &= g_{t\beta}f_{\beta\alpha}, \\ 0 &= g_{t\beta_1\beta_2\beta_3}f_{\beta_1\alpha_1}f_{\beta_2\alpha_2}f_{\beta_3\alpha_3}+g_{t\beta}f_{\beta\alpha_1\alpha_2\alpha_3} \\ 0 &= g_{t\beta_1\beta_2\beta_3\beta_4\beta_5}f_{\beta_1\alpha_1}f_{\beta_2\alpha_2}f_{\beta_3\alpha_3}f_{\beta_4\alpha_4}f_{\beta_5\alpha_5} \\ &\quad +g_{t\beta_1\beta_2\beta^\varepsilon}f_{\beta_1\alpha_1\alpha_2\alpha_3}f_{\beta_2\alpha_4}f_{\beta_3\alpha_5} \\ &\quad +g_{t\beta_1\beta_2\beta_3}f_{\beta_1\alpha_1}f_{\beta_2\alpha_2\alpha_3\alpha_4}f_{\beta_3\alpha_5} \\ &\quad +g_{t\beta_1\beta_2\beta_3}f_{\beta_1\alpha_1}f_{\beta_2\alpha_2}f_{\beta_3\alpha_3\alpha_4\alpha_5} \\ &\quad +g_{t\beta}f_{\beta\alpha_1\alpha_2\alpha_3\alpha_4\alpha_5}, \end{aligned} \qquad \dots\dots \quad \text{(J')}$$

the corresponding Fourier relations being:

$$\begin{aligned} 1 &= G(\omega)F(\omega), \\ 0 &= G(\omega_1,\omega_2,\omega_3)F(\omega_1)F(\omega_2)F(\omega_3)+G(\omega_1+\omega_2+\omega_3)F(\omega_1,\omega_2,\omega_3), \\ 0 &= G(\omega_1,\omega_2,\omega_3,\omega_4,\omega_5)F(\omega_1)F(\omega_2)\dots F(\omega_5) \\ &\quad +G(\omega_1+\omega_2+\omega_3,\omega_4,\omega_5)F(\omega_1+\omega_2+\omega_3)F(\omega_4)F(\omega_5) \\ &\quad +G(\omega_1,\omega_2+\omega_3+\omega_4,\omega_5)F(\omega_1)F(\omega_2+\omega_3+\omega_4)F(\omega_5) \\ &\quad +G(\omega_1,\omega_2,\omega_3+\omega_4+\omega_5)F(\omega_1)F(\omega_2)F(\omega_3+\omega_4+\omega_5) \\ &\quad +G(\omega_1+\omega_2+\omega_3+\omega_4+\omega_5)F(\omega_1,\omega_2,\omega_3,\omega_4,\omega_5), \end{aligned}$$

which may be solved to give:

$$\begin{aligned} G(\omega) &= \frac{1}{F(\omega)}, \\ G(\omega_1,\omega_2,\omega_3) &= \frac{-F(\omega_1,\omega_2,\omega_3)}{F(\omega_1+\omega_2+\omega_3)F(\omega_1)F(\omega_2)F(\omega_3)}, \\ G(\omega_1,\omega_2,\dots,\omega_5) &= \frac{1}{F(\omega_1+\omega_2+\dots+\omega_5)\prod_{i=1}^{5}F(\omega_i)} \\ &\quad \times\{F(\omega_1+\omega_2+\omega_3,\omega_4,\omega_5)+F(\omega_1,\omega_2+\omega_3+\omega_4,\omega_5) \\ &\quad +F(\omega_1,\omega_2,\omega_3+\omega_4+\omega_5)-F(\omega_1,\omega_2,\omega_3,\omega_4,\omega_5)\}. \end{aligned}$$

4.2. *A Servomechanism Problem*

To illustrate the results of the preceding section on the inversion of functionals it will be shown how expressions may be obtained for error and output of a simple servomechanism (James *et al.* 1947) (see diagram). The input, error and output signals are denoted by the letters x, e and y, the functional giving the relation between error and output being denoted by h. h is assumed to be an odd functional of error, as is usually the case in systems of this nature.

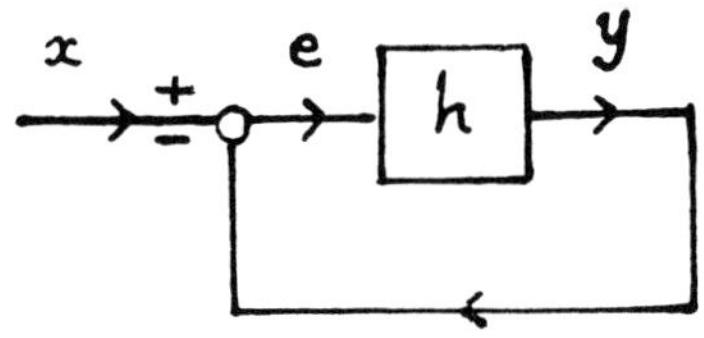

The equations describing the system are then:

$$e(t) = x(t) - y(t), \qquad \text{(A)}$$

$$y(t) = h(t;\, e(t'), t' \leqslant t). \qquad \text{(B)}$$

To study the behaviour of the system it is necessary to consider the relation of error and output to input. From the above equations it follows by elimination of $y(t)$ that:

$$x(t) = e(t) + h(t;\, e(t)', t' \leqslant t). \qquad \text{(C)}$$

It is therefore necessary to invert this functional relation to find $e(t)$ as a functional of input. Once this has been done, it is easy to find $y(t)$ as a functional of input from eqn. (A).

4.2.1. *The linear case*

Here, eqn. (B) may be written:

$$y(t) = \int h(t, t_1) e(t_1)\, dt_1 \qquad \text{(B}'\text{)}$$

and eqn. (C) becomes:

$$x(t) = e(t) + \int h(t, t_1) e(t_1)\, dt_1. \qquad \text{(C}'\text{)}$$

The inversion problem for this equation has already been considered (see § 4.1.1), the solution being:

$$e(t) = x(t) - \int h^{\dagger}(t, t_1) x(t_1)\, dt_1.$$

The output is given by eqn. (A) as:

$$y(t) = \int h^{\dagger}(t, t_1) x(t_1)\, dt_1.$$

Consequently $h^{\dagger}(t, t_1)$ is seen to have a simple interpretation. It is the weighting function (impulse response function) for the whole servomechanism.

In the time-invariant case the method of Fourier transformation may be used. Put $h(t,t_1)=h_1(t-t_1)$, $h^\dagger(t,t_1)=h_1^\dagger(t-t_1)$ and

$$H(\omega)=\int h_1(t)\exp(-i\omega t)\,dt, \quad H(\omega)=\int h_1(t)\exp(-i\omega t)\,dt.$$

Then

$$1-H^\dagger(\omega)=\frac{1}{1+H(\omega)} \qquad \text{cf. § 4.1.1, eqn. (I),}$$

$$H^\dagger(\omega)=\frac{H(\omega)}{1+H(\omega)},$$

a well-known formula. Stability conditions still have to be fulfilled. These are very familiar in linear servomechanism theory; assuming $H(\omega)$ corresponds to a stable system, $H^\dagger(\omega)$ will correspond to a stable system if and only if $1+H(\omega)$ has no zeros in the region $\mathscr{I}(\omega)\leqslant 0$ of the complex ω-plane (James *et al.* 1947).

4.2.2. *Case where h is analytic*

Since h has been assumed odd it will have an expansion:

$$h(t;\,e(t'),t'\leqslant t)=\int h(t;t_1)e(t_1)\,dt_1+\int\int\int h(t;\,t_1,t_2,t_3)e(t_1)e(t_2)e(t_3)\,dt_1\,dt_2\,dt_3 + \dots$$

involving only terms of odd order.

Equation (C) is consequently, in abbreviated notation:

$$\begin{aligned} x_t &= e_t + h_{t\alpha_1}e_{\alpha_1} + h_{t\alpha_1\alpha_2\alpha_3}e_{\alpha_1}e_{\alpha_2}e_{\alpha_3} + \dots \\ &= \delta_{t\alpha_1}e_{\alpha_1} + h_{t\alpha_1}e_{\alpha_1} + h_{t\alpha_1\alpha_2\alpha_3}e_{\alpha_1}e_{\alpha_2}e_{\alpha_3} + \dots \\ &= (\delta_{t\alpha_1} + h_{t\alpha})e_{\alpha_1} + h_{t\alpha_1\alpha_2\alpha_3}e_{\alpha_1}e_{\alpha_2}e_{\alpha_3} + \dots . \end{aligned}$$

It has been seen that the invertibility of this equation is dependent on the invertibility of the linear part, viz.

$$x_t = (\delta_{t\alpha_1} + h_{t\alpha_1})e_{\alpha_1},$$

$$x(t) = e(t) + \int h(t,t_1)e(t_1)\,dt_1.$$

This is the equation of the linear case already considered, the solution being:

$$e(t) = x(t) - \int h^\dagger(t,t_1)x(t_1)\,dt_1;$$

i.e.

$$\begin{aligned} e_t &= x_t - h_{t\alpha}{}^\dagger x_\alpha \\ &= (\delta_{t\alpha} - h_{t\alpha}{}^\dagger)x_\alpha \\ &= g_{t\alpha}x_\alpha . \end{aligned}$$

e may now be calculated as an odd power series in x:

$$e_t = g_{t\alpha}x_\alpha + g_{t\alpha_1\alpha_2\alpha_3}x_{\alpha_1}x_{\alpha_2}x_{\alpha_3} + \dots,$$

where the higher g-kernels may be worked out as in § 4.1.2;

e.g.

$$g_{t\alpha_1\alpha_2\alpha_3} = -g_{tr}h_{r\beta_1\beta_2\beta_3}g_{\beta_1\alpha_1}g_{\beta_2\alpha_2}g_{\beta_3\alpha_3}.$$

If the functional h is time-invariant, Fourier transforms may be used to give the Fourier transforms of the g-kernels in terms of those of the h-kernels.

For example, consider the case where

$$y(t) = \int k_1(t-t_1) f(e(t_1))\, dt_1,$$

$$f(e) = a_1 e + a_3 e^3 + a_5 e^5 + \dots .$$

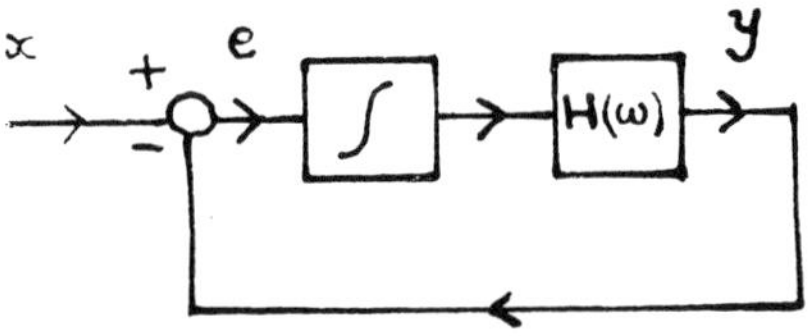

Then

$$y(t) = \int h_1(t-t_1) e(t_1)\, dt_1 + \int\int\int h_1(t-t_1, t-t_2, t-t_3) e(t_1) e(t_2) e(t_3)\, dt_1\, dt_2\, dt_3$$
$$+ \dots$$

on putting

$$\begin{aligned}
h_1(\tau) &= a_1 k_1(\tau),\\
h_1(\tau_1, \tau_2, \tau_3) &= a_3 k_1(\tau_1) \delta(\tau_2 - \tau_1) \delta(\tau_3 - \tau_1),\\
h_1(\tau_1, \tau_2, \dots, \tau_5) &= a_5 k_1(\tau_1) \delta(\tau_2 - \tau_1) \dots \delta(\tau_2 - \tau_1),\\
H(\omega) &= a_1 K(\omega),\\
H(\omega_1, \omega_2, \omega_3) &= a_3 K(\omega_1 + \omega_2 + \omega_3),\\
H(\omega_1, \dots, \omega_5) &= a_5 K(\omega_1 + \dots + \omega_5),
\end{aligned}$$

where $H(\omega)$ is the Fourier transform of $h_1(t)$, etc.

It is found that

$$G(\omega_1, \omega_2, \dots, \omega_5) = \frac{[2 + (3a_3 - a_5) K(\omega_1 + \dots + \omega_5)]}{[1 + a_1 K(\omega_1 + \dots + \omega_5)] \prod_{i=5}^{5} [1 + a_1 K(\omega_i)]},$$

where $G(\omega)$ is the Fourier transform of $g_1(t)$, $g_1(t)$ being defined as $g(t, t_1) = g_1(t - t_1)$ with the same notation for the higher kernels.

§ 5. The Class of Functionals with Finite Mean Square

So far, mainly functionals of analytic type have been considered. The class of analytic functionals includes the linear and polynomial functionals as special cases. A class of functionals will now be introduced which arises naturally in theory based on mean square values of stochastic processes such as occurs in Wiener's theory of least square smoothing and prediction. Analytic functionals are included as special cases as also are certain discontinuous functionals, in particular the simple type of functional (function) occurring in relay systems ($f(x) = +1$, $x > 0$, $f(x) = -1$, $x < 0$).

5.1. *Functionals and Functions with Finite Mean Square*

Consider first the analogous class of functions. Let a bar denote any linear averaging operation of functions, satisfying the axioms:

(i) $\overline{f_1(x)+f_2(x)}=\overline{f_1(x)}+\overline{f_2(x)}$,

(ii) $\overline{kf(x)} \quad =k\overline{f(x)}$ for any constant k,

(iii) $f(x)\geqslant 0$ for all x implies $\overline{f(x)}\geqslant 0$.

Then, relative to this operation, the class of functions of finite mean square will be defined as the class of functions for which $\overline{f(x)^2}$ is finite. A common case is where the averaging operation is given by the formula:

$$\overline{f(x)}=\int_a^b f(x)w(x)\,dx,$$

where $w(x)$ is a non-negative weighting function. If $w(x)=1$, the class of function with finite mean square becomes the class $L^2(a,b)$ of functions of Lebesgue integrable square in (a,b). The case of special interest at present is that in which the bar denotes statistical averaging: this may be written in integral form as:

$$\overline{f(x)}=\int f(x)p(x)\,dx,$$

where $p(x)$ is the probability density function of x. In this case the practical significance of functions of finite mean square is as follows: suppose that the output of a simple instantaneous device is related to the input by

$$y(t)=f(x(t)),$$

$x(t)$ and $y(t)$ representing input and output quantities at time t. Suppose x has a constant one-dimensional distribution with probability density function $p(x)$ and the function f is of finite mean square relative to $p(x)$ as weighting function. Then

$$\int f^2(x)p(x)\,dx,$$

i.e. the mean square output (variance of output) is finite. Hence it may be argued that only functions of this class have physical significance†.

Functionals of finite mean square may be defined similarly, relative to any linear averaging operation.

If $x(t)$ is a stationary ergodic stochastic process, for any functional $f(t;x(t'),t'\leqslant t)$ the mean square value

$$\overline{f(t;x(t'),t'\leqslant t)^2}$$

may be considered. If this is finite, f will be said to be of finite mean square relative to this process. If $x(t)$ is not stationary, the mean square value may be taken at a particular time. Certain difficulties arise in attempting

† Strictly speaking, the process should be ergodic for this argument to apply.

to define an integral form for this averaging operation. In the case where $x(t)$ is a purely random process, the integral has been defined by Wiener (1930).

5.2. *Mean Square Approximation*

The method of mean square approximation may sometimes be used to get approximate solutions of engineering problems where exact methods fail. Thus Booton *et al.* (1953) were able to obtain results about the behaviour of a non-linear servomechanism by using a mean square approximation method which linearized the system. Applications to filtering and prediction theory are given below.

Consider the problem of finding the best mean square approximation to a function $f(x)$ of finite mean square by a polynomial of degree n. This is the problem of minimizing

$$\overline{(f(x)-a_0-a_1x-\ldots-a_nx^n)^2} \quad . \quad . \quad . \quad . \quad . \quad . \quad (A)$$

by variation of the coefficients $a_0, a_1, \ldots, a_n$.

This is a special case of the more general problem of finding the best mean square approximation to a function $f(x)$ of finite mean square by linear combinations of given functions $f_0(x), \ldots f_1(x)., \ldots, f_n(x)$, i.e. of minimizing

$$\overline{(f(x)-a_0f_0(x)-a_1f_1(x)-\ldots-a_nf_n(x))^2} \quad . \quad . \quad . \quad . \quad (B)$$

by variation of the parameters a_0, $a_1, \ldots, a_n$. For the expression to be finite, it is necessary that all functions appearing should be of finite mean square. It may be assumed without any loss of generality that the approximating functions are linearly independent, i.e. no one of them is expressible as a linear sum of the others†. This condition may be shown equivalent to the condition that

$$\begin{vmatrix} \overline{f_0^2} & \overline{f_0f_1} & \overline{f_0f_n} \\ \overline{f_1f_0} & \overline{f_1^2} & \overline{f_1f_n} \\ & & \\ \overline{f_nf_0} & \overline{f_nf_1} & \overline{f_n^2} \end{vmatrix} \neq 0. \quad . \quad . \quad . \quad . \quad . \quad (C)$$

Here $\overline{f_if_j}$ is a contracted notation for $\overline{f_i(x)f_j(x)}$. This notation will be retained.

The calculation of the minimizing values of the parameters $a_0, a_1, \ldots, a_n$ is straightforward.

Denote the expression (B) by $F(a_0, a_1, \ldots, a_n)$:

$$F(a_0, a_1, \ldots, a_n) = \overline{f^2} - 2\sum_{i=0}^{n} a_i\overline{ff_i} + \sum_{i,j=0}^{n} a_ia_j\overline{f_if_j}. \quad . \quad . \quad (D)$$

A necessary condition that $F(a_0, a_1, \ldots, a_n)$ should be a minimum is that

$$\frac{\partial F}{\partial a_0} = \frac{\partial F}{\partial a_1} = \ldots = \frac{\partial F}{\partial a_n} = 0. \quad . \quad . \quad . \quad . \quad (E)$$

† More precisely, it is required that the set of values of x for which any one of the functions is equal to a linear combination of the others should have zero probability measure.

Now

$$\frac{\partial F}{\partial a_k} = -2\overline{f_k f} + 2\sum_{i=0}^{n} a_i \overline{f_i f_k}, \quad k = 0, 1, \ldots, n$$

$$= 2\left\{\sum_{i=0}^{n} a_i \overline{f_i f_k} - \overline{f_k f}\right\}. \qquad \text{(F)}$$

∴ eqns. (E) are:

$$\left.\begin{array}{l} a_0\overline{f_0^2} + a_1\overline{f_0 f_1} + \ldots + a_n\overline{f_0 f_n} = \overline{f_0 f} \\ a_0\overline{f_1 f_0} + a_1\overline{f_1^2} + \ldots + a_n\overline{f_1 f_n} = \overline{f_1 f} \\ \ldots \quad \ldots \quad \ldots = \ldots \\ a_0\overline{f_n f_0} + a_1\overline{f_n f_1} + \ldots + a_n\overline{f_n^2} = \overline{f_n f} \end{array}\right\} \qquad \text{(G)}$$

The determinant of the a's is non-zero by condition (C) and so these equations have a unique solution for the a's. It is easily seen that these values must correspond to a minimum value of $F(a_0, a_1, \ldots, a_n)$. The approximating function, $a_0 f_0(x) + a_1 f_1(x) + \ldots + a_n f(x)$

$$= \begin{vmatrix} 0 & f_0(x) & f_1(x) & \ldots & f_n(x) \\ \overline{f_0 f} & \overline{f_0^2} & \overline{f_0 f_1} & \ldots & \overline{f_0 f_n} \\ \overline{f_1 f} & \overline{f_1 f_0} & \overline{f_1^2} & \ldots & \overline{f_1 f_n} \\ \ldots & \ldots & \ldots & \ldots & \ldots \\ \overline{f_n f} & \overline{f_n f_0} & \overline{f_n f_1} & \ldots & \overline{f_n^2} \end{vmatrix} \div D, \qquad \text{(H)}$$

where D is the determinant on the left-hand side of condition (C).

The value of the expression (A), i.e. the mean square error of the approximation is found as:

$$\begin{vmatrix} \overline{f^2} & \overline{ff_0} & & \overline{ff_n} \\ \overline{f_0 f} & \overline{f_0^2} & & \overline{f_0 f_n} \\ & & & \\ \overline{f_n f} & \overline{f_n f_0} & & \overline{f_n^2} \end{vmatrix} \div D. \qquad \text{(I)}$$

This calculation is the same if the f's represent functionals instead of functions. The expression $f(x)$ may be considered an abbreviation for $f(t; x(t'), t' \leqslant t)$. With functionals the result may be interpreted in engineering terms as follows.

It is required to simulate, or construct a system with functional relation f between output and input when the input has a known distribution. Using a finite number of elements with relations between output and input given by functionals $f_1, f_2, \ldots, f_n$, the arrangement of these elements in parallel which will approximate with least mean square output discrepancy is that shown in the diagram. The multiplication factors have the values found above.

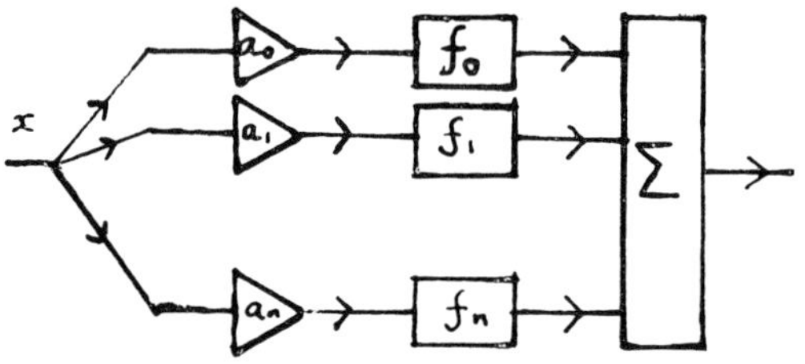

By suitable choice of what might be called coordinate functionals $f_1, f_2, \ldots, f_n$ it should by this means be possible to reproduce quickly to within permissible error a wide range of functional relationships†.

5.3. *The Expansion of a Functional of Finite Mean Square by a Complete Orthonormal System*

Two functionals $f_1(t; x(t'), t' \leqslant t)$ and $f_2(t; x(t''), t'' \leqslant t)$ will be said to be *orthogonal* relative to a stationary ergodic stochastic process if

$$\overline{f_1(t; x(t'), t' \leqslant t) f_2(t; x(t''), t'' \leqslant t)} = 0.$$

If $f_1, f_2, \ldots$ is a finite or infinite set of functionals every two of which are orthogonal, the set will be called an orthogonal set. The norm of a functional relative to a stationary ergodic stochastic process may be defined as

$$\|f\|_x = \sqrt{[\overline{f^2(t; x(t'), t' \leqslant t)}]}.$$

It is thus the root mean square value of the output of any system with functional f when the input is the stochastic process mentioned. If $f_1, f_2, \ldots$ is an orthogonal set of functionals each with norm one relative to some particular stochastic process, the set will be called orthonormal relative to this process.

In this case

$$\overline{f_i(t; x(t'), t' \leqslant t) f_j(t; x(t''), t'' \leqslant t)} = \begin{cases} 1 & i = j, \\ 0 & i \neq j. \end{cases}$$

For finite orthonormal set $f_0, f_1, \ldots, f_n$, the above equations dealing with least mean square approximation take on a very much simpler form.

The determinant

$$D = \begin{vmatrix} 1 & 0 & \ldots & 0 \\ 0 & 1 & \ldots & 0 \\ \ldots & & \ldots & . \\ 0 & 0 & \ldots & 1 \end{vmatrix} = 1.$$

Equations (G) become:

$$\begin{aligned} a_0 \qquad \qquad &= \overline{f_0 f} \\ a_1 \qquad &= \overline{f_1 f} \\ \ldots \qquad & \\ a_n &= \overline{f_n f}, \end{aligned}$$

giving immediately the values of the a's.

† This arrangement has been used by Lampard (1955) to synthesis linear time-invariant filters. See also Lee (1932).

The expression (H) for the approximation becomes:

$$-\begin{vmatrix} 0 & f_0(x) & f_1(x)\ldots & f_n(x) \\ \overline{f_0 f} & 1 & \ldots & \ldots \\ \overline{f_1 f} & & 1 \;\ldots & \ldots \\ \ldots & \ldots & \ldots & \ldots \\ \overline{f_n f} & \ldots & \ldots\;\ldots & 1 \end{vmatrix} = \sum_{i=0}^{n} \overline{ff_i}\, f_i(x)$$

and the minimum mean square value is:

$$\begin{vmatrix} \overline{f^2} & \overline{ff_0} & \ldots & \overline{ff_n} \\ \overline{f_0 f} & 1 & \ldots & 0 \\ \ldots & & \ldots & \ldots \\ \overline{f_n f} & 0 & \ldots & 1 \end{vmatrix} = \overline{f^2} - \sum_{i=1}^{n} (\overline{f_i f})^2.$$

It may happen with an infinite orthonormal set that every functional with finite mean square can be approximated with any prescribed accuracy by finite linear combinations of functionals of this set. Such a set will be said to be *complete*. A complete set of orthonormal functionals for the case where the input stochastic process is completely random has been considered by Cameron and Martin (1947).

In this case it is possible to write:

$$f(x) = \sum \overline{ff_i}\, f_i(x), \qquad \text{(H'')}$$

$$\overline{f^2} = \sum (\overline{ff_i})^2. \qquad \text{(I'')}$$

(I″) thus gives an expansion for any functional of finite mean square in terms of the basic functionals f_i.

5.4. *Formation of Orthogonal and Orthonormal Sets*

Given any linearly independent set of functions or functionals, it is possible to form from these an orthogonal set by first arranging these function(al)s into a sequence:

$$f_0(x), f_1(x), \ldots, f_n(x), \ldots$$

and then forming the sequence (Szegö 1939)†:

$$D_0(x) = f_0(x),$$

$$D_1(x) = \begin{vmatrix} f_1(x) & f_0(x) \\ \overline{f_0 f_1} & \overline{f_0^2} \end{vmatrix},$$

$$D_2(x) = \begin{vmatrix} f_2(x) & f_1(x) & f_0(x) \\ \overline{f_1 f_2} & \overline{f_1^2} & \overline{f_1 f_0} \\ \overline{f_0 f_2} & \overline{f_0 f_1} & \overline{f_0^2} \end{vmatrix}$$

$$\ldots \qquad \ldots \qquad \ldots$$

† The process given here is equivalent to the Gram–Schmidt process.

$$D_n(x) = \begin{vmatrix} f_n(x) & f_{n-1}(x) & \cdots & f_0(x) \\ \overline{f_{n-1}f_n} & \overline{f_{n-1}^2} & \cdots & \overline{f_{n-1}f_0} \\ \cdots & \cdots & \cdots & \cdots \\ \overline{f_0 f_n} & \overline{f_0 f_{n-1}} & \cdots & \overline{f_0^2} \end{vmatrix}$$

$$\cdots = \qquad \cdots \qquad \text{etc.}$$

If any one of these determinants vanishes identically then there will be a linear relation between some of the function(al)s f. Suppose $D_r(x)$ is the first determinant to vanish identically. Then $f_r(x)$ is linearly dependent on its predecessors and may be excluded from the sequence, f_{r+1} being taken in its place.

The determinants $D_n(x)$ are mutually orthogonal: $D_n(x)$ is by its construction, orthogonal to $f_{n-1}(x), \ldots, f_0(x)$. Hence it is orthogonal to any linear combination of these, in particular, to any preceding determinant. Since, of any two determinants, one is of lower order than the other, the result follows; i.e.

$$\overline{D_m(x)D_n(x)} = 0, \quad m \neq n. \qquad \text{(A)}$$

If $m = n$,

$$D_m(x)D_n(x) = D_n^2(x)$$

$$= \begin{vmatrix} f_n(x) & f_{n-1}(x) & \cdots & f_0(x) \\ \overline{f_{n-1}f_n} & \overline{f_{n-1}^2} & \cdots & \overline{f_{n-1}f_0} \\ \cdots & \cdots & \cdots & \cdots \\ \overline{f_0 f_n} & \overline{f_0 f_{n-1}} & & \overline{f_0^2} \end{vmatrix} \times (f_n(x)\Delta_{n-1} + R(x)),$$

where

$$\Delta_{n-1} = \begin{vmatrix} \overline{f_{n-1}^2} & \cdots & \overline{f_{n-1}f_0} \\ \cdots & \cdots & \cdots \\ \overline{f_0 f_{n-1}} & \cdots & \overline{f_0^2} \end{vmatrix}$$

$$\Delta_{-1} = 1.$$

and $R(x)$ is a linear combination of $f_{n-1}(x), \ldots, f_0(x)$.

On taking mean values, the term involving $R(x)$ vanishes by the orthogonality properties of $D_n(x)$.

$$\therefore\ \overline{D_n^2(x)} = \overline{\begin{vmatrix} f_n(x) & f_{n-1}(x) & \cdots & f_0(x) \\ \overline{f_{n-0}f_n} & \overline{f_{n-1}^2} & \cdots & \overline{f_{n-1}f_0} \\ \cdots & \cdots & \cdots & \cdots \\ \overline{f_0 f_n} & \overline{f_1 f_{n-1}} & \cdots & \overline{f_{n-1}^2} \end{vmatrix} \times f_n(x)\Delta_{n-1}}$$

$$= \Delta_n \Delta_{n-1};$$

i.e.

$$\overline{D_n^2(x)} = \Delta_n \Delta_{n-1},\ n = 0, 1, \ldots. \qquad \text{(B)}$$

It follows that the quantities

$$\chi_n(x) = \frac{D_n(x)}{\sqrt{(\Delta_n \Delta_{n-1})}}$$

form an orthonormal set. (It may be shown that $\Delta_n > 0$ and so the square root is always real and positive. This is obvious from (B) since the left-hand side is positive.)

Orthogonal polynomials

Important sets of orthogonal functions (polynomials) are obtained by taking the sequence $f_0(x), f_1(x), \ldots$ to be the sequence of powers of x, viz. $1, x, x^2, \ldots$.

In this case the sequence of D's is:

$$1, \quad \begin{vmatrix} x & 1 \\ \overline{x} & 1 \end{vmatrix}, \quad \begin{vmatrix} x^2 & x & 1 \\ \overline{x^3} & \overline{x^2} & \overline{x} \\ \overline{x^2} & \overline{x} & 1 \end{vmatrix}, \quad \begin{vmatrix} x^3 & x^2 & x & 1 \\ \overline{x^5} & \overline{x^4} & \overline{x^3} & \overline{x^2} \\ \overline{x^4} & \overline{x^3} & \overline{x^2} & \overline{x} \\ \overline{x^3} & \overline{x^2} & \overline{x} & 1 \end{vmatrix}, \ldots .$$

In particular, if the x-distribution is symmetrical with zero mean, all odd moments vanish and this sequence becomes:

$$1, \quad x, \quad \begin{vmatrix} x^2 & 1 \\ \overline{x^2} & 1 \end{vmatrix}, \quad \begin{vmatrix} x^3 & x \\ \overline{x^4} & \overline{x^2} \end{vmatrix}, \ldots,$$

a sequence of alternately even and odd polynomials.

It is convenient to introduce the polynomials:

$$P_n(x) = \frac{D_n(x)}{\Delta_{n-1}}, \qquad n = 0, 1, \ldots$$

which have the orthogonality properties of the D's and, in addition, unit coefficient for the highest power of x.

From eqns. (A) and (B), it follows that:

$$\left.\begin{aligned} \overline{P_m(x)P_n(x)} &= 0, \quad m \neq n, \\ \overline{P_n{}^2(x)} &= \frac{\Delta_n}{\Delta_{n-1}} \end{aligned}\right\} \quad \ldots \ldots \quad \text{(C)}$$

In the case when the x-distribution is normal with unit variance and zero mean, the polynomials $P_n(x)$ become the Hermite polynomials (Cramér 1946) $H_n(x)$ defined by the equation:

$$\frac{d^n}{dx^n} \exp(-x^2/2) = (-1)^n H_n(x) \exp(-x^2/2).$$

5.5. *Multi-dimensional Orthogonal Polynomials: Symmetric Partially Orthogonal Systems*

Multi-dimensional orthogonal polynomials will be mentioned here for they lead towards the definition of the partially orthogonal systems of functionals, which will be used in the applications of the present theory to least square filtering and prediction.

Consider functions in n variables $x_1, x_2, \ldots, x_n$. It is possible, by starting with the sequence:

$$1;\ x_1, x_2, \ldots, x_n;\ x_1^2, x_1, x_2, \ldots, x_1x_n;\ x_2x_1, x_2^2, \ldots, x_2x_n;\ldots$$
$$\ldots x_nx_1 \ldots x_n^2;\ x_0^3, \ldots,$$

and using the preceding method to construct a set of orthogonal or orthonormal polynomials in the n variables $x_1, x_2, \ldots, x_n$. However, this procedure is unsymmetrical in the variables and unsuitable for certain purposes. It is possible to construct a sequence of *symmetrical* polynomials with the property that any two polynomials of different degrees are orthogonal.

The polynomials are, denoting $(x_1, x_2, \ldots, x_n)$ by $\mathbf{x}$:

$$\left.\begin{aligned}
&Q^{(0)}(\mathbf{x}) = 1,\\
&Q_i^{(1)}(\mathbf{x}) = \begin{vmatrix} x_i & 1 \\ \overline{x_i} & 1 \end{vmatrix}, \quad i = 1, 2, \ldots, n,\\
&Q_{ij}^{(2)}(\mathbf{x}) = \begin{vmatrix} x_ix_j & x_1 & x_2 & \ldots & x_n & 1 \\ \overline{x_1x_ix_j} & \overline{x_1^2} & \overline{x_1x_2} & \ldots & \overline{x_1x_n} & \overline{x_1} \\ \ldots & \ldots & \ldots & \ldots & \ldots & \ldots \\ \overline{x_ix_j} & \overline{x_0} & \overline{x_2} & \ldots & \overline{x_n} & 1 \end{vmatrix} i, j, = 1, 2, \ldots\\
&\ldots \quad \ldots \quad \ldots \quad \text{etc.}
\end{aligned}\right\} \qquad \text{(D)}$$

The orthogonality property of polynomials of different degrees is seen as above.

The values of the averaged products of polynomials of the same degree can be evaluated simply by following the previous method. Thus

$$\overline{Q_{ij}^{(2)}(x)Q_{k1}^{(2)}(x)} = \begin{vmatrix} \overline{x_kx_1x_ix_j} & \overline{x_kx_1x_1} & \ldots & \overline{x_kx_1} \\ \overline{x_1x_ix_j} & \overline{x_1^2} & \ldots & \overline{x_1} \\ \ldots & \ldots & \ldots & \ldots \\ \overline{x_ix_j} \ \ldots & \overline{x_1} & \ldots & 1 \end{vmatrix} \times C_2,$$

where (E)

$$C_2 = \begin{vmatrix} \overline{x_1^2} & \overline{x_1x_2} & \ldots & \overline{x_1} \\ \overline{x_2x_1} & \overline{x_2^2} & \ldots & \overline{x_2} \\ \ldots & \ldots & \ldots & \ldots \\ \overline{x_2} & \overline{x_2} & \ldots & 1 \end{vmatrix}.$$

As in the one-dimensional case, it may be convenient to work with polynomials having the coefficient of the term of highest degree unity. These polynomials are:

$$\left.\begin{aligned}
&P^{(0)}(\mathbf{x}) = Q^{(0)}(\mathbf{x}),\\
&P_i^{(1)}(\mathbf{x}) = Q_i^{(1)}(\mathbf{x}), \quad i = 0, 1 \ldots\\
&P_{ij}^{(2)}(\mathbf{x}) = \frac{Q_{ij}^{(2)}(\mathbf{x})}{C_2}, \quad i, j = 0, 1, \ldots\\
&\ldots \quad \ldots \quad \ldots \quad \text{etc.}
\end{aligned}\right\} \quad . \ . \ . \ . \ . \quad \text{(F)}$$

The polynomials $P^{(n)}(x)$ in the case where the variables $x_1, x_2, \ldots, x_n$ are independently and normally distributed with zero mean and unit variance, have been proposed by Grad (1949) as giving a symmetrical set of multi-dimensional Hermite polynomials. In this case, the polynomials of the same degree are orthogonal in the sense that

$$\overline{P_{i_1 i_2}{}^{(n)}{}_{\ldots i_n}(\mathbf{x})\, P^{(n)}{}_{j_1 j_2 \ldots j_n}(\mathbf{x})} = \begin{cases} n! \text{ if } j_1, j_2, \ldots, j_n \text{ is a permutation.} \\ \text{of } i_1, i_2, \ldots, i_n \\ 0 \text{ otherwise.} \end{cases} \qquad \text{(G)}$$

In this case, the first few polynomials have the values (Grad 1949) (changing to Grad's notation):

$$\mathscr{H}^{(0)}(\mathbf{x}) = 1,$$
$$\mathscr{H}^{(1)}(\mathbf{x}) = x_i,$$
$$\mathscr{H}_{ij}{}^{(2)}(\mathbf{x}) = x_i x_j - \delta_{ij},$$
$$\mathscr{H}_{ijk}{}^{(3)}(\mathbf{x}) = x_i x_j x_k - (x_i \delta_{jk} + x_j \delta_{ik} + x_k \delta_{ij}),$$

where δ_{rs} is the Kronecker δ-function: $\delta_{rs} = 0$, $r \neq s$, $\delta_{rs} = 1$, $r = s$; $r, s = 1, 2, \ldots, n$.
In the more general case where the x's may be correlated, the Kronecker δ-function is replaced by the covariance $\overline{x_r x_s}$ between x_r and x_s. Thus

$$\left.\begin{aligned} &\mathscr{H}^{(0)}(\mathbf{x}) = 1, \\ &\mathscr{H}_i{}^{(1)}(\mathbf{x}) = x_i, \\ &\mathscr{H}_{ij}{}^{(2)}(\mathbf{x}) = x_i x_j - \overline{x_i x_j}, \\ &\mathscr{H}_{ijk}{}^{(3)}(\mathbf{x}) = x_i x_j x_k - (x_i \overline{x_j x_k} + x_j \overline{x_k x_i} + x_k \overline{x_i x_j}). \end{aligned}\right\} \qquad \text{(H)}$$

5.6. *Symmetric Partially Orthogonal Systems of Functionals*

The multi-dimensional polynomials just considered are now applied to find a partially orthogonal system of functionals.

As in § 3 (p. 577), a functional may be regarded as a function of a continuous infinity of variables, $x(t)$ where t runs over the appropriate range of time. Thus, to take a definite example, the Hermite polynomials (H) of § 5.5 will become (Friedrichs 1953):

$$\mathscr{H}^{(0)}(x) = 1,$$
$$\mathscr{H}^{(1)}(t_1; x) = x(t_1),$$
$$\mathscr{H}^{(2)}(t_1, t_2; x) = x(t_1)x(t_2) - \overline{x(t_1)x(t_2)},$$
$$\mathscr{H}^{(3)}(t_1, t_2, t_3; x) = x(t_1)x(t_2)x(t_3) - x(t_1)\overline{x(t_2)x(t_3)} - x(t_2)\overline{x(t_3)x(t_1)} - x(t_3)\overline{x(t_1)x(t_2)}.$$

It may be verified that for any values of $s_1, s_2, \ldots, s_m$ and $t_1, t_2, \ldots, t_n$, $\mathscr{H}^{(m)}(s_1, s_2, \ldots, s_m; x)$ and $\mathscr{H}^{(n)}(t_1, t_2, \ldots, t_n; x)$ are orthogonal if $m \neq n$. From this it follows that the functionals

$$\int\!\!\int \cdots \int h(s_1, s_2, \ldots, s_m) \mathscr{H}^{(m)}(s_1, s_2, \ldots, s_m; x)\, ds_1 ds_2 \ldots ds_m$$

$$\int\!\!\int \cdots \int k(t_1, t_2, \ldots, t_n) \mathscr{H}^{(n)}(t_1, t_2, \ldots, t_n; x)\, dt_1\, dt_2 \ldots dt_n$$

are orthogonal for any weighting functions (kernels) h and k. These form the required partially orthonormal system in this case. The general case, for any distribution (stochastic process) will correspond to a system of functionals:

$$\int\int\cdots\int k(t_1, t_2, \ldots, t_n) P^{(n)}(t_1, t_2, \ldots, t_n; x)\, dt_1\, dt_2 \ldots dt_n,$$

where the kernel $k(t_1, t_2, \ldots, t_n)$ is arbitrary and $P^{(n)}(t_1, t_2, \ldots, t_n)$ is the partially orthogonal system of the stochastic process.

The values of the partially orthogonal system of P's does not follow from a simple limiting procedure applied to the polynomials $P_{i_1 i_2 \ldots i_n}{}^{(n)}(\mathbf{x})$ of the multi-nomial case because it is not clear how the determinants will behave, but, taking note of the form of the polynomial P's and the Hermite case just considered, it is possible to write down the expressions $P^{(n)}(t_1, t_2, \ldots, t_n(x)$ as:

$$P^{(0)}(x) = 1,$$

$$P^{(1)}(t_1; x) = x(t_1) - g_{10} P^{(0)}(x),$$

$$P^{(2)}(t_1, t_2; x) = x(t_1)x(t_2) - g_{20}(t_1, t_2) P^{(0)}(x) - \int g_{21}(t_1, t_2; s) P^{(1)}(s; x)\, ds,$$

$$\begin{aligned} P^{(3)}(t_1, t_2, t_3; x) = {} & x(t_1)x(t_2)x(t_3) - g_{30}(t_1, t_2, t_3) P^{(0)}(x) \\ & - \int g_{30}(t_1, t_2, t_3; s) P^{(1)}(s; x)\, ds \\ & - \int\int g_{32}(t_1, t_2, t_3; s_1, s_2) P^{(2)}(s_1, s_2; x)\, ds_1\, ds_2 \\ \ldots \quad \ldots \quad \ldots \quad & \text{etc.,} \end{aligned}$$

and determine the g's by the orthogonality conditions. This leads to a set of simultaneous integral equations for the g's. In the case where the stochastic process is stationary, the kernels g will involve only time differences (see the remark in the next section).

When the x-process is stationary with zero mean and symmetrical distribution the p's become:

$$P^{(0)}(x) = 1,$$

$$P^{(1)}(t_1; x) = x(t_1),$$

$$P^{(2)}(t_1, t_2; x) = x(t_1)x(t_2) - \overline{x(t_1)x(t_2)},$$

$$P^{(3)}(t_1, t_2, t_3; x) = x(t_1)x(t_2)x(t_3) - \int G(s - t_1, s - t_2, s - t_3) x(s)\, ds,$$

$$\ldots \quad \ldots \quad \ldots \quad \ldots \quad \ldots \quad \text{etc.,}$$

where G satisfies

$$\overline{x(t_1)x(t_2)x(t_3)x(t_4)} = \int G(s - t_1, s - t_2, s - t_3) \overline{x(t_4)x(s)}\, ds \ (\text{all } t_1, t_2, t_3 \text{ and } t_4).$$

This only involves time differences and may be solved by taking a 4-dimensional Fourier transform of both sides.

§ 6. Generalized Wiener–Hopf Equations

In this section the orthogonal expansions just discussed will be used to derive a set of equations of the Wiener–Hopf type applicable to the problem of least squares filtering and prediction. In the linear case when the associated stochastic processes are stationary ergodic, the set of equations reduces to one, namely, the Wiener Hopf equation.

The following general problem will be considered ('regression problem').

Given two dependent stochastic processes with values $x(t)$, $\xi(t)$ at time t, what is the best mean square estimate of $\xi(t)$, based on observation o past values of x, i.e. the values $x(t'), t' \leqslant t$? This includes the filtering problem ($\xi =$ signal, $x =$ signal + noise), and the prediction problem ($\xi(t) = x(t+\alpha)$, $\alpha > 0$) and various others, e.g. combined filtering and prediction, approximate differentiation in the presence of noise (Wiener 1949), etc.

This regression problem may be stated in the following way: What Volterra-type functional of finite mean square relative to the x-process minimizes

$$\delta = \overline{(\xi(t) - f(t; x(t'), t' \leqslant t))^2},$$

where the bar denotes ensemble average?

It is not difficult to show (Doob) that there always exists a unique minimizing functional f of the class considered. From the uniqueness property it follows that if the x- and ξ-processes are stationary then the minimizing functional f is time-invariant. For if f minimizes the expression, so do all its time-translates and so these must coincide. In this stationary case, it follows that if the given expression δ is minimized at one time, it is minimized for all time.

If the x- and ξ-processes are stationary ergodic, the ensemble mean may be interpreted as a time-mean

$$\lim_{T\to\infty} \frac{1}{2T}\int_{-T}^{T} (\xi(t) - f(x(t'), t' \leqslant t))^2 \, dt$$

(omitting t from f because f is time-invariant).

6.1. *Use of Orthogonal Expansions*

To derive the above-mentioned equations, f will be developed in a partially orthogonal set of functionals relative to the x-process†. To illustrate the method, the analogous problem will be considered between two real variables x and ξ.

Assuming that x and ξ are real dependent random variables, consider the problem of minimizing $\overline{(\xi - f(x))^2}$ for functions f of finite mean square relative to the x-distribution.

† x is the input to the filter to be optimized.

Expand f by the set of orthogonal polynomials $P_n(x)$ of the x-distribution. Assume this set complete:

$$f(x)=a_0P_0(x)+a_1P_1(x)+\dots \quad \dots+a_nP_n(x)+\dots \qquad \text{(A)}$$

$$\delta=\overline{(\xi-f(x))^2}$$

$$=\overline{\xi^2}-2\sum_{n=0}^{\infty}a_n\overline{\xi P_n(x)}+\sum_{m,n=0}^{\infty}a_ma_n\overline{P_m(x)P_n(x)}$$

$$=\overline{\xi^2}-2\sum_{n=0}^{\infty}a_n\overline{\xi P_n(x)}+\sum_{n=0}^{\infty}a_n^2\overline{P_n^2(x)} \qquad \text{(B)}$$

since $\overline{P_m(x)P_n(x)}=0$ if $m\neq n$. If δ is a minimum, $d\delta/d\alpha^r=0$, $r=0,1,\dots$:

$$\frac{\partial\delta}{\partial a_r}=-2\overline{\xi P_r(x)}+2a_r\overline{P_r^2(x)}$$

$$=0 \quad \text{if}$$

$$\overline{P_r^2(x)}a_r=\overline{\xi P_r(x)}; \qquad \text{(C)}$$

i.e. $\quad$ if $a_r=\dfrac{\overline{\xi P_r(x)}}{\overline{P_r^2(x)}}$ (the denominator does not vanish) $\qquad$ (C)

or

$$f(x)=\sum_{n=0}^{\infty}\frac{\overline{\xi P_n(x)}P_n(x)}{\overline{P_n^2(x)}}. \qquad \text{(D)}$$

This will correspond to the minimum value.

The partial sums of $f(x)$ may be expressed by determinants of the form

$$f(x)=\begin{vmatrix} 0 & 1 & x & \dots & x^m \\ \overline{\xi x} & \overline{x} & \overline{x^2} & \dots & \overline{x^{m'}} \\ \dots & \dots & \dots & \dots & \dots \\ \overline{\xi x^m} & \overline{x^n} & \overline{x^{n-1}} & \dots & \overline{x^{2n}} \end{vmatrix} \div \mathrm{D} \qquad \text{(E)}$$

where D is the leading minor. This follows as in § 5.2.

Now return to the main problem. *It is assumed that there exists a complete set of partially orthogonal functional polynomials* $P^{(0)}(x)$, $P^{(1)}(t;x)$, $P^{(2)}(t_1,t_2;x),\dots$ *relative to the stochastic process* x. Expand f in terms of these:

$$f(t;x)=f(t;x(t')t'\leqslant t)$$

$$=a_0(t)P^{(0)}(x)+\int a_1(t;t_1)P^{(1)}(t_1;x)\,dt_1$$

$$+\iint a_2(t;t_1,t_2,)P^{(2)}(t_1,t_2;x)\,dt_1\,dt_2. \qquad \text{(A}')$$

The problem is to find the optimal values of the kernels a.

$$\delta=\overline{(\xi(t)-f(t;x))^2}$$

$$=\overline{\xi^2(t)}-2\sum_{n=0}^{\infty}\iint\dots\int a_n(t;t_1,t_2,\dots,t_n)\,\overline{\xi(t)P^{(n)}(t_1,t_2,\dots,t_n;x)}\,dt_1\dots dt_n$$

$$+\sum_{n=0}^{\infty}\iint\dots\int a_n(t;s_1,s_2,\dots,s_n)a_n(t;t_1,t_2,\dots,t_n)\,\overline{P^{(n)}(s_1,s_2,\dots,s_n)P^{(n)}(t_1,t_2,\dots,t_n)}\,ds_1\dots ds_1\,dt_1\dots dt_n$$

the cross-terms vanishing as in the dimensional case.

If this expression is minimized by a particular set of kernels a, then the first variation when one of these kernels, say $a_r(t; s_1, s_2, \dots, s_r)$ is varied by $\delta a_r(t; s_1, s_2, \dots, s_r)$ will vanish. This gives, as in the usual Wiener–Hopf derivation, the equation:

$$-2\int\int\dots\int \delta a_r(t; t_1, t_2, \dots, t_n)\Big[\overline{\xi(t)P^{(r)}(t_1, t_2, \dots, t_r; x)}$$

$$-\int\int\dots\int a_r(t; s_1, s_2, \dots, s_r)\overline{P^{(r)}(s_1, s_2, \dots, s_r; x)P^{(r)}(t_1, t_2, \dots, t_r; x)}$$

$$\times\, ds_1\, ds_2 \dots ds_r\Big]\, dt_1\, dt_2 \dots dt_r = 0.$$

$\because$ since $\delta a_r(t; t_1, t_2, \dots, t_r)$ is an arbitrary function in the region $t_1, t_2, \dots, t_r \leqslant t$ it follows that:

$$\int\int\dots\int a_r(t; s_1, s_2, \dots, s_r)\overline{P^{(r)}(s_1, s_2, \dots, s_r; x)P^{(r)})(t_1, t_2, \dots, t_r; x)}\, ds_1\, ds_2 \dots ds_r = \overline{\xi(t)P^{(r)}(t_1, t_2, \dots, t_r; x)}$$

$$\text{for } t_1, t_2, \dots, t_r \leqslant t \qquad r = 0, 1, \dots \qquad \text{(C}')$$

This is the required set of equations to be solved for the kernel a_r:

$P^{(0)}(x) = 1$ Therefore the first equation is

$a_0(t) = \overline{\xi(t)}$ (Compare (C′) with $r = 0$).

$P^{(1)}(x) = x(t)$ Thus the second equation is

$$\int a_1(t; s_1)\overline{x(s_1)x(t_1)}\, ds_1 = \overline{\xi(t)x(t_1)};$$

i.e. the Wiener–Hopf equation in the non-stationary case. This has been considered by Dolph and Woodbury (1952).

In most applications to filtering and prediction problems, the x- and ξ-processes have a symmetrical distribution about $x = 0$, $\xi = 0$. In this case, only the equations corresponding to r odd will remain. For $P^{(r)}(t_1, t_2, \dots, t_r; x)$ is an even polynomial functional of x in this case if r is even.

Therefore $$\overline{\xi(t)P^{(r)}(t_1, t_2, \dots, t_r; x)} = 0$$

by the symmetry of the distribution.

When the x- and ξ-processes are stationary, all the functions occurring in eqns. (C′) depend only on time differences. It would seem likely that these equations could be solved by an extension of the methods used for the Wiener–Hopf equation.

References

Bellman, R., 1954, Rand Corporation Report.

Bochner, S., and Chandrasekharan, K., 1949, *Fourier Transforms.* Annals of Mathematics Studies (Princeton).

Bode, H. W., 1945, *Network Analysis and Feedback Amplifier Design* (D. van Nostrand Co. Inc.).

Bode, H. W., and Shannon, C. E., 1950, *Proc. Inst. Radio Engrs, N.Y.*, **38,** 417.

Booton, R. C., Mathews, M. V., and Seifert, W., 1953, *Report No.* 70, D.A.C.L., M.I.T.

Bourbaki, N., 1952, *Éléments de Mathématique. Actualitiés Scientifiques et Industrielles,* No. 1175 (Paris : Hermann & Cie).

Cameron, R. H., and Martin, W. T., 1947, *Ann. Math.*, **48,** 385.

Cherry, C., 1951, *Phil. Mag.*, **42,** 1161.

Courant, R., and Hilbert, D., 1953, *Methods of Mathematical Physics* (Interscience Publishers).

Cramér, H., 1946, *Mathematical Methods of Statistics* (Princeton).

Dolph, C. L., and Woodbury, M. A., 1952, *Trans. Amer. Math. Soc.*, **72,** 519.

Doob, J. L., *Stochastic Processes* (Chapman & Hall and John Wiley & Sons Inc.).

Friedrichs, K. O., 1953, *Mathematical Aspects of the Quantum Theory of Fields* (Interscience Publishers), p. 52.

Gabor, D., 1946, *J. Instn. elect. Engrs,* **93,** 429.

Goursat, E., 1904, *A Course in Mathematical Analysis* (Hedrick Ginn & Co.).

Grad, H., 1949, *Commun. Pure appl. Math.*, **2,** 325.

Hille, E., 1948, *Functional Analysis and Semi-Groups.* American Mathematical Society Colloquium Publication. Vol. XXXI, Chap. IV.

Ikehara, S., 1951, Report No. 217. Research Laboratory of Electronics, M.I.T.

James, H. M., Nichols, N. B., and Phillips, R. S., 1947, *Theory of Servomechanisms* (McGraw-Hill).

Lafleur, Ch., and Namias, V., 1954, *Proc. U.R.S.I.* (The Hague).

Lampard, D. G., 1955, *Proc. Instn. elect. Engrs,* C, **102,** 34.

Lee, Y. W., 1932, *J. Math. Phys.*, M.I.T., 83.

Lévy, P., 1951, *Problèmes concret's d'Analyse Fonctionelle* (Paris : Gauthier-Villars).

Lovitt, W. V., 1950, *Linear Integral Equations* (Dover).

Millar, W., 1951, *Phil. Mag.*, **42,** 1150.

Schwartz, L., (1950, 1951), *Theories des Distributions.* 2 volumes. Actualitiés Scientifiques et Industrielles, 1091, 1122. Publications de l'institute de mathematique de l'Université de Strasbourg.

Shannon, C. E., and Weaver, W., 1949, *The Mathematical Theory of Communication* (University of Illinois Press).

Szegö, G., 1939, *Orthogonal Polynomials.* American Mathematical Society Colloquium Publication, Vol. XXIII.

Trimmer, J. D., 1950, *The Response of Physical Systems* (Chapman & Hall and John Wiley & Sons Inc.).

Turing, A. M., 1936, *Proc. Lond. Math. Soc.*, **42,** 230.

Volterra, V., 1930, *Theory of Functionals* (Blackie).

Wiener, N., 1930, *Acta Math., Stockh.*, **55,** 117 ; 1942, Report No. 129, Radiation Laboratory, M.I.T. Available as U.S. Department of Commerce report PB-58087 ;
1949, *Extrapolation, Interpolation and Smoothing of Stationary Time Series* John Wiley & Sons Inc. and Technology Press, M.I.T.).

Zadeh, L. S., 1952, *J. Franklin Inst.*, **253,** 310.

Errata

Page 589, lines 7 and 8 from bottom: should read

$$\iiint_{\omega_1+\omega_2+\omega_3+\omega_4=0} \boldsymbol{K}_1^*(\omega_1)\boldsymbol{K}_3^*(\omega_2, \omega_3, \omega_4) \cdots d\omega_1 d\omega_2 d\omega_3 d\omega_4$$

Page 598, lines 5 through 7 after equation (J′): should read

$$\begin{aligned}&+G(\omega_1+\omega_2+\omega_3, \omega_4, \omega_5)\; F(\omega_1, \omega_2, \omega_3)F(\omega_4)F(\omega_5)\\ &+G(\omega_1, \omega_2+\omega_3+\omega_4, \omega_5)\; F(\omega_1)F(\omega_2, \omega_3, \omega_4)F(\omega_5)\\ &+G(\omega_1, \omega_2, \omega_3+\omega_4+\omega_5)\; F(\omega_1)F(\omega_2)F(\omega_3, \omega_4, \omega_5)\end{aligned}$$

Page 598, last two lines: should read

$$\begin{aligned}\times\Bigg\{ &F(\omega_1+\omega_2+\omega_3, \omega_4, \omega_5)\frac{F(\omega_1, \omega_2, \omega_3)}{F(\omega_1+\omega_2+\omega_3)}\\ &+F(\omega_1, \omega_2, +\omega_3+\omega_4, \omega_5)\frac{F(\omega_2, \omega_3, \omega_4)}{F(\omega_2+\omega_3+\omega_4)}\\ &+F(\omega_1, \omega_2, \omega_3+\omega_4+\omega_5)\frac{F(\omega_3, \omega_4, \omega_5)}{F(\omega_3+\omega_4+\omega_5)} - F(\omega_1, \omega_2, \omega_3, \omega_4, \omega_5)\Bigg\}\end{aligned}$$

Page 601, 4th line before §5: should read

It is found that: $G(\omega) = \dfrac{1}{1+a_1K(\omega)}$

$$G(\omega_1, \omega_2, \omega_3) = -a_3\frac{1}{[1+a_1K(\omega_1+\omega_2+\omega_3)]\prod_{i=1}^{3}[1+a_1K(\omega_i)]}$$

Page 601, in the product symbol Π in 3rd line before §5: should read

$$\prod_{i=1}^{5}[1+a_1K(\omega_i)]$$

Page 606, 1st line after equation (I″): should read:

(H″) thus gives . . .

Page 609, 3rd line: should read

$$1;\ x_1, x_2, \ldots, x_n;\ x_1^2, x_1x_2, \ldots, x_1x_n;\ x_2x_1, x_2^2, \ldots, x_2x_n;\ \ldots;\ x_nx_1, \ldots, x_n^2;\ x_1^3, \ldots$$

Page 609, last line in Equation (D): should read

$$\begin{vmatrix} \overline{x_i x_j} & \overline{x_1} & \overline{x_2} & \cdots & \overline{x_n} & 1 \end{vmatrix}$$

Page 609, Equation (E): should read

$$\overline{Q_{ij}^{(2)}(\mathbf{x})\,Q_{kl}^{(2)}(\mathbf{x})} = \begin{vmatrix} \overline{x_k x_l x_i x_j} & & \overline{x_k x_l x_1} & \cdots & \overline{x_k x_l} \\ \overline{x_1 x_i x_j} & & \overline{x_1^{\,2}} & \cdots & \overline{x_1} \\ \cdots & & \cdots & \cdots & \cdots \\ \overline{x_i x_j} & \cdots & \overline{x_1} & \cdots & 1 \end{vmatrix} \times C_2,$$

where

$$C_2 = \begin{vmatrix} \overline{x_1^{\,2}} & \overline{x_1 x_2} & \cdots & \overline{x_1} \\ \overline{x_2 x_1} & \overline{x_2^{\,2}} & \cdots & \overline{x_2} \\ \cdots & \cdots & \cdots & \cdots \\ \overline{x_1} & \overline{x_2} & \cdots & 1 \end{vmatrix}$$

Page 610, line 1: should read

The polynomials $P^{(n)}(\mathbf{x})$ in the case . . .

Page 613, Equation (E): should read

$$f(x) = -\begin{vmatrix} 0 & 1 & x & \cdots & x^m \\ \overline{\xi x} & \overline{x} & \overline{x^2} & \cdots & \overline{x^{m+1}} \\ \cdots & \cdots & \cdots & \cdots & \cdots \\ \overline{\xi x^m} & \overline{x^n} & \overline{x^{n+1}} & \cdots & \overline{x^{2n}} \end{vmatrix} \div \mathrm{D}$$

Page 613, 2nd line from bottom: should read

$$\overline{P^{(n)}(s_1, s_2, \ldots, s_n\,; x)P^{(n)}(t_1, t_2, \ldots, t_n\,; x)}\,ds_1 \cdots ds_n dt_1 \cdots dt_n$$

15

Reprinted from *Proc. IEEE,* **59**(12), 1688–1707 (1971)

The Output Properties of Volterra Systems (Nonlinear Systems with Memory) Driven by Harmonic and Gaussian Inputs

EDWARD BEDROSIAN, SENIOR MEMBER, IEEE, AND STEPHEN O. RICE, FELLOW, IEEE

Abstract—Troublesome distortions often occur in communication systems. For a wide class of systems such distortions can be computed with the help of Volterra series.

Results, both old and new, which will aid the reader in applying Volterra-series-type analyses to systems driven by sine waves or Gaussian noise are presented.

The n-fold Fourier transform G_n of the nth Volterra kernel plays an important role in the analysis. Methods of computing G_n from the system equations are described and several special systems are considered. When the G_n are known, items of interest regarding the output can be obtained by substituting the G_n in general formulas derived from the Volterra series representation. These items include expressions for the output harmonics, when the input is the sum of two or three sine waves, and the power spectrum and various moments, when the input is Gaussian. Special attention is paid to the case in which the Volterra series consists of only the linear and quadratic terms.

PART I: STATEMENT OF RESULTS AND EXAMPLES

I. Introduction

Volterra series were introduced into nonlinear circuit analysis in 1942 by Wiener. Later Wiener [1] extended the theory and applied it in a general way to a number of problems including FM spectra. Since Wiener's early work many reports and papers have dealt with the subject.

Volterra series are particularly useful in calculating small (but troublesome) distortions in communication systems and have been used recently to determine the distortion produced in various types of amplifiers [2]–[6]. The distortion due to filters in an FM system can also be expressed as a Volterra series.

The object of this paper is to present results, both old and new, that will aid the reader in applying Volterra-series-type analyses to systems driven by sine waves or Gaussian noise. The events which led to this paper began when we learned of Mircea's [7] elegant series for the power spectrum of the distortion produced by filters in an FM system and of its extension by Mircea and Sinnreich [5] to systems described by Volterra series. In papers [8]–[10], which we published before we were aware of Mircea's work, we gave the second- and third-order modulation terms in Mircea's FM series. In a subsequent exchange of letters [11] (to the Editor of the Proceedings of the IEEE) with Mircea some new ideas were brought out. The present paper extends these ideas.

In content our paper resembles a 1955 paper by Deutsch [12] and (to a lesser degree) the 1959 report by George [13], in which a number of the topics discussed here are treated somewhat differently.

Volterra series have been described as "power series with memory" which express the output of a nonlinear system in "powers" of the input $x(t)$. A substantial number of the systems encountered in communication problems can be represented as Volterra series. We shall write the series for the typical system as

$$y(t) = \sum_{n=1}^{\infty} \frac{1}{n!}\int_{-\infty}^{\infty} du_1 \cdots \int_{-\infty}^{\infty} du_n g_n(u_1, \cdots, u_n) \prod_{r=1}^{n} x(t - u_r) \quad (1)$$

where $y(t)$ is the output, $x(t)$ the input, and the kernels $g_n(u_1, \cdots, u_n)$ describe the system.[1] It will be noted that the first-order kernel $g_1(u_1)$ is simply the familiar impulse response of a linear network. The higher order kernels can thus be viewed as higher order impulse responses which serve to characterize the various orders of nonlinearity.

The coefficient $1/n!$ in (1) is not used by most writers. We insert it because it simplifies many of our equations. Some authors allow the kernels to be unsymmetric functions of the u's; however, symmetry is necessary for the results presented here. If the response of a system is obtained as a series of the form (1) containing an unsymmetric kernel say γ_n, in place of g_n, a symmetric kernel can be obtained by "symmetrization." This process consists of permuting the subscripts on the u_i in all $n!$ ways, and taking g_n to be $1/n!$ times the sum of the resulting γ_n.

The n-fold Fourier transform

$$G_n(f_1, \cdots, f_n) = \int_{-\infty}^{\infty} du_1 \cdots \int_{-\infty}^{\infty} du_n g_n(u_1, \cdots, u_n) \cdot \exp\left[-j(\omega_1 u_1 + \cdots + \omega_n u_n)\right] \quad (2)$$

where $\omega_i = 2\pi f_i$ plays an important role in the analysis. G_0 is identically zero because our Volterra series starts with $n=1$ (instead of $n=0$, which would imply an active system, i.e., an output without an input). Also, $G_1(f_1)$ will be recognized as the familiar transfer function of a linear network. Thus the transform of the nth-order Volterra kernel is seen to be analogous to an nth-order transfer function. We shall refer to $G_n(f_1, \cdots, f_n)$ as the "nth-order Volterra transfer function." Since $g_n(u_1, \cdots, u_n)$ is a symmetric function of $u_1, \cdots, u_n$, it follows that $G_n(f_1, \cdots, f_n)$ is a symmetric function of $f_1, \cdots, f_n$. As discussed in Section III, in many cases G_n can be obtained without first computing g_n.

Suppose that the G_n, $n=1, 2, \cdots$, for a particular system are known. Suppose further that the input $x(t)$ to the system in (1) consists of 1) one or more sine waves, 2) Gaussian noise, 3) a sine wave plus Gaussian noise, or 4) a random pulse train. Then we can obtain expressions for a number of items of interest regarding the output $y(t)$ by substituting the G_n in formulas derived from the Volterra series for $y(t)$. The leading terms in some of these formulas are listed in Section II. This list is intended to be a guide to the complete formulas derived in the later sections.

The complete formulas are infinite series in which the labor of computing the nth term increases rapidly as n increases. Fortunately, in the study of communication systems it is often possible to neglect

Manuscript received May 12, 1970; revised April 22, 1971.

E. Bedrosian is with the Rand Corporation, Santa Monica, Calif. 90406.
S. O. Rice is with the Bell Telephone Laboratories, Inc., Murray Hill, N. J. 07974.

[1] The arguments of $g_n(u_1, \cdots, u_n)$ and its Fourier transform $G_n(f_1, \cdots, f_n)$ in (2) will occasionally be omitted for brevity when the meaning is clear.

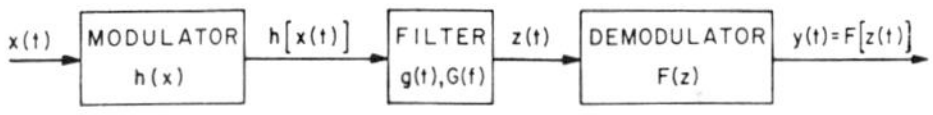

Fig. 1. Modulator–filter–demodulator system.

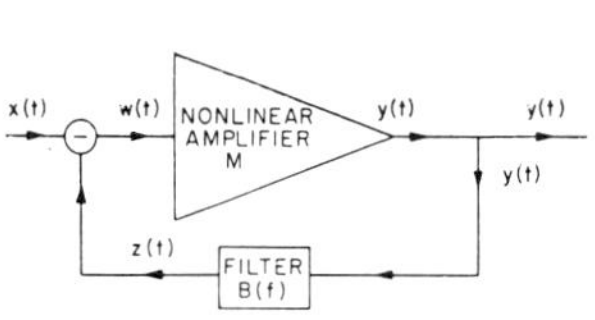

Fig. 2. Feedback system with nonlinear amplifier.

modulation terms (i.e., terms in the Volterra series) of order higher than the second or third.

In practice it appears that Volterra series do not enable us to do anything that cannot be done otherwise. However, a direct attack on modulation problems often leads to a morass of algebra. The Volterra series approach has the virtue that many such problems can be treated in an orderly way by first computing the G_n and then substituting them in the appropriate general formulas.

For convenience, the paper is separated into two major parts: I, Statement of Results and Examples and II, Derivation of Formulas. A summary form of the leading terms of the principal results follows these introductory remarks. Then methods are presented for computing the Volterra transfer functions $G_n(f_1, \cdots, f_n)$ for several types of systems. Finally, a number of examples are worked out to illustrate the use of the formulas. Among these are 1), the general modulator–filter–demodulator system, shown in Fig. 1, which is used to obtain the leading terms in Mircea's series for FM distortion and 2) the nonlinear system with feedback, shown in Fig. 2, recently treated by Narayanan [6].

Part II consists largely of derivations of the various results. These include expressions for the output and its spectrum for a number of inputs. In addition, moments of higher order are considered for the case of a Gaussian input in order to obtain expressions for the cumulants of the output distribution. Formulas are listed which give information about the probability density of $y(t)$ when the cumulants are known. Known results concerning the distribution of quadratic forms are applied to obtain an expression for the probability density of $y(t)$ when $x(t)$ is Gaussian and terms beyond the second are omitted in the Volterra series for $y(t)$.

II. Leading Terms in Formulas

This section lists the leading terms of formulas which give information about the output $y(t)$ for a number of inputs $x(t)$ when the Volterra transfer functions G_n are known. Most of the leading terms listed do not go beyond $G_3(f_1, f_2, f_3)$. Note that G_n is symmetric.

Although the list is intended as a guide to the complete formulas given in the later sections, the leading terms often suffice. In fact, the reader should not expect too much practical help from the complete formulas. Usually only two or three terms beyond those listed in this section can be used with present-day computers because of the rapid increase in complexity.

A. *Sinusoidal Inputs*

When $x(t) = P \cos pt$, (137) gives the complete series

$$y(t) = \sum_{n=1}^{\infty} \sum_{k=0}^{n} \left(\frac{P}{2}\right)^n \frac{\exp[j(2k-n)pt]}{k!(n-k)!} G_{k,n-k}(f_p) \tag{3}$$

where $p = 2\pi f_p$ and $G_{k,n-k}(f_p)$ denote $G_n(f_1, \cdots, f_n)$ with the first k of the f_i equal to f_p and the remaining $n-k$ equal to $-f_p$. For the example of the memoryless nonlinearity in the next section, (23) shows that $G_n(f_1, \cdots, f_n)$ reduces to a constant a_n, and the series (3) may either converge or diverge depending upon P and the a_n. The leading terms in (3) give

$$\begin{aligned} y(t) = &\left[\frac{P^2}{4} G_2(f_p, -f_p) + \cdots\right] \\ &+ e^{jpt}\left[\frac{P}{2} G_1(f_p) + \frac{P^3}{16} G_3(f_p, f_p, -f_p) + \cdots\right] \\ &+ e^{j2pt}\left[\frac{P^2}{8} G_2(f_p, f_p) + \cdots\right] \\ &+ e^{j3pt}\left[\frac{P^3}{48} G_3(f_p, f_p, f_p) + \cdots\right] \\ &+ e^{-jpt}\left[\frac{P}{2} G_1(-f_p) + \frac{P^3}{16} G_3(-f_p, -f_p, f_p) + \cdots\right] \\ &+ e^{-j2pt}\left[\frac{P^2}{8} G_2(-f_p, -f_p) + \cdots\right] \\ &+ e^{-j3pt}\left[\frac{P^2}{48} G_3(-f_p, -f_p, -f_p) + \cdots\right] + \cdots. \end{aligned} \tag{4}$$

Replacing pt by $(pt + \varphi)$ in the exponents in (3) and (4) gives $y(t)$ when $x(t) = P \cos(pt + \varphi)$.

When $x = P \cos pt + Q \cos qt$ where p and q are incommensurable, the complete series for the exp $[j(Np + Mq)t]$ component of $y(t)$ is given by (139). The leading terms in the dc and some of the lower order components of $y(t)$ are

$$\begin{gathered} \left[\frac{P^2}{4} G_2(f_p, -f_p) + \frac{Q^2}{4} G_2(f_q, -f_q)\right] \\ e^{jpt}\left[\frac{P}{2} G_1(f_p) + \frac{P^3}{16} G_3(f_p, f_p, -f_p) + \frac{PQ^2}{8} G_3(f_p, f_q, -f_q)\right] \\ e^{j2pt} \frac{P^2}{8} G_2(f_p, f_p) \qquad e^{j(p+q)t} \frac{PQ}{4} G_2(f_p, f_q) \\ e^{j(2p+q)t} \frac{P^2 Q}{16} G_3(f_p, f_p, f_q). \end{gathered} \tag{5}$$

Changing the signs of q and f_q in the exp $[j(p+q)t]$ component gives the exp $[j(p-q)t]$ component, and so on. When p and q are not incommensurable, some of the components coalesce. For example, if $q = 2p$ and $x(t) = P \cos pt + Q \cos 2pt$, the $2p - q$ and $-2p + q$ terms combine with the dc terms in (5) to give for the leading terms in the new dc component

$$\begin{aligned} &\left[\frac{P^2}{4} G_2(f_p, -f_p) + \frac{Q^2}{4} G_2(2f_p, -2f_p) + \frac{P^2 Q}{16} G_3(f_p, f_p, -2f_p)\right. \\ &\qquad \left. + \frac{P^2 Q}{16} G_3(-f_p, -f_p, 2f_p)\right]. \end{aligned} \tag{6}$$

Similarly, the leading terms in the new exp (jpt) component are given by the sum of the exp (jpt) component in (5) with $f_q = 2f_p$ plus

$$e^{jpt} \frac{PQ}{4} G_2(-f_p, 2f_p) \tag{7}$$

which is the contribution of the $(-p+q)$ term when $q = 2p$. This term is obtained from the $(p+q)$ term in (5) by changing the signs of p and f_p and then setting $q = 2p$ and $f_q = 2f_p$.

When $x(t) = P \cos pt + Q \cos qt + R \cos rt$, the complete series for the exp $[j(Np + Mq + Lr)t]$ component of $y(t)$ is given by (140).

For example, the leading term in the exp $[j(p+q+r)t]$ component of $y(t)$ is

$$e^{j(p+q+r)t}\frac{PQR}{8}G_3(f_p, f_q, f_r). \tag{8}$$

Changing the signs of r and f_r gives the leading term in the exp $[j(p+q-r)t]$ component, and so on.

When $x(t)$ is equal to the sum of an infinite number of sinusoidal components in the sense that it possesses the Fourier transform $X(f)$, i.e.,

$$x(t) = \int_{-\infty}^{\infty} e^{j\omega t}X(f)df, \qquad \omega = 2\pi f \tag{9}$$

then substitution in (1) shows that $y(t)$ and its Fourier transform $Y(f)$ are given by

$$y(t) = \sum_{n=1}^{\infty}\frac{1}{n!}\int_{-\infty}^{\infty}df_1\cdots\int_{-\infty}^{\infty}df_nG_n(f_1,\cdots,f_n) \cdot e^{j(\omega_1+\cdots+\omega_n)t}\prod_{r=1}^{n}X(f_r)$$

$$Y(f) = \frac{1}{1!}G_1(f)X(f) + \frac{1}{2!}\int_{-\infty}^{\infty}df_1G_2(f_1, f-f_1)X(f_1)X(f-f_1) + \frac{1}{3!}\int_{-\infty}^{\infty}df_1\int_{-\infty}^{\infty}df_2G_3(f_1,f_2,f-f_1-f_2)X(f_1)X(f_2) \cdot X(f-f_1-f_2) + \cdots. \tag{10}$$

B. Gaussian Noise Input

In the following, the input $x(t)$ is a zero-mean stationary Gaussian process with the two-sided power spectrum $W_x(f)$. The output $y(t)$ is a stationary process, and the ensemble averages $\langle[y(t)]^l\rangle$ and associated cumulants κ_l do not change with t.

The leading terms in the complete series (147) for $\langle y(t)\rangle$ are

$$\langle y(t)\rangle = \frac{1}{1!2}\int_{-\infty}^{\infty}df_1W_x(f_1)G_2(f_1,-f_1) + \frac{1}{2!2^2}\int_{-\infty}^{\infty}df_1\int_{-\infty}^{\infty}df_2 \, W_x(f_1)W_x(f_2)G_4(f_1,-f_1,f_2,-f_2) + \cdots. \tag{11}$$

The leading terms in the complete series (177) for $\langle y^2(t)\rangle$ are

$$\langle y^2(t)\rangle = \langle y(t)\rangle^2 + \int_{-\infty}^{\infty}df_1W_x(f_1)G_1(f_1)G_1(-f_1) + \int_{-\infty}^{\infty}df_1\int_{-\infty}^{\infty}df_2W_x(f_1)W_x(f_2)[G_1(f_1)G_3(-f_1,f_2,-f_2) + \tfrac{1}{2}G_2(f_1,f_2)G_2(-f_1,-f_2)] + \cdots. \tag{12}$$

The triple integral comprising the third-order term not shown in (12) is given in (180) for the second cumulant $\kappa_2=\langle y^2(t)\rangle-\langle y(t)\rangle^2$. The complete series for $\langle y^2(t)\rangle$ is a special case of the complete series for $\langle y(t+\tau)z(t)\rangle$ given by (152), (156)–(158). Here $z(t)$ is defined by a Volterra series obtained from the series (1) for $y(t)$ by replacing $g_n(u_1,\cdots,u_n)$ by a different kernel $g_n'(u_1,\cdots,u_n)$. Both $y(t)$ and $z(t)$ have the same Gaussian input $x(t)$. The $G_n'(f_1,\cdots,f_n)$ which appears in (157) is the Fourier transform of $g_n'(u_1,\cdots,u_n)$.

The first cumulant for the probability density of $y(t)$ is $\kappa_1=\langle y(t)\rangle$, the second is $\kappa_2=\langle y^2(t)\rangle-\langle y(t)\rangle^2$, and from (180) the leading terms in κ_3 and κ_4 are

$$\kappa_3 = 3\int_{-\infty}^{\infty}df_1\int_{-\infty}^{\infty}df_2W_x(f_1)W_x(f_2)G_1(f_1)G_1(f_2)G_2(-f_1,-f_2)+\cdots$$

$$\kappa_4 = 4\int_{-\infty}^{\infty}df_1\int_{-\infty}^{\infty}df_2\int_{-\infty}^{\infty}df_3W_x(f_1)W_x(f_2)W_x(f_3)G_1(f_1)G_1(f_2) \cdot[G_1(f_3)G_3(-f_1,-f_2,-f_3) + 3G_2(-f_1,f_3)G_2(-f_2,-f_3)] + \cdots. \tag{13}$$

The next terms beyond those shown in (13) are given in (180), but the general forms of the series for κ_3 and κ_4 are not known. The use of the first four cumulants to obtain information about the probability density of $y(t)$ is discussed in Part II and an example is furnished by (84).

The leading terms in the Mircea–Sinnreich [5] series (160) for the two-sided power spectrum $W_y(f)$ of $y(t)$ are shown in

$$W_y(f) = \langle y(t)\rangle^2\delta(f) + W_x(f)\left|G_1(f) + \frac{1}{2}\int_{-\infty}^{\infty}df_1W_x(f_1)G_3(f,f_1,-f_1)+\cdots\right|^2 + \frac{1}{2!}\int_{-\infty}^{\infty}df_1W_x(f_1)W_x(f-f_1)|G_2(f_1,f-f_1)+\cdots|^2 + \frac{1}{3!}\int_{-\infty}^{\infty}df_1\int_{-\infty}^{\infty}df_2W_x(f_1)W_x(f_2)W_x(f-f_1-f_2) \cdot|G_3(f_1,f_2,f-f_1-f_2)+\cdots|^2+\cdots \tag{14}$$

where $\langle y(t)\rangle$ is the dc component of $y(t)$ given by (11) and $\delta(f)$ is the unit impulse function. The right side of (14) shows all of the second-order terms (those which, when written out, contain the product of exactly two W_x) but only some of the third-order terms. All of the third-order terms, and some of the fourth- and fifth-order, would be shown if the double integral containing G_5 and the single integral containing G_4 were added inside the absolute value signs on the second and third lines, respectively, of (14).

When the number of terms in the Volterra series (1) is finite, the series (14) terminates and gives an exact expression for $W_y(f)$. When (14) does not terminate, its application to FM suggests that it may be an asymptotic series [8], [10], [14].

C. Sine Wave Plus Noise Input

In the following, $x(t)=P\cos pt+I_N(t)$, where $I_N(t)$ is a zero-mean stationary Gaussian process with two-sided power spectrum $W_I(f)$.

The ensemble average of $x(t)$ at time t is $\langle x(t)\rangle=P\cos pt$. Similarly, the ensemble average of $y(t)$ consists of a sum of sinusoidal harmonics of cos pt. The complete expression for $\langle y(t)\rangle$ at time t is given by (164) and the leading terms are shown in

$$\langle y(t)\rangle = \left[\frac{P^2}{4}G_2(f_p,-f_p) + \frac{1}{2}\int_{-\infty}^{\infty}df_1W_I(f_1)G_2(f_1,-f_1)+\cdots\right] + e^{jpt}\left[\frac{P}{2}G_1(f_p) + \frac{P^3}{16}G_3(f_p,f_p,-f_p) + \frac{P}{4}\int_{-\infty}^{\infty}df_1W_I(f_1)G_3(f_1,-f_1,f_p)+\cdots\right] + e^{j2pt}\left[\frac{P^2}{8}G_2(f_p,f_p) + \frac{P^2}{16}\int_{-\infty}^{\infty}df_1 \, W_I(f_1)G_4(f_1,-f_1,f_p,f_p)+\cdots\right] + e^{j3pt}\left[\frac{P^3}{48}G_3(f_p,f_p,f_p)+\cdots\right]+\cdots + \{\text{terms with } -p, -f_p \text{ for } p, f_p \text{ in } e^{jkpt}[\cdots], \; k=1,2,\cdots\} \tag{15}$$

where $f_p = p/2\pi$. When $I_N(t)$ is identically zero, $W_I(f)$ is zero, and (15) reduces to (4) for $y(t)$ when $x = P\cos pt$. When P is zero, (15) reduces to (11) for $\langle y(t)\rangle$ when $x(t)$ is Gaussian.

The complete expression for the power spectrum of $y(t)$ is given by (175). The leading terms are shown in

$$\begin{aligned} W_y(f) &= \{\text{spikes due to dc and sine waves in } \langle y(t)\rangle\} \\ &+ W_I(f)\left|G_1(f) + \frac{P^2}{4}G_3(f_p, -f_p, f) \right. \\ &\left. + \frac{1}{2}\int_{-\infty}^{\infty} df_1 W_I(f_1) G_3(f_1, -f_1, f) + \cdots \right|^2 \\ &+ W_I(f - f_p)\left|\frac{P}{2}G_2(f_p, f - f_p) + \cdots\right|^2 \\ &+ W_I(f - 2f_p)\left|\frac{P^2}{8}G_3(f_p, f_p, f - 2f_p) + \cdots\right|^2 + \cdots \\ &+ \{\text{terms with } -f_p \text{ for } f_p \text{ in } W_I(f - kf_p)|\cdots|^2, k = 1, 2, \cdots\} \\ &+ \frac{1}{2!}\int_{-\infty}^{\infty} df_1 W_I(f_1) W_I(f - f_1)|G_2(f_1, f - f_1) + \cdots|^2 \\ &+ \frac{1}{2!}\int_{-\infty}^{\infty} df_1 W_I(f_1) W_I(f - f_1 - f_p) \\ &\cdot \left|\frac{P}{2}G_3(f_1, f_p, f - f_1 - f_p) + \cdots\right|^2 \\ &+ \frac{1}{2!}\int_{-\infty}^{\infty} df_1 W_I(f_1) W_I(f - f_1 + f_p) \\ &\cdot \left|\frac{P}{2}G_3(f_1, -f_p, f - f_1 + f_p) + \cdots\right|^2 \\ &+ \frac{1}{3!}\int_{-\infty}^{\infty} df_1 \int_{-\infty}^{\infty} df_2 W_I(f_1) W_I(f_2) W_I(f - f_1 - f_2) \\ &\cdot |G_3(f_1, f_2, f - f_1 - f_2) + \cdots|^2 + \cdots. \end{aligned} \tag{16}$$

The spikes in $W_y(f)$ due to the dc and sine waves in $\langle y(t)\rangle$ can be computed from (15) for $\langle y(t)\rangle$. The spike due to the component $A_k(f_p, P)\exp(jkpt)$, $k = \cdots -1, 0, \cdots$, is $\delta(f - kf_p)|A_k(f_p, P)|^2$. When P is zero, (16) reduces to (14) for $W_y(f)$ when $x(t)$ is Gaussian. When $I_N(t)$ is identically zero, $W_y(f)$ consists only of the spikes due to the sinusoidal components of $y(t)$.

D. Random Pulse Train Input

Finally we state a result of some interest which has not been thoroughly studied. The input is the pulse train

$$x(t) = \sum_{n=-\infty}^{\infty} a_n \delta(t - nT). \tag{17}$$

When the a_n are identically distributed independent random variables whose probability density is even about $a_n = 0$, and the Volterra series for $y(t)$ stops with the quadratic term, it can be shown that the ensemble average of $y(t)$ consists of the periodic part (of period T)

$$\langle y(t)\rangle = \frac{\langle a^2\rangle}{2T}\sum_{m=-\infty}^{\infty} e^{j2\pi mt/T}\int_{-\infty}^{\infty} df_1 G_2\left(f_1, \frac{m}{T} - f_1\right) \tag{18}$$

and that the power spectrum of $y(t)$ is

$$\begin{aligned} W_y(f) &= \{\text{spikes due to } \langle y(t)\rangle\} + \frac{\langle a^2\rangle}{T}|G_1(f)|^2 \\ &+ \frac{[\langle a^4\rangle - 3\langle a^2\rangle^2]}{4T}\left|\int_{-\infty}^{\infty} df_1 G_2(f_1, f - f_1)\right|^2 \\ &+ \frac{\langle a^2\rangle^2}{2T^2}\sum_{m=-\infty}^{\infty}\int_{-\infty}^{\infty} df_1 G_2(f_1, f - f_1) \\ &\cdot G_2^*\left(f_1 - \frac{m}{T}, f - f_1 + \frac{m}{T}\right). \end{aligned} \tag{19}$$

Here $\langle a^m\rangle$ is the mth moment of a_n, and $G_2(f_1, f_2)$ is assumed to be such that the integrals and the sum converge. The asterisk denotes conjugate complex. Equations (18) and (19) can be proved by using the first two terms in (10) for $y(t)$ and the results

$$\sum_{n=-\infty}^{\infty} e^{-j\omega nT} = T^{-1}\sum_{n=-\infty}^{\infty}\delta(f - nT^{-1}), \qquad \omega = 2\pi f$$

$$W_y(f) = \int_{-\infty}^{\infty} d\tau \int_0^T dt\, T^{-1}\langle y(t + \tau) y^*(t)\rangle e^{-j\omega\tau}.$$

If the pulse shape is $F(t)$ instead of $\delta(t)$, the input is

$$x(t) = \sum_{n=-\infty}^{\infty} a_n F(t - nT) \tag{20}$$

instead of (17), and the corresponding $\langle y(t)\rangle$ and $W_y(f)$ can be obtained by replacing $G_1(f_1)$, $G_2(f_1, f_2)$ in (18) and (19) by $S(f_1)G_1(f_1)$, $S(f_1)S(f_2)G_2(f_1, f_2)$ where

$$S(f) = \int_{-\infty}^{\infty} e^{-j\omega t} F(t)\, dt. \tag{21}$$

III. Computation of Volterra Transfer Functions

Considerable work has been done on the determination of the kernels $g_n(u_1, \cdots, u_n)$ and their Fourier transforms, the Volterra transfer functions $G_n(f_1, \cdots, f_n)$, by measurements made on the system [13], [15], [16]. Here we are principally interested in the calculation of the Volterra transfer functions when the system equations are known and the system can be represented by a Volterra series (which is not always the case).

One of the simplest of these calculations is furnished by the memoryless case

$$y(t) = \sum_{n=1}^{\infty} a_n [x(t)]^n / n! \tag{22}$$

to which the Volterra series reduces when $g_n(u_1, \cdots, u_n)$ is equal to $a_n\,\delta(u_1)\cdots\delta(u_n)$ where $\delta(u)$ is the unit impulse function. Here a_n is a constant and, from definition (2) of the Volterra transfer function,

$$G_n(f_1, \cdots, f_n) = a_n. \tag{23}$$

This case is useful for checking the more complicated results.

This section is divided into three parts. The first describes a summation notation which is convenient in dealing with expressions which hold for general values of n. The second and third parts are concerned with two methods of computing G_n, namely, the "harmonic input" method and the "direct expansion" method.

The harmonic input method is useful in computing $G_n(f_1, \cdots, f_n)$ for the first few values of n and can be used to obtain the expressions for $G_1(f_1)$, $G_2(f_1, f_2)$, and $G_3(f_1, f_2, f_3)$ listed for the examples in this section. The derivations of the expressions listed for $G_n(f_1, \cdots, f_n)$ for arbitrary n are sketched in Sections V-A and V-B and make use of the direct expansion method.

A. Summation Notation

In order to explain the summation notation used to deal with G_n for general values of n, we take for illustration the formula, derived in Section V-A,

$$G_n^{(l)}(f_1,\cdots,f_n) = l! \sum_{(v;l,n)} \sum_N' G_{v_1}(f_1,\cdots,f_{v_1})G_{v_2}(f_{v_1+1},\cdots,f_{v_1+v_2}) \cdots G_{v_l}(f_\mu,\cdots,f_n) \quad (24)$$

for the n-fold Fourier transform of the nth kernel in the Volterra series for $[y(t)]^l$, l being a positive integer, and $1 \le l \le n$. $G_n^{(l)}(f_1,\cdots,f_n)$ is zero for $l>n$ and $G_n^{(n)}(f_1,\cdots,f_n)$ is equal to $n!G_1(f_1)G_1(f_2)\cdots G_1(f_n)$.

In (24) $\mu = v_1+v_2+\cdots+v_{l-1}+1 = n-v_l+1$ and $(v;\ l,\ n)$ beneath the leftmost $\sum$ denotes summation over sets of integers v_i such that

$$v_1 + v_2 + \cdots + v_l = n, \qquad 1 \le v_1 \le v_2 \le \cdots \le v_l. \quad (25)$$

In other words, the summation is taken over those partitions of n which have l parts. The second summation $\sum_N'$ in (24) extends over the N "nonidentical" products that can be obtained by permuting the subscripts on the f's. $G_n(f_1,\cdots,f_n)$ is symmetric, and "identical" is used in the sense that $G_2(f_2,f_1)$ is identical with $G_2(f_1,f_2)$, and $G_1(f_2)G_1(f_1)$ is identical with $G_1(f_1)G_1(f_2)$. The number of terms in $\sum_N'$ is

$$N = n!/v_1!v_2!\cdots v_l!r_1!r_2!\cdots r_k! \quad (26)$$

where r_1 is the number of equal v's in the first run of equalities in the arrangement $v_1 \le v_2 \le \cdots \le v_l$, r_2 the number in the second run, and so on. When the v's are unequal, the r's do not appear.

Sometimes, as in the case of Table I, given later, a table of partitions is of help in writing out the terms in the summation.

When $n=2$ and $l=2$, we have $v_1=v_2=1$, $r_1=2$, $N=2!/(1!1!2!)=1$ and (24) becomes

$$\frac{1}{2!}G_2^{(2)}(f_1,f_2) = G_1(f_1)G_1(f_2). \quad (27)$$

When $n=3$ and $l=2$, we have $v_1=1$, $v_2=2$, $N=3$, and

$$\frac{1}{2!}G_3^{(2)}(f_1,f_2,f_3) = \sum_3' G_1(f_1)G_2(f_2,f_3) = (1)(23) + (2)(13) + (3)(12) \quad (28)$$

in an abbreviated notation. When $n=4$ and $l=2$, there are two 2-part (l-part) partitions of n, $1+3=4$ and $2+2=4$, which give $v_1=1$, $v_2=3$, $N=4$ and $v_1=2$, $v_2=2$; $N=3$, respectively. Hence

$$\frac{1}{2}G_4^{(2)}(f_1,f_2,f_3,f_4) = (1)(234) + (2)(134) + (3)(124) + (4)(123) + (12)(34) + (13)(24) + (14)(23). \quad (29)$$

The number of products of G's in the expressions for $G_2^{(2)}/2!$, $G_3^{(2)}/2!$, $G_4^{(2)}/2!$ are (by counting them in (27)–(29)) 1, 3, 7, respectively. These are the Stirling numbers of the second kind $S(n,2)$ for $n=2,3,4$. In general, the number of products in the sum (24) for $G_n^{(l)}(f_1,\cdots,f_n)/l!$ is $S(n,l)$. This can be shown by taking all of the a_n to be unity in the memoryless case (22) (so that all of the G_n are equal to unity) and using the fact that $l!\,S(n,l)$ is the coefficient of $x^n/n!$ in the expansion of $(e^x-1)^l$.

B. Harmonic Input Method

This method relies on the fact that a harmonic input must result in a harmonic output when (1) holds. Thus, when $x(t)$ is the sum

$$x(t) = \exp(j\omega_1 t) + \exp(j\omega_2 t) + \cdots + \exp(j\omega_n t) \quad (30)$$

where $\omega_i = 2\pi f_i$, $i=1,2,\cdots,n$, and the ω_i are incommensurable, G_n is given by

$$G_n(f_1,\cdots,f_n) = \{\text{coefficient of the } \exp[j(\omega_1+\cdots+\omega_n)t] \text{ term in the expansion of } y(t)\}. \quad (31)$$

This result follows from (1). It enables us to compute $G_1(f_1)$, $G_2(f_1,f_2),\cdots$ in succession. Thus, when we replace $x(t)$ by $\exp(j\omega_1 t)$ in the system equations and assume

$$y(t) = \sum_{k=1}^{\infty} c_k \exp(jk\omega_1 t). \quad (32)$$

$G_1(f_1)$ is equal to c_1. Similarly, $G_2(f_1,f_2)$ is equal to c_{11} where

$$x(t) = \exp(j\omega_1 t) + \exp(j\omega_2 t)$$

$$y(t) = \sum_{k=0}^{\infty}\sum_{l=0}^{\infty} c_{kl}\exp[j(k\omega_1 + l\omega_2)t] \quad (33)$$

$$c_{00} = 0 \qquad c_{10} = G_1(f_1) \qquad c_{01} = G_1(f_2)$$

and $G_3(f_1,f_2,f_3)$ is equal to c_{111} in an analogous triple sum where

$$c_{000} = 0 \qquad c_{110} = G_2(f_1,f_2)$$
$$c_{100} = G_1(f_1) \qquad c_{010} = G_1(f_2),\cdots. \quad (34)$$

To illustrate the use of this method, consider a system described by the equation

$$y(t) = x(t) + \varepsilon[x'(t)]^2 x''(t) \quad (35)$$

which arises in some forms of the quasi-static approximation to filtered FM. Here ε is a constant and the prime denotes differentiation with respect to t. Setting $x(t)$ equal to $\exp(j\omega_1 t)$ gives $G_1(f_1)=1$. Setting $x(t)$ equal to $\exp(j\omega_1 t)+\exp(j\omega_2 t)$ shows that $G_2(f_1,f_2)$ is zero. Setting $x(t)$ equal to the sum of three exponentials and picking out the coefficient of $\exp[j(\omega_1+\omega_2+\omega_3)t]$ in (35) gives G_3. Therefore,

$$G_1(f_1) = 1 \qquad G_2(f_1,f_2) = 0$$
$$G_3(f_1,f_2,f_3) = 2\varepsilon\omega_1\omega_2\omega_3(\omega_1+\omega_2+\omega_3) \quad (36)$$

and G_n is zero for $n>3$. In the Volterra series associated with (35), $g_1(u_1)$ is $\delta(u_1)$ and g_3 is the sum of products of derivatives of impulse functions.

A broader application of the method can be illustrated by considering the system specified by the nonlinear differential equation

$$F(d/dt)y + \sum_{l=2}^{\infty} a_l y^l = x(t) \quad (37)$$

with the condition that $y(t)$ vanish identically when $x(t)$ does. It is assumed that one and only one such solution exists and that the system is stable. $F(d/dt)$ is a polynomial in d/dt, and the coefficients in $F(d/dt)$ and the coefficients a_l are independent of t, x, and y. The first three G_n, derived from (30) through (34), and the recurrence relation derived in Section V-B are

$$G_1(f_1) = 1/F(j\omega_1)$$

$$G_2(f_1,f_2) = -2a_2G_1(f_1)G_1(f_2)/F(j\omega_1+j\omega_2)$$

$$G_3(f_1,f_2,f_3) = -\frac{2a_2\sum_3' G_1(f_1)G_2(f_2,f_3) + 6a_3G_1(f_1)G_1(f_2)G_1(f_3)}{F(j\omega_1+j\omega_2+j\omega_3)}$$

$$G_n(f_1,\cdots,f_n) = -\frac{\sum_{l=2}^{n} a_l G_n^{(l)}(f_1,\cdots,f_n)}{F(j\omega_1+\cdots+j\omega_n)}. \quad (38)$$

The last equation is a recurrence relation because $G_n^{(l)}$ is given by (24) and (for $2 \le l \le n$) the right side of (24) is a combination of some or all (depending on l) of

$$G_1(f_1),\ G_2(f_1,f_2),\cdots,\ G_{n-1}(f_1,\cdots f_{n-1}).$$

As a specific example of the use of these results, let the input $x(t)$ be the voltage applied to a unit inductance connected in series with a slightly nonlinear resistance. The output $y(t)$ is the current through the circuit and is that solution of the Riccati equation

$$\frac{dy}{dt} + \alpha y + \varepsilon y^2 = x(t) \tag{39}$$

which tends to zero when $x(t)$ does. The existence and stability of such a solution is to be expected on physical grounds, provided α and ε are such that the resistance $\alpha + \varepsilon y$ is almost never negative during the operation of the circuit. We regard ε to be so small that εy is almost always small compared to α.

In applying (37) to (39), we take $F(d/dt)$ to be $(d/dt)+\alpha$, $a_2 = \varepsilon$, and obtain from (38), with $\omega_i = 2\pi f_i$,

$$G_1(f_1) = (\alpha + j\omega_1)^{-1}$$

$$G_2(f_1, f_2) = \frac{(-2\varepsilon)[\alpha + j(\omega_1 + \omega_2)]^{-1}}{(\alpha + j\omega_1)(\alpha + j\omega_2)}$$

$$G_3(f_1, f_2, f_3) = \frac{(-2\varepsilon)^2[\alpha + j(\omega_1 + \omega_2 + \omega_3)]^{-1}}{(\alpha + j\omega_1)(\alpha + j\omega_2)(\alpha + j\omega_3)} \cdot \sum_3' \frac{1}{\alpha + j\omega_2 + j\omega_3}$$

$$G_n(f_1, \cdots, f_n) = -\varepsilon[\alpha + j(\omega_1 + \cdots + \omega_n)]^{-1} G_n^{(2)}(f_1, \cdots, f_n) \tag{40}$$

where $G_n^{(2)}$ is given by (24) and in G_3

$$\sum_3' \frac{1}{\alpha + j\omega_2 + j\omega_3} = \frac{1}{\alpha + j\omega_2 + j\omega_3} + \frac{1}{\alpha + j\omega_1 + j\omega_3} + \frac{1}{\alpha + j\omega_1 + j\omega_2}. \tag{41}$$

Another application of the harmonic input method, which is somewhat similar to the nonlinear differential equation system in (37), is concerned with the G_n for the voltage $y(t)$ across the nonlinear device shown in Fig. 3. The voltage $x(t)$ is applied to the series combination of the nonlinear device, defined by

$$I(t) = \sum_{l=1}^{\infty} a_l [y(t)]^l \tag{42}$$

and the linear admittance $H(f)$. As illustrated by the work of Deutsch [12], [17], a knowledge of the series for $y(t)$ is the key to the solution of a number of nonlinear problems. The Volterra transfer functions for $y(t)$ are

$$G_1(f_1) = H(f_1)/[a_1 + H(f_1)]$$

$$G_2(f_1, f_2) = -2a_2 G_1(f_1) G_1(f_2)/[a_1 + H(f_1 + f_2)]$$

$$\vdots$$

$$G_n(f_1, \cdots, f_n) = -\frac{\sum_{l=2}^{n} a_l G_n^{(l)}(f_1, \cdots, f_n)}{a_1 + H(f_1 + \cdots + f_n)} \tag{43}$$

where the last equation is derived in Section V-B. For $n > 1$ these equations differ from (38) only in having the denominator $a_1 + H(f_1 + \cdots + f_n)$ instead of $F(j\omega_1 + \cdots + j\omega_n)$. They furnish a set of recurrence relations for $G_n(f_1, \cdots, f_n)$ which are essentially due to Deutsch [12], [17]. Our $G_n(f_1, \cdots, f_n)$ is $n!$ times the symmetrized version of Deutsch's functions $Q_n(\omega_1, \cdots, \omega_n)$ [with $\omega_i = 2\pi f_i$]. For the circuit of Fig. 3, the multiplier $K(\omega_1)$ appearing in Deutsch's $Q_1(\omega_1)$ is unity.

As a specific example of this application we consider a linear

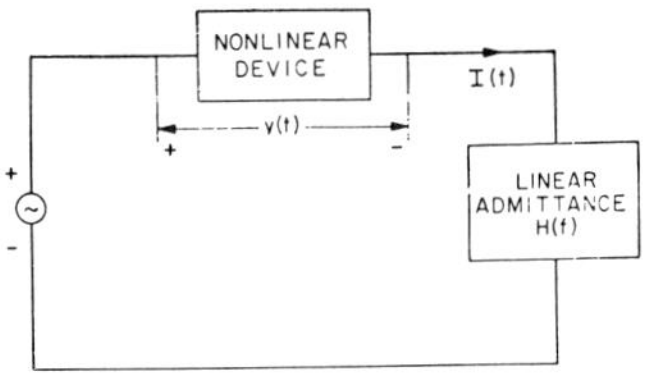

Fig. 3. Nonlinear device in series with linear admittance.

inductance L in series with the square-law device $I(t) = a_2[y(t)]^2$. The circuit equation is

$$\beta \frac{d}{dt} y^2 + y = x(t), \qquad \beta = La_2. \tag{44}$$

This differs somewhat from the Riccati equation (39) for an inductance in series with a nonlinear resistance. The admittance function for the inductance is $H(f) = 1/j\omega L$ and the only coefficient a_l which is not zero is a_2. Substitution in (43) gives, upon omitting the arguments,

$$G_1 = 1$$

$$G_2 = -2a_2 j(\omega_1 + \omega_2)L = -2j\beta(\omega_1 + \omega_2)$$

$$G_3 = 2(-2j\beta)^2(\omega_1 + \omega_2 + \omega_3)^2 \tag{45}$$

where the expression for G_3 is obtained from the recurrence relation

$$G_n = -j\beta(\omega_1 + \cdots + \omega_n) G_n^{(2)}(f_1, \cdots, f_n)$$

and (28) for $G_3^{(2)}$. Although G_4 is a symmetric polynomial of the third degree in $\omega_1, \cdots, \omega_4$, use of (29) for $G_4^{(2)}$ shows that G_4 is not as simple as might be expected.

C. Direct Expansion Method

An alternative to "probing" the system, as in the harmonic input method, is to manipulate the defining system equations until they are brought into the form (1) of a Volterra series expansion. The Volterra transfer functions can then be found by taking the n-fold Fourier transform of the Volterra kernels using (2). This was precisely the technique used in the trivial memoryless case (22). The chief value of the direct expansion method seems to be in the derivation of expressions which hold for general values of n, as in Section V-B. When n is small, it appears simpler to use harmonic inputs.

An example of a nonlinear system that can be analyzed by the direct expansion method is the modulator–filter–demodulator system shown in Fig. 1.

The system output is given by

$$y(t) = F\left\{\int_{-\infty}^{\infty} g(u) h[x(t-u)]\, du\right\}. \tag{46}$$

It is assumed that the modulator and demodulator functions can be expanded in the power series

$$h(x) = \sum_{v=0}^{\infty} h_v x^v / v! \qquad F(z) = \sum_{l=1}^{\infty} F_l (z - z_0)^l / l!$$

$$z_0 = h_0 \int_{-\infty}^{\infty} g(u)\, du \qquad F(z_0) = 0 \tag{47}$$

and the impulse response $g(t)$ of the filter is related to the filter transfer function $G(f)$ by

$$G(f) = \int_{-\infty}^{\infty} e^{-j\omega t} g(t)\, dt. \tag{48}$$

TABLE I

Partition	l	N	Terms in $G_4(f_1, f_2, f_3, f_4)$
4	1	1	$F_1h_4G(f_1+f_2+f_3+f_4)$
1+3	2	4	$F_2h_1h_3\Sigma'_4G(f_1)G(f_2+f_3+f_4)$
2+2	2	3	$F_2h_2^2\Sigma'_3G(f_1+f_2)G(f_3+f_4)$
1+1+2	3	6	$F_3h_1^2h_2\Sigma'_6G(f_1)G(f_2)G(f_3+f_4)$
1+1+1+1	4	1	$F_4h_1^4G(f_1)G(f_2)G(f_3)G(f_4)$

The Volterra transfer functions, i.e., the Fourier transforms of the kernels in the Volterra series for $y(t)$ obtained from (46), are

$$G_1(f_1) = F_1h_1G(f_1)$$

$$G_2(f_1, f_2) = F_1h_2G(f_1 + f_2) + F_2h_1^2G(f_1)G(f_2)$$

$$G_3(f_1, f_2, f_3) = F_1h_3G(f_1 + f_2 + f_3) + F_2h_1h_2 \sum'_3 G(f_1)G(f_2 + f_3) + F_3h_1^3G(f_1)G(f_2)G(f_3)$$

$$G_n(f_1, \cdots, f_n) = \sum_{l=1}^{n} F_l \sum_{(v;l,n)} h_{v_1} \cdots h_{v_l} \sum'_N G(f_1 + f_2 + \cdots + f_{v_1}) G(f_{v_1+1} + \cdots + f_{v_1+v_2}) \cdots G(f_\mu + \cdots + f_n). \quad (49)$$

The last equation is derived in Section V-B. Here, $\sum'_3$ is a sum of the type shown in (28), and the sums denoted by $\sum_{(v;l,n)}$ and $\sum'_N$ are defined in connection with (24)–(29).

We have

$$\sum_{l=1}^{n} \sum_{(v;l,n)} = \sum_{\pi(n)} \quad (50)$$

where $\pi(n)$ beneath the $\sum$ denotes summation over all partitions of n. The number of parts in the partition is l and the v's are the parts. The parts are related by (25). The general form (49) for G_n can be written almost immediately from either 1) the table of partitions given in [18, pp. 831–832], where the values of N are listed in the column labeled M_3, or from 2) the table of Bell polynomials given in [19, p. 125]. Table I illustrates the procedure for $n=4$. The value of G_4 is given by the sum of the terms in the last column.

Let B_n be the number of different products in $G_n(f_1, \cdots, f_n)$ when they are counted in a manner which gives [see (49)] $B_1=1$ for $G_1(f_1)$, $B_2=2$ for $G_2(f_1, f_2)$, $B_3=1+3+1=5$ for $G_3(f_1, f_2, f_3)$, and (see Table I) $B_4=1+4+3+6+1=15$ for G_4. When $G(f)\equiv 1$, $g(u)$ is $\delta(u)$, $y(t)$ is $F\{h[x(t)]\}$, the system is memoryless, and G_n is the coefficient of $[x(t)]^n/n!$ in the expansion of $y(t)$. Setting $h_v=1$ and $F_l=1$ gives $y=\exp[e^x-1]-1$ from (47) and $G_n=B_n$ from (49). Hence B_n is the coefficient of $x^n/n!$ in the expansion of $\exp[e^x-1]$, and from this a recurrence relation for B_n can be obtained. B_n increases rapidly with n. For example, $B_5=52$ and $B_6=203$. The B_n are the Bell numbers [19, p. 192].

A particular case of the modulator–filter–demodulator system is furnished by the phase-modulation system shown in Fig. 4 and discussed further in the next section. Briefly, the input $x(t)$ is used to phase modulate a carrier wave that passes through a filter having a transfer function $K(f)$. The output $\theta(t)$ is taken as the variable portion of the phase angle of the filter output. This corresponds to the system of Fig. 1 with $h(x)=\exp(jx)$, $F(z)=ln\ z$, and $z_0=1$. Then

$$h_v = j^v \qquad F_l = (-1)^{l-1}(l-1)! \quad (51)$$

and substitution in (49) leads to, as shown in Section IV-C, the Volterra transfer functions for $\theta(t)$. For $n=1$ and $n=2$ these are

$$G_{\theta 1}(f_1) = \frac{1}{2}[\Gamma(f_1) + \Gamma^*(-f_1)]$$

Fig. 4. Phase-modulation system.

$$G_{\theta 2}(f_1, f_2) = \frac{j}{2}[\Gamma(f_1 + f_2) - \Gamma(f_1)\Gamma(f_2) - \Gamma^*(-f_1 - f_2) + \Gamma^*(-f_1)\Gamma^*(-f_2)] \quad (52)$$

where the asterisk denotes conjugate complex and $\Gamma(f)=K(f_0+f)/K(f_0)$, f_0 being the carrier frequency $\omega_0/(2\pi)$. The expression (52) for $G_{\theta 2}(f_1, f_2)$ with $\Gamma(f)=\exp(-jbf^2)$ has been used to study the distortion which occurs when an FM wave travels through the ionosphere [9]. The structure of the general $G_{\theta n}(f_1, \cdots, f_n)$ term given here is much the same as the structure of the intermodulation functions introduced by Mircea [7] in his studies of FM distortion.

The other general example of the use of the direct expansion method for determining the Volterra transfer functions is the feedback system shown in Fig. 2.

The system input is $x(t)$ and the output is $y(t)$. The system equations relating $x(t)$ and $y(t)$ are

$$y(t) = \sum_{l=1}^{\infty} \frac{1}{l!} \int_{-\infty}^{\infty} du'_1 \cdots \int_{-\infty}^{\infty} du'_l m_l(u'_1, \cdots, u'_l) \prod_{q=1}^{l} w(t - u'_q) \quad (53)$$

$$w(t) = x(t) - z(t) \quad (54)$$

$$z(t) = \int_{-\infty}^{\infty} b(v)y(t - v)dv \quad (55)$$

where the filter transfer function $B(f)$ [the Fourier transform of $b(v)$] and the symmetric n-fold Fourier transforms $M_n(f_1, \cdots, f_n)$ of the Volterra kernels $m_n(u_1, \cdots, u_n)$ are assumed to be known.

The problem is to determine the Volterra transfer function $G_n(f_1, \cdots, f_n)$, i.e., the n-fold Fourier transform of the nth kernel in the series (1) for $y(t)$. For $n=1$, 2, and 3 the answer can be obtained by the harmonic input method and is

$$G_1(f_1)=[1+M_1(f_1)B(f_1)]^{-1}M_1(f_1)$$

$$G_2(f_1, f_2)=[1+M_1(f_1+f_2)B(f_1+f_2)]^{-1}K_1(f_1)K_1(f_2)M_2(f_1, f_2)$$

$$G_3(f_1, f_2, f_3)=[1+M_1(f_1+f_2+f_3)B(f_1+f_2+f_3)]^{-1} \cdot \{[K_1(f_1)K_2(f_2, f_3)M_2(f_1, f_2+f_3) + K_1(f_2)K_2(f_3, f_1)M_2(f_2, f_3+f_1) + K_1(f_3)K_2(f_1, f_2)M_2(f_3, f_1+f_2)] + K_1(f_1)K_1(f_2)K_1(f_3)M_3(f_1, f_2, f_3)\} \quad (56)$$

where

$$K_1(f_1) = 1 - B(f_1)G_1(f_1)$$

$$K_n(f_1, \cdots, f_n) = -B(f_1 + \cdots + f_n)G_n(f_1, \cdots, f_n), \qquad n > 1. \quad (57)$$

The expression for $G_2(f_1, f_2)$ depends on $G_1(f_1)$ through $K_1(f_1)$; and the expression for $G_3(f_1, f_2, f_3)$ depends on $G_1(f_1)$ and $G_2(f_2, f_3)$ through $K_1(f_1)$ and $K_2(f_1, f_2)$. Therefore the second and third of (56) are recurrence relations.

The three expressions (56) are equivalent to, three given by Narayanan [6]. The $\mu_n(f_1, \cdots, f_n)$, $G_n(f_1, \cdots, f_n)$, and $\beta(f)$ used by Narayanan are equal to, in our notation, $M_n(f, \cdots, f_n)/n!$, $G_n(f_1, \cdots, f_n)/n!$, and $B(f)$, respectively.

The nth recurrence relation for $n>1$, obtained in Section V-B by the direct expansion method, is

$$G_n(f_1, \cdots, f_n) = [1 + M_1(f_1 + \cdots + f_n)B(f_1 + \cdots + f_n)]^{-1}$$
$$\cdot \sum_{l=2}^{n} \sum_{(v;l,n)} \sum_N' K_{v_1}(f_1, \cdots, f_{v_1})K_{v_2}(f_{v_1+1}, \cdots, f_{v_1+v_2})$$
$$\cdots K_{v_l}(f_\mu, \cdots, f_n)M_l(f_1 + \cdots + f_{v_1}, f_{v_1+1} + \cdots$$
$$+ f_{v_1+v_2}, \cdots, f_\mu + \cdots + f_n). \quad (58)$$

Here the integer N and the summation over the sets $(v; l, n)$ of integers v_i are the same as in (24).

The K_n are the n-fold Fourier transforms of the kernels in

$$w(t) = \sum_{i=1}^{\infty} \frac{1}{i!} \int_{-\infty}^{\infty} du_1 \cdots \int_{-\infty}^{\infty} du_i k_i(u_1, \cdots, u_i) \prod_{r=1}^{i} x(t - u_r) \quad (59)$$

which can be obtained by substituting the series (1) for $y(t)$ in (55) to get a series for $z(t)$, and then substituting this series for $z(t)$ in (54), $w(t) = x(t) - z(t)$.

IV. Illustrative Examples

The diversity of the list of formulas in Section II giving properties of the output of nonlinear devices for a variety of inputs makes it desirable to illustrate their use by applying them to practical problems of interest. In this section some of the examples used in Section III to illustrate computation of the Volterra transfer functions will be treated further to obtain output properties of interest for specific input signals.

A. Quasi-Static Filtered FM

The system equation and Volterra transfer functions for this case are, from Section III-B,

$$y(t) = x(t) + \varepsilon[x'(t)]^2 x''(t)$$
$$G_1(f_1) = 1, \qquad G_3(f_1, f_2, f_3) = 2\varepsilon\omega_1\omega_2\omega_3(\omega_1 + \omega_2 + \omega_3) \quad (60)$$

where $\omega = 2\pi f$ and where all of the remaining G_n are zero. When $x(t)$ is Gaussian, (11) shows that $\langle y(t) \rangle$ is zero. From (14) the leading terms in the power spectrum for $y(t)$ are

$$W_y(f) = W_x(f)\left|1 - \varepsilon(2\pi f)^2 \int_{-\infty}^{\infty} df_1 W_x(f_1)(2\pi f_1)^2\right|^2$$
$$+ \frac{1}{3!}\int_{-\infty}^{\infty} df_1 \int_{-\infty}^{\infty} df_2 W_x(f_1)W_x(f_2)W_x(f - f_1 - f_2)$$
$$\cdot |2\varepsilon f_1 f_2(f - f_1 - f_2)f(2\pi)^4|^2. \quad (61)$$

This is also the exact expression for $W_y(f)$ because an examination of the complete expression (160) for $W_y(f)$ shows that all of the remaining terms are zero for this case. When we use the relation $W_{x'}(f) = (2\pi f)^2 W_x(f) = \omega^2 W_x(f)$ between the power spectrum of $x(t)$ and its time derivative $x'(t)$, (61) goes into

$$W_y(f) = W_x(f)\left|1 - \varepsilon\omega^2 \int_{-\infty}^{\infty} df_1 W_{x'}(f_1)\right|^2$$
$$+ \frac{4\varepsilon^2\omega^2}{3!}\int_{-\infty}^{\infty} df_1 \int_{-\infty}^{\infty} df_2 W_{x'}(f_1)W_{x'}(f_2)W_{x'}(f - f_1 - f_2). \quad (62)$$

When $x(t)$ in (60) is equal to $P \cos pt + I_n(t)$, where $I_n(t)$ is Gaussian, (15) and (16) show that the periodic part of $y(t)$ is the ensemble average

$$\langle y(t) \rangle = P\left\{1 - \tfrac{1}{4}P^2\varepsilon p^4 - \varepsilon p^2 \int_{-\infty}^{\infty} df_1 W_{I'}(f_1)\right\}\cos pt$$
$$+ \tfrac{1}{4}P^3\varepsilon p^4 \cos 3pt \quad (63)$$

and that $y(t)$ has the power spectrum

$$W_y(f) = \{\text{Four spikes due to the } \exp(\pm jpt) \text{ and } \exp(\pm j3pt) \text{ components of } \langle y(t) \rangle\}$$
$$+ W_I(f)\left|1 - \tfrac{1}{2}P^2\varepsilon p^2\omega^2 - \varepsilon\omega^2 \int_{-\infty}^{\infty} df_1 W_{I'}(f_1)\right|^2$$
$$+ \tfrac{1}{16}P^4\varepsilon^2 p^4\omega^2[W_{I'}(f - 2f_p) + W_{I'}(f + 2f_p)]$$
$$+ \tfrac{1}{2}P^2\varepsilon^2 p^2\omega^2 \int_{-\infty}^{\infty} df_1 W_{I'}(f_1)[W_{I'}(f - f_1 - f_p)$$
$$+ W_{I'}(f - f_1 + f_p)] + \frac{4\varepsilon^2\omega^2}{3!}\int_{-\infty}^{\infty} df_1 \int_{-\infty}^{\infty}$$
$$\cdot df_2 W_{I'}(f_1)W_{I'}(f_2)W_{I'}(f - f_1 - f_2) \quad (64)$$

where $2\pi f_p = p$. When P is zero, (64) reduces to the expression (62) for $W_y(f)$ when $x(t)$ consists of Gaussian noise alone.

B. Series Inductance and Nonlinear Resistance

Suppose that the voltage $x(t) = P \cos pt + Q \cos qt$ is applied to the series combination of the unit inductance and nonlinear resistance $(\alpha + \varepsilon y)$ described in Section III-B [(39)]. What is the leading term in the $(p-q)$ component of the current $y(t)$ when ε is small? Changing the signs of p and q in (5) for the $(p+q)$ term to obtain the $\pm(p-q)$ terms and substituting $G_2(f_1, f_2)$ from (40) shows that (assuming α and ε real) the desired leading term is

$$2 \operatorname{Re}\left[e^{j(p-q)t} \frac{PQ}{4} \frac{(-2\varepsilon)[\alpha + j(p - q)]^{-1}}{(\alpha + jp)(\alpha - jq)}\right]. \quad (65)$$

When the voltage $x(t)$ applied to the series combination is Gaussian, the leading terms (out to order ε^2) in the power spectrum of the current $y(t)$ are, from (14),

$$W_y(f) = \langle y(t) \rangle^2 \delta(f)$$
$$+ W_x(f)\left|G_1(f) + \frac{1}{2}\int_{-\infty}^{\infty} df_1 W_x(f_1)G_3(f, f_1, -f_1)\right|^2$$
$$+ \frac{1}{2}\int_{-\infty}^{\infty} df_1 W_x(f_1)W_x(f - f_1)|G_2(f_1, f - f_1)|^2 + O(\varepsilon^4) \quad (66)$$

where, from (40) and (11),

$$G_1(f) = (\alpha + j\omega)^{-1}, \qquad \omega = 2\pi f$$
$$|G_2(f_1, f - f_1)|^2 = \frac{4\varepsilon^2(\alpha^2 + \omega^2)^{-1}}{(\alpha^2 + \omega_1^2)[\alpha^2 + (\omega - \omega_1)^2]}$$
$$\langle y(t) \rangle = -\varepsilon\alpha^{-1}\int_{-\infty}^{\infty} df_1 W_x(f_1)/(\alpha^2 + \omega_1^2) \quad (67)$$

and from (40) $G_3(f, f_1, -f_1)$ is $O(\varepsilon^2)$. When ε is small and $x(t)$ is Gaussian, the probability density $p(y)$ of $y(t)$ is almost, but not quite, normal. As discussed in Section VIII-B, the departure of $p(y)$ from normality can be estimated from the values of the cumulants $\kappa_1, \kappa_2, \kappa_3, \kappa_4$. However, when the G_n from (40) are used, the integrals (13) for κ_3, κ_4 are quite complicated and we shall not stop to evaluate them here. A less complicated example is given later in Section IV-D in connection with noise through a square-law device followed by a filter.

C. Filtered Phase Modulation

Here we illustrate the use of (49) for $G_n(f_1, \cdots, f_n)$ by considering a special case of the modulator–filter–demodulator system of Fig. 1, namely, the phase-modulation system shown in Fig. 4. In this system the filter produces undesirable distortion. The input is $x(t)$, the output is $\theta(t)$, and $K(f)$ is the filter transfer function. When

$x(t)$ is zero, the filter output is $a \cos(\omega_0 t + b)$ where [with $\omega_0 = 2\pi f_0$], $K(f_0) = K_0^*(-f_0) = a \exp(jb)$, the envelope factor $R(t)$ in the filter output is unity, and $\theta(t)$ is zero. From phase-modulation theory [8]–[10]

$$\theta(t) = \operatorname{Im} y(t)$$
$$y(t) = \ln\left[\int_{-\infty}^{\infty} du\, \gamma(u) e^{jx(t-u)}\right]$$
$$\gamma(u) = \int_{-\infty}^{\infty} df\, e^{j\omega u}\Gamma(f), \qquad \omega = 2\pi f$$
$$\Gamma(f) = K(f_0 + f)/K(f_0). \tag{68}$$

In this example the system output is $\theta(t)$ instead of $y(t)$. Here $y(t)$ is equal to $\ln R(t) + i\theta(t)$. When the filter is absent, $K(f)$ is independent of the frequency f, $\Gamma(f) = 1$, $\gamma(u) = \delta(u)$, $y(t) = jx(t)$, and $\theta(t)$ is equal to $x(t)$. We are interested in analyzing the difference, usually small, between $\theta(t)$ and $x(t)$ when the filter is present.

In order to apply the formulas listed in Section II to $\theta(t)$, we need the Fourier transforms $G_{\theta n}(f_1, \cdots, f_n)$ of the kernels in the Volterra series for $\theta(t)$. By splitting the Volterra series for $y(t)$ into its real and imaginary parts and using $\theta(t) = \operatorname{Im} y(t)$, it can be shown that

$$G_{\theta n}(f_1, \cdots, f_n) = \left[G_n(f_1, \cdots, f_n) - G_n^*(-f_1, \cdots, -f_n)\right]/(2j) \tag{69}$$

where $G_n(f_1, \cdots, f_n)$ is the Fourier transform of the nth kernel for $y(t)$. Indeed we have the general result that, when $x(t)$ is real, the Fourier transforms of the kernels in the series for the real and imaginary parts, $y_R(t)$ and $y_I(t)$, of $y(t)$ are

$$\left[G_n(f_1, \cdots, f_n) + G_n^*(-f_1, \cdots, -f_n)\right]/2, \qquad \text{for } y_R(t)$$
$$\left[G_n(f_1, \cdots, f_n) - G_n^*(-f_1, \cdots, -f_n)\right]/(2j), \qquad \text{for } y_I(t). \tag{70}$$

Comparison of (68) and (46) for $y(t)$ shows that $g(u) = \gamma(u)$, $h(x) = \exp(jx)$, and $F(z) = \ln z$. Hence $G(f)$ goes into $\Gamma(f)$, the coefficients in the expansion of $h(x)$ are $h_\nu = j^\nu$, and the equation to determine z_0 becomes $z_0 = h_0\Gamma(0) = 1$. Expanding $F(z)$ about $z = z_0 = 1$ gives $F_l = (-1)^{l-1}(l-1)!$ Then the general equations (49) show that the Fourier transforms of the kernels in the Volterra series for $y(t)$ defined by the phase-modulation equations (68) are

$$G_1(f_1) = j\Gamma(f_1).$$
$$G_2(f_1, f_2) = j^2\left[\Gamma(f_1 + f_2) - \Gamma(f_1)\Gamma(f_2)\right].$$
$$G_3(f_1, f_2, f_3) = j^3\left[\Gamma(f_1 + f_2 + f_3) - \Gamma(f_1)\Gamma(f_2 + f_3) - \Gamma(f_2)\Gamma(f_1 + f_3) - \Gamma(f_3)\Gamma(f_1 + f_2) + 2\Gamma(f_1)\Gamma(f_2)\Gamma(f_3)\right].$$
$$\vdots$$
$$G_n(f_1, \cdots, f_n) = j^n \sum_{l=1}^{n} (-1)^{l-1}(l-1)! \sum_{(\nu:l,n)} \sum_N' \Gamma(f_1 + \cdots + f_{\nu_1}) \cdot \Gamma(f_{\nu_1+1} + \cdots + f_{\nu_1+\nu_2}) \cdots \Gamma(f_\mu + \cdots + f_n). \tag{71}$$

The expressions for $G_{\theta 1}$ and $G_{\theta 2}$ obtained by substituting (71) in the general relation (69) for $G_{\theta n}$ are

$$G_{\theta 1}(f_1) = \frac{j}{2j}\left[\Gamma(f_1) + \Gamma^*(-f_1)\right]$$
$$G_{\theta 2}(f_1, f_2) = \frac{j^2}{2j}\left[\Gamma(f_1 + f_2) - \Gamma(f_1)\Gamma(f_2) - \Gamma^*(-f_1 - f_2) + \Gamma^*(-f_1)\Gamma^*(-f_2)\right] \tag{72}$$

as also shown in (52). When $\Gamma^*(-f)$ is equal to $\Gamma(f)$, as it is when the filter transfer function $K(f)$ is "symmetrical" about the carrier frequency f_0, $G_{\theta n}$ is zero if n is even and is equal to $-j\, G_n$ if n is odd.

Now that the $G_{\theta n}$ are known, information regarding $\theta(t)$ for various inputs $x(t)$ can be obtained by replacing $y(t)$ by $\theta(t)$ and $G_n(f_1, \cdots, f_n)$ by $G_{\theta n}(f_1, \cdots, f_n)$ in the formulas listed in Section II. For example, when $x(t) = P \cos pt + Q \cos qt$, the $\exp[j(p-q)t]$ component in $\theta(t)$ has the leading term [upon changing the signs of q and f_q in the $\exp[j(p+q)t]$ term in (5)]

$$e^{j(p-q)t}\frac{PQ}{4} G_{\theta 2}(f_p, -f_q) = e^{j(p-q)t}\frac{PQ}{8} j \cdot \left[\Gamma(f_p - f_q) - \Gamma(f_p)\Gamma(-f_q) - \Gamma^*(-f_p + f_q) + \Gamma^*(-f_p)\Gamma^*(f_q)\right]. \tag{73}$$

Mircea [20] has discussed the case when $x(t) = P \cos pt$, and has given [7] the structure of the general term in the series for the power spectrum $W_\theta(f)$ when $x(t)$ is Gaussian (see also [8]–[11]). The leading terms in the sum of the linear and second-order modulation terms in $W_\theta(f)$ are, upon substitution in the Mircea–Sinnreich series (14),

$$W_x(f)|G_{\theta 1}(f)|^2 + \frac{1}{2}\int_{-\infty}^{\infty} df_1\, W_x(f_1)W_x(f - f_1)|G_{\theta 2}(f_1, f - f_1)|^2 \tag{74}$$

where $G_{\theta 1}$ and $G_{\theta 2}$ are given by (72).

Some insight into the region where the Volterra series approach is useful can be obtained by considering the case $x(t) = P \cos pt$. The usual expression for $\theta(t)$ for this case is, in our notation,

$$\theta(t) = \arctan\left[\frac{\operatorname{Im} S}{\operatorname{Re} S}\right]$$
$$S = \sum_{n=-\infty}^{\infty} j^n J_n(P)\Gamma(nf_p)e^{jnpt} \tag{75}$$

where $J_n(P)$ is the Bessel function of order n. When $\Gamma(f) \equiv 1$, S is equal to $\exp(j P \cos pt)$, and $\theta(t)$ is equal to the input $x(t)$. When the filter bandwidth is large, S remains close to $\exp(j P \cos pt)$. When P is small enough to make $|S - 1| < 1$ for all values of t, $\theta(t)$ can be expanded in a convergent power series in P by using

$$\theta(t) = \operatorname{Im} \ln\left[1 + (S - 1)\right]. \tag{76}$$

On the other hand, the complete series (3) shows that

$$\theta(t) = \sum_{n=1}^{\infty}\left(\frac{P}{2}\right)^n \sum_{k=0}^{n} \frac{\exp\left[j(2k-n)pt\right]}{k!(n-k)!} G_{\theta(k,n-k)}(f_p) \tag{77}$$

where the subscript $\theta(k, n-k)$ on G denotes $G_{\theta n}(f_1, \cdots, f_n)$ with the first k of the f_i equal to f_p and the remaining $n-k$ equal to $-f_p$. Therefore, (77) gives the power series expansion of arctan [Im S/Re S] in powers of P. In an FM system P is the "deviation ratio." This indicates that when the Volterra series analysis is applied to FM systems, it is most useful for systems employing a small deviation ratio—as is the case in many microwave radio relay systems.

D. Filtered Square-Law Detector

The system equation for a square-law device followed by a filter is

$$y(t) = \int_{-\infty}^{\infty} b(u)x^2(t - u)du \tag{78}$$

when $x(t)$ is the input to the square-law device and $y(t)$ is the filter output. We are interested in the probability density $p(y)$ of $y(t)$ when $x(t)$ is a Gaussian noise with power spectrum $W_x(f) = (2\pi)^{-1/2} \exp(-f^2/2)$, and the filter impulse response $b(u)$ and its Fourier transform $B(f)$ are

$$b(u) = \beta(2\pi)^{1/2} \exp(-2\pi^2\beta^2u^2)$$
$$B(f) = \exp[-f^2/(2\beta^2)]. \quad (79)$$

The effective passband of the filter extends from $-\beta$ to $+\beta$ and the effective band of the noise from -1 to $+1$. This is a simpler case than one considered by Slepian [21] in which $x(t)$ is RLC Gaussian noise, $b(u)=1$ for $|u|<T/2$, and $b(u)=0$ for $|u|>T/2$.

First some general remarks. Averaging both sides of (78) and using $\langle x^2(t)\rangle = 1$ shows that $\langle y(t)\rangle = B(0) = 1$, irrespective of the bandwidth β. Since $b(u)\geq 0$, $p(y)$ is zero when $y<0$. When $\beta=\infty$, the filter has no effect, $y(t)=x^2(t)$, and for $y>0$

$$p(y) = (2\pi y)^{-1/2} \exp(-y/2). \quad (80)$$

When $\beta\to 0$, the filter bandwidth tends to zero and we expect [22] $p(y)$ to tend to a normal law of vanishing width centered on the mean value $y=1$:

$$p(y) \to (2\pi\beta)^{-1/2} \exp[-(y-1)^2/(2\beta)]. \quad (81)$$

As β increases from 0 to ∞, $p(y)$ changes from (81) to (80).

Now we apply some of the results mentioned in Section II for Volterra series. The system equation (78) corresponds to a Volterra series with all of its kernels zero except for $n=2$:

$$g_2(u_1, u_2) = 2b(u_1)\delta(u_2 - u_1)$$
$$G_2(f_1, f_2) = 2B(f_1 + f_2). \quad (82)$$

From the more complete series (180) corresponding to (11) and (13) we see that the cumulants $\kappa_1, \kappa_2, \kappa_3, \kappa_4$ for $p(y)$ are, respectively, the integrals of $(W)(1,-1)/2$, $(WW)(1,2)(-1,-2)/2$, $(WWW)(1,2)\cdot(-1,3)(-2,-3)$, $3(WWWW)(1,2)(-1,3)(-2,4)(-3,-4)$. Here (W), (WW), $(1,2),\cdots$ denote $W_x(f_1)$, $W_x(f_1)W_x(f_2)$, $G_2(f_1,f_2),\cdots$ much as in (180). The Gaussian form of the integrands permits the integrations to be performed. It is found that

$$\kappa_1 = 1 \qquad \kappa_2 = 2/c, \qquad c = (4\beta^{-2}+1)^{1/2}$$
$$\kappa_3 = 32/(3c^2+1) \qquad \kappa_4 = 96/[c(c^2+1)]. \quad (83)$$

These values of κ_1 and κ_2 (the mean and variance) agree with those obtained from the limiting forms (80) and (81) of $p(y)$. When β becomes small, $c\to 2/\beta$ and the values of $\kappa_2, \kappa_3, \kappa_4$ approach β, $8\beta^2/3$, $12\beta^3$, respectively. Thus, the variance tends to β, and (181) shows that the skewness γ_1 tends to $(8/3)\beta^{1/2}$ and the excess γ_2 to 12β. The Edgeworth series (182) shows how the normal law (81) is approached:

$$p(y)=\beta^{-1/2}\{Z(u)-\tfrac{4}{9}\beta^{1/2}Z^{(3)}(u)+\beta[\tfrac{1}{2}Z^{(4)}(u)+\tfrac{8}{81}Z^{(6)}(u)]-\cdots\} \quad (84)$$

where $u=(y-1)/\beta^{1/2}$ and $Z(u)$ is equal to $(2\pi)^{-1/2}\exp(-u^2/2)$. From (184) the peak of $p(y)$ occurs at $y_0\approx 1-(4\beta/3)$ where $p(y_0)$ is about $[1+(49\beta/54)]$ times higher than the peak value $(2\pi\beta)^{-1/2}$ of a normal law with the variance β.

At this stage we go to the special results given in Section VIII-C for the two-term Volterra series. For the Gaussian forms of $W_x(f)$ and $G_2(f_1, f_2)$ in our example, it is convenient to work with row 4 of Table II. For our example the integral equation is

$$\lambda\Psi(f) = (2\pi)^{-1/2}\exp(-f^2/2)\int_{-\infty}^{\infty} df_1 2\Psi(f_1) \cdot \exp[-(f-f_1)^2/(2\beta^2)]. \quad (85)$$

The kth eigenvalue and eigenfunction are found to be

$$\lambda_k = \frac{4}{c+1}\left(\frac{c-1}{c+1}\right)^k$$

$$\Psi_k(f) = A_k \exp[-\tfrac{1}{4}(c+1)f^2]H_k\left[f\left(\frac{c}{2}\right)^{1/2}\right] \quad (86)$$

where A_k depends only on k and the Hermite polynomial is given by

$$H_k(x) = e^{x^2}\left(-\frac{d}{dx}\right)^k e^{-x^2} \qquad H_0(x) = 1 \qquad H_1(x) = 2x.$$

That (86) is a solution of (85) can be verified with the help of

$$\int_{-\infty}^{\infty} e^{-px^2+qx}H_k(rx)dx = \left(\frac{\pi}{p}\right)^{1/2}\left(1-\frac{r^2}{p}\right)^{k/2}\exp\left[\frac{q^2}{4p}\right]H_k\left[\frac{qr}{2p}\left(1-\frac{r^2}{p}\right)^{-1/2}\right]. \quad (87)$$

In our example the parameter ξ_k is zero because the first term of the two-term Volterra series is missing, and we need only λ_k in order to calculate $p(y)$ and the cumulants κ_m. The integral (186) for $p(y)$ becomes

$$p(y) = \frac{1}{2\pi}\int_{-\infty}^{\infty} dz e^{-jyz} \prod_{k=0}^{\infty}[1-4jz\rho^k(c+1)^{-1}]^{1/2} \quad (88)$$

where $\rho=(c-1)/(c+1)$. The series in (188) for the mth cumulant gives

$$\kappa_m = \frac{(m-1)!}{2}\sum_{k=0}^{\infty}\lambda_k^m = \frac{(m-1)!2^{2m-1}}{(c+1)^m-(c-1)^m}. \quad (89)$$

For $m=1,2,3,4$ this agrees with (83). Computing κ_m from the integrals in (188) is essentially the same as evaluating the multiple integrals used to obtain (83).

To illustrate the procedure when the first term of the two-term Volterra series is present, we consider

$$y(t) = \int_{-\infty}^{\infty} du\, g_1(u)x(t-u) + \int_{-\infty}^{\infty} du\, b(u)x^2(t-u) \quad (90)$$

where

$$g_1(u)=\alpha\beta_1(2\pi)^{1/2}\exp(-2\pi^2\beta_1^2u^2) \quad G_1(f)=\alpha\exp[-f^2/(2\beta_1^2)]. \quad (91)$$

To calculate $p(y)$ and κ_m, we now need the parameter ξ_k in addition to λ_k [which is still given by (86)]. The orthonormalization relation for $\Psi_k(f)$ given in row 4 of Table II turns out to be that for Hermite polynomials and gives $j^kc^{1/4}(2\pi k!2^k)^{-1/2}$ for the normalization constant A_k in (86). When the normalized $\Psi_k(-f)$ is substituted in the integral for ξ_k given in the last column of Table II and the result evaluated with the help of (87), it is found that when k is odd, ξ_k is zero; and when k is even,

$$\xi_{2n} = \alpha\frac{c^{-1/4}}{n!2^n}\left[\frac{(2n)!2}{a'+1}\right]^{1/2}\left[\frac{a'-1}{a'+1}\right]^n, \qquad a' = (2\beta_1^{-2}+1)/c. \quad (92)$$

We still have $\kappa_1=1$, but from (188) κ_m for $m\geq 2$ is the sum of (89) plus

$$\frac{m!}{2}\sum_{n=0}^{\infty}\xi_{2n}^2\lambda_{2n}^{m-2} = \alpha^2c^{-1/2}\frac{m!}{a'+1}\left(\frac{4}{c+1}\right)^{m-2}\sum_{n=0}^{\infty}\frac{(1/2)_n}{n!} \cdot \left[\left(\frac{a'-1}{a'+1}\right)^2\left(\frac{c-1}{c+1}\right)^{2m-4}\right]^n$$
$$= \alpha^2c^{-1/2}m!4^{m-2}[(a'+1)^2(c+1)^{2m-4} - (a'-1)^2(c-1)^{2m-4}]^{-1/2} \quad (93)$$

where $(\alpha)_0=1$ and $(\alpha)_n=\alpha(\alpha+1)(\alpha+2)\cdots(\alpha+n-1)$ when $n>0$. The series always converges when $m\geq 2$, and for $m=2,3,4$ gives values in agreement with those obtained from the general multiple integrals (180).

When $m=1$, the left side of (93) is of the form $\sum \xi_k^2/(2\lambda_k)$. In Section VIII-C it is pointed out that if this series converges to a value S, and if all of the λ_k are positive, then $y(t)$ is never less than $-S$. Putting $m=1$ in (93) and replacing c and a' by their expressions in terms of the bandwidths β and β_1 shows that the series converges when $\beta^2<2\beta_1^2$ and gives

$$S = \frac{\alpha^2\beta_1^2}{4\beta}(2\beta_1^2 - \beta^2)^{-1/2}. \tag{94}$$

The inequality $y(t)\geq -S$ is a special case of a more general result which we owe to Pollak. Thus, assuming that $b(u)$ is never negative in (90),

$$g_1(u)x(t-u) + b(u)x^2(t-u) \geq -g_1^2(u)/(4b(u))$$

$$y(t) \geq -\int_{-\infty}^{\infty} g_1^2(u)du/(4b(u)). \tag{95}$$

For the $g_1(u)$ and $b(u)$ of our example, the integral converges when $\beta^2<2\beta_1^2$ and again gives $y(t)\geq -S$.

Finally, when $\beta\rightarrow\infty$, the second integral in (90) for $y(t)$ becomes $x^2(t)$, and $c\rightarrow 1$, $\lambda_0\rightarrow 2$, $\lambda_k\rightarrow 0$ for $k>0$. However, the $\Psi_k(f)$ and the ξ_k computed from them do not change markedly. The exponent in the factor $Q(z)$ in the integral (186) for $p(y)$ now contains the sum $\sum \xi_k^2$ which, from (188), is equal to $\kappa_2-\lambda_0^2/2$.

In general, when the second term in a two-term Volterra series is $a_2x(t)^2/2$, analogy with the above example and row 4 of Table II suggests that $\lambda_0=a_2\sigma^2$, $\Psi_0(f)=W_x(f)/\sigma$, $\lambda_k=0$ for $k>0$, and

$$Q(z) = (1-j\lambda_0 z)^{-1/2}\exp\left\{-\frac{z^2}{2}\left[\kappa_2 + j\lambda_0 z\xi_0^2(1-j\lambda_0 z)^{-1}\right]\right\}$$

$$\xi_0 = \int_{-\infty}^{\infty} df\, G_1(f)W_x(f)/\sigma$$

$$\kappa_2 = \int_{-\infty}^{\infty} df\, W_x(f)G_1(f)G_1(-f) \tag{96}$$

where $\sigma^2=\langle x^2(t)\rangle$.

PART II: DERIVATION OF FORMULAS

V. Formulas Associated with the Direct Expansion Method

The direct expansion method is useful in dealing with Volterra transfer functions of arbitrary order. The expansion is usually accomplished with the help of Maclaurin's series and di Bruno's formula for the nth derivative of a function of a function. Frequently, the resulting expansion is a Volterra series with unsymmetrical kernels which must be symmetrized.

Proofs of the general expressions for the Volterra transfer functions listed in Section III are sketched here.

A. Volterra Series for $[y(t)]^l$

This section is devoted to the derivation of an expression for the nth kernel $g_n^{(l)}(u_1,\cdots,u_n)$ in the Volterra series for $[y(t)]^l$, where l is a positive integer.

We introduce the function $H(\zeta)$ which is obtained from the Volterra series (1) for $y(t)$ by replacing the $x(t-u_i)$ by $\zeta x(t-u_i)$. The time t enters $H(\zeta)$ as a parameter, and $H(1)$ is equal to $y(t)$. Let $F(z)$ be z^l. Then

$$[H(\zeta)]^l = F[H(\zeta)] = \sum_{n=1}^{\infty}\frac{\zeta^n}{n!}\left[\frac{d^n}{d\zeta^n}F[H(\zeta)]\right]_{\zeta=0} \tag{97}$$

The nth derivative may be evaluated by di Bruno's formula for the derivative of a function of a function:

$$\frac{d^n}{d\zeta^n}F[H(\zeta)] = \sum_{k=1} F^{(k)}[H(\zeta)]\sum_{(v;k,n)} N(v_1,v_2,\cdots,v_k)$$
$$H^{(v_1)}(\zeta)H^{(v_2)}(\zeta)\cdots H^{(v_k)}(\zeta) \tag{98}$$

where, with k replacing l, the summation notation is the same as in (24) and $N(v_1,\cdots,v_k)$ is the N given by (26).

The kth derivative $F^{(k)}(z)$ is $l(l-1)\cdots(l-k+1)z^{l-k}$. Since $H(\zeta)$ is 0 for $\zeta=0$, the value of $F^{(k)}[H(0)]$ is 0 for $k\neq l$ and is $l!$ for $k=l$. Differentiating the series for $H(\zeta)$ obtained by inserting ζ in the Volterra series (1) for $y(t)$ shows that

$$H^{(v)} = (0)\left[\frac{d^v}{d\zeta^v}H(\zeta)\right]_{\zeta=0} = \int_{-\infty}^{\infty}du_1\cdots\int_{-\infty}^{\infty}du_v g_v(u_1,\cdots,u_v)$$
$$\cdot\prod_{r=1}^{v}x(t-u_r). \tag{99}$$

When these values are substituted in (98), the result is, for $l\leq n$,

$$\left[\frac{d^n}{d\zeta^n}F[H(\zeta)]\right]_{\zeta=0} = l!\sum_{(v;l,n)}N(v_1,\cdots,v_l)\int_{-\infty}^{\infty}du_1\cdots\int_{-\infty}^{\infty}du_n$$
$$\cdot g_{v_1}(u_1,\cdots,u_{v_1})\cdots g_{v_l}(u_\mu,\cdots,u_n)\prod_{r=1}^{n}x(t-u_r). \tag{100}$$

For $l>n$ the right side of (100) is zero.

Substituting (100) in (97), setting $\zeta=1$, and taking the $(v;l,n)$ summation inside the integrations gives a series for $[y(t)]^l$. This series can be converted into a Volterra series by symmetrizing the products

$$g_{v_1}(u_1,\cdots,u_{v_1})\cdots g_{v_l}(u_\mu,\cdots,u_n) \equiv P(u_1,u_2,\cdots,u_n) \tag{101}$$

where, of course, the g_v are symmetric.

The symmetric function formed by permuting the n subscripts in $P(u_1,\cdots,u_n)$ and adding can be reduced to the right side of [see (120)]

$$\frac{1}{n!}\sum_{n!}P(u_1,\cdots,u_n) = \frac{1}{N}\sum_N' P(u_1,\cdots,u_n) \tag{102}$$

where N is given by (26) and is the same as the $N(v_1,\cdots,v_l)$ in (100). As in (24) $\sum_N'$ denotes summation over N nonidentical products. Let u_1 be assigned an arbitrary numerical value, u_2 a different but otherwise arbitrary value, and so on. Then $P(u_1,\cdots,u_n)$ will have a definite numerical value. In Section V-C (102) will be obtained by counting the permutations which leave this value of $P(u_1,\cdots,u_n)$ unchanged.

Thus, the effect of symmetrizing the products (101) is to replace $N(v_1,\cdots,v_l)$ in (100) by the sum $\sum_N'$. This leads to

$$g_n^{(l)}(u_1,\cdots,u_n) = l!\sum_{(v;l,n)}\sum_N' g_{v_1}(u_1,\cdots,u_{v_1})\cdots g_{v_l}(u_\mu,\cdots,u_n). \tag{103}$$

Taking the n-fold Fourier transform of both sides of (103) gives (24) for $G_n^{(l)}(f_1,\cdots,f_n)$:

$$G_n^{(l)}(f_1,\cdots,f_n) = l!\sum_{(v;l,n)}\sum_N' G_{v_1}(f_1,\cdots,f_{v_1})\cdots G_{v_l}(f_\mu,\cdots,f_n). \tag{104}$$

B. Volterra Transfer Functions of Arbitrary Order

The results regarding $[y(t)]^l$ and symmetrization are used in this section to derive the expressions stated in Section III for $G_n(f_1,\cdots,f_n)$ when n is arbitrary.

First consider the differential equation (37). When $x(t)$ is taken to be $\exp(j\omega t)$ and $y(t)$ is assumed to be the series of exponentials (32), substitution in (37) and equating coefficients of exp $(j\omega_1 t)$ gives $c_1 = G_1(f_1) = 1/F(j\omega_1)$. For $n>1$, take $x(t)$ to be the sum of n exponentials and assume that $y(t)$ can be expanded in an n-fold series, analogous to (33) for the case $n=2$, in which the coefficient of exp $[j(\omega_1 + \cdots + \omega_n)t]$ is $G_n(f_1, \cdots, f_n)$. Then $[y(t)]^l$ can be expanded in a similar series in which the coefficient of exp $[j(\omega_1 + \cdots + \omega_n)t]$ is $G_n^{(l)}(f_1, \cdots, f_n)$. Substituting these series in the differential equation (37), equating coefficients of exp $[j(\omega_1 + \cdots + \omega_n)t]$, and noting that $G_n^{(l)}$ is zero when $l > n$ gives

$$F(j\omega_1 + \cdots + j\omega_n)G_n(f_1, \cdots, f_n) + \sum_{l=2}^{n} a_l G_n^{(l)}(f_1, \cdots, f_n) = 0 \quad (105)$$

from which the desired recurrence relation (38) follows.

The system equation for the system shown in Fig. 3, is obtained by equating the series (42) for $I(t)$ to the convolution integral for the current through the admittance $H(f)$:

$$\sum_{l=1}^{\infty} a_l[y(t)]^l = \int_{-\infty}^{\infty} h(\tau)[x(t-\tau) - y(t-\tau)]d\tau \quad (106)$$

where $h(t)$ is the Fourier transform of $H(f)$. Taking $x(t)$ to be exp $(j\omega_1 t)$, assuming the series (32) for $y(t)$, and equating coefficients of exp $(j\omega_1 t)$ in (106) gives

$$a_1 G_1(f_1) = H(f_1) - G_1(f_1)H(f_1). \quad (107)$$

For $n>1$, taking $x(t)$ to be the sum of n exponentials and equating coefficients of exp $[j(\omega_1 + \cdots + \omega_n)t]$ gives

$$a_1 G_n(f_1, \cdots, f_n) + \sum_{l=2}^{n} a_l G_n^{(l)}(f_1, \cdots, f_n) = -G_n(f_1, \cdots, f_n)H(f_1 + \cdots + f_n). \quad (108)$$

The recurrence relations (43) for the circuit of Fig. 3 follow from (107) and (108).

We turn now to Fig. 1. The system equations for the modulator-filter-demodulator system are given by (46) and (47). To start the derivation of the expression (49) for $G_n(f_1, \cdots, f_n)$, we define $H(\zeta)$ by

$$H(\zeta) = \int_{-\infty}^{\infty} g(u)h[\zeta x(t-u)]du \quad (109)$$

where t is regarded as a parameter in $H(\zeta)$. The function $F[H(\zeta)]$ is equal to $y(t)$ when $\zeta = 1$. Also, by direct expansion,

$$F[H(\zeta)] = \sum_{n=0}^{\infty} \frac{\zeta^n}{n!}\left[\frac{d^n}{d\zeta^n} F[H(\zeta)]\right]_{\zeta=0} \quad (110)$$

where we shall evaluate the nth derivative by di Bruno's formula (98). For this we need

$$H^{(\nu)}(0) = h_\nu \int_{-\infty}^{\infty} g(u)[x(t-u)]^\nu du$$

$$H(0) = z_0$$

$$F[H(0)] = F(z_0) = 0$$

$$F^{(l)}[H(0)] = \left[\frac{d^l}{dz^l} F(z)\right]_{z=H(0)} = F_l \quad (111)$$

which follow from (47) and (109). Using di Bruno's formula in (110) and setting $\zeta = 1$ gives

$$y(t) = \sum_{n=1}^{\infty} \frac{1}{n!} \sum_{l=1}^{n} F_l \sum_{(\nu;l,n)} N(\nu_1, \cdots, \nu_l)H^{(\nu_1)}(0) \cdots H^{(\nu_l)}(0). \quad (112)$$

We rewrite the integral for $H^{(\nu)}(0)$ as

$$H^{(\nu)}(0) = h_\nu \int_{-\infty}^{\infty} du_1 \cdots \int_{-\infty}^{\infty} du_\nu \varphi_\nu(u_1, \cdots, u_\nu)x(t-u_1) \cdots x(t-u_\nu)$$

$$\varphi_1(u_1) = g(u_1)$$

$$\varphi_\nu(u_1, \cdots, u_\nu) = g(u_1)\delta(u_2 - u_1) \cdots \delta(u_\nu - u_1), \quad \nu > 1 \quad (113)$$

where $\varphi_\nu(u_1, \cdots, u_\nu)$ can be regarded as a symmetric function possessing the ν-fold Fourier transform $G(f_1 + \cdots + f_\nu)$, $G(f)$ being the Fourier transform (48) of $g(t)$. Then (112) becomes

$$y(t) = \sum_{n=1}^{\infty} \frac{1}{n!} \int_{-\infty}^{\infty} du_1 \cdots \int_{-\infty}^{\infty} du_n \gamma_n(u_1, \cdots, u_n) \prod_{r=1}^{n} x(t-u_r)$$

$$\gamma_n(u_1, \cdots, u_n) = \sum_{l=1}^{n} F_l \sum_{(\nu;l,n)} h_{\nu_1} \cdots h_{\nu_l} N(\nu_1, \cdots, \nu_l)\varphi_{\nu_1}(u_1, \cdots, u_{\nu_1}) \cdots \varphi_{\nu_l}(u_\mu, \cdots, u_n) \quad (114)$$

where $\gamma_n(u_1, \cdots, u_n)$ is usually not symmetric when $n>2$. To convert (114) into a Volterra series, we symmetrize $\gamma_n(u_1, \cdots, u_n)$ with the help of (102). The effect of the symmetrization is to replace $N(\nu_1, \cdots, \nu_l)$ by the sum $\sum_N'$ taken over N nonidentical products. Thus, (114) goes into the Volterra series (1) with the symmetric kernel

$$g_n(u_1, \cdots, u_n) = \sum_{l=1}^{n} F_l \sum_{(\nu;l,n)} h_{\nu_1} \cdots h_{\nu_l} \cdot \sum_N' \varphi_{\nu_1}(u_1, \cdots, u_{\nu_1}) \cdots \varphi_{\nu_l}(u_\mu, \cdots, u_n). \quad (115)$$

This kernel has the n-fold Fourier transform $G_n(f_1, \cdots, f_n)$ stated in (49) as we wished to show.

Finally we sketch the derivation of the general recurrence relation (58) which gives $G_n(f_1, \cdots, f_n)$ for the feedback system of Fig. 2. Replacing t by $t - u_q'$ in the Volterra series (59) for $w(t)$ in "powers" of $x(t)$ gives a series for $w(t-u_q')$. The product $\prod_{q=1}^{l} w(t-u_q')$ in the system equation (53) for $y(t)$ can then be written as an l-fold sum of an $i_1 + \cdots + i_q = n$-fold integral. Changing the order of summation and using

$$\sum_{l=1}^{\infty}\sum_{i_1=1}^{\infty} \cdots \sum_{i_l=1}^{\infty} = \sum_{n=1}^{\infty}\sum_{l=1}^{n} \sum_{i_1+i_2+\cdots+i_l=n}$$

where the summand is of the form $A_{i_1 i_2 \cdots i_l}$, leads to a series for $y(t)$:

$$y(t) = \sum_{n=1}^{\infty} \frac{1}{n!} \int_{-\infty}^{\infty} du_1 \cdots \int_{-\infty}^{\infty} du_n \varphi_n(u_1, \cdots, u_n) \prod_{r=1}^{n} x(t-u_r) \quad (116)$$

where

$$\varphi_n(u_1, \cdots, u_n) = \sum_{l=1}^{n} \frac{1}{l!} \sum_{i_1+\cdots+i_l=n} \frac{n!}{i_1! \cdots i_l!} \int_{-\infty}^{\infty} du_1' \cdots \int_{-\infty}^{\infty} du_l' \cdot m_l(u_1', \cdots, u_l')k_{i_1}(u_1 - u_1', \cdots, u_{i_1} - u_1') \cdots k_{i_l}(\cdots, u_n - u_l'). \quad (117)$$

Comparison of (116) with the Volterra series (1) for $y(t)$ shows that $g_n(u_1, \cdots, u_n)$ is the symmetrized version of $\varphi_n(u_1, \cdots, u_n)$; i.e., $G_n(f_1, \cdots, f_n)$ is the symmetrized version of n-fold Fourier transform $\Phi_n(f_1, \cdots, f_n)$ of φ_n. It turns out that the expression for Φ_n can be obtained from the right side of (117) by replacing the l-fold integral by a product $K_{i_1}K_{i_2} \cdots K_{i_l} M_l$ which has the same form (with ν's replaced by i's) as the product $K_{\nu_1}K_{\nu_2} \cdots K_{\nu_l}M_l$ appearing in the general expression (58) for $G_n(f_1, \cdots, f_n)$. This product (and also the function M_l with arguments $f_1 + \cdots + f_{i_1}$, etc.) has the same type of symmetry as the product $P(f_1, \cdots, f_n; i_1, \cdots, i_l)$ discussed in Section V-C. Symmetrizing Φ_n with the help of (123), setting the result equal to G_n, noting that the term for $l=1$ in Φ_n is the only one containing G_n, and solving for G_n completes the derivation of (58) for G_n.

C. Symmetrization of Products of Symmetric Functions

Let "SV" stand for "symmetrized version of" and denote by $\{\text{SV } F(f_1, f_2, \cdots, f_n)\}$ the function obtained by symmetrizing the arbitrary function $F(f_1, f_2, \cdots, f_n)$:

$$\{\text{SV } F(f_1, \cdots, f_n)\} = \frac{1}{n!} \sum_{n!} F(f_1, \cdots, f_n). \tag{118}$$

Here the subscript $n!$ on $\sum$ denotes that the summation extends over all $n!$ permutations of the subscripts on the f's. Define the product $P(\cdots)$ by

$$P(f_1, \cdots, f_n; i_1, \cdots, i_l) \equiv s_{i_1}(f_1, \cdots, f_{i_1}) s_{i_2}(f_{i_1+1}, \cdots, f_{i_1+i_2}) \cdots s_{i_l}(f_{n-i_l+1}, \cdots, f_n)$$

$$i_1 + i_2 + \cdots + i_l = n, \qquad 1 \le i_q \le n, \qquad q = 1, 2, \cdots, l \tag{119}$$

where the functions $s_i(\cdots)$ are symmetric functions of their arguments.

We first prove that

$$\{\text{SV } P(f_1, \cdots, f_n; v_1, \cdots, v_l)\} = \frac{1}{N} \sum_N' P(f_1, \cdots, f_n; v_1, \cdots, v_l) \tag{120}$$

where n, l, and the set of integers v_i are given. The v's are integers such that, as in (25)

$$v_1 + v_2 + \cdots + v_l = n, \qquad 1 \le v_1 \le v_2 \cdots \le v_l. \tag{121}$$

The sum on the right side of (120) is over all nonidentical products where, as in connection with (26), the number of such products in the sum is

$$N = n!/(v_1! \cdots v_l! r_1! \cdots r_p!). \tag{122}$$

Here r_1 is the number of equal v's in the first run of equal v's in the arrangement $v_1 \le v_2 \le \cdots \le v_l$, r_2 the number in the second run, and so on.

To prove (120) note that for a given set of values of $v_1, v_2, \cdots, v_l$ the value of $P(\cdots)$ is not changed when the f_k in the argument of $s_v(\cdots)$ are permuted. There are $v_1!v_2! \cdots v_l!$ such permutations. Furthermore, the value of $P(\cdots)$ is not changed by permuting the s_v with equal subscripts (equal v's). Permutations of this sort do not violate the condition $v_1 \le v_2 \le \cdots \le v_l$. There are $r_1!r_2! \cdots r_p!$ such permutations. The $n!$ quantities $P(\cdots)$ given by the $n!$ permutations of $f_1, \cdots, f_n$ can be sorted into sets according to the value of $P(\cdots)$. The number of members in each set is the same, namely, $v_1!v_2! \cdots v_l!r_1! \cdots r_p! = M$ and the number of sets is $N = n!/M$. Changing the summation over the $n!$ permutations [see (118)] to a summation over the N sets then gives the desired relation (120).

We next prove that, given n and l,

$$\left\{\text{SV } \frac{1}{l!} \sum_{i_1+\cdots+i_l=n} \frac{n!}{i_1! \cdots i_l!} P(f_1, \cdots, f_n; i_1, \cdots, i_l)\right\} = \sum_{(v;l,n)} \sum_N' P(f_1, \cdots, f_n; v_1, \cdots, v_l) \tag{123}$$

where the summation on the left is taken over the integers $i_1, \cdots, i_l$ such that $i_1 + \cdots + i_l = n$ and $1 \le i_q \le n$ [see (119)]. The $(v; l, n)$ beneath the $\sum$ on the right denotes summation over all sets of integers v which satisfy (121). To prove (123), note that corresponding to each set $v_1, v_2, \cdots, v_l$ in $(v; l, n)$ there are $l!/(r_1! \cdots r_p!)$ sets of $i_1, i_2, \cdots, i_l$ in which the values of the i's are scrambled values of the v's. For any one of these sets of i's we have 1) $i_1! \cdots i_l! = v_1! \cdots v_l!$ and 2)

$$\{\text{SV } P(f_1, \cdots, f_n; i_1, \cdots, i_l)\} = \{\text{SV } P(f_1, \cdots, f_n; v_1, \cdots, v_l)\}. \tag{124}$$

Therefore, the left side of (123) can be written as

$$\begin{aligned} &\frac{1}{l!} \sum_{i_1+\cdots+i_l=n} \frac{n!}{i_1! \cdots i_l!} \{\text{SV } P(f_1, \cdots, f_n; i_1, \cdots, i_l)\} \\ &= \frac{1}{l!} \sum_{v_1+\cdots+v_l=n} \frac{n!}{v_1! \cdots v_l!} \frac{l!}{r_1! \cdots r_p!} \{\text{SV } P(f_1, \cdots, f_n; v_1, \cdots, v_l)\} \\ &= \sum_{(v;l,n)} N\{\text{SV } P(f_1, \cdots, f_n; v_1, \cdots, v_l)\} \\ &= \sum_{(v;l,n)} \sum_N' P(f_1, \cdots, f_n; v_1, \cdots, v_l) \end{aligned} \tag{125}$$

which gives the desired relation (123).

VI. Simple Properties of the Output

Here the Volterra series (1) and the expression (2) for the Volterra transfer function are recast in forms suited to deal with harmonic and Gaussian inputs. The new forms are illustrated by applying them to obtain 1) the general form of the expressions listed in Section II-A for $y(t)$ when $x(t)$ is a sine wave or the sum of two or more sine waves, and 2) the expression for the dc value of $y(t)$ when $x(t)$ is Gaussian. When $x(t)$ is Gaussian, the dc value of $y(t)$ is equal to the expected value, or ensemble average, $\langle y(t) \rangle$.

A. General Relations

The derivation of the formulas whose leading terms are listed in Section II-A is simplified by writing the product $x(t-u_1) \cdots x(t-u_n)$ in (1) as the coefficient of $\alpha_1 \alpha_2 \cdots \alpha_n$ in the expansion of $\exp[\alpha_1 x(t-u_1) + \cdots + \alpha_n x(t-u_n)]$; i.e., as the result of operating on the exponential function by

$$D_\alpha^n \equi \left. \frac{\partial^n}{\partial \alpha_1 \cdots \partial \alpha_n} \right|_{\alpha_1 = \cdots = \alpha_n = 0} \tag{126}$$

Thus

$$x(t-u_1) \cdots x(t-u_n) = D_\alpha^n \exp\left[\sum_{s=1}^n \alpha_s x(t-u_s)\right] \tag{127}$$

$$= D_\alpha^n \left[\sum_{s=1}^n \alpha_n x(t-u_s)\right]^n \Big/ n! \tag{128}$$

and the Volterra series (1) for $y(t)$ can be rewritten in the following two ways:

$$y(t) = \sum_{n=1}^\infty \frac{1}{n!} \int_{-\infty}^\infty du_1 \cdots \int_{-\infty}^\infty du_n g_n(u_1, \cdots, u_n) \cdot D_\alpha^n \exp\left[\sum_{s=1}^n \alpha_s x(t-u_s)\right] \tag{129}$$

$$y(t) = \sum_{n=1}^\infty \frac{1}{n!} \int_{-\infty}^\infty du_1 \cdots \int_{-\infty}^\infty du_n g_n(u_1, \cdots, u_n) \cdot D_\alpha^n \left[\sum_{s=1}^n \alpha_s x(t-u_s)\right]^n \Big/ n!. \tag{130}$$

The series (130) for $y(t)$ and the method of calculating $G_n(f_1, \cdots, f_n)$ by taking $x(t)$ to be the sum of n exponential terms leads to the useful expression

$$G_n(f_1, \cdots, f_n) = \frac{1}{n!} \int_{-\infty}^\infty du_1 \cdots \int_{-\infty}^\infty du_n g_n(u_1, \cdots, u_n) D_\alpha^n \prod_{r=1}^n A_n(f_r) \tag{131}$$

where

$$A_n(f) = \sum_{s=1}^n \alpha_s e^{-j\omega u_s}, \qquad \omega = 2\pi f. \tag{132}$$

The following result [8] will be used in conjunction with the rewritten Volterra series (129) for $y(t)$ when $x(t)$ is Gaussian with the two-sided power spectrum $W_x(f)$. Let L be a linear operator (operating on functions of t) such that

$$L[e^{j\omega t}] = H(f)e^{j\omega t}, \qquad \omega = 2\pi f. \tag{133}$$

Then

$$\langle \exp\{L[x(t)]\}\rangle = \exp\left[\frac{1}{2}\int_{-\infty}^{\infty} df\, W_x(f)H(f)H(-f)\right]. \tag{134}$$

B. Harmonic Input

When $x(t) = P\cos pt$, $p = 2\pi f_p$, the rightmost sum in (130) is

$$\sum_{s=1}^{n} \alpha_s x(t-u_s) = \frac{1}{2}P\sum_{s=1}^{n}\alpha_s(e^{jpt-jpu_s} + e^{-jpt+jpu_s})$$
$$= \frac{1}{2}P[e^{jpt}A_n(f_p) + e^{-jpt}A_n(-f_p)] \tag{135}$$

where A_n is given by (132). From the binomial theorem

$$\left[\sum_{s=1}^{n}\alpha_s x(t-u_s)\right]^n \Big/ n!$$
$$= \left(\frac{P}{2}\right)^n \sum_{k=0}^{n}\frac{\exp[j(2k-n)pt]}{k!(n-k)!}A_n^k(f_p)A_n^{n-k}(-f_p). \tag{136}$$

Substuting this in (130) for $y(t)$ and using (131) for G_n gives

$$y(t) = \sum_{n=1}^{\infty}\left(\frac{P}{2}\right)^n\sum_{k=0}^{n}\frac{\exp[j(2k-n)pt]}{k!(n-k)!}G_{k,n-k}(f_p). \tag{137}$$

Here $G_{k,n-k}(f_p)$ denotes $G_n(f_1,\cdots,f_n)$, with the first k of the f_i equal to $+f_p$ and the remaining $n-k$ equal to $-f_p$.

Selecting the terms in (137) for which $2k-n=N\geq 0$ shows that the $\exp(jNt)$ component of $y(t)$ is

$$e^{jNpt}\sum_{l=0}^{\infty}\frac{(P/2)^{2l+N}}{(N+l)!\,l!}G_{N+l,l}(f_p). \tag{138}$$

The value of $G_{0,0}(f_p)$, which occurs when $N=0$, is zero because $G_0\equiv 0$. Changing the signs of p and f_p in (138) gives the $\exp(-jNpt)$ component of $y(t)$. For $x(t)=P\cos(pt+\varphi)$, $y(t)$ is given by (137) and (138) with pt replaced by $pt+\varphi$. These expressions for the output components are similar to formulas given by Mircea [7].

The same type of argument shows that when $x(t)=P\cos pt + Q\cos qt$, the $\exp[j(Np+Mq)t]$ component in $y(t)$ is, for $M\geq 0$ and $N\geq 0$,

$$e^{j(Np+Mq)t}\sum_{l=0}^{\infty}\sum_{k=0}^{\infty}\frac{(P/2)^{2l+N}(Q/2)^{2k+M}}{(N+l)!\,l!\,(M+k)!\,k!}G_{N+l,l;M+k,k}(f_p,f_q) \tag{139}$$

where $2\pi f_p = p$, $2\pi f_q = q$, and the four subscripts on G mean that it is equal to $G_n(f_1,\cdots,f_n)$ with $n=N+2l+M+2k$ and the first $N+l$ of the f_i equal to f_p, the next l equal to $-f_p$, the next $M+k$ equal to f_q, and the last k equal to $-f_q$. Changing the signs of p and f_p in (139) gives the $\exp[j(-Np+Mq)t]$ component of $y(t)$, and so on.

Similarly, when $x(t)=P\cos pt + Q\cos qt + R\cos rt$, the $\exp[j(Np+Mq+Lr)t]$ component in $y(t)$ is, for $M\geq 0$, $N\geq 0$, and $L\geq 0$,

$$e^{j(Np+Mq+Lr)t}\sum_{l=0}^{\infty}\sum_{k=0}^{\infty}\sum_{i=0}^{\infty}\frac{(P/2)^{2l+N}(Q/2)^{2k+M}(R/2)^{2i+L}}{(N+l)!\,l!\,(M+k)!\,k!\,(L+i)!\,i!}$$
$$G_{N+l,l;M+k,k;L+i,i}(f_p,f_q,f_r) \tag{140}$$

where the order of $G_n(f_1,\cdots,f_n)$ is $n=N+2l+M+2k+L+2i$.

When phase angles appear in the cosine terms in $x(t)$, we replace Npt by $Npt+N\varphi_p$, Mqt by $Mqt+M\varphi_q$, etc., in the exponential terms in (139) and (140).

C. Expected Value of $y(t)$ for Gaussian Input

When $x(t)$ is a zero-mean stationary Gaussian process, the expected value of $y(t)$ is the ensemble average $\langle y(t)\rangle$ obtained by averaging both sides of (129) and using (134) to show that

$$\left\langle \exp\sum_{s=1}^{n}\alpha_s x(t-u_s)\right\rangle = \exp J_{nAA} \tag{141}$$

$$J_{nAA} = \frac{1}{2}\int_{-\infty}^{\infty} df\, W_x(f)A_n(f)A_n(-f). \tag{142}$$

Here the subscript nAA is suggested by (142) and (150), $W_x(f)$ is the two-sided power spectrum of $x(t)$, and $A_n(f)$ is defined by (132). The integral (142) for J_{nAA} is obtained by identifying (141) with (134) and using

$$L[e^{j\omega t}] = e^{j\omega t}\sum_{s=1}^{n}\alpha_s e^{-j\omega u_s} = e^{j\omega t}A_n(f) \tag{143}$$

to show that $H(f)$ goes into $A_n(f)$. Thus the expression (129) for $y(t)$ gives

$$\langle y(t)\rangle = \sum_{n=1}^{\infty}\frac{1}{n!}\int_{-\infty}^{\infty}du_1\cdots\int_{-\infty}^{\infty}du_n g_n(u_1,\cdots,u_n)D_\alpha^n \exp J_{nAA}. \tag{144}$$

The next step is to expand $\exp J_{nAA}$. It is convenient to introduce the operator $Q_k[h(x)]$ which denotes a k-fold integration with respect to $x_1, x_2,\cdots,x_k$, with limits $\pm\infty$. The integrand is $h(x_1)\cdots h(x_k)$ times the function of $x_1,\cdots,x_k$ represented by all of the terms lying to the right of $Q_k[h(x)]$. $Q_0[h(x)]$ denotes the identity operator. When we substitute

$$\exp J_{nAA} = \sum_{\mu=0}^{\infty}\frac{1}{\mu!}J_{nAA}^{\mu}$$
$$= 1 + \sum_{\mu=1}^{\infty}\frac{1}{\mu!\,2^\mu}Q_\mu[W_x(f)]\prod_{r=1}^{\mu}A_n(f_r)A_n(-f_r) \tag{145}$$

in the series (144) for $\langle y(t)\rangle$, the 1 in (145) corresponding to $\mu=0$ contributes nothing because for $n\geq 1$ the value of D_α^n operating on 1 is 0. Interchanging the order of the n and μ summations gives

$$\langle y(t)\rangle = \sum_{\mu=1}^{\infty}\frac{1}{\mu!\,2^\mu}Q_\mu[W_x(f)]\sum_{n=1}^{\infty}\frac{1}{n!}\int_{-\infty}^{\infty}du_1\cdots$$
$$\cdot\int_{-\infty}^{\infty}du_n g_n(u_1,\cdots,u_n)D_\alpha^n\prod_{r=1}^{\mu}A_n(f_r)A_n(-f_r). \tag{146}$$

Since $A_n(f_r)$ is a homogeneous linear function of the α's, the product in (146) is of degree 2μ in the α's; and all of the terms in the n-sum are 0 except the one for $n=2\mu$. Setting $n=2\mu$ and using (131) for G_n leads to

$$\langle y(t)\rangle = \sum_{\mu=1}^{\infty}\frac{1}{\mu!\,2^\mu}Q_\mu[W_x(f)]G_{2\mu}(f_1,-f_1,f_2,-f_2,\cdots,f_\mu,-f_\mu) \tag{147}$$

the first two terms of which have been given in (11).

A somewhat similar expression for $\langle y(t)\rangle$ has been given by Deutsch [12].

VII. Power Spectra

In this section the two-sided power spectrum $W_y(f)$ of $y(t)$ is computed for two cases. In the first, the input $x(t)$ is zero-mean stationary Gaussian noise with power spectrum $W_x(f)$ (the Mircea–Sinnreich case). In the second, the input $x(t)$ is a sine wave plus zero-mean stationary Gaussian noise, $P \cos pt + I_N(t)$. In both cases the ensemble average $\langle y(t+\tau)y^*(t)\rangle$ is computed and then its Fourier transform taken to get $W_y(f)$.

A. *The Expected Value of* $y(t+\tau)z(t)$ *for Gaussian Input*

Let $y(t)$ be given by the Volterra series (1) and $z(t)$ by a similar series with $g'_n(u_1, \cdots, u_n)$ in place of $g_n(u_1, \cdots, u_n)$. Both have the same Gaussian noise input $x(t)$. The steps in calculating the ensemble average $\langle y(t+\tau)z(t)\rangle$ are similar to, but more complicated than, those used to calculate $\langle y(t)\rangle$.

From the rewritten Volterra series (129)

$$\langle y(t+\tau)z(t)\rangle = \sum_{n=1}^{\infty}\sum_{m=1}^{\infty}\frac{1}{n!m!}\int_{-\infty}^{\infty}du_1\cdots\int_{-\infty}^{\infty}du_n\int_{-\infty}^{\infty}dv_1\cdots \cdot\int_{-\infty}^{\infty}dv_m g_n(u_1,\cdots,u_n)g'_m(v_1,\cdots,v_m)D_\alpha^n D_\beta^m \cdot\left\langle \exp\left[\sum_{s=1}^{n}\alpha_s x(t+\tau-u_s)+\sum_{s=1}^{m}\beta_s x(t-v_s)\right]\right\rangle \cdot \quad (148)$$

The ensemble average of the exponential function is again given by (134), but now $H(f)$ is determined by

$$L[e^{j\omega t}] = e^{j\omega t}[e^{j\omega\tau}A_n(f) + B_m(f)] = e^{j\omega t}H(f) \quad (149)$$

where $A_n(f)$ is still given by (132) and $B_m(f)$ by (132) with n, α, u replaced by m, β, v. Therefore, the ensemble average of the exponential function in (148) can be written as $\exp K$, where

$$K = \frac{1}{2}\int_{-\infty}^{\infty} df\, W_x(f)[A_n(f)A_n(-f) + B_m(f)B_m(-f) + 2e^{j\omega\tau}A_n(f)B_m(-f)] = J_{nAA} + J_{mBB} + J_{nmAB}. \quad (150)$$

The evenness of $W_x(f)$ has been used to obtain the term containing $\exp(j\omega\tau)$; J_{nAA} is the integral containing $A_n(f)A_n(-f)$; and so on.

Expanding $\exp J_{nmAB}$ in the same way as was $\exp J_{nAA}$ in (145) gives

$$e^K = \exp[J_{nAA} + J_{mBB}] \cdot\left[1 + \sum_{k=1}^{\infty}\frac{Q_k[W_x(f)]}{k!}\prod_{r=1}^{k} e^{j\omega_r\tau}A_n(f_r)B_m(-f_r)\right]\cdot \quad (151)$$

When this is substituted for the ensemble average in (148), the contribution of $\exp[J_{nAA}+J_{mBB}]$ times 1, the 1 being the term in (151) corresponding to $k=0$, is $\langle y(t)\rangle\langle z(t)\rangle$. This follows from the series (144) for $\langle y(t)\rangle$. The contribution of the remaining portion, i.e., the portion arising from the $k \geq 1$ terms in (151), can be obtained by changing the order of the summations and integrations. After the change the double m, n-sum can be written as the product of the m-sum and the n-sum. Therefore, the substitution of (151) in (148) yields

$$\langle y(t+\tau)z(t)\rangle = a_0 b_0 + \sum_{k=1}^{\infty}\frac{Q_k[W_x(f)]}{k!}e^{j(\omega_1+\cdots+\omega_k)\tau}a_k(f_1,\cdots,f_k)b_k(f_1,\cdots,f_k) \quad (152)$$

where $a_0 = \langle y(t)\rangle$, $b_0 = \langle z(t)\rangle$, and for $k > 0$,

$$a_k(f_1,\cdots,f_k) = \sum_{n=1}^{\infty}\frac{1}{n!}\int_{-\infty}^{\infty}du_1\cdots\int_{-\infty}^{\infty}du_n g_n(u_1,\cdots,u_n)D_\alpha^n e^{J_{nAA}}\prod_{r=1}^{k}A_n(f_r). \quad (153)$$

The function $b_k(f_1, \cdots, f_k)$ is given by an expression obtained by replacing n, u, g_n, α, J_{nAA}, $A_n(f_r)$ in (153) by m, v, g'_m, β, J_{mBB}, $B_m(-f_r)$. For example, if $z(t) \equiv y(t)$, then $b_k(f_1, \cdots, f_k) = a_k(-f_1, \cdots, -f_k)$.

Substituting the series (145) for $\exp J_{nAA}$ in (153) brings in the quantity

$$D_\alpha^n\left[\prod_{q=1}^{\mu}A_n(f'_q)A_n(-f'_q)\right]\left[\prod_{r=1}^{k}A_n(f_r)\right] \quad (154)$$

where r, f, f_r in (145) have been replaced by q, f', f'_q. Since $A_n(f)$ is homogeneous and linear in the α's, (154) is zero unless $n = 2\mu + k$. The double sum for a_k taken over μ and n reduces to a single sum over μ in which the μth term is, with $n = 2\mu + k$,

$$\frac{1}{\mu!2^\mu}Q_\mu[W_x(f')]\frac{1}{n!}\int_{-\infty}^{\infty}du_1\cdots\int_{-\infty}^{\infty}du_n \cdot g_n(u_1,\cdots,u_n)\ \{\text{expression (154)}\}. \quad (155)$$

The expression (131) for $G_n(f_1, \cdots, f_n)$ shows that the multiple integral goes into a G_n function. Consequently, the series (153) for a_k, $k > 0$, becomes

$$a_k(f_1,\cdots,f_k) = \sum_{\mu=0}^{\infty}\frac{Q_\mu[W_x(f')]}{\mu!2^\mu}G_{2\mu+k}(f_1,f_2,\cdots,f_k,f'_1,-f'_1,\cdots,f'_\mu,-f'_\mu)$$
$$= G_k(f_1,\cdots,f_k) + \frac{1}{1!2}\int_{-\infty}^{\infty}df'_1 W_x(f'_1)G_{2+k}(f_1,\cdots,f_k,f'_1,-f'_1) + \frac{1}{2!2^2}\int_{-\infty}^{\infty}df'_1\int_{-\infty}^{\infty}df'_2 W_x(f'_1)W_x(f'_2)G_{4+k}(f_1,\cdots,f_k,f'_1,-f'_1,f'_2,-f'_2) + \cdots. \quad (156)$$

The corresponding expression for $b_k(f_1, \cdots, f_k)$ is obtained by replacing the $G_{2\mu+k}$ functions in (156) by $G'_{2\mu+k}(-f_1, \cdots, -f_k, f'_1, -f'_1, \cdots, f_\mu, -f'_\mu)$ where G'_m is the Fourier transform of the kernel g'_m in the Volterra series for $z(t)$. The signs of $f_1, \cdots, f_k$ are reversed because $B_m(-f_r)$ replaces $A_n(f_r)$ in the analysis. Since the G_m are symmetric, so are $a_k(f_1, \cdots, f_k)$ and $b_k(f_1, \cdots, f_k)$. The series for $b_k(f_1, \cdots, f_k)$ is

$$b_k(f_1,\cdots,f_k) = G'_k(-f_1,\cdots,-f_k) + \frac{1}{1!2}\int_{-\infty}^{\infty}df'_1 W_x(f'_1)G'_{2+k}(-f_1,\cdots,-f_k,f'_1,-f'_1) + \cdots. \quad (157)$$

Writing out the integrals denoted by $Q_\mu[W_x(f)]$ in (152) gives the required expression

$$\langle y(t+\tau)z(t)\rangle = \langle y(t)\rangle\langle z(t)\rangle + \frac{1}{1!}\int_{-\infty}^{\infty}df_1 e^{j\omega_1\tau}W_x(f_1)a_1(f_1)b_1(f_1) + \frac{1}{2!}\int_{-\infty}^{\infty}df_1\int_{-\infty}^{\infty}df_2 e^{j(\omega_1+\omega_2)\tau}W_x(f_1)W_x(f_2)a_2(f_1,f_2)b_2(f_1,f_2) + \cdots \quad (158)$$

where the a_k and b_k are given by the series (156) and (157).

The special case $z(t) \equiv y(t)$ has been considered by Deutsch [12] who outlined a procedure for calculating $\langle y(t+\tau)y(t)\rangle$.

B. *Power Spectrum for Gaussian Input*

The two-sided power spectrum $W_y(f)$ of $y(t)$, for complex $y(t)$ and Gaussian $x(t)$, is the Fourier transform of the function $\langle y(t+\tau)y^*(t)\rangle$ of τ obtained by setting $z(t) = y^*(t)$ in (152). Then 1) $g'_m = g^*_m$, 2)

$$G'_{2\mu+k}(-f_1,\cdots,-f_k,f'_1,-f'_1,\cdots,f'_\mu,-f'_\mu) = G^*_{2\mu+k}(f_1,\cdots,f_k,f'_1,-f'_1,\cdots,f'_\mu,-f'_\mu) \quad (159)$$

and 3) (156) and (157) show that $b_k(f_1, \cdots, f_k)$ is equal to $a_k^*(f_1, \cdots, f_k)$. Consequently, $\langle y(t+\tau)y^*(t)\rangle$ is given by the series (152) with $a_k(f_1, \cdots, f_k)b_k(f_1, \cdots, f_k)$ replaced by $|a_k(f_1, \cdots, f_k)|^2$. Multiplying by $\exp(-j\omega\tau)$ and integrating τ from $-\infty$ to ∞ gives, in our notation, the Mircea–Sinnreich [5] series for the power spectrum of $y(t)$:

$$\begin{aligned} W_y(f) &= |a_0|^2\delta(f) \\ &\quad + \sum_{k=1}^{\infty} \frac{Q_k[W_x(f)]}{k!} \delta(f - f_1 - \cdots - f_k)|a_k(f_1, \cdots, f_k)|^2 \\ &= |a_0|^2\delta(f) + W_x(f)|a_1(f)|^2 \\ &\quad + \frac{1}{2!}\int_{-\infty}^{\infty} df_1 W_x(f_1)W_x(f - f_1)|a_2(f_1, f - f_1)|^2 \\ &\quad + \frac{1}{3!}\int_{-\infty}^{\infty} df_1 \int_{-\infty}^{\infty} df_2 W_x(f_1)W_x(f_2)W_x(f - f_1 - f_2) \\ &\quad \cdot |a_3(f_1, f_2, f - f_1 - f_2)|^2 \\ &\quad + \cdots. \end{aligned} \tag{160}$$

Here $a_0 = \langle y(t)\rangle$ and $a_k(f_1, \cdots, f_k)$ for $k>0$ is given by the series (156). Note that the f in the operator $Q_k[W_x(f)]$ takes on the values $f_1, f_2, \cdots, f_k$. It is not related to the f in $\delta(f - f_1 - \cdots - f_k)$. The series for a_0 is given by the series (147) for $\langle y(t)\rangle$. For $k = 1, 2, 3$, the first few terms in (156) are

$$\begin{aligned} a_1(f) &= G_1(f) + \frac{1}{1!2}\int_{-\infty}^{\infty} df_1' W_x(f_1')G_3(f, f_1', -f_1') + \cdots \\ a_2(\rho, \sigma) &= G_2(\rho, \sigma) + \frac{1}{1!2}\int_{-\infty}^{\infty} df_1' W_x(f_1')G_4(\rho, \sigma, f_1', -f_1') + \cdots \\ a_3(\rho, \sigma, \lambda) &= G_3(\rho, \sigma, \lambda) + \cdots. \end{aligned} \tag{161}$$

The leading terms in (160) for $W_y(f)$ have been given in (14).

When $y(t)$ is complex, it follows from (70) that the power spectrum of the real part of $y(t)$ can be obtained by replacing $G_n(f_1, \cdots, f_n)$ by $[G_n(f_1, \cdots, f_n) + G_n^*(-f_1, \cdots, -f_n)]/2$ in the analysis leading to $W_y(f)$. This is equivalent to replacing $|a_k(f_1, \cdots, f_k)|^2$ by $|a_k(f_1, \cdots, f_k) + a_k^*(-f_1, \cdots, -f_k)|^2/4$ in the series (160) for $W_y(f)$ to get the power spectrum of the real part of $y(t)$. Likewise, the power spectrum of the imaginary part of $y(t)$ is obtained by replacing $G_n(f_1, \cdots, f_n)$ by $[G_n(f_1, \cdots, f_n) - G_n^*(-f_1, \cdots, -f_n)]/(2j)$ in the analysis leading to $W_y(f)$, and $|a_k(f_1, \cdots, f_k)|^2$ by $|a_k(f_1, \cdots, f_k) - a_k^*(-f_1, \cdots, -f_k)|^2/4$ in (160).

C. Power Spectrum for Sine Wave Plus Noise Input

When $x(t) = P\cos pt + I_N(t)$ where $I_N(t)$ is a Gaussian noise having the two-sided power spectrum $W_I(f)$, the ensemble average $\langle y(t)\rangle$ is periodic with period $1/f_p = 2\pi/p$. To obtain $\langle y(t)\rangle$, we proceed as in Section VI-C. Combining (135) and (141) gives

$$\left\langle \exp \sum_{s=1}^{n} \alpha_n x(t - u_s)\right\rangle = \exp\left\{\frac{P}{2}[e^{jpt}A_n(f_p) + e^{-jpt}A_n(-f_p)]\right\}\exp J_{nAA} \tag{162}$$

where J_{nAA} is given by (142) with $W_x(f)$ replaced by $W_I(f)$. If the right side of (162) is expanded with the help of (145), it becomes

$$\sum_{N=0}^{\infty} \frac{1}{N!}\left(\frac{P}{2}\right)^N \sum_{l=0}^{N} \frac{N!e^{j(2l-N)pt}}{l!(N-l)!}[A_n(f_p)]^l[A_n(-f_p)]^{N-l} \cdot\left[1 + \sum_{\mu=1}^{\infty}\frac{1}{\mu!2^\mu}Q_\mu[W_I(f)]\prod_{r=1}^{\mu}A_n(f_r)A_n(-f_r)\right]\cdot \tag{163}$$

When this is substituted in the series of integrals obtained by taking the ensemble average of (129) and changing the order of summation, the operator D_α^n makes all terms zero except those for which $l + (N-l) + 2\mu = N + 2\mu = n$. The result is

$$\begin{aligned} \langle y(t)\rangle = \sum_{N=0}^{\infty}\left(\frac{P}{2}\right)^N \sum_{l=0}^{N} \frac{e^{j(2l-N)pt}}{l!(N-l)!}\sum_{\mu=0}^{\infty}\frac{1}{\mu!2^\mu}Q_\mu[W_I(f)] \\ \cdot G_{N+2\mu}(f_1, -f_1, \cdots, f_\mu, -f_\mu, (f_p)_l, (-f_p)_{N-l}) \end{aligned} \tag{164}$$

where $G_0 \equiv 0$ and $(f_p)_l$ denotes the string of l arguments $f_p, f_p, \cdots, f_p$. If μ or l or $N-l$ are zero, the corresponding arguments in $G_{N+2\mu}$ do not appear. The $G_{N+2\mu}$ term corresponding to $\mu = 0$ in (164), namely, $G_N((f_p)_l, (-f_p)_{N-l})$, is the same as $G_{l,N-l}(f_p)$ in the notation of (137). The f in $Q_\mu[W_I(f)]$ refers only to $f_1, \cdots, f_\mu$, never to f_p. The first few terms of (164) have been given in (15).

The exp $(jnpt)$ component in $\langle y(t)\rangle$ is

$$\begin{aligned} e^{jnpt}\sum_{\sigma=0}^{\infty}\left(\frac{P}{2}\right)^{2\sigma+|n|}\frac{1}{\sigma!(\sigma+|n|)!}\sum_{\mu=0}^{\infty}\frac{1}{\mu!2^\mu}Q_\mu[W_I(f)] \\ \cdot G_{2\sigma+|n|+2\mu}(f_1, -f_1, \cdots, f_\mu, -f_\mu, (f_p s_n)_{\sigma+|n|}, (-f_p s_n)_\sigma) \end{aligned} \tag{165}$$

where $s_n = 1$ for $n \geq 0$ and $s_n = -1$ for $n < 0$. When P is zero, (165) becomes (147), and when $W_I(f)$ is zero, (165) becomes (138) with $N = |n|$.

We now obtain $\langle y(t+\tau)z(t)\rangle$ where $z(t)$ is the same as in Section VII A. For $x(t) = P\cos pt + I_N(t)$,

$$\begin{aligned} &\left\langle \exp\left[\sum_{s=1}^{n}\alpha_s x(t+\tau-u_s) + \sum_{s=1}^{m}\beta_s x(t - v_s)\right]\right\rangle \\ &= \exp\left[\frac{P}{2}\{e^{jpt}[e^{jp\tau}A_n(f_p) + B_m(f_p)] + e^{-jpt}[e^{-jp\tau}A_n(-f_p) \right. \\ &\quad \left. + B_m(-f_p)]\}\right]\exp(J_{nAA} + J_{mBB} + J_{nmAB}) \end{aligned} \tag{166}$$

where J's are defined by the integral (150) with $W_x(f)$ replaced by $W_I(f)$, and $A_n(f)$ and $B_m(f)$ are the same as in (149). The contribution of the first term, i.e., unity, in the expansion of exp J_{nmAB} [see (151)] to the value of the double series for $\langle y(t+\tau)z(t)\rangle$ [obtained by averaging the product of the series (129) for $y(t+\tau)$ and $z(t)$ and using (166)] can be split into the product of the n-sum and m-sum. As in Section VII-A, this contribution is $\langle y(t+\tau)\rangle\langle z(t)\rangle$ in which the averages can be obtained from (164).

To get the contribution of the remaining terms in the expansion of exp (J_{nmAB}), we expand the first exponential on the right in (166) as

$$\sum_{N=0}^{\infty}\frac{1}{N!}\left(\frac{P}{2}\right)^N\sum_{l=0}^{N}\frac{N!e^{j(2l-N)pt}}{l!(N-l)!}\sum_{\lambda=0}^{l}\sum_{\sigma=0}^{N-l}\frac{l!(N-l)!}{\lambda!(l-\lambda)!\sigma!(N-l-\sigma)!}e^{j(\lambda-\sigma)p\tau}A_n^\lambda(f_p)B_m^{l-\lambda}(f_p)A_n^\sigma(-f_p)B_m^{N-l-\sigma}(-f_p). \tag{167}$$

This leads to

$$\begin{aligned} \langle y(t+\tau)z(t)\rangle = \langle y(t+\tau)\rangle\langle z(t)\rangle + \sum_{N=0}^{\infty}\left(\frac{P}{2}\right)^N\sum_{l=0}^{N}e^{j(2l-N)pt}\sum_{\lambda=0}^{l}\sum_{\sigma=0}^{N-l}\frac{e^{j(\lambda-\sigma)p\tau}}{\lambda!(l-\lambda)!\sigma!(N-l-\sigma)!} \\ \cdot\sum_{k=1}^{\infty}\frac{Q_k[W_I(f)]e^{j(\omega_1+\cdots+\omega_k)\tau}}{k!}a_{\lambda,\sigma,k}(f_1, f_2, \cdots, f_k; f_p)b_{l-\lambda,N-l-\sigma,k}(f_1, f_2, \cdots, f_k; f_p) \end{aligned} \tag{168}$$

where

$$a_{\lambda,\sigma,k}(f_1,\cdots,f_k;f_p)=\sum_{v=0}^{\infty}\frac{1}{v!2^v}Q_v[W_I(f')]G_{\lambda+\sigma+k+2v}(f_1',-f_1',\cdots,f_v',-f_v',f_1,\cdots,f_k,(f_p)_\lambda,(-f_p)_\sigma) \quad (169)$$

and $b_{\lambda,\sigma,k}(f_1,\cdots,f_k;f_p)$ is obtained from the right side of (169) by replacing G by G' and $f_1,\cdots,f_k$ by $-f_1,\cdots,-f_k$.

The power spectrum of $y(t)$ is given by

$$W_y(f)=\frac{1}{2\pi}\int_0^{2\pi}d(pt)\int_{-\infty}^{\infty}d\tau\, e^{-j\omega\tau}\langle y(t+\tau)y^*(t)\rangle \quad (170)$$

where the ensemble average is given by (168) with $z(t)=y^*(t)$. The a's are defined by (169), and since $G_n'(f_1,\cdots,f_n)$ is equal to $G_n^*(-f_1,\cdots,-f_n)$,

$$\begin{aligned} b_{\lambda,\sigma,k}(f_1,\cdots,f_k;f_p)&=a^*_{\lambda,\sigma,k}(f_1,\cdots,f_k;-f_p)\\ &=a^*_{\sigma,\lambda,k}(f_1,\cdots,f_k;f_p). \end{aligned} \quad (171)$$

When the modified ensemble average (168) is substituted in (170), the integration with respect to (pt) eliminates all terms in the summation with respect to N and l except those for which $2l=N$, and the integration with respect to τ brings in $\delta(f-f_1-f_2\cdots-f_k-\lambda f_p+\sigma f_p)$. Furthermore, if the expression (164) for $\langle y(t)\rangle$ is written as

$$\langle y(t)\rangle=\sum_{n=-\infty}^{\infty}c_n\exp(jnpt) \quad (172)$$

then the contribution of the product $\langle y(t+\tau)\rangle\langle y^*(t)\rangle$ to the right side of (170) is the series of infinite spikes

$$\sum_{n=-\infty}^{\infty}|c_n|^2\delta(f-nf_p). \quad (173)$$

Combining these results gives

$$\begin{aligned} W_y(f)=&\sum_{n=-\infty}^{\infty}|c_n|^2\delta(f-nf_p)\\ &+\sum_{l=0}^{\infty}\left(\frac{P}{2}\right)^{2l}\sum_{\lambda=0}^{l}\sum_{\sigma=0}^{l}\frac{1}{\lambda!(l-\lambda)!\sigma!(l-\sigma)!}\sum_{k=1}^{\infty}\frac{1}{k!}Q_k[W_I(f)]\\ &\cdot\delta(f-f_1-\cdots-f_k-\lambda f_p+\sigma f_p)a_{\lambda,\sigma,k}(f_1,\cdots,f_k;f_p)\\ &\cdot a^*_{l-\sigma,l-\lambda,k}(f_1,\cdots,f_k;f_p). \end{aligned} \quad (174)$$

Changing the order of summation in the four-fold sum so that the k-summation is the leftmost, and considering the terms in the l,λ,σ-sum for which $\lambda-\sigma$ is equal to a fixed number n, leads to the desired expression

$$\begin{aligned} W_y(f)=&\sum_{n=-\infty}^{\infty}|c_n|^2\delta(f-nf_p)\\ &+\sum_{k=1}^{\infty}\frac{1}{k!}Q_k[W_I(f)]\sum_{n=-\infty}^{\infty}\delta(f-f_1-f_2-\cdots-f_k-nf_p)\\ &\cdot\left|\sum_{\sigma=0}^{\infty}\left(\frac{P}{2}\right)^{2\sigma+|n|}\frac{1}{\sigma!(\sigma+|n|)!}a_{\sigma+|n|,\sigma,k}(f_1,\cdots,f_k;f_ps_n)\right|^2 \end{aligned} \quad (175)$$

where c_n is given by (172) and (165), $a_{\lambda,\sigma,k}(\cdots)$ by (169), the product f_ps_n is equal to f_p when $n\geq 0$, and to $-f_p$ when $n<0$. When P is zero, (175) reduces to (160) for Gaussian $x(t)$. As in (160) the f in $Q_k[W_I(f)]$ takes on only the values $f_1,f_2,\cdots,f_k$, and is not related to the f or the f_p appearing in $\delta(f-f_1-\cdots-f_k-nf_p)$. Also as in (160) the effect of the delta function is to "use up" one of the k integrations denoted by the operator $Q_k[W_I(f)]$ when $k>0$. By noting that (165) is equal to $c_n\exp(jnpt)$ and that $Q_0[W_I(f)]\equiv 1$, we see that the sum of $|c_n|^2\delta(f-nf_p)$ may be regarded as a $k=0$ term, and that (175) can be written as a sum from $k=0$ to $k=\infty$. The first few terms of (175) have been given in (16).

VIII. Higher Moments and Probability Density

The leading terms in the first four cumulants for $y(t)$ when $x(t)$ is Gaussian noise are derived in Section VIII-A. In Section VIII-B formulas are given to show how these cumulants can be used to obtain information about the probability density of $y(t)$. When the terms beyond the second in a Volterra series vanish, the probability density of $y(t)$ can be expressed as an integral containing certain parameters. The values of the parameters can be obtained by solving an integral equation. In Section VIII-C various forms of the integral equation are listed and methods of computing the cumulants are discussed.

A. Cumulants

In this section $x(t)$ is taken to be a real zero-mean stationary Gaussian process with two-sided power spectrum $W_x(f)$. The kernels g_n are assumed to be real so that $y(t)$ is real and $G_n(-f_1,\cdots-f_n)$ is equal to $G_n^*(f_1,\cdots,f_n)$. Since $x(t)$ is stationary, the ensemble averages giving the moments of $y(t)$ do not depend on t.

Substituting $[y(t)]^l$ and $G_{2\mu}^{(l)}$ for $y(t)$ and $G_{2\mu}$ in the series (147) for $\langle y(t)\rangle$ gives a series for the lth moment of $y(t)$,

$$\langle[y(t)]^l\rangle=\sum_{\mu=1}^{\infty}\frac{1}{\mu!2^\mu}Q_\mu[W_x(f)]G_{2\mu}^{(l)}(f_1,-f_1,\cdots,f_\mu,-f_\mu) \quad (176)$$

where $G_{2\mu}^{(l)}$ is given in terms of the G_n by (24). The series obtained by substituting (24) in (176) is not the most desirable one because it can be simplified by making use of the symmetry of the G_n together with $W_x(-f)=W_x(f)$ and appropriate changes of sign in the variables of integration. Unfortunately, a general procedure for simplification is not known. However, a simplified form for $l=2$ (for $l=1$ (176) itself is the simplified form) can be obtained by setting $\tau=0$ and $z(t)=y(t)$ in the series (158) for $\langle y(t+\tau)z(t)\rangle$:

$$\langle y^2(t)\rangle=\langle y(t)\rangle^2+\sum_{k=1}^{\infty}\frac{Q_k[W_x(f)]}{k!}a_k(f_1,\cdots,f_k)a_k(-f_1,\cdots,-f_k). \quad (177)$$

The leading terms in (177) are shown in (180) for $\kappa_2=\langle y^2(t)\rangle-\langle y(t)\rangle^2$.

In the following work we shall be more concerned with the cumulants κ_n of $y(t)$ than with its moments α_n. The cumulants are simpler and appear directly in the discussion of the distribution of $y(t)$ given in Section VIII-B. The first four cumulants are related to the moments by [23, p. 186]:

$$\begin{aligned} \kappa_1&=\alpha_1\\ \kappa_2&=\alpha_2-\alpha_1^2\\ \kappa_3&=\alpha_3-3\alpha_1\alpha_2+2\alpha_1^3\\ \kappa_4&=\alpha_4-3\alpha_2^2-4\alpha_1\alpha_3+12\alpha_1^2\alpha_2-6\alpha_1^4. \end{aligned} \quad (178)$$

The memoryless case (22) in which $y=\sum_1^\infty a_nx^n/n!$ is a useful guide to the general case [the coefficient a_n is unrelated to the $a_k(f_1,\cdots,f_k)$ in (177)]. Here x is a normal random variable with mean 0 and variance σ^2. The nth moment of x is 0 when n is odd and is $1\cdot 3\cdots(n-1)\sigma^n$ when n is even. By first working out the moments for y and then substituting them in (178), we get

$$\kappa_1=\frac{1}{2}a_2\sigma^2+\frac{1}{2!2^2}a_4\sigma^4+\frac{1}{3!2^3}a_6\sigma^6+\cdots$$

$$\kappa_2 = a_1^2\sigma^2 + \left(a_1a_3 + \frac{1}{2}a_2^2\right)\sigma^4 + \left(\frac{1}{4}a_1a_5 + \frac{1}{2}a_2a_4 + \frac{5}{12}a_3^2\right)\sigma^6 + \cdots$$

$$\kappa_3 = 3a_1^2a_2\sigma^4 + \left(\frac{3}{2}a_1^2a_4 + 6a_1a_2a_3 + a_2^3\right)\sigma^6 + \cdots$$

$$\kappa_4 = (4a_1^3a_3 + 12a_1^2a_2^2)\sigma^6 + (2a_1^3a_5 + 18a_1^2a_2a_4 + 36a_1a_2^2a_3 + 12a_1^2a_3^2 + 3a_2^4)\sigma^8 + \cdots. \tag{179}$$

When $y(t)$ is given by the general Volterra series (1), instead of the memoryless power series, the leading terms in the equations corresponding to (179) can be obtained by a similar procedure. The result for κ_1 is given by (176) with $l=1$, i.e., by (147). The result for $\kappa_2=\langle y^2(t)\rangle-\langle y(t)\rangle^2$ is given by (177). The results for κ_3 and κ_4 require much more work and the use of (178). In order to save space in the following list, $W_x(f)df$ and $G_3(f_1, f_2, f_3)$ are written as simply (W) and $(1, 2, 3)$, etc.

$$\kappa_1 = \frac{1}{2}\int_{-\infty}^{\infty} df_1 W_x(f_1)G_2(f_1, -f_1) + \frac{1}{8}\int_{-\infty}^{\infty} df_1 \int_{-\infty}^{\infty} df_2 W_x(f_1)W_x(f_2)G_4(f_1, f_2, -f_1, -f_2) + \cdots$$

$$= \frac{1}{2}\int (W)(1, -1) + \frac{1}{8}\iint (WW)(1, 2, -1, -2) + \cdots$$

$$\kappa_2 = \int (W)(1)(-1) + \iint (WW)\left[(1)(-1, 2, -2) + \frac{1}{2}(1, 2)(-1, -2)\right]$$
$$+ \iiint (WWW)\left[\frac{1}{4}(1)(-1, 2, -2, 3, -3) + \frac{1}{2}(1, 2)(-1, -2, 3, -3) + \frac{1}{4}(1, 2, -2)(-1, 3, -3) + \frac{1}{6}(1, 2, 3)(-1, -2, -3)\right] + \cdots$$

$$\kappa_3 = \iint (WW)3(1)(2)(-1, -2)$$
$$+ \iiint (WWW)\left[\frac{3}{2}(1)(2)(-1, -2, 3, -3) + 3(1)(-1, 2)(-2, 3, -3) + 3(1)(2, 3)(-1, -2, -3) + (1, 2)(-1, 3)(-2, -3)\right] + \cdots$$

$$\kappa_4 = \iiint (WWW)\left[4(1)(2)(3)(-1, -2, -3) + 12(1)(2)(-1, 3)(-2, -3)\right]$$
$$+ \iiiint (WWWW)\Big[2(1)(2)(3)(-1, -2, -3, 4, -4)$$
$$+ \{6(1)(2)(3, 4)(-1, -2, -3, -4) + 12(1)(2)(-1, 3)(-2, -3, 4, -4)\}$$
$$+ \{12(1)(-1, 2)(-2, 3)(-3, 4, -4) + 12(1)(2, -3)(3, 4)(-1, -2, -4) + 12(1)(-1, 2)(3, 4)(-2, -3, -4)\}$$
$$+ \{6(1)(2)(-1, -2, 3)(-3, 4, -4) + 3(1)(2)(-1, 3, 4)(-2, -3, -4) + 3(1)(2)(-1, 3, -3)(-2, 4, -4)\}$$
$$+ 3(1, 2)(-1, 3)(-2, 4)(-3, -4)\Big] + \cdots. \tag{180}$$

The straightforward derivation of the four-fold integral in κ_4 could not be carried through because of its complexity. The expression given in (180) is based upon the conjecture that the only terms occurring in κ_l are those in $\langle y^l(t)\rangle$ which do not separate into products of integrals. The conjecture is supported by the fact that it agrees with the memoryless case results (179) and with the terms in (180) obtained by the earlier method.

If the conjecture is true, it follows from (176) and (24) that the products of G's in the μ-fold integral in the series for κ_l correspond to the l-part partitions of 2μ. For example, consider the product $(1)(2)(-1, -2)$ in the 2-fold integral in κ_3. Here $\mu=2$, $l=3$, and the product corresponds to the 3-part partition $1+1+2$ of $2\mu=4$. This product appears only in the 2-fold integral in κ_3. It does not appear in the 2-fold integrals in κ_1 and κ_2.

B. *Approximate Probability Density*

In this section we review methods of getting information about the probability density $p(y)$ of $y(t)$ from the cumulants κ_n.

The mean and variance of $y(t)$ are κ_1 and κ_2, respectively. The coefficients γ_1 and γ_2 of "skewness" and "excess" used by statisticians to compare the skewness and peakedness of $p(y)$ with a normal curve having the same mean and variance are

$$\gamma_1 = \kappa_3/\kappa_2^{3/2} \qquad \gamma_2 = \kappa_4/\kappa_2^2. \tag{181}$$

When the shape of the central portion of $p(y)$ is known approximately from theoretical considerations, it may be possible to use the first four moments (obtained from the first four cumulants) to fit some appropriate curve, for example a Pearson-type curve.

When the central portion of $p(y)$ is known to be almost normal, the deviation from normality is shown by the Edgeworth-type series [23, pp. 221–232]

$$p(y) = \kappa_2^{-1/2}\left\{Z(u) - \left[\frac{1}{6}\gamma_1 Z^{(3)}(u)\right] + \left[\frac{1}{24}\gamma_2 Z^{(4)}(u) + \frac{1}{72}\gamma_1^2 Z^{(6)}(u)\right] + \cdots\right\} \tag{182}$$

Here u is equal to $(y(t)-\kappa_1)/\kappa_2^{1/2}$ and

$$Z(u) = (2\pi)^{-1/2}\exp(-u^2/2), \qquad Z^{(k)}(u) = (d/du)^k Z(u). \tag{183}$$

The functions $Z^{(k)}(u)$ are tabulated in [18, Table 26.1, pp. 966–973].

When (182) holds, $p(y)$ has its peak at $y=y_0$ where

$$y_0 \approx \kappa_1 - \tfrac{1}{2}\gamma_1\kappa_2^{1/2}$$
$$p(y_0) \approx (2\pi\kappa_2)^{-1/2}\left(1 + \frac{1}{8}\gamma_2 - \frac{1}{12}\gamma_1^2\right). \tag{184}$$

C. *Probability Density of a Two-Term Volterra Series*

An expression for the probability density $p(y)$ of the two-term series

$$y(t) = \frac{1}{1!}\int_{-\infty}^{\infty} du_1 g_1(u_1)x(t-u_1) + \frac{1}{2!}\int_{-\infty}^{\infty} du_1 \int_{-\infty}^{\infty} du_2 g_2(u_1, u_2)x(t-u_1)x(t-u_2) \tag{185}$$

when $x(t)$ is Gaussian [with power spectrum $W_x(f)$ and autocorrelation function $R_x(\tau)$], can be obtained by a method which goes back to [24]. The problem is closely related to that of obtaining the distribution of the second term alone, and hence of quadratic forms of normal variates. Problems of this type have been studied by several authors [17], [21], [24]–[26]. Here we state a method for computing

TABLE II

INTEGRAL EQUATIONS TO DETERMINE λ_n AND ξ_n

$\lambda F(x) = \int k(x, y)F(y)dy$

No.	$F(t)$ or $F(f)$	$k(t, u)$ or $k(f, f_1)$	Orthonormalization	ξ_n
1	$\varphi(t)$	$\iint dv_1 dv_2 g_2(v_1, v_2)a(t-v_1)a(u-v_2)$	$\delta_{mn} = \int dt\, \varphi_m(t)\varphi_n(t)$	$\iint du dv g_1(u)a(v-u)\varphi_n(v)$
2	$\Phi(f)$	$A(f)A(-f_1)G_2(f, -f_1)$	$\delta_{mn} = \int df \Phi_m(f)\Phi_n(-f)$	$\int df G_1(f)A(f)\Phi_n(-f)$
3	$\psi(t)$	$\int dv R_x(t-v)g_2(v, u)$	$\lambda_n\delta_{mn} = \iint du dv g_2(u, v)\psi_m(u)\psi_n(v)$	$\int du g_1(u)\psi_n(u)$
4	$\Psi(f)$	$W_x(f)G_2(f, -f_1)$	$\delta_{mn} = \int df \Psi_m(f)\Psi_n(-f)/W_x(f)$	$\int df G_1(f)\Psi_n(-f)$
5	$\chi(t)$	$\int dv g_2(t, v)R_x(v-u)$	$\lambda_n^2\delta_{mn} = \iint du dv R_x(u-v)\chi_m(u)\chi_n(v)$	$\iint du dv g_1(u)R_x(u-v)\chi_n(v)/\lambda_n$
6	$X(f)$	$W_x(f_1)G_2(f, -f_1)$	$\lambda_n^2\delta_{mn} = \int df W_x(f)X_m(f)X_n(-f)$	$\int df G_1(f)W_x(f)X_n(-f)/\lambda_n$
	$\psi_n(t) = \int du a(u-t)\varphi_n(u)$	$\Psi_n(f) = A(-f)\Phi_n(f)$	$R_x(t-u) = \int dv a(t-v)a(u-v)$	$W_x(f) = \|A(f)\|^2$
	$\chi_n(t) = \int du g_2(t, u)\psi_n(u)$	$X_n(f) = \int df_1 G_2(f, -f_1)\Psi_n(f_1)$	$R_x(t-u) = \Sigma_n\psi_n(t)\psi_n(u)$	$W_x(f)\delta(f-f_1) = \Sigma_n\Psi_n(f)\Psi_n(-f_1)$
	$\lambda_n\psi_n(t) = \int du R_x(t-u)\chi_n(u)$	$\lambda_n\Psi_n(f) = W_x(f)X_n(f)$	$g_2(u, v) = \Sigma_n\lambda_n^{-1}\chi_n(u)\chi_n(v)$	$G_2(f_1, f_2) = \Sigma_n\lambda_n^{-1}X_n(f_1)X_n(f_2)$
	$\lambda_n\delta_{mn} = \int dt \psi_m(t)\chi_n(t)$	$\lambda_n\delta_{mn} = \int df \Psi_m(f)X_n(-f)$	$\iint dv dw g_2(v, w)a(t-v)a(u-w) = \Sigma_n\lambda_n\varphi_n(t)\varphi_n(u)$	$A(f_1)A(f_2)G_2(f_1, f_2) = \Sigma_n\lambda_n\Phi_n(f_1)\Phi_n(f_2)$

Note: $A(f)$, $\Phi(f)$, $\Psi(f)$, $X(f)$ are Fourier transforms of $a(t)$, $\varphi(t)$, $\psi(t)$, $\chi(t)$; $|A(f)|^2 = W_x(f)$, $A(-f) = A^*(f)$, the phase of $A(f)$ is otherwise arbitrary; $\delta(t-u) = \Sigma_n\varphi_n(t)\varphi_n(u) = \Sigma_n\lambda_n^{-1}\chi_n(t)\psi_n(u)$.

$p(y)$ based upon these studies. All of the integrations extend from $-\infty$ to ∞ unless explicitly written otherwise, and $\sum_n$ denotes summation over all eigenstates of an integral equation.

The first problem is to compute a set of eigenvalues λ_n and quantities ξ_n. When these are known, $p(y)$ is given by

$$p(y) = \frac{1}{2\pi}\int_{-\infty}^{\infty} e^{-jyz}Q(z)dz$$

$$Q(z) = \frac{1}{\prod_n (1 - j\lambda_n z)^{\frac{1}{2}}} \exp\left[-\sum_n \frac{\xi_n^2 z^2}{2(1 - j\lambda_n z)}\right] \tag{186}$$

where λ_n and ξ_n are real, and $\arg(1 - j\lambda_n z)^{\frac{1}{2}} = 0$ at $z = 0$. Usually the only practical way to evaluate the integral for $p(y)$ is by numerical integration. When all of the λ_n are positive, moving the path of integration upward to $\text{Im } z = \infty$ shows that $p(y)$ is 0 for y less than $-\sum_n \xi_n^2/(2\lambda_n)$, provided the series converges.

The parameters λ_n and ξ_n are obtained by solving the most convenient [depending on $g_2(u_1, u_2)$, $R_x(\tau)$ and their Fourier transforms $G_2(f_1, f_2)$, $W_x(f)$] one of the six integral equations listed in Table II. These equations are of the form

$$\lambda F(x) = \int k(x, y)F(y)dy. \tag{187}$$

The kernels are listed in the column labeled "$k(t, u)$ or $k(f, f_1)$" and the eigenfunctions in the column labeled "$F(t)$ or $F(f)$." Solving the integral equation gives the λ_n and the eigenfunctions. The ξ_n are then obtained by evaluating the corresponding integral listed in the column labeled "ξ_n." If integral equations 1 or 2 in Table II are selected for solution, some freedom of choice remains in selecting $\arg A(f)$ since $A(f)$ is restricted only by $|A(f)|^2 = W(f)$ and $\arg A(-f) = -\arg A(f)$, ($a(t)$ is the Fourier transform of $A(f)$).

All of the eigenvalues λ_n are real because the kernel shown in row 1 of the table is a symmetric function of t and u.

Table II can be constructed by first taking $x(t)$ to be white noise with $\langle x(t+\tau)x(t)\rangle = \delta(\tau)$ and expanding a typical member $x(t-u)$ of the ensemble as $\sum_n c_n(t)\varphi_n(u)$ where $\varphi_n(u)$ are an as-yet-unspecified orthonormal set. As we go from member to member of the ensemble with t fixed the $c_n(t)$ behave like independent normal random variables with zero mean and unit variance. At this stage $\varphi_n(u)$ is chosen to be the nth eigenfunction of an integral equation having the kernel $g_2(t, u)$. Converting the integral equation into one for which $x(t)$ has the general power spectrum $W_x(f)$ brings in $A(f)$ and leads to 1 in Table II. The arbitrariness associated with $\arg A(f)$ can be removed by introducing $\psi_n(t)$. The eigenfunctions $\psi_n(t)$ and $\chi_m(t)$ are related in essentially the same way, as are the nth modal column ψ_n and mth modal row χ_m of a matrix product Rg where R and g are square symmetric matrices

$$[(I\lambda_n - Rg)\psi_n = 0,\ \chi_m(I\lambda_m - Rg) = 0,\ I = \text{unit matrix}].$$

Corresponding to the three integral equations for $\varphi(t)$, $\psi(t)$, $\chi(t)$ are three more corresponding to their Fourier transforms $\Phi(f)$, $\Psi(f)$, $X(f)$.

The cumulants for the probability density $p(y)$ are proportional to the coefficients in the power series expansion of the characteristic function $Q(z)$. From (186) for $Q(z)$ and Table II it is found that

$$\kappa_1 = \frac{1}{2}\sum_n \lambda_n = \frac{1}{2}\int df\, W_x(f)G_2(f, -f)$$

$$\kappa_2 = \sum_n \left(\frac{1}{2}\lambda_n^2 + \xi_n^2\right)$$

$$= \frac{1}{2}\int df\, W_x(f)G_2'^{(2)}(f, -f) + \int df\, W_x(f)G_1(f)G_1(-f)$$

$$\kappa_m = \sum_n \left[\frac{(m-1)!}{2}\lambda_n^m + \frac{m!}{2}\xi_n^2\lambda_n^{m-2}\right]$$

$$= \frac{(m-1)!}{2}\int df\, W_x(f)G_2'^{(m)}(f, -f) + \frac{m!}{2}\iint df_1 df_2 W_x(f_1) \cdot W_x(f_2)G_1(f_1)G_1(f_2)G_2'^{(m-2)}(-f_1, -f_2). \tag{188}$$

The function $G_2'^{(m)}(f_1, f_2)$ is defined by

$$G_2'^{(1)}(f_1, f_2) = G_2(f_1, f_2)$$

$$G_2'^{(k)}(f_1, f_2) = \int df\, W_x(f)G_2'^{(k-1)}(f_1, f)G_2(-f, f_2), \quad k > 1. \tag{189}$$

Step by step application of orthonormal relation 6 in Table II to the series for $G_2(f_1, f_2)$ in the rightmost column gives

$$G'^{(k)}(f_1, f_2) = \sum_n \lambda_n^{(k-2)}X_n(f_1)X_n(f_2) \tag{190}$$

from which the sum of λ_n^k used in (188) can be obtained.

The values of κ_1, κ_2, κ_3, κ_4 given by (188) agree with those obtained from (180) when $G_n(f_1, \cdots, f_n)$ is zero for $n > 2$.

REFERENCES

[1] N. Wiener, *Nonlinear Problems in Random Theory*. Cambridge, Mass.: Technology Press; and New York: Wiley, 1958.

[2] S. Narayanan, "Transistor distortion analysis using Volterra series representation," *Bell Syst. Tech. J.*, vol. 46, 1967, pp. 991–1023.

[3] R. E. Maurer and S. Narayanan, "Noise loading analysis of a third-

order system with memory," *IEEE Trans. Commun. Technol.*, vol. COM-16, Oct. 1968, pp. 701–712.
[4] K. Franz and K. H. Locherer, "Theorie der nichtlinearen Verzerrungen in einem mehrstufigen Widerstandsverstarker bei schwacher Aussteuerung," *Wiss. Ber. AEG–Telefunken*, vol. 41, 1968, pp. 114–128.
[5] A. Mircea and H. Sinnreich, "Distortion noise in frequency-dependent nonlinear networks," *Proc. Inst. Elec. Eng.*, vol. 116, 1969, pp. 1644–1648.
[6] S. Narayanan, "Application of Volterra series to intermodulation distortion analysis of transistor feedback amplifier," *IEEE Trans. Circuit Theory*, vol. CT-17, Nov. 1970, pp. 518–527.
[7] A. Mircea, "Harmonic distortion and intermodulation noise in linear FM transmission systems," *Rev. Electrotech. Energet.* (Romania), vol. 12, 1967, pp. 359–371.
[8] E. Bedrosian and S. O. Rice, "Distortion and crosstalk of linearly filtered angle-modulated signals," *Proc. IEEE*, vol. 56, Jan. 1968, pp. 2–13.
[9] E. Bedrosian, "Transionospheric propagation of FM signals," Rand Corp., Santa Monica, Calif., RAND Memo. RM-5379-NASA, Aug. 1967; and *IEEE Trans. Commun. Technol.*, vol. COM-18, Apr. 1970, pp. 102–109.
[10] S. O. Rice, "Second and third order modulation terms in the distortion produced when noise modulated FM waves are filtered," *Bell Syst. Tech. J.*, vol. 48, 1969, pp. 87–142.
[11] A. Mircea, E. Bedrosian, and S. O. Rice, "Further comments on 'distortion and crosstalk of linearly filtered, angle-modulated signals,'" *Proc. IEEE* (Lett.), vol. 57, May 1969, pp. 842–844.
[12] R. Deutsch, "On a method of Wiener for noise through nonlinear devices," *IRE Conv. Rec.*, pt. 4, Mar. 1955, pp. 186–192.
[13] D. A. George, "Continuous nonlinear systems," M.I.T. Research Lab. of Electronics, Cambridge, Mass., Tech. Rep. 355, July 24, 1959.
[14] A. Mircea, personal letter to E. Bedrosian and S. O. Rice.
[15] M. Schetzen, "Measurement of the kernels of a nonlinear system of finite order," *Int. J. Contr.*, vol. 1, Mar. 1965, pp. 251–263.
[16] Y. W. Lee and M. Schetzen, "Measurement of the Wiener kernels of a nonlinear system by cross-correlation," *Int. J. Contr.*, vol. 2, 1965, pp. 237–254.
[17] R. Deutsch, *Nonlinear Transformations of Random Processes.* Englewood Cliffs, N. J.: Prentice-Hall, 1962.
[18] M. Abramowitz and I. Stegun, *Handbook of Mathematical Functions.* Washington, D. C.: National Bureau of Standards, AMS no. 55, June 1964.
[19] J. Riordan, *Combinatorial Identities.* New York: Wiley, 1968.
[20] A. Mircea, "FM distortion theory," *Proc. IEEE* (Lett.), vol. 54, Apr. 1966, pp. 705–706.
[21] D. Slepian, "Fluctuation of random noise power," *Bell Syst. Tech. J.*, vol. 37, 1958, pp. 163–184.
[22] S. O. Rice, "Mathematical analysis of random noise," *Bell Syst. Tech. J.*, vol. 24, sec. 4.3, 1945, pp. 46–156.
[23] H. Cramer, *Mathematical Methods of Statistics.* Princeton, N. J.: Princeton Univ. Press, 1966.
[24] M. Kac and A. J. F. Siegert, "On the theory of noise in radio receivers with square-law detectors," *J. Appl. Phys.*, vol. 18, 1947, pp. 383–397.
[25] U. Grenander, H. O. Pollak, and D. Slepian, "The distribution of quadratic forms in normal variates . . . ," *J. SIAM*, vol. 7, 1959, pp. 374–401.
[26] D. Middleton, *Statistical Communication Theory.* New York: McGraw-Hill, 1960.

IV
Basic Papers on Output Spectra of Special Nonlinearities

Editor's Comments on Papers 16, 17, and 18

16 **Bennett:** *Response of a Linear Rectifier to Signal and Noise*

17 **Van Vleck and Middleton:** *The Spectrum of Clipped Noise*

18 **Bennett:** *Spectra of Quantized Signals*

The factor common to all three papers in this part is that they all are basic pioneering efforts published in the mid 1940s, in the early period of development of nonlinear systems analysis with random inputs. Furthermore, they all treat special nonlinearities rather than general ones. However, the special nonlinearities involved are important in applications to practical systems, and the techniques used or the results obtained spurred further developments in the area. Since the papers consider the autocorrelation function or power spectrum of the output of ZNL systems with Gaussian inputs, they may be classified either with the papers on general ZNL systems in Part II, or with the papers in Part V, which are the natural extension of the techniques presented by the papers in this part.

The first paper, by Bennett, derives the output spectrum of a linear rectifier with Gaussian noise. It is a special case which received many later generalizations to quadratic rectifiers and then to νth-law rectifiers, with the addition of signal-and-noise inputs and postfiltering by linear systems. It is one of the first applications of the transform method, which received its full general treatment in Rice's (1944) paper. It is based on a note by Bennett and Rice (1934), which first suggested the techniquc for using the signal frequency components and the transform of the nonlinearity in analyzing the output frequency components of the nonlinearity. The interdependence of the early publications in this subject is evident by the fact that the work reported by Bennett (Paper 16), Van Vleck and Middleton (Paper 17), North (1944), Middleton (1946), and Rice (1944) was carried out approximately simultaneously, and the earlier reports had an impact on Rice's general method.

The paper by Van Vleck and Middleton is a recent version of the original unpublished report by Van Vleck, which dates to 1943. It considers the spectrum of several limiter outputs, both smooth and hard limiters with base-band and narrow-band Gaussian noise. While the direct method is used in some derivations, Rice's transform method is used in others. The reprinted version is enhanced in that it is supplemented by footnotes and later references which place it in the proper historical perspective. This paper provided the impetus for the later publication of several papers on the spectrum of limiters. Several methods were used to derive the spectrum and autocorrelation of the output of smooth limiters, such as Baum (1957), and hard limiters, such as McFadden (1956). Other applications were in the direction of bandpass limiters [Davenport (Paper 20)], with several possible signal inputs: from a single input to several sinusoidal and Gaussian inputs.

The third paper, also by Bennett, treats the spectrum of quantized signals. Quantization is an important aspect of digital processing and digital communications. Therefore, only some aspects of quantization belong in the subject matter of this volume, the behavior of the quantizer as a ZNL system with random inputs.

Bennett's is the first major study of the spectral properties of quantized Gaussian processes. The techniques used are not new to the analysis of ZNL systems with Gaussian inputs. However, the results are important in the analysis and optimization of quantization schemes. It is the principal reference for other papers on optimum quantization, both uniform and nonuniform [see, for example, Smith (1957), Max (1960), and Roe (1964)]. The main results of the paper are asymptotic in nature and were later applied to optimum nonuniform quantizing. They also proved to be useful in the analysis of predictive quantizing. The main results are in the expression for the mean-squared quantizing error as a function of the input probability density, and in the spectral property of the quantizing noise and its asymptotic uncorrelatedness with the signal. Bennett's results in addition to their applications have been directly generalized to the case of sinusoidal signals and Gaussian noise by Barnta (1965) and Hurd (1967).

16

Reprinted from *J. Acoust. Soc. Amer.*, **15**(3), 164–172 (1944)

Response of a Linear Rectifier to Signal and Noise

W. R. Bennett
Bell Telephone Laboratories, New York, New York
(Received November 1, 1943)

When the input to a rectifier contains both signal and noise components, the resultant output is a complicated non-linear function of signal and noise. Given the spectra of the signal and noise input waves, the law of rectification, and the transmission characteristics of the input and output circuits of the rectifier, it should, in general, be possible to describe the spectrum of the resultant output wave. Before discussing the solution of the general problem, we shall derive some results of a simpler nature, which do not require a consideration of the distribution of the signal and noise energies as functions of frequency.

I. DIRECT-CURRENT COMPONENT OF OUTPUT

A quantity of considerable importance is the average value of the output amplitude. This is the quantity which would be read by a direct-current meter. Calculation of the average or d.c. response can be performed in terms of the distribution of instantaneous output amplitudes in time. The distribution of output amplitude can be computed from the distribution of instantaneous input amplitudes and the law of rectification.

As an example, we shall compute the average current obtained from a linear rectifier when the input to the rectifier consists of a sinusoidal signal with random noise superposed upon it. The probability density function of the signal voltage is first determined, and the result given in (3). The corresponding probability density for the voltage of the noise is well known and is given in (4). The distribution of occurrence of the resultant instantaneous amplitudes of the combined noise and signal voltages is then computed by the rules of mathematical probability, and the result is shown in (7). The assumption that the rectifier is linear then leads directly to an integral which yields the average current obtained from the rectifier.

Let the signal voltage E_s be given by

$$E_s = P_0 \cos \omega t. \tag{1}$$

The possible angular values of ωt are uniformly distributed throughout the range 0 to 2π. The range E_s to $E_s + dE_s$ corresponds to the range of values of ωt comprised in the interval.

$$\text{arc cos}\,\frac{E_s}{P_0} < \omega t < \text{arc cos}\,\frac{E_s + dE_s}{P_0}. \tag{2}$$

The angular width of this interval is $(P_0^2 - E_s^2)^{-\frac{1}{2}} dE_s$. There are two such intervals in the range $0 < \omega t < 2\pi$. Values of E_s outside the range $-P_0$ to P_0 do not exist. Hence, the probability that the signal voltage lies in the interval dE_s at any particular E_s is given by

$$\Phi_s(E_s)dE_s$$
$$= \begin{Bmatrix} 0, & |E_s| > P_0 \\ 2(P_0^2 - E_s^2)^{-\frac{1}{2}} dE_s/2\pi, & |E_s| < P_0 \end{Bmatrix} dE_s. \tag{3}$$

Random noise as discussed in this section may be characterized by the fact that the instantaneous amplitudes are normally distributed in time; that is, if $\Phi_n(z)dz$ is the probability that the noise amplitude lies in the amplitude interval of width dz at z,

$$\Phi_n(z) = \frac{1}{\sigma(2\pi)^{\frac{1}{2}}} \exp\left(-z^2/2\sigma^2\right), \tag{4}$$

where σ is the root mean square noise amplitude. The mean noise power dissipated in unit resistance is given by $W_n = \sigma^2$. The corresponding mean signal power is given by $W_s = P_0^2/2$. Let $\Phi_r(z)$ represent the probability density function of the instantaneous sum of the signal and noise amplitudes. Then

$$\Phi_r(z)dz = dz \int_{-\infty}^{\infty} \Phi_s(\lambda)\Phi_n(z-\lambda)d\lambda, \tag{5}$$

or

$$\Phi_r(z) = \frac{1}{\pi\sigma(2\pi)^{\frac{1}{2}}} \int_{-P_0}^{P_0} \frac{\exp\left[-(z-\lambda)^2/2\sigma^2\right]d\lambda}{(P_0^2 - \lambda^2)^{\frac{1}{2}}}. \tag{6}$$

By the substitution $\lambda = P_0 \cos\theta$, we may convert the integral to the form

$$\Phi_r(z) = \frac{1}{\pi\sigma(2\pi)^{\frac{1}{2}}} \int_0^\pi \times \exp\left[-(z-P_0\cos\theta)^2/2\sigma^2\right]d\theta. \quad (7)$$

Suppose we insert a half-wave linear rectifier in series with the source of signal and noise, so that the current I is given in terms of the resultant instantaneous voltage E by

$$I = \begin{Bmatrix} 0, & E<0 \\ \alpha E, & E>0 \end{Bmatrix}. \quad (8)$$

Then the average value of current flowing in the circuit is

$$\bar{I} = \alpha \int_0^\infty z\Phi_r(z)dz = \frac{\alpha}{\pi\sigma(2\pi)^{\frac{1}{2}}} \int_0^\infty zdz \int_0^\pi \times \exp\left[-(z-P_0\cos\theta)^2/2\sigma^2\right]d\theta. \quad (9)$$

The value of this integral is shown in Appendix I to be

$$\bar{I} = \alpha\left(\frac{W_n}{2\pi}\right)^{\frac{1}{2}} \exp(-W_s/2W_n)\left\{I_0(W_s/2W_n) + \frac{W_s}{W_n}\left[I_0\left(\frac{W_s}{2W_n}\right) + I_1\left(\frac{W_s}{2W_n}\right)\right]\right\}. \quad (10)$$

This form is particularly convenient for calculation since Watson's *Theory of Bessel Functions*, Table II, gives $e^{-z}I_0(z)$ and $e^{-z}I_1(z)$ directly.

Limiting forms of this equation may be expressed in terms of series in powers of W_s/W_n when the signal power is small compared with the noise power and in powers of W_n/W_s when the noise power is small compared with the signal power. The ascending series for small signal is:

$$\bar{I} = \alpha\left(\frac{W_n}{2\pi}\right)^{\frac{1}{2}}\left[1 + \frac{1}{2(1!)^2}\frac{W_s}{W_n} + \frac{1(-1)}{2^2(2!)^2}\left(\frac{W_s}{W_n}\right)^2 + \frac{1(-1)(-3)}{2^3(3!)^2}\left(\frac{W_s}{W_n}\right)^3 + \cdots\right]$$
$$= \alpha\left(\frac{W_n}{2\pi}\right)^{\frac{1}{2}} {}_1F_1\left(\frac{-1}{2}; 1; \frac{-W_s}{W_n}\right). \quad (11)$$

The asymptotic series, which is available for computation when the signal is large, is

$$\bar{I} \sim \frac{\alpha(2W_s)^{\frac{1}{2}}}{\pi}\left[1 + \frac{(-1)^2 W_n}{1!4W_s} + \frac{(-1)^2\cdot 1^2}{2!}\left(\frac{W_n}{4W_s}\right)^2 + \frac{(-1)^2\cdot 1^2\cdot 3^2}{3!}\left(\frac{W_n}{4W_s}\right)^3 + \frac{(-1)^2\cdot 1^2\cdot 3^2\cdot 5^2}{4!}\left(\frac{W_n}{4W_s}\right)^4 + \cdots\right]. \quad (12)$$

Curves of $\bar{I}$ have been plotted in three ways. Figure 1 shows the ratio of $\bar{I}$ to $\bar{I}_{s0} = \alpha P_0/\pi$, the

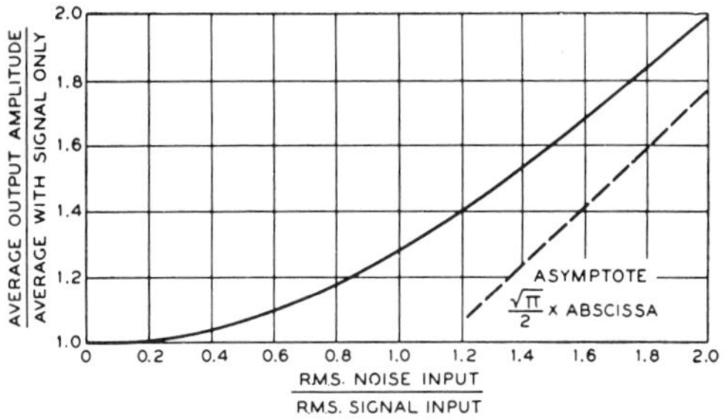

FIG. 1. Variation of direct-current component in response of linear rectifier with ratio of noise input to signal input.

average current in the absence of noise, as a function of ratio of r.m.s. noise input to r.m.s. signal input. Figure 2 shows the ratio of $\bar{I}$ to $\bar{I}_{n0} = \alpha\sigma/(2\pi)^{\frac{1}{2}}$, the average current in the absence of signal, as a function of ratio of r.m.s. signal input to r.m.s. noise input. Figure 3 shows the increment in d.c. power output in decibels as varying amounts of noise expressed in decibels relative to the signal are added. The corresponding result for power addition is given for comparison.

II. SPECTRUM OF OUTPUT

A much more powerful method of attack on this problem is obtained by the use of multiple Fourier series. In this section we shall use Fourier analysis to obtain not only the direct-current output of the rectifier, but also the spectral distribution of the sinusoidal components in the output of the rectifier. We represent the input spectrum by

$$E = P_0 \cos p_0 t + \sum_{n=1}^{N} P_n \cos p_n t. \quad (13)$$

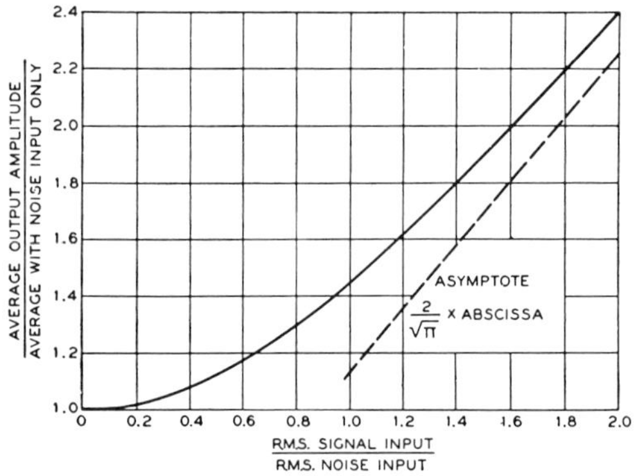

FIG. 2. Variation of direct-current component in response of linear rectifier with ratio of signal input to noise input.

This representation is more general than that given by (4) in that a frequency spectrum as well as an amplitude distribution is defined; it may be shown that the probability density for the sum of N sinusoidal waves with incommensurable frequencies approaches (4) when N is large. The first term represents the sinusoidal signal; the mean power which would be dissipated by this signal in unit resistance is

$$W_s = P_0^2/2. \quad (14)$$

The noise is represented by a large number N of sinusoidal components with incommensurable frequencies (or commensurable frequencies with random phase angles) distributed along the frequency range f_1 to f_2 in such a way that the mean noise power in band width Δf is:

$$w(f)\Delta f = \tfrac{1}{2} \sum_{k=\nu(f-f_1)}^{\nu(f+\Delta f-f_1)} P_n^2 = \nu \Delta f P^2(f)/2. \quad (15)$$

Here ν is the number of components per unit band width and $P(f)$ represents the amplitude of a component in the neighborhood of frequency f. Note also that the mean total noise input power W_n is given by

$$W_n = \int_0^\infty w(f)df = \frac{\nu}{2}\int_0^\infty P^2(f)df. \quad (16)$$

The linear rectifier is specified by the current-voltage relationship (8), which is equivalent to

$$I = -\frac{\alpha}{2\pi}\int_C e^{iEz}\frac{dz}{z^2}, \quad (17)$$

where C is an infinite contour going from $-\infty$ to $+\infty$ with an indentation below the pole at the origin. We may expand I in the multiple Fourier series[1]

$$I = \sum_{m_0=0}^{\infty}\sum_{m_1=0}^{\infty}\cdots\sum_{m_N=0}^{\infty} a_{m_0 m_1 \cdots m_N}$$
$$\times \cos m_0 x_0 \cos m_1 x_1 \cdots \cos m_N x_N, \quad (18)$$

where

$$x_k = p_k t, \quad k = 0, 1, 2, \cdots N; \quad (19)$$

$$a_{m_0 m_1 \cdots m_N} = \frac{\epsilon_{m_0}\epsilon_{m_1}\cdots\epsilon_{m_N}}{\pi^{N+1}}\int_0^\pi dx_0 \int_0^\pi dx_1 \cdots$$
$$\times \int_0^\pi I \cos m_0 x_0$$
$$\times \cos m_1 x_1 \cdots \cos m_N x_N dx_N; \quad (20)$$

$$\epsilon_j = \begin{Bmatrix} 2, & j \neq 0 \\ 1, & j = 0 \end{Bmatrix}. \quad (21)$$

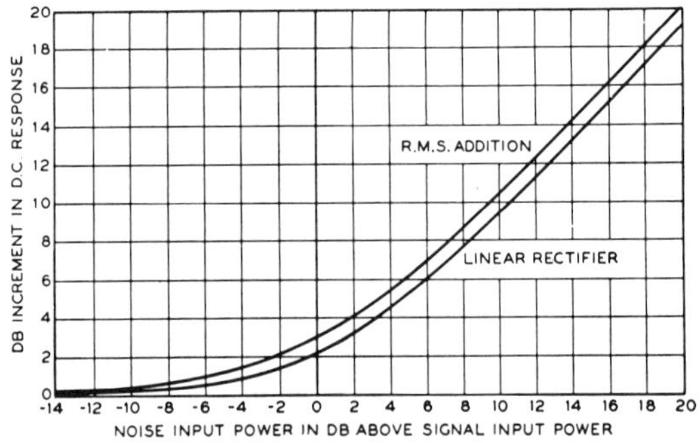

FIG. 3. Variation of direct-current component expressed in decibels, showing comparison between linear rectification and power addition of signal and noise.

The response of the rectifier is thus seen to consist of all orders of modulation products of signal and noise. In a typical case of interest the band of input frequencies is relatively narrow and

[1] W. R. Bennett and S. O. Rice, Phil. Mag. [7], **18**, 422–424 (1934). The present application represents an extension to N variables of the theory there given for two.

centered about a high frequency while the output band includes only low frequencies. In such a case the important components in the output are the beats between signal and noise components and between noise components. The d.c. component is present in the output only if the pass band of the system actually includes zero frequency; we have already computed its value in Section I, but we will derive it again by the method used here as a check.

The amplitude of the d.c. component is in fact:

$$a_{00\cdots 0} = -\frac{\alpha}{2\pi}\int_C \frac{J_0(P_0 z)\prod_{n=1}^{N} J_0(P_n z)}{z^2}dz, \quad (22)$$

on substitution of the expression for E in the integral representation of I, substituting the result in (20), and interchanging the order of integration. When N is large P_n is small; hence the principal contribution to the integral occurs near small values of z, where $J_0(P_n z)$ is nearly equal to unity, since the product of a large number of factors, all less than unity, will be small indeed unless each factor is only slightly less than unity. We therefore replace $J_0(P_n z)$ by a function which coincides with it near $z=0$ and goes rapidly to zero as we depart from this region. Such an approximation (Laplace's process[2]) is

$$J_0(P_n z) \doteq \exp(-P_n^2 z^2/4), \quad (23)$$

which is correct for the first two terms in the Taylor series expansion near $z=0$. Therefore, when P_n approaches zero as N approaches infinity,

$$a_{00\cdots 0} = \bar{I} = -\frac{\alpha}{2\pi}\int_C J_0(P_0 z) \times \exp\left(-\sum_{n=1}^{N} P_n^2 z^2/4\right)\frac{dz}{z^2}$$

$$= -\frac{\alpha}{2\pi}\int_C J_0(P_0 z)\exp(-W_n z^2/2)\frac{dz}{z^2}. \quad (24)$$

The contour integral cannot be replaced by a real integral directly because the integrand goes to infinity at the origin. However, since

$$\frac{J_0(u)}{u^2} = -\frac{J_1(u)}{u} - \frac{d}{du}\frac{J_0(u)}{u}, \quad (25)$$

$$\frac{J_0(Pz)}{z^2} = -\frac{J_1(Pz)}{P^2z^2} - \frac{d}{d(Pz)}\frac{J_0(Pz)}{Pz}$$

$$= -\frac{J_1(Pz)}{P^2z^2} - \frac{1}{P^2}\frac{d}{dz}\frac{J_0(Pz)}{z}, \quad (26)$$

we can substitute (26) in the integral and perform an integration by parts to give the result.

$$\bar{I} = \frac{\alpha}{\pi}\int_0^\infty e^{-W_n z^2/2}\left[\frac{P_0 J_1(P_0 z)}{z} + W_n J_0(P_0 z)\right]dz$$

$$= \alpha\left(\frac{W_n}{2\pi}\right)^{\frac{1}{2}}\left[{}_1F_1\left(\frac{1}{2};1;-\frac{W_s}{W_n}\right) + \frac{W_s}{W_n}{}_1F_1\left(\frac{1}{2};2;-\frac{W_s}{W_n}\right)\right] \quad (27)$$

by Hankel's formula.[3] But it may be shown that (see Appendix II)

$${}_1F_1(\tfrac{1}{2};1;-u) = e^{-u/2}I_0\left(\frac{u}{2}\right), \quad (28)$$

$${}_1F_1(\tfrac{1}{2};2;-u) = e^{-u/2}[I_0(u/2)+I_1(u/2)]. \quad (29)$$

Hence,

$$\bar{I} = \alpha\left(\frac{W_n}{2\pi}\right)^{\frac{1}{2}}\exp(-W_s/2W_n)\left\{I_0(W_s/2W_n) + \frac{W_s}{W_n}[I_0(W_s/2W_n)+I_1(W_s/2W_n)]\right\}, \quad (30)$$

which is identical with the result of Section I, noting that $\sigma = (W_n)^{\frac{1}{2}}$. We point out that a resistance-capacity coupled amplifier will not pass this component since there is no transmission at zero frequency.

The amplitude of the typical difference product

[2] G. N. Watson, *Theory of Bessel Functions* (Cambridge, 1922), p. 421.

[3] See G. N. Watson, reference 2, p. 393. As pointed out by Watson, in a footnote, the difficulty with singularities at the origin could be avoided by expressing Hankel's formula in terms of a contour integral instead of an ordinary integral along the real axis. This procedure would lead directly to the hypergeometric function given in (11).

between the signal and the nth noise component is

$$A_{sn}=\tfrac{1}{2}a_{100\cdots010\cdots0}$$

$$=\frac{\alpha}{\pi}\int_C dz\frac{J_1(P_0z)J_0(P_1z)J_0(P_2z)\cdots\times J_1(P_nz)\cdots J_0(P_Nz)}{z^2}. \quad (31)$$

Using the same process as before, we replace $J_1(P_nz)$ by

$$J_1(P_nz)\doteq\frac{P_nz}{2}\exp(-P_n^2z^2/8) \quad (32)$$

and obtain in the limit as N becomes indefinitely large

$$A_{sn}=\frac{\alpha P_n}{\pi}\int_0^\infty\frac{J_1(P_0z)}{z}\exp(-W_nz^2/2)dz$$

$$=\frac{\alpha P_n}{2}\left(\frac{W_s}{\pi W_n}\right)^{\frac{1}{2}}{}_1F_1\left(\frac{1}{2};2;-\frac{W_s}{W_n}\right)$$

$$=\frac{\alpha P_n}{2}\left(\frac{W_s}{\pi W_n}\right)^{\frac{1}{2}}\exp(-W_s/2W_n)\times\left[I_0\left(\frac{W_s}{2W_n}\right)+I_1\left(\frac{W_s}{2W_n}\right)\right]. \quad (33)$$

Relations between the ${}_1F_1$ function and Bessel functions are discussed in Appendix II.

The shape of the spectrum of the beats between p_0 and the noise input evidently consists of the superposition of the noise spectra above and below p_0, so that if we write $w_{sn}(f)\Delta f$ for the mean energy from this source in that part of the filter output lying in the band of width Δf at f,

$$w_{sn}(f)\Delta f=\frac{\nu\Delta f}{2}[(\overset{+}{A}_{sn})^2+(\overset{-}{A}_{sn})^2], \quad (34)$$

$$\overset{+}{A}_{sn}=[A_{sn}]_{p_n=p_0+2\pi f}, \quad (35)$$

$$\overset{-}{A}_{sn}=[A_{sn}]_{p_n=p_0-2\pi f}, \quad (36)$$

$$P_n=\left[\frac{2w(f_n)}{\nu}\right]^{\frac{1}{2}}, \quad (37)$$

$$w_{sn}(f)=\frac{\alpha^2W_s}{4\pi W_n}\exp(-W_s/W_n)\times\left[I_0\left(\frac{W_s}{2W_n}\right)+I_1\left(\frac{W_s}{2W_n}\right)\right]^2\times[w(f_0+f)+w(f_0-f)]. \quad (38)$$

The total noise from this source in the output of a particular filter of transfer admittance $Y(f)$ is obtained by integrating $w_{sn}(f)\,Y(f)df$ throughout the band of the filter. In the particular case in which the original band of noise is symmetrical about f_0 and occupies the range f_0-f_a to f_0+f_a and an ideal low pass filter cutting off at $f=f_a$ is used in the rectifier output, the total noise output from beats between signal and noise is

$$W_{sn}=2\int_0^{f_a}w_{sn}(f)df=\frac{\alpha^2W_s}{4\pi}\exp(-W_s/W_n)\times[I_0(W_s/2W_n)+I_1(W_s/2W_n)]^2. \quad (39)$$

Next we shall calculate the spectrum of the energy resulting from beats between individual noise components. We write

$$A_{nn}=\tfrac{1}{2}a_{00\cdots010\cdots010\cdots0}$$

$$=\frac{\alpha}{\pi}\int_C dz\frac{J_0(P_0z)J_0(P_1z)\cdots J_1(P_rz)\cdots\times J_1(P_sz)\cdots J_0(P_Nz)}{z^2}$$

$$=\frac{\alpha P_rP_s}{2\pi}\int_0^\infty J_0(P_0z)\exp(-W_nz^2/2)dz$$

$$=\frac{\alpha P_rP_s}{2(2\pi W_n)^{\frac{1}{2}}}{}_1F_1\left\{\frac{1}{2};1;-\frac{W_s}{W_n}\right\}$$

$$=\frac{\alpha P_rP_s}{2(2\pi W_n)^{\frac{1}{2}}}\times\exp(-W_s/2W_n)I_0(W_s/2W_n). \quad (40)$$

To find the resulting spectrum $w_{nn}(f)df$ produced at f by the resultant of all such components, we note that we may sum over all components by beating each component of the primary band with the frequency f above it and adding the resultant power values. The result is

$$w_{nn}(f)=\frac{\alpha^2}{4\pi W_n}\exp(-W_s/W_n)I_0^2(W_s/2W_n)\times\int_0^\infty w(\lambda)w(\lambda+f)d\lambda. \quad (41)$$

In the particular case of a flat band of energy

extending from f_1 to f_2,

$$\int_0^\infty w(\lambda)w(\lambda+f)d\lambda = \int_{f_1}^{f_2-f} \frac{W_n^2}{(f_2-f_1)^2} d\lambda$$

$$= \frac{f_2-f_1-f}{(f_2-f_1)^2} W_n^2, \quad 0<f<f_2-f_1, \quad (42)$$

$$w_{nn}(f) = \frac{\alpha^2(f_2-f_1-f)W_n}{4\pi(f_2-f_1)^2} \times \exp(-W_s/W_n)I_0^2(W_s/2W_n),$$
$$0<f<f_2-f_1. \quad (43)$$

The total mean power of this type lying in the band 0 to f_b is

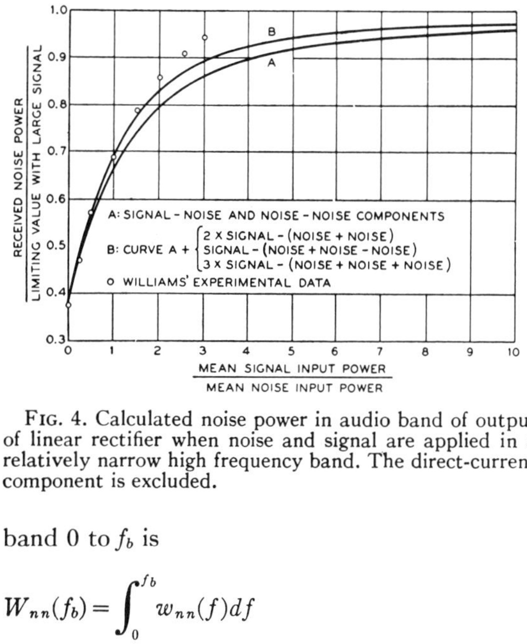

FIG. 4. Calculated noise power in audio band of output of linear rectifier when noise and signal are applied in a relatively narrow high frequency band. The direct-current component is excluded.

$$W_{nn}(f_b) = \int_0^{f_b} w_{nn}(f)df$$

$$= \frac{\alpha^2 W_n(f_2-f_1-f_b/2)f_b}{4\pi(f_2-f_1)^2} \times \exp(-W_s/W_n)I_0^2(W_s/2W_n), \quad (44)$$

provided $f_b<f_2-f_1$. The spectrum is confined to the region $0<f<f_2-f_1$. If f_b is equal to f_2-f_1 so that the output filter passes all the noise of this type, we have

$$W_{nn}(f_2-f_1) = W_{nn}$$

$$= \frac{\alpha^2 W_n}{8\pi} \exp(-W_s/W_n)I_0^2(W_s/2W_n). \quad (45)$$

This result seems to hold approximately for a considerable range of input spectra. For example, if we assume that the original noise is shaped like an error function about f_0, i.e.,

$$w_n(f) = W_n(a/\pi)^{\frac{1}{2}} \exp[-a(f-f_0)^2] \quad (46)$$

with f taken from $-\infty$ to $+\infty$ with small error for large f_0,

$$\int_{-\infty}^\infty w(\lambda)w(\lambda+f)d\lambda$$

$$= W_n^2(a/2\pi)^{\frac{1}{2}} \exp(-af^2/2) \quad (47)$$

$$\int_0^\infty df \int_{-\infty}^\infty w(\lambda)w(\lambda+f)d\lambda = W_n^2/2, \quad (48)$$

which is in agreement with (45).

The output of a half-wave linear rectifier contains fundamental components and all even order modulation products. In general, the amplitudes of the higher order products are small compared with the lower order. In a particular problem some consideration of where the principal products fall in the frequency band is required. The products just considered give a fair approximation for the problem of detection of a radio-frequency band of signal and noise followed by audio amplification. Certain other products should also be added to obtain higher accuracy. We have calculated the products of order zero and two; the next ones of importance are the fourth order, since the third-order products vanish in a perfectly linear rectifier. The fourth-order products in this case which fall in the audio band are of frequency $2p_0-p_r-p_s$, $p_0+p_q-p_r-p_s$, and $p_n+p_q-p_r-p_s$, where the subscripts n, q, r, s refer to the original noise component frequencies. The latter is, however, less important than the sixth-order product $3p_0-p_q-p_r-p_s$, which involves only three noise components. Expressions for the contributions from these products are given in Appendix III.

Figure 4 shows computed curves for the noise produced in an audio band by the various components. Curve A is $W_{sn}+W_{nn}$ and includes what are usually regarded as the principal contributors, the difference frequencies between signal and noise, and between individual noise components. Curve B is obtained by adding to curve A, the contribution from the fourth-order products $2p_0-p_r-p_s$ and $p_0+p_q-p_r-p_s$ and the sixth-order products $3p_0-p_q-p_r-p_s$. Thus all products which include three or less noise fundamental components are included. The curves are plotted in terms of fraction of noise power received compared to the limiting noise when the mean signal input power is made indefinitely large compared to the mean input noise power. Some experimental points given by Williams[4] are shown for comparison. Williams gives the intercept at zero signal power as 35 percent; the theoretical value deduced here is $\pi/8$ or 39.27 percent. It will be noted that the inclusion of the higher order products improves the agreement between experimental and theoretical curves, even though the value of the intercept is unaffected by them. It should also be stressed that our analysis applies strictly to purely resistive networks. The conventional radio detector circuit (which was used by Williams), in which a condenser is shunted across a resistance in series with a diode, departs from the conditions here assumed because of the reactive element, the condenser. The customary approximation made in treating this circuit is that the condenser has infinite impedance in the audiofrequency range and zero impedance at the radiofrequencies. This leads to a bias on the detector which depends on the signal. The methods given here may be applied, but the resulting formulas are much more difficult from the standpoint of numerical computation.

A recent paper by Ragazzini[5] gives an approximate solution based on expanding the envelope of the input wave by the binomial theorem and retaining only the first two terms. The validity depends on the noise amplitude being small compared with the sum of signal and noise, and hence the result should agree with our solution in the neighborhood of $W_n/W_s=0$, which it does. When W_s/W_n is small, the error is appreciable. Ragazzini's result [Eq. (15) of the paper] expressed in our notation is

$$W_{sn}+W_{nn}\doteq\frac{\alpha^2}{\pi^2}\frac{W_n(1+\frac{1}{2}W_n/W_s)}{1+W_n/W_s}. \tag{49}$$

It will be seen by comparing the limiting values for $W_s/W_n=0$ with that of $W_s/W_n=\infty$ from (49) that the intercept of the curve of Fig. 4 would be 50 percent instead of our value of 39.27 percent.

The results given in the present paper have been compiled from unpublished memoranda and notes by the author extending back as far as 1935. Discussions with colleagues have been of great aid, and in particular acknowledgment is made to Messrs. S. O. Rice and R. Clark Jones for many helpful suggestions.

APPENDIX I

Evaluation of Integral for $\bar{I}$

Interchanging the order of integration in (9), we have

$$\bar{I}=\frac{\alpha}{\pi(2\pi W_n)^{\frac{1}{2}}}\int_0^{\pi}d\theta \times\int_0^{\infty}\exp\left[-(z-P_0\cos\theta)^2/2W_n\right]zdz. \tag{50}$$

By substituting $z=P_0\cos\theta+u(2W_n)^{\frac{1}{2}}$, we may evaluate the second integral in terms of the error function, obtaining

$$\begin{aligned}\bar{I}&=\frac{\alpha}{\pi^{\frac{1}{2}}}\int_0^{\pi}d\theta\int_{-P\cos\theta/(2W_n)^{\frac{1}{2}}}^{\infty}\exp(-u^2)\\&\qquad\times\left[u(2W_n)^{\frac{1}{2}}+P_0\cos\theta\right]du\\&=\frac{\alpha}{\pi}\frac{(W_n)^{\frac{1}{2}}}{2\pi}\int_0^{\pi}\exp(-P_0^2\cos^2\theta/2W_n)d\theta\\&\quad+\frac{\alpha P_0}{2\pi}\int_0^{\pi}\operatorname{erf}\left[P_0\cos\theta/(2W_n)^{\frac{1}{2}}\right]\cos\theta d\theta\\&=\frac{\alpha}{\pi}\left(\frac{W_n}{2\pi}\right)^{\frac{1}{2}}\exp(-P_0^2/4W_n)\int_0^{\pi}\exp(-\cos 2\theta/4W_n)d\theta\\&\quad+\frac{\alpha P_0}{2\pi}\int_0^{\pi}\frac{d}{d\theta}\left[\operatorname{erf}\left(\frac{P_0\cos\theta}{(2W_n)^{\frac{1}{2}}}\right)\sin\theta\right]d\theta\\&\quad-\frac{\alpha P_0}{2\pi}\int_0^{\pi}\sin\theta\frac{d}{d\theta}\left[\operatorname{erf}\frac{P_0\cos\theta}{(2W_n)^{\frac{1}{2}}}\right]d\theta\\&=\frac{\alpha}{2\pi}\left(\frac{W_n}{2\pi}\right)^{\frac{1}{2}}\exp(-W_s/2W_n)\\&\quad\times\int_0^{2\pi}\exp(-W_s\cos\Phi/2W_n)d\Phi\\&\quad+\frac{\alpha W_s}{2\pi(2\pi W_n)^{\frac{1}{2}}}\int_0^{2\pi}\exp(-W_s\cos\Phi/2W_n)\\&\quad\times(1-\cos\Phi)d\Phi\\&=\alpha\left(\frac{W_n}{2\pi}\right)^{\frac{1}{2}}\exp(-W_s/2W_n)\Big\{I_0(W_s/2W_n)\\&\qquad+\frac{W_s}{W_n}\left[I_0(W_s/2W_n)+I_1(W_s/2W_n)\right]\Big\}.\end{aligned} \tag{10}$$

---

[4] F. C. Williams, J. Inst. Elec. Eng. **80**, 218–226 (1937).
[5] John R. Ragazzini, Proc. I. R. E. **30**, 277–288 (1942).

In the above we have made use of the relations:

$$\operatorname{erf} z=\frac{2}{(\pi)^{\frac{1}{2}}}\int_0^z \exp(-z^2)dz, \tag{51}$$

$$\frac{d}{dz}\operatorname{erf} z=\frac{2}{(\pi)^{\frac{1}{2}}}\exp(-z^2), \tag{52}$$

$$\int_0^{2\pi} e^{-z\cos\Phi}\cos m\Phi d\Phi=(-)^m 2\pi I_m(z). \tag{53}$$

APPENDIX II

Relations Between Hypergeometric and Bessel Functions

The modulation coefficients appearing in the linear rectification of noise are expressible in compact form in terms of the hypergeometric function:

$$\begin{aligned} {}_1F_1(a;c;-z)&=1-\frac{a}{c}\frac{z}{1!}+\frac{a(a+1)}{c(c+1)}\frac{z^2}{2!}-\cdots \\ &=\frac{\Gamma(c)}{\Gamma(a)}\sum_{m=0}^{\infty}\frac{\Gamma(a+m)}{\Gamma(c+m)m!}(-z)^m. \end{aligned} \tag{54}$$

The ${}_1F_1$ function is a limiting case of the more familiar Gaussian hypergeometric function ${}_2F_1(a, b; c; z)$, *viz.*,

$${}_1F_1(a;c;z)=\operatorname*{Limit}_{b=\infty}{}_2F_1(a,b;c;z/b). \tag{55}$$

In certain special cases this function may be expressed in terms of exponential and Bessel functions. For example, by a formula given by Campbell and Foster,[6] we may show that

$${}_1F_1(\nu+\tfrac{1}{2};2\nu+1;-z)=\frac{2^{2\nu}\Gamma(\nu+1)e^{-z/2}}{(-z)^\nu}I_\nu(-z/2), \tag{56}$$

or setting $\nu=0$

$${}_1F_1(\tfrac{1}{2};1;-z)=e^{-z/2}I_0(z/2), \tag{57}$$

which is one of the functions appearing in our work.

We have also encountered the function ${}_1F_1(\frac{1}{2};2;-z)$ which is not directly reducible by the above formula. The reduction may be effected in a number of ways. By making use of the relation obtained from (56) by setting $\nu=1$,

$${}_1F_1(\tfrac{3}{2};3;-z)=\frac{4}{z}e^{-z/2}I_1(z/2) \tag{58}$$

and noting that

$$\begin{aligned} &{}_1F_1(\tfrac{1}{2};2;-z)-{}_1F_1(\tfrac{1}{2};1;-z) \\ &=\frac{1}{\Gamma(1/2)}\sum_{m=0}^{\infty}\frac{\Gamma(m+1/2)}{m!(m+1)!}(-z)^m \\ &\qquad -\frac{1}{\Gamma(1/2)}\sum_{m=0}^{\infty}\frac{\Gamma(m+1/2)}{(m!)^2}(-z)^m \\ &=\frac{-1}{\Gamma(1/2)}\sum_{m=0}^{\infty}\frac{\Gamma(m+1/2)m}{m!(m+1)!}(-z)^m \\ &=\frac{z}{\Gamma(1/2)}\sum_{m=0}^{\infty}\frac{\Gamma(m+3/2)}{(m+2)!m!}(-z)^m \\ &=\frac{z}{4}{}_1F_1(\tfrac{3}{2};3;-z), \end{aligned} \tag{59}$$

we find that[7]

$${}_1F_1(\tfrac{1}{2};2;-z)=e^{-z/2}[I_0(z/2)+I_1(z/2)]. \tag{60}$$

It may also be verified by integrating the series directly that

$$\int_0^z {}_1F_1(\tfrac{1}{2};1;-z)dz=z\,{}_1F_1(\tfrac{1}{2};2;-z). \tag{61}$$

Combining this relation with (57) and (60) above, we deduce the indefinite integrals

$$\begin{aligned} \int e^x I_0(x)dx&=xe^x[I_0(x)-I_1(x)], \\ \int e^{-x} I_0(x)dx&=xe^{-x}[I_0(x)+I_1(x)], \\ \int e^x I_1(x)dx&=e^x[(1-x)I_0(x)+xI_1(x)], \\ \int e^{-x} I_1(x)dx&=e^{-x}[(1+x)I_0(x)+xI_1(x)]. \end{aligned} \tag{62}$$

These integrals may be derived by differentiating the right-hand members, and could, therefore, serve as a basis for an alternate derivation of (60).

In addition it was noted in Eq. (11) that the constant term in the modulation spectrum could be expressed in terms of ${}_1F_1(-\frac{1}{2};1;-z)$; from the equations given, it follows that we must have the relation:

$${}_1F_1(-\tfrac{1}{2};1;-z)=e^{-z/2}[(1+z)I_0(z/2)+zI_1(z/2)]. \tag{63}$$

Another interesting set of formulas which can be obtained as a by-product from (62) by setting $x=iy$ is:

$$\begin{aligned} \int J_0(y)\cos y dy&=y[J_0(y)\cos y+J_1(y)\sin y], \\ \int J_0(y)\sin y dy&=y[J_0(y)\sin y-J_1(y)\cos y], \\ \int J_1(y)\cos y dy&=yJ_1(y)\cos y-J_0(y)(y\sin y-\cos y), \\ \int J_1(y)\sin y dy&=yJ_1(y)\sin y+J_0(y)(y\cos y-\sin y). \end{aligned} \tag{64}$$

The hypergeometric notation is particularly convenient in determining series expansions for the coefficients to be used for calculation when the variable z is either very small or very large. For small values of z, the form (54) suffices; for large values of z, we may use the general asymptotic expansion formula[8] for the real part of z positive:

$$\begin{aligned} {}_1F_1(a;c;-z)&=\frac{\Gamma(c)}{\Gamma(c-a)z^a}{}_2F_0(a,1+a-c;1/z) \\ &=\frac{\Gamma(c)}{\Gamma(c-a)z^a}\left[1+\frac{a(1+a-c)}{1!z}\right. \\ &\qquad \left.+\frac{a(a+1)(1+a-c)(2+a-c)}{2!z^2}+\cdots\right]. \end{aligned} \tag{65}$$

---

[6] G. A. Campbell and R. M. Foster, "Fourier integrals for practical application," Bell System Monograph B-584, p. 32. See also reference 2, p. 191.

[7] The relation (60) was brought to the attention of the author by Mr. R. M. Foster.

[8] E. T. Copson, *Functions of a Complex Variable* (Oxford, 1935), pp. 264–265.

The series expansions required here could also be obtained from the appropriate series for Bessel functions. It will be noted, however, that the typical modulation coefficient can be expressed in terms of either a single ${}_1F_1$ function or several Bessel functions, so that manipulations must be performed on the series for the latter to give the final result. The Bessel functions, on the other hand, are more convenient for numerical computations because of the excellent tables available.

Reduction formulas for certain other hypergeometric functions are needed in evaluating the higher order products. They are:

$$ {}_1F_1(3/2;1;-z)=e^{-z/2}[(1-z)I_0(z/2)+I_1(z/2)], \tag{66}$$

$$ {}_1F_1(3/2;2;-z)=e^{-z/2}[I_0(z/2)-I_1(z/2)], \tag{67}$$

$$ {}_1F_1(5/2;4;-z)=\frac{4}{z}e^{-z/2}\left[\left(\frac{4}{z}+1\right)I_1(z/2)-I_0(z/2)\right]. \tag{68}$$

Derivation of these is facilitated by the use of the easily demonstrated relations:

$$ {}_1F_1(a;1;-z)=\frac{d}{dz}[z\,{}_1F_1(a;2;-z)], \tag{69}$$

$$ 2z\,{}_1F_1(a;2;-z)=\frac{d}{dz}[z^2\,{}_1F_1(a;3;-z)], \tag{70}$$

$$ {}_1F_1(3/2;3;-z)-{}_1F_1(3/2;2;-z)=\frac{z}{4}\,{}_1F_1(5/2;4;-z). \tag{71}$$

APPENDIX III

Higher Order Products

The methods described in Section II may be applied to calculate the general expression for the general modulation coefficient. The result is for the amplitude of the term $\cos mp_0t \cos pn_1t \cos pn_2t \cdots \cos pn_Mt$:

$$a_{mM}=\frac{(-)^{(m+M/2)+1}P_{n_1}P_{n_2}\cdots P_{n_M}}{\pi(W_n/2)^{(M-1)/2}m!}\Gamma\left(\frac{m+M-1}{2}\right)\left(\frac{W_s}{W_n}\right)^{m/2} \times {}_1F_1\left[\frac{m+M-1}{2};m+1;\frac{-W_s}{W_n}\right]. \tag{72}$$

The coefficient of the term $\cos(mp_0 \pm pn_1 \pm pn_2 \pm \cdots pn_M)^t$ is a_{mM} divided by $2^{M-1}\epsilon_m$. The number of terms of a particular type falling in a particular frequency interval can be calculated by a method previously described by the author.[9] Under the assumed conditions that the original noise spectrum is either flat throughout a limited range, or falls off like an error function, and that the audio amplifier passes all the difference components in question, we find the following results:

$2p_0-p_r-p_s$:

$$W_{2s,nn}=\frac{\alpha^2W_n}{8\pi}\exp(-W_s/W_n)I_1^2(W_s/2W_n); \tag{73}$$

$p_0+p_q-p_r-p_s$:

$$W_{snnn}=\frac{\alpha^2W_s}{32\pi}\exp(-W_s/W_n)\times[I_0(W_s/2W_n)-I_1(W_s/2W_n)]^2; \tag{74}$$

$3p_0-p_q-p_r-p_s$:

$$W_{3s,nnn}=\frac{\alpha^2W_s}{32\pi}\exp(-W_s/W_n)\times[(1+4W_n/W_s)I_1(W_s/2W_n)-I_0(W_s/2W_n)]^2. \tag{75}$$

This includes all beats containing not more than three noise fundamentals. The reductions of hypergeometric functions to exponential and Bessel functions given in Appendix II have been used in deriving the above results.

---

[9] W. R. Bennett, Bell Sys. Tech. J. **19**, 587–610 (1940).

Errata

Page 164, Eq. (3): Delete dE_s after the right-hand brace.
Page 167, Eq. (26), first line: Change the denominator on the left-hand side to P^2z^2 and the denominator of the first term on the right-hand side to Pz.
Page 167, Eq. (26), second line: Change the denominator of the first term to Pz.
Page 170, Eq. (10), fourth line from bottom of page: Insert $\exp(-W_s/2W_n)$ in the numerator of the fraction preceding the integral.
Page 172, Eq. (66): Insert z before $I_1(z/2)$ on the right-hand side.
Page 172, first line under Eq. (72): Drop the exponent t to the line (i.e., it is a multiplying factor, not an exponent).

17

Reprinted from *Proc. IEEE,* **54**(1), 2–19 (1966)

The Spectrum of Clipped Noise

J. H. VAN VLECK AND DAVID MIDDLETON, FELLOW, IEEE

HISTORICAL INTRODUCTION

IT IS RISKING an engineering platitude to observe that noise, whether it be electronic, acoustical, seismic, optical, etc., in origin, is a critical and ultimately limiting factor in all communication processes. The study of such noise and its mathematical description have become an integral part of the sophisticated statistical communication theory of the present era. As is well known, an important element in the analytical description of noise is its power, or intensity, spectrum. In many practical applications one is interested in what happens to the noise spectrum when the noise is passed through both linear and nonlinear devices. One such problem is the subject of the historical paper presented below, namely, the calculation of the power spectrum of normal, or Gaussian, noise after it has been rectified by a "clipper," or zero-memory rectifier, that chops off the extreme values of the noise wave. In particular, it was desired to find out how much of the total available power in the noise wave was redistributed into spectral components away from the original, central region of the input spectrum. The application at the time was to enhance and effect the interference with radar and communication systems over broad spectral regions.

"The Spectrum of Clipped Noise" was written by J. H. Van Vleck and presented as Report No. 51 on July 21, 1943, by the Radio Research Laboratory of Harvard University, operating under the supervision of the Office of Scientific Research and Development. This work was a part of a much larger experimental and theoretical study carried out at the time, during World War II, of the factors which governed the electronic jamming of the radar and communication systems of that period. Although completely declassified at the end of the War, this report was not readily available to the engineering community at large, as it remained unpublished because of post-war adjustments and changes in the original author's area of research. Also, he did not realize that the report would excite so much interest. A short summary of some of the principal results were, however, presented in volume 24 of the M.I.T. Radiation Laboratory Series, but in most cases this was not enough for detailed application. During the period from 1946 to the present, many requests for the original report were received.

Because of this continuing interest, because this work is an illustrative forerunner of the now more highly developed approaches in current use, because a considerable portion of the material is still new, i.e., unpublished, and finally because the work itself appears to have some intrinsic historical importance, the authors have submitted it in its entirety to the readership, feeling (with apologies to Dumas) at the same time a little like the aging d'Artagnan and wondering whether this paper should perhaps better be entitled, "The Spectrum of Clipped Noise—Twenty (three) Years After"! The authors note, in conclusion, that apart from very minor editorial changes and the inclusion of references to subsequent work, the present paper is the original Report, with the addition of a short Appendix of pertinent material (prepared in 1944 by the second writer, who was the principal author's research assistant at Radio Research Laboratory during the war years). A certain amount of technical editing, in the form of footnotes, has also been carried out, in order to relate this earlier work to later developments and to embed it more effectively in the technical history of the field. All footnotes (and the Appendix), accordingly, represent additions made by the second author at the present time. The original Report gives new material not yet published in the literature, viz., the details of the calculations for superclipping of normal, narrow-band noise with a rectangular power spectrum, numerical results for the Gaussian and "optical" (Lorentzian) spectral distributions, and the treatment of Case (B), e.g., carrier amplitude-modulated by superclipped noise.

Manuscript received October 6, 1965.
J. H. Van Vleck is with the Department of Physics, Harvard University, Cambridge, Mass.
David Middleton is a Consulting Physicist located at 23 Park Lane, Concord, Mass.

***Abstract*—The present report calculates in some detail the (intensity) spectrum to be expected for clipped (also called "limited") noise. Two cases are considered: (A), the clipping of an unmodulated noise band (DINA)[1] and (B) a carrier modulated by clipped noise. The computations are made for various shapes of noise bands before clipping, viz., 1) a uniform or rectangular structure, 2) a Gaussian distribution, 3) an "optical" (Lorentzian) shape factor of the type $1/[(\nu-\nu_0)^2+\delta^2]$. The simplest type of calculation to make is that for what we term "extreme clipping," wherein the limiting amplitude is very small compared to the rms amplitude before clipping. The mathematical theory for this is given in Section III, while Section IV develops the theory for clipping at an arbitrary level. The basic mathematical method, which is rather general and is useful, we believe, for a variety of noise problems, is presented in Section II and consists in utilizing a relation between the correlation function and the normal surface, along lines suggested by Rice [1].**

The results of the calculation are discussed in Section I and are displayed in Figures 4–10 and Tables I and II. If the clipping is not down to more than the rms level before limiting (equivalent to clipping at about 1.4 times the rms level after clipping), there is practically no distortion of the spectrum. Even in the case of extreme clipping the wastage of power due to spoiling of the spectrum's uniformity is small, amounting to only 31 percent in (A) and 24 percent in (B). Of the 31 percent loss in (A), 19 percent is due to production of harmonics of the central frequency. Corresponding harmonics are absent in (B). *Clipping is beneficial for jamming purposes in either (A) or (B) since it reduces the peak power requirements. In addition, in (B) it materially diminishes the wastage of power in the carrier frequency.* These facts are demonstrated particularly clearly by *Tables I and II.* For instance, Table II shows us that in (B) the ratio of the energy in the noise sideband to that in the carrier is only 0.23 when the clipping level is twice the original rms noise level, but increases to 0.52 when these two levels are the same, and to 1.0 for extreme clipping.

It is to be cautioned that the present report calculates only the (intensity) spectrum of the clipped noise, and does not deal with its effectiveness on a receiver, which we hope to discuss later from a quantitative standpoint.[2] We can, however, say qualitatively that if the receiver breadth is very small compared to the noise band, the received disturbance will have the same type of Gaussian fluctuation, and hence the same effectiveness as unclipped noise with the same spectral distribution. On the other hand, if the receiver is comparable with the noise band in width, there will be, due to the clipping, a tendency for a "ceiling" in the resultant deflection of the recording device, and under these conditions the utility of clipped noise for jamming is materially diminished.

I. Descriptive Survey and Results

When we speak of a clipped noise we may mean either of two things: (A) a disturbance due entirely to noise (DINA), which we may suppose confined to some frequency band and whose maximum amplitude is limited, or (B) a carrier wave which is noise modulated and whose envelope of modulation is limited. Effect (A) is produced by filtering and then limiting a pure noise source, while (B) may be generated, for instance, by impressing a biased, clipped noise on the grid of a vacuum tube whose plate circuit contains a sinusoidal RF. The Case (B) will also be produced even if the noise on the grid is unclipped, provided the tube amplifies linearly over a certain range of grid voltage, with perfect blocking and saturation at the lower and upper limits of this range, and provided the bias is at the center of this range. The distinction between (A) and (B) is illustrated in Fig. 1, in which the unclipped disturbance is indicated by dotted lines. In studying (B) we assume throughout that the phase of the carrier is undisturbed by the interruptions; this point is discussed more fully at the end of Section II.

It is often illuminating to study particularly what we shall term "extreme clipping." By this we mean that all but a very small portion of the dotted curve in Fig. 1 is shaved off. The resulting wave is then practically rectangular in (A), and its envelope rectangular in (B), as shown in Fig. 2. Usually we are interested in an (intensity) spectrum whose frequency bandwidth before clipping is small compared to the carrier or mean frequency. Then in (A) the instantaneous frequency will deviate only slightly from that at the center of the band. On the other hand, in (B) the noise does not disturb the carrier but gives very irregular discontinuities in the envelope. Correspondingly, in Fig. 2 we have drawn the graph for (A) with more or less equal segments, but that for (B) with a more varied distribution of lengths. In a practical case the deviations from equality in (A) would be even smaller than in Fig. 2, as we have exaggerated them here to make them visible. (It should be mentioned that the two parts of Fig. 1 and also of 2 are drawn to different scales.) The frequency of occurrence of the "teeth" in (A) is approximately the same as the central frequency of the DINA noise band; the carrier frequency in (B) is of the same order as this, whereas the interruptions in (B) represent a modulation frequency which is much lower and which is of the order of the noise bandwidth.

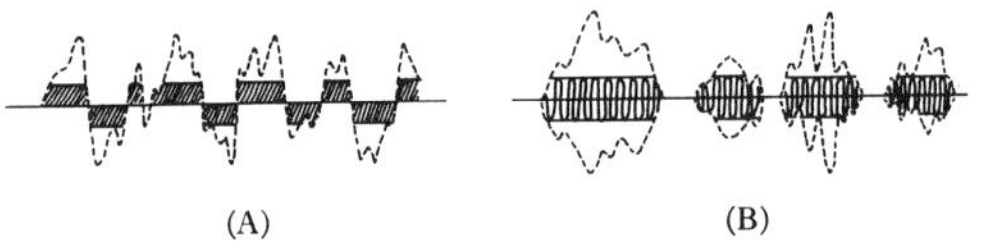

Fig. 1. The general case of clipping at an arbitrary level—schematic illustration of the difference between Cases (A) and (B).

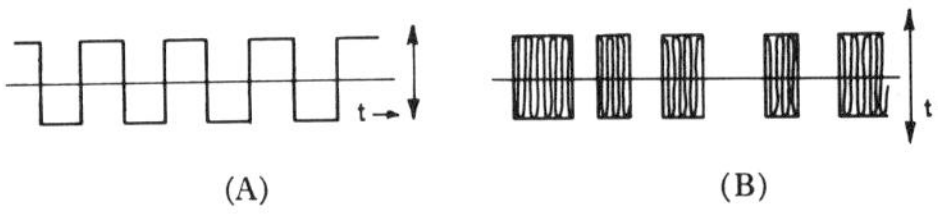

Fig. 2. Examples of extreme clipping for Cases (A) and (B).

It may be objected that the introduction of extreme clipping is too much of an idealization, since it can never be reached in practice. However, the hypothesis of complete clipping greatly simplifies the analytical work, as comparison of Section III (Mathematical Theory of Extreme Clipping) and Section IV (Clipping at an Arbitrary Level) will show. As the extreme model overaccentuates any clipping effects by introducing the maximum possible, it shows in a nutshell what is the general effect of clipping. It turns out that even ex-

[1] This is an acronym for "direct noise amplification"; DINA is usually produced physically as amplified photomultiplier noise.
[2] This analysis was never carried out.

treme clipping does not have a great deal of effect on the spectrum, and so we may feel confident that in no case does clipping do much harm *as far as spectral distribution is concerned.* The present report does not, however, attempt to treat the more delicate question of whether or not clipped and unclipped energies of the same spectral distribution are equally effective in jamming a receiver. We hope to treat this later.[3] Without any quantitative analysis it is apparent that the response of a receiver to clipped noise will have a ceiling, above which a pip is easily spotted, unless the bandwidth of the receiver is very narrow compared with the spectral width of the noise. With a very narrow receiver the noise pattern on the receiver will fluctuate in the same fashion as without clipping, the reason being that the limitation in amplitude due to clipping applies only when all Fourier components are added, and when the receiver selects only a small portion of the noise spectrum this limitation is lost. The quantitative question which we reserve for a later paper[4] is the study of exactly how the ceiling disappears when the receiver width is gradually narrowed. Instead, we shall examine at present simply the spectrum emitted by a transmitter equipped with a clipper, and see how much energy is lost by being spilled outside the frequency range which it is desired to cover with the barrage.

The mathematical theory of the spectrum of a clipped noise[5] will be developed in Sections II–IV; it is based on a statistical method suggested by Rice [1]. We are much indebted to him for stimulating discussions. In mathematical terms, the effect of extreme clipping turns out to be to replace the original correlation function $r(t)$ before clipping with a new correlation function which has the value $(2/\pi)\sin^{-1} r(t)$ [cf., our later (17)]. By the correlation function is meant the mean value of the product of the amplitude at times t_0 and t_0+t, as explained more fully in Section II.

There is an important difference between clipping in Case (A), (DINA), and in (B), (noise-modulated carrier). In (A) part of the energy is located after clipping in bands which are harmonics of the original fundamental band present before clipping, while in (B) the energy is confined to the fundamental. This distinction is illustrated in Fig. 3. The amount of energy located in each of the harmonics in (A) is precisely the same as that in a square wave devoid of any noise effects. The Fourier analysis of such a wave is well known, viz., even harmonics are wanting, and the fractions of the total energy associated with the harmonics $3\omega_c$, $5\omega_c$, $7\omega_c \cdots$ of the fundamental angular frequency ω_c are, respectively, $8/9\pi^2$, $8/25\pi^2$, $8/49\pi^2$, $\cdots$ of the total. All told, a fraction $1-8/\pi^2$ of the total energy is converted into harmonics, and so the energy wasted in harmonics is 19 percent of the total. In practical cases, the wastage is probably less than this estimate would imply, as the resonant amplifying devices of the transmitter are usually not tuned to the harmonics and so dissipate but little energy in the latter. However, it should be mentioned that the filtering of the harmonics increases the fluctuations in the energy, and so does not particularly help if avoidance of high peak power is the important criterion. For instance, it will destroy the equality of rms and peak amplitude characteristic of extreme clipping, as this behavior is secured only when the fundamental and harmonic Fourier components are superposed. In general, removal of the harmonics will make the ratios of peak to average power somewhat greater than the values to be shown in Table I, though not nearly as large as before clipping.

The different behavior of (A) and (B) as regards harmonics is what one might guess from one's physical intuition. Namely, if the noise band is exceedingly narrow, then the disturbance portrayed in Fig. 2(a) is very closely a square wave, as the "caterpillar effect," whereby the successive segments vary a little in length,

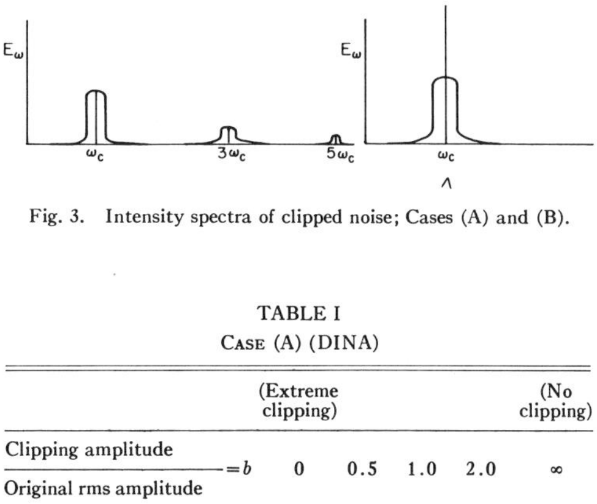

Fig. 3. Intensity spectra of clipped noise; Cases (A) and (B).

TABLE I

CASE (A) (DINA)

	(Extreme clipping)				(No clipping)
$\dfrac{\text{Clipping amplitude}}{\text{Original rms amplitude}} = b$	0	0.5	1.0	2.0	∞
$\dfrac{\text{Clipping amplitude}}{\text{rms amplitude after clipping}}$	1	1.17	1.39	2.09	∞
Bottleneck factor	0.69	0.86	0.95	0.97	1
Fraction of power in harminics $=\gamma$	0.19	0.08	0.03	0.005	0
$\dfrac{\text{Effective power}}{\text{Total rms power}}$	0.69	0.86	0.95	0.97	1
$\dfrac{\text{Effective power}}{\text{Peak power}}$	0.69	0.63	0.49	0.22	0

[3] This was never done.
[4] This was never written.
[5] Subsequent work by Middleton [2], [3], Davenport [4], and Price [5] has extended the theory. For a short account, with associated experimental results, see also Lawson and Uhlenbeck [6]. An extensive treatment of these and related topics involving noise through nonlinear devices may be found in Middleton [7]. We remark that the studies in [2]–[5] are concerned with Case (A) only. A theory for Case (B), including νth-law rectifier and an additive background noise, is developed in [7], Sec. 13.4-4, pp. 591–594.

is small. Hence, under these conditions one is tempted to use the ordinary Fourier analysis of a square wave to determine the ratios of the energies in the various harmonics and the more exact treatment of Section III shows that these ratios are not altered as long as the bandwidth of the noise is small compared with its central frequency. On the other hand, in (B) the interruptions do nothing to the carrier, and so harmonics of the latter do not occur.

Neither in Case (A) nor (B) can the spectrum of the fundamental after clipping, which is the thing of main interest, be determined without statistical and mathematical analysis, which is given in Sections II–IV. Our mathematical method would also enable us to compute the shapes of the harmonic bands in (A) but this does not really interest us, as the harmonics represent spectral domains of far higher frequency than the barrage to be covered. Hence we have not calculated the shapes of the harmonics in detail; without detailed numerical computations the mathematical analysis in Part III shows that the shapes of the harmonics are not the same as the fundamental, and that the diffuseness becomes greater the higher the degree of the harmonic.[6]

The spectral curves for the fundamental which are furnished by the calculations of Section III are shown in Figs. 4 and 5 for extreme clipping of Types (A) and (B), respectively, applied to a noise band which was rectangular before clipping. The ordinate scale is in each case so chosen that the total power in the spectrum is normalized to unity both before and after clipping so as to enable us to compare spectra of equal power content. We use the letter λ for the dimensionless ratio $\lambda = (\omega - \omega_c)/\omega_a$, where ω_c is the central angular frequency in (A) or carrier in (B) and $\omega_c - \omega_a$, $\omega_c + \omega_a$ are the limits of the noise spectrum before clipping. Reference to Figs. 4 and 5 shows that the effect of clipping is to "spill" a certain amount of energy outside the limits $\lambda = \pm 1$, and also to make the spectrum cease to be quite uniform within these limits. It is rather remarkable that even with clipping there is a finite discontinuity at the original boundary $\lambda = \pm 1$. The amount of discontinuity in either (A) or (B) amounts to $2/\pi$ of the height of the curve before clipping. The energy spilled outside $\lambda = \pm 1$ is practically wasted, as it is too diffuse to be of much value; a little may be picked up by receivers near the sides of the barrage, but not much. The fact that the curve for (A) is lower than that for (B) is easily understandable, as in (A) 19 percent of the energy is located in harmonics not shown in the figure. It should be mentioned that the curves for (A) and (B) have somewhat different shapes, i.e., they do not differ merely in ordinates; physically this is because the clipping processes are not quite the same. Apart from the harmonic losses, rather less energy is spilled outside the original band in (A) than in (B).

[6] See [7], Secs. 5.1, 5.2.

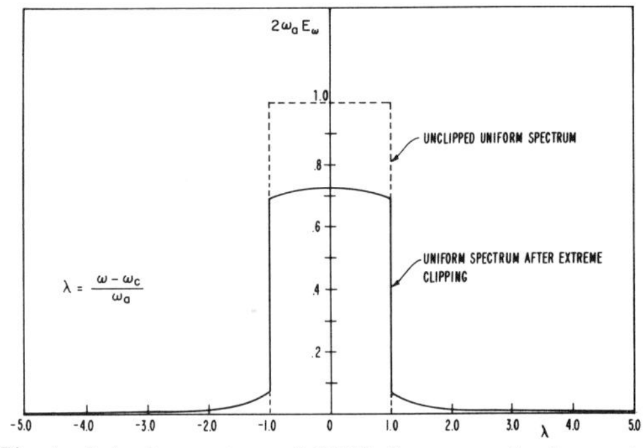

Fig. 4. Intensity spectrum of DINA (i.e., narrow-band normal noise) after extreme clipping, when the original spectrum is rectangular [Case (A)].

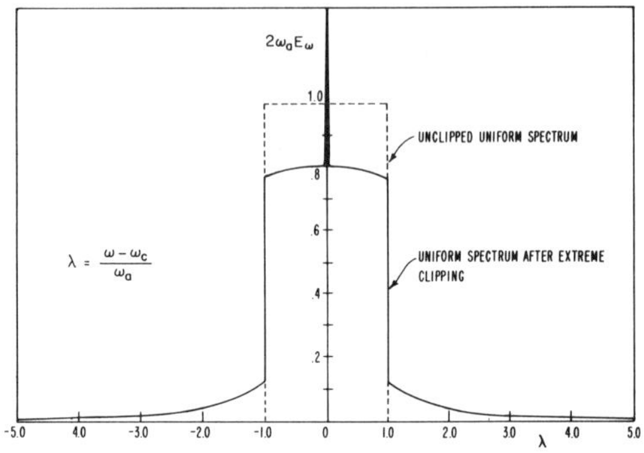

Fig. 5. Intensity spectrum after extreme clipping of a carrier modulated by narrow-band normal noise with a rectangular power spectrum [Case (B)].

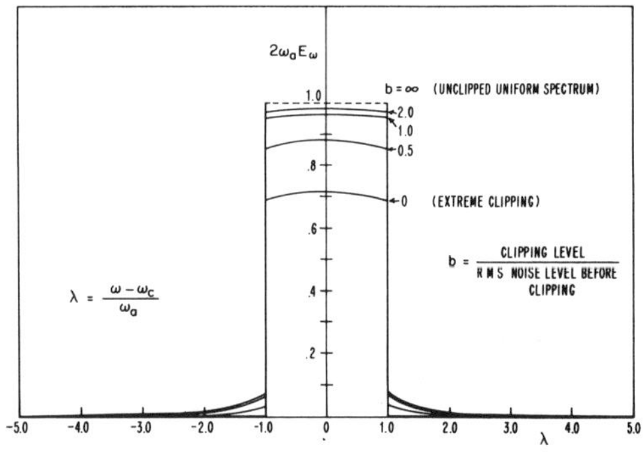

Fig. 6. Same as Fig. 4, now with various levels of clipping.

The effective bottleneck that determines the amount of power needed is the height of the curve at the edges $\lambda = \pm 1$, since a certain minimum power is required to jam any station, and this will be less effective at $\lambda = 1$ than at $\lambda = 0$ because the spectral curve slopes away from its maximum achieved at $\lambda = 0$. (Apart from this, there is the fact that because of the finite bandwidth of the receiver, the barrage generated by the sending station must be somewhat wider than that which it is wished to cover, since one must have power available over most of the bandwidth of the receiver. However, corrections for this effect will be approximately the same with and without clipping since the band is so nearly rectangular in either case.) If we call unity the height of the rectangle before clipping, then the height of the curve at $\lambda = 1$, which we shall call the "bottleneck factor," is 0.69 in Fig. 4 and 0.76 in Fig. 5. Hence, we see that in (A) 31 percent of the energy is wasted, due to the nonrectangular shape of the curve, and 24 percent in (B). Of the 31 percent in (A), 19 percent is due to harmonics and, as already mentioned, this kind of loss can often be avoided through filtering, though at the expense of greater fluctuation.

The mathematical theory for clipping at an arbitrary level, instead of extreme clipping, is developed in Section IV. Figure 6 shows the resulting curves for Case (A) (DINA) for clipping at amplitudes 0.5, 1.0, and 2.0 times the original rms level. As previously, it is assumed that the noise band is uniform or rectangular before clipping, and that the total power is normalized to unity after clipping (as well as before). We have not drawn the corresponding curves for Case (B) (noise-modulated carrier), as the general trend as one goes from no clipping to extreme clipping is the same in (B) as in (A): namely, the distortions in the spectrum occasioned by clipping vary in an exponential fashion, as they are given by formulas whose most significant factor is e^{-b^2}, where b is the ratio of the clipping to rms amplitude. This mode of dependence on the amount of limiting, taken in conjunction with the fact that the distortion is not great even with extreme clipping, shows that if, say, we clip at twice the rms value, the disturbance of the spectrum will be almost negligible, and even at the rms level it is slight.

In comparing the results with clipping at various levels, it is essential in Case (B) to include the energy dissipated in the carrier, which has not been included when in each case we normalize the output to unity. With extreme clipping, the amount of power in the carrier is just equal to that in the noise sidebands or, in other words, the area under the "delta-function" indicated schematically by the vertical straight lines at $\lambda = 0$ in Fig. 5 is just equal to the area under the rest of the curve: namely, unless one goes to the trouble of having a balanced modulator, the transmitting apparatus will have an output only if the grid potential is positive and the noise pattern can be reproduced symmetrically only by using a bias. We assume throughout 100 percent modulation, so as to minimize the carrier wastage. (In other words, we presuppose the minimum bias necessary in order for the tube to function even when the noise voltage on the grid has its maximum negative value.) Because of the bias, instead of having to deal with the spectrum of a function which oscillates between the two values -1, $+1$, we have really a function f whose two possible values are 0, $\sqrt{2}$ (the normalization in each case we take such that $\overline{f^2} = 1$). The amount of energy in the carrier is $(\bar{f})^2$ and that in the sidebands is $\overline{f^2} - (\bar{f})^2$; hence, if f spends on the average equal times at $f = 0$ and $f = \sqrt{2}$, one has $(\bar{f})^2 = (1/2)\overline{f^2}$, so that the energies in the sidebands and in the carrier are just equal. With clipping not of an extreme type, an even greater portion of the energy is wasted in the carrier. The formula for this wastage is given in the final paragraph of Part IV. With unclipped noise, the amount of energy in the carrier would, strictly speaking, be infinitely large compared with that in the noise sidebands, as an infinitely large bias would be required to cope with the largest conceivable fluctuation in a Gaussian law. From a practical standpoint, however, clipping at twice the rms level has only a very small effect on the spectrum.

Table I compares the results of clipping at various levels for Case (A) (DINA), while Table II shows analogous figures for Case (B) (noise-modulated carrier). In the first and second rows of the table we give, respectively, the ratio of the limiting amplitude to the rms amplitude before and after clipping; the latter is obviously the greater of the two ratios, as clipping reduces the mean amplitude. The column labeled with zero ratio of clipping to original rms amplitude corresponds to the theory for extreme clipping. The bottleneck factor is the ordinate of the curves in Figs. 4–6 at the edges $\lambda = \pm 1$, as previously mentioned. The ratio "effective power/total power" is obtained by dividing this factor by $1 + \rho$, where ρ is the ratio of carrier to sideband energy. It is the proper measure of efficiency if the average power generated after clipping is the critical factor governing the expense of the apparatus. If in Case (A), however, the harmonics can be suppressed by filtering, then the estimate of the effective power should be increased by $1/(1-\gamma)$, or approximately $1+\gamma$, where γ is the fraction of energy in harmonics. Often peak power is the restricting factor in the design of apparatus, and then, clearly, the ratio of effective to peak power is that of interest. It is seen that a high degree of clipping improves the efficiency quite materially in either (A) or (B) if the peak criterion is used, and also in (B), even if, instead, the total generated power is employed as a measure. It is to be cautioned that all the figures in the tables make no allowance for the difference between clipped and unclipped noise as regards the confusion

TABLE II

CASE (B) (NOISE-MODULATED CARRIER)

Clipping amplitude / Original rms amplitude $= b$	0	0.5	1.0	2.0	∞
Clipping amplitude / rms amplitude after clipping	1	1.17	1.39	2.09	∞
Bottleneck factor	0.76	0.88	0.95	0.98	1
Ratio of sideband to carrier power $= 1/\rho$	1.00	0.74	0.52	0.23	0
Effective power / Total rms power	0.38	0.37	0.32	0.18	0
Effective power / Peak power	0.19	0.16	0.12	0.06	0

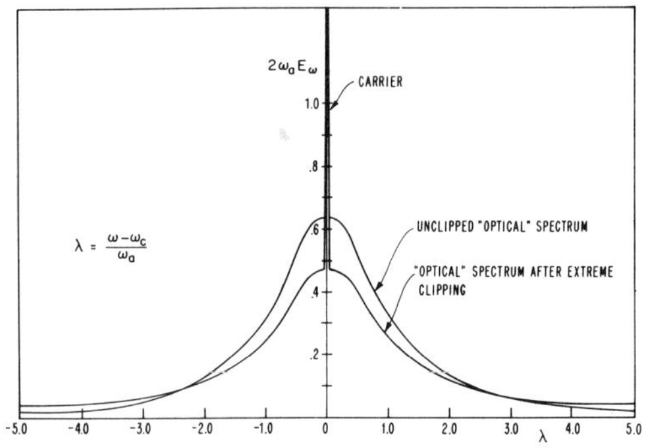

Fig. 9. Intensity spectrum after extreme clipping of a carrier modulated by narrow-band normal noise with an optical (i.e., Lorentzian) spectrum [Case (B)].

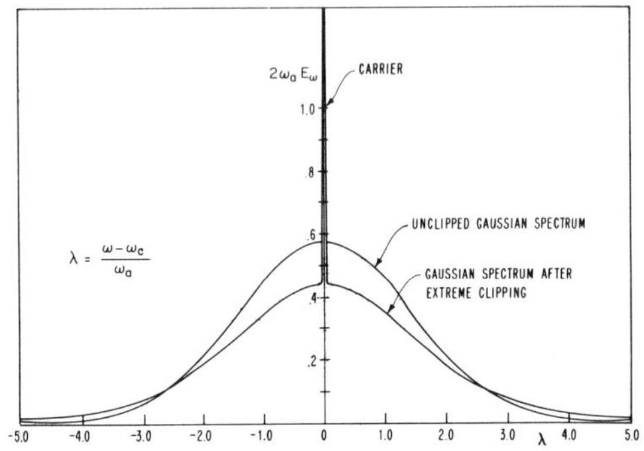

Fig. 7. Intensity spectrum after extreme clipping of a carrier modulated by narrow-band normal noise with a Gaussian spectrum [Case (B)].

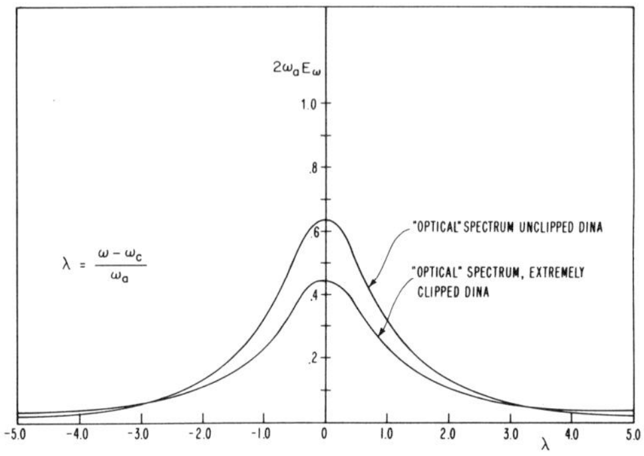

Fig. 10. Same as Fig. 9, but for narrow-band noise (DINA) alone [Case (A)].

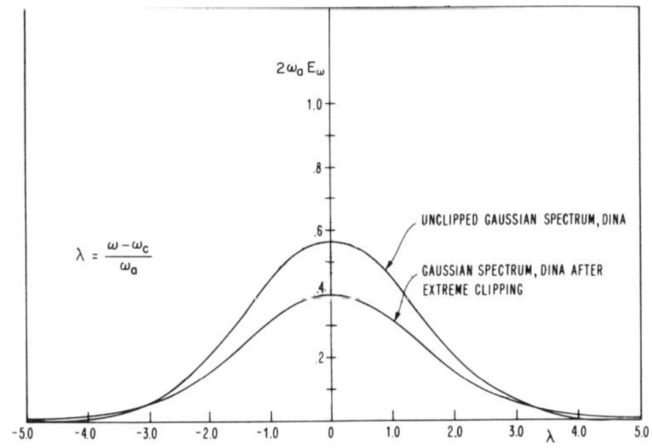

Fig. 8. Same as Fig. 7, but for narrow-band noise (DINA) alone [Case (A)].

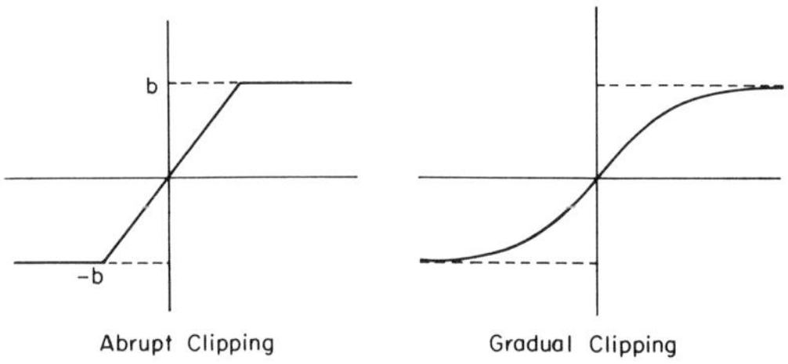

Fig. 11. Dynamic characteristics of amplitude limiters with abrupt and gradual clipping.

produced in the receiver pattern. Hence, unless the receiver width is very small compared with that of the barrage, the advantage of going to extreme clipping will not be as great as is indicated by the tables.

We have also made calculations for the effect of extreme clipping on noise bands which before clipping have a Gaussian shape $e^{-(\omega-\omega_c)^2/\omega_a^2}$, and on bands which have a shape factor $1/[\delta^2+(\omega-\omega_c)^2]$ like that of an optical absorption line or a simple LRC circuit [2]. The results appear in Figs. 7, 8 for clipping of types (B) and (A) for the Gaussian case, and in Figs. 9 and 10 for the optical one. It is seen that in each case the clipping tends to diffuse out the spectrum and lower it in the center, although in the aggregate the distortion is not large. The lowering is greater for (A) than for (B), due to dissipation in the harmonics in (A). The energy wasted in the harmonics in (A) has the same value, 19 percent, as in the case of a uniform band, since this fraction is independent of the shape of the band,[7] provided only that it is narrow compared with the central frequency.

Hitherto we have assumed that the clipping sets in abruptly at some given level. Another type of distortion is what we may term gradual clipping, in which the saturation commences gradually rather than suddenly. The distinction between the two is shown in Fig. 11. A convenient analytical representation of weak clipping is obtained by assuming that the response curve of the amplifier, instead of being linear, is of the form[8] tanh (V/C), where V is the grid voltage measured relative to some appropriately chosen bias, and C is the ratio of the amplitude at extreme saturation to the rms amplitude which would result were there no saturation corrections. The (intensity) spectrum can be readily computed, as is shown at the end of Section II, provided C is large compared with the original rms noise level V_0. The deviations from uniformity, however, are small as long as C is large enough to make the convergence adequate. For instance, if C equals 4, the departure from uniformity amounts to less than one percent, so that gradual clipping of this type is practically without effect on the spectrum. The saturation does, however, reduce the mean square amplitude even when it does not distort the spectrum, lowering it about 10 percent, for example, in the case of $C=4$ (unless it is renormalized to unity as in the procedure followed in Figs. 4–10).

II. The Correlation Function and The Normal Surface

It is perhaps well to begin the mathematical analysis by discussing the role of the so-called correlation function. By the latter is meant the expression

$$R(t) = \overline{f(t_0)f(t_0+t)}, \tag{1}$$

where $f(t)$ is the function being studied. The meaning of the bars requires particular comment. It means an average over all the fluctuations of the system, or, more technically, an average over an ensemble of similarly prepared systems differing only in phase.[9] Under these circumstances, $R(t)$ is by definition independent of the initial time t_0. Also $R(t)$ will be even in t. If there is a carrier frequency ν_c, then the correlation function will be of the form $\psi(t)\cos 2\pi\nu_c t$. The factor cos $\omega_c t$, $(\omega_c = 2\pi\nu_c)$, is of no interest, as it will be ironed out in the video part of the detector. Clearly, $R(t)$, or its envelope in case there is a carrier, is a measure of how rapidly the disturbance tends to fluctuate on the screen of a (radar) receiver.[10]

The use of the correlation function as the measure of the quality of receiver response, however, requires caution. The value of t at which $R(t)$ drops to say half its maximum value can be employed as a gauge of the width of the "blades of grass" only if a), there is no periodicity to the disturbance and b), it does not have any "holes" through which the signal can be seen.

In case a) is not satisfied, the physical reception screen does not really average over the different phases, i.e., $R(t)$ is a periodic function of t, and the disturbance will have a discernible periodicity. For instance, consider a *uniform* square wave, for which f is zero half the time and unity the rest. The uniformity means that all the segments in Fig. 2 are to be considered of the same length. Then $R(t)$ is a periodic function of t and does not approach a definite limit at $t=\infty$. If the reciprocal of the bandwidth of the receiver corresponds to a time interval short compared with the length of the segment, the pattern on the receiver will still be a well-defined square wave with holes and peaks alternating regularly. If, however, the reciprocal of the bandwidth corresponds to a time interval long compared with that represented by the segment length, then the result really will be a uniform smear. The fact that in this example the correlation function does not go to zero as $t\to\infty$ shows that there is a recurrent periodicity. In fact, the behavior of the correlation function at infinity enables us to see whether or not the disturbance is truly irregular. With order, one has $R(\infty)\neq 0$, while with randomness $R(\infty)=0$.

With regard to b), the average in (1) is over all phases, and so includes both holes and peaks. Therefore, the correlation function does not give (an) indication as to the existence of "holes" in the pattern. Even though the decay time of the correlation function is so short that holes of corresponding length would be too brief to be discernible, there can actually be holes of such great length

[7] See [1], Sec. 4.9, Eq. (4.9-19); also [7], Eq. (5.29) and subsequent remarks.

[8] Other rectifier laws, in particular the symmetrical νth-law devices, are discussed in [7], Sec. 13.4-2.

[9] Strictly speaking, we have here an (auto-) covariance function for an assumed stationary process, which, of course, is an ensemble average.

[10] For a detailed discussion of the correlation and covariance functions of a process, and the spectral concepts associated with them, see [7], Secs. 3.1 and 3.2.

as to be troublesome. Let us suppose, for instance, that the disturbance consists of a series of irregularly spaced pulses, so that the width of a pulse is small compared with the spacing. The correlation function is then the same as that for one pulse considered separately, and yields an apparent fluctuation time corresponding to the pulse width, but there will be obvious holes in the pattern. These will disappear when the interval between pulses becomes comparable to the receiver width. A quantitative criterion for determining whether or not the hole difficulty is serious may be established by examining whether $\overline{F(t)^4}$ greatly exceeds $[\overline{F(t)^2}]^2$ or is of the same order of magnitude as the latter: with a disturbance which functions only a small fraction of the time, the former will be much the greater of the two. Here, $F(t)$ specifies the time dependence of the receiver response, and can be identified with the form $f(t)$ of the impinging wave only if the receiver breadth is wide compared with the mean fluctuation frequency of the disturbance.

From the above it is seen that the correlation function[11] must be used with some caution as a gauge of the fluctuation width. The same remarks, however, apply even more strongly to the usual attempts to judge effectiveness by looking at the Fourier components, as the latter obviously give no information concerning the existence of holes. In the last analysis, it is rather futile to say whether the study of Fourier components or correlation function gives the greater information, because the two are so intimately connected that they should be taken in conjunction rather than separately. Namely, if we denote by $E_\omega\, d\omega$ the energy[12] per unit time in the angular frequency interval ω, $\omega+d\omega$, then there are the following very simple relations between E_ω and the correlation function:

$$E_\omega = \frac{1}{\pi}\int_{-\infty}^{+\infty} R(t)\cos\omega t dt,$$

$$R(t) = \int_0^{\infty} E_\omega \cos\omega t d\omega. \tag{2}$$

We express our results throughout in terms of the angular frequency ω rather than the true frequency $\nu=(\omega/2\pi)$ as this will enable us to avoid factors 2π in many places. The corresponding energy densities are connected by the relation $E_\nu = 2\pi E_\omega$.

Equation (2) shows that *except for a proportionality factor, the correlation function is the Fourier transform of the spectral energy density, and vice versa.*[13]

To prove (2) we first write down the Fourier integrals

$$f(t) = \int_{-\infty}^{+\infty} A(\omega)e^{i\omega t}d\omega, \quad A(\omega) = \frac{1}{2\pi}\int_{-\infty}^{+\infty} f(t)e^{-i\omega t}dt, \tag{3}$$

where $A(-\omega)$ is the complex conjugate of $A(\omega)$, since $f(t)$ is real. We use the imaginary exponential rather than the trigonometric form of the integral, as it makes the manipulations easier. From Plancherel's theorem we have:

$$\int_{-\infty}^{+\infty} f(t)^2 dt = 2\pi\int_{-\infty}^{+\infty} | A(\omega) |^2 d\omega. \tag{4}$$

We shall now imagine the disturbance confined to a very long period $2T$ such that this interval is so long that the cutoff at the beginning and end has no bearing on the shape of the Fourier spectrum, but does make the integrals converge. Then we see that

$$\overline{f(t)^2} = \frac{\pi}{T}\int_{-\infty}^{+\infty} | A(\omega) |^2 d\omega. \tag{5}$$

The left side gives the mean energy, and the right side shows how the various frequencies contribute. Since E_ω is to be identified with both the terms in $+\omega$ and $-\omega$, we have immediately the result that

$$E_\omega = \frac{2\pi}{T} | A(\omega) |^2. \tag{6}$$

Now let us compute E_ω in another way. We can write

$$| A(\omega) |^2 = \frac{1}{4\pi^2}\int_{-\infty}^{+\infty} f(t')e^{-i\omega t'}dt' \int_{-\infty}^{+\infty} f(t'')e^{i\omega t''}dt''. \tag{7}$$

For a chaotic (or random) disturbance, if we average over all phases, $\overline{f(t')f(t'')}$ is a function only of $t=t''-t'$, and is even in this variable. Also, we shall imagine the disturbance confined to a finite, but very large interval $2T$. Then, on changing variables to $t=t''-t'$, we have

$$\overline{| A(\omega) |^2} = \frac{1}{4\pi^2}\int_{-\infty}^{+\infty} R(t)e^{i\omega t}dt \int_{-T}^{T} dt'$$

$$= \frac{T}{2\pi^2}\int_{-\infty}^{+\infty} R(t)e^{i\omega t}dt. \tag{8}$$

The first relation of (2) follows immediately on combining (6) and (8) and noting that $R(t)$ is even; the second part of (2) is merely the Fourier transform of the first.[14]

The normal surface. In order to compute the correlation function[15] in statistical problems, it is convenient to use a formula of probability theory known as the normal surface. Let X and Y be two linear functions

[11] This term is used throughout in the sense of a covariance function, e.g., an ensemble average.

[12] Or *power*, in the angular interval of width $d\omega$; ($\omega=2\pi\nu f$); thus E_ω is the power/unit bandwidth, or intensity density, of the process in arbitrary units. The term "energy density" used here refers to energy *per unit time*, per (angular) frequency interval, or, equivalently, a power/(angular) c/s.

[13] This, of course, is the celebrated Wiener-Khintchine theorem ([7], Sec. 3.2-2).

[14] The bar on the left number of (7) has been added subsequently. The "proof" above is strongly heuristic. As is well known now, when $f(t)$ is a random process rather subtle conditions (in addition to the usual ones for analytic functions) must be obeyed in the definition of the intensity spectrum and its relations to the covariance and correlation functions. See, in particular, [7] Secs. 3.2-1 and 3.2-2; also (3) of Sec. 3.2-3, and the footnote on page 152. However, since appropriate statistical averages have been taken in the subsequent treatment here, the results are correct.

[15] E.g., the covariance function.

$$X = \sum_{i=1}^{N} a_i x_i \qquad Y = \sum_{i=1}^{N} b_i x_i$$

of a set of (random) variables $x_1, x_2, \cdots, x_N$, each of which is distributed independently according to the normal error (Gaussian) law with zero mean value ($\bar{x}_i = 0$) and unit-mean-square or standard deviation. In other words, the probability that x_i falls in the interval x_i, $x_i + dx_i$ is taken to be $(1/2\pi)^{1/2} e^{-x_i^2/2} dx_i$. Then X and Y each also have a Gaussian distribution, *though not in general independent of each other*, and we shall assume that their amplitude factors are so normalized that each has a unit standard deviation. This supposition involves no loss of generality, and implies that

$$\sum_{i=1}^{N} a_i^2 = \sum_{i=1}^{N} b_i^2 = 1.$$

The probability that X falls in X, $X+dX$, and that Y also simultaneously falls in Y, $Y+dY$ is[16]

$$\frac{1}{2\pi(1 - r^2)^{\frac{1}{2}}} e^{-(X^2+Y^2-2rXY)/2(1-r^2)} dX dY, \tag{9}$$

where r is the average value of the product XY. The expression (9) is closely related to what is called the normal surface in statistical theory; namely the equation $Z = [2\pi^2(1-r^2)]^{-1/2} \exp[-(X^2+Y^2-2rXY)/2(1-r^2)]$ gives rise to a locus called the normal surface in a three-dimensional plot in which the probability density is taken as the third variable. We owe to S. O. Rice [1] the suggestion of using (9) in problems of the type which we are considering.

A derivation of (9) is found in many books on probability theory but, without consulting them, we may see that (9) is readily established by utilizing the fact that the Gaussian distribution is left invariant under a unitary transformation. The transformation from $x_1, x_2, \cdots, x_N$ to any set of variables inclusive of X, Y cannot be unitary, as X, Y are not orthogonal except in the trivial case $r=0$. However, the variables X, Y' are orthogonal and normalized to unity if we take $Y' = (Y - rX)/(1-r^2)^{1/2}$, and we can, by a well-known procedure (cf., Courant-Hilbert, [8] p. 19), find $N-2$ other variables $x_3', x_4', \cdots, x_N'$ such that the transformation from $x_1, x_2, x_3, \cdots, x_N$ to $X, Y', x_3', \cdots, x_N'$ is unitary. Then the Gaussian distribution law holds in the $X, Y', x_3', \cdots, x_N'$ space, and if we integrate over $x_3', \cdots, x_N'$ we simply get unity and so obtain the Gaussian distribution law $(1/2\pi) \exp[-(X^2+Y'^2)/2]$ applied to the coordinates X, Y' just as though $x_3', \cdots, x_N'$ did not exist. On transforming from the X, Y' to the X, Y space we immediately get (9).

In noise problems we can take X and Y to be the unclipped amplitude at, respectively, $t=0$ and $t=t$. Then $r(t)$ is the correlation function for the unclipped noise.

We are justified in applying statistical reasoning of the type underlying (9) to noise problems because the total disturbance is a linear function of the amplitudes of the individual Fourier components, and the latter are known to obey a Gaussian distribution law for ideal noise.

If we have a distorting apparatus, e.g., a clipper, such that after distortion the amplitude is $f(X)$ rather than X, then the resulting correlation function is by (9):[17]

$$R(t) = \frac{1}{2\pi(1 - r^2)^{\frac{1}{2}}} \int_{-\infty}^{+\infty} \int_{-\infty}^{+\infty} f(X)f(Y) \cdot e^{-(X^2+Y^2-2rXY)/2(1-r^2)} dX dY. \tag{10}$$

It is often convenient to have the distorted amplitude normalized to unity; in this case the right side of (10) should be divided by

$$\frac{1}{(2\pi)^{\frac{1}{2}}} \int_{-\infty}^{+\infty} f(X)^2 e^{-X^2/2} dX.$$

Do not confuse the correlation function $R(t)$ *after* distortion with that, $r(t)$, before. One readily verifies that for the linear amplifier, characterized by $f(X) = X$, the right side of (10) reduces to $r(t)$, as one would expect.

As an example of the use of (10), consider an amplifier whose characteristic is $f(X) = C \tanh (X/C)$. It is easiest to compute the integral by using the coordinate system X, X' described above in the proof of (9). In this system we have

$$R(t) = \frac{C^2}{2\pi} \int_{-\infty}^{+\infty} \int_{-\infty}^{+\infty} \tanh\left(\frac{X}{C}\right) \cdot \tanh\left(\frac{rX + Y'(1 - r^2)^{\frac{1}{2}}}{C}\right) e^{-(X^2+Y'^2)/2} dX dY'. \tag{11}$$

If C is large, i.e., if the saturation corrections are not great at the rms level, then we may evaluate (11) by series expansion

$$\tanh \theta = \theta - \frac{\theta^3}{3} + \frac{2}{15}\theta^5 + \cdots$$

of the hyperbolic tangent. We thus obtain

$$R(t) = r + \frac{1}{C^2}(-2r) + \frac{1}{C^4}(5r + \tfrac{2}{3}r^3) + \cdots. \tag{12}$$

Results based on (12) have already been referred to at the end of Section I.

In closing this section, it may be mentioned that the problem of determining the intensity spectrum of a disturbance subject to extreme clipping is closely related to that of finding the zeros (crossing points of the axis)[18] in Fig. 2(A) and of those of the envelope in Fig. 2(B). If we can determine the distribution of the zeros it is possible to find the correlation function, and hence the spectrum. However, it is much easier to find the correla-

[16] [7], Sec. 7.2, Eq. (7.13a), $\psi = 1$.

[17] [7], Sec. 13.1.

[18] See, for example, [7], Sec. 9.4.

tion (function) by means of (10) than to find the zeros, which is more intricate,[19] and which has been examined to some extent by Rice.[20] One's first guess might be that the zeros of the envelope in Fig. 2(B) obey a random law, like the emission of alpha particles from a radioactive material. This model is exceedingly simple, and has been studied by G. W. Kenrick [9]. At one time we thought it might be a sufficient approximation to the behavior of noise, but actually it is not. It gives a spectrum of type $1/[\delta^2+(\omega-\omega_c)^2]$ like the response curve for a circuit with LRC, and this is very materially different from our results in Figs. 4 and 5, with their pronounced discontinuities at $\lambda=1$. Actually, as Rice's theory[21] shows, the zeros do not obey a (normal) random law, and instead, after one zero, another one is unlikely to occur for a while.

Throughout the report it is assumed that in (B) (modulation by clipped noise) the phase of the carrier persists or, in other words, is unaffected by the blocking of the transmission of the tube when the grid potential is unfavorable. This assumption is usually pretty well warranted. The case that the phase of the carrier has no memory of its past and acquires a new random value each time the grid potential is favorable is much more difficult to treat as the statistical formulas based on the normal surface cannot be used. Without persistence in phase, there will be no coherence between the contributions of the successive wave trains generated by the periods of favorable grid potentials, and the problem is the same as that of determining the average spectrum of an ensemble of wave trains whose distribution in lengths is the same as that of the periods between consecutive zeros. We have already mentioned that this is not an easy problem, and we may note particularly that the determination of the distribution of the intervals between consecutive zeros is more difficult than that of finding the probability that a zero occur a certain time after a selected zero, as the two zeros correlated may not be consecutive. In fact no one appears yet[22] to have found the distribution law for consecutive zeros. Without quantitative analysis it can be said that if there is no persistence in phase or, in other words, if the carrier is reborn for each favorable grid sequence, then the "delta function" or extreme concentration of energy at exactly the carrier frequency will be absent, but the spectrum will otherwise be less uniform than for the case of persistence. It will slope away sharply from a maximum at the center, and will not have the discontinuity at $\lambda=\pm 1$ shown in Fig. 5.

III. Extreme Clipping

With extreme clipping the function $f(X)$ involved in (10) has the form

$$f(X) = +1, \ (X > 0); \quad f(X) = -1 \ (X < 0). \tag{13}$$

We have here assumed a normalization such that after clipping the mean-square amplitude is unity, or in other words that the ordinates of the horizontal straight lines in Fig. 2 are ± 1. The expression (10) becomes

$$R(t) = \frac{1}{2\pi(1-r^2)^{\frac{1}{2}}}\left[\int_0^\infty\int_0^\infty e^{-\alpha}dXdY + \int_{-\infty}^0\int_{-\infty}^0 e^{-\alpha}dXdY - \int_0^\infty\int_{-\infty}^0 e^{-\alpha}dXdY - \int_{-\infty}^0\int_0^\infty e^{-\alpha}dXdY\right] \tag{14}$$

where

$$\alpha = (X^2 + Y^2 - 2rXY)/2(1 - r^2).$$

We can simplify (14) by using the relation

$$\frac{1}{2\pi(1-r^2)^{\frac{1}{2}}}\int_{-\infty}^{+\infty}\int_{-\infty}^{+\infty} e^{-\alpha}dXdY = 1, \tag{15}$$

which is readily verified mathematically and which is also obvious from the fact that the correlation would be unity if instead of (13) we had $f(X)=1$ for all values of the argument. Also, it is convenient to introduce polar coordinates $X=\rho\cos\phi$, $Y=\rho\sin\phi$. It is thus found that (14) can be written

$$R(t) = 4\int_0^{\pi/2} d\phi \int_0^\infty \frac{1}{2\pi(1-r^2)^{\frac{1}{2}}} \cdot e^{-\rho^2(1-r\sin 2\phi)/2(1-r^2)}\rho d\rho - 1. \tag{16}$$

Integration gives

$$R(t) = \frac{2(1-r^2)^{\frac{1}{2}}}{\pi}\int_0^{\pi/2}\frac{d\phi}{1-r\sin 2\phi} - 1 = \frac{2}{\pi}\sin^{-1}(r). \tag{17}$$

We thus have the rather simple and elegant result that the effect of extreme clipping is to make the correlation function $2/\pi$ times the arc sine of the original correlation function before clipping.[23] In case there is a carrier of angular frequency ω_c [Case (B) of Fig. 2], the right side of (17) must be multiplied by $\cos\omega_c t$ to give the true correlation function to be used in (2); namely, the effect of the carrier is to introduce an extra factor $B\cos\omega_c t_0 \cos\omega_c(t+t_0)$ whose mean value is $(\frac{1}{2})B\cos\omega_c t$ on averaging over all phases t_0 and the proportionality factor B is 2 if $R(0)=1$, i.e., if the mean-square amplitude after clipping is taken equal to unity.

[19] See, for example, [7], Sec. 9.4-1.
[20] [1], Secs. 3.3, 3.4.
[21] [1], Secs. 3.3, 3.4.
[22] No precise analytic theory is available to this day. See [7], Sec. 9.4, for a discussion and further references.
[23] This is a special case in the theory of symmetrical limiters, *vide* [7] Sec. 13.4-2(2) and Eq. (13.8 1a,b). See also *ibid.*, Secs. 13.1 and 13.1-1.

The Form of the Correlation Function

Before proceeding further it is necessary to specify the correlation function. The simplest assumption to make in Case (A), (DINA), is that before clipping there is a uniform noise band extending from $\omega_c-\omega_a$ to $\omega_c+\omega_a$. If the total energy[24] before clipping is normalized to unity, so that $r(0)=1$, then the energy density before clipping has the value $E_\omega=1/2\omega_a$ for $\omega_c-\omega_a<\omega<\omega_c+\omega_a$ and vanishes outside this interval. By (2) the correlation function before clipping is now

$$r(t) = \frac{1}{2\omega_a}\int_{\omega_c-\omega_a}^{\omega_c+\omega_a}\cos\omega t\,d\omega = \frac{\sin\omega_a t}{\omega_a t}\cos\omega_c t. \quad (18)$$

The corresponding correlation function after clipping is by (17)

$$R(t) = \frac{2}{\pi}\sin^{-1}\left[\left(\frac{\sin\omega_a t}{\omega_a t}\right)\cos\omega_c t\right]. \quad (19)$$

In Case (B) (modulation by clipped noise), the uniform noise band is low frequency and is used only to modulate the carrier. So if the noise band is assumed to be uniform, it may be taken as extending from $\omega=0$ to $\omega=\omega_a$ with an energy density $1/\omega_a$. The original correlation function is then

$$r(t) = \frac{1}{\omega_a}\int_0^{\omega_a}\cos\omega t d\omega = \frac{\sin\omega_a t}{\omega_a t}, \quad (20)$$

while after clipping it becomes

$$R(t) = \frac{2}{\pi}\cos\omega_c t\sin^{-1}\left[\frac{\sin\omega_a t}{\omega_a t}\right], \quad (21)$$

the factor $\cos\omega_c t$ being added for reasons previously given.

From examination of the preceding equations, we see that the mathematical difference between clipping of Type (A) and that of Type (B) is that in the former the factor $\cos\omega_c t$ appears inside the argument of the arc sine, while in the latter it appears outside.

We shall also give the correlation functions for the case in which the noise band before clipping has a Gaussian spectral distribution, and for the case in which it has what we may term an "optical" distribution. By the latter we mean that it has a structure factor $1/[\omega_a^2+(\omega-\omega_c)^2]$ of the same form as an optical (i.e. Lorentzian) absorption band or as the response curve of a simple circuit with L, R, C (provided[25] $R/2L\ll(LC)^{-1/2}$). The results are given in (22)–(25). The expressions $r(t)$ and $R(t)$ are, respectively, the correlation function before and after clipping, and $\mathcal{E}_\omega$ is the energy density or power spectrum before clipping, which is not to be confused with the energy spectrum E_ω after clipping, to be calculated later as our goal. We assume the energy (e.g., power) normalized to unity, so that

$$\int_0^\infty \mathcal{E}_\omega d_\omega = 1.$$

Thus, we have explicitly:

Gaussian distribution, clipping of Type (A):

$$\mathcal{E}_\omega = \frac{1}{\omega_a}\left(\frac{1}{2\pi}\right)^{\frac{1}{2}} e^{-(\omega-\omega_c)^2/2\omega_a^2};$$

$$r(t) = e^{-\frac{1}{2}\omega_a^2t^2}\cos\omega_c t,$$

$$R(t) = \frac{2}{\pi}\sin^{-1}\left[e^{-\frac{1}{2}\omega_a^2t^2}\cos\omega_c t\right]. \quad (22)$$

Gaussian distribution, clipping of Type (B):

$$\mathcal{E}_\omega = \frac{1}{\omega_a}\left(\frac{2}{\pi}\right)^{\frac{1}{2}} e^{-\omega^2/2\omega_a^2}, \qquad r(t) = e^{-\frac{1}{2}\omega_a^2t^2},$$

$$R(t) = \frac{2}{\pi}\cos\omega_c t\sin^{-1}\left[e^{-\frac{1}{2}\omega_a^2t^2}\right]. \quad (23)$$

Optical distribution, clipping of Type (A):

$$\mathcal{E}_\omega = \frac{\omega_a}{\pi[(\omega-\omega_c)^2+\omega_a^2]}, \qquad r(t) = e^{-\omega_a|t|}\cos\omega_c t,$$

$$R(t) = \frac{2}{\pi}\sin^{-1}\left[e^{-\omega_a|t|}\cos\omega_c t\right]. \quad (24)$$

Optical distribution, clipping of Type (B):

$$\mathcal{E}_\omega = \frac{2\omega_a}{\pi(\omega^2+\omega_a^2)}, \qquad r(t) = e^{-\omega_a|t|}$$

$$R(t) = \frac{2}{\pi}\cos\omega_c t\sin^{-1}\left[e^{-\omega_a|t|}\right]. \quad (25)$$

Spectrum for Originally Rectangular Distribution Subject to Clipping of Type (A)

We now proceed to study the spectrum of unmodulated clipped noise (DINA) which before clipping had a uniform or rectangular distribution, so that (18) and (19) are applicable. By (2) and (19) the energy distribution in the spectrum after clipping is given by the expression

$$E_\omega = \frac{4}{\pi^2}\int_0^\infty \cos\omega t\sin^{-1}\left[\frac{\cos\omega_c t\sin\omega_a t}{\omega_a t}\right]dt. \quad (26)$$

The problem of finding the spectrum is hence reduced to the evaluation of an integral. Since the integral (26) is certainly not evaluable in closed form, one immediately tries expanding the arc sine in a Taylor's series

$$\sin^{-1} r = r + \frac{r^3}{6} + \frac{3r^5}{40} + \frac{5r^7}{112} + \cdots. \quad (27)$$

This gives

[24] E.g. power/unit bandwidth; see remark, footnote following (2).[13]

[25] See [7], Types 3, 4, p. 169.

$$E_\omega = \frac{2}{\pi^2}\int_0^\infty \left\{ [\cos(\omega-\omega_c)t]\left(\frac{\sin\omega_a t}{\omega_a t}\right) + \frac{1}{6}\left[\frac{\cos(\omega-3\omega_c)t + 3\cos(\omega-\omega_c)t}{4}\right]\left(\frac{\sin\omega_a t}{\omega_a t}\right)^3 \right.$$
$$+ \frac{3}{40}\left[\frac{\cos(\omega-5\omega_c)t + 5\cos(\omega-3\omega_c)t + 10\cos(\omega-\omega_c)t}{16}\right]\left(\frac{\sin\omega_a t}{\omega_a t}\right)^5$$
$$+ \frac{5}{112}\left[\frac{\cos(\omega-7\omega_c)t + 7\cos(\omega-5\omega_c)t + 21\cos(\omega-3\omega_c)t + 35\cos(\omega-\omega_c)t}{64}\right]$$
$$\left. \times \left(\frac{\sin\omega_a t}{\omega_a t}\right)^7 + \cdots \right\} dt. \qquad (28)$$

Here and elsewhere we omit terms in $\cos(\omega+\omega_c)t$, $\cos(\omega+3\omega_c)t$, etc. which involve the sum rather than the difference of two high frequencies, as such terms oscillate so rapidly that their contributions to the integrals are negligible. Ordinarily we are interested only in the distribution in the main band, rather than in the harmonics which center about $3\omega_c$, $5\omega_c$, etc. Then in (28) we need retain only the terms having a factor $\cos(\omega-\omega_c)t$. By expressing $\sin^m \omega_a t \cos(\omega-\omega_c)t$ as sums of terms of the type $\sin(\omega-\omega_c+k\omega_a)t$, $-m \leqq k \leqq m$, as can be done by means of elementary trigonometric identities, and then by integrating by parts [cf. the Appendix], the expression (28) can be reduced to the sums of integrals of the form

$$\int_0^\infty [(\sin pt \cos qt)/t]\cdot dt,$$

with p and q real, but not necessarily integral. Since

$$\int_0^\infty \frac{\sin pt \cos qt}{t}\,dt = \frac{\pi}{2}, \qquad (p > q > 0),$$

$$\int_0^\infty \frac{\sin pt \cos qt}{t}\,dt = 0, \qquad (q > p > 0),$$

the analytical expression for E_ω obtained in this fashion has different forms for different domains of ω. It is convenient to introduce the notation $\lambda = (\omega-\omega_c)/\omega_a$. On evaluating the integral in the manner just described (see the Appendix), we obtain expressions for E_ω which are zoned as follows, according to the values of λ:

Zone I: $(|\lambda| < 1)$

$$E_\omega = \frac{1}{\pi\omega_a}\{1.134 - 0.0410\lambda^2 + 9.552\cdot10^{-4}\lambda^4 - 1.060\cdot10^{-5}\lambda^6\}, \qquad (29\text{ I})$$

Zone II: $(1 < |\lambda| < 3)$

$$E_\omega = \frac{1}{\pi\omega_a}\{0.1799 - 0.0890\lambda - 0.00120\lambda^2 + 0.00451\lambda^3 + 1.26\cdot10^{-5}\lambda^4 - 1.113\cdot10^{-4}\lambda^5 + 7.89\cdot10^{-6}\lambda^6\}, \qquad (29\text{ II})$$

Zone III: $(3 < |\lambda| < 5)$

$$E_\omega = \frac{1}{\pi\omega_a}\{0.0808 - 0.04492\lambda + 0.00261\lambda^2 + 0.003208\lambda^3 - 8.809\cdot10^{-4}\lambda^4 + 8.91\cdot10^{-5}\lambda^5 - 3.18\cdot10^{-6}\lambda^6\}, \qquad (29\text{ III})$$

Zone IV: $(5 < |\lambda| < 7)$

$$E_\omega = \frac{1}{\pi\omega_a}\{0.00439 + 0.01613\lambda - 0.01571\lambda^2 + 0.00565\lambda^3 - 1.003\cdot10^{-3}\lambda^4 + 8.91\cdot10^{-5}\lambda^5 - 3.18\cdot10^{-6}\lambda^6\}, \qquad (29\text{ IV})$$

Zone V: $(7 < |\lambda| < \infty)$

$E_\omega = 0$ (unless more than four terms of the series are used). (29 V)

In order to form an estimate of the error involved in using our result (29), etc., based on series expansions, we make use of the identity

$$\int_{-\infty}^{+\infty} G_{\omega'}d\omega' = 2\int_0^\infty G_{\omega'}d\omega' = f(0)$$
$$\text{if } G_{\omega'} = \frac{1}{\pi}\int_0^\infty f(t)\cos\omega t\,dt. \qquad (30)$$

Equation (30) is essentially a special case of (3) obtained by taking r even, identifying $A(\omega)$ with $G_{\omega'}$, and setting $t=0$ in the first relation of (3). We now apply (30) term by term to (28), correlating ω' with $\omega - k\omega_c$ ($k = 1, 3, \cdots$); the lower limit $-\infty$ on ω' may be considered equivalent to 0 on ω, as ω_c is large and in our cases $G_{\omega'}$ is negligible except in the vicinity of $\omega'=0$. In this fashion we obtain

$$\int_0^\infty E_\omega d\omega = \frac{2}{\pi}\left[1 + \frac{1}{6} + \frac{3}{40} + \frac{5}{112}\right] = 0.82. \qquad (31)$$

Actually the integral should be unity, since our total energy is normalized to unity. Hence we have lost 18 percent of the energy by stopping the development (27) after the first four terms, as was done in obtaining (28). The situation is not much improved by carrying the development a few terms further, as the series (27) converges very slowly for $r=1$. At first sight it thus appears that our use of the series is entirely unjustified.

However, most of the lost energy is in the harmonic bands, rather than the fundamental in which we are interested. If we consider only the latter, we insert factors 1, 3/4, 5/8, 35/64 corresponding to the coefficients of $\cos(\omega-\omega_c)t$ in (28), and so we have for the total energy in the fundamental, if four terms of the series are used

$$\frac{2}{\pi}\left[1+\frac{1}{6}\left(\frac{3}{4}\right)+\frac{3}{40}\left(\frac{5}{8}\right)+\frac{5}{112}\left(\frac{35}{64}\right)\right]=0.76. \tag{32}$$

The true value is $8/\pi^2=0.81$, as explained in the next paragraph. Hence the error, obtained by using the series, is about 6 percent if four terms are employed, and hence is not too serious.[26] Most of the energy lost by neglecting the higher terms is in the region well removed from the interval $|\lambda|<1$ covered by the noise band before clipping, and is of only minor interest since it will be well outside the barrage.

The theorem (31) also furnished a proof of our statement in Section I that the distribution in energy between the various harmonics and the main band is just the same as though we clipped an ordinary monochromatic sinusoidal wave of angular frequency ω_c, instead of a noise centering around ω_c. Namely, the expansion (30), carried out arbitrarily far, gives an expression of the form

$$E_\omega=\frac{2}{\pi}\int_0^\infty\{a_1g_1(t)\cos(\omega-\omega_c)t+a_3g_3(t)\cos(\omega-3\omega_c)t+a_5g_5(t)\cos(\omega-5\omega_c)t+\cdots\}dt, \tag{33}$$

where the a_n are numerical factors and g_n are functions of the time which reduce to unity equally well in the limit $t=0$ and in the limit $\omega_a=0$. In virtue of the theorem (30), the amount of energy in the main band is a_1, and the amounts in the various successive harmonics are respectively $a_3, a_5, \cdots$. The coefficients $a_1, a_3, a_5, \cdots$, are the same as those in the expansion of $(2/\pi)\sin^{-1}[\cos\omega_c t]=(2/\pi)[\frac{1}{2}\pi-|\omega_c t|]$ in a Fourier series $\sum_n a_n \cos n\omega_c t$, viz: $a_1=8/\pi^2$, $a_3=(8/9)\pi^2$, $a_i=(8/25)\pi^2, \cdots$. These are the same as half the squares of the coefficients of the expansion of a square wave. The square (of the coefficients in the square wave development) corresponds to the fact that we are dealing with the energy, and hence with the square of the amplitude of the disturbance itself, and the factor $\frac{1}{2}$ with the fact that the mean value of the square of the cosine is $\frac{1}{2}$.

[26] This is in good agreement with Price's more precise results [5] for Case (A). He shows that 10 percent of the power located about the original spectral zone after clipping lies outside the original (rectangular) spectral region. Our calculations give 31 percent −19 percent = 12 percent, cf., abstracted calculation from (29 I), where the 19 percent represents losses in the other harmonic zones, not of interest here. The results for the Gaussian and optical spectra are somewhat more accurate.

Spectrum for Originally Rectangular Distribution Subject to Clipping of Type (B).

We now shall examine Case (B), where a carrier is subject to modulation by clipped noise. If the latter had a uniform distribution before clipping, then by (2) and (21), the formula for the spectrum after clipping is

$$E_\omega=\frac{4}{\pi^2}\int_0^\infty\cos\omega t\cos\omega_c t\sin^{-1}\left[\frac{\sin\omega_a t}{\omega_a t}\right]dt. \tag{34}$$

If we apply the series expansion (27) we obtain an expression similar to (28) except that all the factors [ ] of (28) are replaced by $\cos(\omega-\omega_c)t$. In consequence no energy is dissipated in harmonics of the carrier, as already mentioned in Part I. Integration of this series expansion gives for the various zones

Zone I: $(|\lambda|<1)$

$$E_\omega=(1/\pi\omega_a)[1.1928-0.05785\lambda^2+1.579\cdot10^{-3}\lambda^4-1.94\cdot10^{-5}\lambda^6], \tag{35 I}$$

Zone II: $(1<|\lambda|<3)$

$$E_\omega=(1/\pi\omega_a)[0.2533-0.1174\lambda-0.00659\lambda^2+0.007138\lambda^3+1.334\cdot10^{-4}\lambda^4-2.035\cdot10^{-4}\lambda^5+1.454\cdot10^{-5}\lambda^6], \tag{35 II}$$

Zone III: $(3<|\lambda|<5)$

$$E_\omega=(1/\pi\omega_a)[0.13012-0.06820\lambda+0.00055\lambda^2+0.0064\lambda^3-1.637\cdot10^{-3}\lambda^4+1.63\cdot10^{-4}\lambda^5-5.814\cdot10^{-6}\lambda^6], \tag{35 III}$$

Zone IV: $(5<|\lambda|<7)$

$$E_\omega=(1/\pi\omega_a)[0.00802+0.02950\lambda-0.02875\lambda^2+0.010310\lambda^3-1.832\cdot10^{-3}\lambda^4+1.63\cdot10^{-4}\lambda^5-5.814\cdot10^{-6}\lambda^6], \tag{35 IV}$$

Zone V: $(|\lambda|>7)$

$E_\omega=0$ (unless more than four terms of the series are used). (35 V)

However, (35) is not as good an approximation as was the corresponding result (29) for clipping of Type (A), since in the present case (B) the energy in the fundamental is gauged by (31) (compared to unity) rather than (32) (compared to 0.81), and so only 82 percent of the energy is included if we stop with the first of our four terms of (27), as we have done. The number of terms required to reduce the omitted energy to only a few percent is intractably large. The harm involved in terminating the development with a few terms is, however, not really as bad as appears from our criterion of missing energy, if one is interested only in the energy distribution in the interval $|\lambda|<1$ corresponding to the width of the noise band before clipping, for most of the omitted energy is located outside the band. In other words, the percentage error made by using the series

(27) is least for $\lambda=0$, and becomes progressively larger as one goes to higher values of λ. Usually one is interested primarily in the interval $|\lambda|<1$, as the energy outside it is too sparse to be of much value for jamming purposes. The value of E_ω calculated from (35) for $\lambda=0$ is about 6 percent too low, as is found by comparison with the more exact method given below.

Since the convergence of the series is only mediocre for $\lambda=0$, and is definitely unsatisfactory when λ becomes large compared with unity, it is advisable to find some alternative method for handling the integral (34). Another procedure is as follows. (This scheme, on the other hand, will not work for Case (A), and so it is fortunate that the convergence of the series was better there.) The idea is to split the integral up into two parts, one from 0 to some value t' of t, and the other from t' to ∞. Let the corresponding contributions to E_ω be denoted by E_{ω_1}, E_{ω_2}. We can take t' so large that $x=r(t)$ is small compared to unity in region II, and hence the convergence of (27) is good here. Hence one term or at most two terms of (27) will suffice for an adequate evaluation of E_{ω_2}. With one term, for instance, we have

$$E_{\omega_2} = \frac{2}{\pi^2}\int_{t'}^{\infty} \cos(\omega-\omega_c)t\left[\frac{\sin\omega_a t}{\omega_a t}\right]dt$$

$$= \frac{-1}{\pi^2\omega_a}\left[\mathrm{Si}(1+\lambda)\omega_a t' + \mathrm{Si}(1-\lambda)\omega_a t'\right],$$

where

$$\mathrm{Si}(x) = -\,\mathrm{Si}(-x)$$

$$= -\int_x^{\infty}\frac{\sin\theta}{\theta}d\theta, \quad (x>0). \qquad \lambda \equiv \frac{\omega-\omega_c}{\omega_a}.$$

On the other hand, for the portion E_{ω_1}, where $r(t)$ is nearly unity, we do not use the series (27) at all, and instead develop arc sin $r(r=\sin\omega_a t/\omega_a t)$ as a Taylor's series in t. This is possible, for although $d(\sin^{-1}r)/dr$ has an infinite derivative at $r=1$, on the other hand dr/dt vanishes at $r=1$, and it turns out that $d(\sin^{-1}r)/dt$ has a finite value at $t=0$. One thus finds that

$$R(t) = (2/\pi)\cos\omega_c t\sin^{-1}r$$

$$= \cos\omega_c t\left[1 + 2\pi^{-1}\sqrt{3}(-\theta/3+\theta^3/270\right.$$

$$+ 7.93\cdot 10^{-5}\theta^5 + 8.7\cdot 10^{-7}\theta^7$$

$$\left.- 2.73\cdot 10^{-8}\theta^9 + \cdots)\right]. \quad (\theta=\omega_a t) \qquad (36)$$

With this approximation for $R(t)$ one may immediately evaluate in an elementary manner

$$\int_0^{t'}\cos\omega t R(t)dt.$$

For small values of $\lambda=(\omega-\omega_c)/\omega_a$ it is easiest also to expand cos $(\omega-\omega_c)t$ in a series before integrating, as the exact formula for

$$\int_0^{t'} t^n\cos\omega' t dt$$

involves a good many terms which nearly cancel if $\omega' t'$ is small, and so its use makes numerical accuracy difficult. Figure 5 has been constructed by using partly the calculations of $E_{\omega_1}+E_{\omega_2}$ just described, and partly the expression (35) with corrections extrapolated from comparison with $E_{\omega_1}+E_{\omega_2}$ at the values of λ at which this sum has been computed.

It is to be noted that after clipping the discontinuity at $\lambda=1$ in both Figs. 4 and 5 is $2/\pi$ times the discontinuity at this point before clipping. This result is exact, as in the integral (28) or (34) all the discontinuity in E_ω comes from the first term of the development (27). Higher terms give only discontinuities in derivatives. In consequence, there are no discontinuities in E_ω at the zone boundaries II-III, III-IV, or IV-V in (29) or (35).

Analogous methods can be used to compute the integrals for E_ω when the correlation function has the forms (22)–(25) appropriate to the Gaussian or to the optical distribution before clipping. The results are shown in Figs. 7–10.

IV. Clipping at an Arbitrary Level

We shall now develop the somewhat intricate, but rather interesting, mathematical theory for the case in which the clipping is of the ordinary rather than extreme type, i.e., of the form shown in Fig. 1 rather than Fig. 2. We shall assume that the clipper limits the amplitude to a fraction b of its rms value before clipping, which we shall take as unity. Then in (10) we have

$$f(X) = X, \ (|X|<b), \qquad f(X) = b, \ (X>b),$$
$$f(X) = -b, \ (X<-b). \qquad (37)$$

The linearity of f for arguments of modulus less than b means that the disturbance is assumed unaffected by the clipping instrument unless it reaches the critical value b. With (37), the integral (10) cannot be conveniently evaluated by using polar coordinates, as was done in Section III for extreme clipping. Instead there is an artifice which has been introduced by Bennett [10], Rice [1], and others [2], [3], and which consists in representing the function f defined by (37) as a contour integral.[27] Namely, $f(X)$ can be expressed as

$$f(X) = \frac{1}{2\pi}\int_C \frac{e^{i(X-b)z}-e^{iXz}}{z^2}dz$$
$$-\frac{1}{2\pi}\int_{C'}\frac{e^{i(X+b)z}-e^{iXz}}{z^2}dz, \qquad (38)$$

where C is a path of integration extending along the real axis from $-\infty$ to $+\infty$ except that it is indented downwards near the origin. The path C' is similar to C but

[27] See [7], Secs. 2.3, 5.1, chapters 12, 13 and references therein cited, for a full account, with many illustrations of the method.

is indented upwards near the origin. To prove the legitimacy of the representation (38) for (37) we utilize the fact that by the residue theorem we have

$$\int z^{-2}e^{ikz}dz = -2\pi k$$

for a closed counterclockwise path including the origin, while

$$\int z^{-2}e^{ikz}dz = 0$$

for a closed path not including the origin. The paths C and C' may be closed without affecting the value of the integral by introducing a semicircle of infinitely large radius in the upper or lower half plane, depending on the sign of the exponent.

Now quite generally if

$$f_1(X) = \int_D g(z)e^{iXz}dz, \qquad f_2(Y) = \int_{D'} g(\eta)e^{iY\eta}d\eta$$

where D and D' are arbitrary contours we have

$$(1/2\pi)(1-r^2)^{-\frac{1}{2}}\int_{-\infty}^{+\infty}\int_{-\infty}^{+\infty} f_1(X)f_2(Y) \cdot e^{-(X^2+Y^2-2rXY)/2(1-r^2)}dXdY$$
$$= \int_D\int_{D'} g(z)g(\eta)e^{-(z^2+\eta^2+2rz\eta)/2}dzd\eta$$

as is readily established by interchanging the order of integration. Hence

$$R(t) = \frac{1}{4\pi^2}\Bigg[\int_C \frac{(e^{-ibz}-1)}{z^2}\int_C \frac{(e^{-ib\eta}-1)}{\eta^2}e^{\beta(z,\eta)}dzd\eta$$
$$+\int_{C'} \frac{(1-e^{+ibz})}{z^2}\int_C \frac{(e^{-ib\eta}-1)}{\eta^2}e^{\beta(z,\eta)}dzd\eta$$
$$+\int_C \frac{(e^{-ibz}-1)}{z^2}\int_{C'} \frac{(1-e^{+ib\eta})}{\eta^2}e^{\beta(z\eta)}dzd\eta$$
$$+\int_{C'} \frac{(1-e^{+ibz})}{z^2}\int_{C'} \frac{(1-e^{+ib\eta})}{\eta^2}e^{\beta(z,\eta)}dzd\eta\Bigg] \tag{39}$$

where

$$\beta(z,\eta) = -(z^2+\eta^2+2rz\eta)/2.$$

The expression for the correlation function thus takes a simpler form in the $z\eta$ than in the X, Y space, at least as regards the dependence on r. For the case of extreme clipping, the use of the X, Y space was, to be sure, more straightforward, as the integration went through most simply in polar coordinates.

For clipping at an arbitrary level, however, the polar coordinate scheme ceases to be feasible, and we resort to a series integration which is most simply performed in z, η space. If we develop (39) with respect to r, it is readily seen that no even powers of r are involved, so that

$$R(t) = a_1r + a_3r^3 + \cdots.$$

On working out the coefficients explicitly, we find

$$R(t) = \frac{1}{\pi^2}\Bigg[r\left\{\int_{-\infty}^{+\infty}\frac{\sin bz}{z}e^{-z^2/2}dz\right\}^2$$
$$+\frac{r^3}{6}\left\{\int_{-\infty}^{+\infty} z\sin bze^{-z^2/2}dz\right\}^2$$
$$+\frac{r^5}{120}\left\{\int_{-\infty}^{+\infty} z^3\sin bze^{-z^2/2}dz\right\}^2+\cdots. \tag{40}$$

Here we can omit the indentures near the origin, as each integrand in (40) is finite at $z=0$. Now

$$\frac{1}{\pi}\int_{-\infty}^{+\infty}\frac{\sin bz}{z}e^{-z^2/2}dz$$
$$=\frac{1}{\pi}\int_0^b\int_{-\infty}^{+\infty}\cos bze^{-z^2/2}dzdb$$
$$=\int_0^b\left(\frac{2}{\pi}\right)^{\frac{1}{2}}e^{-b^2/2}db = \Theta(b/\sqrt{2}),$$

where $\Theta(y)$ is the error function[28]

$$(2/\pi^{\frac{1}{2}})\int_0^y e^{-y^2}dy.$$

Also when n is an odd positive integer other than unity we have

$$\int_{-\infty}^{+\infty} z^{n-2}\sin bze^{-z^2/2}dz$$
$$=(-1)^{\frac{1}{2}(n-1)}\frac{d^{n-2}}{db^{n-2}}\int_{-\infty}^{+\infty}\cos bze^{-z^2/2}dz$$
$$=(-1)^{\frac{1}{2}(n-1)}(2\pi)^{\frac{1}{2}}\frac{d^{n-2}}{db^{n-2}}e^{-b^2/2}$$
$$=(-1)^{\frac{1}{2}(n+1)}(2\pi)^{\frac{1}{2}}e^{-b^2/2}H_{n-2}(b),$$

where $H_n(y)$ is the Hermitian polynomial[28]

$$e^{y^2/2}(-1)^n\frac{d^n}{dy^n}(e^{-y^2/2}).$$

Hence

$$R(t) = r\{\Theta(b/\sqrt{2})\}^2 + \sum_{n=3,5,\cdots}^{\infty}\frac{r^n}{n!}[H_{n-2}(b)e^{-b^2/2}]^2$$
$$= r\{\Theta(b/\sqrt{2})\}^2 + \frac{2}{\pi}\Bigg[\frac{r^3b^2e^{-b^2}}{6}+\frac{r^5}{120}(b^3-3b)^2e^{-b^2}$$
$$+\frac{r^7}{5040}(b^5-10b^3+15b)^2e^{-b^2}+\cdots\Bigg]. \tag{41}$$

[28] See [7], Appendix A.1.1, pp. 1071–1073.

As a check on the work it may be noted that (41) reduces properly in limiting cases; viz: for $b=\infty$ it reduces to r, while for b very small it becomes

$$R(t) = \frac{2b^2}{\pi}\left[r + \frac{r^3}{6} + \frac{3r^5}{40} + \frac{5r^7}{112} + \cdots\right]$$

$$= \frac{2}{\pi} b^2 \sin^{-1} r. \tag{42}$$

The result (42) agrees with (17) except that (42) contains an extra factor b^2. This is because in obtaining (41) it was assumed that the mean-square amplitude was normalized to unity before clipping, whereas in (17) it was supposed normalized to unity after clipping. Obviously, clipping will reduce the mean-square amplitude, unless a new normalization is made. If, in general, we wish to renormalize the mean-square amplitude to unity after clipping, we must divide (41) by a constant factor $R(0)$, where

$$R(0) = b^2 - (2/\pi)^{\frac{1}{2}} b e^{-b^2/2} + (1 - b^2)\Theta(b/\sqrt{2}). \tag{43}$$

This division removes the factor b^2 in (42) for the limit of small b.

The integral

$$E_\omega = (2/\pi)\int_0^\infty R(t)\cos \omega t dt$$

may now be evaluated in series by the same general method as described in Section III, and so we shall not give numerical details here. As explained in Sections I and III, the situation is different for Cases (A) and (B), as in the former harmonics of the central frequency appear. In consequence, the convergence in (A) is improved when only the energy in the fundamental band is considered, and is then reasonably satisfactory. The convergence is only mediocre for (B). However, the curves for (B) can be corrected by extrapolation from the case of extreme clipping, where an alternative method was available which was described at the end of Section III, and which was more accurate than the power series in r. (A similar alternative method could probably be developed in the present case, but does not warrant the labor.) Curves based on (41) have already been incorporated in Section I. The important thing is the value of b at which the deviations from the unclipped spectrum become appreciable. This clearly takes place when b becomes comparable with unity. The effect of clipping will hence fade out very rapidly when the limiting amplitude exceeds the rms noise level and, even when the two are equal, the effect is small.

If the tube conducts only for positive values of the grid potential, then in Case (B) a bias of b is necessary if it is to amplify the noise disturbance throughout the amplitude range $-b$ to $+b$. The peak power, the power in the noise sidebands, and the power in the carrier are then proportional, respectively, to $4b^2$, $R(0)$, and b^2 where $R(0)$ is given by (43). The ratio ρ of carrier to sideband power is thus $b^2/R(0)$. In Case (A) there is no carrier, and the ratio of the peak power to the rms noise power is $b^2/R(0)$, whereas in (B) the ratio of the peak power to mean noise sideband power is $4b^2/R(0)$.

Appendix

Evaluation of a Definite Integral

An important definite integral needed in the calculation of the spectrum of superclipped normal noise is

$$\Phi_n(\lambda \mid a, b) \equiv \frac{2}{\pi}\int_0^\infty \left(\frac{\sin\theta}{\theta}\right)^n \cos\lambda\theta\, d\theta$$

$$= \Phi(-\lambda \mid a, b), \tag{44}$$

where (a, b) denotes various ranges of values of $\lambda(=(\omega-\omega_c)/\omega_a)$, a normalized angular frequency, and n is a positive integer. The intervals, or "zones" (a, b), are the various regions throughout which Φ_n assumes different analytic forms. In all cases (except for $n=0$ and 1), Φ_n is continuous over the boundaries of adjacent zones.

Several methods of evaluating (44) suggest themselves: we may use contour integration, noting that

$$\Phi_n = \frac{1}{\pi}\int_C \left(\frac{e^{iz} - e^{-iz}}{2iz}\right)^n e^{i\lambda z}\, dz, \tag{45}$$

where C extends along the real axis, with a downward indentation about the origin. Or we may employ a somewhat more direct approach, using integration by parts. Let us consider the latter: we begin first by writing

$$\Phi_n = \lim_{\epsilon\to 0}\int_\epsilon^\infty \frac{2}{\pi}\left(\frac{\sin\theta}{\theta}\right)^n \cos\lambda\theta\, d\theta \tag{46}$$

and examining the case when n is even, observing that

$$\sin^n\theta = 2^{1-n}\left\{-{}_nC_{n/2}/2 + \sum_{j=0}^{n/2}(-1)^{j+\frac{1}{2}n}{}_nC_j\cos(n-2j)\theta\right\},$$

$$n = 0, 2, \cdots, \tag{47}$$

where ${}_nC_j$ is the usual binomial coefficient $n!/(n-j)!j!$. Integrating by parts, we find that

$$\int_\epsilon^\infty \frac{\cos\lambda\theta}{\theta^n}\, d\theta = \frac{(-1)^{n/2}\lambda^{n-1}}{(n-1)!}\frac{\pi}{2}1_\lambda + \mathcal{R}(\epsilon),$$

$$n = 2, 4, \cdots, \tag{48}$$

where $\mathcal{R}(\epsilon)$ is the (divergent) remainder as $\epsilon\to 0$. The symbol 1_λ is a discontinuity factor which has the value unity when $\lambda>0$ and -1 when $\lambda<0$. Substituting (47) into (46) we see that the *sum* of all the remainders $\mathcal{R}(\epsilon)$ vanishes (for $\epsilon\to 0$), leaving the expected finite result. We obtain finally

$$\Phi_n(\lambda \mid a, b) = \frac{1}{2^n(n-1)!}\Big\{(-1)^{\frac{1}{2}(n+2)}{}_nC_{n/2}\lambda^{n-1}1_\lambda + \sum_{j=0}^{n/2}(-1)^j{}_nC_j\big[(n-2j+\lambda)^{n-1}1_{n-2j+\lambda} + (n-2j-\lambda)^{n-1}1_{n-2j-\lambda}\big]\Big\},$$

$$n = 2, 4, \cdots, \quad (49)$$

When n is odd we observe in a similar way that

$$\sin^n \theta = 2^{1-n}\sum_{j=0}^{(n-1)/2}(-1)^{\frac{1}{2}(n-1)+j}{}_nC_j \sin(n-2j)\theta,$$

$$n = 1, 3, 5, \cdots. \quad (50)$$

We obtain, as before,

$$\int_\epsilon^\infty \frac{\sin \lambda\theta}{\theta^n}\,d\theta = \frac{(-1)^{\frac{1}{2}(n-1)}}{(n-1)!}\,\frac{\pi}{2}\cdot\lambda^{n-1}1_\lambda + \mathcal{R}(\epsilon),$$

$$n = 1, 3, 5, \cdots, \quad (51)$$

on integration by parts, where again $\lim_{\epsilon\to 0}\mathcal{R}(\epsilon)$ is the (divergent) remainder at the lower limit. Applying (50), (51) to (46) now yields

$$\Phi_n(\lambda \mid a, b) = \frac{1}{2^n(n-1)!}\sum_{j=0}^{\frac{1}{2}(n-1)}(-1)^j{}_nC_j \cdot\big\{(n-2j+\lambda)^{n-1}1_{n-2j+\lambda} + (n-2j-\lambda)^{n-1}1_{n-2j-\lambda}\big\},$$

$$n = 1, 3, 5, \cdots. \quad (52)$$

As in the case of even n, the divergent remainders (as $\epsilon \to 0$) all *sum* to zero at $\theta = 0$.

A number of recurrence relations may also be deduced on differentiation with respect to λ. We have, for example,

$$\lambda\Phi_n'(\lambda) = (n-1)\Phi_n(\lambda) - \frac{n}{2}\Phi_{n-1}(\lambda+1) - \frac{n}{2}\Phi_{n-1}(\lambda-1), \quad n \geq 1; \quad (53a)$$

$$\Phi_n(0) = \left(\frac{n}{(n-1)}\right)\Phi_{n-1}(1); \quad n \geq 2. \quad (53b)$$

The first nine ($n=0$ included) functions Φ_n are tabulated below, and plotted in Fig. 12 (for $n \geq 1$). We have specifically

$$\Phi_0(\lambda) = 2\delta(\lambda - 0). \quad (54)$$

Also, we have

$$\begin{cases}\Phi_1(\lambda \mid 0, 1) = 1; \\ \Phi_1(\lambda \mid 1, \infty) = 0.\end{cases} \quad (55)$$

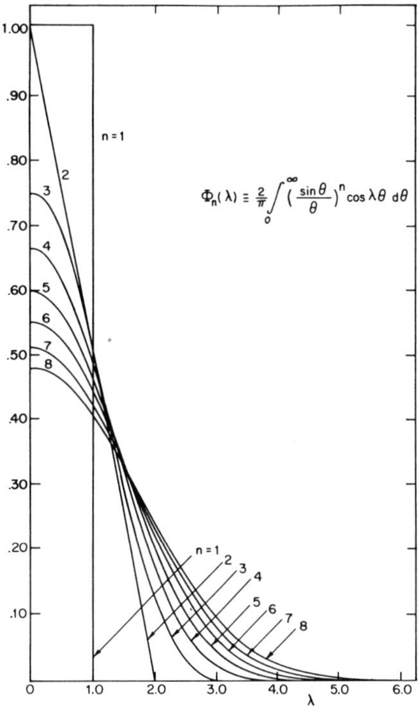

Fig. 12. The function $\Phi_n(\lambda)$ for $0 \leq \lambda$ and $1 \leq n \leq 8$.

$$\Phi_2(\lambda \mid 0, 2) = 1 - \lambda/2;$$
$$\sigma_2(\lambda \mid 2, \infty) = 0. \quad (56)$$

$$\Phi_3(\lambda \mid 0, 1) = (6 - 2\lambda^2)/8;$$
$$\Phi_3(\lambda \mid 1, 3) = (9 - 6\lambda + \lambda^2)/8;$$
$$\Phi_3(\lambda \mid 3, \infty) = 0. \quad (57)$$

$$\Phi_4(\lambda \mid 0, 2) = (32 - 12\lambda^2 + 3\lambda^3)/48;$$
$$\Phi_4(\lambda \mid 2, 4) = (64 - 48\lambda + 12\lambda^2 - \lambda^3)/48;$$
$$\Phi_4(\lambda \mid 4, \infty) = 0. \quad (58)$$

$$\Phi_5(\lambda \mid 0, 1) = (230 - 60\lambda^2 + 6\lambda^4)/384;$$
$$\Phi_5(\lambda \mid 1, 3) = (220 + 40\lambda - 120\lambda^2 + 40\lambda^3 - 4\lambda^4)/384;$$
$$\Phi_5(\lambda \mid 3, 5) = (625 - 500\lambda + 150\lambda^2 - 20\lambda^3 + \lambda^4)/384;$$
$$\Phi_5(\lambda \mid 5, \infty) = 0. \quad (59)$$

$$\Phi_6(\lambda \mid 0, 2) = (2112 - 480\lambda^2 + 60\lambda^4 - 10\lambda^5)/\Delta_6;$$
$$\Delta_6 = 3840;$$
$$\Phi_6(\lambda \mid 2, 4) = (1632 + 1200\lambda - 1680\lambda^2 + 600\lambda^3 - 90\lambda^4 + 5\lambda^5)/\Delta_6;$$

$$\Phi_6(\lambda \mid 4, 6) = (7776 - 6480\lambda + 2160\lambda^2 - 360\lambda^3 + 30\lambda^4 - \lambda^5)/\Delta_6;$$

$$\Phi_6(\lambda \mid 6, \infty) = 0. \tag{60}$$

$$\Phi_7(\lambda \mid 0, 1) = (23\,548 - 4620\lambda^2 + 420\lambda^4 - 20\lambda^6)/\Delta_7; \quad \Delta_7 = 46\,080;$$

$$\Phi_7(\lambda \mid 1, 3) = (23\,548 - 210\lambda - 4095\lambda^2 - 700\lambda^3 + 945\lambda^4 - 210\lambda^5 + 15\lambda^6)/\Delta_7;$$

$$\Phi_7(\lambda \mid 3, 5) = (8274 + 30\,408\lambda - 29\,610\lambda^2 + 10\,640\lambda^3 - 1890\lambda^4 + 168\lambda^5 - 6\lambda^6)/\Delta_7;$$

$$\Phi_7(\lambda \mid 5, 7) = (117\,649 - 100\,842\lambda + 36\,015\lambda^2 - 6860\lambda^3 + 735\lambda^4 - 42\lambda^5 + \lambda^6)/\Delta_7;$$

$$\Phi_7(\lambda \mid 7, \infty) = 0. \tag{61}$$

$$\Phi_8(\lambda \mid 0, 2) = (309\,248 - 53\,760\lambda^2 + 4480\lambda^4 - 280\lambda^6 + 35\lambda^7)/\Delta_8; \quad \Delta_8 = 645\,120;$$

$$\Phi_8(\lambda \mid 2, 4) = (316\,416 - 25\,088\lambda - 16\,128\lambda^2 - 31\,360\lambda^3 + 20\,160\lambda^4 - 4704\lambda^5 + 504\lambda^6 - 21\lambda^7)/\Delta_8;$$

$$\Phi_8(\lambda \mid 4, 6) = (-142\,336 + 777\,728\lambda - 618\,240\lambda^2 + 219\,520\lambda^3 - 42\,560\lambda^4 + 4704\lambda^5 - 280\lambda^6 + 7\lambda^7)/\Delta_8;$$

$$\Phi_8(\lambda \mid 6, 8) = (2\,097\,152 - 1\,835\,008\lambda + 688\,128\lambda^2 - 143\,360\lambda^3 + 17\,920\lambda^4 - 1344\lambda^5 + 56\lambda^6 - \lambda^7)/\Delta_8;$$

$$\Phi_8(\lambda \mid 8, \infty) = 0. \tag{62}$$

References

[1] S. O. Rice, at that time (1943) unpublished notes, which appeared later as "Mathematical analysis of random noise," *Bell Sys. Tech. J.*, vol. 23, p. 282, July 1944, vol. 24, p. 46, January 1945. See especially Part IV.

[2] D. Middleton, "The response of biased, saturated linear and quadratic rectifiers to random noise," *J. Appl. Phys.*, vol. 17, p. 778, 1946, cf., in particular, Sec. VI.

[3] D. Middleton, "Some general results in the theory of noise through nonlinear devices," *Quart. Appl. Math.*, vol. 5, p. 445, 1948, cf., in particular, Eqs. (7.15).

[4] W. B. Davenport, Jr., "Signal-to-noise ratios in bandpass limiters," *J. Appl. Phys.*, vol. 24, p. 720, 1953; cf., Section III, B.

[5] R. Price, "A note on the envelope and phase-modulated components of narrow-band Gaussian noise," *IRE Trans. on Information Theory*, vol. IT-1, pp. 9–13, September 1955. Besides giving a short history of Case (A) here, this paper is particularly interesting for its numerical results and a study of the spectral behavior of clipped noise bands well away from the central frequency.

[6] J. L. Lawson and G. E. Uhlenbeck, *Threshold Signals*, vol. 24, MIT Radiation Laboratory Series, New York, McGraw-Hill, 1950. See pp. 57–59; also Sec. 12.5, pp. 354–358.

[7] D. Middleton, *An Introduction to Statistical Communication Theory*. New York: McGraw-Hill, 1960. See, in particular, Sec. 9.1-2, and pp. 405–408. For various generalizations, see also Problems (9.1), (9.3), (13.15), (13.16), and Sec. 13.4-4 for Case (B).

[8] R. Courant and D. Hilbert, *Methoden der Mathematischen Physik*. Berlin: Springer, 1931.

[9] G. W. Kenrick, "The analysis of irregular motions with applications to the energy frequency spectrum of static and of telegraph signals," *Phil. Mag.*, vol. VIII, ser. 7, pp. 176–196, January 1929.

[10] W. R. Bennett and S. O. Rice, "Note on methods of computing modulation products," *Phil. Mag.*, vol. 18, ser. 7, pp. 422–424, September 1934.

18

Reprinted with permission from *Bell Syst. Tech. J.*, **27,** 446–472 (July 1948)

Spectra of Quantized Signals

By W. R. BENNETT

1. Discussion of Problem and Results Presented

SIGNALS which are quantized both in time of occurrence and in magnitude are in fact quite old in the communications art. Printing telegraph is an outstanding example. Here, time is divided into equal divisions, and the number of magnitudes to be distinguished in any one interval is usually no more than two, corresponding to the closed or open positions of a sending switch. It is only in recent years, however, that the development of high speed electronic devices has progressed sufficiently to enable quantizing techniques to be applied to rapidly changing signals such as produced by speech, music, or television. Quantizing of time, or time division, has found application as a means of multiplexing telephone channels.[1] The method consists of connecting the different channels to the line in sequence by fast moving switches synchronized at the transmitting and receiving ends. In this way a transmission medium capable of handling a much wider band of frequencies than required for one telephone channel can be used simultaneously by a group of channels without mutual interference. The plan is the same as that used in multiplex telegraphy. The difference is that ordinary rotating machinery suffices at the relatively low speeds employed by the latter, while the high speeds needed for time division multiplex telephony can be realized only by practically inertialess electron streams. Also the widths of frequency band required for multiplex telephony are enormously greater than needed for the telegraph, and in fact have become technically feasible only with the development of wide-band radio and cable transmission systems. As far as any one channel is concerned the result is the same as in telegraphy, namely that signals are received at discrete or quantized times. In the limiting case when many channels are sent the speech voltage from one channel is practically constant during the brief switch closure and, in effect, we can send only one magnitude for each contact or quantum of time. The more familiar word "sampling" will be used here interchangeably with the rather formidable term "quantizing of time".

Quantizing the magnitude of speech signals is a fairly recent innovation. Here we do not permit a selection from a continuous range of magnitudes but only certain discrete ones. This means that the original speech signal is to be replaced by a wave constructed of quantized values selected on a minimum error basis from the discrete set available. Clearly if we assign the quantum values with sufficiently close spacing we may make the quantized wave indistinguishable by the ear from the original. The purpose of quantization of magnitudes is to suppress the effects of interference in the transmission medium. By the use of precise receiving instruments we can restore the received quanta without any effect from superposed interference provided the interference does not exceed half the difference between adjacent steps.

By combining quantization of magnitude and time, we make it possible to code the speech signals, since transmission now consists of sending one of a discrete set of magnitudes for each distinct time interval.[2,3,4,5,6,7] The maximum advantage over interference is obtained by expressing each discrete signal magnitude in binary notation in which the only symbols used are 0 and 1. The number which is written as 4 in decimal notation is then represented by 100, 8 by 1000, 16 by 10,000; etc. In general, if we have N digit positions in the binary system, we can construct 2^N different numbers. If we need no more than 2^N different discrete magnitudes for speech transmission, complete information can be sent by a sequence of N on-or-off pulses during each sampling interval. Actually a total of $2^N!$ different coding plans (sets of one-to-one correspondences between signal magnitudes and on-or-off sequences) is possible. The straightforward binary number system is taken as a representative example convenient for either theoretical discussion or practical instrumentation. We assume that absence of a pulse represents the symbol 0 and presence of a pulse represents the symbol 1. The receiver then need only distinguish between two conditions: no transmitted signal and full strength transmitted signal. By spacing the repeaters at intervals such that interference does not reach half the full strength signal at the receiver, we can transmit the signal an indefinitely great distance without any increment in distortion over that originally introduced by the quantizing itself. The latter can be made negligible by using a sufficient number of steps.

To determine the number of quantized steps required to transmit specific signals, we require a knowledge of the relation between distortion and step size. This problem is the subject of the present paper.* We divide the problem into two parts: (1) quantizing the magnitude only and (2) combined quantizing of magnitude and time. The first part can be treated by a simple model: the "staircase transducer", which is a device having the instantaneous ouput vs. input curve shown by Fig. 1. Signals impressed on the stair-

* Other features of the quantizing and coding theory are discussed in forthcoming papers by Messrs. C. E. Shannon, J. R. Pierce, and B. M. Oliver.

case transducer are sorted into voltage slices (the treads of the staircase), and all signals within plus or minus half a step of the midvalue of a slice are replaced in the output by the midvalue. The corresponding output when the input is a smoothly varying function of time is illustrated in Fig. 2. The output remains constant while the input signal rcmains within the boundaries of a tread and changes abruptly by one full step when the signal crosses the boundary. It is not within the scope of the present paper to discuss the internal mechanism of a staircase transducer, which may have many different physical embodiments. We are concerned rather with the distortion produced by such a device when operating perfectly.

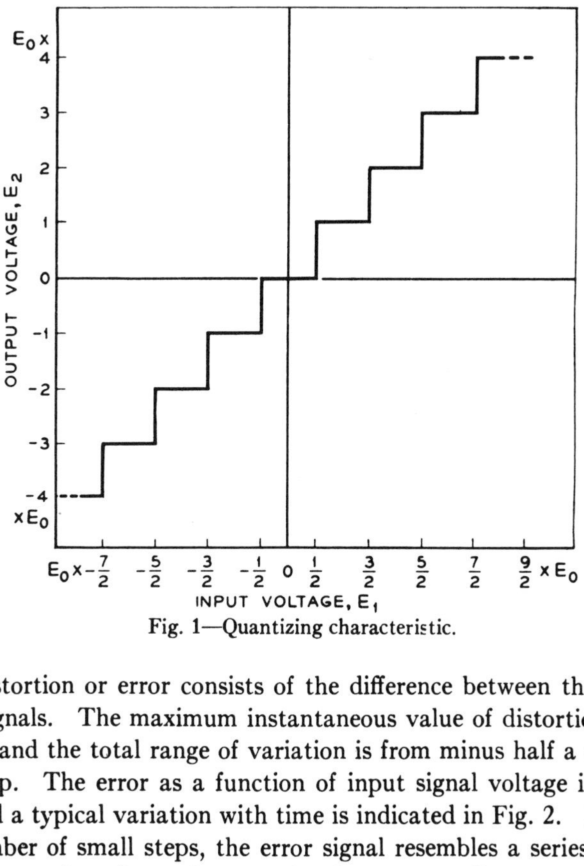

Fig. 1—Quantizing characteristic.

The distortion or error consists of the difference between the input and output signals. The maximum instantaneous value of distortion is half of one step, and the total range of variation is from minus half a step to plus half a step. The error as a function of input signal voltage is plotted in Fig. 3 and a typical variation with time is indicated in Fig. 2. If there is a large number of small steps, the error signal resembles a series of straight lines with varying slopes, but nearly always extending over the vertical interval between minus and plus half a step. The exceptional cases occur when the signal goes through a maximum or minimum within a step. The limiting condition of closely spaced steps enables us to derive quite simply an approximate value for the mean square error, which will later be shown to be sufficiently accurate in most cases of practical importance. This approximation consists of calculating the mean square value of a straight line going from minus half a step to plus half a step with arbitrary slope. If

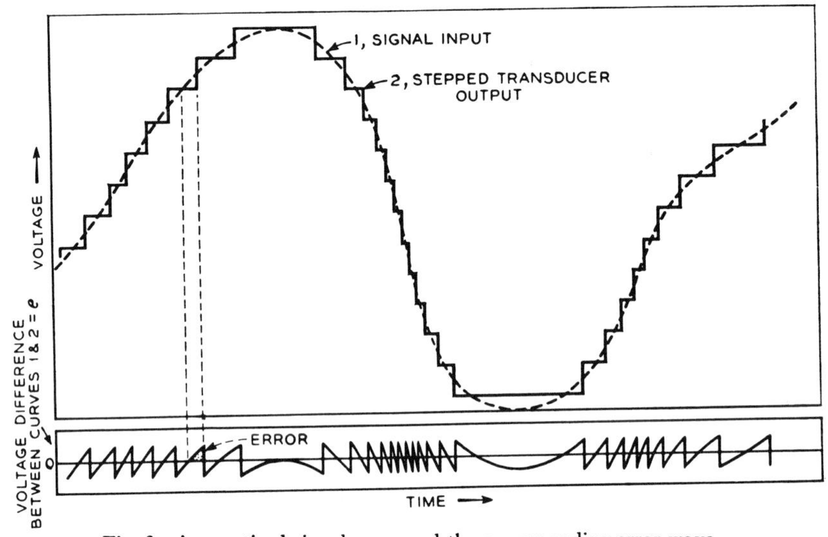

Fig. 2—A quantized signal wave and the corresponding error wave.

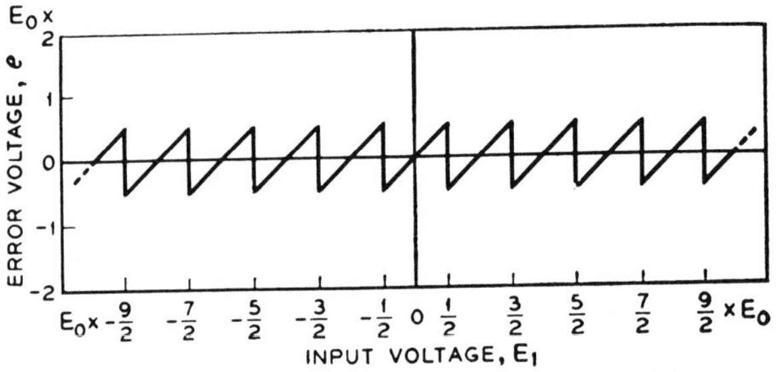

Fig. 3—Characteristic of the errors in quantizing.

E_0 is the voltage corresponding to one step, and s is the slope, the equation of the typical line is:

$$\epsilon = st, \qquad -\frac{E_0}{2s} < t < \frac{E_0}{2s} \tag{1.0}$$

where ϵ is the error voltage and t is the time referred to the midpoint as origin. Then the mean square error is

$$\overline{\epsilon^2} = \frac{s}{E_0} \int_{-E_0/2s}^{E_0/2s} \epsilon^2 \, dt = \frac{E_0^2}{12}, \tag{1.1}$$

or one twelfth the square of the step size.

Not all the distortion falls within the signal band. The distortion may be considered to result from a modulation process consisting of the application of the component frequencies of the original signal to the non-linear staircase characteristic. High order modulation products may have frequencies quite remote from those in the original signal and these can be excluded by a filter passing only the signal band. It becomes of importance, therefore, to calculate the spectrum of the error wave. This we shall do in the next section for a generalized signal using the method of correlation, which is based on the fact that the power spectrum of a wave is the Fourier cosine transform of the correlation function. The result is then applied to a particular kind of signal, namely one having energy uniformly distributed throughout a definite frequency band and with the phases of the components randomly distributed. This is a particularly convenient type of signal because it in effect averages over a large number of possible discrete frequency components within the band. Single or double-frequency signal waves are awkward for analytical purposes because of the ragged nature of the spectra produced. The amplitudes of particular harmonics or cross-products of discrete frequency components are found to oscillate violently with magnitude of input. The use of a large number of input components smooths out the irregularities.

The type of spectra obtained is shown in Fig. 4. Anticipating binary coding, we have shown results in terms of the number of binary digits used. The number of different magnitudes available are 16, 32, 64, 128, and 256 for $N = 4, 5, 6, 7$ and 8 digits, respectively. Here a word of explanation is needed with respect to the placing of the scale of quantized voltages. A signal with a continuous distribution of components along the frequency scale is theoretically capable of assuming indefinitely great values of instantaneous voltage at infrequent instants of time. An actual quantizer (staircase transducer) has a finite overload value which must not be exceeded and hence can have only a finite number of steps. This difficulty is resolved here by the experimentally observed fact that thermal noise, which has the type of spectrum we have assumed for our signal, has never been observed to exceed appreciably a voltage four times its root-mean-square value. Hence we have placed the root-mean-square value of the input signal at one-fourth the overload input to the staircase. This fixes the relation between step size and the total number of steps. In the actual calculation the number of steps is taken as infinite; the effect of the assumed additional steps beyond 2^N is negligible because of the rarity of excursion into this range.

The curves of Fig. 4 are drawn for the case in which the signal band starts at zero frequency. The original signal band width is represented by one unit on the horizontal scale. The relatively wide spread of the distortion spectrum is clearly shown. As the number of digits (or steps) is increased

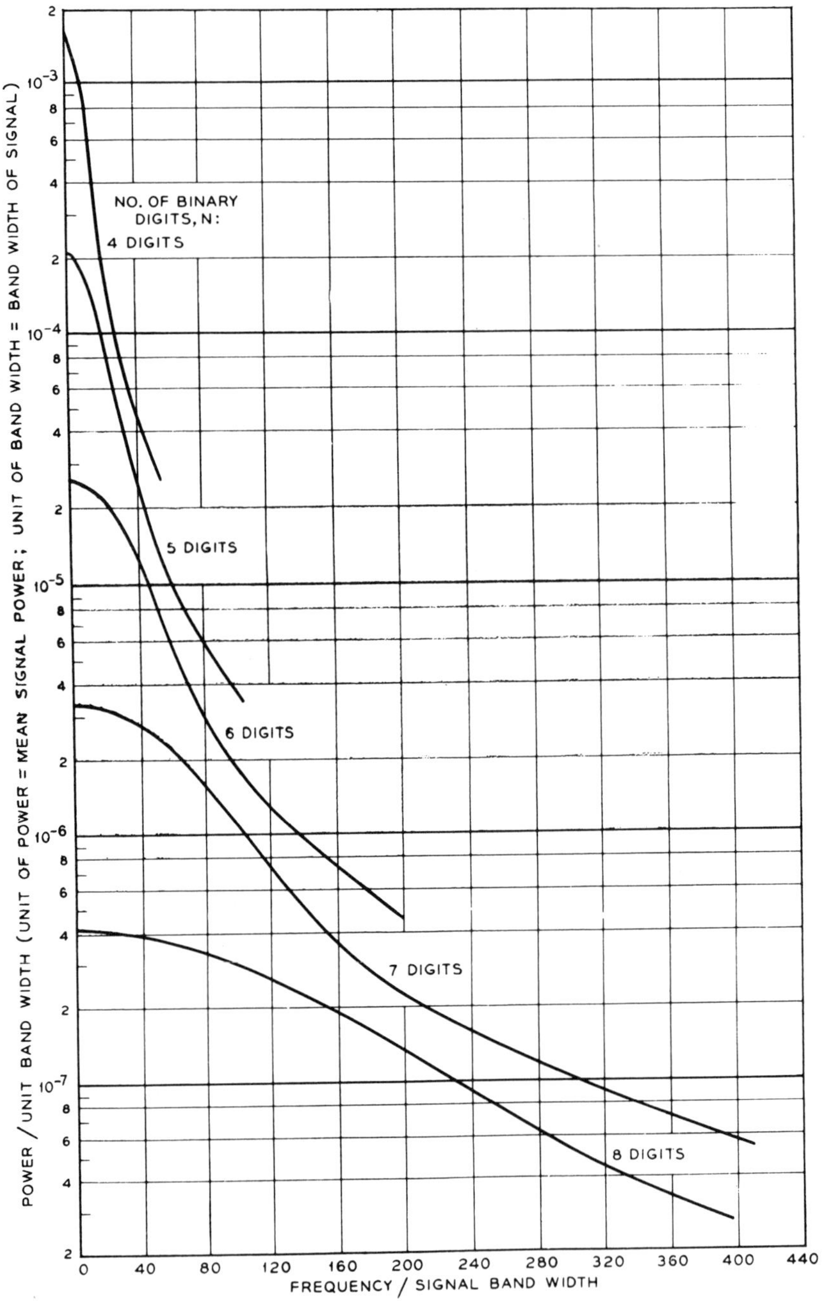

Fig. 4—Spectrum of distortion from quantizing the magnitudes of a random noise wave. Full load on the quantizer is reached by peaks 12 db above the r.m.s. value of input.

the spectrum becomes flatter over a wider range, but with a smaller maximum density. The area under each curve represents the total mean power in the corresponding error wave and is found to agree quite accurately with the approximate result of Eq. (1.1). The distortion power falling in the signal band is represented by the area included under the curve from zero to unit abscissa.

Quantizing the magnitude only is not a technically attractive method of transmission because of the wide frequency band required to preserve the discrete values of the quanta. Thus in a 128-step system, a full load sinusoidal signal passes through 64 different steps each quarter cycle and hence would require transmitting 256 successively different magnitudes during each period of the signal frequency. We therefore consider the second problem—that of sampling the quantized magnitudes.

The theory of periodic sampling of signals is a limiting case of commutator modulation theory as previously shown by the author.[1] We may think of a periodically closed switch in series with the line and source as producing a multiplication of the signal by a switching function. The switching function has a finite value during the time of switch closure and is zero at other times. It may be expanded in a Fourier series containing a term of zero frequency, the repetition frequency of switch closure, and all harmonics of the latter. Multiplication of the signal by the Fourier series representing the constant component of the switching function gives a term proportional to the signal itself. Multiplication of the signal by the fundamental component of the switching function gives upper and lower sidebands on the repetition frequency. Likewise multiplication by the harmonics gives sidebands on each harmonic. The signal is separable from the sidebands on a frequency basis if the signal band does not overlap the lower sideband on the repetition frequency. This leads to the condition for no distortion in time division: the highest signal frequency must be less than one-half the repetition frequency.

To apply the above theory to instantaneous sampling we let the duration of switch closure in one period approach zero. We then approach the condition of one signal value in each period, so that the repetition frequency now becomes the sampling frequency. Clearly the sampling frequency must slightly exceed twice the highest signal frequency. We also note that as the contact time tends toward zero, the switching function approaches a periodically repeated impulse. The important terms of the Fourier series representing the switching function accordingly become a set of harmonics of equal amplitude with a constant component equal to half the amplitude of the typical harmonic. On multiplication of this series by the signal, we get a set of sidebands of equal amplitude including the one corresponding to the original signal itself, the sideband on zero frequency.

These results may be applied to the staircase transducer. The output may be resolved into the input signal plus the error. The sampling frequency is assumed to exceed its minimum required value of twice the top signal frequency. The component of the output that is equal to the original signal can therefore be separated at the receiver by a filter passing the original signal band. A similar statement cannot be made for the error component, for it has been found to extend over a vastly greater range than the original signal. To calculate the total distortion received in the signal band, we can multiply the distortion spectrum by the switching function and sum up all sideband contributions to the original signal band. Each harmonic of the switching function makes such contributions by beating with a band of the error spectrum above and below the frequency of the harmonic. These contributions add as power when the sampling frequency is independent of the individual frequencies contained in the signal. The total error power accepted by the signal band filter decreases as the sampling frequency is increased because each harmonic of the sampling frequency is thereby pushed upward into a less dense portion of the error spectrum. In the limit as the sampling frequency is made indefinitely large, we return to the non-sampled case, that of the staircase transducer only.

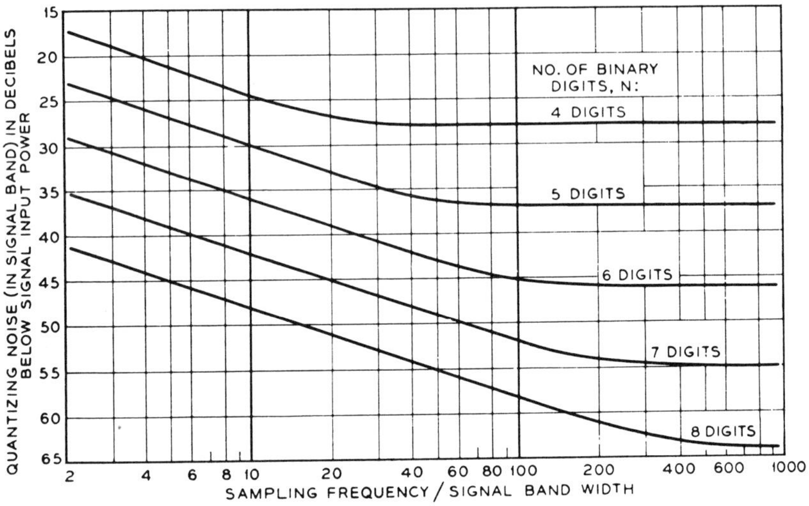

Fig. 5—Total distortion in signal band from quantizing and sampling a random noise wave. Full load on the quantizer is 12 db above the r.m.s. value of input.

Figure 5 shows the calculated curves of distortion in the signal band plotted as a function of ratio of sampling frequency to signal band width. The curves have downward slopes approaching asymptotes corresponding to the area from zero to unity under the corresponding curves of Fig. 4.

The initial points at the minimum sampling rate are determined on the other hand by the total area under the curves of Fig. 4, since the accepted sidebands on the harmonics in this case exactly fill out the entire error spectrum. These initial points are therefore given quite accurately by Eq. (1.1), which, as pointed out before, is a good approximation for the total areas. We can also give a direct demonstration of the applicability of Eq. (1.1) to the initial points of the curves of Fig. 5 by means of the following theorem:

Theorem I. The mean square value of the response of an ideal low-pass filter to a train of unit impulses multiplied by instantaneous samples occurring at double the cutoff frequency is equal to the mean square value of the samples provided no harmonic of the sampling frequency is equal to twice the frequency of one component or equal to the sum or difference of two component frequencies of the sampled signal. Proof of the theorem is given in Appendix I. To apply it here we resolve the input into two components: the true signal and the error. The former is reproduced with fidelity in the output because it contains only frequencies below half the sampling rate. The error component in the output represents the response of the low-pass filter to the error samples. Except for very special types of signals, the error samples are uniformly distributed throughout the range from minus half a step to plus half a step. Calculation of the mean square value of such a distribution gives Eq. (1.1).

We have tacitly assumed above that the sampled values applied to the filter in the output of the system are infinitesimally narrow pulses of height proportional to the samples. In actual systems it is found advantageous to hold the sampled values constant in the individual receiving channels until the next sample is received. This means that the input to the channel filter is a succession of rectangular pulses of heights proportional to the samples. The resulting magnitude of recovered signal is much larger than would be obtained if very short pulses of the same heights were used; stretching the pulses in time produces in effect an amplification. The amplification is obtained, however, at the expense of a variation of channel transmission with signal frequency. Infinitesimally short pulses have a flat frequency spectrum, while pulses of finite duration do not. The frequency characteristic introduced by lengthening the pulses is easily calculated by determining the steady state admittance function of a network which converts impulses to the actual pulses used. The general formula for this admittance when a unit impulse input is converted into an output pulse $g(t)$ is easily shown to be:

$$Y(i\omega) = f_s \int_{-\infty}^{\infty} g(t) e^{-i\omega t} \, dt \tag{1.2}$$

where f_s is the repetition frequency and ω is the angular signal frequency.

We shall call this Theorem II and give the proof in Appendix II. This relation is similar to that found in television and telephotography for the "aperture effect", or variation of transmission with frequency caused by the finite size of the scanning aperture. The pulse shape $g(t)$ is analogous to a variation in aperture height $g(x)$, where x is distance along the line of scanning. Hence it has become customary to use the term "aperture effect" in the theory of restoring signals from samples. The aperture effect associated with rectangular pulses lasting from one sample to the next amounts to an amplitude reduction of $\pi/2$ or 3.9 db at the top signal frequency (one half the sampling rate) compared to a signal of zero frequency. There is also a constant delay introduced equal to half the sampling period. The latter does not cause any distortion and the amplitude effect can be corrected by properly designed equalizing networks.

The fact that many pulse spectra can be simply expressed in terms of a flat spectrum associated with sharp pulses and an aperture effect caused by the particular shape of pulse used does not appear to have been recognized in the recent literature, although applications were made by Nyquist in a fundamental paper[8] of 1928. Premature introduction of a specific finite pulse not only complicates the work, but also restricts the generality of the results.

Distortion caused by quantizing errors produces much the same sort of effects as an independent source of noise. The reason for this is that the spectrum of the distortion in the receiving filter output is practically independent of that of the signal over a wide range of signal magnitudes. Even when the signal is weak so that only a few quantizing steps are operated, there is usually enough residual noise on actual systems to determine the quantizing noise and mask the relation between it and the signal. Eq. (1.1) yields a simple rule enabling one to estimate the magnitude of the quantizing noise with respect to a full load sine wave test tone. Let the full load test tone have peak voltage E; its mean square value is then $E^2/2$. The total range of the quantizer must be $2E$ because the test signal swings between $-E$ and $+E$. The ratio $2E/E_0 = r$ is a convenient one to use in specifying the quantizing; it is the ratio of the total voltage range to the range occupied by one step. The ratio of mean square signal to mean square quantizing noise voltage is

$$\frac{E^2/2}{E_0^2/12} = \frac{6E^2}{4E^2/r^2} = \frac{3r^2}{2} \tag{1.3}$$

Actual systems fail to reproduce the full band $f_s/2$ because of the finite frequency range needed for transition from pass-band to cutoff. If we introduce a factor κ to represent the ratio of equivalent rectangular noise band to $f_s/2$, the actual received noise power is multiplied by κ. Then the signal-to-noise ratio in db for a full load test tone is

$$D = 10 \log_{10} \frac{3r^2}{2\kappa} \text{ db} \tag{1.4}$$

In practical applications the value of κ is about 3/4 which gives the convenient rule:

$$D = 20 \log_{10} r + 3 \text{ db} \tag{1.5}$$

In other words, we add 3 db to the ratio expressed in db of peak-to-peak quantizing range to the range occupied by one step. For various numbers of binary digits the values of D are:

TABLE I

Number of Digits	D
3	21
4	27
5	33
6	39
7	45
8	51

From Table I we can make a quick estimate of the number of digits required for a particular signal transmission system provided that we have some idea of the required signal-to-noise ratio for a full load test tone. The latter ratio may be expressed in terms of the full load test tone which the system is required to handle and the maximum permissible unweighted noise power at the same level point. Since quantizing noise is uniformly distributed throughout the signal band, its interfering effect on speech or other program material is probably similar to that of thermal noise with the same mean power. Requirements given in terms of noise meter readings must be corrected by the proper weighting factor before applying the table. If the signal transmitted is itself a multiplex signal with channels allotted on a frequency division basis, the noise power falling in each channel is the same fraction of the total noise power as the band width occupied by the signal is of the total band width of the system.

We have thus far considered only the case in which the quantized steps are equal. In actual systems designed for transmission of speech it is found advantageous to taper the steps in such a way that finer divisions are available for weak signals. For a given number of total steps this means that coarser quantization applies near the peaks of large signals, but the larger absolute errors are tolerable here because they are small relative to the bigger signal values. Tapered quantizing is equivalent to inserting complementary nonlinear transducers in the signal branch before and after the quantizer. In the usual case, the transducer ahead of the quantizer is of the "compressing" type in which the loss increases as the signal increases. If the full load signal just covers all the linear quantizing steps, a weak signal gets a bigger share of the steps than it would if the transducer were linear. The transducer after the quantizer must be of the "expanding" type which gives decreased loss to the large signals to make the overall combination linear.

On the basis of the theory so far discussed, we can say that the error spectrum out of the linear quantizer is virtually the same whether or not the signal input is compressed. The operation of the expandor then magnifies the errors produced when the signal is large. When weak signals are applied, the mean square error is given by Eq. (1.1), as before, but when the signal is increased an increment in noise occurs. The mean square value of noise voltage under load may be computed from the probability density of the signal values and the output-vs-input characteristic of the expandor, or its inverse, the compressor. A first order approximation, valid when the steps are not too far apart, replaces (1.1) by:

$$\overline{\epsilon^2} = \frac{E_0^2}{12} \int_{Q_2}^{Q_1} \frac{p_1(E_1)\, dE_1}{[F'(E_1)]^2} \tag{1.6}$$

where Q_1 and Q_2 are the minimum and maximum values of the input signal voltage E_1, $p_1(E_1)$ is the probability density function of the input voltage, and $F'(E_1)$ is the slope of $F(E_1)$, the compression characteristic.

Some experimental results obtained with a laboratory model of a quantizer are given in Figs. 6–9. Figs. 6–7 show measurements on the third harmonic associated with 6-digit quantizing. As mentioned before, the amplitude of any one harmonic oscillates with load. The calculated curves shown were obtained by straightforward Fourier analysis. In the measurements it was convenient to spot only the successive nulls and peaks.

In Fig. 6 the bias was set to correspond to the stair-case curve of Fig. 1, while in Fig. 7 the origin is moved to the point $(E_0/2, E_0/2)$, i.e., to the middle of a riser instead of a tread. The peaks of ratio of harmonic to fundamental decrease steadily as the amplitude of the signal is increased to full load, which is just opposite to the usual behavior of a communication system. It is difficult to extrapolate experience with other systems to specify quality in terms of this type of harmonic distortion.

Figure 8 shows measurements of the total distortion power falling in the signal band when the signal is itself a flat band of thermal noise. The technique of making such measurements has been described in earlier articles.[9,10] Measurements are shown for quantizing with both equal and tapered steps. The particular taper used is indicated by the expandor characteristic of Fig. 9. The compression curve is found by interchanging horizontal and vertical scales. The measurements were made on a quantizer with 32, 64, and 128 steps, and a sampling rate of 8,000 cycles per sec-

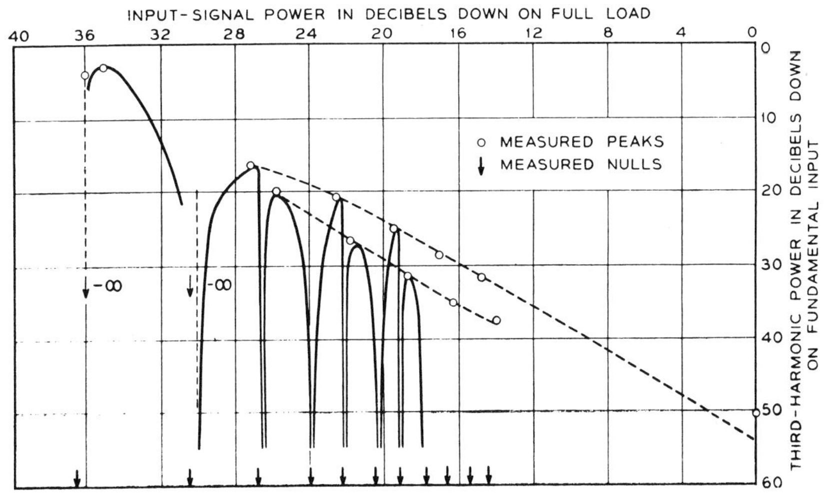

Fig. 6—Third harmonic in 64-step quantized output with bias at mid tread. The smooth curves represent computed values.

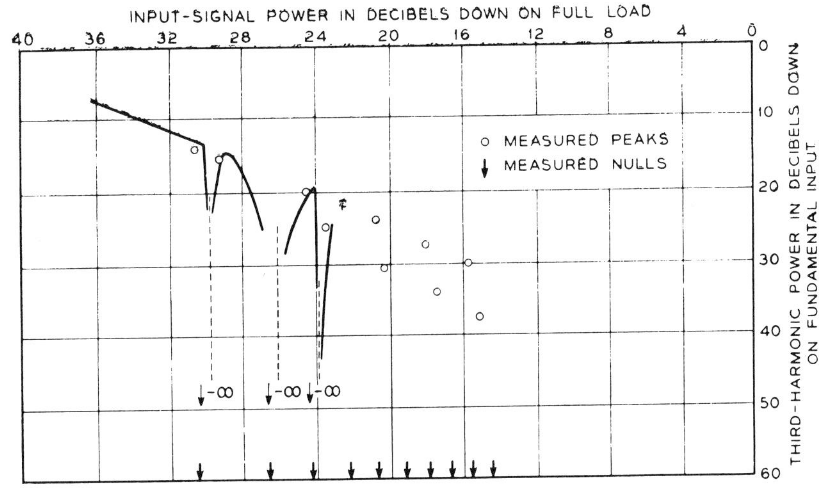

Fig. 7—Third harmonic in 64-step quantized output with bias at mid-riser. The smooth curves represent computed values.

ond. The applied signal was confined to a range below 4,000 cycles per second. With equal steps the distortion power is practically independent of load as shown by the db-for-db straight lines. With tapered steps, the distortion is less for weak signals, and only slightly greater for large signals.

The vertical line designated "full load random noise input" represents the value of noise signal power at which peaks begin to exceed the quantizing

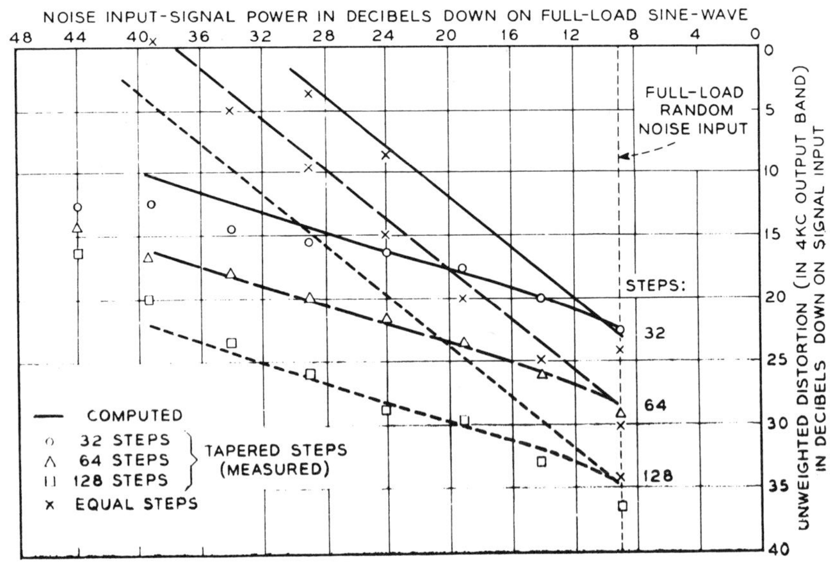

Fig. 8—Total distortion in signal band from quantizing with equal and tapered steps.

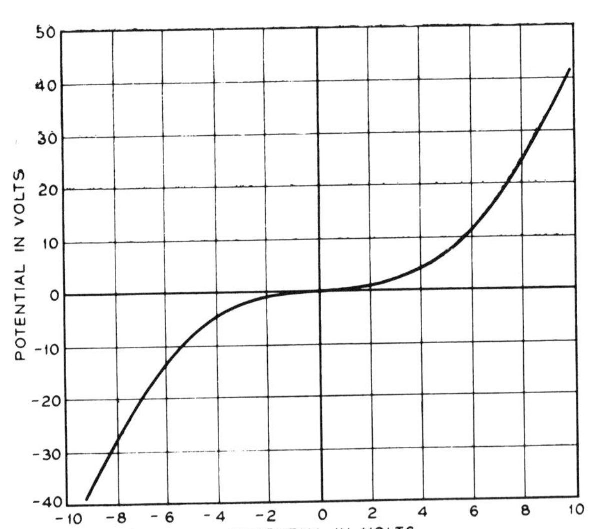

Fig. 9—Expanding characteristic applied to noise in tapered steps of Fig. (8).

range. This occurs when the rms value of input is 9 db below the rms value of the sine wave which fully loads the quantizer.

Flatness of the distortion spectrum with frequency within the signal band is demonstrated by Fig. 10. Two kinds of input were used here—a flat band of thermal noise and a set of 16 sine waves with frequencies distributed throughout the band. Results in the two cases were practically the same. The theoretical levels of distortion power for the band widths of the measuring filters (95 cps) are shown by the horizontal lines.

In the experimental results given here use has been made of laboratory studies by Messrs. A. E. Johanson, W. A. Klute, and L. A. Meacham.

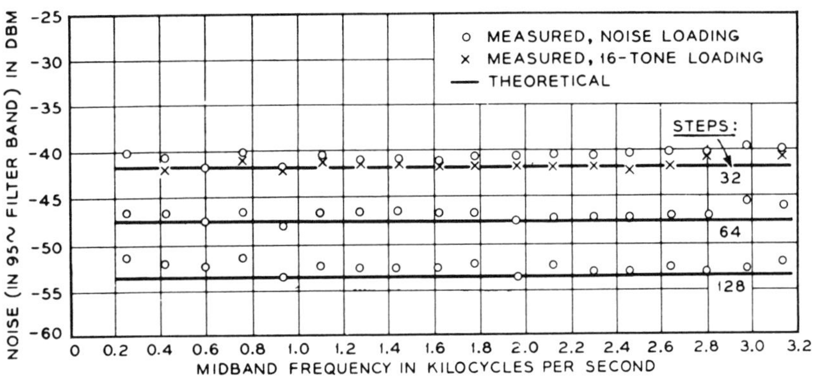

Fig. 10—Spectral density of distortion in signal band from quantizing and sampling. The quantizing steps were equal and the quantizer was fully loaded by a random noise or 16-tone input signal with mean power = −2.5 dbm.

2. Theoretical Analysis

The correlation theorem discovered by N. Wiener[11] may be stated as follows: Let ψ_τ represent the average value of the product $I(t)I(t + \tau)$, where $I(t)$ is the value of a variable such as current or voltage at time t, and $I(t + \tau)$ is the value at a time τ seconds later. Mathematically:

$$\psi_\tau = \overline{I(t)I(t + \tau)} = \lim_{T \to \infty} \frac{1}{T} \int_0^T I(t)I(t + \tau)\, dt \tag{2.0}$$

From analogy with statistical theory, ψ_τ is called the correlation of $I(t)$ with itself, or the autocorrelation function of the signal. Since we shall not deal here with the correlation of two signals, we shall shorten our terms and call ψ_τ simply the correlation of $I(t)$. Let $w_f\, df$ represent the mean power in the output of an ideal bandpass filter of width df centered at f. We assume that the ideal filter is designed to work between resistances of one ohm each and that the input signal $I(t)$ is delivered to the filter from a source with internal resistance of one ohm. (The use of unit resistances does not restrict the generality of the results, since equivalent transmission performance of any linear electrical circuit is obtained by multiplying all impedances by a constant factor. All voltages are multiplied and all currents divided by the same factor. By assuming unit values of resistance we are able to use

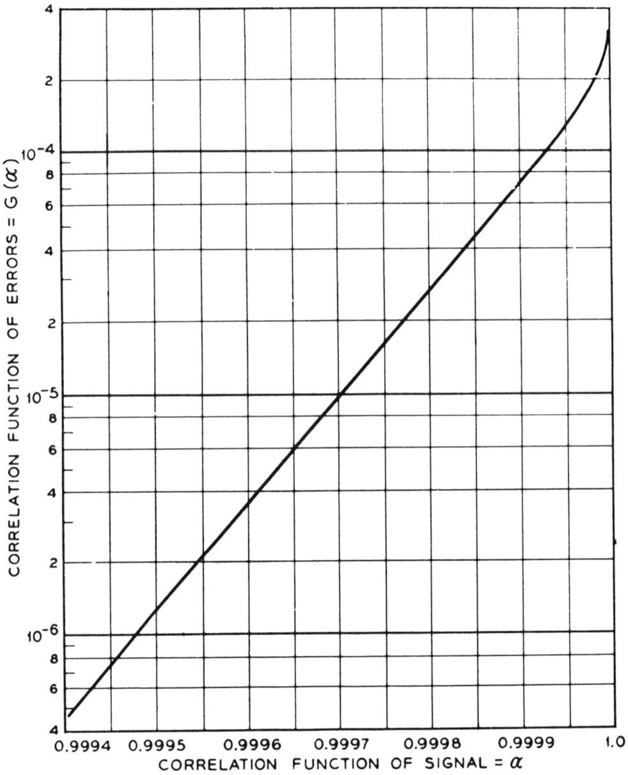

Fig. 11—Correlation function of 7-digit quantizing errors.

squared values of voltages and currents to represent power.) The theorem states that w_f and ψ are related by the equation:

$$w_f = 4 \int_0^\infty \psi_\tau \cos 2\pi f\tau \, d\tau \tag{2.1}$$

Proof may be found in the references cited. When the signal contains periodic components, the integral in (2.1) becomes divergent in the ordinary or Riemann sense, but this difficulty may be overcome by either applying the theory of divergent integrals or replacing Riemann by Stieltjes integration. We shall not require these modifications here because we shall base our analysis on signals with a continuous spectrum. We note that ψ_0 is the mean square value of the signal itself. We also point out that the inversion formula for the Fourier integral enables us to express ψ_τ in terms of w_f, thus:

$$\psi_\tau = \int_0^\infty w_f \cos 2\pi\tau f \, df \tag{2.2}$$

It also may be shown that the ratio ψ_τ/ψ_0 cannot have values outside the interval from -1 to $+1$.

The correlation theorem furnishes a powerful analytical tool for the solution of modulation problems because the calculation of the average ψ_τ is often a straightforward process, while direct calculation of w_f may be a very devious one. Once ψ_τ has been obtained, Eq. (2.1) brings the highly developed theory of Fourier integrals to bear on the computation of w_f.

We shall give the derivation of w_f for quantizing noise making use of the correlation function. In the analysis we shall apply a number of other needed theorems with appropriate references given for proof.

Our first problem is that of calculating the spectrum of the output of the staircase transducer, Fig. 1, when the spectrum of the input signal is given. Let w_f represent the power spectrum of the input signal and ψ_τ the autocorrelation function. The two quantities are related by (2.1) and it is sufficient to express our results in terms of either one. If the instantaneous value of the input signal is represented by E_1, and that of the output by E_2, the staircase function may be defined mathematically by:

$$E_2 = mE_0, \qquad \frac{2m-1}{2} E_0 < E_1 < \frac{2m+1}{2} E_0, \qquad m = 0, \pm 1, \pm 2, \cdots \tag{2.3}$$

The error is the difference between E_1 and E_2 and may be written as

$$\epsilon(t) = E_1 - E_2 = E_1 - mE_0, \qquad \frac{2m-1}{2} E_0 < E_1 < \frac{2m-1}{2} E_0 \tag{2.4}$$

The error characteristic is plotted in Fig. 3.

One approach depends on a knowledge of the probability density function $p(V_1, V_2)$ of the variables $V_1 = E_1$ at time t and $V_2 = E_2$ at time $t + \tau$. The definition of this function is that $p(V_1,V_2)\,dV_1dV_2$ is the probability that V_1 and V_2 lie in a rectangle of dimensions dV_1 and dV_2 centered on the point V_1, V_2 of the V_1V_2-plane. The function $p(V_1,V_2)$ has been calculated for certain types of signals and in theory could be computed for any signal by standard methods. If it is assumed known, we may determine the correlation function of the error. Let

$$F(V_1, V_2) = \epsilon(t)\epsilon(t + \tau) = (V_1 - mE_0)(V_2 - nE_0),$$

$$\frac{2m-1}{2} E_0 < V_1 < \frac{2m+1}{2} E_0, \quad \frac{2n-1}{2} E_0 < V_2 < \frac{2n+1}{2} E_0, \quad m, n = 0, \pm 1, \pm 2, \cdots \tag{2.5}$$

Eq. (2.5) defines $F(V_1, V_2)$ as a definite constant value in each square of width E_0 in the V_1V_2-plane. By elementary statistical theory, the correlation function ξ_τ of the error wave is now

$$\xi_\tau = \overline{F(V_1, V_2)} = \int_{-\infty}^{\infty}\int_{-\infty}^{\infty} F(V_1, V_2)p(V_1, V_2)\, dV_1\, dV_2 \tag{2.6}$$

The correlation may therefore be calculated since F and p are known functions. The power spectrum Ω_f of the error wave is then equal to the right-hand member of (2.1) with ξ_τ substituted for ψ_τ.

We are interested in the case in which the signal voltage has a smoothly varying spectrum over a specified band. This is a property of a random noise function which has a normal distribution of instantaneous voltages. The two-dimensional probability density function of such a wave is known[12]. It is

$$p(V_1, V_2) = \frac{1}{2\pi\sqrt{\psi_0^2 - \psi_\tau^2}} \exp\left[\frac{\psi_0(V_1^2 + V_2^2) - 2\psi_\tau V_1 V_2}{2(\psi_0^2 - \psi_\tau^2)}\right]. \tag{2.7}$$

By inserting this value and that of $F(V_1, V_2)$ from (2.5) in (2.6), making the change of variable:

$$\left.\begin{aligned} V_1 - mE_0 &= E_0x/2 \\ V_2 - nE_0 &= E_0y/2 \end{aligned}\right\} \tag{2.8}$$

and adopting the notation,

$$k = E_0^2/\psi_0, \alpha = \psi_\tau/\psi_0, G(\alpha) = \xi_\tau/\psi_0, \tag{2.9}$$

we obtain the following integral determining ξ_τ,

$$G(\alpha) = \frac{k^2}{32\pi(1 - \alpha^2)^{1/2}} \cdot \int_{-1}^{1}\int_{-1}^{1} xyH(x, y) \exp\frac{-k(x^2 + y^2 - 2\alpha xy)}{8(1 - \alpha^2)}\, dx\, dy \tag{2.10}$$

$$H(x, y) = \sum_{m=-\infty}^{\infty}\sum_{n=-\infty}^{\infty} \cdot \exp\frac{-k[m^2 + m(x - \alpha y) + n^2 + n(y - \alpha x) - 2\alpha mn]}{2(1 - \alpha^2)} \tag{2.11}$$

The power density spectrum of the errors is, from (2.1),

$$\Omega_f = 4\int_0^\infty \xi_\tau \cos 2\pi f\tau \, d\tau$$

$$= 4\psi_0 \int_0^\infty G(\alpha) \cos 2\pi f\tau \, d\tau \tag{2.12}$$

If the signal band is flat from $f = 0$ to $f = f_0$, with no energy outside this band,

$$\alpha = \frac{1}{f_0}\int_0^{f_0} \cos 2\pi\tau f \, df = \frac{\sin 2\pi f_0 \tau}{2\pi f_0 \tau} \tag{2.13}$$

Letting $\gamma = f/f_0$,

$$\Omega_0(\gamma) = \frac{f_0 \, \Omega_f}{\psi_0} = \frac{2}{\pi}\int_0^\infty G\left(\frac{\sin z}{z}\right) \cos \gamma z \, dz, \tag{2.14}$$

To complete the calculation, we must evaluate the integral (2.10). The first step is to transform the double summation (2.11) into products of single sums by the change of indices:

$$\begin{pmatrix} m + n = m' \\ m - n = n' \end{pmatrix} \quad \text{or} \quad \begin{pmatrix} m = \dfrac{m' + n'}{2} \\ n = \dfrac{m' - n'}{2} \end{pmatrix} \tag{2.15}$$

The rearrangement is permissible because the double series is absolutely convergent. The new indices m' and n' also run from minus to plus infinity, but must be either both even or both odd because $m' \pm n'$ is even. On dropping the primes after the substitution is completed, we find

$$\begin{aligned} H(x, y) = \sum_{m=-\infty}^{\infty} \exp \frac{-k[2m(x + y) + 4m^2]}{4(1 + \alpha)} \sum_{n=-\infty}^{\infty} \\ \cdot \exp \frac{-k[2n(x - y) + 4n^2]}{4(1 - \alpha)} + \sum_{m=-\infty}^{\infty} \\ \cdot \exp \frac{-k[(2m + 1)(x + y) + (2m + 1)^2]}{4(1 + \alpha)} \sum_{n=-\infty}^{\infty} \\ \cdot \exp \frac{-k[2n + 1)(x - y) + (2n + 1)^2]}{4(1 - \alpha)} \end{aligned} \tag{2.16}$$

A further simplification results from a change of the variables of integration to eliminate the terms in xy. This is done by setting

$$\begin{pmatrix} x = u + v \\ y = u - v \end{pmatrix} \quad \text{or} \quad \begin{pmatrix} u = (x+y)/2 \\ v = (x-y)/2 \end{pmatrix} \tag{2.17}$$

By calculating the Jacobian of the transformation, we find $dx\,dy = 2\,du\,dv$. The region of integration in the uv-plane is a rhombus bounded by the lines $u \pm v = \pm 1$. We then have:

$$G(\alpha) = \frac{k^2}{16\pi(1-\alpha^2)^{1/2}}\left[\int_{-1}^{0}\int_{-1-u}^{1+u} dv + \int_{0}^{1} du \int_{u-1}^{1-u} dv\right](u^2 - v^2)$$
$$\exp\left[-\frac{k}{4}\left(\frac{u^2}{1+\alpha} + \frac{v^2}{1-\alpha}\right]\sum_{m=-\infty}^{\infty}\exp\frac{-2mk(2u+2m)}{4(1+\alpha)}$$
$$\sum_{n=-\infty}^{\infty}\exp\frac{-2nk(2v+2n}{4(1-\alpha)} + \sum_{m=-\infty}^{\infty}\exp\frac{-(2m+1)k(2u+2m+1)}{4(1+\alpha)}$$
$$\sum_{n=-\infty}^{\infty}\exp\frac{-(2n+1)k(2v+2n+1)}{4(1-\alpha)} \qquad (2.18)$$

If we substitute $u = -x$ in the first double integral, $m = -m'$ in the first series, and $m = -m' - 1$ in the third series, we see that the two double integrals are equal. We therefore drop the first double integral and multiply the second by two. The inner integral may then be split into parts with limits from $v = 0$ to $v = 1 - u$ and $v = u - 1$ to $v = 0$. Substituting $v = -y$ in the second part and treating the series as before, we find that the two parts give equal contributions, so that the bracketed integral terms become

$$4\int_0^1 du \int_0^{1-u} dv$$

applied to the integrand.

The series in (2.18) may be written as Theta Functions, and the imaginary transformation of Jacobi then used as an aid in reduction. We may proceed in a more direct manner, however, by applying Poisson's Summation Formula:[13]

$$\sum_{n=-\infty}^{\infty}\varphi(2\pi n) = \frac{1}{2\pi}\sum_{m=-\infty}^{\infty}\int_{-\infty}^{\infty}\varphi(\tau)e^{-im\tau}\,d\tau \qquad (2.19)$$

We thereby show that

$$\sum_{m=-\infty}^{\infty}\exp\left[-am(x+2m)\right] = \sqrt{\frac{\pi}{2a}}\,e^{ax^2/8}\left[1 + 2\sum_{m=1}^{\infty}e^{-m^2/\pi^2 2a}\cos\frac{m\pi x}{2}\right] \qquad (2.20)$$

$$\sum_{m=-\infty}^{\infty}\exp\left[-a(2m+1)(x+2m+1)\right] = \frac{1}{2}\sqrt{\frac{\pi}{a}}\,e^{ax^2/4}\left[1 + 2\sum_{m=1}^{\infty}(-)^m e^{-m^2\pi^2/4a}\cos\frac{m\pi x}{2}\right] \qquad (2.21)$$

When the series in (2.18) of type corresponding to the left-hand members of (2.20) and (2.21) are replaced by the equivalent righthand members, positive exponents containing the squared variables of integration are introduced which cancel the negative exponents already present in the integrand. The resulting integral may be written:

$$G(\alpha) = \frac{k}{4}\int_0^1 du \int_0^{1-u} (u^2 - v^2)[f_1(1+\alpha, u)f_1(1-\alpha, v) + f_2(1+\alpha, u)f_2(1-\alpha, v)]\, dv, \quad (2.22)$$

where

$$f_1(a, x) = 1 + 2\sum_{m=1}^{\infty} \exp\frac{-m^2\pi^2 a}{2k} \cos\frac{m\pi x}{2} \quad (2.23)$$

$$f_2(a, x) = 1 + 2\sum_{m=1}^{\infty} (-)^m \exp\frac{-m^2\pi^2 a}{2k} \cos\frac{m\pi x}{2} \quad (2.24)$$

The integrations may now be performed without difficulty. The complete result, which as we shall immediately show is hardly ever necessary to use in full is:

$$G(\alpha) = \frac{k}{\pi^2}\sum_{n=1}^{\infty}\frac{1}{n^2}\exp\left(-\frac{4n^2\pi^2}{k}\right)\sinh\frac{4n^2\pi^2\alpha}{k} + \frac{k}{\pi^2}\sum_{m=1}^{\infty}\sum_{n=1}^{\infty}(m \neq n)\frac{1}{(m^2-n^2)}\exp\frac{-4(m^2+n^2)\pi^2}{k}$$

$$\sinh\frac{4(m^2-n^2)\pi^2\alpha}{k} - \frac{k}{\pi^2}\sum_{m=1}^{\infty}\sum_{n=1}^{\infty}(m \neq n)\frac{1}{(m-\frac{1}{2})^2-(n-\frac{1}{2})^2}$$

$$\exp\frac{-4[(m-\frac{1}{2})^2+(n-\frac{1}{2})^2]\pi^2}{k}\sinh\frac{4[(m-\frac{1}{2})^2-(n-\frac{1}{2})^2]\pi^2\alpha}{k} \quad (2.25)$$

An alternative derivation of (2.25), subsequently suggested by Mr. S. O. Rice, is based on the fact that $\epsilon(t)$ as defined by (2.4) or Fig. 3 is a periodic function of E_1 which can be expanded in a Fourier series with period E_0. Substituting the series in (2.5) leads to an expression for $\epsilon(t)\ \epsilon\ (t+\tau)$ as the product of two Fourier series. After proof that it is permissible to write this product as a double series and to calculate the average sum as the sum of the averages of the individual terms the problem is reduced to a double series in which the typical term is proportional to the average value of exp $i(uV_1 + vV_2)$ where u and v are constants depending on the position of the term in the series. Rice has shown[12] that the average value of such a term is exp $[-(u^2+v^2)\psi_0/2 - uv\psi_\tau]$. Summation of these terms leads again to (2.25).

From the defining equation (2.9) we note that k is a small quantity when more than a very few steps are used in the quantizer so that exponentials with exponent containing the factor $-1/k$ are very small except when the factor is multiplied by a number near zero. It will be seen that this can only happen in the first series and then only when α approaches the value unity. We recall that α lies in the range -1 to $+1$ and it is apparent from (2.25) that $G(\alpha)$ is an odd function of α. We thus need consider only positive values of α very slightly less than unity. Only the component of the sin$h$ with positive exponent is then significant, and we write the very accurate approximation for $G(\alpha)$:

$$G(\alpha) \doteq \frac{k}{2\pi^2} \sum_{n=1}^{\infty} \frac{1}{n^2} \exp \frac{-4n^2\pi^2(1-\alpha)}{k} \tag{2.26}$$

A typical curve of $G(\alpha)$ vs. α for a fixed value of k is shown in Fig. 11. The rapidity with which it falls away at the left of the point $\alpha = 1$ is such that the curve can only be plotted by greatly expanding the scale of α in this region. The physical significance of the spike-shaped curve is that $G(\alpha)$ is a measure of the correlation of the errors as a function of the correlation of the applied signal. When there are many steps there is virtually no correlation between errors in successive samples except when there is complete correlation of successive signal values.

Use of the approximation (2.26) enables us to derive a convenient formula for the spectral density of the errors in a flat band input signal. Substituting (2.26) in (2.14) we obtain:

$$\Omega_0(\gamma) \doteq \frac{k}{\pi^3} \sum_{n=1}^{\infty} \frac{1}{n^2} \int_0^{\infty} \exp\left[\frac{-4n^2\pi^2}{k}\left(1 - \frac{\sin z}{z}\right)\right] \cos \gamma z \, dz \tag{2.27}$$

The integrand is negligible except when z is near zero, and in this region we may replace $(\sin z)/z$ by the first two terms of its power series expansion. We then find

$$\begin{aligned}\Omega_0(\gamma) &\doteq \frac{k}{\pi^3} \sum_{n=1}^{\infty} \frac{1}{n^2} \int_0^{\infty} \exp\left(\frac{-2n^2\pi^2z^2}{3k}\right) \cos \gamma z \, dz \\ &= \frac{k}{2\pi^3} \sqrt{\frac{3k}{2\pi}} \sum_{n=1}^{\infty} \frac{1}{n^3} \exp\left(\frac{-3k\gamma^2}{8n^2\pi^2}\right).\end{aligned} \tag{2.28}$$

Only one set of calculations from the infinite series need be made since we may define a function of one variable

$$B(z) = \sum_{n=1}^{\infty} \frac{e^{-z/n^2}}{n^3}. \tag{2.29}$$

Then

$$\Omega_0(\gamma) \doteq \frac{k}{2\pi^3} \sqrt{\frac{3k}{2\pi}}\, B\left(\frac{3k\gamma^2}{8\pi^2}\right). \tag{2.30}$$

The curves of Fig. (4) were obtained in this way. The relation between k and the number of digits N is based on the assumption of the rms value of signal reaching one-fourth the instantaneous overload voltage of the quantizer. Since zero signal voltage is in the middle of the quantizing range $2^N E_0$, the overload signal measured from zero is $2^{N-1}E_0$. The mean square signal input is ψ_0. Therefore

$$2^{N-1}E_0 = 4\sqrt{\psi_0} \tag{2.31}$$

or from (2.9)

$$k = 1/4^{N-3} \tag{2.32}$$

We thus have obtained the spectrum of the quantizing errors without sampling. To apply our results to the sampling case we sum up all contributions from each harmonic of the sampling rate beating with the noise spectrum from quantizing only. The resulting power spectrum is given by

$$A_f = \Omega_f + \sum_{n=1}^{\infty} (\Omega_{nf_s-f} + \Omega_{nf_s+f}), \qquad 0 \leq f \leq f_s/2. \tag{2.33}$$

If y is the ratio of sampling frequency to signal band width and $A_0(y)$ is the ratio of quantizing power received in the signal band to the applied signal power,

$$A_0(y) = \Omega_0(1) + \sum_{n=1}^{\infty} [\Omega_0(ny + 1) + \Omega_0(ny - 1)]. \tag{2.34}$$

This is the equation used in calculating the curves of Fig. (5).

APPENDIX I

Relation Between Mean Squares of Signal and Its Samples

We have already shown that there is a unique relationship between a signal occupying the band of all frequencies less than f_c, and the sampled values of the signal taken at a rate $f_s = 2f_c$. If we are given the signal wave, we can obviously determine the samples; and if we are given the samples, we can determine the signal wave since it is the response of an ideal low-pass filter of cutoff frequency f_c to unit impulses multiplied by the samples. If we apply samples of a signal containing components of frequency greater than f_c, the output of the filter is a new signal with frequencies confined to the band from zero to f_c and yielding the same sampled values as the original wideband signal.

We now consider the problem of determining the mean square value of the samples of an arbitrary function $f(t)$. Let the samples be taken at $t = nT$, $n = 0, \pm 1, \pm 2, \cdots$, where $T = 1/2f_c = 1/f_s$.

We may write an expression for the squared samples as a limit of the product of the squared signal and a periodic switching function of infinitesimal contact time, thus

$$f^2(nt_0) = \lim_{\tau \to 0} f^2(t)S(\tau, t) \tag{I—1}$$

where:

$$S(\tau, t) = \begin{pmatrix} 1, & -\tau/2 < t < \tau/2 \\ 0, & \tau/2 < t < T - \tau/2 \end{pmatrix} \tag{I—2}$$

$$S(\tau, t + T) = S(\tau, t), n = 0, \pm 1, \pm 2, \cdots \tag{I—3}$$

By straightforward Fourier series expansion:

$$S(\tau, t) = \frac{\tau}{T} + \sum_{m=1}^{\infty} \frac{2 \sin m\pi\tau/T}{m\pi} \cos 2m\pi f_s t. \tag{I—4}$$

The mean square value of the samples is the limit of the average value of f^2S taken over the contact intervals of duration τ. The average value of f^2S taken over all time, including the blank intervals, is in the limit a fraction τ/T of the average over the contact intervals only. Therefore

$$\overline{f^2(nt_0)} = \lim_{\tau \to 0} \overline{\frac{T}{\tau} f^2(t)S(\tau, t)}$$

$$= \overline{\lim_{\tau \to 0} f^2(t) + \sum_{m=1}^{\infty} \frac{2T \sin m\pi\tau/T}{m\pi\tau} f^2(t) \cos 2m\pi f_s t} \tag{I—5}$$

$$= \overline{f^2(t)} + \lim_{\tau \to 0} \sum_{m=1}^{\infty} \frac{2T \sin m\pi\tau/T}{m\pi\tau} \overline{f^2(t) \cos 2m\pi f_s t}.$$

Now the long time average value of $f^2(t) \cos 2m\pi f_s t$ must vanish unless $f^2(t)$ contains a component of frequency mf_s. This could not happen except where $f(t)$ itself contains a component of frequency $mf_s/2$ or two components f_1 and f_2 such that

$$|f_1 \pm f_2| = mf_s \tag{I—6}$$

When no such relation of dependency exists:

$$\overline{f^2(nt_0)} = \overline{f^2(t)}. \tag{I—7}$$

As pointed out before if $f(t)$ contains no frequencies above f_c, the response of the ideal low-pass filter to the samples is $f(t)$, and $f(nt_0)$ represents the samples of $f(t)$. If $f(t)$ does contain frequencies exceeding f_c, the response of the filter is $\phi(t)$, where $\phi(t)$ is wholly confined to the band 0 to f_c and yields the same samples as $f(t)$, i.e.,

$$\phi(nt_0) = f(nt_0), n = 0, \pm 1, \pm 2, \cdots \tag{I—8}$$

Eq. (I—7) applied to $\phi(t)$ gives the result:

$$\overline{\phi^2(nt_0)} = \overline{\phi^2(t)}. \tag{I—9}$$

By combining (I—8) and (I—9), we obtain

$$\overline{f^2(nt_0)} = \overline{\phi^2(t)}. \tag{I—10}$$

APPENDIX II

Fundamental Theorem on Aperture Effect in Sampling

If we sample the wave $Q \cos qt$ at a rate f_s, and multiply each sample by a short rectangular pulse of unit height and duration τ centered at the sampling instants, we obtain by reference to Eq. (I–4) replacing $2\pi f_s$ by ω_s,

$$F(t) = Q \cos qt\, S(\tau, t) = \frac{\tau}{T} Q \cos qt + Q \sum_{m=1}^{\infty} \frac{\sin m\pi\tau/T}{m\pi} [\cos (m\omega_s + q)t + \cos (m\omega_s - q)t]. \tag{II—1}$$

The fact that pulse modulation is similar to the more familiar carrier modulation processes is brought out by this equation; the sampling frequency is in fact the carrier. The writer has found that the method of calculation he published in 1933,[14] in which the signal and carrier frequencies are taken as independent variables, is ideally suited for calculations of pulse-modulated spectra. Artificial and cumbersome devices such as assuming the signal and sampling frequencies to be harmonics of a common frequency are thereby avoided.

A unit impulse $\delta(t)$ has zero duration and unit area; hence we may write:

$$\delta(t) = \operatorname*{Lim}_{\tau \to 0} \frac{S(\tau, t)}{\tau}. \tag{II—2}$$

A train of samples in which each sample is multiplied by a unit impulse may therefore be written as

$$\sum_{n=-\infty}^{\infty} Q \cos qt\, \delta(t - t_n) = \operatorname*{Lim}_{\tau \to 0} \left[\frac{Q}{T} \cos qt + Q \sum_{m=1}^{\infty} \frac{\sin m\pi\tau/T}{m\pi\tau} [\cos (m\omega_s + qt) + \cos (m\omega_s - q)t] \right]. \tag{II—3}$$

Suppose we apply the train of waves (II–3) to a linear electrical network which delivers the response $g(t)$ when the input is a unit impulse $\delta(t)$. The steady state admittance of the network is given by[15]

$$Y_0(i\omega) = \int_{-\infty}^{\infty} g(t) e^{-i\omega t}\, dt \tag{II—4}$$

and the response of the network to (II–3) is therefore:

$$I(t) = \frac{Q}{T} \mid Y_0(iq) \mid \cos [qt + ph\, Y_0(iq)]$$
$$+ \frac{Q}{T} \sum_{m=1}^{\infty} (\mid Y_0(im\omega_s + iq) \mid \cos [(m\omega_s + q)t \qquad \text{(II—5)}$$
$$+ ph\, Y_0(im\omega_s + iq)] + \mid Y_0(im\omega_s - iq) \mid \cos [(m\omega_s - q)t$$
$$+ ph\, Y_0(im\omega_s - iq)]).$$

But $I(t)$ evidently represents a train of pulses in which the pulse occurring at $t = nT$ is equal to the *nth* sample multiplied by $g(t - nT)$. We have thus obtained the spectrum of a set of samples in which the pulse representing a unit sample is the generalized wave form $g(t)$. Furthermore if the signal frequency q is less than $\omega_s/2$, an ideal low-pass filter with cutoff at $\omega_s/2$ responds only to the first component of (II—5).

The "aperture effect" or variation of transfer admittance with signal frequency is thus given by

$$Y(iq) = \frac{1}{T} Y_0(iq) = f_s Y_0(iq). \qquad \text{(II—6)}$$

This is Theorem II. Since the system is linear when the signal frequency does not exceed half the sampling frequency, the principle of superposition may be applied to composite signals. In the case of distortion from quantizing errors the aperture effect applies to the error component delivered by the low-pass output filter. For an imperfect low-pass filter in the output we multiply the aperture admittance function by the actual transfer admittance of the filter.

A theorem equivalent to the above has been derived by a different method in a recent paper[16] published after completion of the above work.

REFERENCES

1. W. R. Bennett, Time Division Multiplex Systems, *Bell Sys. Tech. Jour.*, Vol. 18, pp. 1–31; Jan. 1939.
2. H. S. Black, Pulse Code Modulation, *Bell Lab. Record*, Vol. 25, pp. 265–269; July, 1947.
3. W. M. Goodall, Telephony by Pulse Code Modulation, *Bell Sys. Tech. Jour.*, Vol. 26, pp. 395–409; July, 1947.
4. D. D. Grieg, Pulse Count Modulation System, *Tele-Tech.*, Vol. 6, pp. 48–50, 98; Sept. 1947; also *Elect. Comm.*, Vol. 24, pp. 287–296; Sept. 1947.
5. A. G. Clavier, P. F. Panter, and D. D. Grieg, PCM Distortion Analysis, *Elec. Engg.*, Vol. 66, pp. 1110–1122; Nov. 1947.
6. H. S. Black and J. O. Edson, PCM Equipment, *Elec. Engg.*, Vol. 66, pp. 1123–1125; Nov. 1947.
7. L. A. Meacham and E. Peterson, An Experimental Pulse Code Modulation System of Toll Quality, *Bell Sys. Tech. Jour.*, Vol. 27, pp. 1–43; Jan., 1948.
8. H. Nyquist, Certain Topics in Telegraph Transmission Theory, *A. I. E. E. Trans.*, pp. 617–644; April, 1928.
9. E. Peterson, Gas Tube Noise Generator for Circuit Testing, *Bell Lab. Record*, Vol. 18, pp. 81–83; Nov. 1939.

10. W. R. Bennett, Cross-Modulation in Multichannel Amplifiers, *Bell Sys. Tech. Jl.*, Vol. 19, pp. 587–610; Oct. 1940.
11. N. Wiener, Generalized Harmonic Analysis, *Acta. Math.*, Vol. 55, pp. 177–258; 1930.
12. S. O. Rice, Mathematical Analysis of Random Noise, *Bell Sys. Tech. Jour.*, Vol. 24, p. 50; Jan. 1945.
13. R. Courant and D. Hilbert, Methoden der Mathematischen Physik, Vol. 1, p. 64; Berlin, 1931.
14. W. R. Bennett, New Results in the Calculation of Modulation Products, *Bell Sys. Tech. Jour.* Vol. 12, pp. 228–243; April 1933.
15. G. A. Campbell, The Practical Application of the Fourier Integral, *Bell Sys. Tech. Jour.*, Vol. 7, pp. 639–707; Oct. 1928.
16. S. C. Kleene, Analysis of Lengthening of Modulated Repetitive Pulses, *Proc. I. R. E.*, Vol. 35, pp. 1049–1053; 1947.

Errata

Page 454, 2nd paragraph, 3rd line: Insert "$1/f_s$ times" between "equal to" and "the mean square."
Page 463, Eq. (2.7): Insert minus sign ahead of fraction within brackets.
Page 465, Eq. (2.18), 3rd line: Insert) after $(2v + 2n$.
2nd line: Insert large) before right-hand bracket.
Page 465, Eq. (2.20): Transpose / and π^2 in last exponent.
Page 468, Appendix I, 5th line: Insert "f_s times" between "since it is" and "the response."
Page 469, 2nd line under Eq. (I-7): Insert $/f_s$ after $f(t)$.
Page 469, 4th line under Eq. (I-7): Insert $/f_s$ after first $\phi(t)$.

V
Bandpass Nonlinear Systems

Editor's Comments on Papers 19 and 20

19 Middleton: *Some General Results in the Theory of Noise Through Non-linear Devices*

20 Davenport: *Signal-to-Noise Ratios in Band-Pass Limiters*

This section contains two papers whose subject is the power spectrum, signal-to-noise ratio, and other general output properties of ZNL systems followed by linear systems with memory. The linear systems considered are either low-pass or bandpass, depending on the application. The input signals are either noise alone or signal and noise, where in general Gaussian and sinusoidal inputs are considered. Since the input spectra are also assumed to be low-pass or bandpass, the systems involved may therefore be assumed to have the form given in Figure 2.

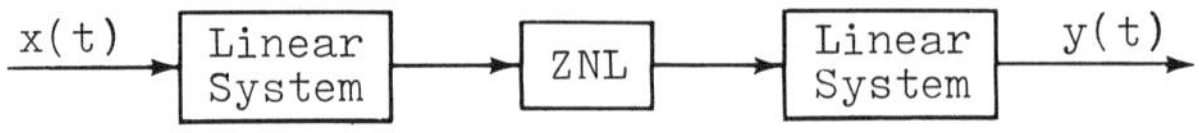

Figure 2

In addition to the fact that the systems and signals are of the same type with various levels of generality, another common factor in both papers is the application of Rice's general method for the analysis of the nonlinearity. Middleton's paper (19) ranks with Rice's (1944) as one of the major contributions to the analysis of nonlinear systems with random signals. It is based on his earlier paper, Middleton (1946), in which linear and quadratic rectifiers were analyzed with noise alone. The paper is comprehensive in nature and generalizes earlier results by Bennett (Paper 16), Rice (1944), and Van Vleck and Middleton (1946) in several directions. The second page of the introduction in the paper provides an itemized description of its major advances. It is the first general treatment of the νth-law device, which is now included in textbooks on random signals and noise [see, for example, Davenport and Root (1958)]. In addition to the derivation of spectra and signal-to-noise ratios, the paper considers also the output densities and other statistical properties of importance. This paper and Rice's are the most quoted references in the analysis of bandpass nonlinearities or of ZNL systems with random signals and noise. Subsequently, many special cases, applications, or generalizations of Middleton's results have been published. Notable among these is Middleton's (1948) paper on the applications of the results to the evaluation of AM receivers. Applications to FM receivers are considered in Middleton (1949a, 1950). Davenport (Paper 20) applies the results to vth-law bandpass limiters, which is an excellent example of the type of papers based on the Rice–Middleton method. Other papers have not been included in this collection both for lack of space and because Davenport's is a good, typical example of such papers. The following is a summary of further contributions.

Campbell (1956) generalized the rectifier results to the case of two signals plus noise inputs. Fellows and Middleton (1956) present an experimental verification of the spectrum of the rectifier case. Other generalizations or simplified expressions for special cases are given by Cahn (1961), Doyle (1962), Doyle and Reed (1964),

Berglund (1964), and Blachman (1964). Notable extensions of the bandpass limiter case to more than one signal are given by Jones (1963) for two input signals plus noise, and by Shaft (1965) for more than two signals.

In summary, Middleton's paper, together with its application to limiters by Davenport, provides a comprehensive treatment of the νth-law device, and spurred the publications of a large number of papers that explore the various aspects of the problem.

19

Reprinted from *Quart. Appl. Math.*, **5**(4), 445–498 (1948)

SOME GENERAL RESULTS IN THE THEORY OF NOISE THROUGH NON-LINEAR DEVICES*

BY

DAVID MIDDLETON

Harvard University

1. Introduction. Because of the great prevalence of noise in almost all electronic processes, a study of the nature and properties of such noise seems desirable. The term "properties of noise" is used here to indicate such measurable quantities as the average and the mean-square voltages and currents, the mean power spectrum, the power or energy associated with the wave, and the power and the correlation function of the disturbance, or of part of it, when noise or a signal and noise is modified by passage through non-linear apparatus. From an analytical point of view, the theory of noise is intrinsically related to that of the Brownian motion so that the results in the discussion of the one may bear rather closely upon the other. There are two different but equivalent lines of attack on the over-all problem: in noise theory we are primarily interested in what happens to random noise waves (with or without an accompanying signal) when they are passed through non linear devices, such as second detectors or mixers in radio receivers, for example, or amplifiers in which cutoff and/or overloading is present. Here the *Fourier series method of Rice*,[1,2] is the more natural approach and is the one followed in the present paper. In the study of Brownian motion and fluctuation phenomena in general, where the variations in the system are described by a diffusion process, the second method of *Fokker-Planck*, or *the diffusion equation method*, is used. We shall not consider this approach here; an excellent discussion has recently been given by Wang and Uhlenbeck,[3] and less recently, an interesting treatment of somewhat similar subjects by Chandrasekhar.[4] We mention only in passing that the two methods can be shown to yield identical results.[3]

Rice† and others[5–9] have used the Fourier series method in the solution of special

---

* Received Jan. 29, 1947.

[1] S. O. Rice, *Mathematical analysis of random noise*, Bell Sys. T. J. **23**, 282 (1944).

[2] S. O. Rice, loc. cit. **24**, 46 (1945). Since this paper was written, Professor Brillouin has kindly called the author's attention to the interesting work of Dr. Blanc-Lapierre along somewhat similar lines, in particular, a thesis: *Sur certaines fonctions aléatoires stationnaires. Application à l'étude des fluctuations dues à la structure électronique de l'électricité*, and *Effet Schottky. Fluctuations dans les amplificateurs linéaires et dans les détecteurs*, Bull. de la Soc. fr. des Elec. (6) **5**, No. 53, Nov. 1945.

[3] M. C. Wang and G. E. Uhlenbeck, *On the theory of the brownian motion II*, Rev. Mod. Phys. **17**, 323 (1945). This paper and reference 4 contain a considerable number of references to previous work along these lines.

[4] S. Chandrasekhar, *Stochastic problems in physics and astronomy*, Rev. Mod. Phys. **15**, No. 1, 1, 1943.

† See part IV of ref. 2.

[5] K. Fränz, Zeits. f. Hoch. u. Elek. **57**, 146 (1941).

[6] W. R. Bennett, *The response of a linear rectifier to signal and noise*, J. Amer. Acous. Soc. **15**, 165 (1944).

[7] D. O. North, *The modification of noise by certain non linear devices*, a synopsis of which was presented at the Jan. 1944 winter technical meeting of the I.R.E.

[8] J. H. Van Vleck and D. Middleton, *A theoretical comparison of the visual, aural, and meter reception of pulsed signals in the presence of noise*, J. Appl. Phys. **17**, 940 (1946).

[9] D. Middleton, *The response of biased, saturated linear and quadratic rectifiers to random noise*, Jour. A. Phys. **17**, 778 (1946).

problems involving the rectification of noise, or of a signal and noise. It is the purpose of this paper to generalize some of the results of previous work on this topic and to obtain original results for a number of unsolved problems in the analysis of noise through non-linear devices. Specifically, it is believed that the treatment of the following topics is new and of interest.

(a) Passage of a *modulated* signal in the presence of noise through a general non-linear apparatus. The case of a sinusoidally modulated carrier (Sec. 2), is examined and attention is also given to the case of narrow-band noise, symmetrically distributed in frequency about the carrier (Sec. 3).

(b) The biased νth-law rectifier, for modulated and unmodulated carriers (Sec. 4). Limiting cases of large noise or signal voltages are also discussed. Reference 9 gives a detailed discussion of this problem for linear and quadratic rectifiers where noise alone is rectified.

(c) The problem of a modulated signal and narrow-band noise, with a determination of the various probability densities associated with the envelope of the wave (Sec. 5). Section (c) and (d) offer alternative solutions to some of the problems discussed in (a) and (b).

(d) The correlation function and mean power associated with the envelope of signal and noise. Attention is given to the low frequency output of the half-wave νth-law device (Sec. 6).

(e) The νth-law, half-wave rectification of noise alone, the results of which are of interest in the measurement of noise by meters, spectrum analyzers, etc., and in the detection of pulse signals in the presence of noise[8,9] (Sec. 7). This work is a generalization of reference 9 in that ν can take on any positive value, but is less general in that only half-wave detection is treated.

(f) A general "small-signal" theory, in which the speak values of the ncoming wave, whether noise or a signal and noise, are sufficiently small that overloading and cut-off do not occur. Rectification takes place because of the curvature of the dynamic characteristic of the device in question.

Not all the material in the present paper is original, it is realized, but in the discussion of (a)–(f) it has been necessary for clarity of treatment to bring together and extend, when necessary, a number of results previously derived in probability theory which are fundamental and hence unavoidable in the analysis of problems of this type. These results include the generalized, s-dimensional random-walk problem,[4] from which in turn one may obtain the characteristic function, with the distinctive property of being the Fourier transform of the probability density, and finally the central-limit theorem,[10] which in the limit of a very large number of events can be shown to yield the s-dimensional Gaussian distribution characteristic of all random processes fulfilling certain rather elastic conditions with regard to the separate distributions of the various events. Some of the details are available in Appendix I. Furthermore, although the concept of the correlation function and its relation to the mean power spectrum has been examined in varying detail elsewhere, we include a brief treatment in Appendix II, along with some of the more significant properties of the correlation function which are necessary in our work. Rigor has not been preserved at all costs, *vide* the use of the Dirac delta-function; the physical significance

[10] H. Cramèr, *Random variables and probability distributions*, Cambridge Tract No. 36 (1937), Chapters VI and X. References 1 and 2 also contain references to this problem; see Secs. 2.9 and 2.10,

of the results is used to check any possible weakness in the rigor. Finally, in Appendix III some of the more unfamiliar special functions that appear in our results are briefly considered. It should be kept in mind that the present discussion applies only when the dynamic path of the non linear device is a one-valued function of the input disturbance. When the path is multi-valued, the theory breaks down. For example, when the plate-load of a rectifying tube is a pure resistance, or at worst, is primarily resistive, the dynamic path is one-valued, or nearly so. Figure 1 illustrates a typical tube characteristic under these conditions. However, when there is apprecia-

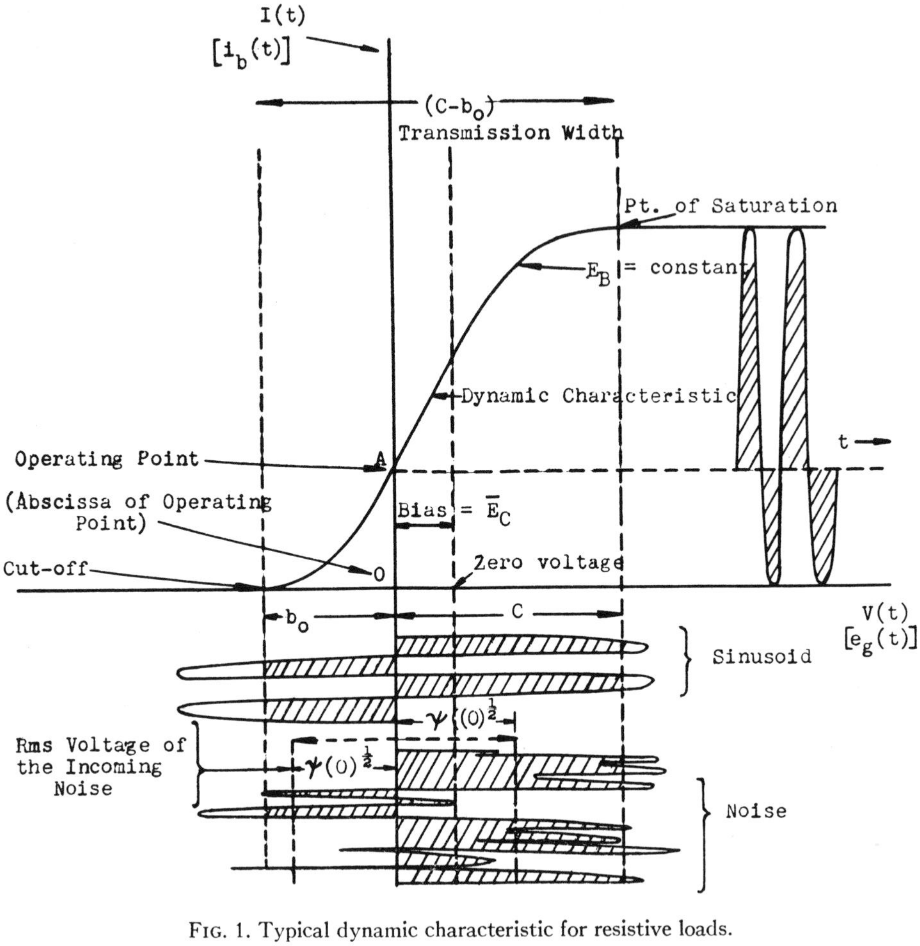

FIG. 1. Typical dynamic characteristic for resistive loads.

ble reactance X_p in comparison with the resistance R_p, the problem is not tractable with the present methods.

At this point it is well to mention what we mean by a random process, and with what class of such processes problems of the present kind deal. We consider $X(t)$ to be a random process when $X(t)$ does not depend in a definite way on the independent variable t, but instead may be specified by an aggregate of different functions $X(t)$, of which experimentally only probability distributions are observable. Here $X(t)$ may be the displacement of the electron beam on the screen of a cathode ray oscilloscope, or, say, the velocity of a particle in Brownian motion. The quantity $X(t)$ may also represent a combination of two or more such stochastic variables, in which case we may speak of a two- or more-dimensional random process. In what follows, we shall consider formally that X and t are continuous variables; for our work, based as it is on the problem of the random walk (see Appendix I), this a plausible assumption. Now, as Wang and Uhlenbeck[3] have shown, it is possible to describe a random process completely by means of a set of distributions determining the probability that X lie in the range X, $X+dX$, at time, t, that X_1 and X_2 fall in the intervals X_1, X_1+dX_1 and X_2, X_2+dX_2, at times t_1 and t_2, respectively, and so on. The classification is considerably simplified when the initial times of the observations do not enter; processes of this kind are *stationary*. Their statistical properties are independent of when the measurements are made, and depend only on the duration of the observations. There is an important class in this group known as *Markoff processes*, which are completely described by the second order density $W_2(X_1, X_2; t)$:

$W_2(X_1, X_2; t)$ = joint probability of obtaining X_1 in the range X_1, X_1+dX_1 and X_2 in the range X_2, X_2+dX_2 at a time t later. (1.1)

Since the present paper deals exclusively with distributions that are stationary and Markoffian, or if not originally the latter, can be extended to a more complex Markoff process by the introduction of suitable additional random variables,[11] we find it helpful to mention the properties of W_2. We have

$$W_2(X_1, X_2; t) = W_1(X_1)P_2(X_1 \mid X_2; t), \tag{1.2}$$

where following Wang and Uhlenbeck,[3] $P_2(X \mid X_2; t)$ denotes the *conditional* probability that, given X_1, we find X_2 in the interval X_2, X_2+dX_2 a time t later. Here $W_1(X_1)$ is simply the first order distribution, giving the probability of locating X_1 in the interval X_1, X_1+dX_1; the time does not enter because the process is assumed to be stationary. We have also the relations

$$\begin{aligned} P_2(X_1 \mid X_2; t) \geqq 0; \qquad \int dX_2 P_2(X_1 \mid X_2; t) = 1; \\ W_1(X_2) = \int W_1(X_1)P_2(X_1 \mid X_2; t)dX_1, \end{aligned} \tag{1.3}$$

where the region of integration includes all possible values of the variables. The further condition

$$\begin{aligned} P_n(X_1, t_1; X_2, t_2; \cdots ; X_{n-1}, t_{n-1} \mid X_n; t_n) \\ = P_2(X_{n-1}, t_{n-1} \mid X_n; t_n), \qquad t_n > t_{n-1} > \cdots > t_1 \end{aligned} \tag{1.4}$$

shows that (1.1), or P_2, completely determines the Markoff process, for once W_2 is known, we may obtain all distributions W_n, $n \geqq 2$, from the above. It should be observed that $P_2 \to W_1(X_2)$ as $t \to \infty$, provided X_2 is purely random and does not contain periodic components. As has been previously pointed out,[3] W_2 or P_2 cannot be selected arbitrarily, for it must satisfy the fundamental equation

[11] See, for example, sec. **3** (c), of ref. 3.

$$P_2(X_1 \mid X_2; t) = \int dX P_2(X_1 \mid X; t_0) P_2(X \mid X_2; t - t_0), \qquad 0 \leqq t_0 \leqq t. \tag{1.5}$$

which is due to Smoluchowski. Equations (1.1)–(1.5) have been written for one- and two-dimensional systems; the extension to a greater number follows at once without modification of the concepts.

The successful solution of non-linear problems involving noise by the Fourier series method of Rice depends, then, on the following definitions, devices, and assumptions.

(1) It is *assumed* that the random process is stationary and Gaussian, i.e., X_1, X_2, etc., all obey a Gaussian distribution law. In the completely Gaussian case, all components of X_1, X_2, $\cdots$, as well, are postulated to have a normal distribution of amplitudes. For our purposes, however, this is not necessary, for by the central limit theorem X_1, X_2, etc., will be Gaussian, irrespective of the distribution law of their components as long as the latter are sufficiently numerous, under conditions easily fulfilled physically in most instances. For convenience, we shall assume that our random processes are initially entirely random in the normal sense. In either case $W_2(X_1, X_2; t)$ is given by (A1.3).

(2) Our choice of input spectra will determine whether or not the system is Markoffian.[12,13] If it is, then W_2 completely describes it; however in the present paper it is not usually critical that we be able to classify the process thoroughly; the chief concern is with the amplitude distribution and with the mean power spectrum, which may always be found when W_1 and W_2 are known, and here it is sufficient that the system be stationary.

(3) The mean power spectrum $W(f)$ may be obtained directly from the correlation function $R(t)$, defined by Eq. (A2.3), since by a well-known result[14–16] they are each other's cosine Fourier transforms:

$$W(f) = 4\int_0^\infty R(t) \cos \omega t dt; \quad \omega = 2\pi f; \quad R(t) = \int_0^\infty W(f) \cos \omega t df. \tag{1.6}$$

The correlation function, in turn, follows from the definition (A2.3) and with the help of W_2 is seen to be

$$R(t) = \iint_{-\infty}^{\infty} X_1 X_2 W_2(X_1, X_2; t) dX_1 dX_2. \tag{1.7a}$$

If the correlation function for the random wave $X(t)$ *after* passage through a non linear device is desired (1.7a) becomes

$$R(t) = \iint_{-\infty}^{\infty} g(X_1) g(X_2) W_2(X_1, X_2; t) dX_1 dX_2, \tag{1.7b}$$

---

[12] J. L. Dobb, Ann. Amer. Stat. **15**, 229 (1944).

[13] Reference 3, Sec. 7, and note II of the appendix for details.

[14] N. Wiener, Acta Math. **55**, 117 (1930).

[15] A. Khintchine, Math. Ann. **109**, 604 (1934).

[16] G. I. Taylor, Proc. Lond. Math. Soc. Sec. 2, **20**, 196 (1920), and Proc. Royal Soc. **164**, 476 (1938) gives applications to the theory of turbulence. Rice's interesting papers [1,2] treat a large number of noise problems with the help of (1–6), while ref. 3 puts more emphasis on the Brownian motion.

where $g(X)$ is the dynamic characteristic of the apparatus in question, and X_1 is the incoming disturbance at some initial time and X_2 at time t later.

(4) The explicit evaluation of the integral (1.7b) is often very difficult in this form because discontinuities in the characteristic g appear as finite limits in the integration. Then it is convenient, almost mandatory, to introduce the *Fourier transform* $f(iz)$ of the dynamic path, so that the output wave is given in terms of the input by[17]

$$g(X) = \frac{1}{2\pi}\int_{\mathbf{C}} f(iz)e^{iXz}dz. \tag{1.8}$$

The contour **C** extends from $-\infty$ to $+\infty$ along the real axis and is indented downward about a pole or branch point at the origin. The evaluation of (1.8) is effected when the contour is extended in an infinitely large semi-circle in a counterclockwise (positive) or clockwise (negative) sense, depending on whether the coefficient of iz in the exponent is positive or negative, respectively. When **C** is traversed in a positive direction, the residue at $z=0$ yields the output as the desired function of the input, while for a negative circuit of the contour, $g(X)$ vanishes. Other contours are also possible, and combinations of such paths may be used for complicated characteristics. In this way, we are able to distinguish between the transmission and the cutoff states of the apparatus. It should be mentioned that when the contour representation (1.8) is employed, the results for the correlation function (1.7b) will be expressed in terms of *the characteristic function* associated with W_2, namely Eq. (A1.1), in the general s-dimensional case, rather than by W_2 itself. This is not surprising, for it is easily shown that the characteristic function is the Fourier transform of the corresponding probability density. See Eqs. (2.13) to (2.16), in the next section.

(5) When a signal, modulated or unmodulated, is introduced along with the noise, the above concepts and artifices hold, with slight modification. For instance, (1.7a, 7b) now represent only the correlation function due to the noise components in the wave; a further average or averages are necessary to account for the signal and modulation components. Here the average over the phases of these components must be performed. In the next section this extension of the method is outlined and illustrated (see also references 2, 6, and 8).

2. Rectification of modulated signals in the presence of noise. Before determining what happens when a modulated wave is detected in the presence of noise, we must represent the incoming disturbance analytically. Let us consider first the signal component, which we shall denote by $V_s(t, t')$. Here the variable t refers to the time-variations of the carrier and t' to those of the modulation. Physically, t and t' both represent the same instant in time, our choice of symbols being merely one of mathematical convenience. For the type of signal of greatest interest here we may therefore write

$$V_s(t, t') = A_0(t') \cos \omega_0 t, \qquad \omega_0 = 2\pi f_0, \qquad A_0(t') \geqq 0, \tag{2.1}$$

where f_0 is the carrier frequency and $A_0(t')$ is some function of the time, which is only properly called the modulation when it varies slowly compared with $\cos \omega_0 t$. In general the carrier and modulation are not commensurable and hence are uncorre-

[17] W. R. Bennett and S. O. Rice, Phil. Mag. **18,** 422 (1934). For an extensive application of the contour representation, see reference 9.

lated; t and t' are consequently independent variables. On the other hand, there are cases when $A_0(t')$ and $\cos \omega_0 t$ bear a commensurable relation to each other, correlation exists, and t and t' are then functionally related. Such instances may occur, for example, when the carrier is over-modulated.

Now let us consider briefly the noise portion of the disturbance entering the non linear device. Several satisfactory representations are possible,[18] of which we choose the Fourier series form

$$V_N(t) = \sum_{n=1}^{N} \{a_n \cos \omega_n t + b_n \sin \omega_n t\}, \qquad \omega_n = 2\pi f_n = 2\pi n/T, \tag{2.2}$$

where the interval of expansion lies between 0 and T, and N is very much greater than unity. The quantities a_n and b_n are independent random variables, having the following properties:

$$\overline{a_n} = \overline{b_n} = 0; \quad \overline{a_n b_m} = 0; \quad \overline{a_n a_m} = \overline{b_n b_m} = w(f_n)\Delta f \delta_n^m = \{|S(f)|^2/T\}\delta_n^m, \tag{2.3}$$

this last from (A2.4). The bar indicates the statistical average over the various random quantities (see Appendix II); here δ_n^m is the familiar Kronecker delta, which has the values $\delta_n^m = 0$, $m \neq n$; $\delta_n^n = 1$. We assume the random process describing the noise to be stationary; this assumption can be verified experimentally. The time average for $T \to \infty$ and the ensemble average, in which the average of an (indefinitely) large number of finite intervals or "strips" corresponding to separate observations is taken, then yield the same results. Under these circumstances the distribution of the random variables a_n and b_n is unaltered. It is convenient to assume also that this distribution is Gaussian, with the standard deviation $[w(f_n)\Delta f]^{1/2}$, but as the central limit theorem shows, [see Appendix I (A1.1, A1.2, A1.3)], any other distribution with the same average and second moments [Eq. (2.3)] leads to identical expressions for the noise wave. Equations (2.3) may then be interpreted as follows: the first relation shows that $\overline{V_N(t)} = 0$, the second shows that a_n and b_n are independent, and the third apportions the mean power in the frequency range $f_n, f_n + \Delta f_n$ dissipated by a current flowing in a unit resistance when a potential difference $V_N(t)$ is maintained between the terminals. In the limit of a very long time, or what is the same thing, of a very large number of "strips" in the ensemble, $\Delta f(=1/T)$ becomes infinitesimal, and the summation may be replaced by an integral from $f = 0$ to $f = \infty$.

With the aid of Appendix I, we proceed to determine $W_1(X)$ and $W_2(X_1, X_2; t)$, cf. (1.1), where we identify V_N at some time t_0 with X_1 and some time t later with X_2. Referring to Eq. (A1.2), we find from (2.2) and (2.3) that on letting $X_{(j=n)} = a_n \cos \omega_n t + b_n \sin \omega_n t$,

$$\begin{aligned} \overline{X} &= \nu = \sum_{n=1} (\bar{a}_n \cos \omega_n t + \bar{b}_n \sin \omega_n t) = 0; \\ \mu_{11} &= \overline{X^2} = \sum_{n=1}^{N} (\overline{a_n^2} \cos^2 \omega_n t + \overline{b_n^2} \sin^2 \omega_n t) = \sum_{n=1}^{N} (\overline{a_n^2} + \overline{b_n^2})/2 \to \int_0^\infty w(f)df \equiv \psi(0), \end{aligned} \tag{2.4}$$

with $\psi(0) = \overline{X^2}$ the mean power in the wave. Clearly here the matrix $\boldsymbol{\mu}$ associated with W_2, (see A1.3), has the single element $\psi(0)$ and $|\mu| = \psi^{1/2}$, $\mu^{11} = 1$, so that the first order distribution for $V_N = X$ is, from (A1.3),

[18] See reference 1, Sec. 2.8, and reference 6.

$$W_1(X) = [2\pi\psi(0)]^{-1/2} \exp [- X^2/2\psi(0)]. \tag{2.5}$$

In a similar fashion the joint probability density $W_2(X_1, X_2; t)$ follows; we have

$$\overline{X}_1 = \overline{X}_2 = \nu_1 = \nu_2 = 0, \quad \text{and}$$

$$\begin{aligned} \mu_{11} &= \sum_{n=1}^{N} (\overline{a_n^2} \cos^2 \omega_n t_0 + \overline{b_n^2} \sin^2 \omega_n t_0) = \overline{X_1^2} = \psi(0) \\ \mu_{22} &= \sum_{n=1}^{N} (\overline{a_n^2} \cos^2 \omega_n (t_0 + t) + \overline{b_n^2} \sin^2 \omega_n (t_0 + t)) = \overline{X_1^2} = \psi(0), \end{aligned} \tag{2.6}$$

from (A1.2) and (A1.4), as a consequence of the conditions (2.3). For the off-diagonal moments one may write

$$\begin{aligned} \mu_{12} = \mu_{21} = \overline{X_1 X_2} &= \sum_{n=1}^{N} [\overline{a_n^2} \cos \omega_n t_0 \cos \omega_n (t_0 + t) + \overline{b_n^2} \sin \omega_n t_0 \sin \omega_n (t_0 + t)] \\ &= \sum_{n=1}^{N} (\overline{a_n^2} \cos^2 \omega_n t_0 + \overline{b_n^2} \sin^2 \omega_n t_0) \cos \omega_n t \\ &\to \int_0^\infty w(f) \cos \omega t df \equiv \psi(t). \end{aligned} \tag{2.7}$$

The quantity $\psi(t)$ is the correlation function of the input noise, and by (1.6),[19] the Fourier transform of the input spectrum $w(f)$. For the two-dimensional distribution $\boldsymbol{\mu}$ becomes

$$\begin{aligned} &= \begin{bmatrix} \mu_{11} & \mu_{12} \\ \mu_{21} & \mu_{22} \end{bmatrix} = \begin{bmatrix} \psi(0) & \psi(t) \\ \psi(t) & \psi(0) \end{bmatrix}; \quad |\mu| = \psi^2(1 - r^2(t)), \quad \text{and} \\ &\mu^{11} = \mu^{22} = \psi(0), \qquad \mu^{12} = \mu^{21} = - \psi(t); \end{aligned} \tag{2.8}$$

$r(t)$ is the normalized correlation function of the incoming disturbance, Eq. (A2.7). With the aid of (A1.3) the joint distribution is easily observed to take the familiar form[20]

$$W_2(X_1, X_2; t) = \frac{1}{2\pi\psi(1 - r^2)^{1/2}} e^{-(X_1^2 + X_2^2 - 2rX_1X_2)/2\psi(1-r^2)}. \tag{2.9}$$

The incoming wave is simply the sum of (2.1) and (2.2), with the various properties of the two components outlined above:

$$V_{in} = X + V_s(t', t). \tag{2.10}$$

Accordingly, if $g(V_{in})$ is the output of the device [see (1.7b)] the correlation function for the noise components is

$$R_N(t_0, t_0'; t) = \iint_{-\infty}^{\infty} g(X_1 + V_s[t_0, t_0'])g(X_2 + V_s[t_0 + t, t_0' + t]) W_2(X_1, X_2; t) dX_1 dX_2. \tag{2.11}$$

[19] See Appendix II for details.

[20] Henceforth we abbreviate $\psi(0)$ by ψ.

The complete correlation function, which includes the average over the phases of the signal as well as over the noise, is

$$R(t) = T_0'^{-1} \int_0^{T_0'} dt_0' \int_0^{T_0} T_0^{-1} R_N(t_0, t_0'; t) dt_0, \tag{2.12}$$

where T_0 and T_0' are the respective periods of the carrier and the modulation. The double integration in (2.12) reduces to a single operation in case t_0 and t_0' are functionally related, i.e., when carrier and modulation have a constant phase difference. Since W_2 is given explicitly [Eq. (2.9)] from our assumptions regarding the character of the noise, (2.11) and (2.12) represent the formal solution to our problem, in as much as the mean power spectrum follows immediately from (1.6). However, as these equations in their present form can be handled only for a few special cases, chiefly in the instance of "small-signal" rectification, where the distortion of the dynamic path arising from cut-off and saturation is gradual, we make use of the device of contour integration, (1.8), for the large and important class of problems in which cut-off and saturation take place abruptly. Then the outputs at times t_0 and t_0+t are respectively

$$g(V_1) = g(V_s(t_0, t_0') + X_1) = \frac{1}{2\pi} \int_{\mathbf{C}} f(iz) e^{iz(X_1 + V_s[t_0', t_0'])} dz, \tag{2.13a}$$

$$g(V_2) = g(V_s[t_0 + t, t_0' + t] + X_2) = \frac{1}{2\pi} \int_{\mathbf{C}'} f(i\xi) e^{i\xi(X_2 + V_s[t_0 + t, t_0 + t])} d\xi, \tag{2.13b}$$

where $\mathbf{C}'$ is a contour similar to $\mathbf{C}$ but in the ξ-plane. Equations (2.11) and (2.12) become finally

$$\begin{aligned} R(t) &= \frac{1}{4\pi^2} \int_{\mathbf{C}} f(iz) dz \int_{\mathbf{C}'} f(i\xi) d\xi \left\{ \iint_{-\infty}^{\infty} W_2(X_1, X_2; t) e^{izX_1 + i\xi X_2} dX_1 dX_2 \right\} \\ &\quad \times \left\{ \frac{1}{T_0 T_0'} \int_0^{T_0} \int_0^{T_0'} dt_0' \, dt_0 e^{izV_s(t_0', t_0) + i\xi V_s[t_0 + t, t_0' + t]} \right\} \\ &= \frac{1}{4\pi^2} \int_{\mathbf{C}} f(iz) dz \int_{\mathbf{C}'} f(i\xi) F_N(z, \xi, t) F_S(z, \xi; t) d\xi, \end{aligned} \tag{2.14}$$

in which $F_S(z, \xi; t)$ is by definition the characteristic function of the signal, given here by

$$F_S(z, \xi; t) = \frac{1}{T_0 T_0'} \int_0^{T_0'} dt_0' \int_0^{T_0} dt_0 e^{izV_s(t_0, t_0') + i\xi V_s(t_0 + t, t_0' + t)}, \tag{2.15}$$

and $F_N(z, \xi; t)$ is similarly seen to be the characteristic function for the noise, and may be written

$$\begin{aligned} F_N(z, \xi; t) &= \iint_{-\infty}^{\infty} W_2(X_1, X_2; t) e^{izX_1 + i\xi X_2} dX_1 dX_2 \\ &= \exp\left[-\tfrac{1}{2}\psi(z^2 + \xi^2) - \psi(t) z\xi\right]. \end{aligned} \tag{2.16}$$

This last result is obtained from (A1.1) with the help of (2.6)–(2.9). From (2.14), it appears that the characteristic function for a modulated signal and noise is merely the product of the separate characteristic functions, for physically the signal and noise components are uncorrelated, and hence their distributions must be independent.

For the modulated carrier Eq. (2.1) may be substituted directly in (2.15). The more usual case of no carrier-modulation correlation gives us

$$F_s(z, \xi; t) = \frac{1}{T_0'}\int_0^{T_0'} dt_0' \left\{\frac{1}{T_0}\int_0^{T_0} e^{izA_0(t_0')\cos\omega_0 t_0 + i\xi A_0(t_0'+t)\cos\omega_0(t_0+t)}dt_0\right\}$$

$$= \frac{1}{T_0'}\int_0^{T_0'} dt_0' \left\{\sum_{m=0}^{\infty}\sum_{p=0}^{\infty} \epsilon_m\epsilon_p i^{m+p} J_m(A_0[t_0']z)J_p(A_0[t_0'+t]\xi)\right.$$

$$\left.\times \frac{1}{T_0}\int_0^{T_0} dt_0 \cos\omega_0 m t_0 \cos p\omega_0(t_0+t)\right\}, \tag{2.17}$$

where we have used the familiar expansion of exp ($i\alpha \cos \beta t$) in a Fourier series,[21] and where $\epsilon_0 = 1$, $\epsilon_m = 2$, $m \geqq 1$. The trigonometric integral has the value $(\cos m\omega_0 t)\delta_m^p/\epsilon_m^p$, so that (2.17) may be written finally

$$F_s(z, \xi; t) = \sum_{m=0}^{\infty}(-1)^m\epsilon_m \cos m\omega_0 t\left\{\frac{1}{T_0'}\int_0^{T_0'} dt_0' J_m(A_0[t_0']z)J_m(A_0[t_0'+t]\xi)\right\}. \tag{2.18}$$

When (2.16) and (2.18) are substituted in the expression for the complete correlation function $R(t)$, we may expand exp $[-\psi(t)z\xi]$, obtaining

$$R(t) = \sum_{m=0}^{\infty}\sum_{n=0}^{\infty}\frac{(-1)^{m+n}\epsilon_m\psi(t)^n \cos m\omega_0 t}{n!}[H_{mn}(t_0')H_{mn}(t_0'+t)]_{\text{av.}}, \tag{2.19}$$

where

$$\left.\begin{aligned} H_{mn}(t_0') &= \frac{1}{2\pi}\int_C z^n f(iz)J_m(zA_0[t_0'])e^{-\psi z^2/2}dz,\\ H_{mn}(t_0'+t) &= \frac{1}{2\pi}\int_{C'} \xi^n f(i\xi)J_m(\xi A_0[t_0'+t])e^{-\psi\xi^2/2}d\xi. \end{aligned}\right\} \tag{2.20}$$

The average indicated above in (2.19) is the time average over the phases of the modulation. The mean power spectrum of the output wave follows from (1.6) and (2.19), and is

$$W(f) = \sum_{m=0}^{\infty}\sum_{n=0}^{\infty}(-1)^{m+n}\frac{\epsilon_m}{n!}C_{mn}(f), \tag{2.21}$$

where $C_{mn}(f)$ is given by

$$C_{mn}(f) = 4\int_0^{\infty}\psi(t)^n \cos m\omega_0 t \cos \omega t[H_{mn}(t_0')H_{mn}(t_0'+t)]_{\text{av.}}dt, \quad \omega = 2\pi f. \tag{2.22}$$

The quantity $[H_{mn}(t_0')H_{mn}(t_0'+t)]_{\text{av.}}$ is, in fact, a kind of correlation function, by formal comparison with (A2.1), except that here no statistical averages appear because $A_0(t)$ is periodic. We can also write formally

[21] Watson, *Theory of Bessel Functions*, Cambridge Univ. Press, 1944, sec. 2.22.

$$[H_{mn}(t_0')H_{mn}(t_0'+t)]_{\text{av.}}$$

$$= \frac{1}{4\pi^2}\int_C z^n f(iz)e^{-\psi z^2/2}dz \int_{C'} \xi^n f(i\xi)e^{-\psi\xi^2/2}[J_m(zA_0[t_0'])J_m(\xi A_0[t_0'+t])]_{\text{av.}}d\xi$$

$$= \sum_{k=0}^{\infty} \epsilon_k h_{kmn}^2 \cos k\omega_A t, \tag{2.23}$$

where $f_A(=\omega_A/2\pi)$ is the fundamental frequency of $A_0(t)$ and h_{kmn}^2 is a mean-square amplitude, which depends on ψ, the amplitude of $A_0(t)$, and on the dynamic characteristic of the apparatus in question.[22] Equation (2.23) is the result of developing the Bessel functions in series, or the expressions integrated first over z and ξ, and arranging the result as a series in cos $k\omega_A t$. The explicit evaluation of h_{kmn} is in general a difficult task. In some cases the work is simplified by taking the time average first, in others, by integrating over z and ξ initially. There seems to be no simple way of handling the integration problem presented by (2.23), but usually the average over the phases of the modulation must be performed last. Some illustrative examples are given later in this section.

From (2.23) we observe that (2.22) becomes

$$C_{mn}(f) = \sum_{k=0}^{\infty} \epsilon_k h_{kmn}^2 c_{kmn}(f), \tag{2.24a}$$

where

$$c_{kmn}(f) = 4\int_0^{\infty} \psi(t)^n \cos m\omega_0 t \cos k\omega_A t \cos \omega t dt; \tag{2.24b}$$

the mean output spectrum (2.21) finally takes the form

$$W(f) = \sum_{m=0}^{\infty}\sum_{n=0}^{\infty}\sum_{k=0}^{\infty} (-1)^{m+n}\frac{\epsilon_m\epsilon_k}{n!} h_{kmn}^2 c_{kmn}(f). \tag{2.25}$$

The effect of the rectifier or similar non-linear device is to "mix" or cross-modulate the noise and signal components with one another so that the (unfiltered) output contains the following three classes of modulation product: (a) *noise*×*noise*, which gives rise to noise, no longer random with Gaussian properties, (b) *noise*×*signal*, which results from the beating of the signal components with the noise wave and which in turn also yields non-Gaussian noise, and finally, (c) *signal*×*signal*, produced by the cross-modulation of the various signal components: this last is entirely periodic and free from noise. A d-c component is often present, but not when the dynamic characteristic is symmetrical, for then the average of the output, as well as the input, vanishes. Further subdivision is possible when there is modulation, as we may then distinguish between modulation products generated by the separate (incommensurable) components of the carrier and modulation, but essentially the three main classifications (a)–(c) still hold.

The different contributions to (a)–(c) may be distinguished in (2.25), since for (a), exclusive of the d-c, we may write, with $m=k=0$, $n\geq 1$,

[22] It is also possible to express the average in terms of the moments $\overline{A_{01}^m A_{02}^p}$, on expanding the Bessel functions with the help of the relation on page 148 of reference 21.

$$W(f)_{\text{noise}\times\text{noise}} = \sum_{n=1}^{\infty} \frac{(-1)^n}{n!} h_{00n}^2 c_{00n}(f), \tag{2.26a}$$

and for (b) we have $m \geqq 1$, $n \geqq 1$, $k \geqq 1$, giving

$$W(f)_{\text{noise}\times\text{signal}} = 4 \sum_{m=1}^{\infty} \sum_{n=1}^{\infty} \sum_{k=1}^{\infty} \frac{(-1)^{m+n}}{n!} h_{kmn}^2 c_{kmn}(f). \tag{2.26b}$$

Type (c) occurs only for $m \geqq 1$, $k \geqq 0$, $n = 0$ or $m \geqq 0$, $k \geqq 1$, $n = 0$:

$$W(f)_{\text{signal}\times\text{signal}} = \sum_{\binom{m=1}{=0}}^{\infty} \sum_{\binom{k=0}{=1}}^{\infty} \epsilon_k (-1)^m h_{km0}^2 \{\delta(f - mf_0 - kf_A) + \delta(f - mf_0 + kf_A)\}, \tag{2.26c}$$

where we have used the delta function to indicate the "discrete" nature of these components, located in frequency at $mf_0 \pm kf_A$. The spectral distributions (a) and (b), on the other hand, representing noise, are continuous. There remains only the d-c, which is specified by $m = n = k = 0$:

$$W_{\text{d-c}} = h_{000}^2 \, 4 \int_0^\infty df \left(\int_0^\infty \cos \omega t dt \right) = h_{000}^2 \, 2 \int_0^\infty \delta(f - 0) df = h_{000}^2. \tag{2.26d}$$

The parts of the correlation function which contribute to (a)–(c) are easily found from (2.19), (2.21), and (1.6).

We turn now to a number of interesting special cases:

I. Unmodulated carrier. This is the first and simplest case involving a signal; A_0 is a constant quantity representing the amplitude of the carrier wave. It is immediately clear that the only non-vanishing terms require $k = 0$ (m, $n \geqq 0$), so that the correlation function reduces to the form given by Eqs. (4.9-7) of reference 2, while $h_{mn} (\equiv h_{mn0})$ is expressed by Eq. (4.9-6). For *noise alone* entering the non linear device, the expressions (2.19)–(2.26d) are still simpler, reducing to the forms given by Eqs. (3.1)–(3.6) of reference 9.

II. Carrier modulated by sine wave.[23] For this important case we have for the signal (2.1)

$$V_s = A_0(1 + \lambda \cos \omega_A t) \cos \omega_0 t, \quad A_0(t_0') = A_0(1 + \lambda \cos \omega_A t_0'), \quad 0 \leqq \lambda \leqq 1, \tag{2.27}$$

where λ is the modulation index and A_0 is the (peak) amplitude of the carrier component. For the moment no restriction is set on f_A and f_0, save that they be incommensurable, which is another way of stating that $A_0(t)$ and $\cos \omega_0 t$ are uncorrelated. Of course, for $A_0(t)$ to be spoken of as a modulation in the usual sense, the frequency f_A must be much less than f_0.

We require the value of $[H_{mn}(t_0')H_{mn}(t_0' + t)]_{\text{av.}}$. Expansion of the Bessel functions in (2.20) yields various moments of the modulation, viz. $[A_0^{\alpha_1}(t_0')A_0^{\alpha_2}(t_0' + t)]_{\text{av.}}$. The average over t_0' follows in the present instance, *cf.* (2.27), with the aid of the relation

[23] For some earlier work on this problem, see J. R. Ragazzini, Proc. I.R.E. **30**, 227, 1942. See also sections 4.1, 4.2, 4.10 of ref. 2. A theoretical discussion of this problem in the case of half-wave rectification has been given by the author in a paper submitted to the I.R.E. A companion paper by Fubini and Johnson verifies the results experimentally.

$$I_{a1,a2}(\phi) = \frac{1}{2\pi}\int_0^{2\pi} (1 + \lambda \cos \theta)^{a_1}(1 + \lambda \cos [\theta + \phi])^{a_2} d\theta, \qquad \phi = \omega_A t,$$

$$= \sum_{j=1}^{a_2} 2^{1-2j}\lambda^{2j}\, {}_{a_2}C_j(-1)^j \sum_{i=0}^{j/2,(j-1)\frac{1}{2}} {}_jC_i\, {}_j\Delta_i[-a_1]_{j-2i} \frac{(2/\lambda)^{2i}}{(j-2i)!} \tag{2.28}$$

$$\times \cos (j-2i)\phi\, {}_2F_1[\tfrac{1}{2}(j - 2i - a_1), \tfrac{1}{2}(j - 2i - a_1 + 1); 1 + j - 2i; \lambda^2], \tag{2.29}$$

where a_1 and a_2 are integers, the C's are the usual binomial coefficients, and ${}_j\Delta_i = \frac{1}{2}$ when $j = 2i$, and ${}_j\Delta_i = 1, j \neq 2i$. The limits on the second summation apply accordingly as j is even or odd, and $[\alpha]_\beta = \alpha(\alpha+1) \cdots [\alpha+\beta-1]$, $[\alpha]_0 = 1$. The expression (2.29) follows from the development of $(1+\lambda \cos [\theta+\phi])^{a_2}$ in a series in $\cos \theta$ and the result[24]

$$\frac{1}{2\pi}\int_0^{2\pi} (1 + \lambda \cos \theta)^l \cos k\theta d\theta$$

$$= \frac{(-l)_k(-\lambda)^k}{2^k k!}\, {}_2F_1[\tfrac{1}{2}(k - l), \tfrac{1}{2}(k - l + 1); k + 1; \lambda^2]. \tag{2.30}$$

From this it is also easily seen that when $\phi = 0$

$$I_{a1,a2}(0) = {}_2F_1[-\tfrac{1}{2}(a_1 + a_2), \tfrac{1}{2}(-a_1 - a_2 + 1); 1; \lambda^2]. \tag{2.31}$$

Equation (2.31) is helpful if one desires to determine the mean power in the wave. A short table of $I_{a_1,a_2}(\phi)$ is given below.

TABLE I

	$a_2=0$	1	2	3	4	5
$a_1=0$	1	1	$1+\lambda^2/2$	$1+3\lambda^2/2$	$1+3\lambda^2/2+3\lambda^4/8$	$1+5\lambda^2$ $+15\lambda^4/8$
1	1	$1+(\lambda^2/2)\cos\phi$	$1+\lambda^2(\frac{1}{2}+\cos\phi)$	$1+(3\lambda^2/2)(1+\cos\phi)$ $+(3\lambda^4/8)\cos\phi$	$1+\lambda^2(3+2\cos\phi)$ $+(3\lambda^4/8)(1+4\cos\phi)$	
2	$1+\lambda^2/2$	$1+\lambda^2(\frac{1}{2}+\cos\phi)$	$1+\lambda^2(1+2\cos\phi)$ $+(\lambda^4/8)(2+\cos 2\phi)$	$1+\lambda^2(2+3\cos\phi)$ $+(3\lambda^4/8)(2+2\cos\phi+\cos 2\phi)$		
3	$1+3\lambda^2/2$	$1+(3\lambda^2/2)(1+\cos\phi)$ $+(3\lambda^4/8)\cos\phi$	$1+\lambda^2(2+3\cos\phi)$ $+(3\lambda^4/8)(2+2\cos\phi+\cos 2\phi)$			
4	$1+3\lambda^2/2$ $+3\lambda^4/8$	$1+\lambda^2(3+2\cos\phi)$ $+(3\lambda^4/8)(1+4\cos\phi)$				
5	$1+5\lambda^2$ $+15\lambda^4/8$					

Once the correlation function has been found it is comparatively easy to determine the mean power in the output by setting $t=0$ in the appropriate expressions for $R(t)$. As shown by (A2.6) and as discussed more fully in Appendix II, this is equivalent to integrating over all spectral components, "discrete" or continuous. It is also evident that the *mean total power* W_T in the output wave *is independent of spectral shape* of the input distribution provided all significant components are included in our calculation. Thus from (2.19) and (2.23) we may write, setting $t=0$,

[24] Reference 2, Eq. (4.2.17).

$$W_\tau = R(0) = \sum_{m=0}^{\infty}\sum_{n=0}^{\infty} \frac{(-1)^{m+n}}{n!} \epsilon_m \psi(0)^n [H_{mn}(t_0')^2]_{\text{av.}}$$

$$= \sum_{m=0}^{\infty}\sum_{n=0}^{\infty}\sum_{k=0}^{\infty} \frac{(-1)^{m+n}\epsilon_m\epsilon_k\psi(0)^n}{n!} h_{kmn}^2 \tag{2.32}$$

for a general, modulated carrier rectified in the presence of random noise. The power in the continuum W_τ is readily obtained from (2.32) and (2.26d), viz., $W_\tau = W_\tau - W_{d-c} = W_\tau - h_{000}^2$, or by omitting from the sum the term in $k=m=n=0$ in (2.32). In a similar fashion, to determine the portion of the total power output attributable to modulation products formed by (a), *noise*$\times$*noise*, (b), *noise*$\times$*signal*, and (c), *signal* $\times$*signal*, we have only to consider in (2.32) those terms for which $m=k=0$, $n\geqq 1$, $m\geqq 1$, $n\geqq 1$, $k\geqq 1$, and $m\geqq 1$, $k\geqq 0$, $n=0$ or $m\geqq 0$, $k\geqq 1$, $n=0$, respectively.

One may also calculate the mean power in the output without knowledge of the correlation function, for if $W_1(V_{in})$ is the first order probability density for the input wave (2.10), then by definition the mean total power dissipated by the rectified disturbance in a unit resistance is

$$W_\tau = \overline{g(V_{in})^2} = \int_{-\infty}^{\infty} g(V_{in})^2 W_1(V_{in}) dV_{in}, \tag{2.33a}$$

with a corresponding expression for the d-c component:

$$W_{\text{d-c}} = [\overline{g(V_{in})}]^2 = \left\{\int_{-\infty}^{\infty} g(V_{in}) W_1(V_{in}) dV_{in}\right\}^2. \tag{2.33b}$$

The quantity $W_1(V_{in})$ has been derived independently by Rice[25] and Bennett[26] for the case of the simple sine wave and noise. For noise alone it is given by (2.5); the distribution is more complicated when a signal is present. The drawback to (2.33a) as a method of determining the power output, however, lies in the difficulty of performing the integrations when g is discontinuous, for instance in the half-wave linear rectifier, where $g>0$, $V_{in}>0$; $g=0$, $V<0$. Equation (2.33a) is useful only when the characteristic g exhibits no abrupt variations, cf., "small-signal" detection, in which the incoming disturbance has a mean square amplitude small compared with the curvature of the dynamic path (see Sec. 8), or when noise alone is subject to rectification, see Ref. 9. In such cases (2.33a) is to be preferred because it yields results in terms of a finite number of (tabulated) functions, whereas the corresponding expression for W_τ obtained from $R(0)$ is given as an infinite series which does not converge too rapidly. On the other hand, use of the correlation function has the advantage that from it can be predicted the order and nature of particular modulation products or groups of products, i.e., whether the contribution arises from *noise*$\times$*noise* or *noise*$\times$*signal*, etc. This, as we shall see in the next section, is an important property of the correlation function, especially useful when the incoming wave is spectrally narrow compared with its central frequency, for then the output is composed of separate harmonic "zones," any one of which may in principle be isolated with a suitable filter and examined.

3. Narrow band waves. So far, in discussing spectra and correlation functions no restrictions have been placed on the spectral nature of the input wave, which is com-

[25] Reference 2, Eq. (3.10.6).
[26] Reference 6, Eq. (3.6).

posed, in its general form, of periodic or of a set of periodic components and of a noise term. Moreover, in determining the mean power spectrum of the output disturbance, no additional modification of the wave has been assumed other than that attributable to the non linear $i_p - e_g$ (output current vs. applied voltage) path of the rectifying device. In somewhat different language this means that the circuits of our apparatus contain only filters whose amplitude response is uniform (for all frequencies) and whose phase shifts vary in linear fashion with the frequency. This means that we restrict ourselves to non linear elements whose instantaneous output does not affect the time dependence of the incoming wave, i.e., to resistive loads, no feedback, or at worst to plate (and grid) circuits whose reactive components are small compared with their resistive ones—all of which is required for essentially one-valued dynamic paths (Fig. 1)—then we may introduce the effect of filters into the analysis, observing that now they simply modify the *frequency* spectrum. In the case of filtering the input the spectral shape is altered, but the quality of the wave is not altered; it is still a Gaussian random process. On the other hand, if the output is filtered, not only may the original, i.e., the unfiltered output, be changed spectrally, but also the quality of

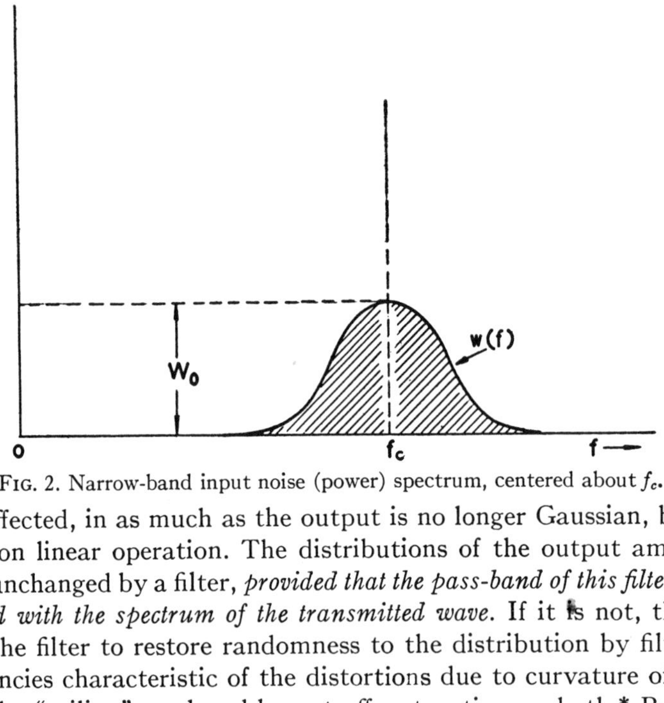

FIG. 2. Narrow-band input noise (power) spectrum, centered about f_c.

the wave is affected, in as much as the output is no longer Gaussian, because of the intervening non linear operation. The distributions of the output amplitude, however, remain unchanged by a filter, *provided that the pass-band of this filter is sufficiently wide compared with the spectrum of the transmitted wave.* If it is not, then there is a tendency of the filter to restore randomness to the distribution by filtering out the higher frequencies characteristic of the distortions due to curvature of the dynamic path and to the "ceiling" produced by cut-off, saturation, or both.* But we can still determine the mean power and mean power spectrum, although the distributions $W_1, W_2, \cdots, W_n$ are at present beyond our powers, by observing that

$$W(f)_{\text{filter in output}} = W(f)_{\text{inf. filter in output}} \cdot |G(f)|^2,$$

where $|G(f)|$ is the modulus of the output filter response.

* *Note added in proof:* The difficult problem of determining the distribution of the amplitudes after rectification and *arbitrary* filtering has been solved by M. Kac and A. J. F. Siegert, J. Appl. Phys., **18**, 383 (1947) for a quadratic detector.

When the spectral width, measured between points at which the mean power spectrum is but a few percent of its maximum value, is small compared to the central or "resonant" frequency, we term the wave narrow-band. A precise definition, of course, requires a knowledge of the spectral shape, whether it has one or more maxima, where we measure width, etc., but here we shall assume that the spectrum is a simple, symmetrical distribution about a single maximum, as shown in Fig. 2. Now an important property of narrow band noise waves is that their correlation functions may be expressed as the product of the correlation functions of a slowly and of a rapidly varying term, corresponding to an envelope or "modulation," and to a "carrier," having the central or resonant frequency f_c of the disturbance. (This is shown to hold for symmetrical noise spectra in Appendix II (A2.8); for a signal the same demonstration may be made, except now, of course, we have a series of discrete components distributed about a true carrier, rather than a continuum about a central frequency.) The correlation function and the spectrum of the output follow directly from (2.14) and (1.6) as before, but instead of expanding exp $[-\psi(t)z\xi]$ at once to obtain the form (2.19), we may use this property of the input correlation, viz:

$$\psi(t) = \psi(0)r_0(t) \cos \omega_c t \tag{A2.8}$$

to show that the outgoing wave is composed of an infinite number of spectral bands centered about harmonics of the central or of the carrier frequency, or of their modulation products. Generally speaking, these bands will overlap to varying degrees, if f_0 and f_c are different; if $f_0 = f_c$, which is the important case in practice, the modulation products involving f_0 and f_c will coincide exactly. The different behaviors are apparent when we attempt to examine a given spectral region, say one centered about the carrier f_0. In the former instance $(f_0 \neq f_c)$ our filter* will exclude part of the spectral zone due to the one and include some of the zone associated with the other, while for the latter $(f_c = f_0)$, our filter passes contributions symmetrically disposed about f_0. For the harmonic regions of higher spectral order, i.e., lf_0, lf_c, $l \geqq 2$, the former effect becomes more pronounced, since now the maxima of the distributions due to the noise and signal are separated by approximately $l|f_0 - f_c|$ cycles. Further there is a "smearing-out" of the spectra, because all orders of *noise*$\times$*noise*, *noise* $\times$*signal*, etc., components are generated. This accounts for the broadening and filling-in of the various output spectra; smearing of the carrier and modulation is also observed. When $l = 0$, we obtain the low-frequency output, or envelope, distorted or not depending on the nature of the tube's dynamic path. This spectral region is usually of principal interest in reception, while in transmission the zone $l = 1$ is of chief concern, where now our rectifier assumes the rôle of mixer (though with a different characteristic, of course). Figure 3 illustrates this discussion.

Now with the help of (A2.8) and the expansion

$$e^{-\psi r_0(t)\cos \omega_0 t} = \sum_{p=0}^{\infty} \epsilon_p(-1)^p I_p(z\xi\psi r_0(t)) \cos p\omega_c t \tag{3.1}$$

we may express the correlation function (2.14) as

* For this purpose all our filters are assumed, to be ideal, i.e., to have a uniform response over a frequency range wide enough to include (almost) all the components in a given zone. We say almost, because the non linear device produces some frequencies which will lie outside the finite pass-band (see the end of Appendix II).

$$R(t) = \sum_{m=0}^{\infty}\sum_{p=0}^{\infty} \epsilon_m\epsilon_p(-1)^{m+p}\cos p\omega_c t\cos m\omega_0 t\left\{\frac{1}{4\pi^2}\int_C f(iz)J_m(zA_0[t_0'])e^{-\psi z^2/2}dz\right.$$

$$\left.\times\int_{C'} f(i\xi)J_m(\xi A_0[t_0'+t])e^{-\psi\xi^2/2}d\xi\cdot I_p(z\xi r\psi)\right\}_{\text{av.}} \quad (3.2)$$

Equation (3.2) is general for all f_0 and f_c. However, in the case of principal interest carrier and central frequency coincide, so that $f_0=f_c$, and (3.2) reduces to a series of

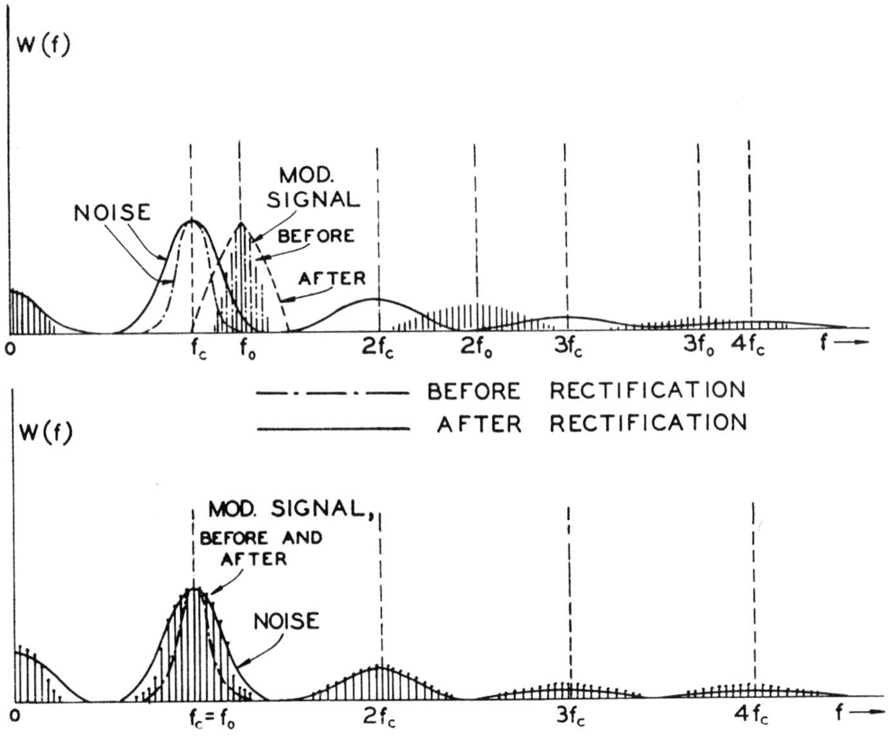

FIG. 3. Narrow-band noise and modulated signal spectra, before and after rectification. The upper of the two applies when $f_c \neq f_0$, where the spectral width is much less than either f_c or f_0. The lower illustrates the spectra when $f_c=f_0$. The normalization is arbitrary.

distinct harmonic regions distributed about lf_0, $l=0, 1, 2, 3 \cdots$, the harmonics of the carrier. Thus, from the fact that replacing $\sum_{m=0}^{\infty}\sum_{p=0}^{\infty}\epsilon_m\epsilon_p(\;)$ by $\sum_{m=-\infty}^{\infty}\sum_{p=-\infty}^{\infty}(\;)$, since $J_m(\alpha)J_m(\beta)=J_{-m}(\alpha)J_{-m}(\beta)$ and $I_p(\gamma)=I_{-p}(\gamma)$, and collecting all terms of $\cos p\omega_0 t \cos m\omega_0 t$ which yield $\cos l\omega_0 t$, we observe that the correlation function for the lth band is

$$R_l(t) = \frac{(-1)^l \cos l\omega_0 t}{4\pi^2}\sum_{m=0}^{\infty}\epsilon_{|m-l|}\left\{\int_C f(iz)J_m(A_0(t_0']z)e^{-\psi z^2/2}dz\right.$$

$$\left.\times\int_{C'} f(i\xi)J_m(A_0[t_0'+t]\xi)I_{|m-l|}(\psi\cdot z\xi r_0)e^{-\psi\xi^2/2}d\xi\right\}_{\text{av.}}$$

$$= (-1)^l\cos l\omega_0 t\sum_{m=0}^{\infty}\frac{\epsilon_{|m-l|}\psi^{|m-l|}}{2^{|m-l|}}\sum_{q=0}^{\infty}\frac{\psi^{2q}r_0(t)^{2q+|m-l|}}{2^{2q}q!(q+|m-l|)!}$$

$$\times\left[H_{m,|m-l|+2q}(t_0')\cdot H_{n,|m-l|+2q}(t_0'+t)\right]_{\text{av.}}, \quad (3.3)$$

on expanding the Bessel function $I_{|m-l|}$. The complete correlation function is, of course, $\sum_{l=0}^{\infty} R_l(t)$. By (A2.6) and (2.23) the power in any zone is

$$R_l(0) = (-1)^l \sum_{m=0}^{\infty} \frac{\psi^{|m-l|}\epsilon_{|m-l|}}{2^{|m-l|}} \sum_{q=0}^{\infty} \frac{\psi^{2q} 2^{-2q}}{q!(q + |m-l|)!} \sum_{k=0}^{\infty} \epsilon_k h^2_{k,m,|m-l+2q|}, \tag{3.4}$$

since $r_0(0) = 1$. Here also we see an illustration of the theorem stated previously, cf. (2.32) and Appendix II, namely, that the power in a spectral region, here an harmonic zone rather than the entire output spectrum, is independent of the spectral shape of the input noise and also of the signal spectrum. This naturally simplifies the calculation of the power, once $R_l(t)$ has been found.

4. The biased νth-law rectifier. In the preceding section certain general aspects of the theory of modulated signals and noise passed through non linear devices have been outlined. Specific results in each instance depend on the evaluation of integrals of the form (2.20). Here attention is directed to the important case of the biased νth-law detector, whose dynamic characteristic $g(V)$ is given by

$$\begin{aligned} I = g(V) &= \beta(V - b_0)^\nu, \qquad V > b_0, \qquad \nu > 0 \\ &= 0, \qquad\qquad\qquad V < b_0, \end{aligned} \tag{4.1}$$

and β has suitable dimensions for the output to be a current when the incoming wave is a voltage; b_0 represents the cut-off voltage, measured from the operating point in the manner of Fig. 1. The Fourier transform of g is readily shown to be

$$f(iz) = \int_0^\infty e^{-(iz)V} g(V)dV = \frac{\beta\Gamma(\nu + 1)e^{-izb_0}}{(iz)^{\nu+1}}. \tag{4.2}$$

With the help of (4.2) the integrals (2.20) may now be determined, in a variety of ways leading to different, but equivalent results, the form of either depending on whether exp $(-ib_0z)$ or $J_m(A_0z)$ is expanded in a series, followed in each instance by termwise integration. The former approach yields a series in $b \equiv b_0/\psi^{1/2}$, where the coefficients of the various powers of b are themselves series in the parameter $p(t) \equiv [A_0(t)]^2/2\psi$, while the latter is expressible as a series in $p(t)$, the coefficient of whose general term likewise is an expansion in b. The two developments, while valid for all values of the parameters b, $p(t)$, and ψ, vary in their usefulness for different ranges of b, $p(t)$, and ψ.

Let us consider first the case where exp $(-izb_0)$ is given as a series. The general integral (2.20) becomes accordingly, with the aid of (4.2),

$$H_{mn}(t) = \frac{\beta\Gamma(\nu + 1)}{2\pi} \int_C z^n e^{-izb_0} J_m(zA_0(t)) e^{-\psi z^2/2} dz/(iz)^{\nu+1} \tag{4.3}$$

$$\begin{aligned} &= \frac{\beta\Gamma(\nu+1)}{2\pi} i^{-\nu-1} \sum_{k=0}^{\infty} \frac{(-i)^k b_0^k}{k!} \int_C z^{k+n+\nu-1} J_m(zA_0(t)) e^{-\psi z^2/2} dz \\ &= \frac{\beta\Gamma(\nu+1)e^{-(m+n)\pi i/2}}{2m!} \left(\frac{\psi}{2}\right)^{(\nu-n)/2} p(t)^{m/2} \\ &\quad \cdot \sum_{k=0}^{\infty} \frac{(-1)^k 2^{k/2} b^k {}_1F_1[(k+m+n-\nu)/2; m+1; -p(t)]}{k!\Gamma[(2+\nu-k-m-n)/2]}, \end{aligned} \tag{4.4}$$

this last from Appendix III, Eq. (A3.15), where also are mentioned some of the properties of the confluent hypergeometric function ${}_1F_1$. For an unmodulated carrier $p(t)$ is merely a constant $p=A_0^2/2\psi$. Observe also that when $(2+\nu-k-m-n)$ is zero or an even *negative* integer the modulation products contributed by such terms vanish, in virtue of the poles of $\Gamma[(2+\nu-k-m-n)/2]$. Further, when ν is integral, only odd values of k appear, $m+n$ even, and only even values of k when $m+n$ is odd (except for $2+\nu>k+m+n>0$). Equation (4.4) is best suited for computation when $b=b_0/\psi^{1/2}$ is of the order of unity or less, even for a wide range of values of $p(t)$. Values of $b>1$ yield too slow a convergence, and the alternative development is then needed.

The expansion of the Bessel function in (4.3) gives us

$$H_{mn}(t) = \frac{\beta\Gamma(\nu+1)}{2\pi} i^{-\nu-1} \sum_{k=0}^{\infty} \frac{(-1)^k A_0(t)^{2k+m}}{k!(k+m)!2^{2k+m}} \int_C z^{2k+m+n-\nu-1} e^{-izb_0-\psi z^2/2} dz,$$

and from Eq. (A3.17) we obtain finally

$$H_{mn}(t) = \frac{\beta\Gamma(\nu+1)}{2} e^{-\pi i(m+n)/2} \left(\frac{\psi}{2}\right)^{(\nu-n)/2} p(t)^{m/2} \sum_{k=0}^{\infty} \alpha_{mnk} p(t)^k, \tag{4.5}$$

where

$$\alpha_{mnk} = \frac{1}{k!(k+m)!} \left\{ \frac{{}_1F_1[(2k+m+n-\nu)/2; 1/2; -b^2/2]}{\Gamma[(2+\nu-2k-m-n)/2]} - \sqrt{2}\, b \frac{{}_1F_1[(2k+m+n-\nu-1)/2; 3/2; -b^2/2]}{\Gamma[(1+\nu-2k-m-n)/2]} \right\}. \tag{4.6}$$

Here large values of b are clearly more easily handled, especially when the asymptotic series for ${}_1F_1$, cf. (A3.3), can be used. Also equations (4.5) and (4.6) in general offer a more satisfactory form from which to determine the time average over the phases of the modulation, viz:

$$[H_{mn}(t_0')H_{mn}(t_0'+t)]_{\text{av.}} = \frac{\beta^2\Gamma(\nu+1)^2(-1)^{m+n}}{4} \left(\frac{\psi}{2}\right)^{\nu-n} \sum_{k_1=0}^{\infty} \sum_{k_2=0}^{\infty} \alpha_{mnk_1}\alpha_{mnk_2} [p(t_0')^{m/2+k_1} p(t_0'+t)^{m/2+k_2}]_{\text{av.}}, \tag{4.7}$$

when the spectrum of the output is desired, cf. Eqs. (2.21), (2.22), since a double series is sufficient, whereas if (4.4) were used, a fourfold infinite series would result for the average. A similar superiority of (4.7) over (4.4) is noted in the limiting case of $p(t)\ll 1$. Observe also from (A3.9) that for integral values of $\nu(>0)$ the various α_{mnk} reduce to the functions defined in Appendix III (A3.4), and are thus expressible as Hermitian polynomials.

A detailed discussion of the above in the case of the biased linear rectifier or mixer is reserved for a later paper.

Limiting cases are of interest: We consider first:

Case I: $b_0 \to 0$, $A_0(t)$, $\psi \neq 0$ *and finite*. Equations (4.4) and (4.7) reduce, equivalently, to

$$H_{mn}(t)_{b_0\to 0} = \frac{\beta\Gamma(\gamma+1)}{2m!} e^{-(m+n)\pi i/2} \left(\frac{\psi}{2}\right)^{(\nu-m)/2} p(t)^{m/2} \left\{ \frac{{}_1F_1[(n+m-\nu)/2; m+1; -p(t)]}{\Gamma[(2+\nu-m-n)/2]} - b\sqrt{2} \frac{{}_1F_1[(1+m+n-\nu)/2; m+1; -p(t)]}{\Gamma[(1+\nu-m-n)/2]} + \cdots \right\}. \tag{4.8}$$

From (4.8) the expressions for half-wave rectification, $b=0$, follow immediately.

Case II: $b_0 \to \pm\infty$, $A_0(t)$, ψ *finite*. Here it is convenient to use the second form (4.5) of $H_{mn}(t)$ and apply the asymptotic series (A3.3) for ${}_1F_1$. It then may be shown that

$$H_{mn}(t)_{b_0 \to \infty} \cong 0. \tag{4.9}$$

since $\alpha_{mnk} \to 0$ for *all* terms in the asymptotic development. This is to be expected physically, since with an indefinitely large cut-off voltage the finite signal amplitudes and even the possibly infinite noise peaks are not transmitted, the latter because they arise a vanishingly small fraction of the time, since the mean power in the input signal and noise waves is finite. On the other hand, when $b_0 \to -\infty$, we find from (A3.3) and (4.5) that

$$H_{mn}(t)_{b_0 \to -\infty} \cong \frac{\beta\Gamma(\nu+1)e^{-(m+n)\pi i/2}\sqrt{\pi}\, 2^{n-\nu} \left| b_0 \right|^{-m-n+\nu} A_0(t)^m}{m!\Gamma[(1+\nu-m-n)/2]\Gamma[(2+\nu-m-n)/2]}$$
$$\times \left\{ 1 + \frac{\psi}{2b_0^2}\left[1 - \frac{A_0^2(t)}{2\psi(m+1)}\right](m+n-\nu+1)(m+n-\nu) + \cdots \right\}, \tag{4.10}$$

which shows that for $\nu > m+n$ the output contains increasingly large terms, while for $\nu < m+n$ all such contributions become negligibly small. As an example, consider the linear rectifier, $\nu = 1$. Here only the terms $(m, n) = (0, 0), (0, 1), (1, 0)$ are significant:

$$H_{00} = \beta \left| b_0 \right|; \quad H_{01} = -i\beta; \quad H_{10}(t) = -i\beta A_0(t)/2; \quad H_{mn}(t) \to 0,\ m+n > 1. \tag{4.10a}$$

The first relation becomes infinite, since infinite (negative) cut-off voltage must be supplied when $b_0 \to -\infty$; the second result, when substituted in (2.21) and (2.22), gives the mean power spectrum of the output noise, which is easily seen to be the same as the input spectrum except for the constant factor β, and finally, the third expression is observed to be the mean power spectrum of the input signal, undistorted because now the linear rectifier is essentially a linear amplifier. In general (4.10) may be said to hold for a small-signal νth-law detector (see Sec. 7).

Case III: $A_0(t) \to 0$, b_0, ψ *finite*. Here the modulation or carrier contribution is allowed to become small. Then the principal term in $H_{mn}(t)$ obtains when $m=0$, and correction terms when $m=0, k=1, m=1, k=0$, and $m=2, k=0$. Either (4.4) or (4.5) reduces to the following

$$H_{0n}(t)_{p(t)\to 0} = \frac{\beta\Gamma(\gamma+1)e^{-\pi n i/2}}{2}\left(\frac{\psi}{2}\right)^{(\nu-n)/2}\{\alpha_{0n0} + \alpha_{0n1}p(t) + \cdots\},$$
$$H_{1n}(t) = -\frac{i\beta\Gamma(\nu+1)e^{-\pi n i/2}}{2}\left(\frac{\psi}{2}\right)^{(\nu-n)/2}\{p(t)^{1/2}\alpha_{1n0} + \cdots\}, \tag{4.11}$$
$$H_{2n}(t) = -\frac{\beta\Gamma(\gamma+1)e^{-\pi n i/2}}{2}\left(\frac{\psi}{2}\right)^{(\nu-n)/2}\{p(t)\alpha_{2n0} + \cdots\}.$$

For noise alone we have

$$H_{0n}(t) = h_{0n} = \frac{\beta\Gamma(\nu+1)e^{-\pi n i/2}}{2}\left(\frac{\psi}{2}\right)^{(\nu-n)/2}\left\{\frac{{}_1F_1[(n-\nu)/2; 1/2; -b^2/2]}{\Gamma[(2+\nu-n)/2]} - \right.$$

$$- b\sqrt{2}\frac{{}_1F_1[(n-\nu+1)/2;3/2;-b^2/2]}{\Gamma[(1+\nu-n)/2]}\Bigg\},\quad (4.12)$$

and in the special cases $\nu=1$ and $\nu=2$ we obtain with the aid of (A3.9) the equations of Reference 9 when there is no saturation.

Case IV: $A_0(t)\to\infty$, b_0, ψ *finite*. The signal is taken to be very large, and again the asymptotic relation (A3.3) is in order, this time applied to (4.4) as $p(t)\to\infty$. The result is

$$H_{mn}(t)_{A_0(t)\to\infty} \cong \frac{\beta\Gamma(\nu+1)(-1)^{(m+n)/2}2^{n-\nu-1}A_0(t)^{\nu-n}}{2\Gamma[(2+\nu-n+m)/2]}\Bigg\{\frac{1}{\Gamma[(2+\nu-n-m)/2]}$$
$$-\frac{2b_0\Gamma[(2+\nu-n+m)/2]}{A_0(t)\Gamma[(1+\nu-n+m)/2](1+\nu-n-m)/2}+\cdots\Bigg\},\quad (4.13)$$

and including terms in $A_0(t)^{-1}$ we may write

$$H_{m0}(t) \cong \beta\Gamma(\nu+1)2^{-\nu-1}(-i)^mA_0(t)^\nu\Bigg\{\frac{1}{\Gamma[(2+\nu+m)/2]\Gamma[(2+\nu-m)/2]}$$
$$-\frac{2b_0\Gamma(2+\nu+m)}{A_0(t)\Gamma[(1+\nu+m)/2]\Gamma[(1+\nu+m/2)]}+\cdots\Bigg\},$$
$$H_{m1}(t) \cong i\beta\Gamma(\nu+1)2^{-\nu}(-i)^mA_0(t)^\nu$$
$$\times\Bigg\{\frac{1}{\Gamma[(1+\nu+m)/2]\Gamma[(1+\nu-m)/2]A_0(t)}+\cdots\Bigg\},\ \text{etc.}\quad (4.14)$$

This shows the complete suppression of the noise when $A_0(t)\gg\psi^{1/2}$. Further, since the cut-off and r-m-s noise voltages are comparable in that they are far smaller than the signal, they, too, are negligible in their effect, and we have essentially half-wave rectification of a modulated signal alone. In the special instance of the low-frequency output of the half-wave linear detector, we have from (4.14), $m=0$, $H_{00}(t)\cong\beta A_0(t)/\pi$, in agreement with Eq. (40) of reference 8.

Case V: $\psi\to\infty$, b_0, $A_0(t)$ *finite*. This corresponds to the case in which the noise overwhelms the signal and much exceeds the cut-off voltage. Equations (4.4) or (4.5) reduce to

$$H_{mn}(t)_{\psi\to\infty} \cong \frac{\beta\Gamma(\nu+1)}{2m!}(-1)^{(m+n)/2}\psi^{(\nu-n-m)/2}2^{(n-\nu-m)/2}A_0(t)^m$$
$$\cdot\Bigg\{\frac{{}_1F_1[(n+m-\nu)/2;m+1;-p(t)]}{\Gamma[(2+\nu-m-n)/2]}$$
$$-\frac{b_0\sqrt{2}}{\psi^{1/2}}\frac{{}_1F_1[(n+m+1-\nu)/2;m+1;-p(t)]}{\Gamma[(1+\gamma-m-n)/2]}$$
$$+\frac{b_0^2}{\psi}\frac{{}_1F_1[(m+n+2-\nu)/2;m+1;-p(t)]}{\Gamma[(\nu-m-n)/2]}+\cdots\Bigg\},\quad (4.15)$$

and the significant terms are

$$H_{00}(t) \cong \frac{\beta\Gamma(\nu+1)\psi^{\nu/2}2^{-\nu/2}}{2\Gamma[(2+\nu)/2]}\left\{1+\frac{b_0\sqrt{2}\,\Gamma[(2+\nu)/2]}{\psi^{1/2}\Gamma[(1+\nu)/2]} + \frac{1}{\psi}\left[\frac{\nu A_0^2(t)}{4}+\frac{b_0^2\Gamma[(2+\nu)/2]}{\Gamma(\nu/2)}\right]+\cdots\right\},$$

$$H_{01}(t) \cong -\frac{i\beta\Gamma(\nu+1)\psi^{(\nu-1)/2}2^{-(\nu+1)/2}}{\Gamma[(\nu+1)/2]}\left\{1-\frac{b_0\sqrt{2}}{\psi^{1/2}}\frac{\Gamma[(1+\nu)/2]}{\Gamma(\nu/2)}+\cdots\right\},$$

$$H_{10}(t) \cong -\frac{i\beta\Gamma(\nu+1)\psi^{\nu/2}2^{-(\nu+3)/2}}{\Gamma[(1+\nu)/2]}\left\{\psi^{-1/2}-\frac{b_0\sqrt{2}}{\psi}\frac{\Gamma[(1+\nu)/2]}{\Gamma(\nu/2)}+\cdots\right\},$$

$$\vdots \qquad\qquad \vdots$$

$$H_{02}(t) \cong -\beta\Gamma(\nu+1)\psi^{(\nu/2)-1}2^{-\nu/2}/\Gamma(\nu/2);$$

$$H_{20}(t) \cong -\beta\Gamma(\nu+1)\psi^{\nu/2}\cdot 2^{-(\nu/2)-3}A_0(t)^2/\psi\Gamma(\nu/2), \text{ etc.} \tag{4.16}$$

Observe that the terms containing pure-signal or cross-term contributions, $n=0$, or m, $n\neq 0$, vanish at least as $\psi^{-1/2}$, and only noise is left. Thus when the input noise voltage becomes sufficiently great in comparison with the signal, the latter is surpressed. In particular, for the half-wave linear and quadratic rectifier, the low-frequency amplitude of the output ($m=n=0$) becomes from (4.16)

$$\left.\begin{aligned} H_{00}(t) - (d-c) &\\ = \frac{\beta\Gamma(\nu+1)\psi^{\nu/2}}{2^{(\nu/2)+3}\Gamma[(\nu+2)/2]}\left(\frac{\nu A_0^2}{\psi}\right) &= \frac{\beta\psi^{1/2}}{2\sqrt{2\pi}}\,[A_0^2(t)/2\psi]_{\nu=1,b_0=0},\quad p(t)\ll 1\\ &= \beta[A_0^2(t)/2]_{\nu=2,b_0=0},\end{aligned}\right\} \tag{4.17}$$

the first of which is Eq. (39) of reference 8, and the second, Eq. (29).

Case VI: $\psi\to 0$, b_0, $A_0(t)$ *finite*. Here it is convenient to distinguish two cases in the limit, one where $|b_0| > A_0(t) \geqq 0$ and the other when $A_0(t) > |b_0 \geqq 0$ For the first we use (4.5) and (4.6) as $b\to\infty$, along with (A3.3) to obtain the not unexpected result that

$$H_{nn}(t)_{\psi\to 0} \cong 0, \qquad b_0 > A_0(t) \geqq 0, \tag{4.18}$$

since the cut-off voltage exceeds the input signal (envelope), so that the wave is not passed. However, when $|b_0| > A_0(t)$, $b_0<0$, we have a different situation, where now the incoming disturbance, which is essentially pure signal (hence $n=0$ in the limit $\psi=0$), is transmitted without distortion due to cut-off or saturation effects, albeit with distortion due to the non linear nature of the dynamic path, $\nu\neq 1$. The expression for the amplitude is accordingly

$$H_{m0}(t) = \frac{\beta\Gamma(\nu+1)i^{-\nu-1}}{2\pi}\int_C e^{-izb_0}J_m(A_0(t)z)dz/dz^{\nu+1}. \tag{4.19}$$

When $b_0<0$, $|b_0| > A_0(t)$ we may use (A3.8) in conjunction with (A3.19) to get finally

$$H_{m0}(t) = \frac{\beta\Gamma(\nu+1)e^{-\pi mi/2}\sqrt{\pi}\,|\,b_0\,|^2}{2^\nu\Gamma[(2+\nu-m)/2]\Gamma[(1+\nu-m)/2]m!}\left[\frac{A_0(t)}{|\,b_0\,|}\right]^m {}_2F_1(m-\nu, m-\nu+1;$$
$$m+1; A_0^2(t)/b_0^2), \qquad b_0 < 0, \ |\,b_0\,| > A_0(t), \ \psi = 0. \tag{4.20}$$

On the other hand, when $|b_0| < A_0(t)$ we obtain from (4.4), including a correction term,

$$H_{mn}(t)_{\psi\to 0} = \beta\Gamma(\nu+1)e^{-(m+n)\pi i/2}2^{n-\nu-1}A_0(t)^{\nu-n}$$
$$\cdot \sum_{k=0}^{\infty} \frac{(-1)^k 2^k b_0^k\{1+\psi(m+n+k-\nu)(n-m+k-\nu)/2A_0(t)^2}{k!A_0(t)^k\Gamma[(2+\nu-k-m-n)/2]\Gamma[(2+\nu-k+m-n)/2]} + \cdots\bigg\}, \tag{4.21}$$

and when $\psi=0$ we may sum (4.21), $n=0$, or use (A3.19) to get

$$H_{m0}(t)_{\psi=0} = \frac{\beta\Gamma(\nu+1)}{2^{\nu+1}} e^{-m\pi i/2}A_0(t)^\nu \left\{\frac{{}_2F_1[(m-\nu)/2, (-m-\nu)/2; \tfrac{1}{2}; b_0^2/A_0(t)^2]}{\Gamma[(2+\nu+m)/2]\Gamma[(2+\nu-m)/2]}\right.$$
$$\left. - \frac{2b_0}{A_0(t)} \frac{{}_2F_1[(m-\nu+1)/2, (-m-\nu+1)/2; \tfrac{3}{2}; b_0^2/A_0(t)^2]}{\Gamma[(1+\nu+m)/2]\Gamma[(1+\nu-m)/2]}\right\}, \tag{4.22}$$
$$0 \leqq |\,b_0\,| < A_0(t).$$

For half-wave detection we have the simple result

$$H_{m0}(t)_{b_0=0,\psi=0} = \frac{\beta\Gamma(\nu+1)}{2^{\nu+1}} \cdot \frac{e^{-m\pi i/2}A_0(t)^\nu}{\Gamma[(2+\nu+m)/2]\Gamma[(2+\nu-m)/2]}. \tag{4.23}$$

Power and spectra may be obtained in the usual manner (see Sec. 2).

5. Probability density of the envelope and phase of a modulated signal and noise. In handling the problem of passage of noise or a signal and noise through a non linear device, it is sometimes convenient first to determine the various first and second order probability densities associated with the incoming wave, and then with their help derive the expressions for the mean power output and the correlation function associated with the transmitted disturbance, after it has been modified by a rectifier or similar non linear apparatus, cf. Eq. (1.7b), for example. Now in particular when the noise and signal are narrow-band, *vide* Sec. 3, this method, alternative to the use of the characteristic function [Eqs. (2.10)–(2.16)] suggests itself. The purpose of this section and of section 6 following is to obtain explicit expressions for the probability density of the envelope and phase of the general type of modulated carrier in the presence of narrow-band noise. Only in such circumstances may one properly speak of an envelope or of phase—i.e., when the part of the incoming wave due to noise and modulation is essentially slowly varying in comparison with the carrier frequency f_0 and the central frequency f_c of the noise band. After rectification it is usually this envelope, the low-frequency part of the disturbance, that is observed.

The input is

$$V(t) = A_0(t)\cos\omega_0 t + V(t)_N$$
$$= (A\cos\omega_d t + V_c)\cos\omega_c t + (V_s - A\sin\omega_d t)\sin\omega_c t, \tag{5.1}$$

where f_d is the difference frequency $f_d=f_c-f_0$ and $A=A_0(t)$, from (2.1) and (2.2). Here V_c is the component of the noise "in phase" with $\cos\omega_c t$ and V_s is the component "in phase" with $\sin\omega_c t$, e.g.:

$$V_c = \sum_{n=1}^{\infty} (a_n \cos\omega_n' t + b_n \sin\omega_n' t),$$
$$V_s = \sum_{n=1}^{\infty} (-a_n \sin\omega_n' t + b_n \cos\omega_n' t); \qquad \omega_n' = \omega_n - \omega_c, \tag{5.2}$$

respectively. By the envelope is meant

$$E(t) = [(A\cos\omega_d t + V_c)^2 + (V_s - A\sin\omega_d t)^2]^{1/2}, \tag{5.3}$$

which is a slowly varying function of the time provided $f_d \ll f_0, f_c$ i.e., the centers of the noise and signal spectra are not too far apart in frequency. The quantities A, V_c, V_s are, of course, relatively low-frequency disturbances.

Our first task is to obtain the joint probability density of E_1 and E_2, where E_1 is the envelope at time t_0 and E_2 its value at some later time t_0+t. We derive the distribution of V_{c1}, V_{c2}, V_{s1}, V_{s2}, and V_{s2} initially in rectangular coördinates. Letting

$$X_1 = V_{c1} = \sum_{n=1}^{\infty} (a_n \cos\omega_n' t_0 + b_n \sin\omega_n' t_0);$$
$$X_3 = V_{c2} = \sum_{n=1}^{\infty} [a_n \cos\omega_n' (t_0+t) + b_n \sin\omega_n' (t_0+t)]$$
$$X_2 = V_{s1} = \sum_{n=1}^{\infty} (-a_n \sin\omega_n' t_0 + b_n \cos\omega_n' t_0);$$
$$X_4 = V_{s2} = \sum_{n=1}^{\infty} [-a_n \sin\omega_n' (t_0+t) + b_n \cos\omega_n' (t_0+t)], \tag{5.4}$$

with the help of (2.3), (2.4), and the generalized Gaussian distribution (A.3), we observe that

$$\overline{X_1^2} = \overline{X_2^2} = \overline{X_3^2} = \overline{X_4^2} = \int_0^{\infty} w(f)\,df \equiv \psi(0) = \mu_{11} = \mu_{22} = \mu_{33} = \mu_{44}, \tag{5.5}$$

$w(f)$ as before being the mean power spectrum of the noise; for $\overline{X_1X_3}$ and $\overline{X_2X_4}$ we have

$$\overline{X_1X_3} = \overline{X_2X_4} = \int_0^{\infty} w(f)\cos(\omega-\omega_c)t\,df \equiv \psi_0(t) = r_0(t)\psi = \mu_{13} = \mu_{24}; \tag{5.6}$$

(see Appendix II). It is also evident that $\overline{X_1X_2}=\overline{X_3X_4}=0$, as there is no correlation between the "in-" and "out-of-phase" components of the noise. There remains finally

$$\overline{X_2X_3} = -\overline{X_1X_4} = \int_0^{\infty} w(f)\sin(\omega-\omega_c)t\,df \equiv \lambda_0(t)\psi = \mu_{23} = -\mu_{14}. \tag{5.7}$$

Accordingly, if we let $Y_1 = A\cos\omega_d t + V_c$, $Y_2 = -A\sin\omega_d t + V_s$, etc., the various Y_1, Y_2,

Y_3, Y_4 represent random variables distributed about the averages $A_0(t_0') \cos \omega_d t_0$, $-A_0(t_0') \sin \omega_d t_0$, $A_0(t_0'+t) \cos \omega_d(t_0+t)$, and $-A_0(t_0'+t) \sin \omega_d(t_0+t)$, where no correlation between the modulation and the carrier, as well as the noise, is assumed. Thus $A_1 = A_0(t_0')$ represents the envelope modulation at a time t_0', not analytically related to t_0, and $A_2 = A_0(t_0'+t)$ is the same modulation a time t later. Equations (5.4) become

$$\begin{aligned} X_1 &= Y_1 - A_1 \cos \omega_d t_0; & X_2 &= Y_2 + A_1 \sin \omega_d t_0; \\ X_3 &= Y_3 - A_2 \cos \omega_d(t_0 + t); & X_4 &= Y_4 + A_2 \sin \omega_d(t_0 + t). \end{aligned} \tag{5.8}$$

Equation (A1.3), where $s=4$, gives us the joint distribution $W_2(X_1X_2: X_3X_4; A_1A_2; t)$, for which the fundamental matrix is

$$\boldsymbol{\mu} = \|\mu_{kl}\| = \begin{bmatrix} \psi & 0 & r_0\psi & -\lambda_0\psi \\ 0 & \psi & \lambda_0\psi & r_0\psi \\ r_0\psi & \lambda_0\psi & \psi & 0 \\ -\lambda_0\psi & r_0\psi & 0 & \psi \end{bmatrix}. \tag{5.9}$$

From (5.5)–(5.7) and from (5.9) the determinant $|\mu|$ and the cofactors μ^{kl} follow at once. Since (5.1) may be written $V(t) = Y_1 \cos \omega_c t + Y_2 \sin \omega_c t$, the envelopes of the wave at the two times t_0' and $t_0'+t$ are respectively

$$E_1 = (Y_1^2 + Y_2^2)^{1/2} \quad \text{and} \quad E_2 = (Y_3^2 + Y_4^2)^{1/2}, \tag{5.10}$$

and following the example of Rice[2] and others, we transform to polar coördinates with the help of

$$Y_1 = E_1 \cos \theta_1; \quad Y_2 = E_1 \sin \theta_1; \quad Y_3 = E_2 \cos \theta_2; \quad Y_4 = E_2 \sin \theta_2, \tag{5.11}$$

for which the Jacobian is easily shown to be E_1E_2.

The probability density W_2 becomes finally

$$\begin{aligned} &W_2(E_1E_2; \theta_1\theta_2; A_1A_2; t) \\ &\quad = \frac{E_1E_2}{4\pi^2\psi(1 - \lambda_0^2 - r_0^2)} \exp\Bigg[-\frac{1}{2\psi(1 - \lambda_0^2 - r_0^2)} \{E_1^2 + E_2^2 + A_1^2 + A_2^2 \\ &\quad - 2r_0E_1E_2 \cos(\theta_2 - \theta_1) - 2\lambda_0E_1E_2 \sin(\theta_2 - \theta_1) \\ &\quad - 2r_0A_1A_2 \cos \omega_d t + 2\lambda_0A_1A_2 \sin \omega_d t \\ &\quad + 2E_2 \sin \theta_2[A_2 \sin \omega_d t_2 - r_0A_1 \sin \omega_d t_1] \\ &\quad + 2E_1 \sin \theta_1[A_1 \sin \omega_d t_1 - r_0A_2 \sin \omega_d t_2] \\ &\quad + 2\lambda_0[E_2A_1 \sin \theta_2 \cos \omega_d t_1 - E_1A_2 \sin \theta_1 \cos \omega_d t_2] \\ &\quad + 2\lambda_0[E_2A_1 \cos \theta_2 \sin \omega_d t_1 - E_1A_2 \cos \theta_1 \sin \omega_d t_2] \\ &\quad - 2E_1 \cos \theta_1[A_1 \cos \omega_d t_1 - r_0A_2 \cos \omega_d t_2] \\ &\quad - 2E_2 \cos \theta_2[A_2 \cos \omega_d t_2 - r_0A_1 \cos \omega_d t_1]\}\Bigg], \end{aligned} \tag{5.12}$$

where $t_1 = t_0$, $t_2 = t_0 + t$. It should be pointed out that W_2 as given by (5.12) is not purely a probability density, in as much as averages over the phases of the modulation and over the phases of the difference frequency terms involving $\omega_d t_1$, $\omega_d t_2$ remain to be taken. Since the modulation and difference terms are not correlated, these averages are independent and, for example, may be determined in the manner of (2.12). Although periodic disturbances are not, strictly speaking, random, they may be treated as such on assuming a random distribution of the time origin with respect to an hypothetical observer. In this way the phase may be considered a random variable uniformly distributed between 0 and 2π, and a statistical average accordingly may be performed for the function containing this random variable. See section 3.10 of reference 2 for a more detailed discussion.

It is evident at once from (5.12) that the presence of a modulated signal introduces considerable mathematical complexity, and comparison with the results of Secs. 2 and 4 would indicate that this form of treatment is perhaps in most cases not so expeditious. Further, it is restricted to narrow-band disturbances, while the approach of Sections 2 to 4 is more general. However, a considerable simplification in (5.12) is possible if we observe that for $\omega' = \omega - \omega_0$, Eq. (5.7) becomes essentially zero, i.e., $\lambda_0(t) \doteq 0$, since the input spectrum is narrow-band and symmetrically distributed about $f = f_c$ and f_c heavily exceeds the effective bandwidth of the noise, as now the contribution from 0 to $-\omega_c$ nearly cancels the integral from 0 to ∞. It follows that all terms in (5.12) containing $\lambda_0(t)$ may then be safely discarded.

The case of greatest practical interest occurs when the carrier and central noise frequencies coincide; then $\omega_d = 0$, and with the usual condition $\lambda_0(t) = 0$ we have for W_2

$$W_2 = \frac{E_1 E_2}{4\pi^2\psi^2(1 - r^2)} \exp - (A_1^2 + A_2^2 - 2r_0 A_1 A_2)/2\psi(1 - r_0^2)$$
$$\cdot \exp - [E_1^2 + E_2^2 - 2r_0 E_1 E_2 \cos(\theta_2 - \theta_1)$$
$$- 2E_1(A_1 - r_0 A_2)\cos\theta_1 - 2E_2\cos\theta_2(A_2 - r_0 A_1)]/2\psi(1 - r_0^2). \tag{5.13}$$

Notice that when $t \to \infty$ we obtain the square of the first-order density W_1. Then since $r_0(\infty) = 0$, $\lambda_0(\infty) = 0$, Eq. (5.12) transforms in the more general case where $\omega_d \neq 0$ to

$$\lim_{t\to\infty} (W_2) = W_1^2 = \left\{\frac{E}{2\pi\psi} \exp - [A_0(t)^2 + E^2 - 2EA_0(t)\cos\phi]/2\psi\right\}^2, \tag{5.14}$$

in which $\phi = \theta + \omega_d t$ is a new phase angle. The probability density of the envelope E is found by integrating over all phases θ, or ϕ, between 0 and 2π. The result is

$$W_1(E, A_0(t)) = (E/\psi) \exp - I_0(EA_0(t)/\psi)[A_0(t)^2 + E^2]/2\psi, \tag{5.15}$$

a generalization of a result derived independently by Goudsmit,[27] North,[7] Rice,[2] and others, when the carrier is not modulated, i.e., when $A_0(t) = A_0$. The complete first-order probability density requires the average over the phases of the modulation, in the manner explained following (5.12). When there is no signal we have the well-known expression $W_1(E) = E\psi^{-1} \exp(-E^2/2\psi)$. Figure 4b illustrates some typical distributions of W_1 for various values of the ratio $A_0/\psi^{1/2}$, or $\sqrt{2}p^{1/2}$. As we would

[27] S. A. Goudsmit, *Comparison between signal and noise*, M.I.T. Rad. Lab. Report 43-21, Jan. 29, 1943.

expect, the presence of the signal shifts the average and the most probable values of the distribution to larger values of the envelope voltage, such that for sufficiently strong signals the most probable value of E coincides with the peak amplitude A_0.

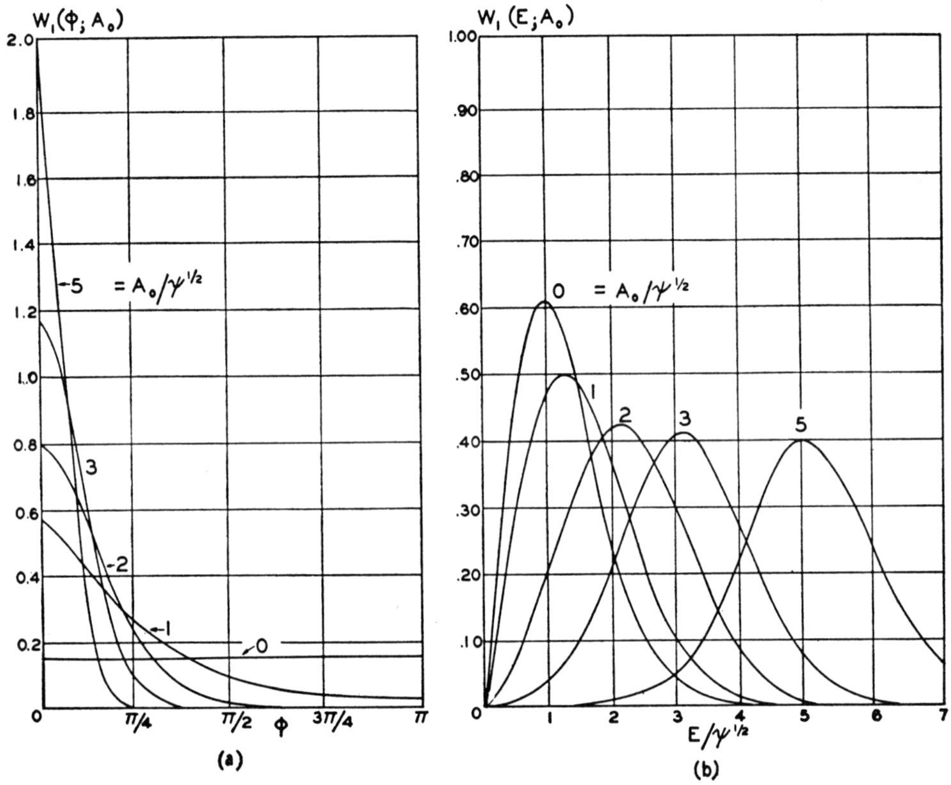

FIG. 4. Curves (a): the probability density of the phase ϕ when the carrier is unmodulated. Curves (b): the probability density of the envelope E of an unmodulated carrier and narrow-band noise, for various ratios of peak signal to r-m-s noise amplitudes, as in the case of (a).

The probability density $W_1(\theta, A_0(t))$ for the phases follows in a similar manner. From (5.14) we have after integration over E with the help of (A3.2) and (A3.7)

$$W_1(\theta, A_0(t)) = \frac{1}{2\pi} e^{-(A_0(t)\sin\phi)^2/2\psi} \left\{ \left(\frac{\pi}{2\pi}\right)^{1/2} A_0(t) \cos\phi + {}_1F_1[-\tfrac{1}{2}; \tfrac{1}{2}; -(A_0(t)\cos\phi)^2/2\psi] \right\}, \tag{5.16}$$

and again to determine the pure distribution $W_1(\theta)$ we must average over the phases of the modulation $A_0(t)$ as described above. The expression $\phi = \theta + \omega_d t$ introduced first in (5.14) is a new phase variable, shifted from the origin of θ by an amount that increases linearly with the time; Eq. (5.16) might be considered a periodically fluctuating probability density. However, the statistics of the phase θ or ϕ are not radically affected; the density functions are the same in either case, only the average values of θ and ϕ are different in the interval $0-2\pi$. Note that when there is no signal

$W_1(\theta) = 1/2\pi$, as we would expect. Figure 4a illustrates the variation of the phase density for an unmodulated carrier where the ratio $A_0/\psi^{1/2}$ takes the same values as in the example of Fig. 4b for the envelope. For no signal the distribution is uniform; as the signal strength is increased relative to the r-m-s noise voltage the phases are grouped progressively closer about the 0°-phase of the carrier, until for no noise, or what is the same thing here, for an overwhelming signal, the density becomes a delta-function at $\phi = 0$. Strong signals thus dominate the distribution of the phases for the noise, and strong signals tend to obliterate the random phasing. The curves of Fig. 4a are symmetrical about $\phi = 0$, the complete interval being $-\pi < \phi \leqq \pi$ or $0 < \phi \leqq 2\pi$.

The second-order probability density W_2 for the envelopes E_1 and E_2, or the phases θ_1 and θ_2 may be determined in the same way that W_1 was. We have finally[28]

$$W_2(E_1E_2; A_1A_2; t)$$

$$= \frac{E_1E_2}{\psi^2(1 - r_0^2)} \exp - [E_1^2 + E_2^2 + A_1^2 + A_2^2 - 2A_1A_2r_0]/2\psi(1 - r_0^2)$$

$$\times \sum_{m=0}^{\infty} \epsilon_m I_m\left(\frac{r_0E_1E_2}{\psi(1 - r_0^2)}\right) I_m\left(\frac{A_1 - r_0A_2}{\psi(1 - r_0^2)} E_1\right) I_m\left(\frac{A_2 - r_0A_1}{\psi(1 - r_0^2)} E_2\right). \tag{5.17}$$

A similar but more complex result obtains from (5.12) if $\omega_d \neq 0$. When there is no modulation (5.17) may be more simply written

$$W_2(E_1E_2; t) = \frac{E_1E_2}{\psi^2(1 - r_0^2)} \exp\left[\frac{-A_0^2}{\psi(1 + r_0)} - (E_1^2 + E_2^2)/2\psi(1 - r_0^2)\right]$$

$$\cdot \sum_{m=0}^{\infty} \epsilon_m I_m\left(\frac{r_0E_1E_2}{\psi(1 - r_0^2)}\right) I_m\left(\frac{A_0E_1}{\psi(1 + r_0)}\right) I_m\left(\frac{A_0E_2}{\psi(1 + r_0)}\right), \tag{5.18}$$

and for no signal at all we have

$$W_2(E_1E_2; t) = \frac{E_1E_2}{\psi^2(1 - r_0^2)} \exp - (E_1^2 + E_2^2)/2\psi(1 - r_0^2)\, I_0\left(\frac{r_0E_1E_2}{\psi(1 - r_0^2)}\right), \tag{5.19}$$

a result originally derived by Uhlenbeck[29] and used by him to determine the correlation function $R(t) = \overline{E_1E_2}$ for a half-wave linear rectifier when random noise alone is detected.

The probability density for the phases θ_1 and θ_2, $\omega_d = 0$, follows in the same way from (5.13), the final result taking the form

$$W_2(\theta_1\theta_2; A_1A_2; t)_{\omega_d=0} = [4\pi^2(1 - r_0^2)]^{-1}$$

$$\exp[-(A_1^2 + A_2^2 - 2r_0A_1A_2)(1 - r_0^2 - \cos^2\theta_1 - \cos^2\theta_2)/2\psi(1 - r_0^2)]$$

$$\times \sum_{m=0}^{\infty} \left[\frac{2r_0 \cos(\theta_2 - \theta_1)}{1 - r_0^2}\right]^m \frac{1}{m!}\left\{a_1(2\psi)^{1/2}\Gamma\left(\frac{m+3}{2}\right) {}_1F_1\left(\frac{-m}{2}; \tfrac{3}{2}; -a_1^2/4\right)\right.$$

---

[28] The result (5.17) has been obtained independently by Rice (communication to the author), for the less general case of an unmodulated carrier, $A_1 = A_2 = A_0$.

[29] G. E. Uhlenbeck, *Theory of the random process*, M.I.T. Laboratory Report #454, Oct. 15, 1943.

$$+ \Gamma\left(\frac{m}{2}+1\right){}_1F_1\left(-\frac{m+1}{2};\tfrac{1}{2};-a_1^2/4\right)\Bigg\}$$

$$\times \left\{a_2(2\psi)^{1/2}\Gamma\left(\frac{m+3}{2}\right){}_1F_1\left(-\frac{m}{2};\tfrac{3}{2};-a_2^2/4\right)\right.$$

$$\left. + \Gamma\left(\frac{m}{2}+1\right){}_1F_1\left(-\frac{m+1}{2};\tfrac{1}{2};-a_2^2/4\right)\right\}, \tag{5.20}$$

$a_1=b_1 \cos \theta_1$, $a_2=b_2 \cos \theta_2$, (for b_1, b_2 see Sec. 6).

6. The correlation function and mean power for the envelope of a signal and noise. The results of the preceding section may be used in (1.7b) to give us an expression for the correlation function and the mean power associated with the *envelope* of an incoming signal and noise wave. Our primary interest is with the low-frequency output, or detected envelope, but the general treatment will first be outlined in brief fashion below, before returning to a more detailed examination of the former.

For the general non linear device we follow the suggestion of Rice (ref. 2, Sec. 4.3) and use the Fourier transform (1.8) to represent our output current $I=g(V)$ as a function of the input disturbance $V=E \cos (\omega_c t-\theta)$. Expanding the exponential in a series of Bessel functions gives us

$$I = \sum_{l=0}^{\infty} I_l = \sum_{l=0}^{\infty} B_l(E) \cos l(\omega_c t - \theta), \tag{6.1}$$

where $B_l(E)$ represents the envelope of the lth output band, viz:

$$B_l(E) = \frac{i^l \epsilon_l}{2\pi}\int_C f(iz)J_l(Ez)dz, \tag{6.2}$$

which in principle may be observed when all other contributions are eliminated by an appropriate band-pass filter, centered about the lth zone and followed by a linear half-wave rectifier or "envelope tracer," as it is sometimes called. If $W_1(E)$ is the probability density of the input envelope, the density function for the lth region is

$$D_{1l}(E) = B_l(E)W_1(E) = W_1(B_l)B_l(E)dB_l/dE, \tag{6.3}$$

since

$$W_1(B_l) = W_1(E)dE/dB_l, \tag{6.4}$$

where $W_1(B_l)$ is the density function of B_l. This latter quantity, $W_1(B_l)$, may be found from (5.15) and from (6.2) by differentiation. For higher order densities we have

$$W_2(B_{1l}, B_{2l}; t) = W_2(E_1E_2; t)\Bigg/\left|\frac{\partial(B_{1l}B_{2l})}{\partial(E_1E_2)}\right|, \text{ etc.} \tag{6.4a}$$

An important consequence of (6.4) and (6.4a) is that these results enable us to determine the probability density of the envelope of a distribution after successive non linear operations, *provided*, of course, that in each operation the concept of the envelope remains, i.e., we have a narrow band wave undergoing rectification and not one whose mean frequency is comparable with the carrier.

The correlation function and the mean power may be written in a similar manner. We have from (2.12), (5.12), and (6.1) for the complete correlation

$$R(t) = \sum_{l=0}^{\infty}\sum_{m=0}^{\infty} T_0'^{-1}\int_0^{T_0'} dt_0' \int_0^{T_0} T_0^{-1}dt_0 \int_0^{\infty} dE_1 \int_0^{\infty} dE_2 \int_0^{2\pi} d\theta_1 \int_0^{2\pi} d\theta_2 B_l(E_1)B_m(E_2)$$
$$W_2(E_1E_2;\ \theta_1\theta_2;\ A_1A_2;\ t) \cos l(\omega_c t_0 - \theta_1) \cos m[\omega_c(t_0 + t) - \theta_2], \tag{6.5}$$

which simplifies considerably when the noise band is symmetrical, i.e., $\lambda_0 \doteq 0$, the carrier is unmodulated, and $\omega_d = 0$. The spectrum may be found from (1.6). The mean power associated with the lth harmonic region is given when $t=0$ in (6.5), but an alternative and sometimes simpler expression may be obtained with the help of $W_1(E)$ and (2.33a), applied to (6.1), (6.2) after averaging over the phases of the zone central frequencies $l\omega_c$.

Specifically, let us examine the interesting case of the νth-law half-wave rectifier when a modulated signal and narrow-band noise together enter the apparatus. This is essentially the receiver problem, as we are concerned only with the low-frequency output—namely, the modified envelope. Then it is in keeping with the problem to set $\omega_d = \lambda_0(t) = 0$, the latter on the assumption of a symmetrical spectrum, and we find for the noise contribution to the correlation function as indicated by the subscript N, *cf.* 2(.11), from (5.17) in (6.5),

$$R(t)_N = \int_0^{\infty} dE_1 \int_0^{\infty} dE_2 B_0(E_1)B_0(E_2)W_2(E_1E_2;\ A_1A_2;\ t)$$
$$= \gamma^2 f(A_1A_2;\ r_0)\sum_{m=0}^{\infty}\epsilon_m \int_0^{\infty} dE_1 \int_0^{\infty} dE_2 (E_1E_2)^{\nu+1} I_m\left(\frac{E_1E_2 r_0}{\psi(1-r_0^2)}\right) I_m(b_1E_1)I_m(b_2E_2)$$
$$\times\ e^{-(E_1^2+E_1^2)/2\psi(1-r_0^2)} \tag{6.6}$$

where $f(A_1, A_2;\ r_0)$ is given by

$$f(A_1, A_2;\ r_0) = \psi^2(1 - r_0^2)]^{-1} \exp\ [(-A_1^2 - A_2^2 + 2r_0A_1A_2)/2\psi(1 - r_0^2)], \quad \text{and}$$

$$b_1 = \frac{A_1 - r_0A_2}{\psi(1 - r_0^2)}; \qquad b_2 = \frac{A_2 - r_0A_1}{\psi(1 - r_0^2)}; \qquad B_0(E_1) = \gamma E_1^{\nu},\ \text{etc.} \tag{6.6a}$$

The scale factor γ is related to the tube factor β used throughout this paper by

$$\gamma = \frac{\beta}{2\pi}\int_0^{\pi} \sin^{\nu}\theta d\theta = \frac{\beta}{2\sqrt{\pi}}\,\frac{\Gamma[(\nu+1)/2]}{\Gamma[(\nu/2+1)]}\,. \tag{6.7}$$

This follows because in (6.6) we are dealing with the *envelope* E rather than the amplitude V of the wave. Now the output current I consists of the positive halves of the oscillations of βV^{ν}, while the envelope of I is the same as that of βV^{ν}. But the area under the loops of I is to the area under βE^{ν} as the area under a loop of $\sin^{\nu}\theta$ is to an area of unit height and length 2π, so that as far as the low-frequency portion of the output is concerned the loops of I are "smeared" together into a current which varies as γE^{ν}, with γ given by (6.7) above. Compare the appropriate results of the present section with those of the next, where the problem is treated from the point of view of the instantaneous amplitude.

There appear to be three principal ways of evaluating $R(t)_N$, each yielding results different in form and usefulness. The first is achieved with the aid of the transformation

$$E_1 = [\psi(1 - r_0^2)]^{1/2} z^{1/2} e^{\phi/2}, \qquad E_2 = [\psi(1 - r_0^2)]^{1/2} z^{1/2} e^{-\phi/2}, \tag{6.8}$$

followed by the successive application of the expression for the product of two Bessel functions (*cf.* (2), p. 148, ref. 21), the integral form of the modified Bessel function of the second kind ((7), p. 182, ref. 21), of argument $n-2k$, and finally, by ((11), p. 410, ref. 21). The second form follows after expanding $I_m(E_1E_2r_0/\psi(1-r_0^2))$ and using termwise integration, with the help of (A3.7), and the third employs a contour integral representation of the Bessel functions in (6.6) applied to the addition formula for $I_0[(x^2+y^2+2xy\cos\phi)^{1/2}]$, along with a reversal of the order of integration. We give the final results:

For the first method we have

$$R(t)_N = \gamma^2 2^\nu \psi^\nu (1 - r_0^2)^{\nu+1} e^{-[A_1^2+A_2^2-2r_0A_1A_2]/2\psi(1-r_0^2)} \sum_{m=0}^{\infty}\sum_{n=0}^{\infty} \frac{\epsilon_m r_0^m c_1^{m+2n} c_2^m}{2^{m+n} m!}$$
$$\times \sum_{k=0}^{n} \frac{(c_1^{-1}c_2)^{2k}\Gamma(\nu/2 + k + m + 1)\Gamma(\nu/2 - k + m + n + 1)}{k!(k + m)!(n - k)!(m + n - k)!}$$
$$\times {}_2F_1\left(\frac{\nu}{2} + k + m + 1;\ \frac{\nu}{2} - k + m + n + 1;\ m + 1;\ r_0^2\right), \tag{6.9}$$
$$c_1 = b_1[\psi(1-\nu_0^2)]^{1/2}, \quad c_2 = b_2[\psi(1-\nu_0^2)]^{1/2}.$$

In the case of modulation we must now apply (2.12) to determine the average over the phases $\omega_A t_0'$. As one can readily see from (6.9) the effort is formidable, even for the simplest modulations. However, a less general but nonetheless important case arises when $A_1 = A_2 = A_0$, a constant. With the assistance of (A3.20b) we obtain for the complete low-frequency correlation function

$$R(t) = \gamma^2 2^\nu \psi^\nu e^{-A_0^2/\psi(1+r_0)} \sum_{m=0}^{\infty}\sum_{n=0}^{\infty} \frac{\epsilon_m r_0^m p^{m+n}}{(1 + r_0)^{2m+2n}}$$
$$\times \sum_{k=0}^{n} \frac{\Gamma(\nu/2 + k + m + 1)\Gamma(\nu/2 - k + m + n + 1)}{k!(k + m)!(n - k)!(m + n - k)!}$$
$$\times {}_2F_1\left(k - \frac{\nu}{2} - n,\ -k - \frac{\nu}{2};\ m + 1;\ r_0^2\right), \quad p \equiv A_0^2/2\psi. \tag{6.10}$$

For noise alone the low-frequency correlation is at once found from (6.10), since only the terms for which $m = n = 0$ contribute:

$$R(t)_{\text{noise}} = \gamma^2 2^\nu \psi^\nu \Gamma\left(\frac{\nu}{2} + 1\right)^2 {}_2F_1\left(-\frac{\nu}{2},\ -\frac{\nu}{2};\ 1;\ r_0^2\right)$$
$$= \frac{\psi^\nu \beta^2 2^{2\nu-2}}{\pi}\, \Gamma\left(\frac{\nu + 1}{2}\right)^2 {}_2F_1\left(-\frac{\nu}{2},\ -\frac{\nu}{2};\ 1;\ r_0^2\right), \tag{6.11}$$

the latter from (6.7). For the second method we write

$$R(t)_N = \gamma^2 2^\nu (1 - r_0^2)^{\nu+1}\psi^\nu \exp\left\{-(A_1^2 + A_2^2 - 2r_0A_1A_2)/2\psi(1 - r_0^2)\right\}$$
$$\times \sum_{m=0}^{\infty}\sum_{n=0}^{\infty} \frac{\epsilon_m r_0^{2n+m}\Gamma(m+n+\nu/2+1)^2(c_1c_2)^m}{(m!)^2 n!(n+m)!2^m}$$
$$\times\, {}_1F_1\left(\frac{\nu}{2}+m+n+1;\, m+1;\, \frac{c_1^2}{2}\right) {}_1F_1\left(\frac{\nu}{2}+m+n+1;\, m+1;\, \frac{c_2^2}{2}\right), \quad (6.12)$$

equivalent to (6.9), while the counterpart of (6.10) is

$$R(t) = \gamma^2 2^\nu \psi^\nu (1 - r_0^2)^{\nu+1} e^{-p} \sum_{m=0}^{\infty} \frac{\epsilon_m r_0^m}{(m!)^2}\left(\frac{1-r_0}{1+r_0}\right)^m p^m \sum_{n=0}^{\infty} \frac{r_0^{2n}\Gamma(m+n+1+\nu/2)^2}{n!(n+m)!}$$
$$\times\, {}_1F_1\left[-n-\frac{\nu}{2};\, m+1;\, -p\frac{(1-r_0)}{1+r_0}\right]^2 \quad (6.13)$$

from (A3.2). When there is no signal it is easy to verify that (6.13) reduces to (6.11), again with the help of (A3.20b).

There remains the third approach. This enables us to write finally, for the correlation function

$$R(t)_N = \gamma^2 2^\nu \psi^{\nu+2}(1 - r_0^2)^{\nu+2} f(A_1A_2; r_0)\Gamma(\nu/2+1)^2 \sum_{m=0}^{\infty} \frac{(\nu/2+1)_m^2}{(m!)^2}$$
$$\times \frac{1}{2\pi i}\int_{\mathbf{D}} \frac{\xi^{\nu+1}e^{\xi}[\xi r_0 + c_1c_2/2]^{2m}d\xi}{[\xi - c_1^2/2]^{\nu/2+1}[\xi - c_2^2/2]^{\nu/2+1}}. \quad (6.14)$$

The path $\mathbf{D}$ of integration in the ξ-plane is taken to be sufficiently large so that the series in the integrand converge satisfactorily. The contour also includes any branch points and poles of the integrand of (6.14). Expanding the denominator of (6.14) in series and using the binomial development of $(\xi r_0 + c_1c_2/2)^{2m}$, gives, after termwise integration for the case when there is no modulation, viz., $A_1 = A_2 = A_0$, and $c_1 = c_2 = c$, $c^2/2 = p(1-r_0)/(1+r_0)$,

$$R(t) = 2^\nu\psi^\nu\gamma^2\Gamma\left(\frac{\nu}{2}+1\right)^2 (1 - r_0^2)^{\nu+1}e^{-p} \sum_{m=0}^{\infty} \frac{(\nu/2+)_m^2(2m)!}{m!^2}$$
$$\times \sum_{n=0}^{2m} \frac{r_0^{2m-n} p^n(1-r_0)^n}{n!^2(2m-n)!(1+r_0)^n}$$
$$\times\, {}_1F_1[-2m-\nu-1+n;\, n+1;\, -p(1-r_0)/(1+r_0)]. \quad (6.15)$$

A somewhat different form of the same expression is found by making the substitution $z = \xi - c^2/2$, and again expanding $(\xi r_0 + c^2/2)^{2m}$ in a binomial series, followed by termwise integration of the resulting series. The low-frequency correlation is then

$$R(t) = 2^\nu\psi^\nu\gamma^2\Gamma\left(\frac{\nu}{2}+1\right)(1 - r_0^2)^{\nu+1}e^{-p} \sum_{m=0}^{\infty} \frac{(\nu/2+1)_m^2(2m)!}{m!^2} \sum_{n=0}^{2m} \frac{(1-r_0)^n r_0^{2m-n} p^n}{(2m-n)!n!}$$
$$\times \sum_{k=0}^{\infty} \frac{1}{k!}\left[\frac{p(1-r_0)}{1+r_0}\right]^{1+\nu-k} \frac{\Gamma(\nu+2)}{\Gamma(\nu-k+2)\Gamma(\nu+n-k+2)}. \quad (6.16)$$

Notice that when ν is integral, the series in k terminates after $\nu+2$ terms.

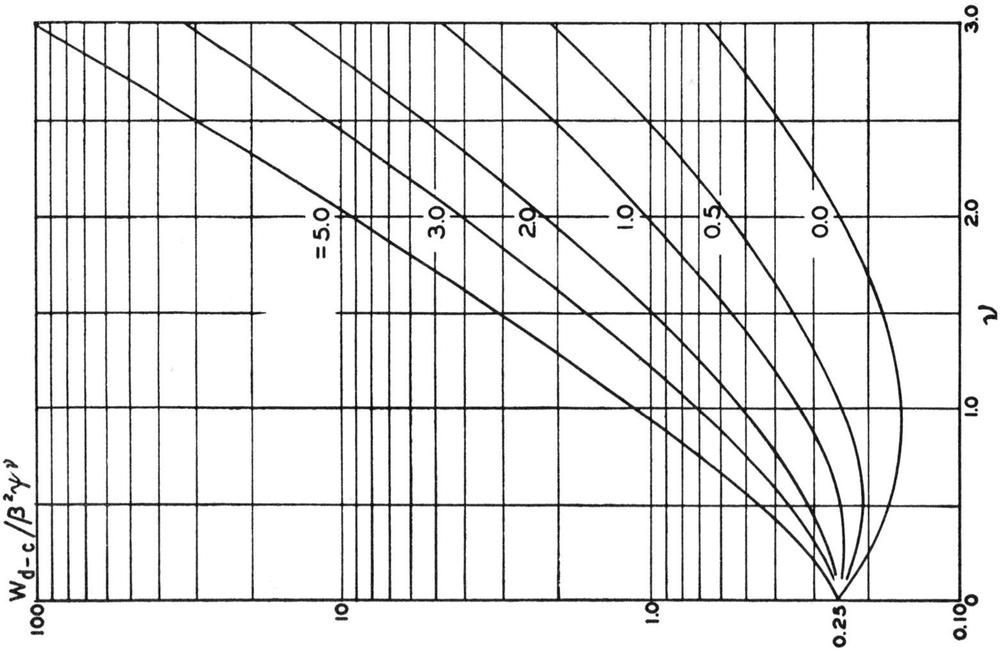

FIG. 6. Mean power output in the d-c component following a half-wave νth-law rectifier for various values of the parameter $p = A_0^2/2\psi$.

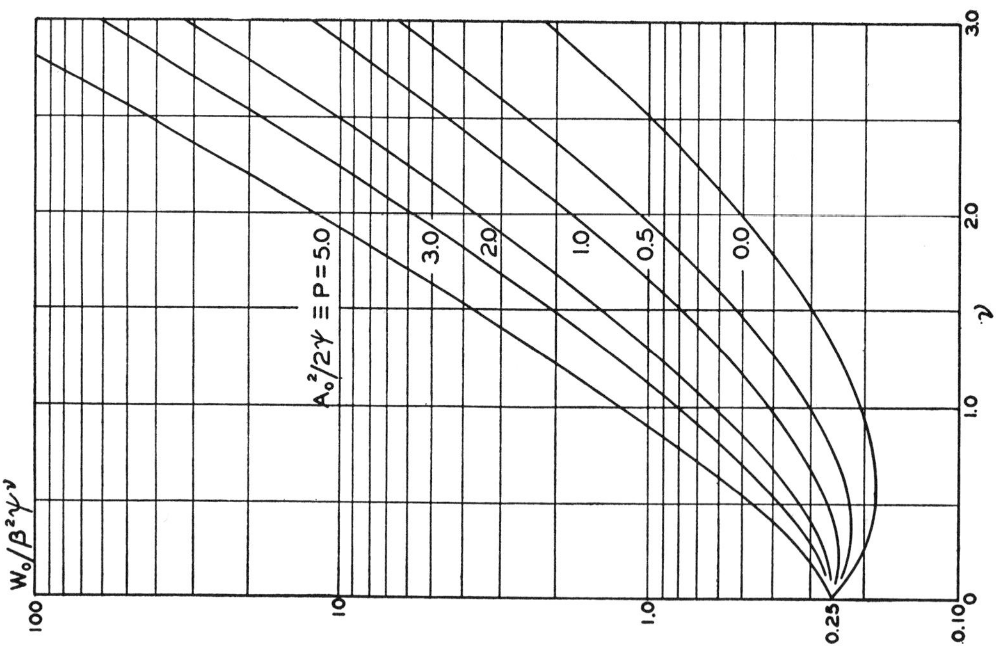

FIG. 5. Mean power output in the low-frequency components following a half-wave νth-law rectifier, for a number of values of the parameter $p = A_0^2/2\psi$.

The mean total low-frequency power $(W_T)_0$ and the power in the $d-c$ component W_{d-c} may be obtained respectively for the various cases considered above on setting $t=0$ and $t=\infty$ in $R(t)$. However, a simpler way of obtaining the same results is to use (6.6) or (6.7) with (5.15) directly. Application of (A3.7) then gives us

$$(W_T)_0 = \overline{I_0^2} = \gamma^2 2^\nu \psi^\nu \Gamma(\nu + 1)\,{}_1F_1(-\nu; 1; -p), \qquad A_1 = A_2 = A_0, \tag{6.17}$$

and for the $d-c$

$$W_{d-c} = (\bar{I}_0)^2 = \gamma^2 2^\nu \psi^\nu \Gamma\left(\frac{\nu}{2} + 1\right)^2 {}_1F_1\left(-\frac{\nu}{2}; 1; -p\right)^2. \tag{6.18}$$

When the carrier is modulated we have only to average the various $p(t_0')^j$ that are present in the above. Curves illustrating $(W_T)_0$ and W_{d-c} as a function of the law of the detector are included in Figs. 5 and 6, for a number of values of $p=A_0^2/2\psi$. It is interesting to note that the output powers actually decrease as ν increases in the interval $0 \backsim \nu \backsim 1.0$ when the signal is weak relative to the noise. Strong signals suppress the noise, and the output is then proportional to p^ν for both total-, low-frequency, and $d-c$ powers, namely,

$$(W_T)_0 \cong \gamma^2 2^\nu \psi^\nu p^\nu(1 + \nu^2/p + \cdots) \text{ and } W_{d-c} \cong \gamma^2 2^\nu \psi^\nu p^\nu (1 + \nu^2/2p + \cdots), \tag{6.19}$$

from (A3.3).

7. General half-wave rectification of random noise. A relatively simple case of considerable interest arises when random noise alone enters a non linear device which passes only the positive amplitudes of the incoming disturbance. The current-voltage characteristic assumes the following general form:

$$\begin{aligned} I &= \beta V^\nu, \qquad V > 0, \qquad \nu > 0 \\ &= 0, \qquad V < 0, \end{aligned} \tag{7.1}$$

where we restrict ourselves to unsaturated cases, corresponding in practice to (relatively) small-signal rectification. Here V represents the instantaneous input noise voltage, and I is the output current.

The analysis of the present section is more general than that of Sections 5 and 6 in that it is capable of handling broad-band noise, where the concept of the envelope is no longer meaningful; it is less general in that signals are excluded from the discussion. Also, a more detailed study of the higher order spectral regions is included, for the case of narrow-band noise. In fact, Section 7 is in itself a study of a special case of the analysis of Sections 3 and 4, when there is no bias and no carrier.

(*a*) *Broad-band Noise.* We consider first the case of wide-band noise, whose central frequency, spectrally speaking, is comparable to the bandwidth of the disturbance. See Fig. 7(a) or (b). From (2.9), (2.11), and (7.1) the correlation function of the output in the absence of a signal is

$$\begin{aligned} R(t) &= \frac{\beta^2\psi^\nu}{2\pi(1-r^2)^{1/2}} \int_0^\infty dX_1 \int_0^\infty dX_2 (X_1X_2)^\nu \\ &\times \exp\left[-(X_1^2 + X_2^2 - 2rX_1X_2)/2\psi(1-r^2)\right] \end{aligned} \tag{7.2}$$

where $r(t)$ is the normalized input correlation, cf. Eq. (A2.7). The integration of (7.2) may be effected directly by using polar coördinates: $x=\rho\cos\theta$, $y=\rho\sin\theta$. The latter approach gives us for (7.2)

$$R(t) = \frac{\beta^2\psi^\nu 2^{-\nu-1}}{[2\pi(1-r^2)^{1/2}]}\int_0^\pi \sin^\nu\phi d\phi \int_0^\infty \rho^{2\nu+1}e^{-a\rho^2}d\rho,$$

$$a = (1-r\sin\phi)/2(1-r^2)$$

$$= \frac{\beta^2\psi^\nu\Gamma(\nu+1)(1-r^2)^{\nu+1/2}}{4\pi}\int_0^\pi \frac{\sin^\nu\phi d\phi}{(1-r\sin\phi)^{\nu+1}}, \tag{7.3}$$

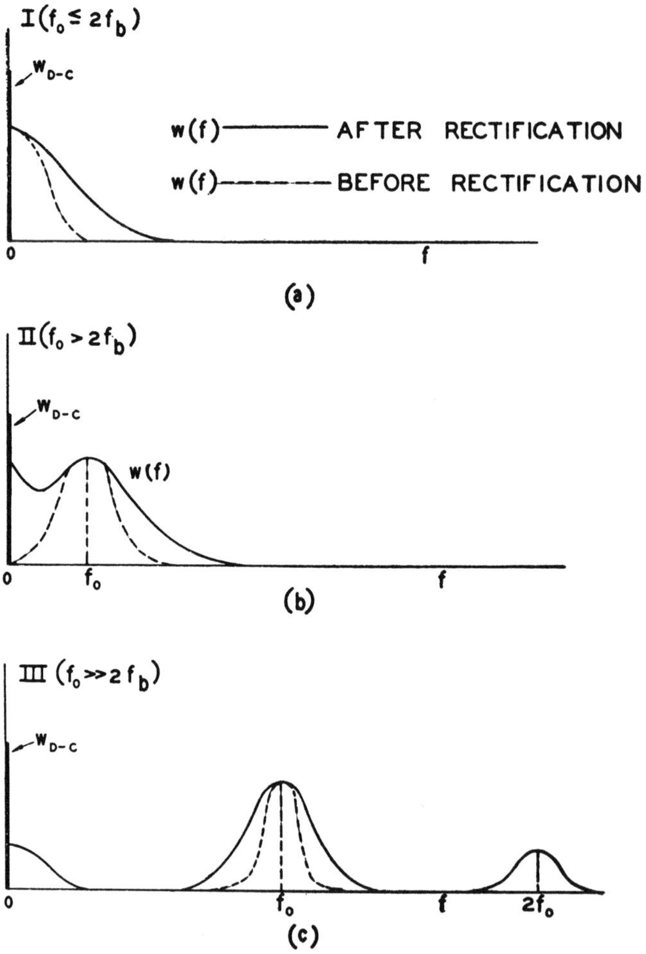

FIG. 7. Examples of narrow-band and broad-band noise before and after rectification. The normalization is arbitrary.

where the substitution $2\theta=\phi$ has been made. Expansion of the denominator and termwise integration (since $r\sin\phi\leqq 1$), with the help of

$$\sin^\nu \phi(1 - r \sin \phi)^{-(\nu+1)} = \sum_{n=0}^{\infty} \frac{r^n (\sin \phi)^{\nu+n} \Gamma(\nu + n + 1)}{n! \Gamma(\nu + 1)} \quad \text{and}$$

$$\int_0^\pi \sin^{\nu+n} \phi d\phi = \frac{\sqrt{\pi}\, \Gamma[(\nu + n + 1)/2]}{\Gamma[(\nu + n)/2 + 1]} \tag{7.4}$$

yields

$$\begin{aligned} R(t) &= \frac{\beta^2 \psi^\nu (1 - r^2)^{\nu+1/2}}{4\sqrt{\pi}} \sum_{n=0}^{\infty} \frac{r^n \Gamma[(\nu + n + 1)/2] \Gamma(\nu + n + 1)}{n! \Gamma[(\nu + n)/2 + 1]} \\ &= \frac{\beta^2 2^{\nu-2} \psi^\nu}{\pi} (1 - r^2)^{\nu+1/2} \left\{ \Gamma\left(\frac{\nu + 1}{2}\right)^2 {}_2F_1\left(\frac{\nu + 1}{2}, \frac{\nu + 1}{2}; \tfrac{1}{2}; r^2\right) \right. \\ &\quad \left. + 2r\Gamma\left(\frac{\nu}{2} + 1\right)^2 {}_2F_1\left(\frac{\nu}{2} + 1, \frac{\nu}{2} + 1; \tfrac{3}{2}; r^2\right) \right\} \\ &= \frac{\beta^2 2^\nu \psi^\nu}{4\pi} \left\{ \Gamma\left(\frac{\nu + 1}{2}\right)^2 {}_2F_1\left(-\frac{\nu}{2}, -\frac{\nu}{2}; \tfrac{1}{2}; r^2\right) \right. \\ &\quad \left. + 2r\Gamma\left(\frac{\nu}{2} + 1\right)^2 {}_2F_1\left(\frac{1 - \nu}{2}, \frac{1 - \nu}{2}; \tfrac{3}{2}; r^2\right) \right\}, \end{aligned} \tag{7.5}$$

this last with the help of (A3.20b). The series form of (7.5) is more convenient when the spectrum is desired; $R(t)$ may be written accordingly

$$R(t) = \frac{\beta^2 2^\nu \psi^\nu}{4\pi} \sum_{n=0}^{\infty} \left[\frac{(-\nu/2)_n^2 \Gamma[(\nu + 2)/2]^2 2^{2n} r^{2n}}{(2n)!} + \frac{[(1 - \nu)/2]_n^2 \Gamma[\nu/2 + 1]^2 2^{2n+1} r^{2n+1}}{(2n + 1)!} \right]. \tag{7.6}$$

A number of interesting special cases follow from (7.6). For half-wave linear rectifiers ($\nu = 1$) we obtain

$$\begin{aligned} R(t)_{\nu=1} &= \frac{\beta^2 \psi}{2\pi} \left({}_2F_1(-\tfrac{1}{2}, -\tfrac{1}{2}; \tfrac{1}{2}; r^2) + \frac{\pi r}{2} \right) \\ &= \frac{\beta^2 \psi}{2\pi} \left\{ r\left(\sin^{-1} r + \frac{\pi}{2}\right) + (1 - r^2)^{1/2} \right\}, \end{aligned} \tag{7.7}$$

a result derived independently by Van Vleck, North, and Rice (see Sec. 4.7 of reference 2). Half-wave quadratic detectors ($\nu = 2$) give us

$$\begin{aligned} R(t)_{\nu=2} &= \frac{\beta^2 \psi^2}{2\pi} \left(\frac{\pi}{2} (1 + 2r^2) + 4r\, {}_2F_1(-\tfrac{1}{2}, -\tfrac{1}{2}; \tfrac{3}{2}; r^2) \right) \\ &= \frac{\beta^2 \psi^2}{2\pi} \left\{ \left(\frac{\pi}{2} + \sin^{-1} r \right) (1 + 2r^2) + 3r(1 - r^2)^{1/2} \right\}. \end{aligned} \tag{7.8}$$

We may continue in this fashion when ν is integral and write $R(t) = f(r, (1 - r^2)^{1/2},$

$\sin^{-1}r$) with the help of the recurrence relations for the hypergeometric function.

The total mean output power W_T, the *d-c* power W_{d-c}, and the mean total *a-c* power W_{a-c} all follow at once from (2.26d), (2.33), (7.5), and (A3.20a). We have finally

$$W_T = \frac{\beta^2\psi^\nu 2^{\nu-1}}{\sqrt{\pi}}\Gamma(\nu+\tfrac{1}{2}); \quad W_{d-c} = \frac{\beta^2\psi^\nu 2^{\nu-2}}{\pi}\Gamma\left(\frac{\nu+1}{2}\right)^2; \quad W_{a-c} = W_T - W_{d-c} \quad (7.9)$$

Curves of W_T, W_{d-c}, and W_{a-c} are shown in Fig. 8. It is evident that characteristics for which ν is large (>2) exhibit outputs which are chiefly *a-c*. The spectra associated with the output may be found from (1.6) and (7.6), where integrals of the type

$$c_{0,n}(f) = 4\int_0^\infty r(t)^n \cos\omega t dt, \qquad \omega = 2\pi f,$$

must be considered. Observe that when ν is an even integer, the first series in (7.6) terminates after the term for which $n=\nu/2$, and when ν is odd, after $n=(\nu-1)/2$. In particular, the mean output spectrum for the linear rectifier is

$$W(f)_{\nu=1} = \frac{\beta^2\psi}{2\pi}\left\{2\delta(f-0) + \frac{\pi}{2}c_{0,1}(f) + \sum_{n=1}^{\infty}\frac{(2n)!c_{0,2n}(f)}{n!^2 2^{2n}(2n-1)^2}\right\}, \quad (7.10)$$

and for a quadratic detector

$$W(f)_{\nu=2} = \frac{\beta^2\psi^2}{2\pi}\left\{\pi\delta(f-0) + 4c_{0,1}(f) + \pi c_{0,2}(f) + 4\sum_{n=1}^{\infty}\frac{(2n)!c_{0,2n+1}(f)}{[n!2^{2n}(2n-1)]^2(2n+1)}\right\}, \quad (7.11)$$

where $\delta(f-0)$ is the familiar delta-function. Curves illustrating (7.10) and (7.11) when the input spectrum is Gaussian, Eq. (A2.9), $\omega_c=0$, are given in reference 9.

(*b*) *Narrow-band Noise.* The output now consists, as explained above in more detail in section 3, of bands of noise located about harmonics of the central frequency f_c. The spectra associated with these harmonic zones are "smeared out," or distorted from their original shape, as Fig. 7(c) indicates. Our result (7.6) is expressed in a form convenient for obtaining the correlation functions associated with the various harmonic regions lf_c, $l=0, 1, 2, 3, \cdots$, and hence the spectral distribution of the mean power and the mean total power itself pertaining to these regions.

Here the input correlation function is given by (A2.8). When this is substituted into (7.6) we obtain series involving $(\cos\omega_c t)^{2n}$ and $(\cos\omega_c t)^{2n+1}$. The expansion in multiples of $\omega_c t$ is then used, and the correlation function of the lth zone follows from (7.6) on taking those values of n such that $j=(2n-l)/2(\geq 0)$ is integral for even l, and $j=(2n-l+1)/2(\geq 0)$ for odd values of $l(>0)$. We may write finally

$$R_l(t) = \pi^{-1}\beta^2\psi^\nu 2^{\nu-2}\Gamma\left(\frac{\nu+1}{2}\right)^2 \epsilon_l \cos l\omega_c t \sum_{n=l/2}^{\infty}\frac{(-\nu/2)_n^2 r_0(t)^{2n}}{(n-l/2)!(n+l/2)!}, \quad l = 0, 2, 4, \cdots, \quad (7.12a)$$

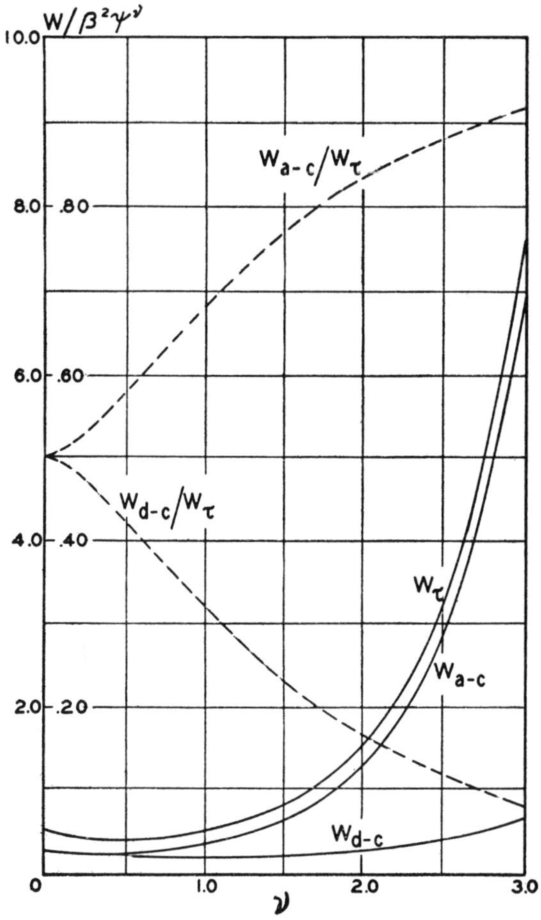

FIG. 8. The mean total, a-c, and d-c power outputs of a half-wave νth-law detector when the incoming disturbance is broadband noise, as a function of ν. The right-hand scale applies for the ratios.

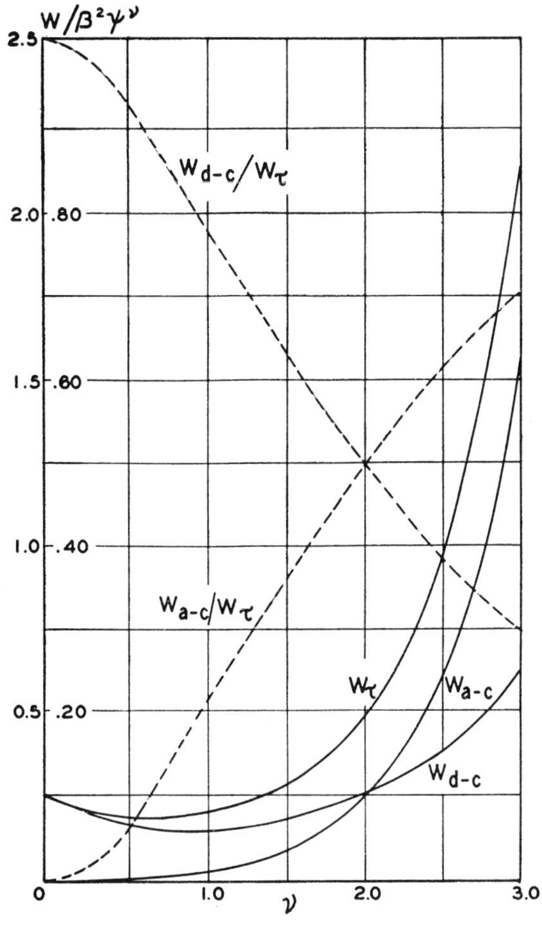

FIG. 9. The mean total and a-c powers associated with the low-frequency output when the incoming random noise is narrow-band, as a function of ν. The d-c power output is the same as that in Fig. 8, and the right-hand scale applies in the case of the ratios.

and

$$R_l(t) = \pi^{-1}\beta^2\psi^\nu 2^{\nu-1}\Gamma\left(\frac{\nu}{2}+1\right)^2 \cos l\omega_c t \sum_{n=(l-1)/2}^{\infty} \frac{[(1-\nu)/2]_n^2 r_0(t)^{2n+1}}{[(2n-l+1)/2]![(2n+l+1)/2]!},$$

$$l = 1, 3, 5, \cdots . \qquad (7.12b)$$

We observe that contributions to the spectrum about the frequency lf_c arise from terms for which $n \geqq l/2$, $(l-1)/2$; values of n less than this do not appear. Now (7.12a) and (7.12b) may be summed; for even values of l we set $n - l/2 = m$, $m = 0, 1, 2, \cdots$, and find that

$$\sum_{n=l/2}^{\infty} \frac{[-\nu/2]_n^2 r_0^{2n}}{(n-l/2)!(n+l/2)!} = r_0^l \sum_{m=0}^{\infty} \frac{[-\nu/2]_{m+l/2}^2 r_0^{2m}}{m!(m+l)!}$$

$$= \frac{r_0^l(-\nu/2)_{l/2}^2}{l!} \sum_{m=0}^{\infty} \frac{[(l-\nu)/2]_m^2 r_0^{2m}}{m!(l+1)_m}$$

$$= \frac{r_0^l[-\nu/2]_{l/2}^2}{l!} \, {}_2F_1\left(\frac{l-\nu}{2}, \frac{l-\nu}{2}; l+1; r_0^2\right), \qquad (7.13)$$

and the correlation function appropriate to the even harmonics is then

$$R(t) = \frac{\beta^2\psi^\nu 2^{\nu-2}}{\pi} \epsilon_l \left(-\frac{\nu}{2}\right)_{l/2}^2 \frac{\Gamma[(\nu+1)/2]^2}{l!} r_0^l \cos l\omega_c t$$

$$\times \, {}_2F_1\left(\frac{l-\nu}{2}, \frac{l-\nu}{2}; l+1; r_0^2\right), \qquad l = 0, 2, 4, \cdots . \qquad (7.14)$$

A similar procedure for odd values of l, letting $n-(l-1)/2=m$, $m=0, 1, 2 \cdots$ in (7.12b) yields

$$R_l(t) = \frac{\beta^2\psi^\nu 2^{\nu-1}}{\pi} \left(\frac{1-\nu}{2}\right)_{(l-1)/2}^2 \frac{\Gamma[\nu/2+1]^2}{l!} r_0^l \cos l\omega_c t$$

$$\times \, {}_2F_1\left(\frac{l-\nu}{2}, \frac{l-\nu}{2}; l+1; r_0^2\right), \qquad l = 1, 3, 5, \cdots . \qquad (7.15)$$

Notice that when the detector characteristic is an even power of the incoming wave, only those zones for which l is even and equal to or less than ν (even) are produced, while all *odd* harmonic zones ($l=1, 3, 5, 7 \cdots$) appear in the output. A like situation is encountered in odd-powered characteristics: only the zones for which $0<l\leqq\nu$ (l and ν odd) are generated, whereas all even regions exist. The above is analogous to the half-wave rectification of a sine wave if we identify the narrow-band input noise with a sinusoid having the frequency $f_0=f_c$. The amplitudes, and hence the powers, of the zones about lf_c corresponding to the sinusoidal components $lf_0=lf_c$ are *not* equal, as a Fourier analysis readily shows, but their number and location in the spectrum are.

Of particular interest in practice is the low-frequency output $l=0$. The complete low-frequency correlation function is seen from (7.14) to be just (6.11), which in the instance $\nu=1$ may also be expressed in terms of the complete elliptic integrals E and K of modulus r_0:

$$R_0(t)_{\nu=1} = \frac{\beta^2\psi}{\pi^2}\left\{E(r_0) - \tfrac{1}{2}(1-r_0^2)K(r_0)\right\}. \qquad (7.16)$$

For the quadratic rectifier the form of $R_0(t)_{\nu=2}$ is quite simple:

$$R_0(t)_{\nu=2} = \beta^2\psi^2(1+r_0^2)/4. \qquad (7.17)$$

When ν is even we obtain a polynomial in r_0, viz:

$$R_0(t)_{\nu=2n} = \beta^2\psi^{2n}\left[\frac{(2n)!}{2^{n+1}n!}\right]^2 {}_2F_1(-n, -n; 1; r_0^2)$$

$$= \left[\frac{\beta\psi^n(2n)!}{2^{n+1}n!}\right]^2 \sum_{j=0}^{n} {}_nC_j^2 r_0^{2(n-j)}. \tag{7.18}$$

The *d-c* component may be found when j is set equal to n.

Again we may determine the various powers by setting $t=0$ or $t=\infty$ in (7.14) and (7.15) and using (A3.20a). The mean total power in the low-frequency region and in the *d-c* are respectively

$$(W_T)_0 = R_0(0) = \frac{\beta^2\psi^\nu 2^{\nu-2}}{\pi} \frac{\Gamma(\nu+1)\Gamma[(\nu+1)/2]^2}{\Gamma(\nu/2+1)^2};$$

$$W_{d-c} = R_0(\infty) = \frac{\beta^2\psi^\nu 2^{\nu-2}}{\pi} \Gamma\left(\frac{\nu+1}{2}\right)^2, \tag{7.19}$$

the latter agreeing with our previous result (7.9), as we would expect in virtue of the general theorem stated at the end of Appendix II. The higher ($l>0$) spectral regions are pure *a-c*. The mean *a-c* power as in general

$$(W_{a-c})_l = R_l(0 = \frac{\beta^2\psi^\nu 2^{\nu-2}}{\pi} \epsilon_l \left(-\frac{\nu}{2}\right)_{l/2}^2 \frac{\Gamma[(\nu+1)/2]^2\Gamma(\nu+1)}{\Gamma[(\nu+l)/2+1]^2},$$
$$l = 2, 4, \cdots \tag{7.20a}$$

and

$$(W_{a-c})_l = \frac{\beta^2\psi^\nu 2^{\nu-1}}{\pi}\left(\frac{1-\nu}{2}\right)_{(l-1)/2}^2 \frac{\Gamma(\nu/2+1)^2\Gamma(\nu+1)}{\Gamma[(\nu+l)/2+1]^2}, \quad l = 1, 3, 5, \cdots. \tag{7.20b}$$

The contributions from all harmonic zones to the total output power $\sum_l R_l(0)$ is simply the first equation in (7.9). Figures 9 and 10 show W_l as a function of ν for several different values of l. It is interesting to note from Fig. 9 that when $\nu<2$ the *d-c* power exceeds the low-frequency *a-c*, while for $\nu>2$ the reverse is true. Contrast this with the behavior of broad-band noise, Fig. 8. We see also in Fig. 10 that over the usual range of ν, i.e., $1<\nu<2$, the power associated with the higher zones ($l\geq 3$) is quite negligible for most purposes compared with that in the zones 0, 1, and 2. In the instance of narrow-band noise the fundamental ($l=1$) spectral region appears to have the greatest relative mean power. Examples of spectra for regions 0 and 1 are to be found in reference 9.

8. Small-signal detection and full-wave rectification. The method of section 7 is particularly well suited to the study of the important practical case of small-signal detection. This process is characterized by input amplitudes sufficiently small so that cut-off, corresponding to large grid-voltage swings in the negative direction, and saturation, arising from excessively large positive swings, are both avoided. For periodic disturbances or in general for waves limited in amplitude, it is possible to achieve this condition. For noise we must modify the "small-signal" concept by specifying that the instantaneous voltage amplitude of the input does not enter the regions of saturation and cut-off (*vide* Fig. 1) more than a given percentage of the time. In our analysis we assume that this percentage is small enough to permit us to replace the physically bounded dynamic characteristic by one unaffected by cut-off or saturation.

We may represent the small-signal dynamic path, or dynamic transfer characteristic, as it is sometimes called,[30] by the following

$$I = \alpha_0 + \alpha_1 V + \alpha_2 V^2 + \cdots + \alpha_k V^k + \cdots = \sum_{k=0}^{\infty} \alpha_k V^k, \tag{8.1}$$

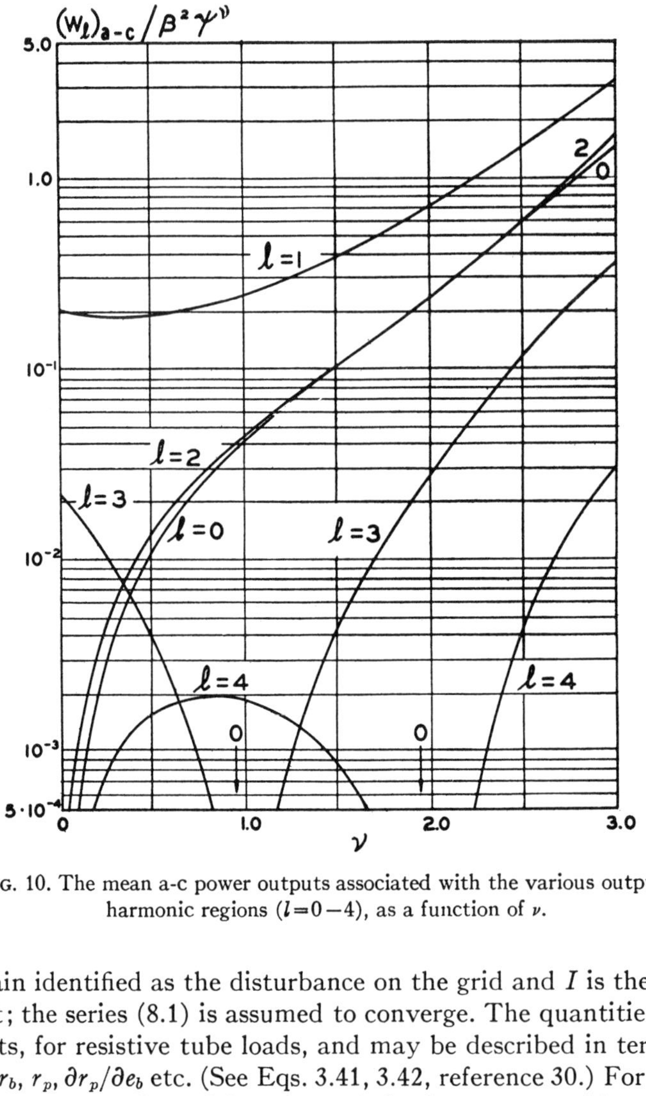

FIG. 10. The mean a-c power outputs associated with the various output harmonic regions ($l=0-4$), as a function of ν.

where V is again identified as the disturbance on the grid and I is the instantaneous current output; the series (8.1) is assumed to converge. The quantities α_k are dimensional constants, for resistive tube loads, and may be described in terms of the tube parameters μ, r_b, r_p, $\partial r_p/\partial e_b$ etc. (See Eqs. 3.41, 3.42, reference 30.) For reactive loads, however, the α_k are complicated functions of the frequency and hence of the input and output waves. As before, our study is restricted to cases where the plate loads are primarily resistive.

[30] H. J. Reich, *Theory and application of electron tubes*, McGraw-Hill (1944) pp. 74–77.

(*a*) *Noise Alone.* When the applied voltage $V(t)$ is random noise, the correlation function of the output is from (2.9) and 1.7b) in normalized form

$$R(t)_N = [2\pi(1 - r^2)^{1/2}]^{-1} \int_{-\infty}^{\infty} dx \int_{-\infty}^{\infty} dy I(x\psi^{1/2}) I(y\psi^{1/2}) e^{-(x^2+y^2-2xyr)/2(1-r^2)}. \tag{8.2}$$

The substitution of (8.1) in (8.2) gives

$$R(t)_N = [2\pi(1 - r^2)^{1/2}]^{-1} \sum_{j=0}^{\infty} \sum_{k=0}^{\infty} \alpha_j \alpha_k \psi^{(j+k)/2} \int_{-\infty}^{\infty} x^j dx \int_{-\infty}^{\infty} y^k dy e^{-(x^2+y^2-2xyr)/2(1-r^2)}. \tag{8.3}$$

The integral in (8.3) may be evaluated with the help of polar coördinates, the method of section 7, or by expanding exp. $(xyr/(1-r^2))$ and applying (A3.20b). The result in either case is

$$\begin{aligned} Q_{j,k}(r) &= [2\pi(1 - r^2)^{1/2}]^{-1} \int_{-\infty}^{\infty} e^{-x^2/2(1-r^2)} x^j dx \int_{-\infty}^{\infty} y^k e^{-y^2/2(1-r^2)} e^{xyr/(1-r^2)} dy \\ &= 0, \quad j + k \text{ odd}, \\ &= 2r \cdot \frac{2^{(j+k)/2}}{\pi} \Gamma\left(\frac{j}{2} + 1\right) \Gamma\left(\frac{k}{2} + 1\right) \\ &\quad \times {}_2F_1\left(\frac{1-j}{2}, \frac{1-k}{2}; \frac{3}{2}; r^2\right), \quad j, k \text{ both odd}, \\ &= \frac{2^{(j+k)/2}}{\pi} \Gamma\left(\frac{j+1}{2}\right) \Gamma\left(\frac{k+1}{2}\right) \\ &\quad \times {}_2F_1\left(-\frac{j}{2}, -\frac{k}{2}; \frac{1}{2}; r^2\right), \quad j, k \text{ both even}. \end{aligned} \tag{8.4}$$

Observe that the hypergeometric functions terminate. Equation (8.2) may be written finally

$$R(t)_N = \sum_{j=0}^{\infty} \sum_{k=0}^{\infty} \alpha_j \alpha_k \psi^{(j+k)/2} Q_{j,k}(r), \quad j + k \text{ even}. \tag{8.5}$$

Only when $j+k$ is even is there a non vanishing output. This quite naturally follows from the basic assumption for our random noise that it contains no *d-c* and consequently has the average value zero, i.e., $\bar{x} = \bar{y} = 0$, and so $\overline{x^j}$ and $\overline{y^k}$ also vanish (j, k odd) for the infinite time average. Then it is clear that $\overline{x^j y^k}$ *on the average* must be zero provided $j+k$ is odd, as borne out by (8.4). This also follows at once from (A1.4) et seq. when $\nu_k = 0$.

The output spectrum is obtained from (8.5) and (1.6), and the total output power is found as before on setting $t = 0$ in $R(t)_N$. With the aid of (A3.20a) we have

$$W_T = R(0) = \sum_{j=0}^{\infty} \sum_{k=0}^{\infty} \alpha_j \alpha_k \left(\frac{\psi}{2}\right)^{(j+k)/2} (j + k)! \Big/ \left(\frac{j+k}{2}\right)!, \quad j + k \text{ even}. \tag{8.6}$$

In practice only the first few terms are generally significant, because α_j and α_k approach zero rapidly for larger values of j and k. The *d-c* output is easily derived on setting $t=\infty$ in (8.5). The expression is

$$W_{d-c} = R(\infty) = \left\{ \sum_{j=0}^{\infty} \alpha_j \left(\frac{\psi}{2}\right)^{j/2} j! \Big/ \left(\frac{j}{2}\right)! \right\}^2, \qquad j = 0, 2, 4, \cdots. \tag{8.7}$$

Equation (8.7) shows that only the even terms in the dynamic characteristic contribute to the *d-c*. The mean power associated with the continuum is now to be obtained from (8.6)–(8.7) and the fundamental relation $W_c = W_T - W_{d-c}$.

When the input noise is broadband there is no separation of the output into spectral bands centered about harmonics of the central frequency as explained in section 3. The spectrum must be calculated directly from (8.5). On the other hand, when the noise is narrow-band, a resolution into harmonics is possible, for which the various correlation functions and hence the powers and spectra may be determined. We note again that the input correlation is given by (A2.8).

Let us consider first the low-frequency output, as it is perhaps of greatest practical interest. Now we observe from (8.4) that only when j and k are both even does $r(t)$ exist in even powers and therefore contributes to the *d-c* and low-frequency continuum. Our procedure for selecting the terms that contribute is similar to that outlined in section 7. Application of the expansion of $(\cos \omega_c t)^{2n}$ to the hypergeometric function in (8.4) yields

$$\begin{aligned} &{}_2F_1\left(-\frac{j}{2}, -\frac{k}{2}; \frac{1}{2}; r^2\right) \\ &= \sum_{n=0}^{j/2,k/2} \frac{(-j/2)_n(-k/2)_n r_0^{2n}}{(\frac{1}{2})_n n! 2^{2n}} \sum_{q=0}^{n} \epsilon_{n-q}\, {}_{2n}C_q \cos 2(n-q)\omega_c t, \quad j, k \text{ even}, \end{aligned} \tag{8.8}$$

where either limit applies on the first summation, according to whichever is the lesser, $j/2$ or $k/2$. Only those terms for which $q=n$ contribute to the low frequency and *d-c*. We have then

$$\begin{aligned} &{}_2F_1\left(-\frac{j}{2}, -\frac{k}{2}; \frac{1}{2}; r^2\right)_0 \\ &= \sum_{n=0}^{j/2,k/2} \frac{(-j/2)_n(-k/2)_n r_0^{2n}\, {}_{2n}C_n}{(\frac{1}{2})_n n! 2^{2n}} = {}_2F_1\left(-\frac{j}{2}, -\frac{k}{2}; 1; r_0^2\right) \quad j, k \text{ even}, \end{aligned} \tag{8.9}$$

since $(1/2)_n = (2n)!/n!2^{2n}$. Equation (8.5) becomes finally

$$\begin{aligned} R_0(t)_N &= \frac{1}{\pi} \sum_{j=0}^{\infty} \sum_{k=0}^{\infty} \alpha_j \alpha_k (2\psi)^{(j+k)/2} \Gamma\left(\frac{j+1}{2}\right) \Gamma\left(\frac{k+1}{2}\right) \\ &\quad \times {}_2F_1\left(-\frac{j}{2}, -\frac{k}{2}; 1; r_0^2\right), \qquad j, k \text{ even}. \end{aligned} \tag{8.10}$$

The low-frequency continuum is calculated from (8.10) when all the constant terms in $\sum {}_2F_1$ corresponding to the *d-c* are removed. The mean total power for the harmonic zone $l=0$ follows from (8.10) and (A3.20a), when $t=0$. The *d-c* power output is seen to be precisely (8.7), as we would expect, and the continuum power $(W_c)_0$ follows from $(W_T)_0 - W_{d-c} = (W_c)_0$.

The above procedure may be applied to the higher spectral zones $l \geqq 1$ where, following the analysis of Sec. 7(b), we may derive the interesting partial sum results

$$
\begin{aligned}
{}_2F_1\left(-\frac{j}{2}, -\frac{k}{2}; \frac{1}{2}; r_0^2 \cos^2 l\omega_c t\right)
&= \sum_{l=0,2,\cdots}^{\infty} \epsilon_{l/2} \cos l\omega_c t \frac{r_0^l}{l!} (-j/2)_{l/2}(-k/2)_{l/2} \\
&\times {}_2F_1\left(\frac{l-j}{2}, \frac{l-k}{2}; l+1; r_0^2\right), \qquad j, k \text{ even}, \qquad (8.11)
\end{aligned}
$$

and

$$
\begin{aligned}
r_0 \cos \omega_c t \, {}_2F_1\left(\frac{1-j}{2}, \frac{1-k}{2}; \frac{3}{2}; r_0^2 \cos^2 \omega_c t\right)
&= \sum_{l=1,3,\cdots}^{\infty} \cos l\omega_c t \frac{r_0^l}{l!} \left(\frac{1-j}{2}\right)_{(l-1)/2} \left(\frac{1-k}{2}\right)_{(l-1)/2} \\
&\times {}_2F_1\left(\frac{l-j}{2}, \frac{l-k}{2}; l+1; r_0^2\right), \qquad j, k \text{ odd}. \qquad (8.12)
\end{aligned}
$$

From (8.11) and (8.12) substituted into (8.5) one may write at once for the correlation functions of the l respective output bands

$$
\begin{aligned}
R_l(t)_N = \epsilon_{l/2} \frac{r_0^l \cos l\omega_c t}{l!} \sum_{j=0}^{\infty} \sum_{k=0}^{\infty} \alpha_j \alpha_k (2\psi)^{(j+k)/2} (-j/2)_{l/2} (-k/2)_{l/2} \Gamma\left(\frac{j+1}{2}\right) \Gamma\left(\frac{k+1}{2}\right) \\
\times {}_2F_1\left(\frac{l-j}{2}, \frac{l-k}{2}; l+1; r_0^2\right), \qquad j, k \text{ even} \qquad (8.13)
\end{aligned}
$$

for even values of l, and when l is odd we obtain

$$
\begin{aligned}
R_l(t)_N = \frac{2r_0^l \cos l\omega_c t}{l!} \sum_{j=0}^{\infty} \sum_{k=0}^{\infty} \alpha_j \alpha_k (2\psi)^{(j+k)/2} \left(\frac{1-j}{2}\right)_{(l-1)/2} \left(\frac{1-k}{2}\right)_{(l-1)/2} \\
\times \Gamma\left(\frac{j}{2}+1\right) \Gamma\left(\frac{k}{2}+1\right) {}_2F_1\left(\frac{l-j}{2}, \frac{l-k}{2}; l+1; r_0^2\right), \; j, k \text{ odd}. \qquad (8.14)
\end{aligned}
$$

From these relations it is observed that for given values of j and k it is not possible to have contributions to zones for which l exceeds j or k, whichever is the lesser. Equation (8.13) and (8.14) also enable us to determine the various mean *a-c* powers in the continuum of the output, on setting $t = 0$ as before, and with the aid of (A3.20a).

(*b*) *Signal and Noise.* Instead of the input voltage being merely random noise, it now consists of a mixture of noise and a signal S, where the latter may or may not be modulated. Since the noise and signal are independent, there can be no correlation between them. Further, if the signal is modulated, correlation may or may not exist between the modulation and the carrier, but in any case there is still no correlation with the noise; (see the first part of section 2). Accordingly, in determining the correlation function that part, R_N, is obtained by a suitable average over the random noise components independently of the contribution arising from the signal. The complete correlation follows from (2.12).

Let S_1 and S_2 be the input signal voltages at times separated by an interval t. Then the total input voltage may be represented in normalized form by

$$v_1 = V_1/\psi^{1/2} = (S_1 + X)/\psi^{1/2} = s_1 + x;$$
$$v_2 = V_2/\psi^{1/2} = (S_2 + Y)/\psi^{1/2} = s_2 + y. \tag{8.15}$$

The contribution to the correlation, attributable to the noise, follows from (8.2) where now the input is given by (8.15). We find then that

$$R_N = \sum_{j=0}^{\infty}\sum_{k=0}^{\infty} \alpha_j\alpha_k\psi^{(j+k)/2}\int_{-\infty}^{\infty}(x + s_1)^j dx \int_{-\infty}^{\infty}(y + s_2)^k W_2(x, y; t)dy, \tag{8.16}$$

with $W_2(x, y; t)$ given by (2.9) after suitable normalization. We note several methods of evaluating these integrals. Unfortunately, none of these methods yields very simple results, although they appear to be the best available. However, in practice the dynamic characteristic (8.1) may be expressed with reasonable accuracy when only the first few terms are considered; the higher coefficients α_j, α_k, j, $k \geqq 3$ are in many cases negligible, and the complexity of the results is consequently much reduced. One approach is that used in evaluating (8.3). Applying it to (8.16) one obtains for the coefficient of $\alpha_j\alpha_k\psi^{(j+k)/2}$:

$$I_{j,k} = \iint_{-\infty}^{\infty}(x + s_1)^j(y + s_2)^k W_2(x, y; t)dxdy$$
$$= j!k!2^{-(j+k)/2}\sum_{p=0}^{j}\sum_{q=0}^{k}\frac{2^{(p+q)/2}s_1^p s_2^q}{p!q![(j-p)/2]![(k-q)/2]!}$$
$$\times\ {}_2F_1\left(\frac{p-j}{2}, \frac{q-k}{2}; \frac{1}{2}; r^2\right), \qquad j-p,\ k-q \text{ even},$$
$$= 2rj!k!2^{-(j+k)/2}\sum_{p=0}^{j}\sum_{q=0}^{k}\frac{2^{(p+q)/2}s_1^p s_2^q}{p!q![(j-p-1)/2]![(k-q-1)/2]!}$$
$$\times\ {}_2F_1\left(\frac{1-j+p}{2}, \frac{1-k+q}{2}; \frac{3}{2}; r^2\right), \quad j-p,\ k-q \text{ odd}, \tag{8.17}$$

and when $j+k-(p+q)(\geqq 0)$ is odd the integrals vanish. We may also derive (8.17) with the aid of the transformation to polar coördinates employed in section 7.

Now let us assume that the modulation A_0 and the carrier $\cos \omega_0 t$ are uncorrelated. Then the input signal S may be written

$$S_1(t_0, t_0') = A_0(t_0')\cos\omega_0 t_0, \qquad S_2(t_0 + t, t_0' + t) = A_0(t_0' + t)\cos\omega_0(t_0 + t). \tag{8.18}$$

The complete correlation may be found from (2.12) on averaging over the phases t_0 and t_0', and from (2.12) the mean power spectrum follows at once, the mean power on setting $t = 0$. As an example, consider the general small-signal quadratic rectifier α_j, $\alpha_k \neq 0$, $j \leqq 2$, $k \leqq 2$. With the help of (8.17)[31] we find for R_N

[31] See Part II of reference 8, section II b.

$$R_N = \alpha_0^2 + \psi^{1/2}\alpha_0\alpha_1(s_1 + s_2) + \psi\alpha_0\alpha_2(2 + s_1^2 + s_2^2) + \psi\alpha_1^2(r + s_1s_2) + \psi^{3/2}\alpha_1\alpha_2(s_1 + s_2)(s_1s_2 + 2r + 1) + \psi^2\alpha_2^2(1 + 2r^2 + s_1^2 + s_2^2 + 4rs_1s_2 + s_1^2s_2^2). \tag{8.19}$$

If the noise is narrow-band, $r = r_0 \cos \omega_c t$ and further, if the carrier is tuned to the center of the noise band, as is the case in receivers or transmitters, then $\omega_c = \omega_0$. Thus, when (8.19) is substituted into the expression for the complete correlation function (2.12) and when (8.18) is used, we observe that the terms involving $s_1, s_2, s_1^2, s_2^2, s_1^2s_2, s_1s_2^2, s_1s_2, s_1r, s_2r$, but *not* s_1s_2r, contribute only to the *d-c* and to frequencies in the neighborhood of f_c and $2f_c$. The low-frequency output, exclusive of direct current, is from (8.19) in conjunction with (2.12).

$$R(t)_{L.F.} = \psi^2\alpha_2^2\left\{r_0(t)^2 + r_0(t)T_0'^{-1}\int_0^{T'}(A_{01}A_{02})_{L.F.}dt_0' + (4T_0')^{-1}\int_0^{T'}(A_{01}^2A_{02}^2)_{L.F.}dt_0'\right\}, \tag{8.20}$$

where we have written A_{01} (for $A_0(t_0')$) and A_{02} for $A_0(t_0'+t)$, and the subscript $L-F$ indicates that *d-c* components are to be removed. Observe that the low-frequency output is composed of three contributions: the first in (8.20) is that arising from the input noise alone, the second represents the cross-modulation of the noise and signal components, and the third is the detected signal envelope, squared, of course, since in this instance the dynamic characteristic is quadratic. We remark that only the term in α_2^2, cf. (8.1), is capable of rectifying; hence the low-frequency output alone results from it. Special cases of (8.19) and (8.20) have been given by Rice, ref. 2, Eqs. (4.10.1) and (4.10.3), when the incoming wave is noise alone, or consists of noise and an unmodulated carrier.

As before, *vide* (8.6) and (8.7), it is possible to deduce (from 8.16) and (8.17) certain general relations for the output, total continuum, and *d-c* power.

(*c*) *Full-Wave Rectification.* By full-wave rectification it is meant that the dynamic characteristic is such that $I_{out} = \beta|V_{in}|^\nu$. When the incoming wave is noise alone, we have merely to multiply the results obtained in section 7 by 2^ν. However, if the input contains a signal as well, Eq. (7.2) multiplied by 2^ν still applies, provided we write for the lower limits of integration $-s_1$ and $-s_2$ respectively, in place of $x = y = 0$. Analytically the problem is now best handled by the methods of section 2 and 3.

We wish to thank Mr. Rice and Prof. J. H. Van Vleck with whom we have discussed these problems from time to time, and also Prof. L. Brillouin, for their helpful criticism of this paper.

Appendix I. In this appendix are briefly summarized the analytical results obtained when the solution to the generalized, *s*-dimensional problem of random flights is used, with the aid of the characteristic function and central limit theorem,[10,32] to derive the multivariate Gaussian distribution law, so fundamental in all problems of the type considered in this paper.

The generalized random-walk problem,[4] resolved by a method due originally to

[32] J. V. Uspensky, *Introduction to mathematical probability*, McGraw-Hill (1937).

Markoff,[33] leads to an expression for the characteristic function $F_N(\boldsymbol{\xi})$, whose associated probability density $W_N(\mathbf{R})$ it is our task to find: viz.

$$F_N(\xi) = \exp\left\{ i \sum_{k=1}^{s} \nu_k \xi_k - \frac{1}{2} \sum_{k=1}^{s} \sum_{l=1}^{s} \mu_{kl} \xi_k \xi_l \right\}, \qquad N \gg 1, \tag{A1.1}$$

where the quantities ν_k and μ_{kl} are given by

$$\nu_k = \sum_{j=1}^{N} \bar{x}_{jk}; \qquad \mu_{kl} = \sum_{j=1}^{N} \overline{x_{jk} x_{jl}}, \qquad N \gg 1. \tag{A1.2}$$

Here $\mathbf{r}_j$ is the jth displacement vector, with components $r_{jk} = x_{jk}$ $(k = 1, 2, \cdots, s)$, and the range of values of x_j extends from $-\infty$ to $+\infty$; N is the total number of displacements, here indefinitely large, and $\mathbf{R}$ is the resultant displacement. Since F_N is the Fourier transform of W_N, we have finally, by the usual methods,[34] the familiar result that[35]

$$W_N(R) = \frac{\exp - \frac{1}{2}(\tilde{\mathbf{X}} - \tilde{\boldsymbol{\nu}})\mathbf{M}(\mathbf{X} - \boldsymbol{\nu})}{(2\pi)^{s/2} \left| \mu \right|^{1/2}}$$

$$= \left[(2\pi)^s \left| \mu \right| \right]^{-1/2} \exp\left\{ -\frac{1}{2} \sum_{k=1}^{s} \sum_{l=1}^{s} \frac{\mu^{kl}}{\left| \mu \right|} (X_k - \nu_k)(X_l - \nu_l) \right\}, \tag{A1.3}$$

where X_k is the kth component of $\mathbf{R}$ and $\overline{X}_k = \nu_k$; here $\mathbf{M}$ is an (sxs) matrix reciprocal to the matrix $\boldsymbol{\mu}$; $|\mu|$ is the determinant of $\boldsymbol{\mu}$, and μ^{kl} is the cofactor of μ_{kl}. Since the average of $\exp\,(i \sum_k \xi_k X_k)$ is, by definition, the characteristic function, it is a simple matter to show that

$$\overline{X}_k = \nu_k = \sum_{j=1}^{N} \bar{x}_{jk}; \quad \overline{X_{k_1} X_{k_2}} = \mu_{k_1 k_2} + \nu_{k_1} \nu_{k_2} = \sum_{j=1}^{N} (\overline{x_{jk_1} x_{jk_2}} + \overline{x_{jk_1}} \cdot \overline{x_{jk_2}}) \tag{A1.4}$$

by expanding this and (A1.1) and equating coefficients of ξ_k, $\xi_{k_1}\xi_{k_2}$, etc. In the same way higher moments may be found, it being noted that if $\nu_k = 0$, all odd-order moments vanish.

Appendix II: Some Remarks on Correlation Functions, Spectra, and Power. The fundamental relationship between the mean power—or mean square amplitude spectrum—and the correlation function, defined below, is well known. We mention it briefly here along with a short discussion of some of the more significant and useful properties of the relation in our work, some of which do not seem to have been treated previously. There appear to be several approaches, yielding similar results, one, for example, through the use of Fourier series,[1] the other with the help of Plancherel's theorem.[36]

Now let us consider $g(t)$ to represent a suitable function of the time, and let us

---

[33] A. Markoff, *Wahrscheinlichkeitsrechnung* (Leipzig, 1912).

[34] See ref. 3; also sec. 10:16 of Margenau and Murphy, *The mathematics of physics and chemistry* (D. Van Nostrand, 1943) and ref. 2, sec. 3.5.

[35] $\mathbf{X}$ and $\boldsymbol{\nu}$ are column matrices, and $\smile$ indicates the transposed matrix.

[36] See E. C. Kemble, *Fundamental principles of quantum mechanics* (McGraw-Hill, p. 36, 1937); also M. Plancherel, Rend. di Palermo **30**, 289 (1910). The theorem is also known by Parseval's name.

require that $g(t)$ be in general different from zero in a long time interval T, but be zero outside this interval. With this in mind we may consider the *correlation function* for $g(t)$, which is defined as

$$R(t) \equiv \lim_{T\to\infty} \frac{1}{T}\int_{-\infty}^{\infty} \overline{g(t_0)g(t_0+t)}dt_0 = [\overline{g(t_0)g(t_0+t)}]_{\text{av.}}, \qquad \text{(A2.1)}$$

where now $g(t_0)$ vanishes outside the interval $0<t<T$. The average in (A2.1) is to be performed over all phases of the disturbance, and the bar indicates the average computed over any random variables in the wave: for it often happens that $g(t)$ is a stochastic or random function of the time t, or at least that some component of $g(t)$ is randomly distributed. Analytically such randomness is introduced by treating the function as involving a certain number of parameters, and then taking these parameters to be random variables, distributed according to a certain law. Now for any given set of parameters there will be a definite correlation function $R(t)$,and as we shall see, a definite mean-square amplitude spectrum also. However, it is the *average* of this set of correlation functions and spectra that is important for our work. These averages may be obtained by averaging over the ranges of the parameters, with the help of their distributions, as indicated by the superscript bar on (A2.1) and elsewhere. We remark further that should $g(t)$ comprise two or more incommensurable periodic disturbances, the definition (A2.1) is easily extended to include the separate averages over the respective phases of the additional components, inasmuch as the correlation function of such a wave is simply the product of the correlation functions of the separate parts. An important example of this type of function is the modulated carrier, *vide* Sec. 2, where no correlation between carrier and modulation exists.

Letting

$$g_1 = g(t_0), \quad (0 < t_0 < T); \qquad g_2^* = g(-t_0 + t), \quad (t_0 - T < t < t_0), \qquad \text{(A2.2)}$$

and with the help of the Fourier transforms of g_1 and g_2^* and Plancherel's theorem, we obtain finally for (A2.1) the correlation function

$$R(t) = \lim_{T\to\infty} \int_{-\infty}^{\infty} \frac{\overline{|S(f)|^2}}{T} e^{-i\omega t}df = \int_0^{\infty} W(f)\cos\omega t df, \qquad f > 0, \qquad \text{(A2.3)}$$

where $W(f)$ is the mean spectral density defined as

$$W(f) = \lim_{T\to\infty} \frac{\overline{2|S(f)|^2}}{T}, \qquad f > 0, \qquad \text{(A2.4)}$$

and $S(f)$ is the amplitude spectrum of $g(t)$. The inversion of (A2.3) gives us at once the other relation

$$W(f) = 4\int_0^{\infty} R(t)\cos\omega t dt. \qquad \text{(A2.5)}$$

The limit in (A2.4) is assumed exist. When it does not, the physical significance of (A2.4) is this: it represents the power at some discrete frequency f_0 rather than a spectral distribution over a continuous range of frequencies. Then analytically in the limit $W(f)$ is observed to become infinite at $f=f_0$ and be zero elsewhere, such that $\int_0^\infty W(f)df$ converges. The spectral density $W(f)$ accordingly exhibits the properties of a delta-function.

On setting $t=0$ we find from (A2.1) and (A2.3) that

$$R(0) = \lim_{T\to\infty} \int_{-\infty}^{\infty} \frac{\overline{g(t_0)^2}}{T} dt_0 = [\overline{g(t_0)^2}]_{\text{av.}} = \int_0^\infty W(f)df, \tag{A2.6}$$

showing that the mean total power may be obtained from the correlation function by putting t equal to zero in the latter; this is in agreement with our definition (A2.4) of the spectral density. Furthermore, *the mean total power is observed* from (A2.6) *to be independent of the shape of the spectrum*, depending only on the integral $\int_0^\infty W(f)df$. It should also be mentioned that the spectrum (A2.4) can never give us the time-variation of the wave, since information about the phases is always wanting. Hence an infinite number of different functions $g(t)$ may be combined to give the same value of $W(f)$.

It is interesting to observe what happens after very long times. For a purely stochastic disturbance, which from (A2.1) is seen to be independent of the average over T, $R(\infty)$ becomes zero: there ceases to be any correlation at all between an event at time t_0 and one at t_0+t, $t>\infty$ later. But for periodic components $R(\infty)$ approaches no definite limit, since this part of the disturbance is indefinitely repeated and can never be said ultimately to die down to zero in time. By considering the oscillatory or constant parts of the expression for the correlation function in the limit $t\to\infty$, we can determine the contribution to the total power arising from the periodic part of the wave, for with the help of (A2.5) and the delta-function the result is seen to be (the sum of) the mean powers in the respective components. These quantities may also be identified as the coefficients of the constant or of the trigonometric parts of the correlation function $R(t)$. The constant part corresponds to *d-c*, the others, to the various discrete frequencies. Thus, in turn, the power in the continuum is $R(0)-R(\infty)$, and may be obtained from that portion of the correlation function which vanishes at $t=\infty$. Examples are given in Sections 2 and 7.

It is convenient to use the *normalized* correlation function $r(t)$, where

$$r(t) \equiv \psi(t)/\psi(0) = \int_0^\infty w(f) \cos \omega t df \Big/ \int_0^\infty w(f)df; \qquad \rho(t) \equiv R(t)/R(0). \tag{A2.7}$$

Here $\psi(t)$ is chosen to represent the correlation of a wave entering some non linear device and $w(f)$ is its mean power spectrum; $\rho(t)$, correspondingly, is the normalized correlation of the output, when $R(t)$ is so distinguished. It follows from (A2.7) that $\rho(0)=r(0)=1$, which is the maximum value of $\rho(t)$ and $r(t)$.

The important special case in which the incoming noise is confined to a symmetrical band of frequencies narrow compared with the central frequency f_c, i.e., $w=w(f-f_c)$ (see Fig. 7c), leads to

$$\begin{aligned} r(t) &= \psi(0)^{-1} \int_0^\infty w(f-f_c) \cos \omega t df \doteq \left\{\psi(0)^{-1} \int_{-\infty}^\infty w(f') \cos \omega' t df'\right\} \cos \omega_c t \\ &\equiv r_0(t) \cos \omega_c t, \end{aligned} \tag{A2.8}$$

on the change of variable $f-f_c=f'$, where we have ignored the spectral "tail" at $f=0$, inasmuch as $w(-f_c)$ is assumed to be very much smaller than $w(0)$. As an example, consider the Gaussian spectrum $w(f)=W_0 \exp[-(f-f_c)^2/f_b^2]$, $f_c/f_b \gg 1$; for this $\psi(t)$ and $\psi(0)$ are readily shown to be

$$\psi(t) = \psi(0) r_0(t) \cos \omega_c t = \left(\frac{W_0 \omega_b}{2\pi^{1/2}}\right) e^{-\omega_b^2 t^2/4} \cos \omega_c t; \quad \psi(0) = \frac{W_0 \omega_b}{2\pi^{1/2}}; \tag{A2.9}$$

W_0 is the maximum spectral density. In a similar fashion correlation functions for other input spectral distributions are easily determined with the help of (A2.7) or (A2.8).

One may expand the output correlation function $R(t)$ in a power series in $\psi(t)$ (or $r(t)$), the input correlation, and hence as a function of $\cos l\omega_c t$ ($l=0, 1, 2, \cdots$), for the case of narrow-band noise, by virtue of (A2.8) and the resolution of $(\cos \omega_c t)^n$ into harmonics.[9] The series then becomes

$$R(t) = \sum_{l=0}^{\infty} G_l(t) \cos l\omega_c t = \sum_{l=0}^{\infty} R_l(t), \tag{A2.10}$$

where now $R_l(t)$ is the correlation of the lth harmonic zone generated in the output. The spectrum of these bands follows at once from (A2.5), showing also that the various "resolved" spectra $W_l(f)$ are distributed about the harmonics lf_c ($l=0, 1, 2 \cdots$). Examination of one such region $W_l(f)$ shows that here, too as in (A2.6), the mean power in any given band is independent of the spectral distribution of the band. Further, since the input spectrum enters only through $\psi(t)$ and $\psi(0)$ (= the mean input power) *the mean power in the l-th band is also independent of the original spectral shape* $w(f)$ of the incoming wave. This is, strictly speaking, only practically true, not rigorously so, as distortion of the input always spreads its spectrum and thus spectral "tails" from one region overlap those of another. This overlapping is quite insignificant most of the time, as long as the band is narrow, the criterion of narrowness depending on what is considered a negligible spectral ordinate.

Appendix III: Special Functions and Integrals. A quantity that appears often in our analysis, cf. Secs. 4 or 5, is the confluent hypergeometric function

$${}_1F_1(\alpha; \beta; x) = 1 + \frac{\alpha x}{\beta 1!} + \frac{\alpha(\alpha+1)}{\beta(\beta+1)} \frac{x^2}{2!} + \cdots \frac{(\alpha)_n x^n}{(\beta)_n n!} + \cdots, \tag{A3.1}$$

where $(\alpha)_n = \alpha(\alpha+1) \cdots (\alpha+n-1)$, and $(\alpha)_0 = 1$, as usual. This function has the important property, known as Kummer's transformation,[21] that

$${}_1F_1(\alpha; \beta; x) = e^x \, {}_1F_1(\beta - \alpha; \beta; -x), \tag{A3.2}$$

and it may be shown that the asymptotic development of ${}_1F_1$ takes the form,[37]

$${}_1F_1(\alpha; \beta; -x) \simeq \frac{\Gamma(\beta)}{\Gamma(\beta-\alpha)} x^{-\alpha} \left\{ 1 + \frac{\alpha(\alpha-\beta+1)}{x 1!} + \frac{\alpha(\alpha+1)(\alpha-\beta+1)(\alpha-\beta+2)}{x^2 2!} + \cdots \right\}, \quad R(x) > 0. \tag{A3.3}$$

[37] See, for example, Whittaker and Watson, *Modern analysis* (Cambridge Univ. Press, 1940), Chapter XVI.

The expression (A3.3) is useful in determining the limiting forms of spectra and power distribution when the root-mean-square noise voltage $\psi^{1/2}$ is much less than either the cut-off voltage b_0 or the amplitude of the carrier; see Sec. 4. The confluent hypergeometric functions of both negative and positive argument may also be expressed in terms of the modified Bessel functions of the first kind, for certain combinations of values of α and β. Examples are given in Appendix II of Bennett's paper[6] and also in ref. 2.

Another function of considerable interest is

$$\phi^{(j)}(b) = \left.\frac{d^{(j)}}{dx^j}\frac{e^{-x^2/2}}{\sqrt{2\pi}}\right|_{x=b} \equiv (-1)^j H_j(b)e^{-b^2/2}/(2\pi)^{1/2}, \qquad j = 0, 1, 2, \cdots . \tag{A3.4}$$

Here the $H_j(b)$ are Hermitian polynomials[38] of order j. It is not difficult to show, by successive differentiations with respect to b_0 when $n=0$, that the ϕ-functions may be given as the following infinite integrals

$$\int_0^\infty z^{2n}\cos b_0 z e^{-\psi z^2/2}dz = \left(\frac{\pi}{2}\right)^{1/2}(-1)^n\psi^{-n-1/2}e^{-b_0^2/2\psi}H_{2n}(b_0/\psi^{1/2})$$
$$= \pi(-1)^n\psi^{-n-1/2}\phi^{(2n)}(b_0/\psi^{1/2})$$
$$\int_0^\infty z^{2n-1}\sin b_0 z e^{-\psi z^2/2}dz = \left(\frac{\pi}{2}\right)^{1/2}(-1)^{n+1}\psi^{-n}e^{-b_0^2/2\psi}H_{2n-1}(b_0/\psi^{1/2})$$
$$= \pi(-1)^n\psi^{-n}\phi^{(2n-1)}(b_0/\psi^{1/2}), \quad n = 0, 1, 2, \cdots, \tag{A3.5}$$

and in this connection

$$\phi^{(-1)}(b_0/\psi^{1/2}) = \frac{1}{\sqrt{2\pi}}e^{-b_0^2/2\psi}H_{-1}(b_0/\psi^{1/2}) = \tfrac{1}{2}\Theta(b_0/\sqrt{2\psi}), \text{ where } \Theta(x) = \frac{2}{\sqrt{\pi}}\int_0^x e^{-y^2}dy$$

is the familiar error function, tabulated, for example, in Jahnke and Emde and in Pierce's Tables. Tables of $\phi^{(n)}$ for $n=0, \cdots, 6$ are given in T. C. Fry's *Probability and its engineering uses* (D. Van Nostrand, 1928, pp. 456, 457). Additional values may be obtained from the recurrence relation

$$\phi^{(n+1)}(b) = -\left\{b\phi^{(n)}(b) + n\phi^{(n-1)}(b)\right\}, \qquad n = 0, 1, 2, \cdots . \tag{A3.6}$$

We may also express $\phi^{(n)}$ in terms of the confluent hypergeometric function ${}_1F_1(\alpha;\beta;-x)$. To do this we need Hankel's exponential integral[39]

$$\int_0^\infty J_\nu(az)z^{\mu-1}e^{-q^2z^2}dz = \frac{\Gamma[(\nu+\mu)/2]}{2q^\mu\Gamma(\nu+1)}\left(\frac{a}{2q}\right)^\nu {}_1F_1\left(\frac{\nu+\mu}{2}; \nu+1; -\frac{a^2}{4q^2}\right),$$
$$R(\mu+\nu) > 0, \qquad |\arg q| < \pi/4, \tag{A3.7}$$

which is readily established by expanding the Bessel function and integrating termwise with help of the Γ-function. Since

$$e^{\pm ib_0 z} = (\pi b_0 z/2)^{1/2}\left\{J_{-1/2}(b_0 z) \pm iJ_{1/2}(b_0 z)\right\}, \tag{A3.8}$$

---

[38] W. Kapteyn, Proc. Royal Acad. Amster. **16,** 1191 (1914), and G. Szegö, Amer. Math. Soc. Colloq. Pub. **23,** 101–104 (1939).

[39] Reference 21, **13.3.**

it follows at once for (A3.6) when this is substituted into (A3.7) that

$$\left.\begin{aligned} \phi^{(2n)}(b_0/\psi^{1/2}) &= \frac{(-1)^n(2n)!}{2^n n!(2\pi)^{1/2}}\,{}_1F_1\left(\frac{2n+1}{2};\ \frac{1}{2};\ -\frac{b_0^2}{2\psi}\right),\\ &\text{and}\\ \phi^{(2n-1)}(b_0/\psi^{1/2}) &= \frac{(-1)^n(2n)!}{2^n n!(2\pi)^{1/2}}\,\frac{b_0}{\psi^{1/2}}\,{}_1F_1\left(\frac{2n+1}{2};\ \frac{3}{2};\ -\frac{b_0^2}{2\psi}\right),\\ &\qquad n = 0, 1, 2, \cdots. \end{aligned}\right\} \tag{A3.9}$$

We notice also from (A3.1), (A3.4), and (A3.9) that the Hermitian polynomials may be written

$$\begin{aligned} H_{2n}(b) &= \frac{(-1)^n(2n)!}{2^n n!}\,{}_1F_1\left(-n;\ \frac{1}{2};\ \frac{b^2}{2}\right);\\ H_{2n-1}(b) &= \frac{(-1)^{n+1}(2n)!}{2^n n!}\,b\,{}_1F_1\left(1-n;\ \frac{3}{2};\ \frac{b^2}{2}\right), \qquad n = 0, 1, 2, \cdots. \end{aligned} \tag{A3.10}$$

From these results it is evident that

$$\begin{aligned} \phi^{(2n)}(0) &= \frac{(-1)^n(2n)!}{2^n n!(2\pi)^{1/2}} = H_{2n}(0)/(2\pi)^{1/2};\\ (2\pi)^{1/2}\phi^{(2n-1)}(0) &= -H_{2n-1}(0) = 0. \end{aligned} \tag{A3.11}$$

Now in the theory of the νth-law non linear device, when $\nu(>0)$ is not necessarily an integer, the integrand of the fundamentals integrals (2.20) contains a branch point at the origin, rather than a simple pole when ν is integral. Accordingly, results like (A3.5) and (A3.7) must be extended to include these more general cases. The first integral to be established is

$$I_1 = \int_{\mathbf{C}} e^{-c^2z^2} z^{2\mu-1}dz = \frac{i\pi e^{-\mu\pi i}}{c^{2\mu}\Gamma(1-\mu)} = ic^{-2\mu}\Gamma(\mu)e^{-\pi\mu i}\sin\pi\mu, \qquad |\arg c| < \pi/4, \tag{A3.12}$$

where the contour $\mathbf{C}$ is the usual one of Eq. (1.8), extending from $-\infty$ to $+\infty$ along the real axis and is indented downward in an infinitesimal semi-circle about $z=0$. Here the argument of z is zero on the positive portion of $\mathbf{C}$ and is $-\pi$ on the negative part. In the neighborhood of the origin the contribution over the semicircle vanishes in the limit, provided that for the moment we require $R(\mu)>0$. Then we may set $z=e^{-\pi i}t$, $0\leqq t\leqq\infty$, $z<0$; $z=t$, $z>0$, so that finally

$$I_1 = (1-e^{-2\pi\mu i})\int_0^\infty t^{2\mu-1}e^{-c^2t^2}dt = (1-e^{-2\pi\mu i})\Gamma(\mu)/2c^{2\mu} = \pi i e^{-\pi\mu i}/c^{2\mu}\Gamma(1-\mu),$$
$$R(\mu) > 0, \qquad |\arg c| < \pi/4. \tag{A3.13}$$

Now $I_1(\mu)$ is certainly analytic for all values of μ, and therefore by analytic continuation we may extend the domain of I_1 to include all values of μ for which $I_1(\mu)$ is finite. This removes the restriction $R(\mu)>0$ and gives us (A3.12).

The generalization of (A3.7) follows from (A3.12). We have only to expand the Bessel function to obtain

$$I_2 = \int_C z^{\mu-1} J_\nu(az) e^{-q^2 z^2} dz = \left(\frac{a}{2}\right)^\nu \sum_{n=0}^{\infty} \frac{(a/2)^{2n}(-1)^n}{n!\Gamma(\nu+n+1)} \int_C z^{\nu+2n-\mu-1} e^{-q^2 z^2} dz, \qquad \text{(A3.14)}$$

termwise integration being allowed because of the absolute convergence of $J_\nu(az)$. Application of (A3.12) with the aid of

$$\Gamma\left(1 - n - \frac{\nu}{2} - \frac{\mu}{2}\right) = \Gamma\left(1 - \frac{\nu+\mu}{2}\right) \Big/ (-1)^n \left(\frac{\nu+\mu}{2}\right)_n$$

gives us finally

$$I_2 = \frac{\pi i^{1-\nu-\mu}(a/2q)^\nu}{q^\mu \Gamma(\nu+1)\Gamma[1-(\nu+\mu)/2]}\, {}_1F_1\left(\frac{\mu+\nu}{2};\ 1+\nu;\ -\frac{a^2}{4q^2}\right),$$
$$|\arg q| < \pi/4. \qquad \text{(A3.15)}$$

Equation (A3.15) enables us to evaluate the generalized version of (A3.5), namely,

$$I_3 = \int_C z^\mu e^{+izb_0 - c^2 z^2} dz, \qquad |\arg c| < \pi/4. \qquad \text{(A3.16)}$$

The substitution of (A3.8) for $e^{\pm izb_0}$ and application of I_2 gives

$$I_3 = \frac{\pi i^{-\mu}}{c^{\mu+1}} \left\{ {}_1F_1\left(\frac{\mu+1}{2};\ \frac{1}{2};\ -x^2\right) \Big/ \Gamma\left(\frac{1-\mu}{2}\right) \right.$$
$$\left. \pm\, 2x\, {}_1F_1\left(\frac{\mu+2}{2};\ -\frac{3}{2};\ -x^2\right) \Big/ \Gamma\left(-\frac{\mu}{2}\right) \right\}, \qquad \text{(A3.17)}$$

where $x = b_0/2c$ and $|\arg.\ c| < \pi/4$, for convergence at infinity. Note that only when μ is an integer can I_3 be expressed as a ϕ-function, cf., Eq. (A3.9).

In the limit of vanishing noise voltage, $\psi \to 0$, or infinite signal or bias and finite noise power, it is necessary to extend the Weber-Schafheitlin integral and evaluate

$$I_4 = \int_C J_\alpha(az) J_\beta(bz) dz / z^\gamma. \qquad \text{(A3.18)}$$

Again, as in the case of the generalized Γ-function, Eq. (A3.12), analytic continuation can be used in precisely similar fashion to give us finally[40]

$$I_4 = \frac{\pi e^{-\pi i(\alpha+\beta-\gamma)/2} b^\beta\, {}_2F_1\left[\tfrac{1}{2}(\alpha+\beta-\gamma+1),\ \tfrac{1}{2}(\beta-\gamma-\alpha+1);\ \beta+1;\ b^2/a^2\right]}{2^{\gamma-1} a^{\beta-\gamma+1} \Gamma(\beta+1) \Gamma[(1+\gamma-\alpha-\beta)/2] \Gamma[(1+\gamma+\alpha-\beta)/2]},\ 0 \leqq b \leqq a, \qquad \text{(A3.19)}$$

subject only to the restrictions that $R(\gamma) > 0$. The function ${}_2F_1$ has two useful properties, needed in the present work, which we list below:

---

[40] Reference 21, Sec. **13.4** gives the value of $\int_0^\infty J_\alpha(az) J_\beta(bz) dz / z^\gamma$.

$$
{}_2F_1(\alpha, \beta; \gamma; 1) = \frac{\Gamma(\gamma)\Gamma(\gamma - \alpha - \beta)}{\Gamma(\gamma - \alpha)\Gamma(\gamma - \beta)}, \quad \begin{array}{l} R(\gamma) \neq 0, -1, -2, \cdots \\ R(\gamma - \alpha - \beta) \neq 0, -1, -2, \cdots, \end{array} \tag{A3.20a}
$$

and

$$
{}_2F_1(\alpha, \beta; \gamma; x) = (1 - x)^{\gamma-\alpha-\beta}\, {}_2F_1(\gamma - \beta, \gamma - \alpha; \gamma; x). \tag{A3.20b}
$$

(See Chapter XIV of reference 37 for a general treatment.)

20

Reprinted from *J. Appl. Phys.*, **24**(6), 720–727 (1953)

Signal-to-Noise Ratios in Band-Pass Limiters*

W. B. DAVENPORT, JR.
Research Laboratory of Electronics, Massachusetts Institute of Technology, Cambridge, Massachusetts

(Received August 26, 1952)

A general analysis is made of the relations between output signal and noise powers and input signal and noise powers for band-pass limiters having odd symmetry in their limiting characteristics. Specific results are given for the case where the limiter has an nth root characteristic, and they include the ideal symmetrical limiter (or clipper) as a limiting case. This analysis shows that, for the band-pass limiter, the output signal-to-noise power ratio is essentially directly proportional to the input signal-to-noise power ratio for all values of the latter. This result is due to the band-pass characteristics rather than to the symmetrical limiting action.

I. INTRODUCTION

SATURATION, or limiting, may often take place in the band-pass amplifier stages of a radio receiver. Sometimes this limiting is inadvertent (as in the case of a much larger than usual input signal which overdrives an amplifier), while sometimes the band-pass stages are deliberately designed to limit (as in the case of most FM receivers). In any case, whether the limiting is inadvertent or deliberate, it is of interest to know quantitatively the action of the band-pass limiter.

The purpose of this report is to determine the relations between the output signal and noise and the input signal and noise for the case of a band-pass limiter for all values of the input signal-to-noise ratio. Previous studies have either considered the effects of noise alone as a limiter input;[1] or, when a signal-plus-noise input was considered, either the action of the limiter was studied only in combination with that of a discriminator[2–4] or results were obtained only for large values of the input signal-to-noise ratio.[5]

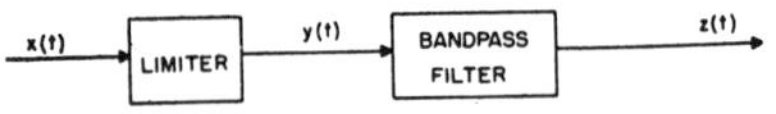

FIG. 1. Block diagram of the band-pass limiter.

The system to be considered here consists of a limiter followed by a band-pass filter, as shown in Fig. 1. The input to this system is assumed to consist of an amplitude-modulated sine wave plus a noise wave

$$x(t)\equiv P(t)\cos pt+N(t). \tag{1}$$

The input noise $N(t)$ is assumed to be Gaussian in nature and to have a narrow band spectrum centered in the vicinity of the signal carrier frequency $p/2\pi$. The spectrum of the limiter output $y(t)$ will consist of signal and noise terms centered on the angular frequencies $\pm mp$, where $m=0, 1, 2, \cdots$. The band-pass filter is assumed to have an ideal rectangular pass-band transfer characteristic which is centered on the fundamental angular frequency p. The filter pass band is assumed to be wide enough to pass all of the limiter output spectrum centered about $\pm p$, but narrow enough to reject those parts of the spectrum centered on $\pm mp$ (where $m\neq 1$).

* This work has been supported in part by the U. S. Signal Corps, the Air Materiel Command, and the U. S. Office of Naval Research.

1 D. Middleton, J. Appl. Phys. **17**, 778 (1946).
2 N. M. Blachman, J. Appl. Phys. **20**, 38 (1949).
3 D. Middleton, J. Appl. Phys. **20**, 334 (1949).
4 D. Middleton, Quart. Appl. Math. **7**, 129 (1949), and **8**, 59 (1950).
5 D. G. Tucker, Wireless Engineer **29**, 128 (1952).

In the analysis to follow, we will obtain expressions for the autocorrelation function at the input to the band-pass filter, as well as expressions for the signal power and noise power at the output of the band-pass filter. From these expressions we will be able to determine the relation between the output signal-to-noise power ratio and the input signal-to-noise power ratio.

II. GENERAL ANALYSIS

Let us first consider the problem of determining the autocorrelation function of the limiter output $y(t)$. This function is defined as the statistical average of $y(t)y(t+\tau)$, i.e.,

$$R_y(\tau)\equiv\langle y(t)y(t+\tau)\rangle_{Av} \tag{2}$$

and has been shown by Wiener[6] to be the Fourier transform of the spectral density.

Rice[7] and Middleton[8] have shown that if the output of a nonlinear device may be expressed as a unique function of its input

$$y(t)=g[x(t)], \tag{3}$$

and if the input to that device is an amplitude-modulated sine wave plus gaussian noise, as in Eq. (1), then the autocorrelation function of the output of the nonlinear device may be expressed as

$$R_y(\tau)=\sum_{m=0}^{\infty}\sum_{k=0}^{\infty}\frac{\epsilon_m}{k!}\langle h_{mk}(t)h_{mk}(t+\tau)\rangle_{Av}R_N{}^k(\tau)\cos mp\tau, \tag{4}$$

where the ϵ_m are the Neumann numbers

$$\epsilon_0=1,\quad \epsilon_m=2 \quad \text{for } m\geqslant 1, \tag{5}$$

where $R_N(\tau)$ is the autocorrelation function of the input

6 N. Wiener, Acta Math. **55**, 117 (1930).
7 S. O. Rice, Bell System Tech. J. **24**, 46 (1945).
8 D. Middleton, Quart. Appl. Math. **5**, 445 (1948).

noise, and where the function $h_{mk}(t)$ is defined by

$$h_{mk}(t)\equiv\frac{j^{m+k}}{2\pi}\int_{-\infty}^{\infty} f(ju)e^{-[\sigma^2(N)\,u^2]/2}u^k J_m[uP(t)]du. \quad (6)$$

$\sigma^2(N)$ is the variance of the input noise, J_m is the mth order Bessel function of the first kind, and $f(ju)$ is the Fourier transform of the transfer characteristic of the nonlinear device

$$f(ju)\equiv\int_{-\infty}^{\infty} g(x)e^{-jux}dx. \quad (7)$$

In our present study we wish to specify that the limiter transfer characteristic $g(x)$ is a nondecreasing odd function of its argument

$$g(x)\equiv\begin{cases} g_+(x) & \text{for } x>0 \\ -g_+(-x) & \text{for } x<0. \end{cases} \quad (8)$$

Because of convergence difficulties, the Fourier transform $f(ju)$ must be replaced in this case by the bilateral Laplace transform[9,10] $f(w)$:

$$f(w)=f_+(w)+f_-(w), \quad (9)$$

where w is the complex variable $v+ju$, and where $f_+(w)$ and $f_-(w)$ are the unilateral Laplace transforms of $g(x)$ in the intervals $(0, +\infty)$ and $(-\infty, 0)$, respectively. The unilateral Laplace transforms $f_+(w)$ and $f_-(w)$ have different regions of convergence in the w plane, and, as we shall see later, each may have a singularity at the origin of that plane. Because of these different regions of convergence for $f_+(w)$ and $f_-(w)$, we must in general employ two inversion integrals with separate integration contours in order to return to $g(x)$ from $f(w)$. For this reason, the single integral expression for the function $h_{mk}(t)$ must be replaced by the sum of two contour integrals in this study

$$h_{mk}(t)=\frac{1}{2\pi j}\int_{C+} f_+(w)e^{[\sigma^2(N)w^2]/2}w^k I_m[wP(t)]dw$$
$$+\frac{1}{2\pi j}\int_{C-} f_-(w)e^{[\sigma^2(N)w^2]/2}w^k I_m[wP(t)]dw, \quad (10)$$

where $C+$ is the contour along the imaginary axis of the w plane with a possible indentation to the right of the origin, and $C-$ is the contour along the imaginary axis with a possible indentation to the left of the origin.

Because of the odd character of $g(x)$, it follows that

$$f_-(w)=-f_+(-w) \quad (11)$$

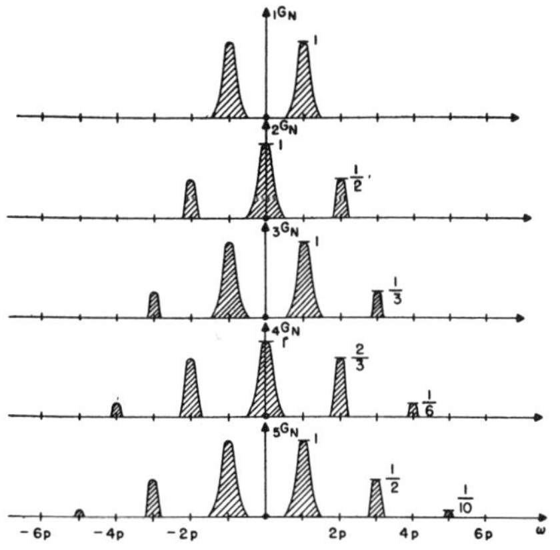

FIG. 2. Plots of the spectral densities $_kG_N(\omega)$, showing the relative values of the various spectral contributions.

and therefore that

$$h_{mk}(t)=\begin{cases} 0 & \text{for } m+k \text{ even} \\ 2\cdot\dfrac{1}{2\pi j}\displaystyle\int_{C+} f_+(w)e^{[\sigma^2(N)w^2]/2}w^k I_m[wP(t)]dw & (12) \\ & \text{for } m+k \text{ odd.}\end{cases}$$

Thus we see that, because of our assumed odd symmetry for the limiter transfer characteristic, the functions $h_{mk}(t)$ vanish whenever the sum of the indices $m+k$ is even. Using this extended definition for $h_{mk}(t)$, we may now use Eq. (4) to determine the autocorrelation function of the limiter output.

From this point on, we will for convenience assume that the input signal is unmodulated. That is, we will assume that $P(t)$ is a constant. In this case, the functions $h_{mk}(t)$ are not functions of t, and Eq. (4) for the limiter output autocorrelation function simplifies to

$$R_y(\tau)=\sum_{\substack{m=0\\(m+k \text{ odd})}}^{\infty}\sum_{k=0}^{\infty}\frac{\epsilon_m h_{mk}^2}{k!}R_N{}^k(\tau)\cos mp\tau, \quad (13)$$

where the coefficients h_{mk} are determined from Eq. (12). It is convenient to expand Eq. (13) as follows

$$R_y(\tau)=2\sum_{\substack{m=1\\(m \text{ odd})}}^{\infty} h_{m0}^2\cos mp\tau+\sum_{\substack{k=1\\(k \text{ odd})}}^{\infty}\frac{h_{0k}^2}{k!}R_N{}^k(\tau)$$
$$+2\sum_{\substack{m=1\\(m+k \text{ odd})}}^{\infty}\sum_{k=1}^{\infty}\frac{h_{mk}^2}{k!}R_N{}^k(\tau)\cos mp\tau. \quad (14)$$

The first set of terms (sum over m) is periodic and consists of the signal output terms representing the interaction of the input signal with itself ($S\times S$ terms). The remaining terms are the limiter output noise terms.

[9] E. C. Titchmarsh, *Fourier Integrals* (Clarendon Press, Oxford, 1948), second edition.

[10] B. van der Pol and H. Bremmer, *Operational Calculus* (Cambridge University Press, Cambridge, 1950).

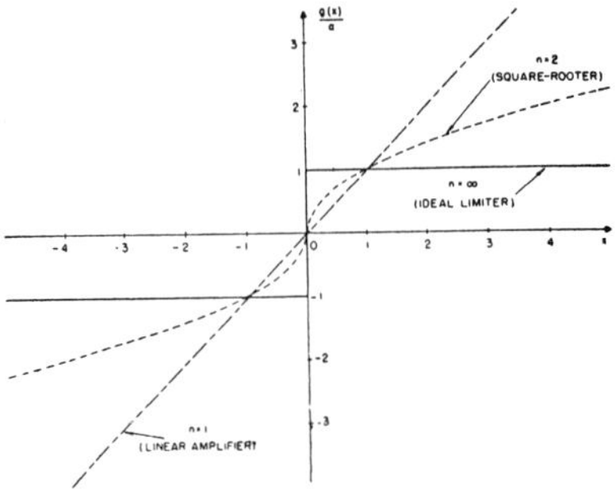

FIG. 3. Plots of the transfer characteristics of several cases of the band-pass rooter.

The second set (sum over k) represents the interaction of the input noise with itself ($N\times N$ terms), while the last set (sum over m and k) represents the interaction of the input signal and noise ($S\times N$ terms).

The spectral density of the limiter output is simply the Fourier transform of the autocorrelation function $R_y(\tau)$, i.e.,

$$G_y(\omega)=\frac{1}{2\pi}\int_{-\infty}^{\infty}R_y(\tau)e^{-i\omega\tau}d\tau, \tag{15}$$

where $G_y(\omega)$ is the so-called "two-sided" spectral density containing both positive and negative frequencies. The Fourier transform of Eq. (14) may be written as

$$G_y(\omega)=\sum_{\substack{m=1\\(m\text{ odd})}}^{\infty}h_{m0}^2\cdot 2\pi[\delta(\omega-mp)+\delta(\omega+mp)] + \sum_{\substack{k=1\\(k\text{ odd})}}^{\infty}\frac{h_{0k}^2}{k!}\,{}_kG_N(\omega)+\sum_{\substack{m=1\\(m+k\text{ odd})}}^{\infty}\sum_{k=1}^{\infty}\frac{h_{mk}^2}{k!} \times[{}_kG_N(\omega-mp)+{}_kG_N(\omega+mp)], \tag{16}$$

where $\delta(\omega)$ is the unit impulse function[10] (Dirac delta-function), and where ${}_kG_N(\omega)$ is the Fourier transform of $R_N{}^k(\omega)$. Successive applications of the convolution theorem[10] shows that ${}_kG_N(\omega)$ may be expressed as the $(k-1)$-fold convolution of $G_N(\omega)$ with itself:

$${}_kG_N(\omega)=\frac{1}{(2\pi)^{k-1}}\int_{-\infty}^{\infty}\cdots\int_{-\infty}^{\infty}G_N(\omega_{k-1}) \times G_N(\omega_{k-2}-\omega_{k-1})\cdots G_N(\omega-\omega_1)d\omega_{k-1}\cdots d\omega_1, \tag{17}$$

where $G_N(\omega)$ is the spectral density of the input noise. One may obtain an idea of the form of the various ${}_kG_N$ by specifying the form of the spectral density of the input noise, and then using Eq. (17) to construct plots of the ${}_kG_N$ up to any desired value of k. Plots were obtained in this manner for several values of k, and are shown in Fig. 2, along with the relative heights of the various spectral contributions. It follows from Eq. (17) that each of the spectral regions in a particular ${}_kG_N$ has the same shape, although this shape may differ from that obtained with a different value of k.

Let us now consider the system output. Because of the band-pass filter, only those terms in Eqs. (14) and (16) which correspond to energy in the vicinity of $\pm p$ will appear at the system output. The output signal autocorrelation function is therefore given by

$$R_{S0}(\tau)=2h_{10}^2\cos p\tau \tag{18}$$

and the signal output spectral density is

$$G_{S0}(\omega)=h_{10}^2\cdot 2\pi[\delta(\omega-p)+\delta(\omega+p)]. \tag{19}$$

The output signal power may be obtained by setting τ equal to zero in Eq. (18)

$$S_0\equiv R_{S0}(0)=2h_{10}^2. \tag{20}$$

The noise terms in the filter output may also be obtained by picking out those terms in Eqs. (14) and (16) which contribute only to the spectral region in the vicinity of $\pm p$. Examination of Eqs. (14) and (16) shows that only those ($N\times N$) terms corresponding to odd values of k appear. From Fig. 2, we then see that all of these terms contribute to the noise output from the band-pass filter. We may determine the filter output noise power due to these terms by setting τ equal to zero in the appropriate terms in Eq. (14), and then multiplying each term in the resultant series by a factor representing the fraction of that term that appears at the filter output. From such a process, we find that the filter output noise power caused by the interaction of the input noise with itself is given by

$$N_{0(N\times N)}=1\cdot\frac{h_{01}^2}{1!}R_N(0)+\frac{3}{4}\cdot\frac{h_{03}^2}{3!}R_N{}^3(0) +\frac{5}{8}\cdot\frac{h_{05}^2}{5!}R_N{}^5(0)+\frac{35}{64}\cdot\frac{h_{07}^2}{7!}R_N{}^7(0)+\cdots. \tag{21}$$

Plots of the spectra of the various ($S\times N$) noise terms at the limiter output may easily be constructed from the plots of Fig. 2 by translating each plot in that figure by an amount $\pm mp$ along the ω axis. A study of Fig. 2 shows that there is a maximum value of the shift mp which will allow a given term to contribute to the spectral region in the vicinity of p. From such a study, we see that for a given k, the only significant values of m are those in the range $(1, k+1)$, such that $(m+k)$ is odd. Therefore, the only terms in Eq. (14) that can contribute to the filter output noise are those given by

$$R'_{y(S\times N)}(\tau)=2\sum_{\substack{m=1\\(m+k\text{ odd})}}^{k+1}\sum_{k=1}^{\infty}\frac{h_{mk}^2}{k!}R_N{}^k(\tau)\cos mp\tau. \tag{22}$$

We may then find the filter output noise power due to the ($S\times N$) terms by setting τ equal to zero in Eq. (22) and multiplying each term by an appropriate factor corresponding to the fraction of that term appearing at the filter output. The result is

$$N_{0(S\times N)}=\frac{1}{2}\cdot 2\frac{h_{21}^2}{1!}R_N(0)+\left(\frac{3}{4}\cdot 2\frac{h_{12}^2}{2!}+\frac{1}{4}\cdot 2\frac{h_{32}^2}{2!}\right)R_N^2(0)$$
$$+\left(\frac{1}{2}\cdot 2\frac{h_{23}^2}{3!}+\frac{1}{8}\cdot 2\frac{h_{43}^2}{3!}\right)R_N^3(0)$$
$$+\left(\frac{5}{8}\cdot 2\frac{h_{14}^2}{4!}+\frac{5}{16}\cdot 2\frac{h_{34}^2}{4!}+\frac{1}{16}\cdot 2\frac{h_{54}^2}{4!}\right)R_N^4(0)+\cdots. \quad (23)$$

The total filter output noise power is then given by the sum of the $(N\times N)$ terms and the $(S\times N)$ terms, and the filter output signal-to-noise power ratio is defined as the ratio of the filter output signal power S_0 to the total filter output noise power N_0. In order to proceed further, we will have to assume a specific form for the limiter transfer characteristic $g(x)$. From this transfer characteristic, we may then determine the coefficients h_{mk} and substitute the result in the above expressions.

III. THE BAND-PASS ROOTER

A. General Results

A convenient form of limiter transfer characteristic is that where the limiter output is proportional to the nth root of its input. We then have

$$g_+(x)\equiv\begin{cases}\alpha x^{1/n} & \text{for } x\geqslant 0\\ 0 & \text{for } x<0,\end{cases} \quad (24)$$

where α is a scaling constant. Plots of $g(x)$ for several values of n are given in Fig. 3. The case $(n=1)$ corresponds to the linear amplifier, while the case $(n=\infty)$ corresponds to the ideal symmetrical limiter.

The Laplace transform $f_+(w)$ of this transfer characteristic is readily determined to be

$$f_+(w)=\alpha\frac{\Gamma\left(1+\frac{1}{n}\right)}{w^{1+(1/n)}}, \quad (25)$$

where Γ is the usual gamma-function. In general, this transform has a branch point at the origin of the w plane, and the contour $C+$ must consequently have an indentation to the right of the origin.

The coefficient h_{mk} is given by substitution of Eq. (25) in Eq. (12) so that

$$h_{mk}=\begin{cases}0 & \text{for } m+k \text{ even}\\ 2\alpha\Gamma\left(1+\frac{1}{n}\right)\cdot\frac{1}{2\pi j}\int_{C+} e^{[\sigma^2(N)w^2]/2} & \\ \quad\times w^{k-1-(1/n)}I_m(Pw)dw & \text{for } m+k \text{ odd.}\end{cases} \quad (26)$$

This contour integral is essentially the same as that required in the study of the νth law detector (if $1/n$ is replaced by ν). Paralleling the evaluation of the corresponding integral for the νth law detector,[6,7] one can readily determine that

$$h_{mk}=\begin{cases}0 & \text{for } m+k \text{ even}\\ \dfrac{\alpha\Gamma\left(1+\frac{1}{n}\right)\left[\frac{P^2}{2\sigma^2(N)}\right]^{m/2}{}_1F_1\left[\frac{m+k-(1/n)}{2};m+1;-\frac{P^2}{2\sigma^2(N)}\right]}{\left[\frac{\sigma^2(N)}{2}\right]^{[k-(1/n)]/2}\Gamma(m+1)\Gamma\left[1-\frac{m+k-(1/n)}{2}\right]} & \text{for } m+k \text{ odd,}\end{cases} \quad (27)$$

where ${}_1F_1$ is the confluent hypergeometric function:[11]

$${}_1F_1[a;c;-z]=1-\frac{a}{c}\frac{z}{1!}+\frac{a(a+1)}{c(c+1)}\frac{z^2}{2!}-\cdots. \quad (28)$$

Reference to Sec. II shows that the coefficient h_{mk} always occurs in combination with $R_N(\tau)$ in the form $h_{mk}^2R_N^k(\tau)$. Now

$$R_N^k(0)=\sigma^{2k}(N), \quad (29)$$

as the input noise has a zero mean. Then by using this result, and the fact that the input signal-to-noise power ratio is given by

$$\left(\frac{S}{N}\right)_i\equiv\frac{S_i}{N_i}=\frac{P^2/2}{\sigma^2(N)}, \quad (30)$$

[11] W. Magnus and F. Oberhettinger, *Formulas and Theorems for the Special Functions of Mathematical Physics* (Chelsea Publishing Company, New York, 1949).

we may write

$$h_{mk}^2 R_N^k(0) = \begin{cases} 0 & \text{for } m+k \text{ even} \\ \dfrac{\alpha^2 2^{k-(1/n)} \Gamma^2\left(1+\frac{1}{n}\right) \sigma^{2/n}(N) \left(\frac{S}{N}\right)_i^m {}_1F_1^2\left[\frac{m+k-(1/n)}{2}; m+1; -\left(\frac{S}{N}\right)_i\right]}{\Gamma^2(m+1)\Gamma^2\left[1-\frac{m+k-(1/n)}{2}\right]} & \text{for } m+k \text{ odd.} \end{cases} \tag{31}$$

This result may now be substituted in expressions (20), (21), and (23) in order to obtain the output signal and noise powers.

A partial check on our results may be obtained by considering the case $(n=1)$. When we substitute $(n=1)$ into Eq. (31), we find that the only nonvanishing coefficients are h_{10} and h_{01}, and that the output signal and noise powers are simply α^2 times the corresponding input powers. These results are comforting in view of the fact that our limiter here is a linear amplifier with a power gain α^2.

B. The Ideal Symmetrical Limiter

As may be seen from Fig. 3, the ideal symmetrical limiter may be represented as a limiting case (for $n=\infty$) of the rooter. Substitution of this limiting value of n in Eq. (31) gives

$$h_{mk}^2 R_N^k(0) = \begin{cases} 0 & \text{for } m+k \text{ even} \\ \dfrac{\alpha^2 2^k \left(\frac{S}{N}\right)_i^m {}_1F_1^2\left[\frac{m+k}{2}; m+1; -\left(\frac{S}{N}\right)_i\right]}{\Gamma^2(m+1)\Gamma^2\left(1-\frac{m+k}{2}\right)} & \text{for } m+k \text{ odd.} \end{cases} \tag{32}$$

If we use this result in our previously obtained expression (20) for the output signal power, we obtain

$$S_0 = \frac{2\alpha^2}{\pi} \cdot \left(\frac{S}{N}\right)_i {}_1F_1^2\left[\frac{1}{2}; 2; -\left(\frac{S}{N}\right)_i\right]. \tag{33}$$

This output signal power has been plotted in Fig. 4 as a function of the input signal-to-noise power ratio $(S/N)_i$.

An expression for the output noise power may be obtained by using Eq. (32) in our previously determined expressions (21) and (23). The result is

$$N_0 = \frac{2\alpha^2}{\pi} \left\{ \begin{aligned} & {}_1F_1^2\left[\frac{1}{2}; 1; -\left(\frac{S}{N}\right)_i\right] + \frac{1}{8} {}_1F_1^2\left[\frac{3}{2}; 1; -\left(\frac{S}{N}\right)_i\right] \\ & + \frac{3}{64} {}_1F_1^2\left[\frac{5}{2}; 1; -\left(\frac{S}{N}\right)_i\right] + \frac{75}{3072} {}_1F_1^2\left[\frac{7}{2}; 1; -\left(\frac{S}{N}\right)_i\right] + \cdots \\ & + \frac{1}{16}\left(\frac{S}{N}\right)_i^2 {}_1F_1^2\left[\frac{3}{2}; 3; -\left(\frac{S}{N}\right)_i\right] + \frac{3}{8}\left(\frac{S}{N}\right)_i {}_1F_1^2\left[\frac{3}{2}; 2; -\left(\frac{S}{N}\right)_i\right] \\ & + \frac{1}{128}\left(\frac{S}{N}\right)_i^3 {}_1F_1^2\left[\frac{5}{2}; 4; -\left(\frac{S}{N}\right)_i\right] + \frac{3}{32}\left(\frac{S}{N}\right)_i^2 {}_1F_1^2\left[\frac{5}{2}; 3; -\left(\frac{S}{N}\right)_i\right] \\ & + \frac{25}{24\,576}\left(\frac{S}{N}\right)_i^4 {}_1F_1^2\left[\frac{7}{2}; 5; -\left(\frac{S}{N}\right)_i\right] + \cdots \end{aligned} \right\}. \tag{34}$$

A plot of this output noise power is given in Fig. 4 as a function of $(S/N)_i$.

Even though we have in Fig. 4 plots of S_0 and N_0 as functions of $(S/N)_i$, it is desirable also to obtain approximate expressions for these powers which are valid in the regions of very large (or very small) values of the input signal-to-noise power ratio.

From the power series expansion about the origin of

the confluent hypergeometric function ${}_1F_1[a;c;-z]$ (Eq. 28), we see that

$$ {}_1F_1[a;c;-z] \to 1 \quad \text{as } z \to 0. \tag{35}$$

Using the limiting result (35) in Eq. (33), we obtain for the output signal power

$$ S_0 \approx \frac{2\alpha^2}{\pi}\left(\frac{S}{N}\right)_i \quad \text{for } \left(\frac{S}{N}\right)_i \to 0. \tag{36}$$

Referring to the numerical calculations used to obtain Fig. 4, we see that the error involved in Eq. (36) is less than ten percent when values of $(S/N)_i$ are less than about two-tenths.

By using the limiting result (35) in Eq. (34), we obtain for the output noise power

$$ N_0 \approx \frac{2\alpha^2}{\pi}\frac{4}{\pi} \quad \text{for } \left(\frac{S}{N}\right)_i \to 0. \tag{37}$$

A study of Eq. (34) shows that in the region of small $(S/N)_i$, the dominant output noise is that due to direct feedthrough of the noise input to the limiter.

From Eqs. (36) and (37) we obtain

$$ \left(\frac{S}{N}\right)_0 \approx \frac{\pi}{4}\left(\frac{S}{N}\right)_i \quad \text{for } \left(\frac{S}{N}\right)_i \to 0. \tag{38}$$

Thus we see that, for the ideal, symmetrical, band-pass limiter, the output signal-to-noise power ratio is directly proportional to the input signal-to-noise power ratio in the region of very small $(S/N)_i$. This result differs radically from the familiar square-law behavior of detectors[7,8] in the region of small $(S/N)_i$. The present result is due primarily to the fact that the output terms of interest in the present case are those in the vicinity of the input frequencies, while in the detector case, the output terms of interest are those in the vicinity of zero frequency.

From the series expansion about infinity[11] of the confluent hypergeometric function ${}_1F_1[a;c;-z]$, we see

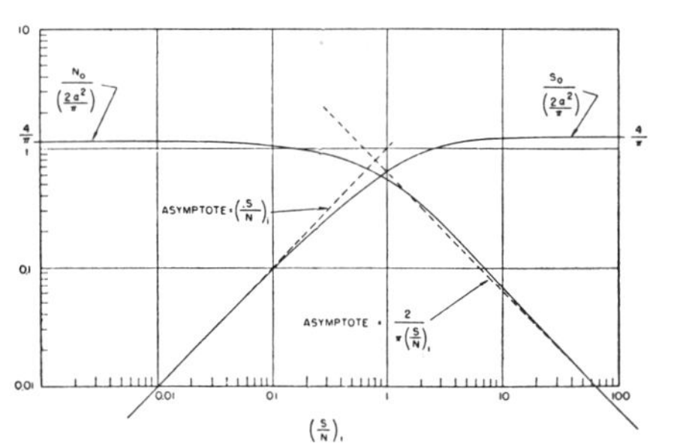

FIG. 4. Output signal power S_0 and output noise power N_0 plotted as functions of the input signal-to-noise power ratio $(S/N)_i$ for the case of the ideal, symmetrical, band-pass limiter.

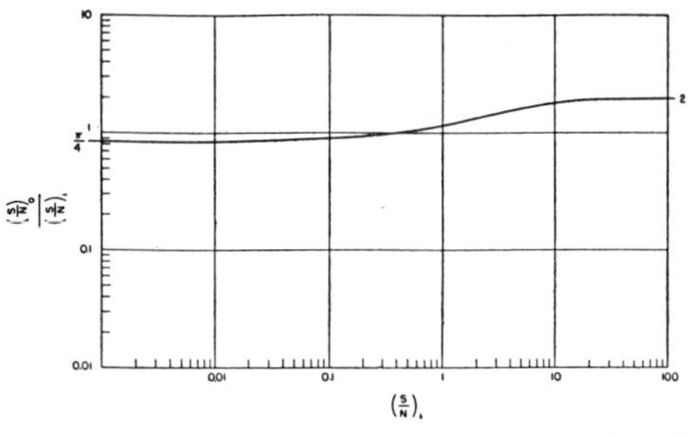

FIG. 5. The ratio of output signal-to-noise power ratio to the input signal-to-noise power ratio as a function of the latter, for the case of the ideal, symmetrical, band-pass limiter.

that

$$ {}_1F_1[a;c;-z] \approx \frac{\Gamma(c)}{\Gamma(c-a)z^a} \quad \text{for } z \to \infty. \tag{39}$$

If we now use the limiting result (39) in Eq. (33), we obtain for the output signal power

$$ S_0 \approx \frac{2\alpha^2}{\pi}\left(\frac{4}{\pi}\right) \quad \text{for } \left(\frac{S}{N}\right)_i \to \infty. \tag{40}$$

Application of the limiting expression (39) to Eq. (34) gives for the output noise power

$$ N_0 \approx \frac{2\alpha^2}{\pi}\cdot\frac{2}{\pi(S/N)_i} \quad \text{for } \left(\frac{S}{N}\right)_i \to \infty. \tag{41}$$

A study of Eq. (34) shows that the dominant noise output terms in this case are (a) the direct feed-through noise term, and (b) the noise resulting from the interaction of the input noise with the second harmonic of the input sine-wave signal.

The output signal-to-noise power ratio becomes

$$ \left(\frac{S}{N}\right)_0 \approx 2\left(\frac{S}{N}\right)_i \quad \text{for } \left(\frac{S}{N}\right)_i \to \infty. \tag{42}$$

Thus, for very large values of the input signal-to-noise power ratio, we find that the output signal-to-noise power ratio is twice that of the input. This is the result obtained by Tucker.[5]

From Eqs. (38) and (42), we see that the output signal-to-noise power ratio is directly proportional to $(S/N)_i$ in both the large and small value regions of $(S/N)_i$. In order to determine the behavior of $(S/N)_0$ for intermediate values of $(S/N)_i$, the ratio of $(S/N)_0$ to $(S/N)_i$ was calculated from the data used to plot Fig. 4, and the results were plotted to form Fig. 5. From this plot, we see that in the case of the ideal, symmetrical, band-pass limiter, the output signal-to-noise power ratio is essentially linearly proportional to the input signal-to-noise power ratio for all values of the latter. A review of this analysis shows that this behavior is due primarily to the action of the band-pass filter following the limiter rather than to the symmetry of the limiter.

C. The Square-Rooter

Let us now consider the case of the square-rooter: Substitution of ($n=2$) in the general expression, Eq. (31), gives

$$h_{mk}^2 R_N{}^k(0)=\begin{cases} 0 & \text{for } m+k \text{ even} \\ \dfrac{\alpha^2\pi}{4}\dfrac{\sigma(N)}{\sqrt{2}}\left(\dfrac{S}{N}\right)_i^m \dfrac{2^k\,{}_1F_1{}^2\left[\dfrac{m+k-\frac{1}{2}}{2};m+1;-\left(\dfrac{S}{N}\right)_i\right]}{\Gamma^2(m+1)\Gamma^2\left(1-\dfrac{m+k-\frac{1}{2}}{2}\right)} & \text{for } m+k \text{ odd}\end{cases} \tag{43}$$

for the case of the square-rooter.

Substitution of Eq. (43) into Eq. (20) gives

$$S_0=\frac{\alpha^2}{4\pi}\Gamma^2\left(\frac{1}{4}\right)\frac{\sigma(N)}{\sqrt{2}}\left(\frac{S}{N}\right)_i {}_1F_1{}^2\left[\frac{1}{4};2;-\left(\frac{S}{N}\right)_i\right] \tag{44}$$

as an expression for the output signal power from the band-pass square-rooter.

Substitution of Eq. (43) into Eqs. (21) and (23) gives

$$N_0=\frac{\alpha^2}{4\pi}\Gamma^2\left(\frac{1}{4}\right)\frac{\sigma(N)}{\sqrt{2}}\left\{\begin{aligned} &{}_1F_1{}^2\left[\frac{1}{4};1;-\left(\frac{S}{N}\right)_i\right]+\frac{1}{32}\,{}_1F_1{}^2\left[\frac{5}{4};1;-\left(\frac{S}{N}\right)_i\right] \\ &+\frac{25}{3072}\,{}_1F_1{}^2\left[\frac{9}{4};1;-\left(\frac{S}{N}\right)_i\right]+\cdots \\ &+\frac{1}{64}\left(\frac{S}{N}\right)_i^2 {}_1F_1{}^2\left[\frac{5}{4};3;-\left(\frac{S}{N}\right)_i\right] \\ &+\frac{3}{32}\left(\frac{S}{N}\right)_i {}_1F_1{}^2\left[\frac{5}{4};2;-\left(\frac{S}{N}\right)_i\right] \\ &+\frac{25}{4608}\left(\frac{S}{N}\right)_i^3 {}_1F_1{}^2\left[\frac{9}{4};4;-\left(\frac{S}{N}\right)_i\right] \\ &+\frac{25}{1536}\left(\frac{S}{N}\right)_i^2 {}_1F_1{}^2\left[\frac{9}{4};3;-\left(\frac{S}{N}\right)_i\right] \\ &+\frac{125}{524\,288}\left(\frac{S}{N}\right)_i^4 {}_1F_1{}^2\left[\frac{13}{4};5;-\left(\frac{S}{N}\right)_i\right]+\cdots \end{aligned}\right\} \tag{45}$$

as an expression for the output noise power from the band-pass square-rooter.

Let us now obtain limiting expressions for the output signal and noise powers valid in the region of large (or small) values of input signal-to-noise power ratio. Substitution of the small $(S/N)_i$ limiting expression (35) for the confluent hypergeometric function in Eq. (44) gives for the output signal power

$$S_0\approx\frac{\alpha^2}{4\pi}\Gamma^2\left(\frac{1}{4}\right)\frac{\sigma(N)}{\sqrt{2}}\left(\frac{S}{N}\right)_i \quad \text{for } \left(\frac{S}{N}\right)_i\to 0. \tag{46}$$

Substitution of the expression (35) in Eq. (45) gives for the output noise power

$$N_0\approx\frac{\alpha^2}{4\pi}\Gamma^2\left(\frac{1}{4}\right)\frac{\sigma(N)}{\sqrt{2}}(1.04) \quad \text{for } \left(\frac{S}{N}\right)_i\to 0. \tag{47}$$

From these expressions, we obtain

$$\left(\frac{S}{N}\right)_0\approx 0.96\left(\frac{S}{N}\right)_i \quad \text{for } \left(\frac{S}{N}\right)_i\to 0. \tag{48}$$

Thus, in the region of very small values of the input

signal-to-noise power ratio, the output signal-to-noise power ratio is equal to 96 percent of the input ratio. Substitution of the large $(S/N)_i$ limiting expression (39) for the confluent hypergeometric function in Eq. (44) gives for the output signal power

$$S_0 \approx \frac{\alpha^2}{4\pi}\left(\frac{8}{9\pi^2}\right)\Gamma^2\left(\frac{1}{4}\right)\frac{\sigma(N)}{\sqrt{2}}\left(\frac{S}{N}\right)_i^{1/2} \quad \text{for} \left(\frac{S}{N}\right)_i \to \infty. \tag{49}$$

Substitution of Eq. (39) in Eq. (45) gives for the output noise power

$$N_0 \approx \frac{\alpha^2}{4\pi}\left(\frac{5}{9\pi^2}\right)\Gamma^2\left(\frac{1}{4}\right)\frac{\sigma(N)}{\sqrt{2}}\left(\frac{S}{N}\right)_i^{-(1/2)} \quad \text{for} \left(\frac{S}{N}\right)_i \to \infty. \tag{50}$$

From these expressions, we obtain

$$\left(\frac{S}{N}\right)_0 \approx \frac{8}{5}\left(\frac{S}{N}\right)_i \quad \text{for} \left(\frac{S}{N}\right)_i \to \infty. \tag{51}$$

Thus, in the region of very large values of the input signal-to-noise power ratio, the output signal-to-noise power ratio is again found to be directly proportional to the input ratio. We see then, as in our previous cases, that the output signal-to-noise power ratio is essentially directly proportional to the input signal-to-noise power ratio for the case of the band-pass square-rooter.

IV. CONCLUSIONS

We have presented here an analysis of the relation between the output signal-to-noise power ratio and the input signal-to-noise power ratio for the case of a band-pass limiter. This analysis shows that, for this type of system, the output signal-to-noise power ratio is essentially directly proportional to the input signal-to-noise power ratio for all values of the latter. This type of behavior is due primarily to the fact that, in the systems studied here, the system output has been so filtered that it contains only those frequency components in the immediate vicinity of the input frequencies.

VI
Output Distributions of Zadeh's Class η_1 Functionals

Editor's Comments on Papers 21 Through 26

21 **Kac and Siegert:** *On the Theory of Noise in Radio Receivers with Square Law Detectors*

22 **Meyer and Middleton:** *On the Distributions of Signals and Noise After Rectification and Filtering*

23 **Darling and Siegert:** *A Systematic Approach to a Class of Problems in the Theory of Noise and Other Random Phenomena: Part I*

24 **Siegert:** *A Systematic Approach to a Class of Problems in the Theory of Noise and Other Random Phenomena: Part II. Examples*

25 **Siegert:** *A Systematic Approach to a Class of Problems in the Theory of Noise and Other Random Phenomena: Part III. Examples*

26 **Grenander, Pollak, and Slepian:** *The Distribution of Quadratic Forms in Normal Variates: A Small Sample Theory with Applications to Spectral Analysis*

The output distributions of ZNL systems with random inputs may be derived in a relatively straightforward manner; however, if the ZNL system is followed by a linear system, the problem of deriving the output distribution becomes rather difficult. For such systems it is easier in principle to obtain the output autocorrelation or power spectrum, which is the subject of the papers in the earlier sections. While Gaussian inputs in linear systems result in Gaussian outputs, the preprocessing of the Gaussian inputs by a ZNL system complicates the matter of the output distribution. The output distributions of such systems are important in evaluating detection probabilities and other nonlinear communication systems. All the papers in this section are concerned with the output distributions of a special class of nonlinear systems, namely, an extension of Zadeh's class η_1 to the vector case:

$$y(t) = \int_{t_0}^{t} K\left[\mathbf{x}(\tau), \tau\right] \,|d\tau$$

The major contributor to this area is A. J. F. Siegert, who coauthored four of the six papers in this section. The papers consider the distribution of the output $y(t)$ for several special and general cases of the functional described above for Gaussian or Markov inputs.

The first contribution to this subject was made by Kac and Siegert (Paper 21), who treated the output distribution of a square-law device followed by a linear system with memory. The resulting distribution involves the solution for the eigenvalues and eigenfunction of a homogeneous linear integral equation. They also considered explicit or approximate solutions for special cases. Their paper also discussed application to the discrete case, a subject that has received considerable attention. The paper in its treatment of a special nonlinearity established the method, which has been used by later generalizations, and prompted many extensions.

The first extension is provided by Meyer and Middleton in Paper 22; here the basic technique of Kac and Siegert is applied to the quadratic nonlinearity for the derivation of nth-order distributions. Emerson (1953) suggested a series expansion for the resulting output probability density considered by Kac and Siegert, and provided a method for deriving the cumulants of the distribution. Lampard (1956) further generalized the problem in considering the output distribution of a multiplier with two inputs followed by a linear system. He thus extended the problem from one signal in a quadratic device to two signals.

The major generalization appeared in a series of three papers by Darling and Siegert (Paper 23) and Siegert (Papers 24 and 25). The preliminary basis for the general theory appeared in an earlier paper by Siegert (1954), in which the integral equation for the output distribution for Markov inputs is reduced to a Fokker–Planck type of differential equation. The general result of Darling and Siegert derives the differential equations for the output distribution of the functional mentioned above, for the case where the input $\mathbf{x}(t)$ is a vector Markov process. The result is applicable only to the case where the Markov input conditional density function satisfies a Fokker–Planck equation. In the second paper in the series, Siegert applies the general method in Paper 23 to the square-law device considered in Kac and Siegert (Paper 21) and shows directly the equivalence of the two approaches. Finally, in the last paper in the series (Paper 25) the general method is applied by Siegert to the quadratic case where the form of the functional kernel is given as

$$K(\mathbf{x}, \tau) = \mathbf{x}^{\mathrm{T}}\mathscr{K}(\tau)\mathbf{x}$$

where $\mathscr{K}(\tau)$ is a square matrix. The problem in this case is shown to require the solution of initial condition differential equations only. The importance of this special case lies in the fact that it includes the correlator or multiplier considered by Lampard (1956) as a special case, and is directly related to the last paper in this part.

While most of the results and applications are given for the continuous case, they apply also to the discrete case of the distribution of quadratic forms in n-variates. Such an application has already been mentioned in the case of Kac and Siegert (Paper 21). However, a parallel but similar development on the distributions of discrete quadratic forms has been published in the statistical literature. The first paper on this subject is by Robbins (1948) and considers the distribution of a definite quadratic form. Extensions to definite and indefinite quadratic forms have been given by Gurland (1953, 1955). The results are similar in nature to Kac and Siegert in that they also require the solution of a linear integral eigenvalue problem. An excellent summary of these earlier results together with approximate methods for the explicit solution is given in the paper by Grenander, Pollak, and Slepian (Paper 26). In addition, the paper contains applications to spectral estimation, which is one of the major applications of both the quadratic forms method and that of Kac and Siegert. The extent of the application of Kac and Siegert to correlators may be inferred from the following papers all dealing with the output distribution of correlators: Jacobson (1963), Cooper (1965), Brown and Piper (1967), and Andrews (1973). Other applications to special nonlinearities, special linear systems, or special input characteristics have also been published in the 1960s, such as: Doyle, McFadden, and Marx (1962), Henry and Schultheiss (1962), and Pawula and Tsai (1969).

Finally, it should be noted that the Fokker–Planck method used for the dynamic systems analysis case plays a central role in the general result of Darling and Siegert. Consequently, the subject of this part is strongly related not only to Part I, which considers Markov processes satisfying the cross-correlation property, but also to the analysis of dynamic nonlinear systems with random inputs, a subject not included in this collection.

21

Reprinted from *J. Appl. Phys.*, **18**(4), 383–397 (1947)

On the Theory of Noise in Radio Receivers with Square Law Detectors*

Mark Kac
Department of Mathematics, Cornell University, Ithaca, New York

AND

A. J. F. Siegert**
Radiation Laboratory, Massachusetts Institute of Technology, Cambridge, Massachusetts

(Received September 13, 1946)

For the video output V of a receiver, consisting of an i-f stage, a quadratic detector, and a video amplifier, the probability density $P(V)$ has been obtained for noise alone and for noise and signal. The results are expressed in terms of eigenvalues and eigenfunctions of the integral equation

$$\int_0^\infty K(t)\rho(s-t)f(t)dt=\lambda f(s),$$

where $\rho(\tau)$ is the i-f correlation function (i.e., the Fourier transform of the i-f power spectrum) and $K(t)$ is the response function of the video amplifier (i.e., the Fourier transform of the video amplitude spectrum). Two special cases are discussed in which the integral equation can be solved explicitly. Approximations for general amplifiers are given in the limiting cases of wide and narrow videos. Some applications of the method to other problems are shown in Sections 7B and 9.

I. THE RECEIVER MODEL

The response of a linear circuit to noise has been investigated by several authors. The probability densities and correlation coefficients of the noise outputs of some types of detectors are also known.[1] To find the probability density of the noise voltage output of the next stage

* The major part of this paper is based on work done for the Office of Scientific Research and Development under contract OEMsr-429 with Cornell University and contract OEMsr-262 with Massachusetts Institute of Technology.

** Now at the Physics Department, Syracuse University, Syracuse, New York.

[1] References to the work in this field can be found in the articles by S. O. Rice, Bell Sys. Tech. J. **23**, 282 (1944); **24**, 46 (1945), and M. C. Wang and G. E. Uhlenbeck, Rev. Mod. Phys. **17**, 323 (1946).

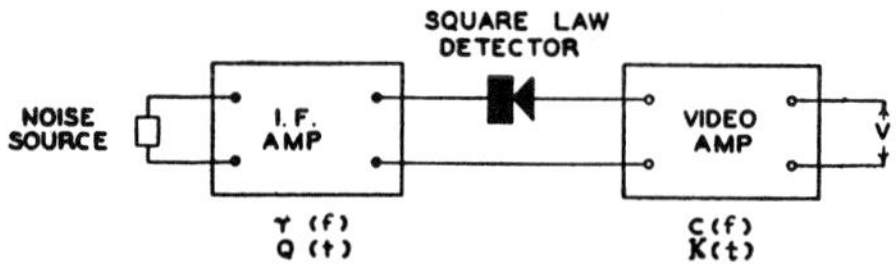

FIG. 1. The receiver model.

(video or audio amplifier) is a problem of a different type, since after the detector, the noise voltage is no longer Gaussianly distributed.

In this paper we have, therefore, calculated the probability density $P(V)$ of the noise voltage V at the end terminals of a receiver consisting of an intermediate-frequency amplifier, a quadratic detector and a video (or audio) amplifier (Fig. 1). The response of both amplifiers is assumed to be linear and characterized by the steady-state response function $\gamma(f)$ and $C(f)$, respectively, or the response functions $Q(t)$ and $K(t)$, such that the output voltages of the amplifiers are $\gamma(f)e^{2\pi ift}$ and $C(f)e^{2\pi ift}$, respectively, if the input voltage is $e^{2\pi ift}$. If, therefore, the video input voltage is $v(t)$, the video output voltage V is given by

$$V(t)=\int_{-\infty}^{\infty}dfC(f)e^{2\pi ift}\int_{-\infty}^{\infty}dt'e^{-2\pi ift'}v(t')$$
$$=\int_{-\infty}^{\infty}dt'K(t-t')v(t') \qquad (1.11)$$

where

$$K(\tau)=\int_{-\infty}^{\infty}dfC(f)e^{2\pi if\tau}. \qquad (1.12)$$

Since $V(t)$ can depend only on the values which $v(t')$ assumes at times $t'\leqq t$, it follows that $K(\tau)=0$ for $\tau<0$, which means that all poles of $C(f)$ lie above the real axis. Since K must be real, we have $C(-f)=C^*(f)$. Both conditions are fulfilled for passive networks. We shall normalize K by

$$\int_0^{\infty}K(t)dt=1.$$

In writing i-f voltages $V_{\text{I.F.}}{}^{(i)}$ and $V_{\text{I.F.}}{}^{(o)}$ for input and output it is generally convenient to consider them as modulated carriers, i.e., we write

$$V_{\text{I.F.}}{}^{(i)}=Re\{u(t)e^{2\pi if_0t}\}$$
$$=x(t)\cos 2\pi f_0t+y(t)\sin 2\pi f_0t,$$

where

$$u(t)=x(t)-iy(t),$$

and f_0 is the carrier frequency. The i-f output voltage is then

$$V_{\text{I.F.}}{}^{(o)}=Re\left\{\int_{-\infty}^{\infty}dfe^{2\pi ift}\gamma(f)\times\int_{-\infty}^{\infty}dt'e^{-2\pi ift'}u(t')e^{2\pi if_0t'}\right\}$$
$$=Re\left\{e^{2\pi if_0t}\int_{-\infty}^{\infty}dt'u(t')\times\int_{-\infty}^{\infty}dfe^{2\pi i(f-f_0)(t-t')}\gamma(f)\right\}$$
$$=Re\{U(t)e^{2\pi if_0t}\}$$

where

$$U(t)=\int_{-\infty}^{\infty}dt'Q(t-t')u(t'),$$

and

$$Q(\tau)=\int_{-\infty}^{\infty}dfe^{2\pi i(f-f_0)\tau}\gamma(f)=\int_{-\infty}^{\infty}df'e^{2\pi if'\tau}\gamma(f'+f_0).$$

We shall normalize γ and Q by demanding that

$$\int_{-\infty}^{\infty}Q^2(\tau)d\tau=1;$$

then we have

$$\int_{-\infty}^{\infty}Q^2(\tau)d\tau=\iiint_{-\infty}^{\infty}d\tau df'df''e^{2\pi i(f'+f'')\tau}\gamma(f'+f_0)\times\gamma(f''+f_0)=1,$$

or

$$\int_{-\infty}^{\infty}df'\gamma(f_0+f')\gamma(f_0-f')=1.$$

The calculations are appreciably simplified by assuming a symmetrical pass band* for the i-f amplifier, i.e.,

$$\gamma(f_0+f')=\gamma^*(f_0-f'). \qquad (1.20)$$

We also put $|\gamma(f_0+f)|=B(f)$. Inserting this in the normalization equation we have

$$\int_{-\infty}^{\infty}|\gamma^2(f)|df=1,$$

i.e., the power spectrum is normalized. The sym-

* See, however, footnote 3, Section IV.

metry of γ implies that Q is real and $V_{\text{I.F.}}^{(o)}$ can now be written as

$$V_{\text{I.F.}}^{(o)}=X(t)\cos 2\pi f_0 t+Y(t)\sin 2\pi f_0 t \quad (1.21)$$

with

$$X(t)=\int_{-\infty}^{\infty}dt' Q(t-t')x(t')$$
$$Y(t)=\int_{-\infty}^{\infty}dt' Q(t-t')y(t'). \quad (1.22)$$

$Q(\tau)$ must of course vanish for $\tau<0$.

The detector is assumed to yield the output

$$X^2(t)+Y^2(t), \quad (1.31)$$

if the input voltage is

$$X(t)\cos 2\pi f_0 t+Y(t)\sin 2\pi f_0 t. \quad (1.32)$$

In order that this be a reasonable model of an actual detector it is necessary to assume

$$\left|\frac{x'(t)}{x(t)}\right|\ll f_0 \quad\text{and}\quad \left|\frac{y'(t)}{y(t)}\right|\ll f_0,$$

which implies that the width of the pass band of the i-f amplifier is small compared to the intermediate frequency f_0.

II. REPRESENTATION OF NOISE

The i-f output noise voltage (omitting the irrelevant d.c. component) is represented[2] as

$$V_{\text{I.F.}}^{(o)}=\frac{\sqrt{N}}{\sqrt{T}}\sum_{k=1}^{\infty}|\gamma(f_k')| \times(X_k'\cos 2\pi f_k't+Y_k'\sin 2\pi f_k't) \quad (2.11)$$

with

$$f_k'=k/T,$$

where X_1', Y_1', X_2', $Y_2'\cdots$ are independent normally distributed random variables each having the probability density

$$\pi^{-\frac{1}{2}}\exp(-u^2).$$

The representation holds only for the time interval $0<t<T$, but this is adequate if T is chosen large compared to the time during which observations are made. We shall later let $T\rightarrow\infty$. The properties of $\gamma(f)$ were discussed above and N denotes the input noise power per unit frequency.

[2] S. O. Rice, "Mathematical analysis of random noise," Bell Sys. Tech. J. **23**, 282 (1944), see p. 328.

According to (1.31) and (1.32), the detector output voltage arising from this i-f input voltage is given by

$$X^2(t)+Y^2(t)=V_D^{(o)},$$

where

$$X(t)=\frac{\sqrt{N}}{\sqrt{T}}\sum_{k=1}^{\infty}|\gamma(f_k')|[X_k'\cos 2\pi(f_k'-f_0)t + Y_k'\sin 2\pi(f_k'-f_0)t]$$
$$Y(t)=\frac{\sqrt{N}}{\sqrt{T}}\sum_{k=1}^{\infty}|\gamma(f_k')|[-X_k'\sin 2\pi(f_k'-f_0) + Y_k'\cos 2\pi(f_k'-f_0)t]. \quad (2.12)$$

Making use of (1.20) and the fact that the band width of the i-f amplifier is small compared to the intermediate frequency f_0, we may write

$$X(t)=\frac{\sqrt{N}}{\sqrt{T}}\sum_{-\infty}^{\infty}B(f_k)(X_k\cos 2\pi f_k t+Y_k\sin 2\pi f_k t),$$
$$Y(t)=\frac{\sqrt{N}}{\sqrt{T}}\sum_{-\infty}^{\infty}B(f_k) \times(-X_k\sin 2\pi f_k t+Y_k\cos 2\pi f_k t), \quad (2.13)$$

where

$$f_k=k/T,\quad B(f_k)=|\gamma(f_k+f_0)|,$$
$$X_k=X'_{k+k_0}\quad\text{and}\quad Y_k=Y'_{k+k_0} \quad (2.14)$$

with $k_0=Tf_0$, assumed to be an integer.

The video output voltage $V(t)$ is given by (1.11) as

$$V(t)=\int_{-\infty}^{\infty}K(t-t')[X^2(t')+Y^2(t')]dt'$$
$$=\int_{0}^{\infty}K(u)[X^2(t-u)+Y^2(t-u)]du. \quad (2.15)$$

In the derivation of Chapter 3 we shall consider the probability distribution of the slightly modified quantity

$$V_T(t)=\int_0^T K(u)[X^2(t-u)+Y^2(t-u)]du, \quad (2.16)$$

and then pass to the limit as $T\rightarrow\infty$.

III. "WHITE NOISE"

The i-f output $X(t)\cos 2\pi f_0 t+Y(t)\sin 2\pi f_0 t$ with $X(t)$, $Y(t)$ given by (2.13) can be described as caused by the "white noise"

$$x(t)\cos 2\pi f_0 t + y(t)\sin 2\pi f_0 t,$$

where $x(t)$ and $y(t)$ are given by

$$x(t) = \frac{\sqrt{N}}{\sqrt{T}} \sum_{k=-L}^{L} (X_k \cos 2\pi f_k t + Y_k \sin 2\pi f_k t)$$

$$y(t) = \frac{\sqrt{N}}{\sqrt{T}} \sum_{k=-L}^{L} (-X_k \sin 2\pi f_k t + Y_k \cos 2\pi f_k t),$$

where L is a number so large that the frequencies $f_0 \pm L/T$ lie outside of the pass band of the i-f amplifier. This "white noise" is the input thermal and shot noise which actually is known to have a constant power spectrum up to very high frequencies.

In the derivation of the next section it will be advantageous to represent $x(t)$ and $y(t)$ in a different orthogonal system, and we will now show that the expansion coefficients in any other complete orthogonal system are Gaussianly distributed and independent of each other, if L is sufficiently large, i.e., if the power spectrum is constant up to sufficiently high frequencies.

Let $\phi_j(t)$ be an orthonormal, complete set in the interval $(0, T)$. Expanding $x(t)$ in this system we have

$$x(t) = \sum_{1}^{\infty} u_j \phi_j(t), \tag{3.1}$$

where

$$u_j = \left(\frac{N}{T}\right)^{\frac{1}{2}} \sum_{k=-L}^{L} \left(X_k \int_0^T \phi_j(t) \cos 2\pi f_k t dt + Y_k \int_0^T \phi_j(t) \sin 2\pi f_k t dt \right). \tag{3.2}$$

The coefficients u_j are thus linear combinations of the Gaussian variables X_k, Y_k and are, therefore, Gaussianly distributed. To prove independence of linear combinations of independent Gaussian variables, it is necessary only to show that $\langle u_l u_j \rangle = 0$ for $l \neq j$. Using the statistical properties of X_k and Y_k, we have

$$\langle u_l u_j \rangle_{\text{Av}} = \frac{1}{2}\frac{N}{T} \int_0^T \int_0^T ds dt \phi_l(t) \phi_j(s) \times \sum_{-L}^{L} \cos 2\pi f_k (s-t). \tag{3.3}$$

For sufficiently large L

$$\frac{1}{T} \sum_{-L}^{L} \cos 2\pi f_k (s-t)$$

can be replaced by $\delta(s-t)$ and, therefore,

$$\langle u_l u_j \rangle_{\text{Av}} \rightarrow \frac{N}{2} \delta_{jl}. \tag{3.4}$$

In the same way we can show, for the coefficients v_j of $y(t)$, that

$$\langle v_l v_j \rangle_{\text{Av}} \rightarrow \frac{N}{2} \delta_{lj} \tag{3.5}$$

and

$$\langle u_k v_l \rangle_{\text{Av}} \rightarrow 0. \tag{3.6}$$

IV. DERIVATION OF PROBABILITY DENSITY FOR NOISE ALONE

To derive the probability density $P(V)$ we use the above representation for the incoming "white noise" writing the i-f input voltage[3] as

$$V_{\text{I.F.}}^{(i)} = x(t)\cos 2\pi f_0 t + y(t) \sin 2\pi f_0 t, \tag{4.11}$$

with

$$x(t) = \sum_j u_j \phi_j(t)$$

and

$$y(t) = \sum_j v_j \phi_j(t). \tag{4.12}$$

Choosing N as unit of power we have

$$\langle u_k u_l \rangle_{\text{Av}} = \langle v_k v_l \rangle_{\text{Av}} = \tfrac{1}{2} \delta_{lk}. \tag{4.13}$$

The i-f output voltate $V_{\text{I.F.}}^{(o)}$ is given by

$$V_{\text{I.F.}}^{(o)} = X(t)\cos 2\pi f_0 t + Y(t) \sin 2\pi f_0 t, \tag{4.14}$$

where

$$X(t) = \int_{-\infty}^{\infty} Q(t-\tau) x(\tau) d\tau,$$
$$Y(t) = \int_{-\infty}^{\infty} Q(t-\tau) y(\tau) d\tau. \tag{4.15}$$

[3] The authors are aware that the results of this section can be obtained in a more elegant way by using the representation (4.12) for $V_{\text{I.F.}}^{(i)}$ rather than for the components $x(t)$ and $y(t)$, and by defining the quadratic detector by

$$V_D^{(o)} = [V_{\text{I.F.}}^{(o)}]^2$$

rather than by Eq. (4.31), provided that the I.F. and video band widths are small compared with f_0 [cf. Phys. Rev. **70**, 449, C6 (1946)]. The simplifying assumption of symmetrical I.F. pass band can then be easily dispensed with. The method presented here was chosen to keep the continuity with the existing literature.

The correlation coefficient

$$\rho(\tau)=\langle X(t)X(t+\tau)\rangle_{\text{Av}}/\langle X^2(t)\rangle_{\text{Av}} \quad (4.21)$$

can be expressed in terms of the function Q as follows:

$$\langle X(t)X(t+\tau)\rangle_{\text{Av}}=\int\int_{-\infty}^{\infty} Q(t-\tau_1)Q(t+\tau-\tau_2)\times\langle X(\tau_1)X(\tau_2)\rangle_{\text{Av}}d\tau_1 d\tau_2$$

$$=\frac{1}{2}\int_{-\infty}^{\infty} Q(t-\tau_1)Q(t+\tau-\tau_1)d\tau_1$$

$$=\frac{1}{2}\int_{-\infty}^{\infty} Q(\theta)Q(\theta+\tau)d\theta, \quad (4.22)$$

$$\langle X^2(t)\rangle_{\text{Av}}=\frac{1}{2}\int_{-\infty}^{\infty} Q^2(\theta)d\theta=\tfrac{1}{2}, \quad (4.23)$$

and therefore

$$\rho(\tau)=\int_{-\infty}^{\infty} Q(\theta)Q(\theta+\tau)d\theta. \quad (4.24)$$

We note that

$$\int_{-\infty}^{\infty} e^{2\pi i f\tau}\rho(\tau)d\tau$$

$$=\int_{-\infty}^{\infty}\int_{-\infty}^{\infty} Q(\theta)Q(\theta+\tau)e^{2\pi i f\tau}d\theta d\tau$$

$$=\int\int\int\int_{-\infty}^{\infty} df'df''\gamma(f'+f_0)\gamma(f''+f_0)\times e^{2\pi i[f'\theta+f''(\theta+\tau)+f\tau]}d\theta d\tau$$

$$=\int\int_{-\infty}^{\infty} df'df''\gamma(f'+f_0)\gamma(f''+f_0)\times\delta(f'+f'')\delta(f''+f)$$

$$=\gamma(f+f_0)\gamma(f_0-f)=|\gamma^2(f+f_0)| \quad (4.25)$$

which is a special case of the Wiener-Khintchine relation.

The detector output voltage $V_D^{(o)}$ is, therefore,

$$V_D^{(o)}=X^2(t)+Y^2(t)$$

$$=\int\int_{-\infty}^{\infty} Q(t-\tau_1)Q(t-\tau_2)\times[x(\tau_1)x(\tau_2)+y(\tau_1)y(\tau_2)]d\tau_1 d\tau_2, \quad (4.31)$$

and the output of the video amplifier

$$V(t)=\int_{-\infty}^{\infty} K(t-\theta)[X^2(\theta)+Y^2(\theta)]d\theta$$

$$=\int_{-\infty}^{\infty} K(t-\theta)d\theta\int\int_{-\infty}^{\infty} Q(\theta-\tau_1)Q(\theta-\tau_2)\times[x(\tau_1)x(\tau_2)+y(\tau_1)y(\tau_2)]d\tau_1 d\tau_2. \quad (4.32)$$

Changing variables to

$$\theta'=t-\theta, \quad \tau_1'=t-\tau_1, \quad \tau_2'=t-\tau_2, \quad (4.33)$$

we have

$$V(t)=\int_{-\infty}^{\infty} K(\theta')d\theta'\int\int_{-\infty}^{\infty} Q(\tau_1'-\theta')Q(\tau_2'-\theta')\times[x(t-\tau_1')x(t-\gamma_2')+y(t-\tau_1')y(t-\tau_2')]d\tau_1'd\tau_2'. \quad (4.34)$$

If the i-f input noise voltage is a homogeneous random function $x(t-\tau)$ and $y(t-\tau)$ are given by

$$x(t-\tau)=\sum_j u_j'\phi_j(\tau),\qquad y(t-\tau)=\sum_j v_j'\phi_j(\tau), \quad (4.35)$$

where u_j' and v_j' are random variables with the same properties as u_j and v_j, and we have—omitting the primes—

$$V(t)=\int_{-\infty}^{\infty} K(\theta)d\theta\int\int_{-\infty}^{\infty} Q(\tau_1-\theta)Q(\tau_2-\theta)\times\sum_{ij}(u_iu_j+v_iv_j)\phi_i(\tau_1)\phi_j(\tau_2)d\tau_1 d\tau_2 \quad (4.36)$$

$$=\sum_{ij}(u_iu_j+v_iv_j)\times\int\int_{-\infty}^{\infty}\Lambda(\tau_1,\tau_2)\phi_i(\tau_1)\phi_j(\tau_2)d\tau_1 d\tau_2 \quad (4.37)$$

with

$$\Lambda(\tau_1,\tau_2)=\int_{-\infty}^{\infty} K(\theta)d\theta Q(\tau_1-\theta)Q(\tau_2-\theta). \quad (4.38)$$

We note that t does not appear anymore on the r.h.s., that means that $V(t)$ is a homogeneous random function. This is caused by the fact that the input noise is itself a homogeneous random function.

The essential step in our calculation is now to choose as the function $\phi_i(\tau)$ the normalized eigenfunctions of the integral equation

$$\int_{-\infty}^{\infty} \Lambda(\tau_1, \tau_2)\phi(\tau_2)d\tau_2 = \lambda\phi(\tau_1). \quad (4.41)$$

Then $V(t)$ is expressed as

$$V = \sum_i (u_i^2+v_i^2)\lambda_i, \quad (4.42)$$

where the quantities λ_i are the eigenvalues of the integral equation. The quantities

$$u_i^2+v_i^2 = Z_i \quad (4.43)$$

are independent and their probability densities $W(Z_i)$ are given by

$$W(Z_i) = \begin{cases} e^{-Z_i} & \text{if } Z_i>0, \\ 0 & \text{if } Z_i<0, \end{cases} \quad (4.44)$$

because u_i and v_i are independent Gaussian variables of spread $\frac{1}{2}$. The probability distribution of the sum representing V is therefore

$$\begin{aligned} P(V)dV &= dV\int_0^{\infty}\cdots\int_0^{\infty}\delta(V-\textstyle\sum_i \lambda_i Z_i)\prod_k W(Z_k)dZ_k \\ &= dV\frac{1}{2\pi}\int_{-\infty}^{\infty} d\xi e^{-iV\xi}\int_0^{\infty}\cdots\int_0^{\infty}\textstyle\prod_k e^{-Z_k(1-i\lambda_k\xi)}dZ_k \\ &= dV\frac{1}{2\pi}\int_{-\infty}^{\infty} d\xi e^{-iV\xi}\frac{1}{\prod_k(1-i\lambda_k\xi)}. \end{aligned} \quad (4.45)$$

For $V>0$ the path of integration is closed in the lower half plane and contracted around all poles $\xi_s = -i/\lambda_s$ with $\lambda_s>0$ and *vice versa*. We thus obtain

$$P(V) = \frac{1}{2\pi}\sum_s{}^{+}\oint^{(1/i\lambda_s,-)}\frac{d\xi}{(-i\lambda_s)\left(\xi-\frac{1}{i\lambda_s}\right)}\frac{e^{-iV\xi}}{\prod'_{k\neq s}(1-i\lambda_k\xi)} = \sum_s{}^{+}\frac{e^{(-V/\lambda_s)}}{\lambda_s\prod'_{k\neq s}\left(1-\frac{\lambda_k}{\lambda_s}\right)} \quad \text{for } V>0 \quad (4.46)$$

where $\sum_s^{+}$ means summation over all values of s for which $\lambda_s>0$. For $V<0$ we obtain

$$P(V) = \frac{1}{2\pi}\sum_n{}^{(-)}\oint\frac{d\xi}{(-i\lambda_n)\left(\xi-\frac{1}{i\lambda_n}\right)}\frac{e^{-iV\xi}}{\prod'_{k\neq n}(1-i\lambda_k\xi)} = \sum_n{}^{(-)}\frac{e^{(-V/\lambda_n)}}{\lambda_n\prod'_{k\neq n}\left(1-\frac{\lambda_k}{\lambda_n}\right)} \quad (4.47)$$

where $\sum_n^{(-)}$ indicates summation over all n for which $\lambda_n<0$. The product is extended always over all k with exception of $k=n$.

It is sometimes more convenient to have the integral equation in a different form, involving the functions K and ρ rather than K and Q. This form is obtained by introducing the new unknown

$$f(\theta) = \int_{-\infty}^{\infty} Q(\tau_2-\theta)\phi(\tau_2)d\tau_2, \quad (4.51)$$

so that

$$\int_{-\infty}^{\infty} d\theta K(\theta)Q(\tau_1-\theta)f(\theta) = \lambda\phi(\tau_1). \quad (4.52)$$

Multiplying both sides with $Q(\tau_1-\theta')$ and integrating over τ_1 we obtain

$$\int_{-\infty}^{\infty} K(\theta)d\theta\int_{-\infty}^{\infty} d\tau_1 Q(\tau_1-\theta') \times Q(\tau_1-\theta)f(\theta) = \lambda f(\theta'). \quad (4.53)$$

We can now write

$$\int_{-\infty}^{\infty} Q(\tau_1-\theta')Q(\tau_1-\theta)d\tau_1 = \int_{-\infty}^{\infty} d\tau Q(\tau)Q(\tau+\theta'-\theta) = \rho(\theta'-\theta), \quad (4.54)$$

and have the integral equation in the form

$$\int_{-\infty}^{\infty} K(\theta)\rho(\theta'-\theta)f(\theta)d\theta = \lambda f(\theta'),$$

or, since $K(\theta)=0$ for $\theta<0$,

$$\int_0^\infty K(\theta)\rho(\theta'-\theta)f(\theta)d\theta=\lambda f(\theta'). \quad (4.55)$$

Finally, by writing $(K(\theta))^{\frac{1}{2}}f(\theta)=\Phi(\theta)$ we can bring the equation into the form which will be derived directly in Section 6:

$$\int_0^\infty [K(\theta')]^{\frac{1}{2}}\rho(\theta'-\theta)[K(\theta)]^{\frac{1}{2}}\Phi(\theta)d\theta=\lambda\Phi(\theta'). \quad (4.56)$$

By means of the expression

$$V=\sum \lambda_j Z_j, \quad (4.61)$$

we can easily verify formulas for the moments. For the first two moments we have for instance

$$\langle V\rangle_{Av}=\sum \lambda_j\langle Z_j\rangle_{Av}=\sum \lambda_j=\int_0^\infty K(s)ds=1, \quad (4.62)$$

$$\begin{aligned}\langle V^2\rangle_{Av}&=\sum \lambda_j^2\langle Z_j^2\rangle_{Av}+\sum_{j\neq k} \lambda_j\lambda_k\langle Z_jZ_k\rangle_{Av}\\ &=\sum_1^\infty \lambda_j^2+(\sum \lambda_j)^2 \quad (4.63)\\ &=1+\sum \lambda_j^2.\end{aligned}$$

For the fluctuation

$$\sigma^2=\langle V^2\rangle_{Av}-\langle V\rangle_{Av}^2. \quad (4.64)$$

We thus obtain

$$\sigma^2=\sum \lambda_j^2=\int_0^\infty\int_0^\infty K(s)K(t)\rho^2(s-t)dsdt, \quad (4.65)$$

this being the trace of the first iterated kernel.

V. DERIVATION OF THE PROBABILITY DENSITY FOR SIGNAL AND NOISE

For the case of signal and noise only one derivation will be given here, since the direct derivation runs parallel to the analogous derivation for noise alone. The i-f input voltage $V_{\text{I.F.}}^{(i)}$ is in this case given by

$$V_{\text{I.F.}}^{(i)}=[x(t)+\alpha(t)]\cos 2\pi f_0t+[y(t)+\beta(t)]\sin 2\pi f_0t, \quad (5.11)$$

where $x(t)$, $y(t)$ are the input white noise amplitudes, and $\alpha(t)$, $\beta(t)$ are the input signal amplitudes. The i-f output voltage is then given by

$$V_{\text{I.F.}}^{(o)}=[X(t)+p(t)]\cos 2\pi f_0t+[Y(t)+q(t)]\sin 2\pi f_0t, \quad (5.12)$$

where X and Y are defined as above and

$$p(t)=\int_{-\infty}^\infty Q(t-\tau)\alpha(\tau)d\tau \quad\text{and}\quad q(t)=\int_{-\infty}^\infty Q(t-\tau)\beta(\tau)d\tau \quad (5.13)$$

are the signal amplitudes after the i-f amplifier. The detector output voltage is then

$$V_D^{(o)}=[X(t)+p(t)]^2+[Y(t)+q(t)]^2 \quad (5.14)$$

and the output of the video amplifier is

$$\begin{aligned}V(t)&=\int_{-\infty}^\infty K(t-\theta)d\theta\{[X(\theta)+p(\theta)]^2+[Y(\theta+q(\theta)]^2\}\\ &=\int_{-\infty}^\infty K(t-\theta)d\theta\iint_{-\infty}^{\infty} Q(\theta-\tau_1)Q(\theta-\tau_2)\{[X(\tau_1)+\alpha(\tau_1)][x(\tau_2)+\alpha(\tau_2)]\\ &\qquad+[y(\tau_1)+\beta(\tau_1)][y(\tau_2)+\beta(\tau_2)]\}d\tau_1d\tau_2. \quad (5.15)\end{aligned}$$

Changing variables to

$$\theta'=t-\theta,\ \tau_1'=t-\tau_1,\ \tau_2'=t-\tau_2, \quad (5.16)$$

we obtain

$$V(t)=\int_{-\infty}^{\infty}K(\theta')d\theta'\iint_{-\infty}^{\infty}Q(\tau_1'-\theta')Q(\tau_2'-\theta')\{[x(t-\tau_1')+\alpha(t-\tau_1')][x(t-\tau_2')+\alpha(t-\tau_2')]$$
$$+[y(t-\tau_1')+\beta(t-\tau_1')][y(t-\tau_2')+\beta(t-\tau_2')]\}d\tau_1'd\tau_2'$$
$$=\int_{-\infty}^{\infty}\Lambda(\tau_1,\tau_2)d\tau_1d\tau_2\{[x(t-\tau_1)+\alpha(t-\tau_1)][x(t-\tau_2)+\alpha(t-\tau_2)]$$
$$+[y(t-\tau_1)+\beta(t-\tau_1)][y(t-\tau_2)+\beta(t-\tau_2)]\}. \quad (5.17)$$

We now expand noise and signal amplitudes in terms of the eigen functions ϕ_j of Λ, writing

$$x(t-\tau)=\sum_j u_j\phi_j(\tau),\quad \alpha(t-\tau)=\sum_j \alpha_j(t)\phi_j(\tau),$$
$$y(t-\tau)=\sum_j v_j\phi_j(\tau),\quad \beta(t-\tau)=\sum_j \beta_j(t)\phi_j(\tau), \quad (5.18)$$

and obtain

$$V(t)=\sum_j \lambda_j\{[u_j+\alpha_j(t)]^2+[v_j+\beta_j(t)]^2\}. \quad (5.19)$$

The probability density of the variables u_j and v_j are $\pi^{-\frac{1}{2}}\exp(-u_j^2)$ and $\pi^{-\frac{1}{2}}\exp(-v_j^2)$, respectively, and the characteristic function for $V(t)$ is, therefore, given by

$$\langle e^{i\xi V}\rangle_{\text{Av}}=\int_{-\infty}^{\infty}\cdots\int\exp\{i\xi\sum_j\lambda_j[(u_j+\alpha_j)^2+(v_j+\beta_j)^2]\}\prod_j\frac{\exp[-(u_j^2+v_j^2)]}{\pi}du_jdv_j$$

$$=\prod_j\frac{1}{\pi}\iint_{-\infty}^{\infty}du_jdv_j\exp\{-[(u_j^2+v_j^2)(1-i\xi\lambda_j)-2i\xi\lambda_j(u_j\alpha_j+v_j\beta_j)]+i\xi\lambda_j(\alpha_j^2+\beta_j^2)\}$$

$$=\prod_j\frac{1}{\pi}\iint_{-\infty}^{\infty}du_jdv_j\exp\left\{\left(-u_j(1-i\xi\lambda_j)^{\frac{1}{2}}-\frac{i\xi\lambda_j\alpha_j}{(1-i\xi\lambda_j)^{\frac{1}{2}}}\right)^2-\left(v_j(1-i\xi\lambda_j)^{\frac{1}{2}}-\frac{i\xi\lambda_j\beta_j}{(1-i\xi\lambda_j)^{\frac{1}{2}}}\right)^2\right\}$$
$$\times\exp\left[-\frac{\lambda_j^2\xi^2(\alpha_j^2+\beta_j^2)}{1-i\xi\lambda_j}+i\xi\lambda_j(\alpha_j^2+\beta_j^2)\right], \quad (5.21)$$

$$\langle e^{i\xi V}\rangle_{\text{Av}}=\prod_j\frac{\exp[i\xi\lambda_j(\alpha_j^2+\beta_j^2)/(1-i\xi\lambda_j)]}{1-i\xi\lambda_j}. \quad (5.22)$$

The Fourier inversion integral of this expression which represents the probability density cannot be evaluated in closed form.

We shall check this expression by deriving the known [4] probability density for the voltage after a quadratic detector. This distribution is obtained in the limit of infinitely wide video band width, which implies that

$$\int_0^{\infty}K(\theta)\psi(\theta)d\theta=\psi(0) \quad (5.31)$$

for any function $\psi(\theta)$. Then $\Lambda(\tau_1\tau_2)$ becomes simply the product $Q(\tau_1)Q(\tau_2)$ and the integral equation reduces to

$$Q(\tau_1)\int_{-\infty}^{\infty}Q(\tau_2)\phi(\tau_2)d\tau_2=\lambda\phi(\tau_1). \quad (5.32)$$

The only eigenfunction of this integral equation is of the form $CQ(\tau_1)$ where C is a constant, and from

[4] S. A. Goudsmit, RL 43-21; K. A. Norton and V. D. Landon, Proc. I.R.E. **30**, 425, Sept. 42; J. C. Slater, RL V-23; D. G. Fink, Report T8; D. O. North, RCA Tech. Rep. PTR6C.

$\int_{-\infty}^{\infty} Q^2(\tau)d\tau = 1$ we have $C=1$ and $\lambda=1$. The characteristic function, therefore, reduces to

$$\frac{\exp i\xi(\alpha^2+\beta^2)/(1-i\xi)}{1-i\xi},$$

where

$$\alpha = \int_{-\infty}^{\infty} \alpha(t-\tau)Q(\tau)d\tau = \int_{-\infty}^{\infty} \alpha(\tau')Q(\tau-\tau')d\tau' = p(t),$$
$$\beta = \int_{-\infty}^{\infty} \beta(t-\tau)Q(\tau)d\tau = \int_{-\infty}^{\infty} \beta(\tau')Q(t-\tau')d\tau' = q(t). \quad (5.33)$$

Calling the signal after the detector $s^2 = s^2(t)$ we have

$$s^2(t) = p^2(t) + q^2(t) = \alpha^2+\beta^2, \quad (5.34)$$

and

$$P(V) = \frac{1}{2\pi}\int_{-\infty}^{\infty} \frac{\exp(-i\xi V + i\xi s^2/(i-i\xi))}{(1-i\xi)} d\xi$$
$$= \frac{1}{2\pi}\oint^{-i,-} \frac{\exp(-i\xi V + i\xi s^2/(1-i\xi))}{1-i\xi} d\xi \quad (5.35)$$

for $V>0$ and zero otherwise. With $u = 1-i\xi$ this becomes

$$P(V) = \exp(-V-s^2)\frac{1}{2\pi i}\int^{0,+} \exp[Vu + s^2/u]du/u$$
$$= \exp(-V-s^2)J_0(2is\sqrt{V}) \text{ for } V>0 \text{ and} \quad (5.36)$$

zero otherwise, where J_0 is the Bessel function of order 0. This result is in agreement with the earlier results.

VI. DIRECT DERIVATION FOR NOISE ALONE

In order to give a direct derivation of the results obtained in Section 4, we investigate the probability density of the function

$$V(t) = \int_0^T K(u)[x^2(t-u)+y^2(t-u)]du \quad (6.11)^*$$

and let, at an appropriate instant, T approach infinity. For the sake of simplicity, we shall restrict ourselves to the case $K(u)>0$. We divide the interval $(0, T)$ into a large number n of equal subintervals and consider the expression

$$\frac{T}{n}\sum_{j=1}^{n} [x^2(t-u_j)+y^2(t-u_j)]K(u_j);$$
$$\frac{j-1}{n}T < u_j < \frac{j}{n}T. \quad (6.12)$$

* For convenience, x and y are used in this section to denote the quantities formerly denoted by X and Y.

The characteristic function of the joint probability density of

$$x_1 = x(t-u_1), \quad x_2 = x(t-u_2), \quad \cdots, \quad x_n = x(t-u_n)$$

defined as the average

$$\left\langle \exp\left[i\sum_1^n \xi_j x(t-u_j)\right]\right\rangle_{\text{Av}}$$

is easily found to be

$$\exp\left[-\frac{1}{4}\sum_{j,k=1}^{n} \rho_T(u_j-u_k)\xi_j\xi_k\right] \quad (6.13)$$

where the correlation coefficients ρ_T for the finite interval T are defined as

$$\rho_T(t) = \frac{1}{T}\sum_{-\infty}^{\infty} A(f_k)\cos 2\pi f_k t, \quad (6.14)$$

and

$$A(f) = B^2(f) = |\gamma^2(f\pm f_0)|. \quad (6.15)$$

Thus the joint probability density of

$$x_1, x_2, x_3, \cdots, x_n$$

is given by the formula

$$\frac{1}{(2\pi)^n}\int_{-\infty}^{\infty}\cdots\int_{-\infty}^{\infty}\exp\left\{-i\sum_1^n \xi_j x_j\right\}$$

$$\times\exp\left\{-\frac{1}{4}\sum_1^n \rho_T(u_j-u_k)\xi_j\xi_k\right\}d\xi_1\cdots d\xi_n$$

$$=\frac{1}{(\sqrt{\pi})^n}\frac{1}{\sqrt{D}}\exp\left\{-\sum_1^n \alpha_{jk}x_jx_k\right\}, \quad (6.16)$$

where the matrix $[\alpha_{jk}]$ is the inverse of the correlation matrix $[\rho_T(u_j-u_k)]$ and D the determinant of the correlation matrix.

We next find the characteristic function of

$$\frac{T}{n}\sum_1^n K(u_j)x^2(t-u_j),$$

i.e., the average

$$\left\langle \exp\left\{\frac{i\xi T}{n}\sum_1^n K(u_j)x^2(t-u_j)\right\}\right\rangle_{\text{Av}}.$$

We have

$$\left\langle \exp\left\{\frac{i\xi T}{n}\sum_1^n K(u_j)x^2(t-u_j)\right\}\right\rangle_{\text{Av}}$$

$$=\frac{1}{(\sqrt{\pi})^n}\frac{1}{\sqrt{D}}\int_{-\infty}^{\infty}\cdots\int_{-\infty}^{\infty}\exp\left\{\frac{i\xi T}{n}\sum_1^n K(u_j)x_j^2\right\}$$

$$\times\exp\left\{-\sum_1^n \alpha_{jk}x_jx_k\right\}dx_1\cdots dx_n. \quad (6.21)$$

Since $K(u_j)>0$ we can introduce the new variables

$$x_j'=[K(u_j)]^{\frac{1}{2}}x_j. \quad (6.22)$$

Obtaining

$$\left\langle \exp\left\{\frac{i\xi T}{n}\sum_1^n K(u_j)x^2(t-u_j)\right\}\right\rangle_{\text{Av}}$$

$$=\frac{1}{(\sqrt{\pi})^n}\frac{1}{\sqrt{}}\frac{1}{\prod_1^n[K(u_j)]^{\frac{1}{2}}}$$

$$\times\int_{-\infty}^{\infty}\cdots\int_{-\infty}^{\infty}\exp\left\{\frac{i\xi T}{n}\sum_1^n x_j'^2\right\}$$

$$\times\exp\left\{-\sum_1^n \beta_{jk}x_j'x_k'\right\}dx_1'\cdots dx_n', \quad (6.23)$$

where

$$\beta_{jk}=\frac{\alpha_{jk}}{[K(u_j)]^{\frac{1}{2}}[K(u_k)]^{\frac{1}{2}}}. \quad (6.24)$$

Let $\mu_1^{(n)}, \mu_2^{(n)}, \cdots, \mu_n^{(n)}$ be the eigenvalues of the matrix $[\beta_{jk}]$; then changing the coordinates in the last integral to "principal axes" of the "ellipsoid"

$$\sum_1^n \beta_{jk}x_j'x_k'=1 \quad (6.25)$$

and noting that $\sum x_j'^2$ remains invariant under the transformation, we get easily

$$\left\langle \exp\left\{\frac{i\xi T}{n}\sum_1^n K(u_j)x^2(t-u_j)\right\}\right\rangle_{\text{Av}}$$

$$=\frac{1}{\sqrt{D}\prod_1^n[K(u_j)]^{\frac{1}{2}}}\frac{1}{\prod_1^n\left(\mu_j^{(n)}-\frac{i\xi T}{n}\right)^{\frac{1}{2}}}. \quad (6.26)$$

Since we have assumed that $A(f)$ is an even function we can show without much difficulty that

$$\sum_1^n K(u_j)x^2(t-u_j) \quad \text{and} \quad \sum_1^n K(u_j)y^2(t-u_j)$$

are independent in the statistical sense of the word. Thus the characteristic function of

$$\frac{T}{n}\sum_1^n K(u_j)\{x^2(t-u_j)+y^2(t-u_j)\} \quad (6.27)$$

is

$$\frac{1}{D\prod_1^n K(u_j)\prod_1^n\left(\mu_j^{(n)}-\frac{i\xi T}{n}\right)}$$

we now observe that the matrix $[\beta_{jk}]$ is the inverse of the matrix

$$[(K(u_j))^{\frac{1}{2}}\rho_T(u_j-u_k)(K(u_k))^{\frac{1}{2}}] \quad (6.28)$$

and that the determinant of (6.28) is

$$D\prod_1^n K(u_j).$$

Denoting by $\lambda_1^{(n)}, \lambda_2^{(n)}, \cdots, \lambda_n^{(n)}$, the eigenvalues of matrix (6.28) we see that the μ's are the inverses of the λ's and that

$$D\prod_1^n K(u_j)=\lambda_1^{(n)}\cdot\lambda_2^{(n)}\cdots\lambda_n^{(n)}.$$

These facts imply that the characteristic function of (6.27) is

$$1\Big/\prod_{1}^{n}\left(1-\frac{iT}{n}\lambda_j^{(n)}\xi\right). \tag{6.29}$$

It follows from Hilbert's treatment of integral equations that the eigenvalues of matrix (6.28) when multiplied by T/n approach, as n approaches infinity, the eigenvalues of the integral equation

$$\int_0^T [K(s)]^{\frac{1}{2}}\rho_T(s-t)[K(t)]^{\frac{1}{2}}f(t)dt=\lambda f(s), \tag{6.31}$$

provided these eigenvalues are simple. Since the distribution of (6.27) approaches that of $V(t)$ we deduce that the characteristic function of $V(t)$ is given by the formula

$$1\Big/\prod_{1}^{\infty}(1-i\lambda_j^{(T)}\xi), \tag{6.32}$$

where $\lambda_1^{(T)}$, $\lambda_2^{(T)}$, $\cdots$, are the eigenvalues of the integral equation (6.31). For large T we can replace the integral equation (6.31) by

$$\int_0^\infty [K(s)]^{\frac{1}{2}}\rho(s-t)[K(t)]^{\frac{1}{2}}f(t)dt=\lambda f(s), \tag{6.33}$$

and the characteristic function of $V(t)$ by

$$1/\prod_j(1-i\lambda_j\xi), \tag{6.34}$$

where λ_1, λ_2, λ_3, $\cdots$ are now the eigenvalues of the integral equation (6.34) which we again assume to be simple. The probability density $P(V)$ of V can now be expressed by means of the inversion formula

$$P(V)=\frac{1}{2\pi}\int_{-\infty}^{\infty}\frac{e^{-iV\xi}}{\prod_j(1-i\lambda_j\xi)}d\xi. \tag{6.35}$$

At the end of Section 4, the integration has been carried out by the method of residues. The justification of the formal integration is not quite simple since one must investigate the behavior of the entire function

$$\phi(Z)=\prod_1^\infty(1-i\lambda_jZ)$$

for large $|Z|$.[5] However, it can be considerably simplified if one assumes that for some $\nu(0<\nu<1)$ the series

$$\sum_1^\infty \lambda_j^\nu$$

converges. On the other hand it is easy to see that the exact probability density $P(V)$ is the limit, as n approaches infinity, of $P_n(V)$ where

$$P_n(V)=\sum_{s=1}^{n}\frac{\exp(-V/\lambda_s)}{\lambda_s\prod_1^{n}{}'(1-\lambda_j/\lambda_s)}; \quad V>0. \tag{6.36}$$

$P_n(V)=0$; $V<0$ since with the assumption $K(u)>0$ negative eigenvalues cannot occur. In fact, the characteristic function corresponding to $P_n(V)$ which is seen to be

$$\prod_1^n\frac{1}{1-i\xi\lambda_j}$$

approaches the characteristic function associated with $P(V)$ thus $P_n(V)$ approaches $P(V)$.

VII. EXPLICIT SOLUTIONS IN SOME SPECIAL CASES

1. Single-Tuned I.F. with Simple Low Pass Video

For the special case of a single tuned i-f amplifier with the video amplifier acting as a simple low pass filter the integral equation can be solved explicitly and the eigenvalues can be expressed in terms of roots of Bessel functions. For one special ratio of video band width to i-f band width it is even possible to obtain $P(V)$ in closed form. For other ratios it is not too difficult to sum the series for $P(V)$ numerically.

The correlation function $\rho(t)$ is in this case

$$\rho(t)=e^{-\alpha|t|} \tag{7.11}$$

which is equivalent to writing

$$A(f)=\frac{\text{Const.}}{1+(2\pi f/\alpha)^2}. \tag{7.12}$$

[5] We note that $\phi(Z)=D(iZ)$ where $D(\mu)$ is the Fredholm Determinant of the above integral equation. It is noteworthy that this relationship between the characteristic function of $P(V)$ and the Fredholm Determinant persists, even in the general case in which $K(t)$ is not assumed positive (in which case the simple product representation need not hold, since $\Sigma\lambda$ need not converge). In the general case one must take the Fredholm Determinant of the kernel $K(t)\rho(s-t)$.

For the video we assume

$$K(u)=\begin{cases}0 & ,\ u>0\\ \beta\exp(-\beta u), & u<0\end{cases}\qquad(7.13)$$

i.e.,

$$C(f)=\frac{1}{1+2\pi if/\beta}.\qquad(7.14)$$

The integral equation (6.33) now assumes the form

$$\beta\int_0^{\infty}e^{-\alpha|s-t|}e^{-(\beta/2)(s+t)}f(t)dt=\lambda f(s).\qquad(7.15)$$

Putting

$$\psi(t)=\exp\left(\frac{\beta}{2}t\right)f(t)\qquad(7.16)$$

we have

$$\int_0^{\infty}\exp(-\alpha|s-t|)\exp(-\beta t)\psi(t)dt=\frac{\lambda}{\beta}\psi(s).\qquad(7.17)$$

This integral equation is solved by[6]

$$\psi(s)=J_{2\alpha/\beta}\left[2\left(\frac{2\alpha}{\beta\lambda}\right)^{\frac{1}{2}}e^{-\beta s/2}\right]\qquad(7.18)$$

and the eigenvalues are determined by

$$J_{(2\alpha/\beta)-1}[2(2\alpha/\beta\lambda)^{\frac{1}{2}}]=0.\qquad(7.19)$$

Thus the eigenvalues of our integral equation are expressible in terms of roots of Bessel functions and are seen to be simple. It is interesting to note that if $2\alpha/\beta=\frac{3}{2}$ we can express $P(V)$ in terms of one of the theta functions. In fact, if $2\alpha/\beta=\frac{3}{2}$, Eq. (7.19) assumes the form

$$J_{\frac{1}{2}}[(6/\lambda)^{\frac{1}{2}}]=0\qquad(7.21)$$

and since

$$J_{\frac{1}{2}}(x)=\left(\frac{2}{\pi}\right)^{\frac{1}{2}}\frac{\sin x}{\sqrt{x}}\qquad(7.22)$$

we see that

$$\lambda_n=\frac{6}{\pi^2}\frac{1}{n^2}\quad(n=1,2,\cdots)\qquad(7.23)$$

and formula (6.36) for $V>0$ becomes

$$P(V)=\frac{\pi^2}{6}\sum_{s=1}^{\infty}\frac{s^2\exp(-\pi^2s^2V/6)}{\prod_j'(1-(s^2/j^2))}.\qquad(7.24)$$

It can be shown quite easily that

$$\prod_j'\left(1-\frac{s^2}{j^2}\right)=-\frac{\cos\pi s}{2}\qquad(7.25)$$

and therefore,

$$P(V)=\frac{\pi^2}{3}\sum_{s=1}^{\infty}(-1)^{s+1}s^2\exp(-\pi^2s^2V/6).\qquad(7.26)$$

Introducing the theta-function notation

$$\theta(q;0)=1-2q+2q^4-2q^9+\cdots\qquad(7.27)$$

we have

$$q\theta'(q;0)=-2\sum_{s=1}^{\infty}(-1)^{s+1}s^2q^{s^2}\qquad(7.28)$$

and finally

$$P(V)=-\frac{\pi^2}{6}\exp\left(\frac{\pi^2V}{6}\right)\theta'(e^{-\pi^2V/6};0).\qquad(7.29)$$

For other values of the ratio α/β we can calculate values of $P(V)$ and plot the corresponding graphs.

2. The Noise Power, Averaged over a Finite Time Interval

In some problems it is of interest to know the distribution of the average

$$M=\frac{1}{\tau}\int_0^{\tau}I^2(t)dt.\qquad(7.31)$$

The derivation of Section 4 is clearly applicable to this case if we put

$$K(u)=\begin{cases}1/\tau, & 0<u<\tau\\ 0 & \text{elsewhere.}\end{cases}\qquad(7.32)$$

The probability density is given by (4.46) replacing V by M except that the λ's are now eigenvalues of the integral equation

$$\frac{1}{\tau}\int_0^{\tau}\rho(s-t)f(t)dt=\lambda f(s).\qquad(7.33)$$

In the simple case

$$\rho(t)=e^{-\alpha|t|}\qquad(7.34)$$

the eigenvalues can be shown to be given by the formula

$$\lambda_m=\frac{2}{g}\frac{1}{1+y_m^2}\qquad(7.35)$$

[6] For the solution and discussion of the integral equation see M. L. Juncosa "An integral equation related to Bessel functions," Duke Math. J. 12, 465–471 (1945).

where y_m is the mth (positive) root of the equation

$$\tan gy = -2y/(1-y^2) \quad (7.36)$$

and $g = \alpha\tau$.

VIII. APPROXIMATIONS FOR $P(V)$

At the end of Section 5, it was shown that, for infinitely wide video, $P(V)$ becomes identical with the well-known probability density of the output voltage of a quadratic detector. We will derive now an approximation for the case that the video band width is very large compared with the i-f band width.

It can be shown that in this case one of the eigenvalues say λ_1, tends to unity while all others approach zero. In order to obtain the dependence of the eigenvalues on the band width ratio we shall write

$$A(f') = \frac{1}{B} a\left(\frac{f'}{B}\right) \quad (8.11)$$

and

$$C(f) = c\left(\frac{f-f_0}{b}\right), \quad (8.12)$$

where

$$\int_{-\infty}^{\infty} a(\nu)d\nu = 1 \quad (8.13)$$

and

$$c(f_0/b) = 1 \quad (8.14)$$

in conformity with our original normalization for $A(f)$ and $K(t)$. The limiting process is carried through for $B/b = \epsilon \to 0$, while the functions $a(\nu)$ and $c(\nu)$ remain unchanged. In terms of A and C our integral equation becomes

$$\int_{-\infty}^{\infty} A(f_1)C(f_2-f_1)g(f_2)df_2 = \lambda g(f_1) \quad (8.21)$$

and traces S_k are, therefore,

$S_1 = 1,$

$$S_2 = \iint_{-\infty}^{\infty} d\nu_1 d\nu_2 a(\nu_1)a(\nu_2) \times c(\epsilon[\nu_2-\nu_1]-f_0/b)c(\epsilon[\nu_1-\nu_2]-f_0/b), \quad (8.22)$$

$$S_3 = \iiint_{-\infty}^{\infty} d\nu_1 d\nu_2 d\nu_3 a(\nu_1)a(\nu_2)a(\nu_3) \times c(\epsilon[\nu_2-\nu_1]-f_0/b)c(\epsilon[\nu_1-\nu_3]-f_0/b) \times c(\epsilon[\nu_3-\nu_2]-f_0/b),$$

etc. Expanding $c(\epsilon\nu - f_0/b)$ in a power series around f_0/b we shall write

$$c(\epsilon\nu - f_0/b) = 1 + C_1\epsilon\nu + C_2\epsilon^2\nu^2 + \cdots. \quad (8.23)$$

Assuming that

$$\int_{-\infty}^{\infty} a(\nu)\nu^4 d\nu$$

is finite we obtain the following approximations.

$$\begin{aligned} S_2 &\cong 1 + \alpha\epsilon^2 + 0(\epsilon^4), \\ S_3 &\cong 1 + 3\alpha\epsilon^2/2 + 0(\epsilon^4), \\ S_4 &\cong 1 + 2\alpha\epsilon^2 + 0(\epsilon^4), \end{aligned} \quad (8.24)$$

where

$$\alpha = 2(2C_2 - C_1^2)\int_{-\infty}^{\infty} a(\nu)\nu^2 d\nu \quad (8.25)$$

and essential use has been made of the fact that $a(\nu) = a(-\nu)$. Noticing that

$$\lambda_1^2 S_2 > S_4 \quad (8.26)$$

(since $S_n = \sum_1^\infty \lambda_k^n$, and λ_1 is by defintion the largest eigenvalue) we obtain

$$\lambda_1^2 > S_4/S_2 = 1 + \alpha\epsilon^2 + 0(\epsilon^4). \quad (8.27)$$

Therefore, $\sum_2^\infty \lambda_k^2$ is of order ϵ^4 or higher, and $\sum_2^\infty \lambda_k^4$ is of order ϵ^8 or higher. Now we notice that

$$(S_4)^{\frac{1}{4}} = \lambda_1 + 0(\epsilon^8) \quad (8.28)$$

so that

$$\lambda_1 = 1 + \alpha\epsilon^2/2 + 0(\epsilon^4). \quad (8.29)$$

Making use of the expression

$$V = \sum_1^\infty \lambda_k Z_k \quad (8.31)$$

[cf. Eqs. (4.42), (4.43)], we write

$$V = V_1 + v \quad (8.32)$$

with

$$V_1 = 1 - \lambda_1 + \lambda_1 Z_1 \quad (8.33)$$

and

$$v = \sum_2^\infty \lambda_k(Z_k - 1). \quad (8.34)$$

If $P_1(V_1)$ is the probability density of V_1, given by

$$P_1(V_1) = 0 \quad \text{for} \quad V_1 < 1 - \lambda_1 \quad (8.35)$$

and

$$P_1(V_1) = \lambda_1^{-1} \exp \frac{V_1 - (1-\lambda_1)}{\lambda_1} \quad \text{for} \quad V_1 > (1-\lambda_1),$$

and $w(v)$ is the probability density of v, we have

for $P(V)$ the exact expression

$$P(V)=\int_{-\infty}^{\infty} dv P_1(V-v)w(v)$$
$$=\int_{-\infty}^{V-1+\lambda_1} dv P_1(V-v)w(v). \quad (8.36)$$

Since the width of $w(v)$ is of order ϵ^2 we can use $P_1(V)$ as the first approximation of $P(V)$. To discuss the reliability of this appproximation we expand $P(V)$ in the form

$$P(V)=P_1(V)-P_1'(V)\int_{-\infty}^{V-1+\lambda_1} vdvw(v)$$
$$+\tfrac{1}{2}P_1''(V)\int_{-\infty}^{V-1+\lambda_1} v^2dvw(v)$$
$$+\frac{1}{3!}\int_{-\infty}^{V-1+\lambda_1} v^3dvw(v)P_1'''(V-\xi) \quad (8.37)$$

where, in the last term,

$$0<\xi<|v|.$$

For any fixed value of $V>0$, the integrals in the first and second term can be replaced by $\bar{v}=0$ and $\langle v^2\rangle=\sum_2^\infty\lambda_k^2$ with an error of order ϵ^4 or smaller, because

$$\int_{V-(1-\lambda_1)}^{\infty} vdvw(v)$$
$$<\int_0^\infty \frac{v^4}{[V-(1-\lambda_1)]^3}w(v)dv=0(\epsilon^8), \quad (8.38)$$
$$\int_{V-(1-\lambda_1)}^{\infty} v^2dvw(v)$$
$$<\int_0^\infty \frac{v^4}{[V-(1-\lambda_1)]^2}w(v)dv=0(\epsilon^8).$$

The last term in (8.37) can be shown to be of order ϵ^6 or higher. Thus we have

$$P(V)=P_1(V)+\tfrac{1}{2}P_1''(V)\langle v^2\rangle_{Av}+0(\epsilon^6) \quad (8.39)$$

for any fixed $V>0$. Substituting

$$\lambda_1=1+\alpha\epsilon^2/2+0(\epsilon^4)$$

and

$$\langle v^2\rangle_{Av}=\sum_{k=2}^{\infty}\lambda_k^2=0(\epsilon^4),$$

we see that for fixed $V>0$, $P_1(V)$ is a good approximation except for terms of order ϵ^4 and higher.

If, in addition, the existence of

$$\int_{-\infty}^{\infty} a(\nu)\nu^6 d\nu \quad (8.41)$$

is assumed the approximation can be improved by including the term

$$\tfrac{1}{2}P_1''(V)\langle v^2\rangle_{Av}$$

where

$$\langle v^2\rangle_{Av}=S_2-(S_6)^{\frac{1}{3}}+0(\epsilon^{12}) \quad (8.42)$$

and λ_1 [which occurs in $P_1(V)$] is calculated from

$$\lambda_1=(S_4)^{\frac{1}{4}}+0(\epsilon^8). \quad (8.43)$$

Substituting these values we obtain an approximation for $P(V)$ valid up to terms of order ϵ^4 inclusive. In the neighborhood of $V=0$, no approximation which utilizes only the traces of a limited number of iterated kernels, can be expected to hold, since there—as can be seen from the general solution—the eigenvalues individually determine the shape.

The considerations of this section cannot be expected to be applicable to the case treated exactly in Section 7 since there not even

$$\int_{-\infty}^{\infty} a(\nu)\nu^2 d\nu$$

exists. In that case one sees directly that the eigenvalues are such that

$$\lambda_1=1-0(\epsilon), \quad \sum_2^\infty \lambda_k^2=0(\epsilon^2).$$

In order to be able to obtain a general approximation for wide videos one must demand that the i-f spectrum falls off sufficiently rapidly at infinity.

In the opposite limit, $b/B\to 0$, similar considerations show that $P(V)$ becomes Gaussian with mean $\langle V\rangle_{Av}=1$ and spread proportional to $(b/B)^{\frac{1}{2}}$.

IX. APPLICATION TO A DISCRETE CASE

For some applications of noise theory it is desirable to have the characteristic function

$$\exp\left\langle i\sum_{k=1}^{n}\xi_k I_k\right\rangle_{Av}=\phi(\xi_1\cdots\xi_n) \quad (9.11)$$

of the joint probability of n successive values $I_1, I_2\cdots, I_n$; $I_k=x^2(t_k)+y^2(t_k)$ of the detector output, from which the averages of products $\langle\prod_k I_k^{m_k}\rangle_{\text{Av}}$ (m_k integer) can be derived. $\phi(\xi_1\cdots\xi_n)$ can be obtained by carrying through the idea of transformation to principal axes for the discrete case, that is, by a slight modification of Section 6. However, we shall derive $\phi(\xi_1\cdots\xi_n)$ from the results of Section 4 since this seems the quicker way.

We simply chose for $K(t)$ the function

$$K(t)=\frac{1}{\xi}\sum_1^n \xi_k\delta(t-t_k) \tag{9.12}$$

where $\xi=\sum_1^n\xi_k$ for the sake of normalization, and the time variable t is chosen in such a way that all $t_k<0$. For this choice of $K(t)$ we have

$$V=\int_0^\infty K(t)[x^2(t)+y^2(t)]dt=\frac{1}{\xi}\sum_1^n \xi_k I_k \tag{9.13}$$

and, provided that the I_k are normalized so that $\langle I_k\rangle_{\text{Av}}=1$, we have

$$\phi(\xi_1\cdots\xi_n)\equiv e^{i\Sigma\xi_k I_k}=\langle e^{i\xi V}\rangle_{\text{Av}}=\frac{1}{\prod_j(1-i\lambda_j\xi)}. \tag{9.14}$$

The integral equation, whose eigenvalues are λ_j degenerates in this case into a system of linear equations for n discrete variables $f(t_k)=f_k$:

$$\frac{1}{\xi}\sum_{k=1}^n \xi_k\rho_{km}f_k=\lambda f_m, \tag{9.15}$$

where

$$\rho_{km}=\rho(t_k-t_m)$$

with

$$\Lambda=\xi\lambda.$$

We write these equations as

$$\sum_{k=1}^n \xi_k\rho_{km}f_k=\Lambda f_m. \tag{9.16}$$

To obtain ϕ, it is not necessary actually to determine the eigenvalues since the product $\prod_j(1-i\Lambda_j)$ can be expressed in terms of the determinant $\det|\xi_k\rho_{km}+i\delta_{km}|$ as follows:
We note that

$\prod_1^n(iu-i\Lambda_j)$ is a polynomial in u which has zeros at $u=\Lambda_j$ and starts with $i^n u^n$. The determinant $\det|\xi_k\rho_{km}-u\delta_{km}|$ is a polynomial in u, which has the same roots and starts with $(-)^n u^n$. We therefore have

$$\prod_1^n(iu-i\lambda_j)=i^{-n}\det|\xi_k\rho_{km}-u\delta_{km}| \tag{9.17}$$

and, specially, for $u=-i$

$$\prod_1^n(1-i\Lambda_j)=i^{-n}\det|\xi_k\rho_{km}+i\delta_{km}|.^7$$

The characteristic function is thus obtained as

$$\langle e^{i\Sigma\xi_k I_k}\rangle_{\text{Av}}=\frac{i^n}{\det|\xi_k\rho_{km}+i\delta_{km}|}. \tag{9.18}$$

While it has not been possible to evaluate the joint probability

$$W_n(I_1\cdots I_n)=\frac{1}{(2\pi)^n}\int\cdots\int d\xi_1\cdots d\xi_n \times\phi(\xi_1\cdots\xi_n)e^{-i\Sigma\xi_k I_k}, \tag{9.19}$$

explicitly (except for $n=1$ and $n=2$, in which cases they are well known) one can obtain averages of the form

$$\langle\prod_k I_k^{m_k}\rangle_{\text{Av}} \text{ for integer values of } m_k,$$

by expanding both sides of (9.18) in power series in the variables ξ_k and by comparing coefficients.

[7] For the general kernel a corresponding relation can be derived between the above product and the Fredholm determinant of the kernel.

22

Reprinted from *J. Appl. Phys.*, **25**(8), 1037–1052 (1954)

On the Distributions of Signals and Noise after Rectification and Filtering

M. A. MEYER, *Laboratory for Electronics, Inc., Boston 14, Massachusetts*

AND

DAVID MIDDLETON, *Cruft Laboratory, Harvard University, Cambridge 38, Massachusetts*

(Received January 9, 1954)

The probability distributions for broad- and narrow-band signals and normal random noise, following a square-law rectifier *and* a video (or audio) filter of arbitrary width, are examined. The present approach is based on a method originally used by Kac and Seigert [J. Appl. Phys. **18**, 383 (1949)] for noise alone (after quadratic rectification but no filter) in the case of the nth-order distribution ($n \geq 2$). The procedure requires an appropriate transformation to express the output waveform in terms of the input disturbance; the statistics of the output are now determined by a suitable transformation with respect to the original, normal statistics. Explicit solutions are obtained for an integral equation involving the autocorrelation function of the noise, the weighting function of the video (or audio) filter, and an orthonormal set of eigenfunctions. For noise alone only the eigenvalues need be determined, but for a signal and noise it is necessary to know the eigenfunctions as well. Among the new results are (1), the calculation of the general, nth-order characteristic function for the filtered output following quadratic rectification for an input noise and a signal *and* noise; (2), a discussion of the second-order probability density W_2 in this case; (3), some examples of practical interest; and (4), the limiting cases of broad and narrow post-detection filters (vis-à-vis the predetection filter), with particular attention to approximations to W_2 for signal and noise for narrow videos and an improved approximation for W_1 for noise alone, as well.

1. INTRODUCTION

In most electronic systems we are concerned with the observation of some physical quantity. This quantity may be random, or it may have a well-defined representation in time and may be accompanied by noise, which can interfere with the measurement. Here we consider as typical inputs to our electronic system either normal random noise, or a normal random signal (the quantity of interest), or a combination of normal random noise and a signal with a definite structure. It is assumed, in addition, that the random processes are stationary, and in some cases, ergodic.

To obtain the desired information from the input, a generalized filter is used which may be active or passive and which may contain both linear and nonlinear filters. One such generalized filter is the combination of an IF amplifier followed by a quadratic detector, which in turn precedes a second, linear ("video") filter. Arrangements of this type are common in radar receivers, for example, and are also basic for power-spectrum analyzers. In the present paper we attempt to determine the statistical properties of the *video filter's* output, since from these statistics the necessary description of the observations can be given, and many of the observed properties of the original wave obtained.[1–3]

A random process can be completely described by a set $W_1, \cdots, W_n$ of probability (density) functions.[4] Here W_n is the nth-order probability density function $W_n(X_1; X_2,t_2; \cdots; X_n,t_n)$ which gives the joint probability of finding an n-tuple of values of X in the ranges $(X_1,X_1+dX_1), \cdots, (X_n,X_n+dX_n)$, at the successive times $t_1(=0), t_2, t_3, \cdots$, etc. Since the process is assumed stationary, only the time differences between pairs of values of X are significant. An important consequence of the normal structure of the input wave is that the set $W_1, \cdots, W_n$ is completely specified if the (co-) variances of the input wave are known. (Here the means are assumed equal to zero.) Thus, if the statistics of the input wave are given, it should be possible in principle to calculate the statistics anywhere else in the system.

The problem considered here is the effect on the statistics of the normal process, when the incoming random wave is subjected to a linear filtering operation, represented by the IF amplifier (Fig. 1), rectified by a quadratic detector, and then passed again through a second linear filter, which we shall call the "video." In particular, we wish the second-order probability density W_2 for the output of this second linear filter—an output which will not in general possess normal properties, because of the intervening nonlinear operation. We use here a generalization of a method first introduced by Kac and Siegert,[5] who obtained the first-order density W_1 for noise and for a signal and noise. Before this, earlier work in the unfiltered (i.e., infinitely wide-video) case[1,2] has given us W_1 and W_2 in a direct way, for all nonlinear elements. Kac and Siegert showed, however, for the first time, that explicit results in the much more difficult situation of arbitrary post-detection filtering could be obtained, at least for the full-wave quadratic rectifier. Their results were subsequently verified by Jastram[6] in certain instances. It turns out

[1] S. O. Rice, Bell System Tech. J. **23**, 282 (1944); **24**, 46 (1945).

[2] D. Middleton, Quart. Appl. Math. **5**, 445 (1948); see also J. Appl. Phys. **22**, 1143, 1153 (1951).

[3] Davenport, Johnson, and Middleton, J. Appl. Phys. **23**, 377 (1952).

[4] M. C. Wang and G. E. Uhlenbeck, Revs. Modern Phys. **17**, 323 (1946).

[5] M. Kac and A. J. F. Siegert, J. Appl. Phys. **18**, 383 (1949).

[6] P. Jastram, "The effect of nonlinearity and frequency distortion on the amplitude distribution for stationary random processes," doctoral dissertation, University of Michigan (1947).

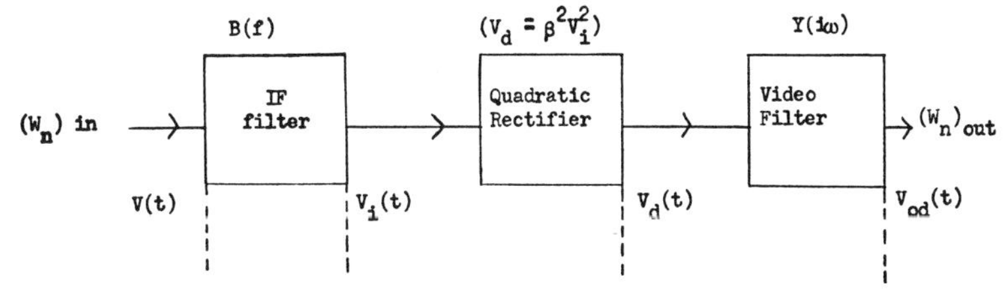

FIG. 1. The generalized filter.

that W_1 can be represented explicitly,[7] provided the eigenvalues of a certain integral equation can be found. For signals and noise both the eigenvalues and the associated eigenfunctions of this integral equation are needed, and, moreover, only the characteristic function of the distribution is expressible in convenient form. Approximations for W_1 have also been developed in the two extremes of video filters wide or narrow compared with the first, or IF response; see Fig. 1.

The choice of a quadratic rectifier is important for several reasons: (1) In receiving systems which must reproduce adequately over extended dynamic ranges, the minimum (or threshold) signal, and therefore the noise, is confined to the lower portions of the rectifier characteristic, which for a wide class of practical detectors approximates a square-law[8] response. (2) Often one desires a quantity proportional to the instantaneous power in a random wave, and this requires usually a quadratic rectifier. (3) From a purely mathematical point of view, the quadratic rectifier is much easier to treat analytically than other nonlinear devices, and in fact, seems at the moment to be the only response for which more or less explicit solutions can be found. (4) Finally, a study of W_2 (as well as W_1) after rectification and filtering provides results useful in the problem of determining, say, the correlation function or spectrum of the output of a *second* nonlinear device, following the video filter. Methods to date have assumed that any filters between successive nonlinear operations are either very broad or very narrow, in which case standard procedures[1,2] may be used.

Kac and Siegert achieved their results in two ways. Their first approach, which we shall call the "expansion method," develops the input random wave in a series using an appropriate set of orthonormal functions. However, this method does not appear possible in higher-order problems (W_2, etc.), since the description by such a set is not general enough.[9] The second, or "direct," derivation employed by Kac and Siegert is the basis of our approach to the higher-order problems. This direct approach requires an appropriate transformation to express the output wave form in terms of the input. The statistics of the output are then determined by suitable additional transformations with respect to the original, input statistics. With this method Kac and Siegert also obtained the nth-order characteristic function for the detector's output in the case of noise alone.

The present paper obtains the following new results:

(1) The general nth-order characteristic function for the filtered output following quadratic rectification, in terms of the eigenvalues (and eigenfunctions) of a certain integral equation, *for signal and noise*, as well as noise alone.

(2) The second-order characteristic function $F_v(\zeta_1,\zeta_2)$ as a special case of (1) for both broad- and narrow-band input waves.

(3) A special video filter which gives the same second-order density W_2 as for an infinitely wide video, in the case of a band-limited IF response.

(4) Approximations for W_2 in the limiting cases of wide-or narrow-videos for noise alone, and for a signal and noise.

These results show, as expected, that the distribution tends to that obtained for an infinitely wide filter when the final filter is wide compared to the IF, and to a normal distribution,[5,9a] when the final filter is narrower than the IF.

The main features of the analysis, which is necessarily quite involved, are outlined in the succeeding sections, along with a summary of special results of interest in the limiting cases. Sufficient detail is maintained to facilitate calculations of any additional cases, if so desired.

2. THE CHARACTERISTIC FUNCTION AFTER RECTIFICATION. NOISE ALONE

Our first major problem is to find the nth-order characteristic function ϕ_n for the output of the quadratic rectifier in terms of the statistics of the input IF wave. We next determine the transformed characteristic function for the output of a linear filter when the input has a given characteristic function. Applying these operations successively gives us the desired charac-

[7] Recent work of R. C. Emerson, J. Appl. Phys. 24, 1168 (1953), discusses this problem from an alternative point of view, using the cumulants (semi-invariants) of the distribution to avoid solving the associated integral equations of Kac and Siegert's and the present approach.

[8] J. H. Van Vleck and D. Middleton, J. Appl. Phys. 17, 940 (1946); see also D. Middleton, Proc. Inst. Radio Engrs. 36, 1467 (1948).

[9] M. A. Meyer, "On Some Distribution Functions in the Theory of Random Noise," doctoral dissertation, Harvard University, June (1952), pp. 2-6, 2-7, and following.

[9a] G. R. Arthur, J. Appl. Phys. 23, 1143 (1952).

teristic function $F_V(\zeta_1,\cdots,\zeta_n)$ for the output of the video filter when the input is the detected IF disturbance.

Two forms of F_V are needed in limiting cases. One of these is expressed in terms of the eigenvalues and eigenfunctions of the integral equation mentioned above; the other is associated with the Fredholm determinant of this integral equation. The quantity F_V is calculated initially in terms of the time-response (i.e., weighting) functions of the IF and video filters. We next transform these expressions of F_V into representations where the frequency-response (i.e., system) functions now appear. From the integral equation and the determinantal expansion in this new form we obtain finally the desired approximations and other results, since a "wide" or "narrow" video has meaning primarily in the frequency domain.

We start with the nth-order normal distribution density and associated characteristic function of a random variable x which has a zero mean:[10]

$$W_n(x_1,\cdots,x_n)=(2\pi)^{-n}|\mathbf{u}|^{-1/2}\times\exp\left\{-(1/2)\sum_{kl}^{n}\frac{\mu^{kl}}{|\mathbf{u}|}x_kx_l\right\},\quad (2.1a)$$

$$\phi(\zeta_1,\cdots,\zeta_n)\equiv\langle\exp(i\sum_k\zeta_kx_k)\rangle=\exp(-(1/2)\sum_{kl}\zeta_k\zeta_l\mu_{kl}),\quad (2.1b)$$

where $\mathbf{u}$ is the matrix of the variances $\langle x_kx_l\rangle$, $\langle x_k,x_l\rangle=0$; μ^{kl} is the cofactor of the kth row, lth column of $\mathbf{u}$, and $\mathbf{u}$ itself is a symmetric matrix, since $\langle x_kx_l\rangle=\langle x_lx_k\rangle$. Furthermore, we assume an ergodic process, so that the statistical averages $\langle x_kx_l\rangle$ are equivalent to the time averages $\langle x_kx_l\rangle_{Av}=R(t_k-t_l)$, $(k,l=1,\cdots,n)$, where the R_{kl} form a set of autocorrelation functions, and $x_1\cdots x_n$ represent the random wave at n successive times $(t_1,\cdots,t_n)$. By the theorem of Wiener and Khintchine[11,12] we can write

$$R(\tau)=\int_0^\infty w(f)\cos 2\pi\tau df,\quad (2.2)$$

where $w(f)$ is the (power) spectral density of the random wave (in this case before it enters the IF filter). Thus, if we know the spectral density, we can completely specify the probability density (2.1) in this instance.

When the input is passed through the IF, one easily shows that the correlation function and spectrum of the output are

$$\langle V_i(t+t_j)V_i(t+t_k)\rangle=R(t_k-t_j)_{IF}=\int_0^\infty w(f)_{IF}\cos 2\pi f(t_k-t_j)df,\quad (2.2a)$$

and

$$w(f)_{IF}=|B(f)|^2w(f),\quad (2.2b)$$

where $B(f)$ is the system function of the IF filter. In particular, if the input wave is white noise of constant spectral intensity D, then $w(f)_{IF}=D|B(f)|^2$; henceforth $V_i(t)$ represents the random input to the quadratic detector.

We now distinguish between various outputs of the IF amplifier on the basis of their spectral character: when the power spectrum (measured, say, between half-intensity points) is narrow compared to the central or "resonant" frequency of the wave, it is called *narrow-band*; when this condition is not satisfied, we have *broad-band* noise. The former, for instance, is typical of white noise through a relatively high-Q filter. It turns out analytically that it is more convenient to discuss first these narrow-band cases, and then to specialize the results to the broad-band situation. Accordingly, we begin by assuming a narrow-band wave possessing a symmetrical spectrum, centered at the frequency f_0, as shown in Fig. 2. The wave $V_i(t)$ can then be represented as a combination of slowly and rapidly varying components in the usual way[1,2]

$$V_i(t)=R(t)\cos[\omega_0t+\Psi(t)]=x(t)\cos\omega_0t+y(t)\sin\omega_0t,\quad (2.3a)$$

$$x(t)=R\cos\Psi;\quad y(t)=R\sin\Psi,\quad (2.3b)$$

where R and Ψ are, respectively, the envelope and phase of the wave.

The quadratic detector through which $V_i(t)$ is next passed yields the instantaneous output

$$V_0(t)=\beta^2V_i(t)^2,\quad (-\infty<V_i<\infty),\quad (2.4a)$$

$$=\frac{\beta^2}{2}[(x^2+y^2)+(1/2)(x^2\cos 2\omega_0t+(2xy-y^2)\sin 2\omega_0t)]\quad (2.4b)$$

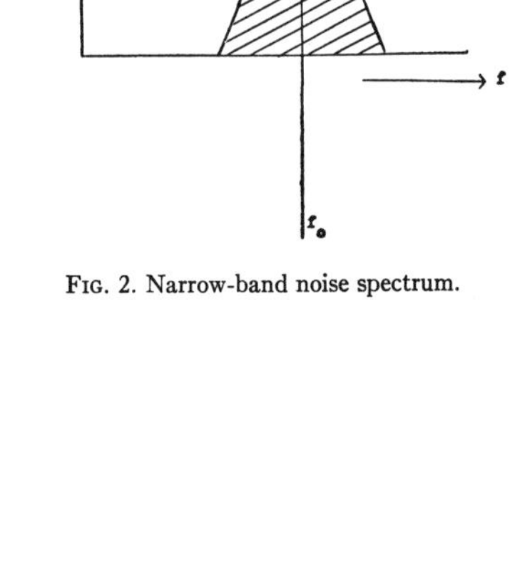

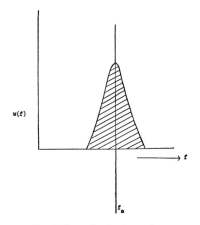

FIG. 2. Narrow-band noise spectrum.

[10] H. Cramér, *Mathematical Methods of Statistics* (Princeton University Press, Princeton, (1946); see Chap. 10.
[11] N. Wiener, Acta Math. **55**, 117 (1930).
[12] A. Khintchine, Math. Ann. **109**, 604 (1934).

for narrow-band waves. (For simplicity, in subsequent work we set $\beta^2=2$ for narrow-band waves and $\beta^2=1$ for broad-band disturbances, the extension to other values being obvious.) The low-frequency output, with which we are concerned here, is at once

$$V_0(t)_{LF}=x^2+y^2;\ \text{at}\ t=t_j,$$
$$(V_{LF})_j=x(t_j)^2+y(t_j)^2=x_j^2+y_j^2, \tag{2.5}$$

this last for the n time points t_j $(j=1\cdots n)$. But since $V_i(t)$ is normally distributed, so also are x and y,[1,2] giving the moments

$$\langle x_j^2\rangle=\int_0^\infty w(f)_{IF}df=\langle y_j^2\rangle\doteq\int_0^\infty w(f-f_0)_{IF}df\equiv\psi/2,$$
$$=\langle V_N^2\rangle, \tag{2.6a}$$

$$\langle x_jx_k\rangle=\langle y_jy_k\rangle\doteq\int_0^\infty w(f-f_0)_{IF}\cos\omega(t_k-t_j)df$$
$$\equiv\frac{\psi}{2}\alpha(t_k-t_j)=\frac{\psi}{2}\alpha_{kj}, \tag{2.6b}$$

$$\langle x_jy_k\rangle=-\langle x_ky_j\rangle\doteq\int_0^\infty w(f-f_0)_{IF}\sin\omega(t_k-t_j)df, \tag{2.6c}$$

this last since the spectrum is narrow-band.

The nth-order characteristic function associated with the nth-order distribution density *of the low-frequency output* of the quadratic detector at n successive times is

$$\phi_N(\zeta_1,\cdots,\zeta_n)\equiv\left\langle\exp i\sum_{j=1}^n\zeta_jV_{LF}(t+t_j)\right\rangle$$
$$=\int_{-\infty}^\infty\cdots\int W_n(x_1\cdots x_n;\,y_1\cdots y_n)$$
$$\times\left[\exp\left\{i\sum_{j=1}^n\zeta_j(x_j^2+y_j^2)\right\}\right]dx_1\cdots dx_ndy_1\cdots dy_n, \tag{2.7}$$

where W_n is the $2n$th-order distribution obtained from (2.1) on letting

$$x_k=x_k,\ (k,l=1,\cdots,n);$$
$$x_k=y_k,\ (k,l=n+1,\cdots,2n)$$

therein. In most instances we can assume a symmetrical pass band, so that $\langle x_jy_k\rangle$, Eq. (2.6c), vanishes, whereupon $W_n(x_1,\cdots,x_n;\,y_1,\cdots,y_n)$ factors into $W_n(x_1\cdots x_n)\times W_n(y_1\cdots y_n)$, each term of which is precisely (2.1). With the notation of Kac and Siegert,[5] *viz.*, (2.6b):

$$\langle x_jx_k\rangle=(1/2)\int_{-\infty}^\infty w(f)_{IF}\cos(\omega-\omega_0)(t_k-t_j)df=\frac{\psi}{2}\alpha_{jk},$$

we obtain finally for independent x and y

$$\phi_N(\zeta_1,\cdots,\zeta_n)=(\psi^n\pi^n|\alpha|)^{-1}\left[\int_{-\infty}^\infty\cdots\int dx_1\cdots dx_n\right.$$
$$\left.\times\exp\left\{-\sum_{k,l=1}^n\left(\frac{\alpha^{kl}}{|\psi\alpha|}-i\zeta_k\delta_{kl}\right)x_kx_l\right\}\right] \tag{2.8}$$

in which δ_{kl} is the familiar Kronecker delta: $\delta_{kl}=1$, $k=l$; $=0$, $k\neq l$.

The next step is to integrate (2.8) by reducing the quadratic form to the sum of squares (see Appendix I). When this is done we obtain a result of Kac and Siegert[5]:

$$\phi_N(\zeta_1,\cdots,\zeta_n)=i^n/|\zeta_k\psi\rho_{kj}+i\delta_{kj}|,\quad\rho_{kj}=\alpha_{kj}, \tag{2.9}$$

where we have replaced α_{kj} by ρ_{kj} to get a more familiar nomenclature. The characteristic function ϕ_N is given in terms of the determinant $|\zeta_k\psi\rho_{kj}+i\delta_{kj}|$, in which the statistical nature of the input noise enters only through the correlation coefficients

$$\psi\rho_{kj}=\int_{-\infty}^\infty w(f-f_0)_{IF}\cos\omega(t_k-t_j)df=2\langle x_jx_k\rangle.$$

An alternative form of (2.9), which is more useful in the discussion to follow, is (see Appendix I):

$$(\textit{narrow-band}):\begin{cases}\phi_N(\zeta_1,\cdots,\zeta_n)=\prod_{j=1}^n(1-i\lambda_j)^{-1}, & (2.10a)\\ \text{with} & \\ \psi\sum_{k=1}^n\zeta_k\rho_{km}f_{jk}=\lambda_jf_{jm}, & (2.10b)\end{cases}$$

i.e. λ_j is the jth eigenvalue of the set of homogeneous linear equations given in (2.10b), and f_{jk} are the components of the jth eigenvector f_j. Note that in this representation λ_j is a function of $\zeta_1,\zeta_2,\cdots,\zeta_n$.

The broad-band case for noise alone follows at once from the observation that now the output of the detector (2.4a) can be written $V_0(t)=\beta^2X^2(t)$, where the moments are

$$\langle X_jX_k\rangle=\psi\rho_{jk}=\int_{-\infty}^\infty w(f)_{IF}\cos\omega(t_j-t_k)df;$$
$$\langle X_j^2\rangle=\langle X_k^2\rangle=\psi,\ \text{etc.} \tag{2.11}$$

Setting $\beta=1$ here, and repeating the procedure outlined above for (2.7)–(2.9), we obtain just the square root of (2.9) and (2.10) for the characteristic function after rectification in this case:

$$(\textit{broad-band}):\phi_N(\zeta_1,\cdots,\zeta_N)=i^{n/2}/|\zeta_k\psi\rho_{kj}+i\delta_{kj}|^{1/2}$$
$$=\prod_{j=1}^n(1-i\lambda_j)^{-1/2}. \tag{2.12}$$

3. CHARACTERISTIC FUNCTION AFTER RECTIFICATION. SIGNAL AND NOISE

As in Sec. 2, we divide the discussion into two parts, considering first the more general analysis of the narrow-band case, where now a signal as well as noise constitutes the input to the IF filter. The noise part of the rectifier's input is given as before by (2.3), while the signal is represented by $M(t')\cos\omega_0t$. Here $M(t')$ is a modulation of the carrier† $\cos\omega_0 t$, and t' rather than t is used to indicate the fact that modulation and (IF) carrier are uncorrelated. There is, of course, no correlation between signal and noise either. The total input to the detector is therefore

$$V_i(t)=[x(t)+M(t')]\cos\omega_0 t+y(t)\sin\omega_0 t, \quad (3.1)$$

and from (2.5) the corresponding low-frequency output of the quadratic detector becomes ($\beta=\sqrt{2}$)

$$V_{LF}(t)=x^2+y^2+2xM+M^2. \quad (3.2)$$

Letting $x(t+t_j)=x_j$, $y(t+t_j)=y_j$, $M(t'+t_j)=M_j(t')$, we find that the characteristic function takes the more general form

$$\phi_{S+N}(\zeta_1,\cdots,\zeta_n)=\int_{-\infty}^{\infty}\cdots\int W_n(x_1,\cdots,x_n)W_n(y_1,\cdots,y_n)$$

$$\times\exp\left(i\sum_{j=1}^{n}\zeta_j[x_j^2+y_j^2+2x_jM_j(t')+M_j(t')^2]\right)dx_1\cdots dy_n, \quad (3.3)$$

which is explicitly, see (2.8),

$$\phi_{S+N}(\zeta_1,\cdots,\zeta_n)=(\psi^n\pi^n|\alpha|)^{-1}\exp\sum_j^n i\zeta_j M_j^2(t')$$

$$\cdot\int_{-\infty}^{\infty}\cdots\int dy_1\cdots dy_n\exp\left(i\sum_k^n\zeta_k y_k^2-\sum_{kl}^n\frac{\alpha^{kl}y_ky_l}{|\alpha|\psi}\right)$$

$$\cdot\int_{-\infty}^{\infty}\cdots\int dx_1\cdots dx_n\exp\left(i\sum_{kl}^n\zeta_k x_k^2-\sum_{kl}^n\frac{\alpha^{kl}x_kx_l}{\psi|\alpha|}+2i\sum_k^n\zeta_k x_k M_k(t')\right). \quad (3.4)$$

Again introducing a suitable principal-axis transformation (see Appendix I) and integrating give us finally

$$(\textit{narrow-band}):\ \phi_{S+N}(\zeta_1\cdots\zeta_n)=\prod_{k=1}^{n}(1-i\lambda_k)^{-1}$$

$$\cdot\prod_{k=1}^{n}\exp\left[i\left(\sum_j^n M_j(t')\zeta_j f_{kj}\right)^2\Big/(1-i\lambda_k)\right], \quad (3.5a)$$

where the eigenvalues $\lambda_j=\lambda_j(\zeta_1,\cdots,\zeta_n)$ and eigenfunctions f_{kj} are defined by (2.10b), with the additional normalizing relation

$$\sum_{l=1}^{n}\zeta_l f_{kl}f_{jl}=\delta_{kl}. \quad (3.5b)$$

The corresponding expression for broad-band waves is easily found. We now let $S(t')$ represent the signal and $X(t)$ the noise, so that the output (2.4a) of the quadratic detector, when $V_i=X+S$ is the input, becomes at the times t_j ($j=1,\cdots,n$)

$$V_j=[X_j+S_j(t')]^2=X_j^2+S_j^2(t')+2X_jS_j(t'),\quad(\beta=1). \quad (3.6)$$

Comparison with (3.2) and (3.4) gives at once the desired characteristic function

$$(\textit{broad-band}):\ \phi_{S+N}(\zeta_1,\cdots,\zeta_n)=\prod_{k=1}^{n}(1-i\lambda_k)^{-1/2}$$

$$\cdot\prod_{k=1}^{n}\exp\left[i\left(\sum_{j=1}^{n}S_j(t')\zeta_j f_{kj}\right)^2\Big/(1-i\lambda_k)\right]. \quad (3.7)$$

Both (3.5a) and (3.7) are generalizations of Kac and Siegert's earlier result (2.12) for noise alone.

4. THE POST-DETECTION FILTER

There remains now to determine the effects of the linear, post-detection ("video") filter on the statistics of the rectified wave. Our first task is to obtain the desired characteristic function after this post-detection filter for the broad and narrow-band cases discussed above. Special limiting cases of interest are reserved for Secs. (5–7) and Appendix II.

If the input to the "video" filter is $V_d(t)$ where $V_d(t)$ is $V_{LF}(t)$, for broad- or narrow-band inputs, and if $V_{0d}(t)$ is the corresponding output of the filter, we can write in the steady state

$$V_{0d}(t)=\int_{-\infty}^{\infty}A(\tau)V_d(t-\tau)d\tau, \quad (4.1)$$

where $A(\tau)$ is the filter's weighting function, which is related to the system function $Y(i\omega)$ by the Fourier transforms

$$A(t)=\int_{-\infty}^{\infty}Y(i\omega)e^{i\omega t}df,\quad Y(i\omega)=\int_{-\infty}^{\infty}A(t)e^{-i\omega t}dt \quad (4.2)$$

for linear stable filters. For all real filters $A(t)=0$, $t<0-$.

The procedure now is to replace the integral (4.1) by a sum which (in the appropriate limit) will again be equal to the integral. The sum is here

$$(V_{0d})_k\equiv V_k=\sum_{l=0}^{L}A_l(V_d)_{(k-l)},$$

and

$$(V_{0d})_{k-m(j)}=\sum_{l=0}^{L}A_l(V_d)_{k-l-m(j)}, \quad (4.3)$$

where the interval over which $A(\tau_l)$ is significant is divided into $L+1$ points, each of which is a distance h apart, and for each $(k-1)h\leq\tau_k\leq kh$; $A(\tau_l)=A_lh$.

† Since we are dealing with stationary waves, one can always choose the origin of time such that $\cos(\omega_0 t'+\phi)=\cos\omega_0 t$.

The nth-order characteristic functions $F_V(\zeta_1',\cdots,\zeta_n')$ for the output $V_{0d}(t)$ of the post-detection filter is

$$F_V(\zeta')=\left\langle\exp\left\{i\sum_{k=1}^{n}\zeta_k V_{0d}(t-\tau_{k-1})\right\}\right\rangle. \tag{4.4}$$

Substituting for $V_{0d}(t-\tau_j)$, etc., from (4.3) into (4.4) and performing the statistical average gives

$$F_V(\zeta')=\int_{-\infty}^{\infty}\cdots\int W_L[(V_d)_0,\cdots,(V_d)_{(-L)}]$$
$$\times\exp\left\{i\sum_{k=1}^{n}\zeta_k\sum_{l=0}^{L}A_l(V_d)_{(k-l-m(k-1))}\right\}$$
$$\times d(V_d)_0\cdots d(V_d)_{-L}, \tag{4.5}$$

where $[m(j)-1]h\leq\tau_j\leq m(j)h$. Here W_L is the $(L+1)$st-order probability density for the $L+1$ inputs to the filter at the successive times $t-\tau_L$ to t. (According to the present notation $m(0)=0$.)

Our next step is to replace‡ the sum

$$\sum_{l=0}^{L}A_l(V_d)_{(k-l-m(j))}\quad\text{by}\quad\sum_{l=0}^{L}A_{l-m(j)}(V_d)_{k-l}.$$

Equation (4.5) is now written with the help of the Fourier transform of W_n, *viz.*, the characteristic function $\phi(\zeta_0,\cdots,\zeta_{-L})$,

$$F_V(\zeta')=\int_{-\infty}^{\infty}\cdots\int d(V_d)_0\cdots d(V_d)_{-L}$$
$$\times\exp\left\{i\sum_{k=1}^{n}\zeta_k'\sum_{l=0}^{L}A_{l-m(k-1)}(V_d)_{k-l}\right\}$$
$$\cdot(2\pi)^{-L}\int_{-\infty}^{\infty}\cdots\int d\zeta_0\cdots d\zeta_{-L}\phi(\zeta_0,\cdots,\zeta_{-L})$$
$$\times\exp\left\{-i\sum_{l}^{n}\zeta_{-l}(V_d)_{-l}\right\}. \tag{4.6}$$

The characteristic function $\phi(\zeta_0,\cdots,\zeta_{-L})$ is associated with the $L+1$ inputs $(V_d)_0,\cdots,(V_d)_{-L}$. Integrating over $(V_d)_{-q}$ $(q=0,\cdots,L)$ we find the first set of integrals to be

$$\sum_{l=0}^{L}\delta\left(\zeta_{-1}-\sum_{k=1}^{n}\zeta_k'A_{l-m(k-1)}\right) \tag{4.6a}$$

so that (4.6) becomes finally

$$F_V(\zeta')=\phi\left(\sum_{k=1}^{n}\zeta_k'A_{-m(k-1)};\cdots;\right.$$
$$\left.\sum_{k=1}^{n}\zeta_k'A_{j-m(k-1)};\cdots;\sum_{k=1}^{L}\zeta_k'A_{L-m(k-1)}\right), \tag{4.7}$$

where we remember that ϕ is here the characteristic function for the *output* of the quadratic rectifier, and $F_V(\zeta')$ is the desired characteristic function of the output of the rectifier plus post-detection filter. A number of specific cases follow:

(a) *Narrow-Band Noise*

We apply (4.7) now, using the results of Secs. 2 and 3. Equations (2.9) and (4.7) give us first one form of the nth-order characteristic function for narrow-band noise after video filtering:

$$F_V(\zeta')_N=i^{L+1}\bigg/\left|\psi\rho_{kj}\left(\sum_{q=1}^{n}\zeta_q'A_{k-m(q)}\right)+i\delta_{kj}\right|,$$
$$(k=1,\cdots,L), \tag{4.8}$$

and the determinant is an $(L\times L)$ one. Corresponding to (2.10) we have the alternative representation

$$F_V(\zeta)_N=\prod_{j=0}^{L}(1-i\lambda_j)^{-1}, \tag{4.9a}$$

where

$$\sum_{k=0}^{L}\psi\rho_{km}\left(\sum_{q=1}^{n}\zeta_q'A_{k-m(q-1)}\right)f_{jk}=\lambda_j f_{jm}, \tag{4.9b}$$

and since $A_l=A(t_l)h$ this sum can be rewritten

$$\left\{\sum_{k=0}^{L}(\psi\rho_{km}f_{jk})\sum_{q=1}^{n}\zeta_q'A(t_k-t_{m(q-1)})\right\}h=\lambda_j f_{jm}. \tag{4.9c}$$

Letting L approach ∞ and h approach 0 we get an integral for the sum over k, and so for the characteristic function of the continuous, filtered wave $V_{0d}(t)$, Eq. (4.1), the result (4.9a) becomes in the limit (dropping the primes on ζ_1, etc.)

$$F_V(\zeta)_N=\prod_{j=1}^{\infty}(1-i\lambda_j)^{-1}, \tag{4.10a}$$

where λ_j are the eigenvalues (here functions of $\zeta_1,\cdots,\zeta_n$ and $\tau_1,\cdots,\tau_{n-1}$) of the integral equation§

$$\psi\int_0^{\infty}dt\left[\sum_{q=1}^{n}\zeta_qA(t-\tau_{q-1})\right]\rho(t,s)f(t)=\lambda f(s). \tag{4.10b}$$

‡ This substitution may be justified by the following argument: if we let $l+m(j)=q$, then

$$\sum_{l=0}^{L}A_l(V_d)_{k-l-m(j)}=\sum_{m(j)}^{L+m(j)}A_{(q-m(j))}(V_d)_{k-q}$$
$$=\sum_{0}^{L+m(j)}A_{(q-m(j))}(V_d)_{k-q},$$

since $A(\tau_j)=0$; $\tau_j<0$. Thus, if L is chosen sufficiently large, so that $A_{(L-m(j))}\to0$, then $A_L\to0$ also, and we can replace the upper limit on the last sum with L only.

§ The eigenvalues of the finite sum are certainly discrete, but it may be questioned whether or not they remain discrete in the limit of the integral equation, since one can always find kernels for which this is not the case. However, for symmetric kernels

In the limit, of course, one gets an infinite determinant, if (4.8) is used.

The examples of chief interest occur for $n=1, 2$. For $n=1$ (first-order probability density and characteristic function) we find that (4.10b) reduces to

$$\psi\zeta_1\int_0^\infty dtA(t)\rho(t,s)f(t)=\lambda f(s)=\lambda(\zeta_1)f(s). \quad (4.11a)$$

Letting $\lambda_j=\lambda_j'\zeta_1\psi$, we see that the integral equation is modified to

$$\int_0^\infty A(t)\rho(t,s)f(t)dt=\lambda' f(s), \quad (4.11b)$$

and the desired characteristic function becomes

$$F_V(\zeta_1)_N=\prod_{j=1}^{\infty}(1-i\zeta\psi\lambda_j')^{-1}, \quad (4.12)$$

which is the result obtained by Kac and Siegert[5] when $\psi=1$.

When $n=2$, we get from (4.10a, b)

$$F_V(\zeta_1,\zeta_2;\tau)_N=\prod_{j=1}^{\infty}[1-i\lambda_j(\zeta_1,\zeta_2;\tau)]^{-1} \quad (4.13a)$$

with the accompanying integral equation

$$\psi\int_0^\infty dt[\zeta_1A(t)+\zeta_2A(t-\tau)]\rho(t,s)f(t)=\lambda f(s). \quad (4.13b)$$

$K(s,t)$, which are quadratically integrable in the sense

$$\int_0^\infty\int K^2(s,t)dsdt=A \text{ (exists)}, \quad (i)$$

the proof of discreteness is easily given. We summarize the main points of the argument: let the eigenfunctions and eigenvalues be defined from

$$\int_0^\infty K(s,t)y_j(t)dt=\lambda_jy_j(s). \quad (ii)$$

Since we are looking for quadratically integrable solutions of (ii), we can choose these from an orthonormal set. Now the Fourier coefficients of $K(s,t)$, with respect to $y_j(t)$, are given by $\lambda_jy_j(s)$. Bessel's inequality yields (for arbitrary n)

$$\sum_{j=1}^{n}\lambda_j^2y_j^2(s)\leq\int_0^\infty K^2(s,t)dt. \quad (iii)$$

Integrating over s we obtain

$$\sum_{j=1}^{n}\lambda_j^2\leq\int_0^\infty\int K^2(s,t)dsdt=A \text{ (exists)} \quad (iv)$$

from (i). Therefore $\sum_{j=1}^{\infty}\lambda_j^2$ converges, and the λ_j's can have no finite point of accumulation. The spectrum is accordingly discrete. In our examples the quadratic integrability may be shown from the fact that $\int_{-\infty}^{\infty}|A(t)|dt$ exists, and the fact that $\rho(s,t)=\rho(|t-s|)$ never has a weaker falling-off for large $(t-s)$ than $|(t-s)|^P\cdot e^{-|s-t|}$, where P is finite (≥ 0). The former applies because we are dealing with linear, stable filters, and the latter, because ρ represents the autocorrelation function of intensity-limited noise after it has been passed through such filters. We are indebted to Professor Ivar Stakgold for calling to our attention this question of the discrete character of the eigenvalues.

The alternative form (4.8) proves useful in limiting cases of wide and narrow video. If we let

$$\Delta=|\delta_{jk}+\gamma hK_{jk}|$$

be the Lth-order determinant (4.8), where $A_k=A(t_k)h$, we may expand Δ as follows[13]

$$\Delta=1-\gamma\sum_{j}^{L}K_{jj}h+\frac{\gamma^2}{2}\sum_{jk}^{L}h^2\begin{vmatrix}K_{jj} & K_{jk}\\ K_{kj} & K_{kk}\end{vmatrix}\cdots. \quad (4.14)$$

In the limits $M\rightarrow\infty$, $h\rightarrow 0$, sums are replaced by integrals, as before, $\gamma=i$, and we get finally

$$F_V(\zeta_1,\zeta_2;\tau)_N=\left\{1-i\int_0^\infty K(t,t)_\tau dt-(1/2!)\right.$$

$$\left.\times\int_0^\infty\int\begin{vmatrix}K(t_1,t_1)_\tau & K(t_1,t_2)_\tau\\ K(t_2,t_1)_\tau & K(t_2,t_2)_\tau\end{vmatrix}dt_1dt_2+\cdots\right\}^{-1}, \quad (4.15a)$$

where

$$K(t_1,t_2)_\tau=\psi\rho(t_2-t_1)[\zeta_1A(t_1)+\zeta_2A(t_1-\tau)]. \quad (4.15b)$$

Applications of this approach are considered in Secs. 5 and 6.

We note finally that if the video filter's weighting function is such that $A(t)\geq 0$, all $t\geq 0$, then only positive eigenvalues can occur, corresponding physically to the fact that the output of such a filter is always positive (when the preceding detector is quadratic). Consequently, there will be a dc term, appearing as a finite average value for the output, i.e., $\langle V_{out}\rangle>0$, as well as an ac fluctuation about this value, but no $\langle V_{out}\rangle<0$. A simple example arises when the video filter is an RC network, with the output taken across the condenser, as then $A(t)=(RC)^{-1}e^{-t/RC}$, $t>0$. However, when $A(t)$ can be negative for some $t(>0)$, a [finite¶] number of negative eigenvalues appear as well: the output can be negative part of the time, although there can still be a positive mean value $\langle V\rangle$ (which vanishes in the limit of a very narrow video). An example of this is provided by a CR post-detection filter, with the output observed now across the resistance, as here

$$A(t)=\delta(t-0)-(RC)^{-1}e^{-t/RC}, \quad (t>0-).$$

These phenomena are specifically illustrated by the results of Secs. 5, 6, see (5.3), (5.4), and (5.22), etc., and the foregoing remarks apply also for (b)–(d) following.

(b) Broad-Band Noise

From (2.11), (2.12) it is clear that the results of Sec. 4 may be used directly for broad-band problems, with the

[13] Lovitt, *Linear Integral Equations* (McGraw-Hill Book Company, Inc., New York, 1924), p. 24.

¶ This number depends on the kernel and on the weighting function $A(t)$, of course, but is finite for realizable networks.

simple modifications of Eqs. (4.8), (4.10), (4.13a), etc., i.e. of replacing the right-hand members by their square roots.

(c) *Narrow-Band Signal and Noise*

A direct modification of (3.5a) in conjunction with the transformation used to replace ζ_{-l}' in obtaining (4.7) permits us to write $\zeta_j(\equiv\zeta_{-j})=\zeta_1'A_j$, $(j=0,\cdots, L\to\infty)$ for the first-order characteristic function where there is a signal. Dropping the prime on ζ_1' gives us

$$F_V(\zeta_1)_{S+N}=\prod_{k=1}^{\infty}(1-i\psi\zeta_1\lambda_k')^{-1}\times\exp\left\{i\zeta_1^2\left(\int_{-\infty}^{\infty}M(t'+t)A(t)f^{(k)}(t)dt\right)\Big/(1-i\zeta_1\psi\lambda_k')\right\},\quad(4.16)$$

where $\lambda_k=\lambda_k'\zeta_1\psi$ as in (4.11a, b). The associated integral equation and normalization of the eigenfunctions $f^{(k)}$ follow, respectively, from

$$\int_{-\infty}^{\infty}A(t)\rho(t,s)f^{(k)}(t)dt=\lambda_k'f^{(k)}(s);$$

$$\zeta_1\int_{-\infty}^{\infty}A(t)f^{(k)}(t)^2dt=1.\quad(4.17)$$

From (4.16) it is clear that in dealing with a signal, as well as noise, one must also know the eigenfunctions; it is not enough to determine the eigenvalues alone, as is true of the case for noise only.

The second-order characteristic function is obtained in the same way. The general transformation relating ζ_{-j} and ζ_k' now reduces to the two terms $\zeta_j=\zeta_1'A_j+\zeta_2'A_{j-m}$ $(j=0,\cdots, L\to\infty)$. Dropping primes again we get finally

$$F_V(\zeta_1,\zeta_2;\tau)_{S+N}=\prod_{k=1}^{\infty}(1-i\lambda_k)^{-1}\times\exp\left\{i\left(\int_{-\infty}^{\infty}M(t'+t)[\zeta_1A(t)+\zeta_2A(t-\tau)]\cdot f^{(k)}(t)dt\right)^2\Big/(1-i\lambda_k)\right\},\quad(4.18)$$

where we remember that now $\lambda_k=\lambda_k(\zeta_1,\zeta_2;\tau)$. The associated integral equation and normalization condition on $f^{(k)}$ are

$$\psi\int_{-\infty}^{\infty}[\zeta_1A(t)+\zeta_2A(t-\tau)]\rho(t,s)f^{(k)}(t)dt=\lambda_kf^{(k)}(s),\quad(4.19a)$$

$$\int_{-\infty}^{\infty}[\zeta_1A(t)+\zeta_2A(t-\tau)]f^{(k)}(t)^2dt=1.\quad(4.19b)$$

A check on these results is given in Appendix II in the special case of an infinitely wide video.

(d) *Broad-Band Signal and Noise*

The expressions (4.16)–(4.19) may be adapted at once to this case if we replace $M(t'+t)$ by $S(t'+t)$ and $\Pi_k(1-i\lambda_k)^{-1}$ by $\Pi_k(1-i\lambda_k)^{-1/2}$. [See Eq. (3.7) for infinitely wide video filter; also Appendix II.]

5. THE NARROW VIDEO FILTER

(a) *Noise Alone*

If we consider now the determinantal relation (4.15a) and expand it in such a form as to produce the coefficients of the various powers of $\zeta_1^j\zeta_2^{j-k}$, $\zeta_2^j\zeta_1^{j-k}$, $(j=1\cdots n;\ k=0\cdots j)$ we find that these coefficients consist of a sum of terms of order $(\Delta f_r/\Delta f_F)$, $(\Delta f_r/\Delta f_F)^2$, $\cdots$, $(\Delta f_r/\Delta f_F)^{j-1}$. When the video band width Δf_r is small compared to the IF band width‖ Δf_F a good approximation may be obtained by keeping terms up to and including $(\Delta f_r/\Delta f_F)^2$. Letting $n\to\infty$ and picking out the significant terms to the desired order, we find finally that

$$F_V(\zeta_1,\zeta_2;\tau)_N=e^{i\psi(\zeta_1+\zeta_2)}\left\{1+\frac{\psi^2}{2}[(\zeta_1^2+\zeta_2^2)q(0)+2\zeta_1\zeta_2q(\tau)]+\frac{i\psi^3}{3}[(\zeta_1^3+\zeta_2^3)p(0)+3\zeta_1^2\zeta_2p(\tau)+3\zeta_1\zeta_2^2p(-\tau)]+0(\Delta f_r/\Delta f_F)^3\right\}^{-1},\quad(5.1)$$

where

$$p(\tau)=\iiint_{-\infty}^{\infty}Y(i\omega)Y(i\omega')Y(i\omega'+i\omega)^*\times H(f'')H(f'+f'')H(f-f'')e^{-i\omega\tau}\cdot dfdf'df''\equiv0(\Delta f_r/\Delta f_F)^2,\quad(5.2a)$$

$$q(\tau)=\iint_{-\infty}^{\infty}H(f)H(f')\,|Y(i\omega+i\omega')|^2e^{i(\omega+\omega')\tau}dfdf'=0(\Delta f_r/\Delta f_F).\quad(5.2b)$$

Our next step is to express (5.1) in terms of an exponential with linear and quadratic arguments, the higher-order terms appearing as a factor of the exponential. Inversion then gives the (asymptotic) expression for the second-order probability density for narrow videos (vis-à-vis the IF):

‖ These band widths are those of the equivalent rectangular filters. For details of the method, see reference 8, Chap. V.

(*narrow-band*)

$$W_2(V_1,V_2;\tau)_N \simeq \left\{1-\frac{\psi^3}{3}\left[\left(\frac{\partial^3}{\partial V_1^3}+\frac{\partial^3}{\partial V_2^3}\right)p(0) + 3\frac{\partial^3}{V_1^2\partial V_2}p(\tau)+\frac{3\partial^3}{\partial V_1\partial V_2^2}p(-\tau)\right] + \frac{\psi^4}{8}\left[\left(\frac{\partial^4}{\partial V_1^4}+\frac{\partial^4}{\partial V_2^4}\right)q(0)+2\frac{\partial^4}{\partial V_1^2\partial V_2^2}(q^2(0)+q^2(\tau)) + \left(\frac{\partial^4}{\partial V_1^3\partial V_2}+\frac{\partial^4}{\partial V_1\partial V_2^3}\right)4q(0)q(\tau)\right] + 0(\Delta f_r^3/\Delta f_F^3)\right\} W_2(V_1,V_2;\tau)_{N,0}, \quad (5.3)$$

where $(W_2)_{N,0}$ is the distribution density defined by

$$W_2(V_1,V_2;\tau)_{N,0}=[2\pi\psi^2q(0)(1-\rho_0^2)^{1/2}]^{-1} \times\exp\left(-\left\{\frac{(V_1-\psi)^2+(V_2-\psi)^2-2(V_1-\psi)(V_2-\psi)\rho_0}{2\psi^2q(0)(1-\rho_0^2)}\right\}\right). \quad (5.4)$$

in which

$$\rho_0(\tau)\equiv\Phi_1(\tau)/\Phi_1(0)=\int_{-\infty}^{\infty}A(t)A(t+\tau)dt \Big/ \int_{-\infty}^{\infty}A^2(t)dt. \quad (5.5)$$

[See Appendix III, Eq. (A3.1).]

For *broad-band noise* one may use (5.3) after inserting a factor $\frac{1}{2}$ before each term in the brackets [ ] and replacing ψ by $\psi/2$, ψ^2 by $\psi^2/2$ in (5.4); ψ itself is then reinterpreted according to the remarks following reference 14.

The desired development of the characteristic function and probability density may also be obtained directly from (4.13a): for the narrow-band case we have formally for small λ_j

$$F_V(\zeta_1,\zeta_2,\tau)_N=\prod_{j=1}^{\infty}[1-i\lambda(\zeta_1,\zeta_2;\tau)]^{-1}=\Delta^{-1} =\exp\left[-\sum_j^{\infty}\log(1-i\lambda_j)\right] =\exp\left(i\sum_j^{\infty}\lambda_j-(1/2)\sum_j^{\infty}(\lambda_j^2+\cdots\right). \quad (5.6)$$

These sums of the eigenvalues are found in the usual way by expressing the symmetrical kernel (4.15b) as a series in the *orthonormal* set of eigenfunctions $f(t)$

$$K(t_1,t_2)=\textstyle\sum_j \lambda_j f^{(j)}(t_1)f^{(j)}(t_2)$$

and integrating over t_1 and t_2. The sums in (5.6) are easily seen to be the trace of the kernel and of the first iterated kernel, respectively, which in the time and frequency domains become

$$\int_{-\infty}^{\infty}K(t,t)_\tau dt=\int_{-\infty}^{\infty}\mathcal{K}(f,f)df=\sum_{j=1}^{\infty}\lambda_j=\psi(\zeta_1+\zeta_2), \quad (5.7)$$

$$\int_{-\infty}^{\infty}\int K(t_1,t_2)_\tau K(t_2,t_1)_\tau dt_1dt_2 =\int_{-\infty}^{\infty}\int \mathcal{K}(f_1,f_2)\mathcal{K}(f_2,f_1)df_1df_2=\sum_{j=1}^{\infty}\lambda_j^2 =(\zeta_1^2+\zeta_2^2)\psi^2q(0)+2\zeta_1\zeta_2\psi^2q(\tau), \quad (5.8)$$

and so on for the second and higher iterations. The frequency form of K is

$$\mathcal{K}(f_1,f_2)=\psi(\zeta_1+\zeta_2e^{i(\omega_2-\omega_1)})H(f_1)Y(i\omega_1-i\omega_2). \quad (5.9)$$

The correspondence between these results and (5.1)–(5.2b) indicates that the λ_j are indeed small for the narrow video (5.2), so that this latter approach may be followed for the corresponding problem of broad-band noise and signal and noise (considered below).

(b) *Signal and Noise*

Here the governing integral equation and the characteristic function for the second-order case ($n=2$) are given by (4.19), (4.18) for narrow-band waves. As mentioned above, when the video is narrow compared to the IF, the eigenvalues are small and the product $\prod_{k=1}^{\infty}(1-i\lambda_k)^{-1}$ (for noise alone) approaches a normal characteristic function. Using an expansion similar to (5.6) we have for the exponent in (4.18)

$$\exp\{i\sum_1^{\infty}c_k^2-\psi\sum_1^{\infty}\lambda_kc_k^2-i\psi^2\sum_1^{\infty}\lambda_k^2c_k^2+\cdots\}, \quad (5.10a)$$

where

$$c_k=\int_{-\infty}^{\infty}M(t'+t)[\zeta_1A(t)+\zeta_2A(t-\tau)]f^{(k)}(\tau)dt. \quad (5.10b)$$

Expanding $M(t'+t)$ in the orthonormal set $f^{(k)}(t)$ gives

$$M(t'+t)=\sum_1^{\infty}b_kf^{(k)}(t), \quad (5.11)$$

and from (5.10b) and the integral equation (4.19a) one sees that $b_k=c_k$. Then, multiplying (5.11) by $M(t'+t)\times[\zeta_1A(t)+\zeta_2A(t-\tau)]$ and integrating gives us an expression for the first term in the exponent of (5.10a):

$$\int_{-\infty}^{\infty}M^2(t'+t)[\zeta_1A(t)+\zeta_2A(t-\tau)]dt=\sum_1^{\infty}c_k^2. \quad (5.12)$$

The other term in (5.10a) may be found by a similar technique: Expansion of $\rho(s,t)$ gives

$$\rho(s,t)=\sum_1^\infty d_k f^{(k)}(t)=\sum_1^\infty \lambda_k f^{(k)}(s) f^{(k)}(t), \quad (5.13)$$

and multiplying by

$$M(t+t')M(s+t') \times[\zeta_1 A(t)+\zeta_2 A(t-\tau)][\zeta_1 A(s)+\zeta_2 A(s-\tau)]$$

and integrating over s,t yields

$$\int_{-\infty}^{\infty}\int \rho(s,t)M(t+t')M(s+t')\{\zeta_1^2 A(s)A(t) +\zeta_2^2 A(s-\tau)A(t-\tau)+2\zeta_1\zeta_2 A(s)A(t-\tau)\}dsdt =\sum_1^\infty \lambda_k c_k^2. \quad (5.14)$$

For the third term in (5.10a) we multiply (5.13) by $[\zeta_1 A(t)+\zeta_2 A(t-\tau)]\rho(x,\tau)$ and with the aid of the integral equation (4.19a) obtain

$$\int_{-\infty}^{\infty} [\zeta_1 A(t)+\zeta_2 A(t-\tau)]\rho(x,t)\rho(s,t)dt =\sum_1^\infty \lambda_k^2 f^{(k)}(x) f^{(k)}(s). \quad (5.15)$$

Multiplying this by the same expression used in deriving (5.14), replacing t by x, we obtain finally

$$\int_{-\infty}^{\infty}\int\int \{\zeta_1 A(x)+\zeta_2 A(x-\tau)\}\{\zeta_1 A(s)+\zeta_2 A(s-\tau)\} \times\{\zeta_1 A(t)+\zeta_2 A(t-\tau)\}M(t'+s)M(t'+x) \times\rho(s,t)\rho(x,t)dxdsdt=\sum_1^\infty \lambda_k^2 c_k^2 \quad (5.16)$$

in the general case.

Because of spatial limitations we now outline only the derivation of the leading term in W_2; the correction terms may be found in straightforward fashion with the help of (5.16) (expressed in frequency form), combined with (5.1), and then developed into a structure of the type (5.3), where, of course, $(W_2)_{N,0}$ is replaced by $(W_2)_{S+N,0}$ derived below. A further simplification (which does not in any way restrict the generality of the *method*) sets the signal equal to a constant: $M(t')=M$. Then (5.12) becomes

$$\sum_{k=1}^\infty c_k^2=M^2(\zeta_1+\zeta_2), \quad (5.17)$$

and the integrals occurring in (5.14) can be expressed in the frequency domain as

$$\int_{-\infty}^{\infty}\int A(s-t_1)A(t-\tau_2)\rho(s,t)dsdt =\int_{-\infty}^{\infty} H(f)|Y)i\omega)|^2 e^{i\omega(\tau_1-\tau_2)}df =H(0)\int_{-\infty}^{\infty} |Y(i\omega)|^2 e^{i\omega(\tau_1-\tau_2)} =H(0)\Phi_1(\tau_1-\tau_2)=H(0)\Phi_1(\tau_2-\tau_1), \quad (5.18)$$

since for the video narrow compared to the IF, $H(f)\doteq H(0)$. We get finally

$$\sum_{k=1}^\infty \lambda_k c_k^2=M^2H(0)\{(\zeta_1^2+\zeta_2^2)\Phi_1(0)+2\zeta_1\zeta_2\Phi_1(\tau)\}. \quad (5.19)$$

The characteristic function is the product of (5.10a) and (5.1), which for the normal term becomes on writing $H(0)=H_0$,

$$F_V(\zeta_1,\zeta_2;\tau)_{S+N,0} =\exp[i(\psi+M^2)(\zeta_1+\zeta_2)-(1/2)\{(\psi^2H_1+2\psi M^2H_0) \cdot((\zeta_1^2+\zeta_2^2)\Phi(0)+2\zeta_1\zeta_2\Phi_1(\tau))\}], \quad (5.20)$$

where we have made the substitution for $q(\tau)$:

$$q(\tau)=\Phi_1(\tau)\int_{-\infty}^{+\infty} H^2(f)df\equiv H_1\Phi_1(\tau), \quad (5.21)$$

which is valid for narrow videos. The corresponding probability density is

$$W_2(V_1,V_2;\tau)_{S+N,0}=[2\pi(\psi^2H_1+2\psi M^2H_0)\Phi_1(0)(1-\rho_0^2)^{\frac{1}{2}}]^{-1} \exp\left[\frac{[V_1-(\psi+M^2)]^2+[V_2-(\psi+M^2)]^2-2[V_1-(\psi+M^2)][V_2-(\psi+M^2)]\rho_0}{(\psi^2H_1+2\psi M^2H_0)\Phi_1(0)(1-\rho_0^2)}\right], \quad (5.22)$$

where ρ_0 is given by (5.5).

$$\left(\text{Note that } H_1=\int_{-\infty}^{\infty} \rho(\tau)^2dt=\int_{-\infty}^{\infty} H(f)^2df\right).$$

We observe in both (a) and (b) here, especially (5.3), (5.4), and (5.22) that in this approximation ($q(0)\neq0$, $\Phi_1(0)\neq0$) the probability densities appear to exist for V_1, $V_2<0$. Of course, this cannot strictly occur if $A(t)\geq0$, $t>0$, since there can be no negative output of the final video filter in such cases; (see remarks at end of (a), Sec. 4). The higher-order terms in these approximations tend to reduce this contribution for V_1, $V_2<0$ to zero.

6. THE WIDE VIDEO FILTER

(a) *Narrow-Band Noise*

The case when the video band width is narrow compared to that of the IF stage has just been treated. We can also obtain expressions for the alternative limiting case where the video band width is large compared to the IF band width. The expected result should approach that of the distribution obtained with an infinitely wide, uniform video response.

If the IF band is limited (practically speaking) to a band** $2B$, and the video band width is b, one makes a Taylor series expansion

$$Y(i\omega)=1+\alpha_1\frac{i\omega}{2\pi B}+\alpha_2\left(\frac{\omega}{2\pi B}\right)^2+\cdots,$$
$$\alpha_1\sim 0(B/b),\quad \alpha_2\sim 0(B/b)^2 \tag{6.1}$$

which certainly converges for $\omega<2\pi b$.

Evaluating the coefficients in the determinantal expansion of $F_V(\zeta_1\zeta_2\tau)$, see (4.15), we obtain

$$D_2(\tau)=\int_{-b}^{b}\int_{-b}^{b}H(f)H(f')\left[1+\frac{(\alpha_1^2+2\alpha_2)}{(2\pi)^2B^2}\times(\omega^2+2\omega\omega'+\omega'^2)\right]e^{i(\omega+\omega')\tau}dfdf'$$
$$+2\int_b^\infty\int_b^\infty H(f)H(f')\,|\,Y(i\omega+i\omega')\,|^2e^{i(\omega+\omega')\tau}dfdf, \tag{6.2}$$

retaining terms through the second order. The contribution of the second integral is negligible, since $H(f)$ falls off at least as $(1/\omega^2)$ and $Y(i\omega)$ at least as $(1/\omega)$ for $\omega>b/2\pi$. Also, if $H(f)$ falls off fast enough, the finite limits in the first integral may be replaced with negligible error by infinite limits. The latter step is not necessary, but is convenient for some cases.

Also we find [see (A.1) and (A3.9)] that

$$D_3(\tau,\tau)=\int_{-\infty}^{\infty}\int_{-\infty}^{\infty}\int_{-\infty}^{\infty}H(f'')H(f'+f'')H(f-f'')\cdot e^{-i\omega\tau}\left[1+(\alpha_1^2+2\alpha_2)\frac{(\omega^2+\omega'^2+\omega\omega')}{(2\pi)^2B^2}\right]dfdf'df'', \tag{6.3}$$

** This is true in many practical networks. In a stagger-tuned amplifier of n stages, for example, the normalized frequency response is

$$|\,Y(i\omega)|=\left[\frac{1}{1+(\omega/\omega_b)^n}\right]^{\frac{1}{2}}$$

The power, proportional to $|\,Y(i\omega)|^2$ is $\frac{1}{2}$ at $\omega=\omega_b$, and is $(1+2^{2n})^{-1}\doteq 2^{-2n}$ at $\omega=4\omega_b$. Thus for a 5-stage amplifier the ratio of power at $\omega=\omega_b$ to $\omega=4\omega_b$ is $512-1$.

which produces the desired approximation for the characteristic function

$$F_V(\zeta_1\zeta_2;\tau)_N\cong\Bigg[F_V(\zeta_1,\zeta_2)_N|_\infty+\left(\frac{\alpha_1^2+2\alpha_2}{\beta^2}\right)\times\left\{\frac{q_1(0)\psi^2}{2}(\zeta_1^2+\zeta_2^2)+\psi^2\zeta_1\zeta_2q_1(\tau)+\frac{i\psi^3}{6}(\zeta_1^3+\zeta_2^3)q_1(0)\right.$$
$$+i\psi^3\left(q_1(\tau)+S'(\tau)-\frac{q_1(0)}{2}\right)\zeta_1^2\zeta_2$$
$$\left.\left.+i\psi^3\left(q_1(\tau)+S'(-\tau)-\frac{q_1(0)}{2}\right)\zeta_1\zeta_2^2\right\}\right]^{-1}, \tag{6.4}$$

where

$$q_1(\tau)=\frac{-\partial^2[\rho^2(\tau)]}{(2\pi)^2\partial\tau^2}, \tag{6.5}$$

$$S'(\tau)=\left(\frac{1}{2\pi}^2\right)\int\int_{-\infty}^{\infty}\int\omega\omega'H(f'')H(f'+f'')\times H(f-f'')e^{-i\omega\tau}dfdf'df'', \tag{6.6}$$

$$\rho(\tau)=\int_{-\infty}^{\infty}H(f)e^{-i\omega\tau}df; \tag{6.7}$$

and $F_V(\zeta_1,\zeta_2;\tau)_N|_\infty$ is the characteristic function for infinite video when noise alone enters the quadratic detector. We get finally

$$F_V(\zeta_1\zeta_2;\tau)_N\cong F_V(\zeta_1\zeta_2;\tau)_N|_\infty\times\left[1-\frac{(2\alpha_2+\alpha_1^2)\psi^2}{B^2}F_V(\zeta_1\zeta_2;\tau)_N\Big|_\infty\theta(\zeta_1,\zeta_2)+\cdots\right], \tag{6.8}$$

where specifically

$$\theta(\zeta_1,\zeta_2)=\left[\frac{q_1(0)}{2}(\zeta_1^2+\zeta_2^2)+\zeta_1\zeta_2q_1(\tau)+\frac{i\psi}{6}(\zeta_1^3+\zeta_2^3)q_1(0)\right.$$
$$+i\psi\left[q_1(\tau)+S'(\tau)-\frac{q_1(0)}{2}\right]\zeta_1^2\zeta_2$$
$$\left.+i\psi\left[q_1(\tau)+S'(-\tau)-\frac{q_1(0)}{2}\right]\zeta_1\zeta_2^2\right]. \tag{6.9}$$

Thus

$$W_2(V_1,V_2;\tau)_N=W_2(V_1V_2;\tau)_{N,\infty}-\left(\frac{2\alpha_2+\alpha_1^2}{B^2}\right)\psi^2\times\theta\left(\frac{i\partial}{\partial V_1},\frac{i\partial}{\partial V_2}\right)W_2^{(1)}(V_1,V_2;\tau)_{N,\infty}, \tag{6.10}$$

where $W_2(V_1,V_2;\tau)_{N,\infty}$ is the distribution for the infinitely wide video, see (A2.16), and $W_2^{(1)}$ is obtained from the convolution

$$W_2^{(1)}(V_1,V_2;\tau)_{N,\infty}=\int_{-\infty}^{\infty}\int W_2(z,y;\tau)_{N,\infty}$$

$$\times W_2(V_1-z,V_2-y;\tau)_{N,\infty}dzdy. \quad (6.11)$$

(b) Wide-Band Noise

The results for wide-band noise follow easily from the work above. We find

$$F_V(\zeta_1\zeta_2;\tau)_N\cong F_V(\zeta_1\zeta_2;\tau)^{\frac{1}{2}}{}_{N,\infty}\left[1-\frac{1}{2}\left(\frac{2\alpha_2+{}_1\alpha^2}{B^2}\right)\right.$$

$$\left.\times\psi^2F_V(\zeta_1,\zeta_2;\tau)_{N,\infty}\theta(\zeta_1,\zeta_2)+\cdots\right]. \quad (6.12)$$

Here $\theta(\zeta_1,\zeta_2)$ is given by (6.9) and the Fourier transform of $F_V(\zeta_1,\zeta_2;\tau)_{N,\infty}{}^{\frac{1}{2}}$ is the second-order probability density $W_2'(V_1,V_2,\tau)_N|_\infty$ obtained by the square-law rectification of *wide-band noise*. $W_2'(V_1,V_2;\tau)_N|_\infty$ can be obtained directly from the normal distribution, that is, from (A2.18). The result then is

$$W_2'(V_1,V_2;\tau)_N=\left[W_2'(W_1,V_2;\tau)_{N,\infty}-\frac{\psi^2}{2}\left(\frac{2\alpha_2+\alpha_1^2}{B^2}\right)\right.$$

$$\left.\times\theta\left(i\frac{\partial}{\partial x_1},i\frac{\partial}{\partial x_2}\right)W_2^{(1)}(x_1x_2;\tau)_N\right], \quad (6.13)$$

where $W_2^{(1)}(V_1,V_2,\tau)_N$ is given here by

$$\int_{-\infty}^{\infty}\int_{-\infty}^{\infty}W_2(z,y;\tau)_{N,\infty}W_2'(x_1-z,\ x_2-y;\tau)_{N,\infty}dzdy. \quad (6.14)$$

(c) First-Order Approximation (Narrow-Band Noise)

As an example, we observe that it is possible to get approximations to $W_1(V)_N$ for narrow-band noise by solving the integral equation for the first-order case, *viz.*,

$$\int_{-\infty}^{\infty}Y(i\omega_1-i\omega_2)H(f_1)g^{(k)}(f_2)df_2=\lambda_kg^{(k)}(f). \quad (A3.11)$$

The present method differs from that used by Kac and Siegert in that it yields any number of eigenvalues, depending on how closely one approximates the video filter by a Taylor series. Certain limitations on the types of filter for which approximations to $W_1(V)$ can be found by the method of Kac and Siegert are removed.

Using the Taylor series expansion (6.1) and substituting into the integral equation above, we obtain

$$H(f_1)\int_{-b}^{b}\left[1+i\alpha\frac{(\omega_1-\omega_2)}{2\pi B}+\frac{\alpha_2(\omega_1-\omega_2)^2}{(2\pi B)^2}\right]g^{(k)}(f_2)df_2$$

$$\text{I } H(f)\int_{b}^{\infty}Y(i\omega_1-i\omega_2)g^{(k)}(f_2)df_2$$

$$+H(f_1)\int_{-\infty}^{b}Y(i\omega-i\omega_2)g^{(k)}(f_2)df_2=\lambda_kg^{(k)}(f). \quad (6.15)$$

Expanding and neglecting the last two integrals, we find that††

$$H(f_1)\left[\int_{-\infty}^{\infty}g^{(k)}(f_2)\left[1+i\frac{\alpha_1\omega_1}{2\pi B}+\frac{\alpha_2\omega_1^2}{(2\pi B)^2}\right]df_2\right.$$

$$+\int_{-\infty}^{\infty}\omega_2g^{(k)}(f_2)\left[-\frac{i\alpha_1}{2\pi B}-\frac{2\alpha_2\omega_1}{(2\pi B)^2}\right]df_2$$

$$\left.+\int_{-\infty}^{\infty}\omega_2^2g^{(k)}(f_2)\frac{\alpha_2}{(2\pi B)^2}df_2\right]=\lambda_kg^{(k)}(f). \quad (6.16)$$

Solving the integral equation, we obtain finally these important eigenvalues:

$$\lambda_1\doteq1+\left(\frac{2\alpha_2+\alpha_1^2}{B^2}\right)\int_{-\infty}^{\infty}f^2H(f)df, \quad (6.17a)$$

which is the same result for λ_1 as obtained by Kac and Siegert.‡‡ We have also

$$\lambda_2\doteq\left(\frac{2\alpha_2+\alpha_1^2}{B^2}\right)\frac{\rho''(0)}{(2\pi)^2} \quad (6.17b)$$

and

$$\lambda_3=\frac{2\alpha_2^3}{(2\alpha_2+\alpha_1^2)B^4}\left[\int_{-\infty}^{\infty}f^2H(f)df-\int_{-\infty}^{\infty}f^4H(f)df\right]. \quad (6.17c)$$

Using Eqs. (4.46), (4.47) of reference 5 and the foregoing, we may write approximately

$$W_1(V)_N=\left[\frac{1-2\alpha}{\psi}\right]\exp[-V(1-\alpha)/\psi]$$

$$-\frac{1}{\psi}\exp\frac{V}{\alpha\psi}-\frac{\gamma}{\alpha\psi}\exp(-V/\gamma\psi),\quad V<0, \quad (6.18)$$

where $1-\lambda_1=\lambda_2=-\alpha$, $\lambda_3=\gamma$; $(\alpha>0,\ \gamma<0)$ and all terms of $0(B/b)^2$ are included. In this case $V>|\alpha|\psi$, as $\alpha\to0$, and we obtain on noting that $|\gamma|\ll(\alpha)$ the further

†† The infinite limits apply in the same sense as the cases considered in Eqs. (6.2) and (6.3).

‡‡ Note that for certain functions $H(f)$ we cannot integrate between infinite limits, but must use $-b$ and $+b$.

approximation

$$W(V)_N \doteq \frac{e^{-V/\psi}}{\psi}\left[1-2\alpha+\alpha\frac{V}{\psi}\right], \quad (V>0). \quad (6.19)$$

(d) A First-Order Approximation. Narrow-Band Signal and Noise

A further example in the case of wide video filters is provided when the input is the sum of signal and noise; the main features of the approximation are outlined below.

We start with (4.16) and the integral equation (4.17). Our approximation is limited to the calculation of a single eigenvalue (λ_1), and a single eigenfunction $f^{(1)}(t)$. Calling

$$\left(\int_{-\infty}^{\infty} M(t+t')A(t)f^{(1)}(t)dt\right)^2 = k^2/\zeta_1,$$

and writing $\lambda_1=1+\alpha$, *viz.* (6.18), we see that Eq. (4.16) becomes

$$F_V(\zeta)_{S+N} \doteq (1-i\psi\zeta_1-\alpha i\zeta_1\psi)^{-1}$$
$$\times\exp[i\zeta_1\psi k^2/(1-i\zeta_1\psi-\alpha i\zeta_1\psi)]. \quad (6.20)$$

Making the approximation for small α

$$(1-i\psi\zeta_1-\alpha i\zeta_1\psi)^{-1}=(1-i\zeta\psi)^{-1}+\alpha i\zeta_1\psi/(1-i\zeta_1\psi)^2$$

and expanding the corresponding term (containing α) in the exponent of (6.20), we obtain finally

$$F_V(\zeta_1)_{S+N}=\frac{1}{(1-i\zeta_1\psi)}\exp\left(\frac{i\zeta_1\psi k^2}{1-i\zeta_1\psi}\right)$$
$$\times[1+i\alpha\zeta_1\psi F_V(\zeta)_{S+N,\infty}-\alpha\psi^2\zeta_1^2k^2F_V^2(\zeta)_{S+N,\infty}], \quad (6.21)$$

where we have retained terms through $0(\alpha)$ only. The Fourier transform of (6.21) is easily found, see (A2.1), (A2.2), and we can write finally

$$W_1(V)_{S+N} \backsimeq W_1(V)_{S+N}|_\infty-\alpha\psi\frac{\partial}{\partial V_1}W_1^{(1)}(V)_{S+N}$$
$$+k^2\alpha\psi\frac{\partial^2}{\partial V_1^2}W^{(2)}(V)_{S+N} \quad (6.22)$$

to this order of approximation; $W_1(V)_{S+N}|_\infty$ is given by (A2.2), and

$$W_1^{(1)}(V)_{S+N}=\int_{-\infty}^{\infty} e^{-x/\psi}W_1(V-x)_{S+N}|_\infty dx/\psi,$$

$$W_1^{(2)}(V)_{S+N}=\int_{-\infty}^{\infty} e^{-x/\psi}W_1^{(1)}(V-x)_{S+N}dx/\psi.$$

7. A SPECIAL VIDEO FILTER

If the IF noise and signal wave are confined to a band $2f_b$, this implies that the slowly varying part has a maximum frequency spread f_b. A square-law detector doubles this spread to $2f_b$, so that the maximum band that the detected wave occupies is $2f_b$. Thus, if the video filter is flat over this band, one would expect the same result as for an infinitely wide filter. The expected result can be shown as follows:

We choose

$$Y(i\omega)=e^{-i\alpha\omega}, \quad 0<|f|<2f_b,$$

$$H(f)=0\begin{cases} f<-f_b \\ f>f_b \end{cases}. \quad (7.1)$$

The condition $H(f)=0$ at frequencies greater than f_b from the center of the IF pass band is often practically met, as for example in a stagger-tuned amplifier with many stages.

Using the frequency form of the integral equation (A3.11), choosing

$$g(f)=[k_1\zeta_1H(f)+k_2\zeta_2e^{-i\omega t}H(f)]e^{i\alpha\omega} \quad (7.2)$$

and substituting this into the integral equation, we obtain finally

$$[\zeta_1\{k_1\zeta_1+k_2\zeta_2\rho(\tau)\}+\zeta_2e^{-i\alpha\tau}\{k_1\zeta_1\rho(\tau)+k_2\zeta_2\}]$$
$$=\lambda(k_1\zeta_1+k_2\zeta_2e^{-i\omega\tau}). \quad (7.3)$$

This relation can be satisfied if we set

$$(\zeta_1-\lambda)k_1+\zeta_2k_2\rho(\tau)=0, \quad (7.4a)$$

$$\zeta_1k_1\rho(\tau)+k_2(\zeta_2-\lambda)=0, \quad (7.4b)$$

which leads to the same values of λ as found in Appendix II (A2.9), for the case of an infinitely wide video.

8. CONCLUDING REMARKS

The preceding analysis has shown that the second- and higher-order distribution functions for noise or signal and noise which have been rectified by a quadratic detector and then filtered can be represented only in closed form in terms of the characteristic functions of these distributions. Even so, such representations imply the solution of a difficult integral equation.

In many instances of practical importance, we are interested in limiting cases, i.e., either a very wide or a very narrow video. For wide video, by expanding the characteristic function we can obtain approximations to the second-order distribution function, with correction terms obtained from the known distribution function for infinitely wide video response. As seen in Sec. 7, a wide video means a video band pass that is of the order of the band width at the IF frequency which contains most of the noise power.

Again, for cases where another nonlinear operation follows video (or audio) filtering, the characteristic function is of value and the approximations developed here can be used to advantage. As the video filter becomes more narrow, terms in the denominator of

$F_V(\zeta_1,\zeta_2;\tau)$ representing powers of (ζ_1,ζ_2) greater than unity, assume more importance. It appears that powers of ζ up to the third yield results which imply a first-order correction to the distribution for the infinitely wide post detection filter.

As may be seen from the approximations developed for $W_1(x)$ in Sec. 6, more terms in the Taylor series approximation to the video filter function must be considered as the video becomes narrower, and more eigenvalues of significance are obtained.

In the extreme of a narrow-video filter one would expect from physical considerations that the distribution becomes normal. The results of the analytical work show to what extent the actual distribution departs from the normal for videos narrow compared to the IF band. Correction terms have coefficients whose magnitudes are of the order of powers of the ratio

$$\left(\frac{\text{video width}=b}{IF\ \text{width}=2B}\right).$$

APPENDIX I. MATRIX EQUATIONS

In Eq. (2.8) the exponent

$$\sum_{k,l=1}^{n}\left(\frac{\alpha^{kl}}{|\alpha|\psi}-i\zeta_k\delta_{kl}\right)x_kx_l$$

is represented in matrix form by $\mathbf{x}'\mathbf{B}\mathbf{x}$, where $\mathbf{x}$ is a column vector, $\mathbf{x}'$ the transpose, and $\mathbf{B}$ is a symmetrical matrix with components $B_{kl}=(\alpha^{kl}/\psi|\alpha|)-i\zeta_k\delta_{kl}$; $|\alpha|$ is the determinant of $\boldsymbol{\alpha}$ and α^{kl} the cofactor of α_{kl} in $\boldsymbol{\alpha}$. An orthogonal matrix can be found such that $\mathbf{P}'\mathbf{B}\mathbf{P}=\boldsymbol{\Lambda}\mathbf{I}$, where $\boldsymbol{\Lambda}$ is diagonal $(\mathbf{I}=\delta_{kl})$. Letting $\mathbf{Z}=\mathbf{P}^{-1}\mathbf{x}$, $\mathbf{x}'\mathbf{B}\mathbf{x}$ becomes $\sum_{j=1}^{n}\lambda_jZ_j^2$, where the λ_j are the components of $\boldsymbol{\Lambda}=[\lambda_j\delta_{jk}]$.

By a simple quadrature one obtains

$$\phi_N(\zeta_1,\cdots\zeta_n)=\psi^{-n}|\alpha|^{-1}\prod_{j=1}^{n}\lambda_j^{-1}=\psi^{-n}|\mathbf{B}\boldsymbol{\alpha}|^{-1}, \quad \text{(A1.1)}$$

since $\prod_{j=1}^{n}\lambda_j=|\boldsymbol{\Lambda}|=|\mathbf{B}|$. This gives (2.9) with a little further manipulation, since

$$\mathbf{B}=\boldsymbol{\alpha}^{-1}\psi^{-1}-i\boldsymbol{\zeta};\quad \boldsymbol{\alpha}^{-1}=[\alpha^{kl}/|\alpha|]=\boldsymbol{\varrho}^{-1}. \quad \text{(A1.2)}$$

Now, by finding a symmetrical matrix $\mathbf{Q}$, such that $\mathbf{Q}(\mathbf{A}-i\mathbf{I})\mathbf{Q}^{-1}=\boldsymbol{\Lambda}'\mathbf{I}$, where

$$[\mathbf{A}]_{k,j}=\zeta_k\psi\rho_{k,j}+i\delta_{k,j}, \quad \text{(A1.3)}$$

inasmuch as $\phi_N(\zeta_1,\cdots\zeta_n)=i^n|\mathbf{A}|^{-1}=|\mathbf{I}-i\boldsymbol{\Lambda}'|^{-1}$, we obtain the result (2.10a), *viz.*,

$$\phi_N(\zeta_1,\cdots\zeta_n)=\prod_{j=1}^{n}(1-\lambda_j')^{-1}. \quad \text{(A1.4)}$$

The equations for the rows of $\mathbf{Q}$ become the equations for the eigenvectors $\mathbf{f}_{(j)}(\mathbf{A}-i\mathbf{I})=\boldsymbol{\lambda}'\mathbf{f}_{(j)}$, or

$$\psi\sum_{k}^{n}\zeta_k\rho_{km}f_{jk}=\lambda_j'f_{jm}, \quad \text{(A1.5)}$$

which is Eq. (2.10b).

APPENDIX II. DISTRIBUTIONS FOR THE INFINITELY WIDE VIDEO FILTER

When the video filter is infinitely wide we expect the general results of Sec. 4 to reduce to the more familiar distributions of the amplitude following square-law rectification alone. The latter may, of course, be obtained directly from (2.10)–(2.12) for noise alone, or from (3.5a), (3.7) for a signal and noise, or, as was originally done in first-order cases, by a direct transformation of variables $V_0=\beta^2V_{\text{in}}^2$ according to the detector's law, applied to the original normal distribution for the noise (see Sec. 2). Accordingly, we now verify that our general results for arbitrary filtering, in the case of the first- and second-order densities, reduce to the expected forms when the video filter is allowed to become indefinitely wide.§§ The first-order densities for signal and noise follow at once.

(I) First-Order Densities

For an infinitely wide video, the filter's weighting function is $A(t)=\delta(t-0)$, and the solution to (4.17) is at once $f(s)=C\rho(s)$ and $\lambda_1'=1$. The normalization condition gives $C=\zeta_1^{-1/2}$, whereupon (4.16) becomes

(*a*) *Narrow-Band Waves.—*

$$F_V(\zeta_1)_{S+N}|_\infty=\phi_{S+N}(\zeta_1)=(1-i\zeta_1\psi)^{-1}\times\exp[i\zeta_1M(t')^2/(1-\psi\zeta_1)]. \quad \text{(A2.1)}$$

The corresponding probability density is found with the help of Pair No. 650.0 [G. A. Campbell and R. M. Foster, *Fourier Integrals for Practical Applications* (D. Van Nostrand Company, Inc., New York, 1947)] to be

$$W_1(V)_{S+N}|_\infty=\psi^{-1}\exp(-(V+M^2)/\psi)I_0(2MV^{\frac{1}{2}}/\psi),\quad V>0;\ =0;\ V<0;\ (\psi=2\langle V_N^2\rangle) \quad \text{(A2.2)}$$

which is a well-known result.[5,14] For noise alone one has

$$W_1(V)|_\infty=\psi^{-1}e^{-V/\psi},\quad V>0;\ =0;\ V<0, \quad \text{(A2.3)}$$

which is a χ^2 density with 2 degrees of freedom, as expected. [From (2.6a) we recall that since $\beta^2=2$ here, $\psi=2\langle V_N^2\rangle$, where $\langle V_N^2\rangle$ is the mean-square noise output of the IF filter.]

(*b*) *Broad-band Waves.—*

Here (A2.1) is appropriately modified, to give us

$$W_1(V)_{S+N}|_\infty=\int_{-\infty}^{\infty}\exp[(-i\zeta_1V+i\zeta_1s^2)/(1-i2\psi\zeta_1)]\times\left(\frac{d\zeta_1}{(1-\zeta_1^2\psi)^{\frac{1}{2}}2\pi}\right)$$

$$=(2\pi V\psi)^{-\frac{1}{2}}\exp[-(s^2+V)/2\psi]\cosh[V^{\frac{1}{2}}s/\psi], \quad \text{(A2.4)}$$
$$V>0;\ =0;\ V<0,$$

where we have again used transform pair No. 650.0. This result has also been obtained by Emerson[15] in a recent article. For noise alone we get a χ^2 distribution with 1 degree of freedom:

$$W_1(V)_N|_\infty=(2\pi V\psi)^{-\frac{1}{2}}e^{-V/2\psi},\quad V>0;\ =0;\ V<0,\ (\psi=\langle V_N^2\rangle). \quad \text{(A2.5)}$$

(II) Second-Order Densities

Here the calculations are more involved. From (4.18) we have for general post-detection filtering

$$F_V(\zeta_1,\zeta_2;\tau)_{S+N}=\prod_{k=1}^{\infty}(1-i\lambda_k)^{-\alpha}\exp\left\{\left(i\int_{-\infty}^{\infty}M(t'+t)\times[\zeta_1A(t)+\zeta_2A(t-\tau)]f^k(t)dt\right)^2\Big/(1-i\lambda_k)\right\}, \quad \text{(4.18)}$$

where $\alpha=1$ for narrow-band waves, and $\alpha=\frac{1}{2}$ for broad-band disturbances, with M replaced by S. The associated integral

§§ The carrier or central-frequency zone in the narrow-band case is, however, not passed by this "infinitely" wide video.

[14] S. A. Goudsmit, RL 43-21; J. C. Slater, RL V-23; D. O. North, RCA Tech. Rept. PTR6C; K. A. Norton and V. D. Landon, Proc. Inst. Radio Engrs. 30, 425 (1942). (RL = Radiation Laboratory, MIT.)

[15] R. C. Emerson, J. Appl. Phys. 24, 1168 (1953). (Here $V=\psi Y$, $S^2=\psi X$, and our $\psi=\langle V_N^2\rangle$, since here $\beta^2=1$.) In general, for arbitrary β we must replace ψ by $\beta^2\psi/2=\beta^2\langle V_N^2\rangle/2$ for narrow-band waves, and $\beta^2\psi=\beta^2\langle V_N^2\rangle$ for broad-band waves, in all results of the present paper.

equation is

$$\psi\int_{-\infty}^{\infty}[\zeta_1 A(t)+\zeta_2 A(t-\tau)]\rho(t,s)f^{(k)}(t)dt=\lambda_k f^{(k)}(s) \quad \text{(A2.6)}$$

and since $A(t)=\delta(t-0)$; $\rho(0,s)=\rho(-s)=\rho(s)$; $\rho(\tau,-s)=\rho(\tau-s)$, etc., Eq. (A2.6) becomes

$$\zeta_1\rho(s)f^{(k)}(0)+\zeta_2\rho(\tau-s)f^{(k)}(\tau)=\lambda_k f^{(k)}(s)\psi^{-1}, \quad \text{(A2.7a)}$$

subject to the normalization conditions

$$\zeta_1 f^{(k)}(0)^2+\zeta_2 f^{(k)}(\tau^2)=1. \quad \text{(A2.7b)}$$

In general, the integral equation (A2.6) is here satisfied by

$$f^{(k)}(s)=C_1^{(k)}(\zeta_1,\zeta_2)\zeta_1\rho(s)+C_2^{(k)}(\zeta_1,\zeta_2)\zeta_2\rho(s-\tau). \quad \text{(A2.8)}$$

Substituting (A2.8) into (A2.7a) (or (A2.6)), rearranging terms, and setting the coefficients of $\rho(s)$ and $\rho(s-\tau)$ equal to zero, since the equations are then satisfied by arbitrary $\rho(s)$, arbitrary s, τ, we obtain

$$C_1^{(k)}\left(\frac{\lambda_k}{\psi}-\zeta_1\right)-C_2^{(k)}\zeta_2\rho(\tau)=0;$$

$$-C_1^{(k)}\zeta_1\rho(\tau)+C_2^{(k)}\left(\frac{\lambda_k}{\psi}-\zeta_2\right)=0. \quad \text{(A2.9)}$$

For nontrivial values of C_1, C_2 we determine from this the eigenvalues λ_k, by setting the determinant of the coefficients C_1, C_2 equal to zero, obtaining two values

$$\lambda_k=\frac{\psi}{2}\{\zeta_1+\zeta_2\pm[\zeta_1^2+\zeta_2^2-4\zeta_1\zeta_2(1-\rho^2)]^{\frac{1}{2}}\}; \quad k=1,2 \quad \text{(A2.10)}$$

and hence *four* coefficients $C_1^{(k)}$, $C_2^{(k)}$. From the two relations (A2.8) in (A2, 7a), and the two equations (A2.7b) for normalization we get the necessary number of expressions to determine the coefficients, and hence the eigenfunctions. Unlike the case of noise alone, the eigenfunctions, as well as the eigenvalues λ_k ($k=1, 2$) are needed when there is a signal, as can be seen from (4.18) above. The desired probability density $W_2(V_1,V_2,\tau)_{S+N}|_\infty$ then follows by inversion, in the usual way.

Since $C_1^{(k)}$, $C_2^{(k)}$ are, however, complicated functions of ζ_1, and ζ_2, the actual inversion is not easy. The expected result may be obtained, none the less, by a direct transformation of W_2, where $V_{\text{out}}=\beta^2 V_{\text{in}}^2=R_{\text{in}}^2$. For narrow-band cases, using a result of Middleton,[16] we have at once

(*narrow-band*):

$$W_2(V_1,V_2;\tau)_{S+N}|_\infty=\{[\psi^2(1-\rho^2)]^{-1}\exp[-(V_1+V_2)/\psi(1-\rho^2)]$$
$$\times\exp[-(M_1^2+M_2^2-2M_1M_2\rho)/\psi(1-\rho^2)]\}$$
$$\cdot\sum_{M=0}^{\infty}\epsilon_m I_m\left(\frac{2\rho(V_1V_2)^{\frac{1}{2}}}{\psi(1-\rho^2)}\right)I_m\left(\frac{M_1-M_2\rho}{\psi(1-\rho^2)}2V_1^{\frac{1}{2}}\right)I_m\left(\frac{M_2-M_1\rho}{\psi(1-\rho^2)}2V_2^{\frac{1}{2}}\right),$$
$$V_1,\ V_2>0;\ V_1,\ V_2<0 \quad \text{(A2.11)}$$

and where $M_1=M(t_1)$; $M_2=M(t_2)$.

In the broad-band situation a direct transformation of (2.9), using (2.10), of reference (16) and the fact that $V_{\text{out}}=\beta^2 V_{\text{in}}^2$ gives us immediately ($\beta^2=1$):

(*broad-band*):

$$W_2(V_1,V_2;\tau)_{S+N}|_\infty=(\{2\pi\psi[V_1V_2(1-\rho^2)]^{\frac{1}{2}}\}^{-1}$$
$$\cdot\exp\{-[(V_1+V_2-2\rho(V_1V_2)^{\frac{1}{2}}]/2\psi(1-\rho^2)\})$$
$$\cdot\exp\{-[S_1^2+S_2^2-2\rho S_1S_2-2(V_1)^{\frac{1}{2}}(S_1-\rho S_2)$$
$$-2(V_2)^{\frac{1}{2}}(S_2-\rho S_1)]/2\psi(1-\rho^2)\},$$
$$V_1;\ V_2>0;\ =0,\ V_1,\ V_2<0. \quad \text{(A2.12)}$$

Where there is no signal, we do not need the eigenfunctions, only the eigenvalues (A2.10). The characteristic function (4.18) reduces now in the narrow-band case to

$$F_V(\zeta_1,\zeta_2;\tau)_N|_\infty=\phi(\zeta_1,\zeta_2)_N=[1-\psi^2\lambda_1\lambda_2-i\psi(\lambda_1+\lambda_2)]^{-1}$$
$$=[1-i\psi(\zeta_1+\zeta_2)-\psi^2\zeta_1\zeta_2(1-\rho^2(\tau))]^{-1}. \quad \text{(A2.13)}$$

[16] D. Middleton, Quart. Appl. Math. 5, 445 (1948), Eq. (5.17).

The second-order density is obtained from the Fourier transform of (A2.13); letting $1-\rho^2=k$, we can write this transform as

$$W_2(V_1,V_2;\tau)_N|_\infty=\psi^{-1}(2\pi)^{-2}\int_{-\infty}^{\infty}\frac{d\zeta_2 e^{-i\zeta_2V_2}}{(1-i\psi k\zeta_2)}\int_{-\infty}^{\infty}d\zeta_1 e^{-i\zeta_1V_1}$$
$$\times\left[\zeta_1+\frac{i}{\psi}\left(\frac{1-i\psi\zeta_2}{1-i\psi\zeta_2 k}\right)\right]^{-1}. \quad \text{(A2.14)}$$

The second integral in (A2.14) has a pole in the lower plane for all real values of ζ_2. For $V_1<0$ the integral is zero, and since (A2.14) is symmetrical in ζ_1V_1, ζ_2V_2, it vanishes also for $V_2<0$. For $V_1>0$, we get for the integral in ζ_1 the value

$$-2\pi i\exp[-V_1(1-i\psi\zeta_2)/\psi(1-i\psi k\zeta_2)].$$

Letting $\mu=k\psi\zeta_2+i=[(V_1/V_2)(1+k)]^{\frac{1}{2}}\eta$, we have

$$W_2=\frac{ie^{-(V_1+V_2)/k\psi}}{2\pi k\psi^2}\int_{-\infty+i}^{\infty+i}[\exp-i(V_1V_2(1+k)/k^2\psi^2)^{\frac{1}{2}}]$$
$$\times[\eta-\eta^{-1}]d\eta/\eta. \quad \text{(A2.15)}$$

This integral is recognized as a modified Bessel function of zeroth order, giving us finally the result

$$W_2(V_1,V_2;\tau)_N|_\infty=[\psi(1-\rho^2)]^{-1}\exp[-(V_1+V_2)/\psi(1-\rho^2)]$$
$$\times I_0\left(\frac{2\rho(V_1V_2)^{\frac{1}{2}}}{\psi(1-\rho^2)}\right),\ V_1,\ V_2>0;\ =0,\ V_1,V_2<0 \quad \text{(A2.16)}$$

for the narrow-band case, which is immediately verified from (A2.11) on setting $M_1=M_2=0$.

The broad-band case may be obtained in a similar way, where now the characteristic function is

$$F_V(\zeta_1,\zeta_2;\tau)_N|_\infty=[1-i\psi(\zeta_1+\zeta_2)-\psi^2\zeta_1\zeta_2(1-\rho^2)]^{-\frac{1}{2}}. \quad \text{(A2.17)}$$

We have finally

(*broad-band*):

$$W_2(V_1,V_2;\tau)_N|_\infty=\{2\pi\psi[V_1V_2(1-\rho^2)]^{\frac{1}{2}}\}^{-1}$$
$$\times\exp\{-[V_1+V_2-2\rho(V_1V_2)^{\frac{1}{2}}]/2\psi(1-\rho^2)\},$$
$$V_1,\ V_2>0;\ =0,\ V_1,V_2<0, \quad \text{(A2.18)}$$

as expected, see (A2.12) when $S_1=S_2=0$.

APPENDIX III. EXPANSION OF THE CHARACTERISTIC FUNCTION $F_v(\zeta)$. TIME AND FREQUENCY REPRESENTATIONS

As we have seen, a form of the characteristic function which is useful in the important limiting cases of wide or narrow video filters is given by the expansion of the reciprocal of the Fredholm determinant associated with the integral equation (4.10b) in the general case, and with (4.13b) for $n=2$. For the latter the characteristic function is Δ^{-1}, see (4.15).

To represent the expansion of $F_V(\zeta_1,\zeta_2;\tau)_N$ compactly in both the time and frequency domains, the following Fourier transforms are needed, relating system-, weighting-, correlation functions, and intensity spectra associated with the IF and video filters. We have for the IF filter

$$\rho(\tau)=\int_{-\infty}^{\infty}H(f)e^{i\omega\tau}df;\quad \rho(0)=1 \quad \text{(A3.1)}$$

$$H(f)=\int_{-\infty}^{\infty}\rho(\tau)e^{-i\omega\tau}d\tau=|B(f\pm f_0)|^2, \quad \text{(A3.2)}$$

$$\int_{-\infty}^{\infty}H(f)df=1;\quad \int_{-\infty}^{\infty}H^2(f)df=\int_{-\infty}^{\infty}\rho^2(\tau)d\tau\equiv H_1. \quad \text{(A3.3)}$$

For the video filter we write

$$Y(i\omega)=\int_{-\infty}^{\infty}A(t)e^{-i\omega t}dt;\quad A(t)=\int_{-\infty}^{\infty}Y(i\omega)e^{i\omega t}df;$$
$$t<0,\ =0 \quad \text{(A3.4)}$$

$$|Y(i\omega)|^2=\int_{-\infty}^{\infty}\Phi_1(t)e^{-i\omega t}dt;\quad \Phi_1(t)=\int_{-\infty}^{\infty}|Y(i\omega)|^2e^{i\omega t}df \quad \text{(A3.5)}$$

$$=\int_{-\infty}^{\infty}A(t')A(t'\pm t)dt'=\Phi_1(\pm t), \quad \text{(A3.5a)}$$

and

$$\Phi_2(\tau_1,\tau_2)\equiv\int_{-\infty}^{\infty}A(t')A(t'+\tau_1)A(t'+\tau_2)dt'$$

$$=\iint_{-\infty}^{\infty} Y(i\omega)Y(i\omega')Y(i\omega+i\omega')^* \exp[i\omega'(\tau_1-\tau_2)-i\omega\tau_2]dfdf'. \tag{A3.6}$$

When the determinantal expansion is carried out through terms of order 3 in (ζ_1,ζ_2), the result is finally

$$F_V(\zeta_1,\zeta_2;\tau)_N=\Big\{1-i\psi(\zeta_1+\zeta_2)$$

$$-\frac{\psi^2}{2!}[(\zeta_1^2+\zeta_2^2)(1-D_2(0))+2\zeta_1\zeta_2(1-D_2(-\tau))]$$

$$+\frac{i\psi^3}{3!}[(\zeta_1+\zeta_2)^3+3(\zeta_1+\zeta_2)\{(\zeta_1^2+\zeta_2^2)D_2(0)+2\zeta_1\zeta_2D_2(-\tau)\}$$

$$+2(\zeta_1^3+\zeta_2^3)D_3(0,0)+2\zeta_1^2\zeta_2\{2D_3(-\tau,0)+D_3(\tau,\tau)\}$$

$$+2\zeta_1\zeta_2^2\{2D_3(\tau,0)+D_3(-\tau,-\tau)\}]+\cdots\Big\}^{-1}, \tag{A3.7}$$

where for broad-band noise $\{\ \}^{-1}$ is replaced by $\{\ \}^{-\frac{1}{2}}$, as usual. The functions D_2, D_3 are specifically

$$D_2(\tau)=\int_{-\infty}^{\infty}\rho^2(t)\Phi_1(t+\tau)dt=\iint_{-\infty}^{\infty} H(f)H(f')$$

$$\times|Y(i\omega+i\omega')|^2\exp[i(\omega+\omega')\tau]dfdf', \tag{A3.8}$$

$$D_3(\tau,\tau)=\iint_{-\infty}^{\infty}\Phi_2(u+\tau,v+\tau)\rho(u)\rho(v)\rho(u-v)dudv$$

$$=\iiint_{-\infty}^{\infty} Y(i\omega)Y(i\omega')Y(i\omega+i\omega')^*H(f-f'')$$

$$\times H(f'+f'')H(f'')e^{-i\omega\tau}dfdf'df''. \tag{A3.9}$$

Further relations of some use are

$$2D_3(-\tau,0)+D_3(\tau,\tau)=3\iiint_{-\infty}^{\infty} Y(i\omega)Y(i\omega')Y(i\omega+i\omega')^*$$

$$\times H(f-f'')H(f'+f'')H(f'')e^{-i\omega\tau}dfdf'df'' \tag{A3.10}$$

and $2D_3(\tau,0)+D_3(-\tau,-\tau)$ is precisely (A3.10) with ω replaced by ω' in the exponent. The integral equation (4.13b) for $n=2$ is transformed in a similar way to

$$\psi H(f)\int_{-\infty}^{\infty}\{\zeta_1+\zeta_2\exp[i(\omega'-\omega)\tau]\}Y(i\omega-i\omega')g^{(k)}(f')df'$$

$$=\lambda_k g^{(k)}(f), \tag{A3.11}$$

where it is assumed that the $f^{(k)}(t)$ possess a transform

$$f^{(k)}(t)=\int_{-\infty}^{\infty}g^{(k)}(f)e^{i\omega t}df;\quad g^{(k)}(f)=\int_{-\infty}^{\infty}f^{(k)}(t)e^{-i\omega t}dt. \tag{A3.12}$$

APPENDIX IV. REMARK ON THE CALCULATION OF MOMENTS

The moments of the distribution for $W_2(V_1,V_2;\tau)$ can be obtained easily from the Taylor series expansion for $F_V(\zeta_1,\zeta_2;\tau)$, as is well known. That is, we can write, when these quantities exist,

$$\langle V_1^kV_2^l\rangle=A_{kl}(-i)^{k+l}/l!k!,$$

where the A_{kl} are the coefficients of the series expansion. One then uses the characteristic function expansion of $F_V(\zeta_1,\zeta_2;\tau)$ to obtain the corresponding coefficients in the Taylor's series.

One result of this calculation is

$$\langle V_1V_2\rangle=\psi^2\Big(1+\iint_{-\infty}^{\infty}H(f)H(f')|Y(f+f')|^2$$

$$\times\exp[-i(\omega+\omega')\tau]dfdf'\Big), \tag{A4.1}$$

the correlation function of the output wave.

23

Reprinted from *IRE Trans. Information Theory,* **IT-3**(1), 32–37 (1957)

A Systematic Approach to a Class of Problems in the Theory of Noise and Other Random Phenomena —Part I*

D. A. DARLING† AND A. J. F. SIEGERT‡

Summary—**The problem of finding the probability of distribution of the functional**

$$\int_{t_0}^{t} \Phi(X(\tau), \tau)\, d\tau,$$

where $X(\tau)$ is a (multidimensional) Markoff process and $\Phi(X, \tau)$ is a given function, appears in many forms in the theory of noise and other random phenomena. We have shown that a certain function from which this probability distribution can be obtained is the unique solution of two integral equations. We also developed a perturbation formalism which relates the solutions of the integral equations belonging to two different functions $\Phi(X, \tau)$. If the transition probability density for $X(\tau)$ is the principal solution of two partial differential equations of the Fokker-Planck-Kolmogoroff type, the principal solution of two similar differential equations is the solution of the integral equations. As an example, we calculated the probability distribution of the sample probability density for a stationary Markoff process.

Introduction

In the theory of noise and similar random phenomena, a small number of special problems have been solved by various special methods. Each of these methods seems to apply only to the particular problem for which it was developed or at best to a rather restricted class of problems. It seemed of interest, therefore, to develop a systematic approach to a wider class of problems, which contains as special cases most of the problems solved before. Even though this approach leads to rather formidable integral or differential equations, so that the number of new problems which can be solved exactly will be small, it leads to a perturbation formalism for problems "in the neighborhood" of those permitting exact solutions.

We consider the problem of finding the probability distribution of the random variable

$$u = \int_{t_0}^{t} \Phi(X(\tau), \tau)\, d\tau \tag{1}$$

where $X(\tau)$ is a Markoff process with components $x_1(\tau)$, $x_2(\tau) \cdots x_n(\tau)$. This problem arose originally as the problem of finding the probability distribution of the noise output of a radio receiver consisting of a linear amplifier, an arbitrary detector, and a second linear amplifier. Let $x(\tau)$ be the output voltage of the first amplifier at a time $\tau \geq 0$ before observation,[1] $\varphi[x(\tau)]$ the output voltage of the detector at the same time, and $K(\tau)$ the output of the second amplifier at the time of observation if a δ-function pulse is applied to it at the time τ. The output voltage V of the second amplifier in response to $x(\tau)$ is then

$$V = \int_{0}^{t} K(\tau)\varphi(x(\tau))\, d\tau \tag{2}$$

if the noise was turned on at time $t \geq 0$ before observation. If the input of the first amplifier is white noise, $x(\tau)$ is a Gaussian random function. This fact has made it possible to reduce the problem for the special case $\varphi(x) \equiv x^2$ to the solution of an integral equation in one variable only. Except for this special form of the detector function, and of course in the trivial case $\varphi(x) \equiv x$, the Gaussian property of $X(\tau)$ does not simplify the problem.

If the first amplifier is equivalent to a network with lumped circuit elements and its input is white noise, $x(\tau)$ is also a component of a Markoff process. This led us to consider the more general problem stated above which has many applications apart from the noise output of radio receivers.

If, for instance, a domain $\mathfrak{D}$ is chosen in X space, and $\Phi(X, \tau)$ is defined by

$$\Phi(X, \tau) = \begin{cases} 1 & \text{when } X \text{ is in } \mathfrak{D} \\ 0 & \text{otherwise} \end{cases}$$

then u is that part of the time $(t - t_0)$ during which X is in $\mathfrak{D}$ in the time interval $(t - t_0)$, and $(t - t_0)^{-1}u$ is an estimate for the probability that X is in $\mathfrak{D}$, obtained from the finite sample. The distribution of u is thus of importance if it is desired to estimate the accuracy with which the probability distribution of time homogeneous processes $X(\tau)$ can be obtained from finite samples.

If, specially, $\mathfrak{D}$ is defined by

$$x_1 < a,$$

with $x_2, x_3 \cdots x_n$ unrestricted, the probability for $u = t - t_0$ is the probability that $x_1(t') < a$ for all t' in $t_0 < t' < t$ (except for a set of measure zero) and, for a continuous function $x_1(\tau)$, this is the cumulative distribution of the absolute maximum of $x_1(t)$ in the interval

* Manuscript received by the PGIT, August 13, 1956. Most of this work was done while the authors were consultants for The RAND Corp., Santa Monica, Calif.

† Dept. of Mathematics, University of Michigan, Ann Arbor, Mich.

‡ Dept. of Physics, Northwestern University, Evanston, Ill.

[1] It is convenient to choose the time scale positive into the past.

$(t . t)$, if it is considered as function of a, or the cumulative probability distribution of the one-sided first-passage time, if it is considered as function of t.

If $\mathfrak{D}$ is defined by

$$b < x_1 < a$$

with $x_2, x_3, \cdots x_n$ unrestricted, one obtains in a similar way for continuous $x_1(\tau)$ the distribution of the two-sided first-passage time (escape time), and the distribution of the range $r \equiv \max_\tau x(\tau) - \min_\tau x(\tau)$, $t_0 < \tau < t$. For one-dimensional Markoff processes the problem of first-passage time, range, and maximum has been solved by an older method.[2]

The problem of finding the distribution of the empirical spectrum or the Fourier coefficients obtained from a sample can also be formulated as a special case of our problem. If $\psi_1(\tau), \psi_2(\tau), \cdots \psi_l(\tau)$ are given functions, for instance trigonometric functions, the characteristic function

$$\left\langle \exp\left\{ i \sum_{k=1}^{l} \zeta_k \int_{t_0}^{t} x_1(\tau)\psi_k(\tau)\, dt \right\} \right\rangle_{\text{av}}$$

for the joint distribution of the Fourier coefficients is also the characteristic function

$$\left\langle \exp\left\{ i\zeta \int_{t_0}^{t} \Phi(x_1(\tau), \tau)\, d\tau \right\} \right\rangle_{\text{av}}$$

for the random variable u defined by (1), if one chooses

$$\Phi(X(\tau), \tau) \equiv \zeta^{-1} x_1(\tau) \sum_{k=1}^{l} \zeta_k \psi_k(\tau). \quad (3)$$

Integral Equations for the Conditional Characteristic Function, Perturbation Formula, and Differential Equation

In the present paper we present a heuristic derivation[3] of two integral equations for the function

$$r(X_0, t_0 \mid X, t; \lambda) \equiv \left\langle \exp\left\{ -\lambda \int_{t_0}^{t} \Phi(X(\tau), \tau)\, d\tau \right\} \middle| \; X(t_0) = X_0, X(t) = X \right\rangle_{\text{av}} \cdot p(X_0, t_0 \mid X, t) \quad (4)$$

where $\langle | \rangle_{av}$ denotes the average of the functional on the left of the vertical bar under the condition written on its right, and where $p(X_0, t_0) \mid X, t) dX$ is the probability that $X(t)$ is in the volume element dX at X, if $X(t_0) = X_0$. The parameter λ will be chosen positive real if Φ is non-negative, and negative imaginary otherwise. The Fourier or Laplace transform of the probability density for the variable u defined by (1) is obviously r/p if initial and end conditions are imposed and $\int r dX$ if only initial conditions are imposed, and $\int W(X_0, t_0) r dX dX_0$ if no conditions are imposed, where $W(X_0, t_0)$ is the probability density for $X(t_0)$.

Consider now $X(\tau)$ as the path of a particle in X space. If, at first, $\Phi(X(\tau), \tau)$ is assumed to be non-negative and λ real and positive, then $\lambda\Phi(X, \tau) d\tau$ can be interpreted as the probability of a "collision" at the point X in the time interval $(\tau, \tau + d\tau)$. A "collision" is thereby understood to be an event which does not affect the path of the particle nor the probability of later collisions, but leaves a mark on the particle so that the number of collisions experienced by the particle can be counted. The functional $\exp[-\lambda \int_{t_0}^{t} \Phi(X(\tau), \tau) d\tau]$ is thus the probability that the particle suffers no collisions on a path $X(\tau)$ which leads from X_0 to X; and $r(X_0, t_0 \mid X, t; \lambda)\, dX$ is the probability of finding the particle at time t in the volume element dX at X without any marks, if it started at X_0 at time t_0. An integral equation for $r(X_0 t_0 \mid X, t; \lambda)$ is obtained by subtracting from $p(X_0, t_0 \mid X, t)\, dX$ the probability that the particle reached some point X' in the volume element dX' at some time t' without collisions, suffered the first collision there in the time interval $(t', t' + dt')$, and went on from there to X suffering an arbitrary and irrelevant number of collisions. One thus has the integral equation

$$r(X_0, t_0 \mid X, t; \lambda) = p(X_0, t_0 \mid X, t) - \lambda \int_{t_0}^{t} dt' \int dX' \cdot r(X_0, t_0 \mid X', t'; \lambda)\Phi(X', t')p(X', t' \mid X, t). \quad (5)$$

Repeating the same argument with the last collision one obtains

$$r(X_0, t_0 \mid X, t; \lambda) = p(X_0, t_0 \mid X, t) - \lambda \int_{t_0}^{t} dt' \int dX' \cdot p(X_0, t_0 \mid X', t')\Phi(X', t')r(X', t' \mid X, t; \lambda). \quad (6)$$

A formal derivation which removes the restrictions $\lambda > 0$, $\Phi(X(\tau), \tau) \geq 0$ is given in Appendix I.

Since the integral equations will in general be difficult to solve in closed form, a perturbation formalism seems of value. Suppose that the solution of the integral equations is $r_1(X_0, t_0 \mid X, t; \lambda)$ for $\Phi(X, t) \equiv \Phi_1(X, t)$. Let now a second scattering medium be added, such that the probability of a collision at X in $(t, t + dt)$ is increased by $\lambda[\Phi_2(X, t) - \Phi_1(X, t)]dt$. By repetition of the argument given above one then obtains two integral equations for the solution $r_2(X_0, t_0 \mid X, t; \lambda)$ of (5) and (6), when $\Phi(X, t) \equiv \Phi_2(X, t)$:

$$r_2(X_0, t_0 \mid X, t; \lambda) = r_1(X_0, t_0 \mid X, t; \lambda) - \lambda \int_{t_0}^{t} dt' \int dX' r_2(X_0, t_0 \mid X', t') \cdot [\Phi_2(X', t') - \Phi_1(X', t')] r_1(X', t' \mid X, t; \lambda) \quad (7)$$

[2] A. J. F. Siegert, "On the first passage time probability problem," *Phys. Rev.*, vol. 81, pp. 617–623; February 15, 1951.
D. A. Darling and A. J. F. Siegert, "The first passage problem for a continuous Markoff process," *Ann. Math. Stat.*, vol. 24, pp. 624–639; December, 1953, and work quoted there.

[3] A rigorous derivation has been given by D. A. Darling and A. J. F. Siegert, "On the Distribution of Certain Functionals of Markoff Processes," The RAND Corp., Rep. P-429; April, 1954. Appeared in abbreviated form, *Proc. Natl. Acad. Sci.*; August, 1956. A short sketch of the method was given by Siegert, "Passage of stationary processes through linear and non-linear devices," IRE Trans., vol. PGIT-3, pp. 4–25; March, 1954.

and

$$r_2(X_0, t_0 \mid X, t; \lambda) = r_1(X_0, t_0 \mid X, t; \lambda) - \lambda \int_{t_0}^{t} dt' \int dX' r_1(X_0, t_0 \mid X', t') \cdot [\Phi_2(X', t') - \Phi_1(X', t')] r_2(X', t' \mid X, t; \lambda). \quad (8)$$

We can thus obtain successive approximations for r_2 if r_1 is known, by the usual iteration procedure.

A formal derivation independent of the restrictions imposed on λ and Φ has been given.[3] This derivation serves, furthermore, to prove the uniqueness of the solutions of (5) and (6).

In many cases of practical interest $p(X_0, t_0 \mid X, t)$ is the principal solution of two partial differential equations of the form

$$\frac{\partial p}{\partial t} = Lp \quad (9)$$

$$-\frac{\partial p}{\partial t_0} = L_0^+ p \quad (10)$$

where L is defined by[4]

$$Lp \equiv \frac{1}{2} \sum_{kl} \frac{\partial^2}{\partial x_k \partial x_l} [B_{kl}(X, t)p] - \sum_k \frac{\partial}{\partial x_k} [A_k(X, t)p] \quad (11)$$

and L_0^+ is the adjoint of this operator with X and t replaced by X_0 and t_0. The physical meaning of (10) is that of a continuity equation

$$\frac{\partial p}{\partial t} = -\text{Div } J \quad (12)$$

where Div is the divergence and J is the probability current in X space. The current J can be interpreted as a diffusion current with components $-\frac{1}{2} \sum_l B_{kl}\, \partial p/\partial x_l$ and a drift current with components $(A_k - \frac{1}{2} \sum_l \partial B_{kl}/\partial x_l)p$ if one wants to keep the form $-\frac{1}{2}B \cdot \text{Grad } p$ for the diffusion current. (If one prefers to retain the drift current in the form $V_{av}p$, where V_{av} is the average velocity one interprets $A_k p$ as the drift current and $-\frac{1}{2} \sum_l \partial/\partial x_l (B_{kl}p)$ as diffusion current.) Formal application of the operator $L - \partial/\partial t$ to (5) yields

$$\left(L - \frac{\partial}{\partial t}\right) r = \lambda \int dX' r(X_0, t_0 \mid X', t; \lambda) \cdot \Phi(X', t) p(X', t \mid Xt). \quad (13)$$

With the initial condition

$$p(X', t \mid X, t) = \delta(X' - X) \quad (14)$$

this becomes

$$\frac{\partial}{\partial t} r(X_0, t_0 \mid X, t; \lambda) = \{L - \lambda\Phi(X, t)\} r(X_0, t_0 \mid X, t; \lambda). \quad (15)$$

The interpretation of (15) is clearly that the rate of particle loss by collisions $\lambda\Phi(X, t)$ has been added to the continuity equation. Formal application of the operator $L_0^+ + \partial/\partial t_0$ to (6) yields in the same way the differential equation

$$-\frac{\partial}{\partial t_0} r(X_0, t_0 \mid X, t, \lambda) = \{L_0^+ - \lambda\Phi(X_0, t_0)\} r(X_0, t_0 \mid X, t; \lambda). \quad (16)$$

We showed[3] that the principal solution of either (15) or (16), if it exists, is actually a solution of (5) and (6) and is, therefore, by virtue of the uniqueness theorem, the solution of these integral equations.

In special cases, (5), (6), (13), and (15) reduce to the integral and differential equations derived by Kac, Rosenblatt,[6] and Fortet.[7] When $X(\tau)$ is taken to be the one-dimensional Wiener function $x(\tau)$ (once integrated "white noise") with $x(0) = 0$, and $\Phi[X(\tau), \tau] = V(x)$, one obtains from (5) the integral equation (3.8) of Kac, and the Laplace transform of (15) reduces to equation (3.14) of Kac.[5] When the components $x_k(\tau)$ of $X(\tau)$ are Wiener functions with $x_k(0) = 0$, (5) reduces to equation (1.9) of Rosenblatt.[6] The differential equation (16) was derived directly by Fortet.[7]

If $X(\tau)$ is a Gaussian Markoff process, and $\Phi(X(\tau), \tau) = K(t)x_1^2(\tau)$ the method of Kac and Siegert[8] can be applied also, and leads to an integral equation in the time variable only. In this case the solution of (15) is an exponential function of a second-degree polynomial in the components of X_0 and X, and (15) leads to first-order nonlinear differential equations for the coefficients. The equivalence of these to the integral equation of Kac and Siegert[8] requires a somewhat lengthy discussion and will be given in Part II of this paper. It seems interesting to note that the present procedure yields some of the results of Kac and Siegert[8] in closed form.

Example: The Distribution of the Sample Probability Density for a Stationary Markoff Process

It is often necessary to infer the probability density for a random process from a sample. If the process is stationary a convenient estimate $w^*(z)$ of the probability density $w(z)$ is the fraction of the sample length during which the value of the random process $X(\tau)$ lies in a small interval or volume element Δ centered on z, divided by Δ.

The calculation will be carried through for the Markoff

[3] M. C. Wang and G. E. Uhlenbeck, "On the theory of the Brownian motion II," *Rev. Mod. Phys.*, vol. 17, pp. 323–342; April–July; 1945.

[4] A. Kolmogoroff, "Über die analytischen methoden in der wahrscheinlichkeitsrechnung," *Math. Ann.*, vol. 104, pp. 415–458; March, 1931.

[5] M. Kac, "On some connections between probability theory and differential and integral equations," *Proc. Second Berkeley Symposium on Mathematical Statistics and Probability*, University of California Press, Berkeley, Calif., pp. 189–215; 1951.

[6] M. Rosenblatt, "On a class of Markov processes," *Trans. Am. Math. Soc.*, vol. 71, pp. 120–135; July, 1951.

[7] A. Blanc-Lapierre and R. Fortet, "Théorie des Fonctions Aléatoires," Masson et Cie, Paris, 321 pp.; 1953.

[8] M. Kac and A. J. F. Siegert, "On the theory of noise in radio receivers with square law detectors," *J. Appl. Phys.*, vol. 18, pp. 383–397; April, 1947. For an improvement of this method, see Siegert, "Passage of stationary processes through linear and nonlinear devices," IRE Trans., vol. PGIT-3, pp. 4–25; March, 195 .

process with one component $x(\tau)$; the generalization to the n-dimensional process is trivial, if one is interested in the joint distribution of the components. [It must be emphasized, however, that the generalization to the sample probability of *one* component of a *multidimensional* Markoff process is not trivial, since the integral equation (6) does not simplify appreciably in that case.] It would be easy, on the other hand, to generalize our calculation to obtain the joint distribution of $w^*(z_1)$, $w^*(z_2)$ $\cdots$ $w^*(z_k)$. This distribution may be useful in obtaining an approximation to the distribution of

$$\int_0^t \Phi(x(\tau))\,d\tau \equiv \int \Phi(z)w^*(z)\,dz,$$

if $\Phi(z)$ and $w^*(z)$ are slowly varying functions so that the last integral can be approximated by $\sum_{i=1}^k \Phi(z_i)w^*(z_i)\Delta_i$.

Since we have restricted our problem to stationary Markoff functions we will use the notation

$$r(x_0 \mid x, t; \lambda) \equiv r(x_0, t_0 \mid x, t_0 + t; \lambda) \tag{17}$$

$$p(x_0 \mid x, t) \equiv p(x_0, t_0 \mid x, t_0 + t). \tag{18}$$

The estimate $w^*(z)$ defined above can be written in the form

$$w^*(z) = u/t$$

where u is defined by (1), with $t_0 = 0$ and Φ a function of x only, which is defined by

$$\Phi(x) = \begin{cases} \Delta^{-1} & \text{if } |x - z| < \Delta/2, \\ 0 & \text{otherwise.} \end{cases} \tag{19}$$

Eq. (6) becomes

$$r(x_0 \mid x, t; \lambda) = p(x_0 \mid x, t) - \frac{\lambda}{\Delta}\int_0^t dt' \int_{z-\Delta/2}^{z+\Delta/2} \cdot p(x_0 \mid x', t')r(x' \mid x, t - t'; \lambda)\,dx'. \tag{20}$$

In the limit $\Delta \to 0$ this equation simplifies to the integral equation

$$r(x_0 \mid x, t; \lambda) = p(x_0 \mid x, t) - \lambda \int_0^t dt' \cdot p(x_0 \mid z, t')r(z \mid x, t - t'; \lambda) \tag{21}$$

which can be solved by taking Laplace transforms.

If r_L and p_L denote the Laplace transforms of r and p, e.g.,

$$r_L(x_0 \mid x, s; \lambda) = \int_0^\infty e^{-st} r(x_0 \mid x, t; \lambda)\,dt, \tag{22}$$

one obtains

$$r_L(x_0 \mid x, s; \lambda) = p_L(x_0 \mid x, s) - \lambda p_L(x_0 \mid z, s) r_L(z \mid x, s; \lambda). \tag{23}$$

To solve (23) for $r(x_0 \mid x, s; \lambda)$ we first put $x_0 = z$ and obtain

$$r_L(z \mid x, s; \lambda) = p_L(z \mid x, s) - \lambda p_L(z \mid z, s) r_L(z \mid x, s; \lambda) \tag{24}$$

and

$$r_L(z \mid x, s; \lambda) = p_L(z \mid x, s)/[1 + \lambda p_L(z \mid z, s)]. \tag{25}$$

Substituting this result into (23) yields

$$r_L(x_0 \mid x, s; \lambda) = p_L(x_0 \mid x, s) - \lambda p_L(x_0 \mid z, s) p_L(z \mid x, s)/[1 + \lambda p_L(z \mid z, s)]. \tag{26}$$

We denote by $\rho(x_0 \mid x, u, t)$ the joint probability density for x and u at t with fixed initial value of $x(\tau)$ ($x(0) = x_0$), so that

$$r(x_0 \mid x, t; \lambda) = \int_0^\infty e^{-u\lambda} \rho(x_0 \mid x, u, t)\,du$$

and we denote by $\rho_L(x_0 \mid x, u, s)$ the Laplace transform of ρ with respect to t. We compute first $\rho_L(x_0 \mid x, u, s)$ by Laplace inversion of (26). To do this we have to split off the term which leads to a delta function in $\rho_L(x_0 \mid x, u, s)$ (unless its coefficient vanishes). This term corresponds to a delta function in $\rho(x_0 \mid x, u, t)$ and the coefficient of the delta function is the probability that $w^*(z) = 0$, which will occur with nonvanishing probability, for instance, if x_0 and x are both smaller or larger than z.

We, therefore, split off those terms in (26) which do not vanish in the limit $\lambda \to \infty$ and write

$$\begin{aligned} r_L(x_0 \mid x, s, \lambda) &= p_L(x_0 \mid x, s) - p_L(x_0 \mid z, s)p(z \mid x, s)/p_L(z \mid z, s) \\ &\quad - \left[\frac{\lambda p_L(x_0 \mid z, s)p_L(z \mid x, s)}{1 + \lambda p_L(z \mid z, s)} - \frac{p_L(x_0 \mid z, s)p_L(z \mid x, s)}{p_L(z \mid z, s)}\right] \\ &= \frac{p_L(x_0 \mid x, s)p_L(z \mid z, s) - p_L(x_0 \mid z, s)p_L(z \mid x, s)}{p_L(z \mid z, s)} \\ &\quad + \frac{p_L(x_0 \mid z, s)p_L(z \mid x, s)}{p_L(z \mid z, s)(1 + \lambda p_L(z \mid z, s))}. \end{aligned} \tag{27}$$

Taking the Laplace inverse with respect to λ we obtain[9]

$$\begin{aligned} \rho_L(x_0 \mid x, u, s) &= \frac{p_L(x_0 \mid x, s)p_L(z \mid z, s) - p_L(x_0 \mid z, s)p_L(z \mid x, s)}{p_L(z \mid z, s)}\,\delta_+(u) \\ &\quad + \frac{p_L(x_0 \mid z, s)p_L(z \mid x, s)}{p_L^2(z \mid z, s)}\,e^{-u/p_L(z \mid z, s)} \end{aligned} \tag{28}$$

for $u \geq 0$ and zero otherwise, where $\delta_+(u)$ is defined by $\delta_+(u) = 0$ for $u \neq 0$ and

$$\int_0^\epsilon \delta_+(u)\,du = 1 \quad \text{for any} \quad \epsilon > 0. \tag{29}$$

[9] Eq. (28) can be checked by comparing the moments of u with those obtained directly. See Appendix I.

The conditional probability of finding $w^*(z)t$ in the interval $(u, u + du)$, if $x(0) = x_0$ and $x(t) = x$ is known, is thus given by $\rho(x_0 \mid x, u, t)du/p(x_0 \mid x, t)$ where $\rho(x_0 \mid x, u, t)$ is to be obtained by Laplace inversion from $\rho_L(x_0 \mid x, u, s)$, given by (28). The (unconditional) probability of finding $w^*(z)t$ in the interval $(u, u + du)$ is given by

$$\rho(u, t) \equiv \iint_{-\infty}^{\infty} w(x_0)\rho(x_0 \mid x, u, t)\, dx_0\, dx$$

and is to be obtained by Laplace inversion of

$$\rho_L(u, s) \equiv \iint_{-\infty}^{\infty} w(x_0)\rho_L(x_0 \mid x, u, s)\, dx_0\, dx$$

$$= \left(\frac{1}{s} - \frac{w(z)}{s^2 p_L(z \mid z, s)}\right)\delta_+(u) + \frac{w(z)}{s^2 p_L^2(z \mid z, s)} e^{-u/p_L(z \mid z, s)}. \qquad (30)$$

This inversion will in general be too complicated to perform exactly, but an asymptotic evaluation for large t can be carried through. We expect that the quantity $(u - wt)/\sqrt{t}$ becomes normally distributed in this limit. We thus consider the distribution

$$F(v, t) \equiv \text{prob}\,\{(u - wt)/\sqrt{t} \le v\}$$

$$= \begin{cases} \displaystyle\int_0^{v\sqrt{t}+wt} du \left\{\frac{1}{2\pi i}\int_{-i\infty}^{+i\infty} \rho_L(u, s)e^{st}\, ds\right\} & \text{for } v \ge -w\sqrt{t} \\ 0 \text{ for } v < -w\sqrt{t} \end{cases} \qquad (31)$$

where the path of integration has to be taken to the right of singularities of ρ_L. Using (30) and interchanging the order of integration yields

$$F(v, t) = \frac{1}{2\pi i}\int_{-i\infty+\gamma}^{+i\infty+\gamma}\left\{\frac{1}{s} - \frac{w}{s^2 p_L} e^{-(v\sqrt{t}+wt)/p_L}\right\}e^{st}\, ds \qquad (32)$$

$$= 1 - \frac{w}{2\pi i}\int_{-i\infty+\gamma}^{+i\infty+\gamma}\frac{ds}{s^2 p_L}$$

$$\cdot \exp\,[(-vs\sqrt{t} + s^2 t(p_L - w/s))/(w + s(p_L - w/s)]$$

for $v \ge -w\sqrt{t}$, where w and p_L stand for $w(z)$ and $p_L(z \mid z, s)$, respectively. With $\zeta = -is\sqrt{t}$ and $\tau(z, s) \equiv p_L - w/s$ we write this result in the form

$$F(v, t) = 1 - \frac{w}{2\pi i}\int_{-\infty}^{\infty}\frac{d\zeta}{\zeta}\,\frac{1}{w + i\zeta\tau(z, i\zeta/\sqrt{t})/\sqrt{t}} \qquad (33)$$

$$\cdot \exp\{[-i\zeta v - \zeta^2\tau(z, i\zeta/\sqrt{t})]/[w + i\zeta\tau(z, i\zeta/\sqrt{t})/\sqrt{t}]\}$$

where the path of integration has to be taken below any singularities of the integrand. If

$$\lim_{s \to 0}\tau(z, s) = \tau(z) \qquad (34)$$

exists at least for Re $s \ge 0$ and is finite and larger than zero and if the limit $t \to \infty$ and the integration can be interchanged, then $F(v, t)$ approaches the normal distribution for $v \ge -w\sqrt{t}$, *i.e.*,

$$F(v, \infty) = [4\pi\tau(z)\cdot w(z)]^{-1/2}\int_{-\infty}^{v} e^{-\eta^2/4w(z)\tau(z)}\, d\eta. \qquad (35)$$

For the existence of the limit $\tau(z)$ it is sufficient that the stationary distribution is approached sufficiently fast so that

$$\int_0^{\infty} \mid p(z \mid z, t) - w(z) \mid dt \quad \text{exists.} \qquad (36)$$

The significance of the condition $\tau(z) > 0$ can be seen in various ways. We note first that the unconditional first and second moments of u are, according to (51) in Appendix II,

$$\bar{u} = wt \qquad (37)$$

$$\langle u^2\rangle_{\text{av}} = 2w\int_0^t dt_2\int_0^{t_2} p(z \mid z, t_2 - t_1)\, dt_1$$

$$= 2w\int_0^t dt_2\int_0^{t_2} p(z \mid z, t')\, dt'$$

$$= 2w\left\{t\int_0^t p(z \mid z, t')\, dt' - \int_0^t t_2 p(z \mid z, t_2)\, dt_2\right\}. \qquad (38)$$

One has thus

$$\langle u^2 - \bar{u}^2\rangle_{\text{av}} = 2w\int_0^t \{p(z \mid z, t') - w\}(t - t')\, dt' \qquad (39)$$

or

$$t^{-1}\langle u^2 - \bar{u}^2\rangle_{\text{av}} = 2w\int_0^t \{p(z \mid z, t') - w\}\left(1 - \frac{t'}{t}\right) dt'. \qquad (40)$$

In the limit $t \to \infty$, the second factor in the integrand merely a convergence creating factor, so that if the integral converges with the first factor alone, *i.e.*, *a fortiori* if the condition (36) is fulfilled

$$\lim_{t\to\infty} t^{-1}\langle u^2 - \bar{u}^2\rangle_{\text{av}} = 2w(z)\int_0^{\infty}\{p(z \mid z, t') - w(z)\}\, dt'$$

$$= 2w(z)\tau(z). \qquad (41)$$

This shows that for $w(z) \ne 0$, $\tau(z)$ must be at least nonnegative.

From (26) or (30) one sees that

$$\int_0^{\infty} e^{-st}\,\text{prob}\,\{u \ne 0\}\, dt = w/s^2 p_L. \qquad (42)$$

We can consider prob $\{u \ne 0\}$ also as the probability that the time ϑ at which the first contribution to u occurs is smaller than t. The first moment $\bar{\vartheta}$ is, therefore,

[10] For a continuous process, this is the first-passage time.

$$\bar{\vartheta} = -\left[\frac{d}{ds}\left(\frac{w}{sp_L}\right)\right]_{s=0}$$
$$= w\left[\frac{p_L + sdp_L/ds}{s^2 p_L^2}\right]_{s=0}$$
$$= w^{-1}[p_L - w/s]_{s=0}$$
$$= w^{-1}\tau(z). \quad (43)$$

This shows that $\tau(z)$ is finite and positive as required, if $\neq 0$ and if the average time at which the first contribution to u occurs is finite and different from zero.

Appendix I

To derive (5) more formally we use the trivial identity

$$\exp\left\{-\lambda\int_{t_0}^{t}\Phi(X(\tau),\tau)\,d\tau\right\} = 1 - \lambda\int_{t_0}^{t} dt'\Phi(X(t'),t') \cdot\exp\left\{-\lambda\int_{t_0}^{t'}\Phi(X(\tau),\tau)\,d\tau\right\}. \quad (44)$$

By averaging both sides with initial and end point of $X(t)$ fixed, and multiplying by $p(X_0, t_0 \mid X, t)$ one obtains from the definition (4)

$$r(X_0, t_0 \mid X, t; \lambda) = p(X_0, t_0 \mid X, t)\Big(1 - \lambda\int_{t_0}^{t} dt' \cdot\Big[\Phi(X(t'),t')\exp\left\{-\lambda\int_{t_0}^{t'}\Phi(X(\tau),\tau)\,d\tau\right\} \mid \cdot X(t_0) = X_0, X(t) = X\Big]_{\text{av}}\Big). \quad (45)$$

In the second term, we now introduce a third condition $X(t') = X'$ and compensate for this by multiplicatiom with $p(X_0, t_0; X, t \mid X', t')dX'$ and integration over X', where $p(X_0, t_0; X, t \mid X', t')dX'$ is defined as the probability that $X(t')$ in dX' at X' for a path with fixed initial and end point $X(t_0) = X_0$ and $X(t) = X$, respectively. The second term thus becomes

$$\int_{t_0}^{t} dt' \int dX'\, p(X_0, t_0; Xt \mid X', t') \Big[\Phi(X(t'),t')\exp\left\{-\lambda\int_{t_0}^{t'}\Phi(X(\tau),\tau)\,d\tau\right\} \mid \cdot X(t_0) = X_0, X(t') = X', X(t) = X\Big]_{\text{av}}. \quad (46)$$

We can now take $\Phi(X(t'), t')$ out of the average symbol as $\Phi(X', t')$, since $X(t') = X'$ is held fixed. We also can omit the condition $X(t) = X$ in the average symbol since, by virtue of the Markoff property, $X(t)$ is statistically independent of values of $X(t'')$ for $t'' < t'$, if $X(t')$ is held fixed. Also from the Markoff property and the definition of conditional probabilities follows for $t_0 \leq t' \leq t$,

$$p(X_0, t_0; Xt \mid X', t') = p(X_0, t_0 \mid X', t') \cdot p(X', t' \mid X, t)/p(X_0, t_0 \mid X, t). \quad (47)$$

We thus have

$$r(X_0, t_0 \mid X, t; \lambda) = p(X_0, t_0 \mid X, t) - \lambda\int_{t_0}^{t'} dt' \int dX' \Big[\exp\left\{-\lambda\int_{t_0}^{t'}\Phi(X(\tau),\tau)\,d\tau\right\} \mid \cdot X(t_0) = X_0, X(t') = X'\Big]_{\text{av}} p(X_0, t_0 \mid X', t') \cdot\Phi(X', t')p(X', t' \mid X, t) \quad (48)$$

and using the definition of $r(X_0, t_0 \mid X', t', \lambda)$ on the right-hand side, we obtain (5). By a similar argument starting with the identity

$$\exp\left\{-\lambda\int_{t_0}^{t}\Phi(X(\tau),\tau)\,d\tau\right\} = 1 - \lambda\int_{t_0}^{t} dt'\,\Phi(X(t'),t') \cdot\exp\left\{-\lambda\int_{t'}^{t}(X(\tau),\tau)\,d\tau\right\} \quad (49)$$

one obtains (6).

Appendix II

Eq. (28) can be checked directly by the method of moments. Since for integer $n \geq 1$

$$u^n = \int_0^t\cdots\int \prod_{j=1}^{n}\left\{dt_j \frac{1}{\Delta}\int_{z-\Delta/2}^{z+\Delta/2}\delta(x(t_j) - z_j)\,dz_j\right\}$$
$$= n!\int_0^t dt_n \int_0^{t_n} dt_{n-1}\cdots\int_0^{t_2} dt_1\,\Delta^{-n} \cdot\prod_{j=1}^{n}\int_{z-\Delta/2}^{z+\Delta/2}\delta(x(t_j) - z_j)\,dz_j \quad (50)$$

one has

$$\langle u^n \mid x(t_0) = x_0, x(t) = x\rangle_{\text{av}}$$
$$= n!\int_0^t dt_n\int_0^{t_n} dt_{n-1}\cdots\int_0^{t_2} dt_1\,\Delta^{-n}\iint_{z-\Delta/2}^{z+\Delta/2} p(x_0 \mid z_1, t_1) \cdot\prod_{j=1}^{n-1} p(z_j \mid z_{j+1}, t_{j+1} - t_j)p(z_n \mid x, t - t_n)\prod_1^n dz_j$$
$$= n!\int_0^t dt_n p(z \mid x, t - t_n)\int_0^{t_n} dt_{n-1} \cdot p(z \mid z, t_n - t_{n-1})\cdots\int_0^{t_2} p(z \mid z, t_2 - t_1) \cdot p(x_0 \mid z, t_1). \quad (51)$$

Taking Laplace transforms one has

$$\int_0^\infty e^{-st}\,dt\,\langle u^n \mid x(t_0) = x_0, x(t) = x\rangle_{\text{av}} = n!\,p_L(x_0 \mid z, s)p_L(z \mid x, s)p_L(z \mid z, s)^{n-1}. \quad (52)$$

This is in agreement with Laplace transforms of the moments obtained by integration with respect to u from (28).

24

Reprinted from *IRE Trans. Information Theory*, **IT-3**(1), 38–43 (1957)

A Systematic Approach to a Class of Problems in the Theory of Noise and Other Random Phenomena —Part II, Examples*

ARNOLD J. F. SIEGERT†

Summary—The method of Part I is applied to the problem of finding the probability distribution of $u \equiv \int_0^t K(\tau)x^2(\tau)\,d\tau$, where $K(\tau)$ is a given function and $x(\tau)$ is the Uhlenbeck process. The earlier methods of Kac and the author yielded the characteristic function of this distribution as the reciprocal square root of the Fredholm determinant D of an integral equation. The present method yields a second-order linear differential equation with initial condition only for D as function of t. For the special cases $K(\tau) = 1$ and $K(\tau) = e^{-\alpha\tau}$ the characteristic function is obtained in closed form.

In Section III, we have verified directly from the integral equation the differential equation for D and some relations between D and the initial and end point values of the Volterra reciprocal kernel which appear in the joint characteristic function for u, $x(0)$ and $x(t)$.

Section I

In previous papers,[1] Darling and the author derived two integral equations for a function closely related to the characteristic function of the probability distribution of the functional $\int_{t_0}^t \phi(X(\tau), \tau)\,d\tau$, where the components $x_j(\tau)$ of $X(\tau)$ form a Markoff process and $\phi(X, \tau)$ is a given function of X and τ. For an important class of Markoff processes, the solution of these integral equations was shown to be the principal solution of a partial differential equation similar to the Fokker-Planck equation. We also derived an integral equation relating two solutions of the problem with two functions, $\phi_2(X, \tau)$ and $\phi_1(X, \tau)$, which can be used for a perturbation calculation to obtain solutions of the problem when the solution for $\phi_1(X, \tau)$ is known, and the solution for a function $\phi_2(X, \tau)$ "in the neighborhood" of $\phi_1(X, \tau)$ is desired.

Since the integral equations as well as the differential equations derived in this previous paper are rather formidable, it seemed of interest to consider first some cases for which the problem could be solved or at least be reduced to an integral equation in a single variable by an older method.[2,3] In the case $\phi(X, \tau) = K(\tau)x^2(\tau)$, where $x(\tau)$ is a Gaussian random function with arbitrary autocorrelation function $\rho(\tau)$, the older method applies and the problem can be reduced to the problem of solving either the homogeneous integral equation[2]

$$\int_{t_0}^t \rho(\tau_1 - \tau)K(\tau)\varphi(\tau)\,d\tau = \lambda\varphi(\tau_1) \tag{1}$$

or the inhomogeneous integral equation[3]

$$g_\lambda(\tau_1, \tau_2) + 2\lambda \int_{t_0}^t \rho(\tau_1 - \tau) \cdot K(\tau)g_\lambda(\tau, \tau_2)\,d\tau = \rho(\tau_1 - \tau_2). \tag{2}$$

If, specially, the function $x(\tau)$ is a component $x_1(\tau)$ of a Markoffian Gaussian process, the method referred to[1] applies and leads to the partial differential equation [(15) of Part I]:

$$(L - \lambda K(t)x_1^2)r = \frac{\partial r}{\partial t} \tag{3}$$

if the transition probability density $p(X_0 \mid X, t)$ is the principal solution of the differential equation

$$Lp = \frac{\partial p}{\partial t} \tag{4}$$

where L operates on the components of X. One sees easily that in the case of Gaussian $p(X_0 \mid X, t)$ the additional term in (3) does not seriously complicate the differential equation and that r is still of the Gaussian form. For the coefficients of this Gaussian, one obtains a system of differential equations of first order and second degree with t as the independent variable with initial conditions only. The equivalence of this system in which t appears as the independent variable and the integral equation (2) in which t appears only parametrically is not trivial even though (2) can be reduced to a differential equation with appropriate boundary conditions when $\rho(\tau)$ is the auto-correlation function of a Markoffian Gaussian process. We have, therefore, in Section II worked out in detail the case of the one-dimensional, Markoffian Gaussian random process (Uhlenbeck process).[4] As special examples, we give in closed form the characteristic functions of $\int_0^t x^2(\tau)d\tau$ and $\int_0^t e^{-\alpha\tau}x^2(\tau)d\tau$. The latter represents the output of a receiver consisting of single tuned IF and audio stage with quadratic detector, wi

* Manuscript received by the PGIT, August 13, 1956. Most of this work was done while the author was a consultant for The RAND Corp., Santa Monica, Calif.

† Dept. of Physics, Northwestern University, Evanston, Ill.

[1] D. A. Darling and A. J. F. Siegert, "On the Distribution of Certain Functionals of Markoff Processes," The RAND Corp., Paper P-429; October 31, 1953, and Part I of this paper (P-738), "A Systematic Approach to a Class of Problems in the Theory of Noise and Other Random Phenomena;" September, 1955. Darling and Siegert, "On the distribution of certain functionals of Markoff chains and processes," *Proc. Natl. Acad. Sci.*, vol. 42, pp. 525–529; August, 1956.

[2] M. Kac and A. J. F. Siegert, "Note on the theory of noise in receivers with square law detector," *Phys. Rev.*, vol. 70, p. 449; September, 1946.

Kac and Siegert, "On the theory of noise in radio receivers with square law detector," *J. Appl. Phys.*, vol. 18, pp. 383–397; April, 1947.

[3] A. J. F. Siegert, "Passage of stationary processes through linear and non-linear devices," IRE Trans., vol. PGIT-3, pp. 4–25; March, 1954. (The RAND Corp., (P-419); October 29, 1953.)

[4] Analogous problems for the Wiener process have been trea·e independently by R. Deutsch, "Piecewise quadratic detecto·, 1956 IRE Convention Record, Part 4, pp. 15–20.

white noise input turned on a time t before observation. In Section III, we have shown for this case how the system of differential equations originally obtained by the method[1] follows directly from (2). The purpose of this derivation was primarily to show that some of the equations derived by the method[1] remain valid when the auto-correlation function of the Uhlenbeck process $[\rho(\tau_1 - \tau) = \exp(-\beta \mid \tau_1 - \tau \mid)]$ is replaced by a general function $h(\tau_1, \tau)$. We found, *e.g.*, a simpler expression for the characteristic function $f = \langle \exp[-\lambda \int_0^t K(\tau)x^2(\tau)d\tau]\rangle_{av}$ where $x(\tau)$ is a Gaussian random function with arbitrary correlation function $\rho(\tau_1 - \tau)$. We had shown[3] that the function f can be expressed in terms of the trace of the solution $g_\lambda(\tau_1, \tau_2)$ of the integral equation (2) (which depends parametrically on t) by

$$f = \exp\left[-\int_0^\lambda d\kappa \int_0^t K(\tau)g_\kappa(\tau, \tau)\, d\tau\right].$$

This equation is suggested by the well-known expression for the Fredholm determinant in terms of the Volterra reciprocal function. We now obtained the simpler expression

$$f = \exp\left[-\lambda \int_0^t K(t')g_\lambda(t', t')\, dt'\right]$$

where $g_\lambda(\tau_1, \tau_2)$ is the solution of (2) with the upper limit replaced by t' so that $g_\lambda(t', t')$ depends on t' implicitly also. [See (58).]

The results[1] and the present paper raise an interesting question for further investigation. In Kac and Siegert,[2] the fact that the problem of the probability distribution of a quadratic integral form of a Gaussian process $x(t)$ could be reduced to the solution of an integral equation involving as variable only the time was clearly a consequence of the fact that the joint probability distribution for $x(t_1), x(t_2) \cdots x(t_n)$ is the exponential function of a quadratic form. The results[1] show that this simplification can also be understood as a consequence of the fact that the differential equation (4) is not essentially complicated by the addition of a quadratic term. One may thus expect to find for Markoff processes other than the Gaussian processes certain functionals for which the problem of finding the characteristic function reduces to differential equations with only the time as independent variable.

Section II

In this section we will apply the method[1] to the problem of evaluating the function

$$r(x_0 \mid x, t, \lambda)$$

$$\left\langle \exp\left[-\lambda \int_0^t K(\tau)x^2(\tau)\, d\tau\right] \middle| x(0) = x_0, x(t) = x \right\rangle_{av} \cdot p(x_0 \mid x, t) \qquad (5)$$

where $x(\tau)$ is the stationary one-dimensional Markoffian Gaussian process (Uhlenbeck process), which is described by the transition probability density

$$p(x_0 \mid x, t) = [2\pi(1 - e^{-2\beta t})]^{-1/2} \exp\left\{-\frac{(x - x_0e^{-\beta t})^2}{2(1 - e^{-2\beta t})}\right\} \qquad (6)$$

with constant β, $K(\tau)$ is a given function, λ is to be chosen positive if $K(\tau) \geq 0$ and negative imaginary if $K(\tau)$ can assume negative values. The symbol $\langle | \rangle_{av}$ denotes the average of the functional to the left of the vertical line under the conditions written to the right of the vertical line. The extension of the result to, *e.g.*, the characteristic function for $\int_0^t K(\tau) \sum_i x_i^2(\tau)\, d\tau$ where $x_i(\tau)$ are independent Uhlenbeck processes is trivial.

This problem can be treated by the method[2] and the result can be brought into a slightly more convenient form by an extension of this method.[3]

It is convenient to work with the function

$$\hat{r}(\eta \mid \zeta, t, \lambda) \equiv (2\pi)^{-1/2} \iint_{-\infty}^{\infty} \exp\left(-\frac{x_0^2}{2} + i\eta x_0 + i\zeta x\right) \cdot r(x_0 \mid x, t, \lambda)\, dx\, dx_0 \qquad (7)$$

$$= \left\langle \exp\left\{i[\eta x(0) + i\zeta x(t)] - \lambda \int_0^t K(\tau)x^2(\tau)\, d\tau\right\}\right\rangle_{av}$$

from which $r(x_0 \mid x, t, \lambda)$ is easily obtained by Fourier inversion and multiplication by $\sqrt{2\pi}\, e^{x_0^2/2}$. Note that the characteristic function for the distribution of $\int_0^t K(\tau)x^2(\tau)d\tau$ without initial and end conditions is simply obtained by choosing $\eta = \zeta = 0$.

The result obtained by using the methods[2,3] is

$$\hat{r}(\eta \mid \zeta, t, \lambda) = \exp\left\{-\int_0^\lambda d\kappa \int_0^t K(\tau)g_\kappa(\tau, \tau)\, d\tau - \frac{1}{2}(\eta^2 g_\lambda(0, 0) + 2\eta\zeta g_\lambda(0, t) + \zeta^2 g_\lambda(t, t))\right\} \qquad (8)$$

where $g_\lambda(\tau_1, \tau_2)$ is the solution of the integral equation

$$g_\lambda(\tau_1, \tau_2) + 2\lambda \int_0^t K(\tau)\rho(\tau_1 - \tau) \cdot g_\lambda(\tau, \tau_2)\, d\tau = \rho(\tau_1 - \tau_2). \qquad (9)$$

(See Appendix II.)

This result is valid for any stationary Gaussian process with auto-correlation function $\rho(\tau)$. The variable τ_2 appears only as a parameter in the integral equation. If $x(\tau)$ represents the output of a network with lumped circuit and with white noise input, the integral equation reduces to a differential equation. For the Uhlenbeck process, one has specially $\rho(\tau) = e^{-\beta|\tau|}$, and one would reduce the integral equation to a second-order linear differential equation with appropriate boundary conditions with τ_1 as the independent variable.

It is interesting to see how the new method leads to a Riccati equation for $g_\lambda(t, t)$, with t as the independent variable. This Riccati equation is of course also equivalent to a second-order linear differential equation, which, however, is not the same as the differential equation obtained for $g_\lambda(\tau_1, \tau_2)$ as function of τ_1.

The new method leads to the differential equation

$$\frac{\partial r}{\partial t} = \beta\left\{\frac{\partial^2 r}{\partial x^2} + \frac{\partial(xr)}{\partial x}\right\} - \lambda K x^2 r \qquad (10)$$

Since p satisfies

$$\frac{\partial p}{\partial t} = \beta\left\{\frac{\partial^2 p}{\partial x^2} + \frac{\partial(xp)}{\partial x}\right\}. \qquad (11)$$

From this one obtains

$$\frac{\partial \hat{r}}{\partial t} = -\beta\left(\zeta\frac{\partial \hat{r}}{\partial \zeta} + \zeta^2 \hat{r}\right) + \lambda K \frac{\partial^2 \hat{r}}{\partial \zeta^2}. \qquad (12)$$

Making the Ansatz

$$\hat{r} = f \exp\left[-\tfrac{1}{2}(\sigma_0\eta^2 + 2\sigma\zeta\eta + \sigma_1\zeta^2)\right] \qquad (13a)$$

one obtains (with dots indicating differentiation with respect to t)

$$\frac{\dot{f}}{f} - \tfrac{1}{2}(\dot{\sigma}_0\eta^2 + 2\dot{\sigma}\zeta\eta + \dot{\sigma}_1\zeta^2) = -\beta[\zeta(-\sigma\eta - \sigma_1\zeta) + \zeta^2]$$
$$+ \lambda K[(\sigma\eta + \sigma_1\zeta)^2 - \sigma_1] \qquad (13b)$$

and by comparing coefficients

$$\frac{d \ln f}{dt} = -\lambda K \sigma_1 \qquad (14a)$$

$$\frac{d\sigma_1}{dt} = 2\beta(1 - \sigma_1) - 2\lambda K \sigma_1^2 \qquad (14b)$$

$$\frac{d\sigma_0}{dt} = -2\lambda K \sigma^2 \qquad (14c)$$

$$\frac{d \ln \sigma}{dt} = -\beta - 2\lambda K \sigma_1. \qquad (14d)$$

Since $r(x_0 \mid x, 0, \lambda) = \delta(x - x_0)$, we have

$$\hat{r}(\eta \mid \zeta, 0, \lambda) = (2\pi)^{-1/2} \iint_{-\infty}^{\infty} \exp\left(-\frac{x_0^2}{2} + i\eta x_0 + i\zeta x\right)$$
$$\delta(x - x_0)\, dx\, dx_0 \qquad (15)$$
$$= (2\pi)^{-1/2} \int_{-\infty}^{\infty} \exp\left(-\frac{x_0^2}{2} + i(\eta + \zeta)x_0\right) dx_0$$
$$= \exp\left(-\tfrac{1}{2}(\eta + \zeta)^2\right).$$

The initial conditions for the differential equations (14a) to (14d) are, therefore,

$$f(0) = \sigma_0(0) = \sigma_1(0) = 1. \qquad (16)$$

We note that σ_0, σ, and f are obtained by quadratures, if σ_1 has been found, which means that $g_\lambda(0, 0)$, $g_\lambda(0, t)$ and $\int_0^\lambda d\kappa \int_0^t K(\tau) g_\kappa(\tau, \tau) d\tau$ are obtained by quadratures from $g_\lambda(t, t)$. It should be remembered that these relations need hold only for the special choice $\rho(\tau) = e^{-\beta|\tau|}$ since a stationary Gaussian process is Markoffian only if it has this special auto-correlation function, and (10) is based on the Markoff property. We will show in the following Section III, however, that (14a) and (14c) are independent of this special form of ρ.

It will generally be more convenient to convert the Riccati equation (10b) into a second-order linear differen- tial equation. With the substitutions

$$dx = 2\lambda K\, dt \qquad (17)$$

and

$$\sigma_1 = u'/u \qquad (18)$$

where prime denotes differentiation with respect to x we get

$$\frac{d\sigma_1}{2\lambda K\, dt} = \frac{\beta}{\lambda K}(1 - \sigma_1) - \sigma_1^2 \qquad (19)$$

$$\left(\frac{u'}{u}\right)' = \frac{\beta}{\lambda K}\left(1 - \frac{u'}{u}\right) - \left(\frac{u'}{u}\right)^2 \qquad (20)$$

or

$$\lambda\beta^{-1}K(t(x))\, u'' + u' - u = 0. \qquad (21)$$

One initial condition is obtained from $\sigma_1(0) = 1$

$$u'(x_0) = u(x_0) \qquad (22a)$$

which simplifies to

$$u''(x_0) = 0 \quad \text{if} \quad K(t(x_0)) \neq 0 \qquad (22b)$$

where x_0 is the value assumed by x for $t = 0$. This de- termines $u(x)$ except for a constant factor which is irrelevant since all results depend only on $\sigma_1 = u'/u$. Of special interest is the characteristic function for the unconditional distribution, f. We obtain from (14a)

$$\frac{d \ln f}{2\lambda K\, dt} = \frac{d \ln f}{dx} = -\frac{1}{2}\sigma_1 = -\frac{1}{2}\frac{d \ln u}{dx} \qquad (23)$$

or

$$f = \sqrt{\frac{u(x_0)}{u(x)}}. \qquad (24)$$

Since $u(x_0)$ is essentially the Wronskian, it can always be written in a convenient form (see Appendix 1).

The present method thus presents the characteristic function in terms of the solution of a differential equation with initial condition rather than through an eigenvalue problem or an inhomogeneous integral equation.

For σ and σ_0 we obtain from (14c), (14d), and (16)

$$d \ln \sigma = -\beta\, dt - \sigma_1\, dx \qquad (24)$$

$$\ln \sigma = -\beta t(x) - \ln u + \text{const} \qquad (25)$$

$$\sigma = e^{-\beta t}\, u(x_0)/u(x) \qquad (26)$$

and

$$\frac{d\sigma_0}{dx} = -\sigma^2 = -e^{-2\beta t(x)}\, u^2(x_0)/u^2(x) \qquad (27)$$

or

$$\sigma_0 = 1 - u^2(x_0) \int_{x_0}^{x} e^{-2\beta t(y)}\, u^{-2}(y)\, dy. \qquad (28)$$

In Kac and Siegert,[2] solutions in terms of infinite products were given for the two cases $K(\tau) = 1$, and $K(\tau) = e^{-\alpha\tau}$ with $t = \infty$. We will give here the two solutions in closed form with the second case for general t.

Case 1: $K(t) = 1$.

$$\lambda\beta^{-1}u'' + u' - u = 0 \tag{29}$$

with

$$x = 2\lambda t. \tag{30}$$

The general solution is

$$u = e^{-\beta x/2\lambda}(ae^{kx} + be^{-kx}) \tag{31}$$

with

$$\kappa = \sqrt{(\beta/2\lambda)^2 + \beta/\lambda} \tag{32}$$

The initial condition (22a) requires

$$\frac{a-b}{a+b} = \kappa^{-1}(1 + \beta/2\lambda). \tag{33}$$

We thus have from (24)

$$f = e^{\beta x/4\lambda}\left(\frac{a+b}{ae^{\kappa x} + be^{-\kappa x}}\right)^{1/2}$$

$$= e^{\beta x/4\lambda}\left(\frac{a+b}{(a+b)\cosh\kappa x + (a-b)\sinh\kappa x}\right)^{1/2}$$

$$= e^{\beta x/4\lambda}(\cosh\kappa x + \kappa^{-1}(1+\beta/2\lambda)\sinh\kappa x)^{-1/2}. \tag{34}$$

With

$$\eta \equiv \sqrt{1 + 4\lambda/\beta} \tag{35}$$

this becomes[5]

$$f = e^{\beta t/2}(\cosh\beta\eta t + \eta^{-1}(1 + 2\lambda/\beta)\sinh\beta\eta t)^{-1/2}. \tag{36}$$

The roots $\lambda^{(m)}$ of f are determined by the roots η_ν of

$$\text{th}\beta\eta t = -\frac{\eta}{1 + 2\lambda/\beta} = -\frac{2\eta}{1+\eta^2} \tag{37}$$

with

$$\lambda^{(m)} = \frac{\beta}{4}(\eta_m^2 - 1). \tag{38}$$

In Kac and Siegert,[2] the meaning of α and β is interchanged and λ_ν is in our present notation given by

$$\lambda_m = -\frac{1}{2\lambda^{(m)}t} \tag{39}$$

so that we have

$$\lambda_\nu = \frac{2}{\beta t(1 - \eta_m^2)}. \tag{40}$$

With $\eta = iy$ and $\eta_m = -iy_m$ we then have agreement with equations (7.35) and (7.36) of Kac and Siegert.[2]

Case 2: $K(t) = e^{-\alpha t}$.

In this case we get from (17)

$$x = 2\lambda\alpha^{-1}e^{-\alpha t} \tag{41}$$

since it turns out to be convenient to choose the integration constant equal to zero. We thus have

$$x_0 = -2\lambda\alpha^{-1} \tag{42}$$

and

$$\lambda K(t(x)) = -1/2\,\alpha x. \tag{43}$$

Eq. (21) becomes

$$u'' - \frac{2\beta}{\alpha x}(u' - u) = 0 \tag{44}$$

and has the solution[6]

$$u = x^{p/2}Z_p(\sqrt{\gamma x}) \tag{45}$$

where $Z_p(y)$ is a Bessel function of order

$$p = 1 + (\gamma/4) \tag{46a}$$

and

$$\gamma = 8\beta/\alpha. \tag{46b}$$

With the aid of the identity[7]

$$\frac{d}{dx}[x^{p/2}Z_p(\sqrt{\gamma x})] = \tfrac{1}{2}\sqrt{\gamma}\,x^{(p-1)/2}Z_{p-1}(\sqrt{\gamma x}) \tag{47}$$

the initial condition is seen to be satisfied by

$$u(x) = x^{p/2}[N_{p-2}(\sqrt{\gamma x_0})J_p(\sqrt{\gamma x}) - J_{p-2}(\sqrt{\gamma x_0})N_p(\sqrt{\gamma x})] \tag{48}$$

where J_p is the ordinary Bessel function and N_p is the Neumann function

$$N_p = (J_p\cos\pi p - J_{-p})/\sin\pi p. \tag{49}$$

The function $u(x_0)$ must simplify since it is essentially a Wronskian and we have using (47)

$$x_0^{(p-2)/2}u(x_0) = \frac{2}{\sqrt{\gamma}}\frac{d}{dx_0}[x_0^{(p-1)/2}N_{p-1}x_0^{p/2}J_p - x_0^{(p-1)/2}J_{p-1}x_0^{p/2}N_p]$$

$$= \frac{2}{\sqrt{\gamma}}\frac{d}{dx_0}\left\{x_0^{p-(1/2)}\frac{2}{\pi\sqrt{\gamma x_0}}\right\} \tag{50}$$

where the second equation follows from an identity.[8] We thus get

$$u(x_0) = 4\gamma^{-1}\pi^{-1}(p-1)x_0^{(p-2)/2} = \pi^{-1}x_0^{(p-2)/2} \tag{51}$$

and from (24) for the unconditional characteristic function

$$f(t) = \{\pi x_0^{(2-p)/2}x^{p/2}[N_{p-2}(\sqrt{\gamma x_0})J_p(\sqrt{\gamma x}) - J_{p-2}(\sqrt{\gamma x_0})N_p(\sqrt{\gamma x})]\}^{-1/2} \tag{52}$$

[5] In Kac and Siegert (footnote 2), f^2 was given in product form, since we treated there the envelope detector.

[6] E. Jahnke and F. Emde, "Tables of Functions," Dover Publications, New York, N.Y., 4th ed., p. 146, sec. 7, second equation; 1945.

[7] *Ibid.*, p. 145, sec. 5, fourth equation.

[8] *Ibid.*, p. 144, sec. 4, third equation.

where x, x_0, p and γ are given by (41), (42), (46a), and (46b).

For $\alpha \to 0$ this must reduce to (36). For $t \to \infty$, $x \to 0$ and the first term in the bracket becomes negligible. We then have

$$x^{p/2}N_p(\sqrt{\gamma x}) \cong -\pi^{-1}(\tfrac{1}{2}\sqrt{\gamma})^{-p}\,\Gamma(p) \quad (53)$$

$$f(\infty) = \{\gamma^{-1}2^p(\gamma x_0)^{(2-p)/2}J_{p-2}(\sqrt{\gamma x_0})\,\Gamma(p)\}^{-1/2}. \quad (54)$$

The product form[2] is then checked by using the Weierstrass product[9]

$$(\gamma x_0)^{(2-p)/2}J_{p-2}(\sqrt{\gamma x_0}) = 2^{2-p}\prod_n\left(1 - \frac{\gamma x_0}{y_n^2}\right)\Big/\Gamma(p-1) \quad (55)$$

which yields

$$f(\infty) = \prod_n\left(1 - \frac{\gamma x_0}{y_n^2}\right)^{-1/2} = \prod_n\left(1 + \frac{4\beta\lambda}{\alpha^2 y_n^2}\right)^{-1/2} \quad (56)$$

where the numbers y_n are the roots of $J_{p-2}(y)$.

In Kac and Siegert,[2] the probability density $P(V)$ was obtained for the random variable $\beta\int_0^\infty e^{-\beta\tau}\,(x_1^2(\tau) + x_2^2(\tau))d\tau$, where $x_1(\tau)$ and $x_2(\tau)$ are two independent Uhlenbeck processes with $\overline{x_1^2} = \overline{x_2^2} = 1/2$ and autocorrelation function $e^{-\alpha|\tau|}$. We get for this the result (after interchanging α and β to conform with the notation[2]):

$$P(V) = \frac{1}{2\pi i}\int_{-i\infty}^{+i\infty} e^{\lambda V/2\beta}f^2(\infty)\,d\lambda/2\beta$$
$$= \frac{1}{2\pi}\int_{-\infty}^{+\infty} e^{-i\zeta V}\,d\zeta\prod_n\left(1 - \frac{8i\alpha\zeta}{\beta y_n^2}\right)^{-1} \quad (57)$$

where y_n is the nth root of $J_{(2\alpha/\beta)-1}(y) = 0$ in agreement with equations (4.45) and (7.19) of Kac and Siegert.[2]

Section III

It seemed to interest to verify the differential equations (14a) to (14d) directly from the integral equation (9) and to show that (14a) and (14c) remain valid when ρ is replaced by a general symmetric kernel $h(\tau_1, \tau_2)$. We will here only outline the derivation; the details are given in a RAND Corp. paper.[10]

From the integral equation

$$g(\tau_1, \tau_2) + 2\lambda\int_0^t h(\tau_1, \tau)K(\tau)g(\tau, \tau_2)\,d\tau = h(\tau_1, \tau_2) \quad (58)$$

one obtains an integral equation for $\partial g(\tau_1, \tau_2)/\partial t$ and an integral equation for $g(\tau_1, t)$. Comparing these two integral equations, whose solutions can be shown to be unique, one obtains

$$\partial g(\tau_1, \tau_2)/\partial t = -2\lambda K(t)g(\tau_1, t)g(t, \tau_2). \quad (59)$$

Since $g(\tau_1, \tau_2)$ is symmetric because of the symmetry of the kernel, (59) yields

$$dg(0, 0)/dt = -2\lambda K(t)g^2(0, t) \quad (60)$$

which proves (14c).

From (87), of Appendix II, follows[11]

$$\ln f = -\int_0^\lambda d\kappa\int_0^t d\tau K(\tau)g_\kappa(\tau, \tau) \quad (61)$$

where $g_k(\tau, \tau)$ is the solution of (58) with λ replaced by k. Differentiation with respect to t and use of (59) yields

$$d\ln f/dt = K(t)\int_0^\lambda [2\kappa g_\kappa^{(2)}(t, t) - g_\kappa(t, t)]\,d\kappa \quad (62)$$

where $g_k^{(2)}$ is defined by

$$g_\kappa^{(2)}(\tau_1, \tau_2) \equiv \int_0^t g_\kappa(\tau_1, \tau)K(\tau)g_\kappa(\tau, \tau_2)\,d\tau. \quad (63)$$

From (58) one obtains integral equations for $g_\kappa^{(2)}(\tau_1, \tau_2)$ and $\partial g_\kappa(\tau_1, \tau_2)/\partial\kappa$. Comparing these one finds

$$g_\kappa^{(2)}(\tau_1, \tau_2) = -\tfrac{1}{2}\partial g_\kappa(\tau_1, \tau_2)/\partial\kappa. \quad (64)$$

Substituting this result in (62) yields

$$d\ln f/dt = -\lambda K(t)g_\lambda(t, t) \quad (65)$$

and proves (14a).

The specific form of the kernel, $h(\tau_1, \tau_2) = e^{-\beta|\tau_1-\tau_2|}$ is needed for the verification of (14b) and (14d). The functions $g(\tau_1, \tau_2)$ and $e^{-\beta|\tau_1-\tau_2|}$ have a discontinuous derivative at $\tau_1 = \tau_2$, but we can obtain $dg(t, t)/dt$ from

$$dg(t, t)/dt = dg^*(t, t)/dt = [\partial g^*(\tau_1, \tau_2)/\partial t]_{\tau_1=\tau_2=t}$$
$$+ [\partial g^*(\tau_1, t)/\partial\tau_1]_{\tau_1=t} + [\partial g^*(t, \tau_2)/\partial\tau_2]_{\tau_2=t} \quad (66)$$

where g^* is defined by

$$g^*(\tau_1, \tau_2) = g(\tau_1, \tau_2) - e^{-\beta|\tau_1-\tau_2|}. \quad (67)$$

The first term in (66) is evaluated by means of (59). The second term contains $[\partial g(\tau_1, t)/\partial\tau_1]_{\tau_1=t}$ which can be evaluated by differentiation of (58); one obtains

$$[\partial g(\tau_1, t)/\partial\tau_1]_{\tau_1=t} = \beta[2 - g(t, t)]. \quad (68)$$

The third term of (66) is equal to the second term by symmetry. Subtracting the corresponding differential quotients of $e^{-\beta|\tau_1-\tau_2|}$ one obtains (14b) with $\sigma_1 = g(t, t)$.

The proof of (14d) starts from

$$dg(0, t)/dt = [\partial g(0, \tau)/\partial t]_{\tau=t} + [\partial g(0, \tau)/\partial\tau]_{\tau=t} \quad (69)$$

and follows very closely the preceding derivation.

[9] G. N. Watson, "A Treatise on the Theory of Bessel Functions," Cambridge University Press, Cambridge, Eng., 2nd ed., p. 498; 1952.

[10] A. J. F. Siegert, "A Systematic Approach to a Class of Problems in the Theory of Noise and Other Random Phenomena. II. Examples," The RAND Corp., Paper P-730, sec, 3; September 1955. In this paper, the function h in (3.8), p. 16, should be replaced by f. The function g in (3.21), p. 19, should be replaced by $g^* \equiv g - \rho$. The right-hand sides of (3.23), p. 19, and (3.24), p. 20, should have positive signs, and the right-hand side of (3.25), p. 20, should be $\beta[2 - g(t, t)]$.

[11] Eq. (61) is essentially the expression for the Fredholm determinant of a kernel expressed in terms of its Volterra reciprocal function. See E. T. Whittaker and G. N. Watson, "A Course of Modern Analysis," Cambridge University Press, Cambridge, Eng., sec. 11.21, example 2, and sec. 11.22; 1940.

Appendix I

Eq. (21) can be written in the form

$$u'' + 2\beta \frac{dt(x)}{dx}(u' - u) = 0. \quad (70)$$

The Wronskian $\omega(x)$ defined by

$$\omega(x) = u_2'u_1 - u_1'u_2$$

can thus be conviently computed as

$$\omega(x) = \omega(a) \exp\{2\beta[t(a) - t(x)]\}. \quad (71)$$

If $u_1(x)$, $u_2(x)$ are two linearly-independent solutions of (70) then the initial condition for $u(x)$ is satisfied by

$$u(x) = [u_2'(x_0) - u_2(x_0)]u_1(x) - [u_1'(x_0) - u_1(x_0)]u_2(x) \quad (72)$$

unless both coefficients vanish. Using (70) once more one obtains

$$u(x) = -\left(2\beta \frac{dt}{dx}\right)_{x_0}^{-1} \{u_2''(x_0)u_1(x) - u_1''(x_0)u_2(x)\} \quad (73)$$

and

$$u(x_0) = -\left(2\beta \frac{dt}{dx}\right)_{x_0}^{-1} d\omega(x_0)/dx_0 = -\omega(a)\, e^{2\beta t(a)} \quad (74)$$

or

$$f = [u(x_0)/u(x)]^{1/2} = [\omega(a)\, e^{2\beta t(a)} \cdot K(0)/2\beta\{u_2''(x_0)u_1(x) - u_1''(x_0)u_2(x)\}]^{1/2}. \quad (75)$$

Appendix II

To obtain (8) for the function $\hat{r}(\eta \mid \zeta, t, \lambda)$ defined by (7), we write the Gaussian random function $x(t)$ in the form

$$x(t) = \sum_\nu c_\nu \sqrt{\lambda_\nu}\, \varphi_\nu(t) \quad (76)$$

where the numbers λ_ν and the functions $\phi_\nu(t)$ are defined as the eigenvalues and eigenfunctions of the integral equation

$$\int_0^t \rho(\tau - \tau')K(\tau')\varphi_\nu(\tau')\, d\tau' = \lambda_\nu \varphi_\nu(\tau) \quad (77)$$

with normalization

$$\int_0^t K(\tau)\varphi_\nu^2(\tau)\, d\tau = 1, \quad (78)$$

and where the random variables c_ν are independent and Gaussian with $\langle c_\nu \rangle_{av} = 0$ and $\langle c_\nu^2 \rangle_{av} = 1$.

We then have

$$\left\langle \exp\left(i[\eta x(0) + \zeta x(t)] - \lambda \int_0^t K(\tau)x^2(\tau)\, d\tau\right)\right\rangle_{AV} = \int \exp\left(i \sum_\nu c_\nu \varphi_\nu - \lambda \sum_\nu \lambda_\nu c_\nu^2\right) \prod_\nu \exp\left(-\frac{c_\nu^2}{2}\right) \frac{dc_\nu}{\sqrt{2\pi}} \quad (79)$$

with

$$\psi \equiv \sqrt{\lambda_\nu}\, (\eta\varphi_\nu(0) + \zeta\varphi_\nu(t)), \quad (80)$$

since

$$\int_0^t K(\tau)x^2(\tau)\, d\tau = \sum_{\mu\nu} \sqrt{\lambda_\nu\lambda_\mu}\, c_\nu c_\mu \cdot \int_0^t K(\tau)\varphi_\nu(\tau)\varphi_\mu(\tau)\, d\tau = \sum_\nu \lambda_\nu c_\mu^2. \quad (81)$$

Evaluation of the integral (79) yields

$$\hat{r}(\eta \mid \zeta, t, \lambda) = \prod_\nu (1 + 2\lambda\lambda_\nu)^{-1/2} \cdot \exp\left(-\frac{1}{2} \sum_\nu \frac{\lambda_\nu[\eta\varphi_\nu(0) + \zeta\varphi_\nu(t)]^2}{(1 + 2\lambda\lambda_\nu)}\right). \quad (82)$$

We define the function $g_\lambda(\tau_1, \tau_2)$ by

$$g_\lambda(\tau_1, \tau_2) = \sum \frac{\lambda_\nu\varphi_\nu(\tau_1)\varphi_\nu(\tau_2)}{1 + 2\lambda\lambda_\nu} \quad (83)$$

and obtain from (77) that

$$g_\lambda(\tau_1, \tau_2) + 2\lambda \int_0^t \rho(\tau_1 - \tau')K(\tau')g_\lambda(\tau', \tau_2) = \sum_\nu \lambda_\nu\varphi_\nu(\tau_1)\varphi_\nu(\tau_2) = \rho(\tau_1 - \tau_2). \quad (84)$$

In terms of $g_\lambda(\tau_1, \tau_2)$ we can now express the exponent in (82):

$$\sum_\nu \frac{\lambda_\nu[\eta\varphi_\nu(0) + \zeta\varphi_\nu(t)]^2}{1 + 2\lambda\lambda_\nu} = \eta^2 g_\lambda(0, 0) + 2\eta\zeta g_\lambda(0, t) + \zeta^2 g_\lambda(t, t). \quad (85)$$

The product is obtained by writing

$$\frac{\partial}{\partial\lambda} \ln \prod_\nu (1 + 2\lambda\lambda_\nu)^{-1/2} = -\sum_\nu \frac{\lambda_\nu}{1 + 2\lambda\lambda_\nu} = -\int_0^t K(\tau)g_\lambda(\tau, \tau)\, d\tau. \quad (86)$$

Since the product is equal to unity when $\lambda = 0$, we get

$$\prod_\nu (1 + 2\lambda\lambda_\nu)^{-1/2} = \exp\left(-\int_0^\lambda d\kappa \int_0^t K(\tau)g_\kappa(\tau, \tau)\, d\tau\right). \quad (87)$$

25

Reprinted from *IRE Trans. Information Theory*, **IT-4**(1), 4–14 (1958)

A Systematic Approach to a Class of Problems in the Theory of Noise and Other Random Phenomena —Part III, Examples*

A. J. F. SIEGERT†

Summary—The method of Part I is applied to the problem of finding the characteristic function for the probability distribution of $\int_0^t \sum_{jk} x_j(\tau)\, K_{jl}(\tau) x_l(\tau)\, d\tau$, where $x_j(\tau)$ denotes the jth component of a stationary n-dimensional Markoffian Gaussian process. The problem is reduced to the problem of solving $2n$ first-order linear differential equations with initial conditions only. For the case of constant K, the explicit solution is given in terms of the eigenvalues and the first $2n - 1$ powers of a constant $2n \times 2n$ matrix. For the case of a symmetric correlation matrix which commutes with K, the problem is reduced to the one-dimensional case treated in Part II. For the case $K_{ij}(t) = \delta_{il}\delta_{jl}e^{-t}$, where the functional represents the output of a receiver consisting of a lumped circuit amplifier, a quadratic detector, and a single-stage amplifier, the solution has been obtained in a form which is more explicit than that provided by the earlier methods.

I. Introduction

In this part, the method described in Part I[1] is applied to the problem of finding the characteristic function for the probability distribution of $\mathfrak{F} \equiv \int_0^t \sum_{i,k} x_j(\tau) K_{jl}(\tau) x_l(\tau) d\tau$, with or without conditions on $x_j(0)$ and $x_j(t)$, where the functions $x_j(\tau)$ are the components of a stationary n-dimensional Markoffian Gaussian process $x(\tau)$.

The older method of attacking this reduces it to the problem of finding the solution of a certain integral equation. Special cases, for instance the probability distribution of the output of a radio receiver with square law detector and the probability distribution of the filtered output of a multiplier,[2] have been treated in this way. Another special case of interest is the joint distribution of the sample cross-correlation coefficients.

In its original form the older method required finding the eigenvalues of a homogeneous integral equation and the calculation of the Fredholm determinant as an infinite product. It is possible to avoid the latter part of the calculation by solving instead an inhomogeneous integral equation, which is essentially the equation for the Volterra reciprocal kernel; for the special case of the Markoff process, the integral equation can be reduced to a differential equation.[3] However, the actual determination of that solution of the differential equation which also solves the integral equation requires a rather tedious procedure of satisfying boundary and matching conditions even if the differential equation can be solved. It is not surprising, therefore, that the number and type of cases which have been solved explicitly have been very restricted.

While there are approximation methods valid in limiting cases, the exploration for cases which are exactly soluble seems to be of importance, since in Part I a perturbation formalism was derived, which permits the calculation of the characteristic function for functionals "in the neighborhood" of one for which the solution can be obtained. For this purpose, one needs not only the characteristic function of $\mathfrak{F}$, but also a function $\hat{r}$ which is similar to a joint characteristic function and defined as

$$\hat{r}(\eta_1, \eta_2 \cdots \eta_n; \xi_1, \xi_2 \cdots \xi_n; t, \lambda) = \left\langle \exp\left\{ i \sum_{k=1}^{n} (\eta_k x_k(0) + \xi_k x_k(t)) - \lambda \mathfrak{F} \right\} \right\rangle_{\text{Av}}. \tag{1}$$

In Part II we studied this function for the Uhlenbeck process $x(t)$ and $\mathfrak{F} = \int_0^t K(\tau)\, x^2(\tau) d\tau$ and found that it is of the form

$$\hat{r}(\eta, \xi; t, \lambda) = f \cdot \exp\{-\tfrac{1}{2}(a\eta^2 + 2b\xi\eta + c\xi^2)\}, \tag{2}$$

where f, a, b, and c are functions of t only, and satisfy a system of first-order differential equations with initial conditions only. The function c specially was found to be the solution of a Riccati equation, which could be converted by standard procedure into a linear second-order differential equation for a function u, again with initial conditions only. If u could be determined, then f, a, and b could be expressed in terms of u, by quadratures or more simply; we found, for instance, that f was given simply by

$$f(t) = \{u(0)/u(t)\}^{1/2}.$$

Since f was expressed by the older method as the reciprocal square root of the Fredholm determinant of the kernel $e^{-\beta|\tau_1 - \tau_2|} K(\tau_2)$, this showed that the Fredholm determinant of this particular kernel satisfies a second-order linear differential equation as a function of the upper limit of the integral equation. Furthermore, the functions

* Manuscript received by the PGIT, August 16, 1957.

† Northwestern University, Evanston, Ill. and consultant for The RAND Corp., Santa Monica, Calif.

[1] D. A. Darling and A. J. F. Siegert, "A Systematic Approach to a Class of Problems in the Theory of Noise and Other Random Phenomena—Part I," The RAND Corp., Santa Monica, Calif., Paper P-738; September 10, 1955, and IRE Trans. on Information Theory, vol. 3, pp. 32–37; March, 1957. See also, A. J. F. Siegert, "Part II, Examples," The RAND Corp., Paper P-730; September, 1955, and IRE Trans. on Information Theory, vol. 3, pp. 38–43; March, 1957.

[2] D. G. Lampard, "The probability distribution for the filtered output of a multiplier whose inputs are correlated, stationary, Gaussian time-series," IRE Trans. on Information Theory, vol. 2, pp. 4–11; March, 1956. References to earlier publications using the older method are given in this paper.

[3] A. J. F. Siegert, "Passage of stationary processes through linear and non-linear devices," IRE Trans. on Information Theory, vol. 3, pp. 4–25; March, 1954.

a, b, and c are essentially the values $g(0, 0)$, $g(0, t)$, and $g(t, t)$ of the Volterra reciprocal kernel $g(\tau_1, \tau_2)$; one thus had the result that these special values of g can be obtained trivially from the Fredholm determinant, for the special form of the kernel,[4] so that it is not necessary to solve the integral equation in order to obtain these special values of g.

In the present paper these results have been generalized to the case of quadratic functionals of a stationary, n-dimensional Markoffian Gaussian random process. The analogs of the functions a, b, and c are now matrices of n rows and columns, which are functions of t. In Section II we obtain two first-order linear differential equations, with initial conditions only, for the matrices $u \equiv b^{-1}$ and $v \equiv cb^{-1}$. The Fredholm determinant of the older method now reduces, but for a trivial factor, to the determinant of u. In Section III an explicit solution is given for the case of constant K, which requires only the finding of the eigenvalues and the first $2n-1$ powers of a certain $2n \times 2n$ matrix with constant elements, or alternatively the eigenvalues and eigenvectors of this matrix. The case where the correlation matrix is symmetric and commutes with K has been reduced in Section IV to the one-dimensional problem. In Section V the principal equation of the present method was derived directly by means of the older method. The characteristic function for the probability distribution of $\int_0^\infty e^{-t}x_1^2(t)dt$ has been obtained in Section VI in an explicit form, as the reciprocal square root of an $n \times n$ determinant whose elements are series of hypergeometric type.

The linear first-order equations, of course, are in general not easy to solve. They seem to be, however, a more economical formulation of the problem, freed of the necessity of solving the integral equation, in a case where one is interested only in the Fredholm determinant and certain special values of the reciprocal kernel. Therefore, they also may provide a more convenient basis for numerical computation.

II. Reduction of the Problem to Linear First-Order Differential Equations

The n-dimensional stationary Gaussian process $X(t) = \{x_1(t), x_2(t), \cdots x_n(t)\}$ with mean zero is described by the correlation matrix $R(\tau)$ with components[5]

$$R_{kl}(\tau) = \langle x_k(t)x_l(t+\tau)\rangle_{\text{Av}}; \qquad \tau \geq 0 \tag{3}$$

$$R(\tau) = \tilde{R}(-\tau) \tag{4}$$

where the tilde denotes transposed matrix. For stationary Markoffian Gaussian process, $R(\tau)$ has the form

$$R(\tau) = e^{Q\tau} \tag{5}$$

for $\tau \geq 0$, if $R(0)$ is the unit matrix.[6]

The characteristic function χ_2 of the joint probability distribution of $X(0)$ and $X(\tau)$ is then

$$\chi_2(\eta_1, \cdots \eta_n, \xi_1, \cdots \xi_n, t) \equiv \left\langle \exp\left[i \sum_k (\eta_k x_k(0) + \xi_k x_k(t))\right]\right\rangle_{\text{Av}} = \exp\left\{-\frac{1}{2}\left[\sum_k (\eta_k^2 + \xi_k^2) + 2\sum_{k,l} \eta_k R_{kl}(t)\xi_l\right]\right\}. \tag{6}$$

The characteristic function χ_2 satisfies the differential equation

$$\frac{\partial \chi_2}{\partial t} = \sum_{i,l}\left(\frac{\partial \chi_2}{\partial \xi_i} + \xi_i\chi_2\right)Q_{il}\xi_l \tag{7}$$

which can be verified directly using (5).

To obtain the characteristic function for the distribution of the functional

$$\mathfrak{F} \equiv \int_0^t \sum_{i,l} x_i(\tau)K_{il}(\tau)x_l(\tau)\, d\tau \tag{8}$$

[with or without conditions on $X(0)$ and $X(t)$] where K is a symmetric matrix, we define the characteristic function for the joint distribution of $X(0)$, $X(t)$, and u,

$$\hat{f}(\eta_1 \cdots \eta_n; \xi_1, \cdots \xi_n, t, \lambda) \equiv \langle \exp\{i \sum_k (\eta_k x_k(0) + \xi_k x_k(t)) - \lambda\mathfrak{F}\}\rangle_{\text{Av}} \tag{9}$$

where λ is taken to be real if $\mathfrak{F} > 0$, and negative imaginary otherwise.

Proceeding as in Parts I and II, we then obtain

$$\frac{\partial \hat{f}}{\partial t} = \sum_{i,l}\left(\frac{\partial \hat{f}}{\partial \xi_i} + \xi_i\hat{f}\right)Q_{il}\xi_l + \lambda \sum_{i,l} K_{il}(t)\frac{\partial^2 \hat{f}}{\partial \xi_i\, \partial \xi_l}. \tag{10}$$

The initial condition is obtained from (9) as

$$\hat{f}_{t=0} = \exp\{-\tfrac{1}{2}\sum_k (\eta_k + \xi_k)^2\}. \tag{11}$$

Eq. (10) is reduced to a set of first-order differential equations with t as the independent variable by the ansatz

$$\hat{f} = fe^{\phi} \tag{12}$$

where f is a function of t alone, and

$$\phi \equiv -\tfrac{1}{2}\sum_{i,k} [\eta_i a_{ik}(t)\eta_k + 2\eta_i b_{ik}(t)\xi_k + \xi_i c_{ik}(t)\xi_k] \tag{13}$$

where a and c are symmetric matrices depending only on t. Eq. (10) then becomes

$$\frac{\dot{f}}{f} + \dot{\phi} = \sum_{i,l}\left(\frac{\partial \phi}{\partial \xi_i} + \xi_i\right)Q_{il}\xi_l + \lambda \sum_{i,l} K_{il}\left(\frac{\partial^2 \phi}{\partial \xi_i\, \partial \xi_l} + \frac{\partial \phi}{\partial \xi_i}\frac{\phi}{\partial \xi_l}\right) \tag{14}$$

[4] Some of these relations could be shown to be independent of the special form of the kernel. Recently, R. Bellman obtained a Riccati type integrodifferential equation for the Fredholm resolvent of a Fredholm integral equation; see, "Functional Equations in the Theory of Dynamic Programming-VII: An Integro-Differential Equation for the Fredholm Resolvent," The RAND Corp., Santa Monica, Calif., Paper P-859; May 7, 1956.

[5] M. C. Wang and G. E. Uhlenbeck, "On the theory of the Brownian Motion II," *Rev. Mod. Phys.*, vol. 17, pp. 323–342; August, 1945.

[6] This can be assumed without loss of generality; see *ibid*, footnote 19, p. 330.

where dots indicate differentiation with respect to t. Using the symmetry of the matrix c, one has

$$\frac{\partial\phi}{\partial\xi_j} = -\sum_i \eta_i b_{ij} - \tfrac{1}{2}\sum_i (\xi_i c_{ij} + c_{ji}\xi_i) = -\sum_i (\eta_i b_{ij} + \xi_i c_{ij}) \qquad (15)$$

and (14) becomes

$$d \ln f/dt - \tfrac{1}{2}\sum_{i,k} (\eta_i \dot{a}_{ik}\eta_k + 2\eta_i \dot{b}_{ik}\xi_k + \xi_i \dot{c}_{ik}\xi_k) \qquad (16)$$

$$= -\sum_{j,l}\Big(\sum_i (\eta_i b_{ij} + \xi_i c_{ij}) - \xi_j\Big)Q_{jl} + \lambda \sum_{j,l} K_{jl}\Big[-c_{lj} + \sum_{ik}(\eta_i b_{ij} + \xi_i c_{ij})(\eta_k b_{kl} + \xi_k c_{kl})\Big]$$

$$= -\lambda \sum_{jl} c_{lj}K_{jl} + \lambda \sum_{i,j,k,l} \eta_i(b_{ij}K_{jl}b_{lk})\eta_k + \sum_{i,k}\eta_i\Big[-\sum_j b_{ij}Q_{jk} + 2\lambda \sum_{jl} b_{ij}K_{jl}c_{lk}\Big]\xi_k + \sum_{i,k}\xi_i\Big[\Big(-\sum_j c_{ij}Q_{jk} + Q_{ik}\Big) + \lambda \sum_{jl} c_{ij}K_{jl}c_{lk}\Big]\xi_k$$

$$= -\lambda \operatorname{tr}(Kc) + \lambda \sum_{i,k}\eta_i(bKb)_{ik}\eta_k + \sum_{ik}\eta_i[-(bQ)_{ik} + 2\lambda(bKc)_{ik}]\xi_k + \sum_{i,k}\xi_i[-(cQ)_{ik} + Q_{ik} + \lambda(cKc)_{ik}]\xi_k$$

$$= -\lambda \operatorname{tr}(Kc) + \frac{\lambda}{2}\sum_{i,k}\eta_i(bKb + \tilde{b}K\tilde{b})\eta_k + \sum_{i,k}\eta_i[-(bQ)_{ik} + 2\lambda(bKc)_{ik}]\xi_k + \tfrac{1}{2}\sum_{i,k}\xi_i(-cQ - \tilde{Q}c + Q + \tilde{Q} + 2\lambda cKc)_{ik}\xi_k$$

where the second and fourth terms have been written in symmetric form in order to facilitate comparison of coefficients, and where tr indicates the trace of a matrix. One thus obtains

$$\frac{d \ln f}{dt} = -\lambda \operatorname{tr}(Kc) \qquad (17a)$$

$$\dot{c} = -(Q + \tilde{Q}) + cQ + \tilde{Q}c - 2\lambda cKc \qquad (17b)$$

$$\dot{a} = -\lambda(bKb + \tilde{b}K\tilde{b}) \qquad (17c)$$

$$\dot{b} = b[Q - 2\lambda Kc]. \qquad (17d)$$

The initial conditions are obtained from (11)–(13) as

$$f(0) = 1 \qquad (18a)$$

$$a(0) = b(0) = c(0) = I, \qquad (18b)$$

where I denotes the unit matrix.

The scalar equation (17a) and the matrix differential equations (17b)–(17d) reduce to (14a)–(14d) of Part II for the one-dimensional case. Eqs. (17a) and (17c) can still be solved by quadratures, but (17d) is no longer as simple as the corresponding (14d) of Part II. It turns out, however, that the solution of (17d) depends on the solution of the same set of linear equations as (17b) and therefore we will start with the latter.

Eq. (17b) is a matrix analog of the scalar Riccati equation, but the familiar substitution leading from the latter to a second-order linear differential equation does not work for the matrix equation. A similar substitution, however, leads to two linear differential equations for two matrices u and v. We write c in the form

$$c = vu^{-1} \qquad (19)$$

and obtain

$$\dot{v}u^{-1} - vu^{-1}\dot{u}u^{-1} = -(Q + \tilde{Q}) + vu^{-1}Q + \tilde{Q}vu^{-1} - 2\lambda vu^{-1}Kvu^{-1} \qquad (20)$$

or

$$\dot{v} = \tilde{Q}v - (Q + \tilde{Q})u + vu^{-1}(\dot{u} + Qu - 2\lambda Kv). \qquad (21)$$

Since we are free to choose one relation between u and v, we can let the factor of vu^{-1} be zero and obtain the two linear differential equations

$$\dot{v} = \tilde{Q}v - (Q + \tilde{Q})u \qquad (22)$$

$$\dot{u} = 2\lambda Kv - Qu. \qquad (23)$$

The initial condition is $u(0) = v(0)$, since $c(0) = I$. The matrix Riccati equation (17b) is thus reduced to a pair of first-order linear differential equations for the matrices u and v, *i.e.*, to a system of $2n$ first-order linear differential equations.

The matrix b turns out to be simply

$$b(t) = u(0)u^{-1}(t), \qquad (24)$$

since (17d), (19), and (23) yield

$$\dot{b} = -b\dot{u}u^{-1} \qquad (25)$$

or

$$\dot{b}u + b\dot{u} = 0 \qquad (26)$$

and, therefore,

$$bu = b(0)u(0) = u(0). \qquad (27)$$

The function f also is expressed simply in terms of the determinant of u

$$\frac{d}{dt}\det u = \sum_{ik}\frac{du_{ik}}{dt}\frac{\partial \det u}{\partial u_{ik}} = \sum_{ik}\frac{du_{ik}}{dt}(u^{-1})_{ik}\det u, \qquad (28)$$

where det u denotes the determinant of u. We thus have

$$\frac{d \ln \det u}{dt} = \operatorname{tr}\dot{u}u^{-1} = \operatorname{tr}(2\lambda Kvu^{-1} - Q) = -\operatorname{tr} Q + 2\lambda \operatorname{tr} Kc \qquad (29)$$

and

$$\frac{d \ln f}{dt} = -\frac{1}{2} \operatorname{tr} Q - \frac{1}{2}\frac{d}{dt} \ln \det u. \tag{30}$$

Since $f(0) = 1$, we thus have

$$f(t) = e^{-t \operatorname{tr} Q/2} \sqrt{\frac{\det u(0)}{\det u(t)}}. \tag{31}$$

The matrix $u(0)$ can be chosen arbitrarily, as long as the matrix $v(0)$ is chosen equal to $u(0)$.

One could have obtained (22)–(24) and (19) directly, using (17d) to obtain

$$\frac{db^{-1}}{dt} = -b^{-1}\dot{b}b^{-1} = -(Qb^{-1} - 2\lambda Kcb^{-1}) \tag{32}$$

and (17b) to obtain

$$\begin{aligned}\frac{d(cb^{-1})}{dt} &= \dot{c}b^{-1} + c\frac{db^{-1}}{dt} \qquad (33)\\ &= -(Q + \tilde{Q})b^{-1} + cQb^{-1} + \tilde{Q}cb^{-1}\\ &\quad - 2\lambda cKcb^{-1} - c(Qb^{-1} - 2\lambda Kcb^{-1})\\ &= -(Q + \tilde{Q})b^{-1} + \tilde{Q}cb^{-1}\end{aligned}$$

so that one has two linear equations for b^{-1} and cb^{-1}.

III. The Case of Constant Matrix K

This case is of interest since the characteristic function of $\mathfrak{F}$ defined by (8), is for constant K, closely related to the characteristic function of the empirical variances and covariances of the components of $X(\tau)$.

The solution then can be reduced to the solution of $2n$ homogeneous linear algebraic equations, by writing (22) and (23) in the form

$$\begin{pmatrix} \dot{v} \\ \dot{u} \end{pmatrix} = \mathfrak{M} \begin{pmatrix} v \\ u \end{pmatrix} \tag{34}$$

where $\mathfrak{M}$ is the $2n \times 2n$ matrix

$$\mathfrak{M} = \begin{pmatrix} \tilde{Q} & -(Q + \tilde{Q}) \\ 2\lambda K & -Q \end{pmatrix}. \tag{35}$$

(Script capitals will be used to denote the $2n \times 2n$ matrices.)

Choosing the initial conditions $u(0) = v(0) = I$, we then have

$$\begin{pmatrix} v \\ u \end{pmatrix} = e^{\mathfrak{M}t} \begin{pmatrix} I \\ I \end{pmatrix} \tag{36}$$

and

$$u = (e^{\mathfrak{M}t})_{21} + (e^{\mathfrak{M}t})_{22} \tag{37}$$

$$e^{\mathfrak{M}t} = \begin{pmatrix} (e^{\mathfrak{M}t})_{11} & (e^{\mathfrak{M}t})_{12} \\ (e^{\mathfrak{M}t})_{21} & (e^{\mathfrak{M}t})_{22} \end{pmatrix}. \tag{38}$$

where $(e^{\mathfrak{M}t})_{21}$ and $(e^{\mathfrak{M}t})_{22}$ are $n \times n$ submatrices of $e^{\mathfrak{M}t}$, obtained by dividing $e^{\mathfrak{M}t}$ in the form

The unconditional characteristic function $f(t)$ for the distribution of $\mathfrak{F}(t)$ is obtained thus from (31) as

$$f(t) = e^{-t \operatorname{tr} Q/2}[\det \{(e^{\mathfrak{M}t})_{21} + (e^{(\mathfrak{M}t)})_{22}\}]^{-1/2}. \tag{39}$$

If the matrix $\mathfrak{M}$ can be diagonalized,

$$\mathfrak{M} = \mathfrak{C}^{-1}\mathfrak{D}\mathfrak{C}, \tag{40}$$

where $\mathfrak{D}$ is a diagonal matrix, the matrix $e^{\mathfrak{M}t}$ can be evaluated by using

$$e^{\mathfrak{M}t} = \mathfrak{C}^{-1}e^{\mathfrak{D}t}\mathfrak{C}, \tag{41}$$

where $e^{\mathfrak{D}t}$ is the diagonal matrix with elements $\exp(\mathfrak{D}_{ii}t)$.

Alternatively, one can evaluate $e^{\mathfrak{M}t}$ by the following procedure, which does not depend on the possibility of diagonalizing $\mathfrak{M}$. Let $D(z)$ be the secular determinant of $\mathfrak{M}$,

$$D(z) = \det (zI - \mathfrak{M}) \tag{42}$$

which is a polynomial of degree $2n$ in z. Define the function $\phi(\mu, z)$ by

$$\phi(\mu, z) = \frac{D(\mu) - D(z)}{\mu - z}. \tag{43}$$

[The explicit form of $\phi(\mu, z)$ is given in (49).]

Since $(\mu - z)$ is a divisor of $D(\mu) - D(z)$, the function $\phi(\mu, z)$ is a polynomial of degree $(2n - 1)$ in both μ and z, and is symmetric in μ and z. We then have

$$e^{\mathfrak{M}t} = \frac{1}{2\pi i}\oint e^{\mu t}\frac{\phi(\mu, \mathfrak{M})}{D(\mu)}\, d\mu \tag{44}$$

where the path of integration encircles all roots of $D(\mu)$, and the integral can be evaluated by the method of residues and appears as a polynomial of degree $2n - 1$ in the matrix $\mathfrak{M}$ with time-dependent coefficients. The fact that $e^{\mathfrak{M}t}$ has this particular form can already be concluded from the Cayley[7] theorem, which states that

$$D(\mathfrak{M}) = 0 \tag{45}$$

and allows expression of $\mathfrak{M}^{2n}$ and all higher powers of $\mathfrak{M}$ in the power series for $e^{\mathfrak{M}t}$ by the first $2n - 1$ powers of $\mathfrak{M}$.

That the coefficients are correctly given by (44) can be seen by differentiating (44). One obtains

$$\begin{aligned}\left(\frac{d}{dt} - \mathfrak{M}\right)e^{\mathfrak{M}t} &= \frac{1}{2\pi i}\oint e^{\mu t}\frac{(\mu - \mathfrak{M})\phi(\mu, \mathfrak{M})}{D(\mu)}\, d\mu \qquad (46)\\ &= \frac{1}{2\pi i}\oint e^{\mu t}\, d\mu = 0.\end{aligned}$$

To complete the proof, one need only to show that (44) yields the unit matrix for $t = 0$. We transform to a new variable $\xi = 1/\mu$ and can contract the path of integration to a small circle around $\xi = 0$, and obtain

$$\frac{1}{2\pi i}\oint^{0,+}\frac{\phi(\mu, \mathfrak{M})}{D(\mu)}\, d\mu = \frac{1}{2\pi i}\oint^{0,+}\frac{\phi(\xi^{-1}, \mathfrak{M})}{\xi^2 D(\xi^{-1})}\, d\xi. \tag{47}$$

[7] R. Courant and D. Hilbert, "Methoden der Mathematischen Physik," Springer Verlag, Berlin, Germany, vol. 1, p. 18; 1931.

Now if
$$D(z) = \sum_{s=0}^{2n} b_s z^s, \quad (48)$$

then
$$\begin{aligned}
\phi(\mu, z) &= \sum_{s=1}^{2n} b_s \frac{\mu^s - z^s}{\mu - z} \\
&= \sum_{s=1}^{2n} b_s \sum_{r=0}^{s-1} \mu^r z^{s-1-r} \\
&= \sum_{r=0}^{2n-1} \mu^r \sum_{s=r+1}^{2n} b_s z^{s-1-r}
\end{aligned} \quad (49)$$

and
$$\begin{aligned}
\oint^{0,+} \frac{\phi(\xi^{-1}, \mathfrak{M})}{\xi^2 D(\xi^{-1})} d\xi &= \oint^{0,+} \frac{\sum_{r=0}^{2n-1} \xi^{-r} \sum_{s=r+1}^{2n} b_s \mathfrak{M}^{s-1-r}}{\xi^2 \sum_{s=0}^{2n} b_s \xi^{-s}} \\
&= \oint^{0,+} \frac{d\xi}{\xi} \frac{b_{2n} I + \xi(b_{2n}\mathfrak{M} + b_{2n-1}I) + \cdots}{b_{2n} + \xi b_{2n-1} + \cdots} \\
&= 2\pi i I.
\end{aligned}$$

IV. REDUCTION OF A SPECIAL CASE TO THE ONE-DIMENSIONAL PROBLEM

If Q is symmetric (which implies real eigenvalues), it can be written in the form
$$Q = \tilde{Q} = S^{-1} \Lambda S \quad (50)$$

where Λ is a diagonal matrix, since by definition Q is real. It turns out that the matrix S need not be known explicitly. Eqs. (22) and (23) then can be reduced to
$$\dot{V} = \Lambda V - 2\Lambda U \quad (51)$$
$$\dot{U} = 2\lambda \bar{K} V - \Lambda U \quad (52)$$

where
$$V = Sv, \quad U = Su, \quad \bar{K} = SKS^{-1}. \quad (53)$$

Especially if K commutes with Q, the matrix S can be chosen so that $\bar{K}$ is diagonal and, if its elements do not vanish identically,[8] (51) and (52) reduce to those of the one-dimensional case. If $\bar{u}$ is defined by
$$\bar{u} = e^{\Lambda t} U, \quad (54)$$

then
$$\dot{\bar{u}} = e^{\Lambda t}(\dot{U} + \Lambda V) = e^{\Lambda t} 2\lambda \bar{K} V. \quad (55)$$

Since $\bar{K}$ was assumed to be diagonal, we then have
$$\begin{aligned}
\left(-\frac{1}{2}\Lambda^{-1}\frac{d}{dt} + 1\right)[(2\lambda\bar{K})^{-1}\dot{\bar{u}}] &= \left(-\frac{1}{2}\Lambda^{-1}\frac{d}{dt} + 1\right)(e^{\Lambda t} V) \\
&= e^{\Lambda t} U.
\end{aligned} \quad (56)$$

---

[8] If some elements of $\bar{K}$ vanish, the corresponding rows of U are obtained directly from (52) and the initial condition as $U_{jk} = e^{\Lambda jj t} \delta_{jk}$, otherwise the calculation is unchanged.

We thus obtain, with
$$\Lambda_{jj} = -\beta_j \quad (57)$$
and
$$\bar{K}_{jj} = \kappa_j(t), \quad (58)$$

the equation
$$\frac{1}{2}\beta_j^{-1}\frac{d}{dt}\left(\frac{1}{2\lambda\kappa_j(t)}\dot{\bar{u}}_{jk}\right) + \frac{1}{2\lambda\kappa_j(t)}\dot{\bar{u}}_{jk} - \bar{u}_{jk} = 0 \quad (59)$$

or with
$$dx_j = 2\lambda\kappa_j(t) \, dt \quad (60)$$

the equation
$$\lambda\beta_j^{-1}k_j(t(x_j)) \, d^2\bar{u}_{jk}/dx_j^2 + d\bar{u}_{jk}/dx_j - \bar{u}_{jk} = 0 \quad (61)$$

which are the same as (21) of Part II, and were solved there explicitly for some special forms of $k_j(t)$.

The initial condition $u(0) = v(0)$ requires only $U(0) = V(0)$ and the initial condition for (61) is, therefore,
$$\left(\frac{d\bar{u}}{dt}\right)_{t=0} = 2\lambda\bar{K}(0)\bar{u}_{t=0}. \quad (62)$$

If $\kappa_j(0)$ does not vanish, this simplifies to
$$\left[\frac{d\bar{u}_{jk}}{dx_j}\right]_{t=0} = [u_{jk}]_{t=0}. \quad (63)$$

Without losing generality, one can choose
$$[\bar{u}_{t=0}] = I; \quad (64)$$

then $\bar{u}$ is a diagonal matrix for all x [provided that $\kappa_j(0) < \infty$] and one has
$$f(t) = \left\{\prod_{j=1}^{n} \bar{u}_{jj}(x(t))\right\}^{-1/2}$$

with $\bar{u}_{jj}(x)$ defined by (61), (63), and (64).

V. DERIVATION OF THE MATRIX RICCATI EQUATION BY THE OLDER METHOD

Since $R_{jj}(\tau)$ is the autocorrelation function of $x_j(t)$, it can be written in the form
$$R_{jj}(\tau) = \int_{-\infty}^{\infty} h_j(\vartheta) h_j(\tau + \vartheta) \, d\vartheta, \quad (65)$$

where one may demand $h_j(\vartheta) = 0$ for $\vartheta < 0$. One then can write $x_j(t)$ in the form
$$x_j(t) = \sum_{\nu} c_{\nu} \int_{-\infty}^{\infty} h_j(t - \vartheta) \psi_{\nu}(\vartheta) \, d\vartheta, \quad (66)$$

where the coefficients c_{ν} are independent Gaussian random variables with mean zero and variance unity, and where the functions $\psi_{\nu}(\vartheta)$ are a complete orthonormal set. Then the correlation matrix is given by
$$R_{kl}(\tau) = \int_{-\infty}^{\infty} h_k(\vartheta) h_l(\tau + \vartheta) \, d\vartheta. \quad (67)$$

Now let the functions $\psi_\nu(t)$ be the eigenfunctions of the integral equation

$$\lambda_\nu \psi_\nu(\vartheta_1) = \int_{-\infty}^{\infty} \Lambda(\vartheta_1, \vartheta_2)\psi_\nu(\vartheta_2)\, d\vartheta_2, \tag{68}$$

where

$$\Lambda(\vartheta_1, \vartheta_2) = \int_0^t \sum_{j,l} h_j(\vartheta - \vartheta_1)K_{jl}(\vartheta)h_l(\vartheta - \vartheta_2)\, d\vartheta; \tag{69}$$

then we have

$$\sum_{jl} \int_0^t x_j(\vartheta)K_{jl}(\vartheta)x_l(\vartheta)\, d\vartheta = \sum_\nu c_\nu^2 \lambda_\nu, \tag{70}$$

using (66), (68), and (69).

One then obtains for the joint characteristic function

$$\begin{aligned} &\left\langle \exp\left[i \sum_j (\eta_j x_j(0) + \xi_j x_j(t)) - \lambda \int_0^t \sum_{j,l} K_{jl}(\tau)x_j(\tau)x_l(\tau)\, d\tau \right]\right\rangle_{Av} \\ &= \left\langle \exp\left[i \sum_\nu c_\nu \Psi_\nu - \lambda \sum_\nu c_\nu^2 \lambda_\nu \right]\right\rangle_{Av} \\ &= \prod_\nu (1 + 2\lambda\lambda_\nu)^{-1/2} \cdot \exp\left[-\frac{1}{2} \sum_\nu \Psi_\nu^2/(1 + 2\lambda\lambda_\nu)\right] \end{aligned} \tag{71}$$

with the abbreviations

$$\Psi_\nu \equiv \sum_j [\eta_j \phi_{j\nu}(0) + \xi_j \phi_{j\nu}(t)] \tag{72}$$

and

$$\phi_{j\nu}(t) \equiv \int_{-\infty}^{\infty} h_j(t - \vartheta)\psi_\nu(\vartheta)\, d\vartheta. \tag{73}$$

The function f and the matrix c introduced by (12) and (13), therefore, are given by

$$f = \prod_\nu (1 + 2\lambda\lambda_\nu)^{-1/2} \tag{74}$$

and

$$c_{kl} = \sum_\nu \frac{\phi_{k\nu}(t)\phi_{l\nu}(t)}{1 + 2\lambda\lambda_\nu}. \tag{75}$$

We now define the matrix function $g(\tau_1, \tau_2)$ by

$$\begin{aligned} g_{kl}(\tau_1, \tau_2) &= \sum_\nu \frac{\phi_{k\nu}(\tau_1)\phi_{l\nu}(\tau_2)}{1 + 2\lambda\lambda_\nu} \\ &= \iint_{-\infty}^{\infty} h_k(\tau_1 - \vartheta_1)h_l(\tau_2 - \vartheta_2)\gamma(\vartheta_1, \vartheta_2)\, d\vartheta_1 d\vartheta_2, \end{aligned} \tag{76}$$

where

$$\gamma(\vartheta_1, \vartheta_2) \equiv \sum_\nu \frac{\psi_\nu(\vartheta_1)\psi_\nu(\vartheta_2)}{1 + 2\lambda\lambda_\nu}. \tag{77}$$

Using (68), one sees that

$$\gamma(\vartheta_1, \vartheta_2) + 2\lambda \int_{-\infty}^{\infty} \Lambda(\vartheta_1, \vartheta)\gamma(\vartheta, \vartheta_2)\, d\vartheta = \delta(\vartheta_1 - \vartheta_2) \tag{78}$$

and, therefore,

$$\begin{aligned} &g_{il}(\tau_1, \tau_2) + 2\lambda \iiint h_i(\tau_1 - \vartheta_1)\Lambda(\vartheta_1, \vartheta)\gamma(\vartheta, \vartheta_2) \cdot h_l(\tau_2 - \vartheta_2)\, d\vartheta\, d\vartheta_1\, d\vartheta_2 \\ &= \int h_i(\tau_1 - \vartheta_1)h_l(\tau_2 - \vartheta_1)\, d\vartheta_1 \\ &= g_{il}(\tau_1, \tau_2) + 2\lambda \int h_i(\tau_1 - \vartheta_1) \int_0^t \sum_{jk} h_j(\tau - \vartheta_1)K_{jk}(\tau) \cdot h_k(\tau - \vartheta)\gamma(\vartheta, \vartheta_2)h_l(\tau_2 - \vartheta_2)\, d\tau\, d\vartheta\, d\vartheta_1\, d\vartheta_2 \\ &= g_{il}(\tau_1, \tau_2) + 2\lambda \int_0^t \sum_{jk} R_{ij}(\tau - \tau_1)K_{jk}(\tau)g_{kl}(\tau, \tau_2) \\ &= R_{il}(\tau_2 - \tau_1), \end{aligned} \tag{79}$$

using (76), (69), and (65).

In matrix form this equation is written as

$$g(\tau_1, \tau_2) + 2\lambda \int_0^t R(\tau - \tau_1)K(\tau)g(\tau, \tau_2)d\tau = R(\tau_2 - \tau_1) \tag{80}$$

and $g(t, t)$ is the matrix $c(t)$.

The iteration solution, written for $\tau_1 = \tau_2 = t$, becomes

$$g(t, t) = I + \sum_1^\infty (-2\lambda)^n \int_0^t H^{(n)}(t, \tau)R(t - \tau)\, d\tau \tag{81}$$

where

$$H^{(1)}(\tau_1, \tau) \equiv H(\tau_1, \tau) = R(\tau - \tau_1)K(\tau) \tag{82}$$

and

$$H^{(n)}(\tau_1, \tau_2) = \int_0^t H^{(n-1)}(\tau_1, \tau)H^{(1)}(\tau, \tau_2)\, d\tau. \tag{83}$$

Differentiation yields

$$\frac{dg(t, t)}{dt} = \sum_1^\infty (-2\lambda)^n \left\{ H^{(n)}(t, t) + \int_0^t \left[\frac{dH^{(n)}(t, \tau)}{dt} R(t - \tau) + H^{(n)}(t, \tau)R(t - \tau)Q \right] d\tau \right\}, \tag{84}$$

where

$$\begin{aligned} \frac{dH^{(n)}(t, \tau)}{dt} &= \frac{d}{dt} \int_0^t \cdots \int_0^t H(t, \tau_1)H(\tau_1, \tau_2) \cdots H(\tau_{n-1}, \tau)\, d\tau_1 \cdots d\tau_{n-1} \\ &= \sum_{k=1}^{n-1} H^{(k)}(t, t)H^{(n-k)}(t, \tau) + \int_0^t \frac{H(t, \tau_1)}{\partial t} H^{(n-1)}(\tau_1, \tau)\, d\tau_1. \end{aligned} \tag{85}$$

Using (82), (4), and (5), one has

$$\frac{\partial H(t, \tau)}{\partial t} = \tilde{Q}H(t, \tau). \tag{86}$$

Substituting in (84) and using (81), one obtains

$$\frac{dg(t, t)}{dt} = \tilde{Q}(g(t, t) - I) + (g(t, t) - I)Q + \sum_{1}^{\infty} (-2\lambda)^n \Big\{ H^{(n)}(t, t) + \sum_{k=1}^{n-1} H^{(k)}(t, t) \int_0^t H^{(n-k)}(t, \tau)R(t - \tau)\, d\tau \Big\}. \tag{87}$$

It is convenient to define

$$\int_0^t H^{(0)}(t, \tau)\phi(\tau)\, d\tau = \phi(t) \tag{88}$$

for any function $\phi(\tau)$; then the last sum can be written as

$$\sum_{n=1}^{\infty} (-2\lambda)^n \sum_{k=1}^{n} H^{(k)}(t, t) \int_0^t H^{(n-k)}(t, \tau)R(t - \tau)\, dt$$
$$= \sum_{k=1}^{\infty} \sum_{l=0}^{\infty} (-2\lambda)^{k+l} H^{(k)}(t, t) \int_0^t H^{(l)}(t, \tau)R(t - \tau)\, dt$$
$$= \sum_{m=0}^{\infty} (-2\lambda)^{m+1} H^{(m+1)}(t, t) \cdot \sum_{l=0}^{\infty} (-2\lambda)^l \int_0^t H^{(l)}(t, \tau)R(t - \tau)\, d\tau$$
$$= -2\lambda \cdot \Big[K(t) + \sum_{m=1}^{\infty} (-2\lambda)^m H^{(m+1)}(t, t) \Big] \cdot \Big[I + \sum_{l=1}^{\infty} (-2\lambda) \int_0^t H^{(l)}(t, t)R(t - \tau)\, d\tau \Big] \tag{89}$$

using (82) and (88). We recognize the second bracket as $g(t, t)$ and the first bracket as $g(t, t)\, K(t)$, since

$$R(t - \tau)K(t) = H(\tau, t) \tag{90}$$

and therefore

$$g(t, t)K(t) = K(t) + \sum_{n=1}^{\infty} (-2\lambda)^n \int_0^t H^{(n)}(t, \tau)H(\tau, t)\, d\tau. \tag{91}$$

Now writing $c(t)$ for $g(t, t)$ we obtain from (87) and (89)

$$\frac{dc(t)}{dt} = -\tilde{Q} - Q + \tilde{Q}c(t) + c(t)Q - 2\lambda c(t)K(t)c(t) \tag{92}$$

in agreement with (17b).

VI. Characteristic Functions for the Probability Distribution of $\int_0^\infty e^{-t}\, x_1^2\, (t)\, dt$

For the calculations of this section it is useful to have (22), (23), and (31) in a different form.

Since

$$e^{-t\,\mathrm{tr}\,Q} = \det e^{-Qt}, \tag{93}$$

(31) can be written as

$$f^{-2}(t) = \det\, [e^{Qt}u(t)] \tag{94}$$

when $u(0) = v(0)$ is chosen to be the unit matrix.

From (23) it follows that

$$\frac{d}{dt}\, [e^{Qt}u(t)] = 2\lambda e^{Qt}K(t)v(t) \tag{95}$$

and

$$e^{Qt}u(t) = I + 2\lambda \int_0^t e^{Qt'}K(t')v(t')\, dt'. \tag{96}$$

This equation is specially useful in the limit $t \to \infty$, since the integral, in this limit, can be expressed in terms of the Laplace transform of $K(t')\, v(t')$ with $-\,Q$ substituted for the Laplace variable.

With

$$V_{hl}(t) \equiv [K(t)v(t)]_{hl} \tag{97}$$

and

$$\hat{V}_{hl}(s) \equiv \int_0^\infty e^{-st}\, V_{hl}(t)\, dt, \tag{98}$$

one has

$$\lim_{t\to\infty} [e^{Qt}u(t)]_{kl} = \delta_{kl} + 2\lambda \sum_h \int_0^\infty (e^{Qt})_{kh} V_{hl}(t)\, dt$$
$$= \delta_{kl} + 2\lambda \sum_h \int_0^\infty [e^{Qt}\, V_{hl}(t)]_{kh}\, dt$$
$$= \delta_{kl} + 2\lambda \sum_h [\hat{V}_{hl}(-Q)]_{kh}. \tag{99}$$

The derivation shows that the last expression is to be evaluated as follows. The matrix elements $\hat{V}_{hl}(s)$, which are ordinary functions of s, become matrices $\hat{V}_{hl}(-\,Q)$ when $-\,Q$ is substituted for s. These matrices have elements $[\hat{V}_{hl}(-\,Q)]_{kh}$, and the summation is to be carried out over the index h, to obtain the (kl) element of the matrix $\lim_{t\to\infty} (e^{Qt}\, u(t) - I)$. If the matrix Q is available in diagonalized form

$$Q_{kh} = -\sum_\gamma S_{k\gamma}\beta_\gamma S_{\gamma h}^{-1}, \tag{100}$$

the above result can be written as

$$\lim_{t\to\infty} [e^{Qt}u(t)]_{kl} = \delta_{kl} + 2\lambda \sum_\gamma S_{k\gamma}(S^{-1}\hat{V}(\beta_\gamma))_{\gamma l} \tag{101}$$

which can also be derived directly from (99). Another form of this result, which is useful in the special case treated below, is obtained from (44), with Q substituted for $\mathfrak{M}$. The integral in (99) then becomes

$$\int_0^\infty e^{Qt}\, V(t)\, dt = \int_0^\infty \frac{1}{2\pi i} \oint e^{\mu t}\, \frac{\varphi(\mu, Q)}{D(\mu)}\, V(t)\, dt$$
$$= \frac{1}{2\pi i} \oint \frac{\varphi(\mu, Q)}{D(\mu)}\, \hat{V}(-\mu)\, d\mu \tag{99a}$$

where the path of integration encircles the roots of

$$D(\mu) \equiv \prod_{\kappa=1}^{n} (\mu + \beta_\kappa) \tag{99b}$$

and where $(-\beta_\kappa)$ are the n eigenvalues of Q. If these are all distinct, one has

$$\int_0^\infty e^{Qt} V(t)\, dt = \sum_{\gamma=1}^{n} \varphi(-\beta_\gamma, Q) \hat{V}(\beta_\gamma) \prod_{\substack{\kappa=1 \\ \kappa \neq \gamma}}^{n} (\beta_\kappa - \beta_\gamma). \tag{99c}$$

The matrix $v(t)$ satisfies the integral equation

$$v(t) = e^{-Qt} + 2\lambda \int_0^t [e^{-Q(t-t')} - e^{\tilde{Q}(t-t')}] K(t') v(t')\, dt'. \tag{102}$$

To check this, one multiplies from the left by $e^{-\tilde{Q}t}$ and differentiates and obtains

$$\frac{d}{dt} [e^{-\tilde{Q}t} v(t)] = \frac{d}{dt} [e^{-\tilde{Q}t} e^{-Qt}] + 2\lambda \frac{d}{dt} \left\{ e^{-\tilde{Q}t} e^{-Qt} \int_0^t e^{Qt'} K(t') v(t')\, dt' \right\} - 2\lambda \frac{d}{dt} \int_0^t e^{-\tilde{Q}t'} K(t') v(t')\, dt'$$
$$= \frac{d}{dt} [e^{-\tilde{Q}t} e^{-Qt}] \cdot \left(I + 2\lambda \int_0^t e^{Qt'} K(t') v(t')\, dt' \right), \tag{103}$$

since the terms arising from differentiation of the integrals cancel. Using (23), one then has

$$\frac{d}{dt} [e^{-\tilde{Q}t} v(t)] = \frac{d}{dt} [e^{-\tilde{Q}t} e^{-Qt}] \cdot e^{Qt} u(t) = e^{-\tilde{Q}t} (Q + \tilde{Q}) u(t) \tag{104}$$

in agreement with (22).

Instead of considering (102) as an integral equation for $v(t)$, one may consider its Laplace transform

$$\hat{v}(s) = (sI + Q)^{-1} + 2\lambda \{ (sI + Q)^{-1} - (sI - \tilde{Q})^{-1} \} \hat{V}(s) \tag{105}$$

as an algebraic relation between $\hat{v}(s)$ and $\hat{V}(s)$ to be used for the determination of $\hat{v}(s)$ and $\hat{V}(s)$, together with

$$\hat{V}(s) = \int_0^\infty e^{-st} K(t) v(t)\, dt, \tag{106}$$

which follows from (97) and (98).

As an example of a case in which this approach is useful, we now consider the special case

$$K(t) = Re^{-t} \tag{107a}$$

where R is the constant matrix with elements

$$R_{kl} = \delta_{k1} \delta_{l1}. \tag{107b}$$

The case $K(t) = Re^{-\alpha t}$ of course, can be reduced to this case by choosing α^{-1} as unit of time. Eq. (106) then reduces to

$$\hat{V}(s) = R\hat{v}(s + 1) \tag{108}$$

so that only the elements of the first row of $\hat{v}(s)$ are needed for the evaluation of (99). For these, (105) reduces to the difference equation

$$\hat{v}_{1l}(s) = a_l(s) + 2\lambda b(s) \hat{v}_{1l}(s + 1) \tag{109}$$

where

$$a_l(s) = (sI + Q)^{-1}_{1l} \tag{110}$$

$$b(s) \equiv \{ (sI + Q)^{-1} - (sI - Q)^{-1} \}_{11}. \tag{111}$$

(The tilde indicating the transposed matrix can be omitted for the diagonal element.)

The functions $a_l(s)$ and $b(s)$ are rational functions of s.

Although there are more elegant methods of solving (109), the most transparent result is obtained by iteration, which yields

$$\hat{v}_{1l}(s) = a_l(s) + \sum_{r=1}^{\infty} (2\lambda)^r \prod_{m=0}^{r-1} b(s + m) \cdot a_l(s + r) \tag{112}$$

and

$$\hat{v}_{1l}(s + 1) = a_l(s + 1) + \sum_{r=1}^{\infty} (2\lambda)^r \prod_{m=1}^{r} b(s + m) a_l(s + 1 + r). \tag{113}$$

This can be written as

$$\hat{v}_{1l}(s + 1) = \sum_{r=0}^{\infty} (2\lambda)^r \prod_{m=1}^{r} b(s + m) a_l(s + 1 + r), \tag{113a}$$

if we adopt the convention that products containing no factors, such as $\prod_{m=1}^{0}$, are to be replaced by unity, in this and all subsequent equations.

The terms of the series can be obtained in a more explicit form by the procedure used in Section III, (42) *et seq.* Let $D(x)$ be the secular determinant of Q

$$D(x) = \det (xI - Q) = \prod_{\kappa=1}^{n} (x + \beta_\kappa) \tag{114}$$

when Q is a matrix of n rows and columns with eigenvalues $(-\beta_\gamma)$. The eigenvalues are assumed to be distinct. Let $\varphi(x, y)$ be the polynomial of degree $(n - 1)$ in both x and y defined by

$$(x - y)\varphi(x, y) = D(x) - D(y). \tag{115}$$

The matrices used in (110) and (111) then can be written in the form of polynomials in s and Q

$$(sI - Q)^{-1} = \varphi(s, Q)/D(s) \tag{116}$$

$$(sI + Q)^{-1} = -(-sI - Q)^{-1} = -\varphi(-s, Q)/D(-s). \tag{117}$$

The function $b(s)$ is thus

$$b(s) = -[\varphi(-s, Q)_{11} D(s) + \varphi(s, Q)_{11} D(-s)]/D(s)D(-s). \tag{118}$$

Since the terms with the highest exponent of x in $\varphi(x, y)$ and $D(x)$ are x^{n-1} and x^n, respectively, the coefficient of s^{2n-1} in the numerator vanishes and the numerator is a polynomial of degree $2(n - 1)$. It could be of lower order only if Q_{11} vanished, which is a case of no practical interest. Only even powers of s can occur, because the numerator is obviously an even function of s.

Let the roots of the numerator in (118) be $\pm\, \alpha_\mu$ ($\mu = 1, 2, \cdots, n-1$). Then the function $b(s)$ can be written in the form

$$b(s) = \text{const} \cdot \prod_{\mu=1}^{n-1} (s^2 - \alpha_\mu^2) \Big/ \prod_{\kappa=1}^{n} (s^2 - \beta_\kappa^2). \tag{119}$$

The constant is easily determined by expanding (111) for large s. One has

$$b(s) = -2s^{-2}Q_{11} + \cdots \tag{120}$$

and, therefore,

$$b(s) = -2Q_{11} \prod_{\mu=1}^{n-1} (s^2 - \alpha_\mu^2) \Big/ \prod_{\kappa=1}^{n} (s^2 - \beta_\kappa^2). \tag{121}$$

In order to obtain the product in (112) explicitly, note that

$$\prod_{m=1}^{r} (s + m - \alpha) = \Gamma(s + r + 1 - \alpha)/\Gamma(s + 1 - \alpha) \tag{122}$$

and, therefore,

$$\prod_{m=1}^{r} b(s + m) = (-2Q_{11})^r \cdot \prod_{\mu=1}^{n-1} \frac{\Gamma(s + r + 1 - \alpha_\mu)\Gamma(s + r + 1 + \alpha_\mu)}{\Gamma(s + 1 - \alpha_\mu)\Gamma(s + 1 + \alpha_\mu)} \cdot \prod_{\kappa=1}^{n} \frac{\Gamma(s + 1 - \beta_\kappa)\Gamma(s + 1 + \beta_\kappa)}{\Gamma(s + r + 1 - \beta_\kappa)\Gamma(s + r + 1 + \beta_\kappa)} \tag{123}$$

where again for $n = 1$, the first product is to be replaced by unity. We need this expression only for $s = \beta_\gamma$, and can take out of the second product the term $\kappa = \gamma$

$$\prod_{m=1}^{r} b(\beta_\gamma + m) = \frac{(-2Q_{11})^r}{r!} \frac{\Gamma(1 + 2\beta_\gamma)}{\Gamma(1 + r + 2\beta_\gamma)} \cdot \prod_{\mu=1}^{n-1} \frac{\Gamma(\beta_\gamma + r + 1 - \alpha_\mu)\Gamma(\beta_\gamma + r + 1 + \alpha_\mu)}{\Gamma(\beta_\gamma + 1 - \alpha_\mu)\Gamma(\beta_\gamma + 1 + \alpha_\mu)} \cdot \prod_{\substack{\kappa=1 \\ \kappa \neq \gamma}}^{n} \frac{\Gamma(1 + \beta_\gamma - \beta_\kappa)\Gamma(1 + \beta_\gamma + \beta_\kappa)}{\Gamma(r + 1 + \beta_\gamma - \beta_\kappa)\Gamma(r + 1 + \beta_\gamma + \beta_\kappa)}. \tag{124}$$

The functions $a_l(\beta_\gamma + 1 + r)$ can be written as

$$\begin{aligned} a_l(\beta_\gamma + 1 + r) &= -\varphi(-\beta_\gamma - 1 - r, Q)_{1l}/D(-\beta_\gamma - 1 - r) \\ &= (-)^{n+1}\varphi(-\beta_\gamma - 1 - r, Q)_{1l} \Big/ \prod_{\kappa=1}^{n} (r + 1 + \beta_\gamma - \beta_\kappa) \end{aligned} \tag{125}$$

by use of (110), (114), and (117). The denominator of this expression combines with the factor

$$r! \prod_{\substack{\kappa=1 \\ \kappa \neq \gamma}}^{n} \Gamma(r + 1 + \beta_\gamma - \beta_\kappa)$$

in the preceding equation.

Eq. (99c) becomes, in the special case defined by (107a)-(108),

$$\int_0^\infty [e^{Qt}V(t)]_{kl}\, dt = \sum_{\gamma=1}^{n} \varphi(-\beta_\gamma, Q)_{k1}\hat{v}_{1l}(\beta_\gamma + 1) \Big/ \prod_{\substack{\kappa=1 \\ \kappa \neq \gamma}}^{n} (\beta_\kappa - \beta_\gamma). \tag{126}$$

Combining (113a) and (124)-(126), one obtains

$$\lim_{t\to\infty} [e^{Qt}u(t)]_{kl} = \delta_{kl} + 2\lambda \sum_{\gamma=1}^{n} \varphi(-\beta_\gamma, Q)_{k1} \cdot \sum_{r=0}^{\infty} \frac{(-4\lambda Q_{11})^r}{(r + 1)!} \frac{\Gamma(1 + 2\beta_\gamma)}{\Gamma(1 + r + 2\beta_\gamma)} \varphi(-\beta_\gamma - 1 - r, Q)_{1l} \cdot \prod_{\mu=1}^{n-1} \frac{\Gamma(\beta_\gamma + r + 1 - \alpha_\mu)\Gamma(\beta_\gamma + r + 1 + \alpha_\mu)}{\Gamma(\beta_\gamma + 1 - \alpha_\mu)\Gamma(\beta_\gamma + 1 + \alpha_\mu)} \cdot \prod_{\substack{\kappa=1 \\ \kappa \neq \gamma}}^{n} \frac{\Gamma(\beta_\gamma - \beta_\kappa)\Gamma(1 + \beta_\gamma + \beta_\kappa)}{\Gamma(r + 2 + \beta_\gamma - \beta_\kappa)\Gamma(r + 1 + \beta_\gamma + \beta_\kappa)} \tag{127}$$

where φ is defined by (115), and the numbers $\pm\, \alpha_\mu$ are the roots of the numerator in (118), and $(-\, \beta_\gamma)$ are the eigenvalues of Q. A product without factors, such as $\prod_{\mu=1}^{n-1}$ with $n = 1$, has to be replaced by unity.

For $n = 1$, this reduces to the result obtained in (54) of Part II. In that case, the matrices reduce to numbers, and

$$Q = Q_{11} = -\beta_1 \tag{128}$$

$$D(x) = (x + \beta_1) \tag{129}$$

$$\varphi(x, y) = \frac{(x + \beta) - (y + \beta)}{x - y} = 1 \tag{130}$$

and, since the products contain no factors, the result reduces in the one-dimensional case to

$$\begin{aligned} f^{-2}(\infty) &= \lim_{t\to\infty} [e^{Qt}u(t)] \\ &= 1 + 2\lambda \sum_{r=1}^{\infty} \frac{(4\lambda\beta_1)^r}{(r + 1)!} \frac{\Gamma(1 + 2\beta_1)}{\Gamma(1 + r + 2\beta_1)} \\ &= \Gamma(2\beta_1) \sum_{r=0}^{\infty} \frac{(4\lambda\beta_1)^r}{r!\Gamma(r + 2\beta_1)} \\ &= \Gamma(2\beta_1)(-4\lambda\beta_1)^{-\beta_1+1/2} J_{2\beta_1-1}(\sqrt{-16\lambda\beta_1}). \end{aligned} \tag{131}$$

With $\beta_1 = \beta/\alpha$ this agrees with the result quoted above.

If the diagonalizing matrix S defined by (100) is available, a matrix slightly simpler than that given in (127) can be obtained, which has the same determinant. Let the matrix F be defined by

$$F_{\sigma\tau} = \sum_{kl} S_{\sigma k}^{-1} \lim_{t\to\infty} [e^{Qt}u(t)]_{kl} S_{l\tau} \tag{132}$$

and note that the determinant is unchanged by this transformation. The same procedure applied to the matrix elements on the right-hand side of (127) results in

$$\sum_k S_{\sigma k}^{-1}\varphi(-\beta_\gamma, Q)_{k1} = \varphi(-\beta_\gamma, -\, \beta_\sigma)S_{\sigma 1}^{-1} \tag{133}$$

and

$$\sum_l \varphi(-\beta_\gamma - 1 - r, Q)_{1l}S_{l\tau} = S_{1\tau}\varphi(-\beta_\gamma - 1 - r, -\, \beta_\tau). \tag{134}$$

From (114) and (115) one obtains

$$\varphi(x, -\beta_\sigma) = D(x)/(x+\beta_\sigma) = \prod_{\substack{\kappa \\ \kappa\neq\sigma}} (x+\beta_\kappa), \quad (135)$$

$$\varphi(-\beta_\gamma, -\beta_\sigma) = \prod_{\substack{\kappa \\ \kappa\neq\sigma}} (\beta_\kappa - \beta_\gamma) = \delta_{\gamma\sigma} \prod_{\substack{\kappa \\ \kappa\neq\sigma}} (\beta_\kappa - \beta_\sigma) \quad (136)$$

and

$$\varphi(-\beta_\gamma - 1 - r, -\beta_\tau) = \prod_{\substack{\kappa \\ \kappa\neq\tau}} (\beta_\kappa - \beta_\gamma - 1 - r). \quad (137)$$

Substitution of these results in (132) yields

$$F_{\sigma\tau} = \delta_{\sigma\tau} + 2\lambda g_{\sigma\tau} S_{\sigma 1}^{-1} S_{1\tau} \quad (138)$$

with

$$\begin{aligned} g_{\sigma\tau} = {} & \sum_{r=0}^{\infty} (-4\lambda Q_{11})^r \\ & \cdot \prod_{\mu=1}^{n-1} \frac{\Gamma(\beta_\sigma - \alpha_\mu + r + 1)\Gamma(\beta_\sigma + \alpha_\mu + r + 1)}{\Gamma(\beta_\sigma - \alpha_\mu + 1)\Gamma(\beta_\sigma + \alpha_\mu + 1)} \\ & \cdot \prod_{\kappa=1}^{n} \frac{\Gamma(\beta_\sigma - \beta_\kappa + 1)\Gamma(\beta_\sigma + \beta_\kappa + 1)}{\Gamma(\beta_\sigma - \beta_\kappa + r + 1)\Gamma(\beta_\sigma + \beta_\kappa + r + 1)} \\ & \cdot (\beta_\sigma - \beta_\tau + r + 1)^{-1}. \end{aligned} \quad (139)$$

The series is of hypergeometric type.[9]

In the Appendix $(\det F)^{-1/2}$ has been computed to second order in λ and the result checked by direct computation of the first and second moment.

VII. Appendix

To check the result given by (138) and (139), we compute $f(\infty)$ to second order in λ from these equations and directly.

Writing the matrix F in the form

$$F = I + \lambda G + \lambda^2 H + \cdots \quad (140)$$

with

$$G_{\sigma\tau} = 2S_{\sigma 1}^{-1} S_{1\tau} (\beta_\sigma - \beta_\tau + 1)^{-1} \quad (141)$$

$$\begin{aligned} H_{\sigma\tau} = {} & 2S_{\sigma 1}^{-1} S_{1\tau} (-4Q_{11}) \prod_{\mu=1}^{n-1} [(\beta_\sigma + 1)^2 - \alpha_\mu^2] \\ & \cdot \prod_{\kappa=1}^{n} [(\beta_\sigma + 1)^2 - \beta_\kappa^2]^{-1} (\beta_\sigma - \beta_\tau + 2)^{-1} \end{aligned} \quad (142)$$

we have, to second order in λ,

$$\begin{aligned} f(\infty) &= (\det F)^{-1/2} = \exp\{-\tfrac{1}{2} \operatorname{tr} \ln F\} \\ &= \exp\{-\tfrac{1}{2} \operatorname{tr} [\lambda G + \lambda^2 (H - \tfrac{1}{2} G^2)]\} \\ &= 1 - \frac{\lambda}{2} \operatorname{tr} G \\ &\quad - \frac{\lambda^2}{2} [\operatorname{tr} H - \tfrac{1}{2} \operatorname{tr} G^2 - \tfrac{1}{4} (\operatorname{tr} G)^2] + \cdots \end{aligned} \quad (143)$$

where tr stands for trace. The individual terms are evaluated as follows

$$\operatorname{tr} G = \sum_\sigma 2S_{\sigma 1}^{-1} S_{1\sigma} = 2. \quad (144)$$

$$\begin{aligned} \operatorname{tr} G^2 &= \sum_{\sigma,\tau} G_{\sigma\tau} G_{\tau\sigma} \\ &= 4 \sum_{\sigma,\tau} S_{\sigma 1}^{-1} S_{1\tau} S_{\tau 1}^{-1} S_{1\sigma} (1 + \beta_\sigma - \beta_\tau)^{-1} \cdot (1 + \beta_\tau - \beta_\sigma)^{-1} \\ &= 2 \sum_{\sigma\tau} S_{\sigma 1}^{-1} S_{1\sigma} S_{\tau 1}^{-1} S_{1\tau} \left[\frac{1}{1 + \beta_\sigma - \beta_\tau} + \frac{1}{1 + \beta_\tau - \beta_\sigma} \right] \\ &= 2 \Big(\sum_\sigma S_{\sigma 1}^{-1} S_{1\sigma} (1 + \beta_\sigma + Q)_{11}^{-1} + \sum_\tau S_{\tau 1}^{-1} S_{1\tau} (1 + \beta_\tau + Q)_{11}^{-1} \Big) \\ &= 4 \sum_\sigma S_{\sigma 1}^{-1} S_{1\sigma} a_1 (1 + \beta_\sigma) \\ &= 4[a_1(I - Q)]_{11} \end{aligned} \quad (145)$$

where $a_1(s)$ is defined by (110).

$$\begin{aligned} \operatorname{tr} H &= -4Q_{11} \sum_\sigma S_{\sigma 1}^{-1} S_{1\sigma} \cdot \prod_{\mu=1}^{n-1} [(\beta_\sigma + 1)^2 - \alpha_\mu^2] \prod_{\kappa=1}^{n} [(\beta_\sigma + 1)^2 - \beta_\kappa^2] \\ &= 2 \sum_\sigma S_{\sigma 1}^{-1} S_{1\sigma} b(\beta_\sigma + 1) \\ &= 2[b(I - Q)]_{11} \end{aligned} \quad (146)$$

where the explicit form of $b(s)$, given in (121), has been used. Substitution of these results in (143) yields

$$\begin{aligned} f(\infty) &= 1 - \lambda - \frac{\lambda^2}{2} \{2[b(I - Q)]_{11} - 2[a_1(I - Q)]_{11} - 1\} + \cdots \\ &= 1 - \lambda + \frac{\lambda^2}{2} \{1 + 2[a_1(I - Q) - b(I - Q)]_{11}\} + \cdots . \end{aligned} \quad (147)$$

The first two moments of

$$\mathfrak{F} \equiv \int_0^\infty e^{-t} x_1^2(t)\, dt, \quad (148)$$

where $x_1(t)$ is the first component of the stationary Gaussian process specified by (3)–(5) are easily obtained by direct computation.

The first moment is

$$\langle \mathfrak{F} \rangle_{\text{Av}} = \int_0^\infty e^{-t} \langle x_1^2(t) \rangle_{\text{Av}}\, dt = 1. \quad (149)$$

The second moment is

$$\begin{aligned} \langle \mathfrak{F}^2 \rangle_{\text{Av}} &= \left\langle \iint_0^\infty e^{-t-t'} x_1^2(t) x_1^2(t')\, dt\, dt' \right\rangle_{\text{Av}} \\ &= 2 \int_0^\infty e^{-t} \int_0^t e^{-t'} \langle x_1^2(t') x_1^2(t) \rangle_{\text{Av}}\, dt'\, dt. \end{aligned} \quad (150)$$

[9] G. N. Watson, "A Treatise on the Theory of Bessel Functions," Cambridge University Press, New York, N. Y., p. 100; 1944.

For a Gaussian random function, one has

$$\langle x(t_1)x(t_2)x(t_3)x(t_4)\rangle_{Av} = \langle x(t_1)x(t_2)\rangle_{Av}\langle x(t_3)x(t_4)\rangle_{Av} + \langle x(t_1)x(t_3)\rangle_{Av}\langle x(t_2)x(t_4)\rangle_{Av} + \langle x(t_1)x(t_4)\rangle_{Av}\langle x(t_2)x(t_3)\rangle_{Av} \quad (151)$$

and, with $t_1 = t_2 = t'$ and $t_3 = t_4 = t$, and $t' \leq t$,

$$\langle x_1^2(t')x_1^2(t)\rangle_{Av} = \langle x_1^2(t')\rangle_{Av}\langle x_1^2(t)\rangle_{Av} + 2\langle x_1(t')x_1(t)\rangle_{Av}^2 = 1 + 2\{(e^{Q(t-t')})_{11}\}^2. \quad (152)$$

Substitution in (150) yields

$$\langle \mathfrak{F}^2\rangle_{Av} = 1 + 4\int_0^\infty e^{-t}\int_0^t e^{-t'}\{(e^{Q(t-t')})_{11}\}^2\,dt' = 1 + 2\int_0^\infty e^{-t}\{(e^{Qt})_{11}\}^2\,dt \quad (153)$$

by virtue of the folding theorem for Laplace transforms.

Using (100), one obtains

$$\langle \mathfrak{F}^2\rangle_{Av} = 1 + 2\sum_{\sigma,\tau} S_{1\sigma}S_{\sigma 1}^{-1}S_{1\tau}S_{\tau 1}^{-1}(1 + \beta_\sigma + \beta_\tau)^{-1} \quad (154)$$

or, with $a_1(s)$ and $b(s)$ defined by (110) and (111)

$$\langle \mathfrak{F}^2\rangle_{Av} = 1 + 2(a_1(I - Q) - b(I - Q))_{11}. \quad (155)$$

We, therefore, have to second order in λ

$$f(\infty) \equiv \langle e^{-\lambda \mathfrak{F}}\rangle_{Av} = 1 - \lambda\langle \mathfrak{F}\rangle_{Av} + \frac{\lambda^2}{2}\langle \mathfrak{F}^2\rangle_{Av} + \cdots = 1 - \lambda + \frac{\lambda^2}{2}\{1 + 2[a_1(I - Q) - b(I - Q)]_{11}\} + \cdots \quad (156)$$

in agreement with (147).

26

Reprinted from *J. Soc. Ind. Appl. Math.*, **7**(4), 374–401 (1959)

THE DISTRIBUTION OF QUADRATIC FORMS IN NORMAL VARIATES: A SMALL SAMPLE THEORY WITH APPLICATIONS TO SPECTRAL ANALYSIS*

ULF GRENANDER, H. O. POLLAK, AND D. SLEPIAN

TABLE OF CONTENTS

Introduction. This paper is concerned with the problem of obtaining satisfactory approximations and computing techniques for the determination of the distribution function of quadratic forms in correlated normal variates. Such quadratic forms are of importance in many applications of the theory of stochastic processes. The estimation of power spectral densities and the determination of total noise power are two particularly important well-studied applications involving these sums. While much of what follows is directed specifically to these two applications, many of our remarks are valid for the more general distribution problem as well.

Although exact expressions for the distribution function of quadratic forms in normal variates are readily available in terms of Fourier integrals or convolutions, these expressions must be regarded as largely of theoretical interest since computations with them are in general impossible even with modern computing machines. When the number of observations is

* Received by the editors October 29, 1958.

large, approximations based on the central limit theorem are available. Much of the extensive literature on power spectrum estimation has this large sample character. The present paper has grown out of an attempt to solve the distribution problem in spectral analysis for moderate sample sizes and hence is concerned with approximations that do not depend upon the presence of a very large number of variables.

In Part 1, we discuss exact expressions for the distribution function and mention some of the methods that have been suggested for their numerical evaluation. A new exact formulation well suited for calculation is presented whereby the distribution function is obtained as the solution of a homogeneous integral equation. Use of this method requires knowledge of the behavior of the distribution function near the origin and the necessary asymptotics are developed.

A number of elementary approximate distribution functions are discussed in the first section of Part 2. The remainder of Part 2 is concerned with an approximation valid for a special but important class of quadratic forms. All the quadratic forms studied in this paper can be transformed to weighted sums of squares of independent identically distributed normal variates. In many applications, these weights are or approximate the eigenvalues of a Toeplitz matrix. Application of the fundamental theorem for Toeplitz matrices governing the behavior of their eigenvalue distrbution yields the small sample approximation theory mentioned.

In Part 3, the notions of the preceding sections are applied specifically to the problem of power spectral density estimation. Numerous examples are treated.

Part 4 is devoted to a brief discussion of the numerical techniques involved in carrying out the integral equation method described in Part 1. Curves are presented which compare exact distributions and various approximations.

Part 5 gives a guide to the use of the approximations in different situations.

Part 1. Exact Distributions

1.1 Exact expression for frequency function. We are concerned with the distribution of the random variable

$$Q = \sum_{i,j=1}^{n} w_{ij}x_ix_j \tag{1.1.1}$$

where the w_{ij} are elements of a nonnegative definite symmetric matrix W, and the x's are normal variates each with mean zero. The elements of the covariance matrix R of the x's are denoted by $r_{ij} = E\{x_ix_j\}, i, j = 1, \cdots, n$.

It is well known (see, e.g., [1] p. 118) that the characteristic function of Q is given by

(1.1.2) $$\varphi(z) = Ee^{izQ} = |I - 2izRW|^{-\frac{1}{2}} = \prod_{\nu=1}^{n} (1 - 2iz\lambda_\nu)^{-\frac{1}{2}}$$

where the λ_ν are the n eigenvalues of the matrix product RW. Since both R and W are nonnegative definite, it follows that their product has nonnegative eigenvalues, although RW is not necessarily a symmetric matrix. For the cumulants of Q, one finds readily

$$\kappa_\nu = (\nu - 1)!2^{\nu-1}\sum_{j=1}^{n} \lambda_j^\nu$$

so that the mean value m and variance σ^2 are

(1.1.3) $$m = \sum_{j=1}^{n} \lambda_j$$

(1.1.4) $$\sigma^2 = 2\sum_{j=1}^{n} \lambda_j^2.$$

Fourier's inversion formula then gives for the frequency function, $g(x)$, of Q,

(1.1.5) $$g(x) = \frac{1}{2\pi} \int_{-\infty}^{\infty} e^{-ixz} \prod_{\nu=1}^{n} (1 - 2iz\lambda_\nu)^{-\frac{1}{2}}\, dz.$$

This integral cannot be explicitly evaluated in general, nor is it well suited for numerical work.

1.2 Some previously suggested methods of evaluating (1.1.5). Since the integrand in (1.1.5) is an analytic function of z except at the branch points $z_\nu = -i/(2\lambda_\nu)$, the contour of integration can be deformed into a set of circles enclosing pairs of branch points (assuming that n is even). By collapsing the circles, one obtains an expression for $g(x)$ as an alternating sum of finite integrals,

(1.2.1) $$g(x) = \frac{1}{\pi} \sum_{\nu=1}^{n/2} (-1)^{\nu+1} \int_{(2\lambda_{2\nu-1})^{-1}}^{(2\lambda_{2\nu})^{-1}} e^{-yx} \prod_{j=1}^{n} (1 - 2y\lambda_j)^{-\frac{1}{2}}\, dy.$$

The singularities of the integrand at the end points of the interval of integration can be removed by an appropriate change of variable. This formula has been used by one of the authors [12], and has proved to be useful for numerical evaluation. If the λ's cluster toward zero, then except near $x = 0$, only the first few terms of the series contribute appreciably to the sum.

The case of (exactly) double eigenvalues is of a certain interest. We then have $\varphi(z) = \prod_1^{n/2}(1 - 2iz\lambda_\nu)^{-1}$, a rational function with simple poles $z_\nu = -i/(2\lambda_\nu)$. The inversion formula (1.1.5) can then be evaluated directly as

$$g(x) = \sum_{\nu=1}^{n/2} e^{-x/2\lambda_\nu} d_\nu$$

with

$$d_\nu = \prod_{\mu\neq\nu} (1 - \lambda_\nu/\lambda_\mu)^{-1}.$$

This observation forms the basis of an evaluation technique for g suggested by Robbins [10]. The end result of his analysis is a power series expansion of g as well as a representation of g as a mixture of χ^2-distributions. Gurland [7] has developed an expansion in terms of Laguerre polynomials. All of these expansions seem to converge slowly.

1.3 A new approach—an integral equation for $g(x)$. The numerical determination of the exact frequency function $g(x)$ consists of two steps: (1) getting the eigenvalues λ_ν and (2) computing the Fourier transform of $\varphi(z)$. The first step in most cases involves a great deal of numerical work and usually becomes almost impossible for large sample sizes, say $n = 40$. The methods described in the preceding section for carrying out the second step are also somewhat cumbersome. We now discuss an alternative approach that seems well adapted for use with a modern computer.

On taking the logarithmic derivative of (1.1.2), one finds

$$\frac{\varphi'(z)}{\varphi(z)} = i \sum_{\nu=1}^{n} \frac{\lambda_\nu}{1 - 2iz\lambda_\nu},$$

or

$$-i\varphi'(z) = \varphi(z) \sum_{\nu=1}^{n} \frac{\lambda_\nu}{1 - 2iz\lambda_\nu}.$$

The inverse Fourier transform yields

$$xg(x) = \int_0^x g(x - y)\, h(y)\, dy \tag{1.3.1}$$

where

$$h(x) = \tfrac{1}{2} \textstyle\sum_{\nu=1}^{n} e^{-x/2\lambda_\nu}. \tag{1.3.2}$$

From (1.3.1), one readily finds a similar integral equation for

$$G(x) = \int_0^x g(s)\, ds,$$

the distribution function of Q. The equation in question is

$$xG(x) = \int_0^x G(x - y)\, H(y)\, dy \tag{1.3.3}$$

where

$$H(y) = 1 + h(y). \tag{1.3.4}$$

Either (1.3.1) or (1.3.3) can be used as the basis of a computational scheme to numerically determine the distribution of Q. For example, on replacing (1.3.1) by a set of linear equations, one obtains

$$\nu g_\nu = \sum_{\mu=1}^{\nu} g_{\nu-\mu} h_\mu \qquad (\nu = 1, 2, \cdots) \tag{1.3.5}$$

or a similar set of equations if a different formula for numerical quadrature is used for the right member. Here we have put $g_\nu = g(\nu\Delta)$ and $h_\nu = h(\nu\Delta)$ where Δ is a mesh size. Since the matrix of coefficients in (1.3.5) is triangular, the solution can be obtained successively by simple operations.

Details of the use of this method will be given in Part 4. Here we pause only to point out a difficulty in the starting procedure for using (1.3.5). The frequency function $g(x)$ will always have a high order zero at $x = 0$ if n is large. This makes it necessary to obtain initial values for the solution of (1.3.5) by some other means. This is readily accomplished, for

$$\begin{aligned} g(x) &= \frac{1}{2\pi i} \int_{-i\infty}^{i\infty} e^{sx} \prod_{1}^{n} (1 + 2\lambda_\nu s)^{-\frac{1}{2}}\, ds \\ &= \frac{2^{-n/2} \Pi\lambda_\nu^{-\frac{1}{2}}}{2\pi i} \int_{-i\infty}^{i\infty} e^{sx} s^{-n/2} \left\{1 - \frac{1}{4s} \sum \lambda_\nu^{-1} + O(s^{-2})\right\} ds \\ &= \frac{x^{(n/2)-1}}{2^{n/2}\, \Pi\lambda_\nu^{\frac{1}{2}}\, \Gamma(n/2)} \left[1 - \frac{x}{4(n/2)} \sum \lambda_\nu^{-1} + O(x^2)\right] \end{aligned} \tag{1.3.6}$$

for small positive values of x.

It is an interesting observation which can be proved easily from (1.3.1) that the first term of the asymptotic formula (1.3.6) is in fact an upper bound. Indeed, let

$$g(x) = x^{(n/2)-1} \eta(x);$$

we shall prove that $\eta'(x) < 0$. The equation is

$$x^{n/2} \eta(x) = \int_0^x y^{(n/2)-1} \eta(y) h(x - y)\, dy$$

which gives after differentiation

$$\frac{n}{2} x^{(n/2)-1} \eta(x) + x^{n/2} \eta'(x) = x^{(n/2)-1} \eta(x) h(0) + \int_0^x y^{(n/2)-1} \eta(y) h'(x - y)\, dy.$$

Remembering that $h(0) = n/2$ and $h(x)$ is decreasing we obtain our result.

The integral equation technique described here has also been applied in several other contexts in which a complicated characteristic function is essentially the exponential of a simpler function. A particular example is contained in a forthcoming paper by Gilbert and Pollak [4] on the amplitude distribution of impulse noise.

1.4 Continuous quadratic forms. All of the foregoing can be extended to continuous quadratic forms in normal variates,

$$Q = \int_0^T dt_1 \int_0^T dt_2\, W(t_1, t_2)x(t_1)x(t_2)$$

where $W(t_1, t_2)$ is a nonnegative definite kernel and $x(t)$ is a sufficiently well-behaved Gaussian process with mean zero and covariance function $R(t_1, t_2) = E\{x(t_1)\, x(t_2)\}$. The characteristic function of Q is now given by an infinite product

$$\varphi(z) = \prod_{\nu=1}^{\infty} (1 - 2iz\lambda_\nu)^{-\frac{1}{2}}$$

where the λ_ν are eigenvalues of the Fredholm equation

$$\lambda\psi(t) = \int_0^T U(t, s)\psi(s)\, ds \tag{1.4.1}$$

and the kernel is defined by

$$U(t, s) = \int_0^T R(t, x)W(x, s)\, dx.$$

In applying the evaluation method described in the first part of 1.2, one must ascertain that the change of integration contour is legitimate. A sufficient condition is that the entire function $\varphi^{-2}(z)$ be of order less than unity. Such a condition is reflected in the rate of decrease of the λ_ν for large ν. For further details, see [12].

The integral equations (1.3.1) and (1.3.3) remain the same. The kernel $h(x)$, however, is now given by an infinite sum which diverges for $x = 0$. For numerical purposes, it is convenient to rewrite the equation as

$$g(x)\left(x - \sum_1^{\infty} \lambda_\nu\right) = -\int_0^x dy g'(x - y) \sum_1^{\infty} \lambda_\nu e^{-y/2\lambda_\nu}$$

which is obtained from (1.3.1) by an integration by parts. Note that

$$\sum_1^{\infty} \lambda_\nu = E\{Q\} < \infty.$$

Part 2. Approximate Distributions

2.1 Some simple approximations. Since the exact distribution of Q is quite difficult to obtain numerically, one is naturally led to ask for approximations. An obvious approximation is the Gaussian distribution with the parameters (1.1.3) and (1.1.4). Since Q has the same distribution as $Q_0 = \sum_1^n \lambda_\nu \xi_\nu^2$, where the ξ_ν are normal independent Gaussian variables, the central limit theorem can be used to prove asymptotic normality of Q for large sample sizes. Unfortunately this approximation is quite poor for small n, which is not surprising since the Gaussian distribution extends over the whole real line while the values of Q are only positive.

A somewhat better approximation is the one suggested by Rice ([9], p. 99), which uses a type III distribution with parameters chosen to give the values (1.1.3) and (1.1.4) for the first two moments,

$$g(x) = \left(\frac{m}{\sigma^2}\right)^r \frac{x^{r-1}}{\Gamma(r)} e^{-mx/\sigma^2}$$

with

$$r = m^2/\sigma^2.$$

Approximations that are also bounds on the distribution function G can be obtained as follows. Let n be even and order the eigenvalues so that $\lambda_1 \leqq \lambda_2 \leqq \cdots \leqq \lambda_n$. Introduce the quadratic form

$$Q' = \sum_1^n \lambda_\nu' \xi_\nu^2$$

where $\lambda_1' = \lambda_2' = \lambda_1$, $\lambda_3' = \lambda_4' = \lambda_3$, etc. Then since $Q' \leqq Q_0$, it follows that

$$G_{Q'}(x) \geqq G_{Q_0}(x)$$

for the corresponding distribution functions. But $F_{Q'}(x)$ can be obtained explicitly as a sum of exponentials since the λ''s are pairwise equal (see (1.2.1)). In an analogous manner a lower bound for $G_{Q_0}(x)$ can be obtained and it is clear that the two bounds will be close if the eigenvalues are not spaced too far apart from each other.

2.2 Approximations associated with Toeplitz forms. In certain applications of the theory of stochastic processes (see Part 3 for a detailed example) it happens that the weights λ_ν of the diagonalized quadratic form

$$Q_0 = \sum_1^n \lambda_\nu \xi_\nu^2 \tag{2.2.1}$$

are the eigenvalues (or approximate the eigenvalues) of a Toeplitz matrix. The following approximation for the distribution of Q_0 is based upon the theory of these matrices. The reader is referred to [6] for detailed statements and proofs.

We suppose that the λ's of (2.2.1) are the eigenvalues of an Hermitian Toeplitz matrix C,

$$C = \begin{bmatrix} c_0 & c_1 & c_2 & \cdot & \cdot & c_n \\ c_{-1} & c_0 & c_1 & \cdot & \cdot & \cdot \\ c_{-2} & c_{-1} & c_0 & \cdot & \cdot & \cdot \\ \cdot & \cdot & \cdot & \cdot & \cdot & \cdot \\ \cdot & \cdot & \cdot & \cdot & \cdot & \cdot \\ c_{-n} & c_{-n+1} & c_{-n+2} & \cdot & \cdot & c_0 \end{bmatrix}$$

where

$$c_\nu = \frac{1}{2\pi}\int_{-\pi}^{\pi} e^{i\nu\lambda}\, dW(\lambda) \qquad (\nu = 0, \pm 1, \pm 2, \cdots)$$

and W is bounded and nondecreasing. For convenience of exposition, we shall suppose that W is absolutely continuous with a density $w(\lambda) = W'(\lambda)$. The fundamental theorem of Szegö states that the distribution of eigenvalues of C (it is discrete consisting of $n + 1$ values) converges to the distribution of values of the function $w(\lambda)$ when λ is uniformly distributed over $(-\pi, \pi)$. More precisely,

$$\lim_{n\to\infty} \frac{\text{no. eigenvalues} \leqq x}{n} = \frac{1}{2\pi} \text{meas}\, [\lambda \mid w(\lambda) \leqq x].$$

We now use this simple asymptotic distribution of the λ's in place of the actual λ values in the formulae of Part 1 to obtain approximations for the distribution of Q.

Let $\varphi(z)$ be the characteristic function associated with Q. Define $\psi(z) = \log \varphi(z)$ where that branch of the logarithm is taken which vanishes for $\varphi = 1$. One has then

$$\psi(z) = -\tfrac{1}{2}\textstyle\sum_1^n \log(1 - 2iz\lambda_\nu),$$

and since the λ_ν behave approximately like the equidistributed ordinates of $w(\lambda)$,

$$\psi(z) \approx \psi_a(z) = -\frac{n}{4\pi}\int_{-\pi}^{\pi} \log\, [1 - 2izw(\lambda)]\, d\lambda.$$

The integration here should be carried out over the set of λ's where $w(\lambda) > 0$. This set of λ's is called the *support* of $w(\lambda)$ and is denoted by S. If it is assumed that $w(\lambda)$ is bounded, $0 \leqq w(\lambda) \leqq M < \infty$, then $\psi_a(z)$ is analytic in the z-plane cut from $-i/2M$ to $-i\infty$. On applying the Fourier inversion formula, we get the approximate frequency function

$$g_a(x) = \frac{1}{2\pi}\int_\Gamma e^{-ixz+\psi_a(z)}\, dz$$

where Γ loops the cut $(-i/2M, -i\infty)$ in the clockwise direction. The integration contour can be deformed in this manner because of the boundedness of $w(\lambda)$. This means that

$$g_a(x) = \frac{1}{2\pi}\int_{1/2M}^{\infty} e^{-xv}\, p(v)\, dv$$

where

$$p(v) = i \lim_{\epsilon \downarrow 0}\, [e^{\psi_a(-iv-\epsilon)} - e^{\psi_a(-iv+\epsilon)}]$$

is the saltus function at the cut line.

Introduce the angle

$$\theta(v) = \text{meas}\left(\lambda \mid w(\lambda) \geqq \frac{1}{2v}\right)$$

and the norm

$$R(v) = \frac{1}{2\pi}\int_S \log \mid 1 - 2\nu w(\lambda) \mid d\lambda$$

where we assume that this integral converges. One has then the approximate frequency function

$$g_a(x) = \frac{1}{\pi}\int_{1/2M}^{\infty} e^{-xv-(n/2)R(v)} \sin\left[\frac{n}{4}\theta(v)\right] dv. \tag{2.2.2}$$

This integral can be evaluated explicitly for a few $w(\lambda)$ of interest, and can be considered as an alternative to (1.2.1) for numerical evaluation.

The approximation suggested here can be used, of course, with the integral approach of §1.3. In place of (1.3.2) one now has the approximate kernel

$$h_a(x) = \frac{n}{4\pi}\int_S e^{-x/2w(\lambda)} d\lambda.$$

Again to start the integral equation (1.3.1) it is necessary to know the behavior of the solution for small x. The asymptotics can be carried out in the following way.

The Laplace transform of the approximate frequency function is

$$\begin{aligned} p(s) &= \varphi_a(is) = e^{\psi_a(is)} \\ &= \exp\left[-\frac{n}{4\pi}\int_S \log\left[1 + 2sw(\lambda)\right] d\lambda\right]. \end{aligned}$$

Assuming that $w(\lambda)$ is bounded away from zero and infinity, in S, one has for large values of s,

$$\begin{aligned} p(s) &= \exp\left[-\frac{n}{4\pi}\int_S \left(\log 2sw(\lambda) + \log\left[1 + \frac{1}{2sw(\lambda)}\right]\right) d\lambda\right] \\ &= \exp\left[-\frac{n}{4\pi}\left(m(S)\log 2s + \int_S \log w(\lambda)\, d\lambda + O(1/s)\right)\right] \qquad (2.2.3) \\ &= (2s)^{-(n/4\pi)m(S)} \exp\left[-\frac{n}{4\pi}\int_S \log w(\lambda)\, d\lambda\right]\left[1 + O(1/s)\right]. \end{aligned}$$

This implies (see [13] p. 192) for small values of x

$$g_a(x) = \frac{x^{\alpha-1}}{2^{n/2}\Gamma(\alpha)} \exp\left[-\frac{n}{4\pi}\int_S \log w(\lambda)\, d\lambda\right]\left[1 + O(x)\right]$$

where $\alpha = (n/4\pi)m(S)$. If $w(\lambda)$ is not bounded away from zero in S but has a finite number of zeros of the type $(\lambda - \lambda_0)^\beta$, the asymptotic expression (2.2.3) still holds with the term $O(x)$ replaced by $O(x/\log x)$ if $\beta < 1$ and $O(x^{1/\beta})$ if $\beta > 1$.

2.3 Approximations with continuous Toeplitz forms. The results of the preceding section can be extended in a rather obvious way to continuous quadratic forms. One is concerned now with Toeplitz integral equations, rather than with Toeplitz matrices (see [6], §8.6). We suppose that (1.4.1) now has a difference kernel $U(t, s) = U(t - s)$ and that

$$U(\tau) = \frac{1}{2\pi} \int_{-\infty}^{\infty} e^{i\lambda\tau}\, dW(\lambda) \tag{2.3.1}$$

with $W(\lambda)$ a bounded and nondecreasing function. We again suppose that W has a density $w(\lambda) = W'(\lambda)$. The fundamental theorem about the behavior of the eigenvalues of the kernel (2.3.1) is concerned with their distribution as T in (1.4.1) becomes large. It states that if the interval $[a, b]$ does not contain the origin and if meas $\{\lambda | w(\lambda) = a\} =$ meas $\{\lambda | w(\lambda) = b\} = 0$, then

$$\lim_{T\to\infty} \frac{\text{no. eigenvalues in } [a, b]}{T} = \frac{1}{2\pi} \text{meas}\,[\lambda \mid a < w(\lambda) < b].$$

The necessary modifications to be made on the equations of §2.2 therefore, consist largely of replacing appropriate n's by T's and we omit the details here with the following exception.

To start the integral equation (1.3.1) in a computational scheme, the behavior of the approximate frequency function $g_a(x)$, must be found for small x. To accomplish this, we consider the function

$$q(s) = \exp\left[-\frac{T}{2\pi}\int_0^\infty \log\,[1 + 2sw(\lambda)]\, d\lambda\right]$$

for large values of s. We shall assume that

(1) $w(\lambda)$ is decreasing for large values of λ,

(2) the inverse function $\lambda(w)$ satisfies, for small w (large λ),

$$\lambda(w) = \gamma_0 w^{-1/\alpha} + \gamma_1 w^{1-1/\alpha} + O(w^{2-1/\alpha}) \qquad (\alpha > 1).$$

One finds then

$$\int_0^\infty \log\,[1 + 2sw(\lambda)]\, d\lambda = -2s\int_0^\infty \frac{\lambda w'(\lambda)}{1 + 2sw(\lambda)}\, d\lambda$$

$$= \int_0^A \frac{\lambda(w)}{w + 1/2s}\, dw,$$

where we have put $w(0) = A$. But this last integral can be written as

$$\int_0^A \frac{\lambda(w) - \gamma_0 w^{-1/\alpha} - \gamma_1 w^{1-1/\alpha}}{w + 1/2s} dw + \int_0^A \frac{\gamma_0 w^{-1/\alpha} + \gamma_1 w^{1-1/\alpha}}{w + 1/2s} dw$$

and

$$\int_0^A \frac{\gamma_0 w^{-1/\alpha} + \gamma_1 w^{1-1/\alpha}}{w + 1/2s} dw = \gamma_0(2s)^{1/\alpha} \int_0^\infty \frac{x^{-1/\alpha}\, dx}{1 + x}$$
$$+ \gamma_1 \int_0^A \frac{w^{1-1/\alpha} dw}{w + 1/2s} - \gamma_0 \int_A^\infty \frac{w^{-1/\alpha}}{w + 1/2s} dw = \frac{\gamma_1 A^{1-1/\alpha}}{1 - 1/\alpha}$$
$$- \gamma_0 \alpha A^{-1/\alpha} + \gamma_0(2s)^{1/\alpha} \frac{\pi}{\sin(\pi/\alpha)} + O(s^{1/\alpha - 1}).$$

Therefore,

$$q(s) = B \exp\left[-2^{1/\alpha - 1} \frac{\gamma_0 T}{\sin(\pi/\alpha)} s^{1/\alpha}\right][1 + O(s^{1/\alpha-1})]$$

where

$$B = \exp\left(-\frac{T}{2}\left[\int_0^A \frac{\lambda(w) - \gamma_0 w^{-1/\alpha} - \gamma_1 w^{1-1/\alpha}}{w} dw + \frac{\gamma A^{1-1/\alpha}}{1 - 1/\alpha} - \gamma_0 \alpha A^{-1/\alpha}\right]\right).$$

The Tauberian relation expressing the local behavior of $g_a(x)$ at $x = 0$ in terms of $q(s)$ for large values of s can be found in [8]. After a few elementary transformations there results finally

$$g_a(x) = \frac{B}{\pi^{\frac{3}{2}}}\left(\frac{T\gamma_0}{2^{2\alpha-1}\alpha \sin(\pi/2\alpha)}\right)^{\alpha/(2\alpha-2)} \left(\cos\frac{\pi}{2\alpha}\right)^{-\alpha/(2\alpha-1)} \cdot (1 - 1/\alpha)^{-\frac{1}{2}} D(x) E(x)[1 + O(1)]$$

where

$$D(x) = x^{-(2\alpha-1)/(2\alpha-2)}$$

and

$$E(x) = \exp\left\{-2^{(2\alpha-1)/(\alpha-1)} \alpha^{1/(1-\alpha)} \left(1 - \frac{1}{\alpha}\right) \left(\cot\frac{\pi}{2\alpha}\right)^{\alpha/(1-\alpha)} (\gamma_0 T)^{\alpha/(\alpha-1)} x^{-1/(\alpha-1)}\right\}.$$

Part 3. Application to Spectral Analysis

3.1 Introduction. One is frequently confronted in applications with a stationary time series, $\cdots x_{-1}, x_0, x_1, \cdots$. We assume that this process has mean zero. Because of the stationarity, the covariance matrix has elements $r_{ij} = E\{x_i x_j\}$ which depend only on the absolute difference $|i - j|$ and we accordingly adopt a single subscript notation $r_j = E\{x_i x_{i+j}\}$, $i, j = 0, \pm 1, \pm 2, \cdots$. It is well known that the covariances have a Fourier-Stieltjes representation

$$r_\nu = \frac{1}{2\pi} \int_{-\pi}^{\pi} e^{i\lambda\nu} \, dF(\lambda) \tag{3.1.1}$$

where $F(\lambda)$ is a bounded, nondecreasing function which can be interpreted as describing the distribution in frequency of the power of the process. In many cases this function, called the spectral distribution function, is absolutely continuous so that

$$F(\lambda) = \int_{-\pi}^{\lambda} f(y) \, dy.$$

One of the important problems in the modern theory of time series analysis is the practical determination of $f(\lambda)$, the spectral density, from observed values of the process.

For this purpose one introduces a quadratic form

$$Q = \sum_{\mu,\nu=1}^{n} w_{\nu-\mu} x_\nu x_\mu . \tag{3.1.2}$$

At first glance it may appear unnecessarily restrictive to assume that the coefficients $w_{\nu-\mu}$ depend only on the difference $\nu - \mu$. However, it has been shown (see [5], p. 123) that little is lost by doing so, and further that we can assume that

$$w_\nu = \frac{1}{2\pi} \int_{-\pi}^{\pi} e^{i\nu\lambda} \, dW(\lambda) \tag{3.1.3}$$

where W is bounded, nondecreasing and of total variation unity. Again, W will usually be taken absolutely continuous with a density $w(\lambda) = W'(\lambda)$ called the spectral weight function or *spectral window*.

The choice of spectral window has been discussed extensively in the literature and the following is known. To estimate the spectral density at $\lambda = \lambda_0$ one should choose a narrow window around this frequency. The narrower the window, the better the estimate resolves the spectrum. At the same time, however, the sampling variability of the estimate increases so that some sort of compromise is necessary. It becomes important, therefore, to be able to compute the distribution of Q in order to obtain confidence limits for the estimates of spectral density made with various windows.

3.2 Applicability of Parts 1 and 2. Frequently in time series analysis it is assumed that the data follow a Gaussian probability law. The quadratic form (3.1.2) then is a special case of (1.1.1) and the results of Part 1 can be used to obtain its distribution. This distribution will apparently depend upon the covariance matrix R assumed for the time series in a complicated manner through the λ_ν, the eigenvalues of the matrix product RW.

Now in general R is unknown to the observer. He has only W or equivalently the spectral window $w(\lambda)$ at his choice. To be of use to him, curves of the theoretical distribution of Q for a given spectral window $w(\lambda)$ must be insensitive to the detailed structure of R or equivalently to the detailed structure of the spectral density $f(\lambda)$. To see that this is indeed the case for windows sufficiently narrow compared to the local changes in $f(\lambda)$ near λ_0, we make use of the Toeplitz nature of R and W as evidenced by (3.1.1) and (3.1.3).

It is true that while R and W are Toeplitz matrices their product RW, which governs the distribution of Q, is not in general a Toeplitz matrix. Nevertheless, the eigenvalues of the product RW are distributed asymptotically like the values of the product of $f(\lambda)w(\lambda)$ when λ is uniformly distributed in $(-\pi, \pi)$. This asymptotic result is shown in [6], Chapter 8. This has an immediate consequence for the spectral estimation problem: if $w(\lambda)$ is sufficiently peaked at some frequency λ_0 compared to local variations of $f(\lambda)$ near λ_0, then the distribution of the eigenvalues of RW will depend to a good first approximation upon the function $f(\lambda)$ only through the value $f(\lambda_0)$. That is, the eigenvalues of RW will be distributed like the values of $f(\lambda_0)w(\lambda)$. To this approximation, then, the distribution of $Q/f(\lambda_0)$, where Q is the estimator (3.1.2), will be independent of the shape of the spectral density $f(\lambda)$.

We can now, therefore, apply the approximation methods of §2.2 using the eigenvalues belonging to the single matrix W. Since the eigenvalue distribution for Toeplitz matrices approaches the limiting distribution quite rapidly a small sample theory is obtained. The range of validity of these and other approximations will be discussed below.

3.3 Some standard spectral windows. Quite a number of spectral windows $w(\lambda)$ have been suggested in the literature. At present one does not know very much about their possible optimum properties, partly because no small sample distribution theory has been available. We will discuss some of them briefly.

One usually starts with a window $u(\lambda)$ symmetric around $\lambda = 0$ and then defines

$$w(\lambda) = \tfrac{1}{2}[u(\lambda - \lambda_0) + u(\lambda + \lambda_0)].$$

If the Fourier coefficients of $u(\lambda)$ are denoted by u_ν, then the weights w_ν in the spectral estimate are given by

$$w_\nu = u_\nu \cos \nu\lambda_0 .$$

1. Perhaps the simplest possible case is obtained by choosing a rectangular window

$$u(\lambda) = \begin{cases} \pi/h & \text{if } |\lambda| < h \\ 0 & \text{otherwise.} \end{cases}$$

The weights are then

$$w_\nu = \frac{\sin h\nu}{h\nu} \cos \nu\lambda_0 .$$

The bandwidth h will be taken as a small positive number.

2. To get weights that decrease faster to zero it has been suggested that one should use

$$w_\nu = \left(1 - \frac{|\nu|}{n}\right)^p \cos \nu\lambda_0 .$$

In order to get a narrow window the value of p should be taken small. There are also other possibilities where u_ν is a polynomial in $|\nu|$.

3. Alternatively one may use

$$w_\nu = \begin{cases} \left(1 - \dfrac{|\nu|}{m}\right) \cos \nu\lambda_0 & \text{if } |\nu| < m \\ 0 & \text{otherwise.} \end{cases}$$

This has the numerical advantage that only m of the n possible product sums need to be computed. Small values of m correspond to a wide window.

4. Choosing $u(\lambda)$ as a Poisson kernel

$$u(\lambda) = \frac{1 - \rho^2}{1 - 2\rho \cos \lambda + \rho^2} \qquad (0 < \rho < 1),$$

we get exponential weights

$$w_\nu = \rho^{|\nu|} \cos \nu\lambda_0 .$$

These weights decrease rapidly for small ρ: a wide window. The simple analytic form of both window and weights may be useful.

5. We can of course let the window consist of spikes instead of letting it be continuous. One such choice is to take the spikes at equidistant points with the spacing $2\pi/n$. We then get estimates of the form

$$Q = \sum c_j I_j$$

where the I_j are ordinates of the usual periodogram

$$I_j = \left(\sum x_\nu \cos \frac{2\pi j\nu}{n}\right)^2 + \left(\sum x_\nu \sin \frac{2\pi j\nu}{n}\right)^2.$$

6. In the classical theory of time series analysis similar quadratic forms have been used for hypothesis testing and estimation. We mention only the lagged product sums

$$S_p = \sum_{\nu=1}^{n-p} x_\nu x_{\nu+p} \qquad (0 \leqq p < n),$$

which correspond to weight functions of the type

$$w(\lambda) = \cos p\lambda.$$

3.4 Some Toeplitz approximations. In certain cases the Toeplitz approximations have been determined explicitly. The simplest case is for the spectral window 1, the rectangular window. Then

$$\psi_a(z) = -\frac{hn}{\pi} \log \left[1 - \frac{2iz\pi}{h}\right],$$

which corresponds to a type III distribution, so that in this case we are led to the same approximation as the one suggested by Rice (see §2.1).

For the spectral window 4 an approximate distribution has been given in [3]; the reader is referred to that paper for the derivation and a numerical discussion.

The spectral window 5 is very easy to deal with. Assuming that the spikes c_j have been chosen so that the bandwidth is small we can consider $f(\lambda)$ as almost a constant $f(\lambda_0)$, just as discussed in §3.2. But this corresponds to independent observations for which it is known that the I_j are also independent and distributed as χ^2-variables with 2 d.f. each (see [2]).

We have not included any windows for time series with a continuous parameter although this may very well be possible. As an example we consider the case of RC noise when

$$f(\lambda) = \frac{1}{\alpha^2 + \lambda^2}$$

and the quadratic form is

$$Q = \int_0^T x^2(t)\, dt.$$

We have to evaluate

$$q(v) = \int_{-\infty}^{\infty} \log \left[1 - \frac{2v}{\alpha^2 + \lambda^2}\right] d\lambda$$

and find

$$q(v) = 2\pi[\sqrt{\alpha^2 - 2v} - \alpha],$$

which has a branch point at $v = \alpha^2/2$. Using the Fourier inversion formula and changing the path of integration in the way that has been described above we get

$$\begin{aligned} g_a(x) &= \frac{1}{2\pi}\int_{-\infty}^{\infty} \exp\left\{-izx - \frac{T}{4\pi}\, q(iz)\right\} dz \\ &= \frac{1}{\pi}\int_{\alpha^2/2}^{\infty} \exp\left\{-vx + \frac{T}{2}\,\alpha\right\} \sin\left(\frac{T}{2}\sqrt{2v - \alpha^2}\right) dv \\ &= \frac{1}{2\pi}\exp\left\{\frac{T}{2}\,\alpha - \frac{\alpha^2}{2}\,x\right\} \int_{-\infty}^{\infty} e^{-xu^2/2} u \sin\frac{T}{2}\,u\, du \\ &= \frac{T}{4\pi x}\exp\left\{\frac{\alpha}{2}\,(T - \alpha x)\right\} \int_{-\infty}^{\infty} e^{-xu^2/2} \cos\frac{T}{2}\,u\, du \\ &= \frac{T}{2\sqrt{2\pi}}\, e^{\alpha T/2} x^{-\frac{3}{2}} e^{-T^2/8x} e^{-\alpha^2 x/2}. \end{aligned}$$

Note the behavior of this frequency function at $x = 0$; there is contact of infinite order between $g_a(x)$ and the x-axis.

3.5 Sharper approximations. It may happen that the Toeplitz approximations are not sufficiently accurate. This is likely to occur when the time of observation, n or T, is too short compared to the time constant associated with the bandwidth of the spectral window. In such a case we may still be able to find the eigenvalues of the W-matrix exactly or at least with considerable accuracy. The method of §1.3 can then be used, for example, to compute the desired distribution function.

The determination of eigenvalues for many of the spectral windows of interest is not as formidable a task as it might seem at first sight. Frequently the solutions of

$$l\varphi_\mu = \sum_{\nu=1}^{n} w_{\mu-\nu}\varphi_\nu \qquad (\mu = 1, 2, \cdots, n) \tag{3.5.1}$$

satisfy linear difference equations with constant coefficients. Consideration of these equations yields a transcendental equation for the eigenvalues. This equation contains n parametrically and the need for dealing with large matrices is obviated. Similar techniques can often be used to solve the continuous analogue of (3.5.1), namely

$$\lambda\varphi(t) = \int_0^1 K(t - s)\varphi(s)\, ds. \tag{3.5.2}$$

Here linear differential equations replace the linear difference equations.

We shall discuss the method of solution only for (3.5.2), the modification for (3.5.1) being evident. A sketch of the method will suffice; the algebra is cumbersome but straightforward. Before doing so, however, we note that (3.5.2) can sometimes be used to obtain approximate eigenvalues for (3.5.1). This approximation will be good if the time constant of W is not small compared to n, for then the sequence w_ν varies slowly as ν increases from 0 to n. The large eigenvalues of (3.5.2) with $K(\nu/n) = w_\nu$ will then be good approximations to the large eigenvalues of W. Of course this will only be of use when the solution of (3.5.2) is more easy to obtain than that of (3.5.1).

For the class of kernels under discussion, it is possible to find differential operators L and M,

$$L = \sum_{\nu=0}^{p} a_\nu \frac{d^\nu}{dt^\nu}, \qquad M = \sum_{\nu=0}^{q} b_\nu \frac{d^\nu}{dt^\nu} \qquad (p > q)$$

such that if

$$h(t) \equiv \int_0^1 K(t - s)\varphi(s)\, ds$$

then

$$Lh(t) = M\varphi(t)$$

for all φ continuous in (0, 1). The solutions φ of (3.5.2) must therefore satisfy

$$(\lambda L - M)\varphi = 0, \tag{3.5.3}$$

a linear differential equation of order p with constant coefficients. The eigenfunctions φ must be linear combinations

$$\varphi = \textstyle\sum_0^p c_\nu f_\nu(t) \tag{3.5.4}$$

of the p linearly independent solutions $f_\nu(t)$ of (3.5.3). Here each $f_\nu(t)$ can be taken of the form $t^j e^{z_i t}$ where z_i is a root (of multiplicity greater than j) of the algebraic equation

$$\lambda \sum a_\nu z^\nu - \sum b_\nu z^\nu = 0.$$

The z_i are thus functions of the parameter λ.

Substitute (3.5.4) into the integral equation (3.5.2) and require that the latter be satisfied identically in t. By equating to zero the coefficients of the p independent terms $t^j e^{z_i t}$, one obtains a set of p equations linear and homogeneous in the p quantities c_ν. On equating to zero the determinant of this system, one obtains a transcendental equation for the determination of the eigenvalues λ of (3.5.2).

It is easy to show that if associated with $K(t - s)$ are operators L and M of order p and q, then associated with the kernel $K(t - s) \cos \lambda_0(t - s)$ there are operators $\hat{L}$ and $\hat{M}$ of order $2p$ and $q' < 2p$ respectively, so that shifting the position of a spectral window causes no trouble. We therefore restrict our attention in what follows to windows centered at the origin.

The method just outlined is clearly applicable when $K(\tau)$ consists of a polynomial in $|\tau|$ or the sum of products of such polynomials with exponentials in $|\tau|$. (For the latter case see [11]. A different method of solution is given in [14].) A slight modification of the method will also treat the case in which $K(\tau)$ is of this form in some interval $|\tau| \leqq T$ and is zero for $|\tau| > T$. The differential equation (3.5.3) then becomes a difference-differential equation. Thus windows 2, 3 and 4 can all be treated in this manner. We now comment briefly on these separate cases.

3.5.1. *Window* 2. Consider kernels of the form

$$K(t) = \textstyle\sum_{\nu=0}^{r} a_\nu |t|^\nu.$$

These include window 2 of §3.3. Let

$$h_\nu(t) = \int_0^1 |t - s|^\nu \varphi(s)\, ds \qquad (0 \leqq \nu \leqq r).$$

If ν is even, then $h_\nu(t)$ is a νth degree polynomial and its $(r + 1)$th derivative vanishes. If ν is odd,

$$\frac{d^{\nu+1}}{dt^{\nu+1}} h_\nu(t) = 2\nu!\varphi(t).$$

Hence for (3.5.3) we have

$$\lambda \frac{d^{r+1}}{dt^{r+1}} \varphi = 2 \sum_{j=1}^{(m+1)/2} (2j - 1)!\, a_{2j-1} \frac{d^{r-2j+1}\varphi}{dt^{r-2j+1}}$$

where m is the largest odd integer for which $a_m \neq 0$. If r is small, the solution outlined above can be readily carried out in full.

3.5.2 *Window* 4. This case is treated in detail in [11]. The matrix case

$$W = \{\rho^{|\nu-\mu|}\} \qquad (\nu, \mu = 1, 2, \cdots, n)$$

is of special interest, however, so we present an alternative method which may be illuminating.

Introduce the determinants

$$\Delta_{n-1}(\lambda) = |\rho^{|\nu-\mu|} - \lambda\delta_{\mu\nu}| \qquad (\mu, \nu = 1, \cdots, n).$$

One readily derives the difference equation

$$\Delta_n(\lambda) = [1 - \lambda - \rho^2(1 + \lambda)]\Delta_{n-1}(\lambda) - \rho^2\lambda^2\Delta_{n-2}(\lambda).$$

Introduce the angle x defined by

$$\lambda = \lambda(x) = \frac{1-\rho^2}{1 - 2\rho \cos x + \rho^2} \qquad (0 \leqq x < \pi).$$

Then

$$\Delta_n(\lambda) = \frac{(-\lambda\rho)^{n+1}}{1-\rho^2}\left[\frac{\sin(n+2)x}{\sin x} - 2\rho\frac{\sin(n+1)x}{\sin x} + \rho^2\frac{\sin nx}{\sin x}\right]$$
$$= \frac{(-\lambda\rho)^{n+1}}{1-\rho^2} P(\cos x)$$

where P is a polynomial of degree $n + 1$. The $n + 1$ zeros x_ν of $P(\cos$
lie in the interval $(0, \pi)$ and the eigenvalues of W are given by

$$\lambda_\nu = \lambda(x_\nu).$$

3.5.3. *Window* 3. For simplicity of exposition, we treat only the case

$$K(t) = \begin{cases} 1 - k\,|\,t\,| & (|\,t\,| \leqq k \\ 0 & (|\,t\,| > k \end{cases}$$

where k is a positive integer. The method can be extended to arbitrar and to kernels that are sums of products of polynomials in $|\,\tau\,|$ with ponentials in $|\,\tau\,|$ in the interval $|\,\tau\,| < k^{-1}$.

On letting $\alpha(t) = \max(0, t - k^{-1})$ and $\beta(t) = \min(1, t + k^{-1})$, one write the integral equation

$$\lambda\varphi(t) = \int_{\alpha(t)}^{t} [1 - k(t-s)]\varphi(s)\,ds + \int_{t}^{\beta(t)} [1 - k(s-t)]\varphi(s)\,d$$

Differentiation yields the equations

$$\lambda\varphi''(t) = \begin{cases} k\varphi(t+k^{-1}) - 2k\varphi(t) & (0 \leqq t \leqq \\ k\varphi(t+k^{-1}) - 2k\varphi(t) + k\varphi(t-k^{-1}) & (k^{-1} \leqq t \leqq 1 - \\ \qquad - 2k\varphi(t) + k\varphi(t-k^{-1}) & (1 - k^{-1} \leqq t \end{cases}$$

(If $k = 1$, only $\lambda\varphi''(t) = -2k\varphi(t)$ results.) These difference-diffe equations play the role of (3.5.3) in the cases treated earlier. They ar conveniently solved by introducing the intervals

$$E_r = \left(\frac{r}{k}, \frac{r+1}{k}\right) \qquad (r = 0, 1, 2, \cdots,$$

and the functions

$$f_r(s - rk^{-1}) = \varphi(s), \qquad ($$
$$f_{-1}(s) = f_{k+1}(s) = 0.$$

We then have

$$\lambda f_r''(x) = kf_{r-1}(x) - 2kf_r(x) + f_{r+1}(x)$$

where we have put $x = t - rk^{-1}$, $0 < x < k^{-1}$.

This system of ordinary second order differential equations can be solved by putting

$$f_r(x) = \sum_\nu a_{r\nu} e^{\omega_\nu x}$$

where the ω_ν and $a_{r\nu}$ must satisfy

$$\lambda\omega_\nu^2 a_{r\nu} = ka_{r-1,\nu} - 2ka_{r\nu} + ka_{r+1,\nu} \qquad (r = 0, 1, \cdots, k-1).$$

If λ is to be an eigenvalue of the integral equation there must be at least one $a_{r\nu} \neq 0$. But then $\lambda\omega_r^2/k$ must be an eigenvalue of the matrix

$$M = \begin{Bmatrix} -2 & 1 & 0 & 0 & \cdot & \cdot & \cdot & 0 \\ 1 & -2 & 1 & 0 & \cdot & \cdot & \cdot & 0 \\ \cdot & \cdot & \cdot & \cdot & \cdot & \cdot & \cdot & \cdot \\ 0 & 0 & 0 & 0 & \cdot & \cdot & \cdot & -2 \end{Bmatrix}$$

which implies

$$\frac{\lambda\omega_r^2}{k} = -2 + 2\cos\frac{\pi\nu}{k+1} \qquad (\nu = 1, 2, \cdots, k).$$

Since $\lambda > 0$ then ω_ν must be purely imaginary,

$$\omega_\nu = \pm i\sqrt{\frac{2k}{\lambda}\left(1 - \cos\frac{\pi\nu}{k+1}\right)}$$

$$= \pm 2i\sqrt{\frac{k}{\lambda}}\sin\frac{\pi\nu}{2(k+1)}$$

and we will number the ω_ν so that $\omega_{-\nu} = -\omega_\nu$, $\nu = \pm1, \pm2, \cdots, \pm k$.

We now have to adjust the constants $A_{r\nu}$ so that

$$\varphi(s) = \sum_{\nu=-k}^{k} A_{r\nu} e^{\omega_\nu s} \qquad (s \in E_r),$$

is an eigenfunction. We use the convention $A_{-1\nu} = A_{k\nu} = A_{r0} = 0$. We then get

$$\lambda\varphi(s) = \alpha_r + \beta_r s + \sum_{\nu=-k}^{k} \gamma_{r\nu} e^{\omega_\nu s} \qquad (s \in E_r),$$

where the constants α_r, β_r, $\gamma_{r\nu}$ can easily be evaluated. We give only $\gamma_{r\nu}$

$$\gamma_{r\nu} = \frac{k}{\omega_\nu^2} e^{-\omega_\nu/k} A_{r-1,\nu} - 2\frac{k}{\omega_\nu^2} A_{r\nu} + \frac{k}{\omega_\nu^2} e^{\omega_\nu/k} A_{r+1,\nu}.$$

It is necessary that $\alpha_r = \beta_r = 0$ and that $\gamma_{r\nu} = \lambda A_{r\nu}$. For fixed ν we introduce the column vector A with components $A_{r\nu}e^{(r/k)\omega_\nu}$ and get

$$\frac{k}{\omega_\nu^2} MA = \lambda A.$$

Hence A must be proportional to that eigenvector of M that corresponds to the eigenvalue $\lambda\omega_\nu^2/k$:

$$(A)_r = b_\nu \sin \frac{(r+1)\nu\pi}{k+1}$$

or

$$A_{r\nu} = b_\nu e^{-rw_\nu/k} \sin \frac{(r+1)\nu\pi}{k+1},$$

and at least one b_ν must differ from zero. The conditions $\alpha_r = \beta_r = 0$ give us $2k$ homogeneous linear equations in the $2k$ unknowns b_ν. Putting the determinant of this system equal to zero gives a transcendental equation in λ determining the eigenvalues of the integral equation.

For a reader interested in nonintegral values of k we point out that the role played here by the points r/k will be taken over by the points r/k and $1 - r/k$ in the interval $(0, 1)$.

3.5.4. *Window* 1. It has been shown in [11] how to find the eigenvalues corresponding to the kernel

$$K(t) = \frac{\sin kt}{kt}.$$

They are given by the expression

$$\lambda_\nu = \left[R_{0\nu}^{(1)}\left(\frac{h}{2}, 1\right)\right]^2$$

where the $R_{0\nu}^{(1)}$ are prolate spheroidal Bessel functions.

For this window the explicit eigenvalues in the corresponding case of discrete time are not known.

PART 4. NUMERICAL EXAMPLES

4.1 The integral equation method. The integral equation (1.3.1), namely

$$xg(x) = \int_0^x g(x-y)h(y)\,dy = \int_0^x h(x-y)g(y)\,dy$$

is singular due to the appearance of x as a factor in the left member. It is not surprising then that there are certain difficulties attendant on its use nor that a detailed analysis of the errors introduced by its numerical solution appears to be quite difficult. Accordingly, we here make a few comments concerning our experience with its numerical solution.

On using a trapezoidal integration formula, one can approximate the equation by

$$(4.1.1)\quad g(j\Delta) = \frac{1}{j - \frac{1}{2}h(0)} \sum_{k=1}^{j=1} g(k\Delta)h[(j - k)\Delta] \quad (j = 2, \cdots),$$

which expresses $g(j\Delta)$ in terms of g for smaller argument. Here we have used the fact that $g(0) = 0$. As mentioned earlier, it is necessary to start off this recurrence with values of g obtained by other means. Unfortunately, in cases of interest, it frequently happens that asymptotic formulae for g near the origin are valid for only an exceedingly small fraction of the total x range of interest. (It is obvious from (1.3.6) that this will happen if at least one λ is very close to zero.) This forces one to choose an initial mesh size Δ so small that continued use of (4.1.1) to investigate the entire range of interest leads to a prohibitively large number of iterations. It becomes necessary, therefore, to increase the mesh size Δ at various stages of the computation.

The scheme used by the authors is as follows. The asymptotic expression for $g(x)$ was used to compute the values $g(j\Delta_1), j = 1, 2, \cdots, j_1$. Equation (4.1.1) was then used for $j = j_1 + 1, j_1 + 2, \cdots, 2j_1$. A new mesh size $\Delta_2 = 2\Delta_1$ was then chosen. Values of $g(j\Delta_2)$, $j = 1, 2, \cdots, j_1$ have already been computed, so that (4.1.1) with $\Delta = \Delta_2$ could be used for $j = j_1 + 1, \cdots, 2j_1$. This process was continued until a large enough mesh size was obtained to allow economical computation out to the largest x value of interest.

It is clear that if j_1 is chosen too small the g values obtained will be distorted in passing from $g(2j_1\Delta_i)$ to $g[(j_1 + 1)\Delta_{i+1}]$. It was found in practice that for $j_1 = 20$ the jumps were still noticeable, though small. For $j_1 = 50$ they could not be detected.

Fig. 1 compares an exact frequency function with one obtained by the integral equation method. The eigenvalues were chosen equal in pairs as indicated. The exact frequency function is given in §1.2. Using values $\Delta_1 = .0002$ and $j_1 = 50$, the integral equation method agreed with the exact curve to three significant figures.

Some qualitative remarks about the frequency function can easily be deduced. In most cases of interest the λ_j approach zero with increasing j. Ignoring the small λ's has very little effect on the shape of g except near the origin. Inclusion of the small λ's causes an increase in the order of contact at the origin and shifts the entire curve slightly to the right. These observations have been verified by many numerical examples.

In general, the integral equation (1.3.1) for the frequency function yields

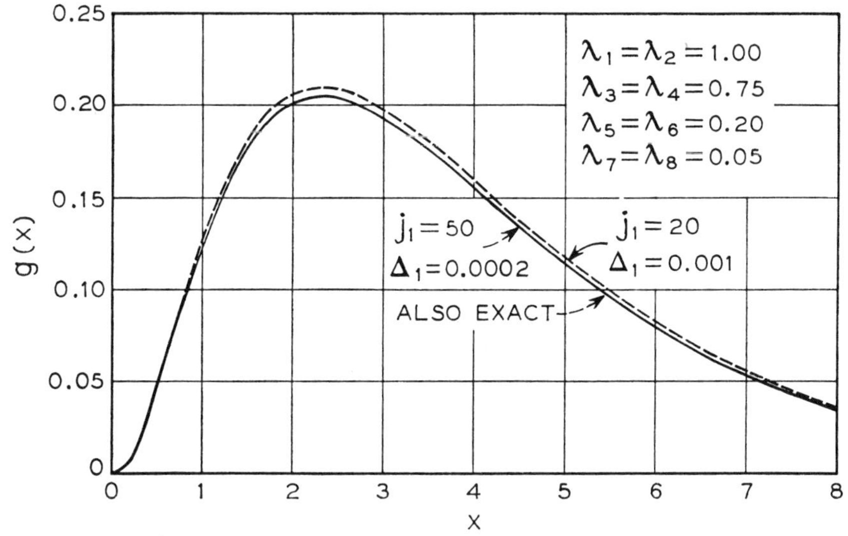

FIG. 1. *Comparison of exact density with solution of integral equation.* See §4.1

better results than (1.3.3) for the distribution function for the same mesh sizes and numbers of computing steps.

Typical curves presented in this section took from 3 to 15 minutes computing time on the IBM 704 computer. While this is not unduly long, it is to be hoped that further study of the integral equation will yield a more efficient numerical technique for obtaining the distribution function.

4.2 Spectral estimation with triangular window. Let $w(\lambda)$ be the "triangular window"

$$w(\lambda) = \begin{cases} \frac{4}{\pi}\left(1 - \frac{4}{\pi}|\lambda|\right) & \left(|\lambda| \leq \frac{\pi}{4}\right) \\ 0 & \left(|\lambda| > \frac{\pi}{4}\right). \end{cases} \tag{4.2.1}$$

The frequency function of

$$Q = \textstyle\sum_1^{20} c_{i-j}x_ix_j$$

where

$$c_\nu = \frac{1}{2\pi}\int_{-\pi}^{\pi} e^{i\nu\lambda}\, w(\lambda)\, d\lambda$$

has been computed in two ways as an illustration of our methods. The resulting frequency functions are shown on Fig. 2.

The curve labeled "exact" was obtained by first computing numerically the 20 eigenvalues of the matrix $| c_{i-j} |$. The largest eigenvalue was 1.094;

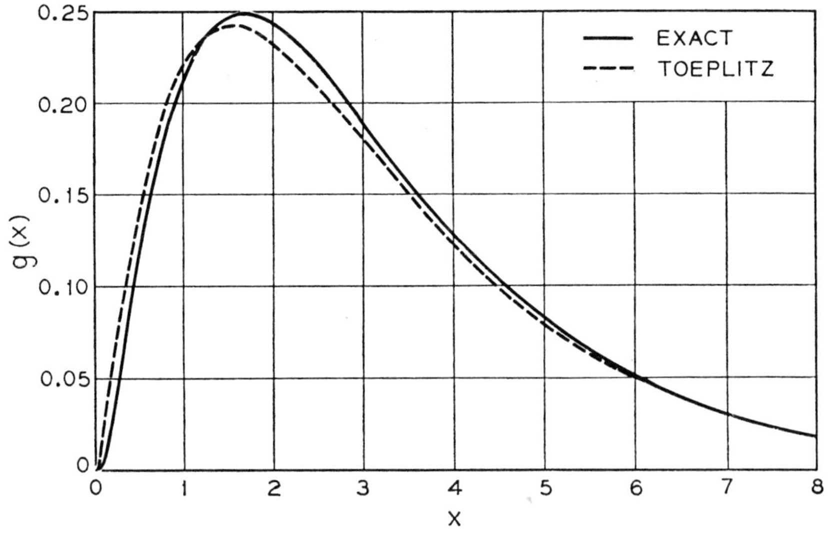

FIG. 2. *Frequency function for Q for triangular window—20 samples.* See §4.2.

the sixth largest was .0519; the seventh was .0075 and the sum of the remaining 13 eigenvalues was .0021. The first six eigenvalues were used to compute $g(x)$ by the integral equation method with h given by (1.3.2).

The curve labelled "Toeplitz approximation" was also obtained by the integral equation method. The kernel h used was obtained from (2.2.2) and (4.2.1). The indicated integral can be expressed in terms of elementary functions and the exponential integral function $\mathrm{Ei}(x)$.

4.3 Toeplitz approximation for *R-C* noise power. In [12] curves are given for the frequency function of

$$Q = \frac{1}{T}\int_0^T [x(t)]^2\,dt \tag{4.3.1}$$

where $x(t)$ is a Gaussian process with mean zero and covariance $r(\tau) = E\{x(t)x(t+\tau)\} = e^{-\alpha|\tau|}$. The Toeplitz approximation of §2.3 is particularly convenient in this case since the indicated integrations can all be performed analytically. The following approximate frequency function for (4.3.1) results

$$g_T(x) = \sqrt{\frac{\beta}{2\pi}}\, e^{\beta}\, x^{-3/2}\, e^{-(\beta/2)(x+1/x)} \tag{4.3.2}$$

where

$$\beta = \alpha T/2.$$

On Fig. 3, curves of the frequency function for $Q = \beta = .5$, 5, and 30 as

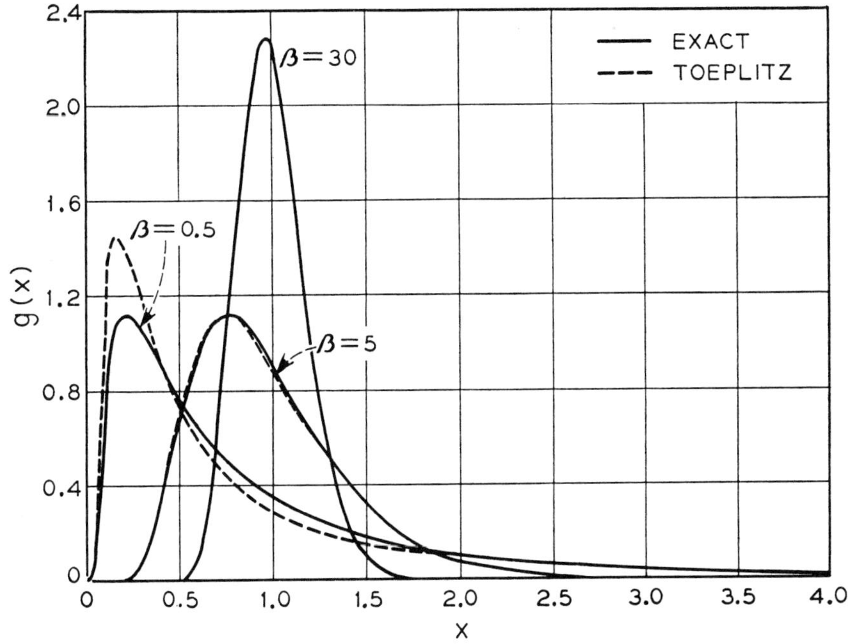

FIG. 3. *RC noise power*. See §4.3.

obtained from [12] are shown. For comparison, plots of (4.3.2) for $\beta = .5$ and 5 are also given. The plot of (4.3.2) for $\beta = 30$ coincides so closely with that obtained from [12] that it could not be plotted separately in Fig. 3. For $\beta = 5$, it is seen that the Toeplitz approximation is already very good.

4.4 Power distribution of band-limited noise. Let Q be given by (4.3.1) where $x(t)$ is a Gaussian process with mean zero and covariance

$$r(\tau) = \frac{\sin 2\pi f_0 \tau}{2\pi f_0 \tau}. \qquad (4.4.1)$$

The corresponding spectrum is

$$w(\lambda) = \begin{cases} \frac{1}{2} f_0 & (|\lambda| \leqq 2\pi f_0) \\ 0 & (|\lambda| > 2\pi f_0). \end{cases}$$

The Toeplitz approximation, $g_T(x)$, to the frequency function of Q can be readily obtained by the methods of §2.3. They yield

$$g_T(x) = \frac{(n/2)^{n/2}}{\Gamma(n/2)} x^{(n/2)-1} e^{-(n/2)x} \qquad (4.4.2)$$

where

$$n = 2f_0 T.$$

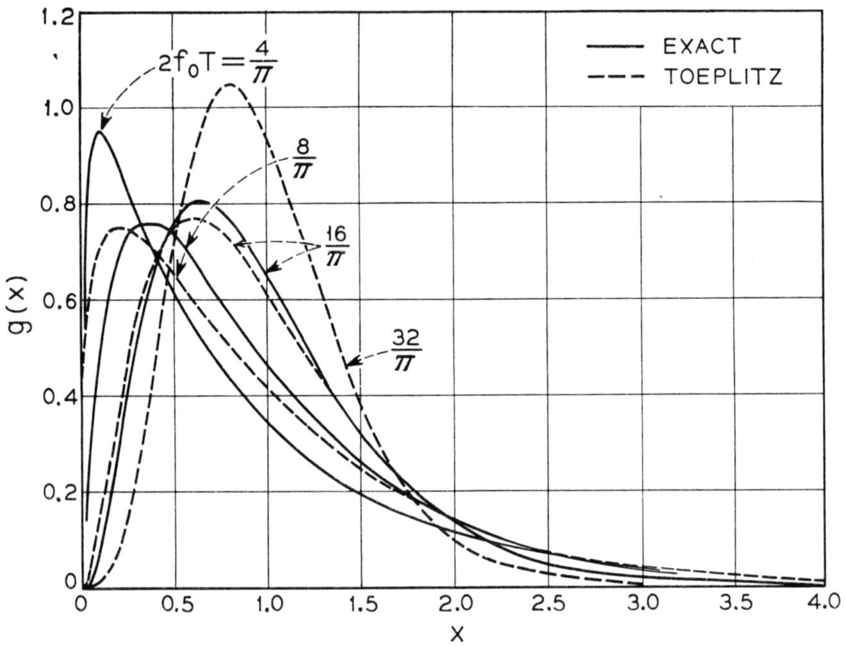

FIG. 4. *Band-limited noise power*. See §4.4.

This formula has a ready interpretation in terms of the sampling theorem for band-limited functions. For large T, approximate the integral (4.3.1) by

$$Q' = \frac{1}{T}\sum_{j=1}^{n} x^2\left(\frac{jT}{n}\right)\frac{T}{n}, \qquad n = 2f_0\,T\,.$$

The variables $x(j(T/n))$ are statistically independent, however, so that Q' has the χ^2 distribution with n degrees of freedom (4.4.2).

Fig. 4 shows plots of (4.4.2) for $n = 8/\pi$, $16/\pi$, $32/\pi$ along with exact values of $g(x)$ for $n = 4/\pi, 8/\pi$ and $16/\pi$. The exact values were determined by solving the integral equation

$$\lambda\varphi(x) = \frac{1}{T}\int_0^T r(x - y)\varphi(y)\,dy$$

with $r(\tau)$ given by (4.4.1). The solution was obtained by the method of [11]. For the values of $n = 2f_0T$ chosen, the eigenvalues drop off rapidly. For the three cases $n = 4/\pi$, $8/\pi$, $16/\pi$, it was necessary to keep only 3, 5 and 7 eigenvalues respectively to assure that the largest eigenvalue neglected was less than .002 times the largest eigenvalue. The integral equation method of §1.3 was then used to obtain the curves.

The agreement between the Toeplitz approximation and the exact curve is quite good already for $2f_0T = 16/\pi$.

Part 5. Discussion of the Approximations

If the sample size is not large enough for the normal or type III approximations to be valid, we can use the theory of Toeplitz forms to get better approximations. This will usually be done in two steps.

The first consists in replacing the eigenvalues of the matrix WR by those of $f(\lambda_0)W$, where $f(\lambda_0)$ is the spectral density evaluated at the center of the spectral window. It is obvious that this approximation is better the narrower the window is. It seems likely that this part of the approximation will be adequate for most practical purposes.

In the second step we replace the distribution of the eigenvalues of W by the continuous distribution given by Szegö's limit theorem for Toeplitz forms. Using this we will be able in a few cases to compute the approximate frequency function $g_a(x)$ analytically. If this is not possible we can determine $g_a(x)$ numerically as the solution of the integral equation (1.3.1), where now the kernel $h(x)$ is given by

$$h_a(x) = \frac{n}{4\pi} \int_{-\pi}^{\pi} e^{-x/2f(\lambda)}\, d\lambda.$$

Alternatively we might use the integral representation (2.2.2), but this has not been tried so far and may not be practical. The approximation in the second step will be adequate if the sample number n is not too small when compared to the time constant τ of the spectral window. If n/τ is small, say less than 5, we first apply the sharper methods developed in §3.5 and obtain the largest eigenvalues. Inserting these into the integral equation of §1.3 we can get $g(x)$ by numerical integration. This will perhaps not be necessary very often, but we have discussed this possibility, because spectral analysis will probably be used more for small sample numbers in the future, and then the need for such approximations will arise.

REFERENCES

1. H. Cramér, *Mathematical Methods of Statistics*, Princeton Univ. Press, 1946.
2. R. A. Fisher, *Tests of significance in harmonic analysis*, Proc. Roy. Soc. London, Ser. A, 125 (1929), pp. 54–59.
3. W. Freiberger and U. Grenander, *Approximate Distributions of Noise Power Measurements*, Quart. Appl. Math., 17 (1959), pp. 271–284.
4. E. N. Gilbert and H. O. Pollak, *On the Amplitude Distribution of Impulse Noise*, to be published.
5. U. Grenander and M. Rosenblatt, *Statistical Analysis of Stationary Time Series*, Wiley, New York, 1957.
6. U. Grenander and G. Szegö, *Toeplitz Forms and Their Applications*, Univ. of Calif. Press, Berkeley, 1958.

7. J. GURLAND, *Distributions of definite and indefinite quadratic forms*, Ann. Math. Statist., 26 (1955), pp. 122–127.
8. YU. V. LINNIK, *On stable probability laws with exponent less than one* (Russian), Dokl. Akad. Nauk SSSR, 94 (1954), pp. 619–621.
9. S. O. RICE, *Mathematical analysis of random noise*, Bell System Tech. J., 23 (1944), pp. 282–332; 24 (1945), pp. 46–156.
10. H. E. ROBBINS, *Distributions of a definite quadratic form*, Ann. Math. Statist., 19 (1948), pp. 266–270.
11. D. SLEPIAN, *Estimation of signal parameters in the presence of noise*, IRE Trans. on Information Theory, PGIT-3, 1954.
12. ——, *Fluctuations of random noise power*, Bell System Tech. J., 37 (1958), pp. 163–184.
13. D. V. WIDDER, *The Laplace Transform*, Princeton Univ. Press, 1941.
14. D. C. YOULA, *A homogeneous Wiener-Hopf integral equation*, IRE Trans. on Information Theory, IT-3, No. 3, (1957), pp. 187–193,.

BROWN UNIVERSITY, (1)
BELL TELEPHONE LABORATORIES, (2) AND (3)

Bibliography

IV† Allison, J. S. (1967), "The Linear Rectification of Two Amplitude-Modulated Signals and Noise," *IEEE Trans. Information Theory,* **IT-13,** pp. 69–75, January.

VI Andrews, L. C. (1973), "The Probability Density Function for the Output of a Cross-Correlator with Band-Pass Inputs," *IEEE Trans. Information Theory,* **IT-19,** pp. 13–19, January.

III Arozullah, M. (1966), "Passage of Stochastic Signals Through a Class of Stochastic Nonlinear Systems," Univ. Ottawa, Dept. Elec. Eng. Tech. Rept. 66–12, October.

II Atherton, D. P. (1962), "The Evaluation of the Response of Single-Valued Nonlinearities to Several Inputs," *Proc. IEE,* **109,** Part C, pp. 146–157, March.

IV Banta, E. D. (1965), "On the Autocorrelation Function of Quantized Signal Plus Noise," *IEEE Trans. Information Theory,* **IT-11,** pp. 114–117, January.

III Barrett, J. F. (1964), "Hermite Functional Expansions and the Calculation of Output Autocorrelation and Spectrum for any Time-Invariant Nonlinear System," *J. Electronics and Control,* **16,** pp. 107–113, February.

II ———, and J. F. Coales (1956), "An Introduction to the Analysis of Nonlinear Control Systems with Random Inputs," *Proc. IEE,* **103,** Part C, pp. 190–199, March.

IV Baum, R. F. (1957), "The Correlation Function of Smoothly Limited Gaussian Noise," *IRE Trans. Information Theory,* **IT-3,** pp. 193–197, September.

III Bédard, G. (1969), "Moments of Gaussian Processes in Terms of Generalized Multidimensional Hermite Polynomials," *IEEE Trans. Information Theory,* **IT-15,** pp. 611–612, September.

II Bello, P. A. (1965), "On the RMS Bandwidth of Nonlinearly Envelope Detected Narrow-Band Gayssian Noise," *IEEE Trans. Information Theory,* **IT-11,** pp. 236–239, April.

IV ———, and W. Higgins (1961), "Effect of Hard Limiting on the Probabilities of Incorrect Dismissal and False Alarm at the Output of an Envelope Detector," *IRE Trans. Information Theory,* **IT-7,** pp. 60–66, April.

IV Bennett, W. R., and S. O. Rice (1934), "Note on Methods of Computing Modulation Products," *Phil. Mag.,* Ser. 7, **18,** pp. 422–424.

V Berglund, C. N. (1964), "A Note on Power-Law Devices and Their Effect on Signal-to-Noise Ratio," *IEEE Trans. Information Theory,* **IT-10,** pp. 52–57, January.

II Blachman, N. M. (1953), "The Output Signal-to-Noise Ratio of a Power Law Device," *J. Appl. Phys.,* **24,** pp. 783–785, June.

IV ——— (1960), "The Effect of a Limiter upon Signals in the Presence of Noise," *IRE Trans. Information Theory,* **IT-6,** p. 52, March.

V ——— (1964), "Band-Pass Nonlinearities," *IEEE Trans. Information Theory,* **IT-10,** pp. 162–164, April.

†This column indicates the part most appropriate to the particular reference. However, references may relate to more than one part; this is especially true for parts II, IV, and V.

IV ——— (1964), "The Correlation Function of Gaussian Noise After Error-Function Limiting," *J. Electronics and Control,* **16,** pp. 509–511, May.

V ——— (1971), "Detectors, Band-Pass Nonlinearities and Their Optimization: Inversion of the Chebyshev Transform," *IEEE Trans. Information Theory,* **IT-17,** pp. 398–404, July.

Blake, I. F., and W. C. Lindsey (1973), "Level-Crossing Problems for Random Processes," *IEEE Trans. Information Theory,* **IT-19,** pp. 295–317, May.

II Bonet, G. (1964), "Nonlinear Zero-Memory Transformations of Random Signals," *Ann. Telecommun.,* **19,** pp. 203–220, September–October.

II Booton, R. C., Jr. (1954), "Nonlinear Control Systems with Random Inputs," *IRE Trans. Circuit Theory,* **CT-1,** pp. 9–18, March.

III Bose, A. G. (1956), "A Theory of Nonlinear Systems," Mass. Inst. Technol. Res. Lab. Electronics Tech. Rept. 309, May.

II Bowen, B. A. (1967), "The Transform Method for Nonlinear Devices with Non-Gaussian Noise," *IEEE Trans. Information Theory,* **IT-13,** pp. 326–328, April.

III Brilliant, M. B. (1958), "Theory of the Analysis of Nonlinear Systems," Mass. Inst. Technol. Res. Lab. Electronics Tech. Rept. 345, March.

I Brown, J. L., Jr. (1957), "On a Cross-Correlation Property for Stationary Random Processes," *IRE Trans. Information Theory,* **IT-3,** pp. 28–31, March.

I ——— (1958), "A Criterion for the Diagonal Expansion of a Second Order Probability Distribution in Orthogonal Polynomials," *IRE Trans. Information Theory,* **IT-4,** p. 172, December.

IV ——— (1965), "Output Correlation Function for a Half-Wave Smooth Limiter," *Intern. J. Control,* **1,** pp. 1–5, January.

II ——— (1967), "A Generalized Form of Price's Theorem and Its Converse," *IEEE Trans. Information Theory,* **IT-13,** pp. 27–30, January.

I ——— (1968), "On the Expansion of the Bivariate Gaussian Probability Density Using Results of Nonlinear Theory," *IEEE Trans. Information Theory,* **IT-14,** pp. 158–159, January.

II ———, and G. P. Patil (1967), "On the Statistical Independence of Joint Gaussian Variables After Nonlinear Transformation," *IEEE Trans. Information Theory,* **IT-13,** pp. 123–124, January.

VI ———, and H. S. Piper, Jr. (1967), "Output Characteristic Function for an Analog Cross-Correlator with Band-Pass Inputs," *IEEE Trans. Information Theory,* **IT-13,** pp. 6–10, January.

II Burgess, R. E. (1951), "The Rectification and Observation of Signals in the Presence of Noise," *Phil. Mag.,* Ser. 7, **42,** pp. 475–503, May.

V Cahn, C. R. (1961), "A Note on Signal-to-Noise Ratio in Band-Pass Limiters," *IRE Trans. Information Theory,* **IT-7,** pp. 39–43, January.

V Campbell, L. L. (1956), "Rectification of Two Signals in Random Noise," *IRE Trans. Information Theory,* **IT-2,** pp. 119–124, December.

Chandrasekhar, S. (1943), "Stochastic Problems in Physics and Astronomy," *Rev. Mod. Phys.,* **15,** pp. 1–89, January. Also in Wax (1954), pp. 3–91.

III Chesler, D. A. (1960), "Nonlinear Systems with Gaussian Inputs," Mass. Inst. Technol. Res. Lab. Electronics Tech. Rept. 366, February.

VI Cooper, D. C. (1965), "The Probability Density Function for the Output of a Correlator with Band-Pass Input Waveforms," *IEEE Trans. Information Theory,* **IT-11,** pp. 190–195, April.

III Crandall, S. H. (1962), "Random Vibration of a Nonlinear System with a Set-up Spring," *J. Appl. Mech.,* **29E,** pp. 477–482, September.

Davenport, W. B., Jr., and W. L. Root (1958), *An Introduction to the Theory of Random Signals and Noise,* McGraw-Hill, New York. Chapters 12–13.

VI Davisson, L. D., and P. Papantoni-Kazakos (1973), "On the Distribution and Moments of *RC*-Filtered Hard-Limited *RC*-Filtered White Noise," *IEEE Trans. Information Theory,* **IT-19,** pp. 411–414, July.

III Deutsch, R. (1955), "On a Method of Wiener for Noise Through Nonlinear Devices," *IRE Conv. Rec.,* Part 4, pp. 186–192, March.

——— (1962), *Nonlinear Transformations of Random Processes,* Prentice-Hall, Englewood Cliffs, N. J.

V Doyle, W. (1962), "Elementary Derivation for Band-Pass Limiters *S/N*," *IRE Trans. Information Theory,* **IT-8,** p. 259, April.

V ———, and I. S. Reed (1964), "Approximate Band-Pass Limiter Envelope Distributions," *IEEE Trans. Information Theory,* **IT-10,** pp. 180–185, July.

VI ———, J. A. McFadden, and I. Marx (1962), "The Distribution of a Certain Nonlinear Functional of an Ornstein–Uhlenbeck Process," *J. Soc. Ind. Appl. Math.,* **10,** pp. 381–393, June.

IV Dukes, J. M. C. (1955), "The Effect of Severe Amplitude Limitation on Certain Types of Random Signals: A Clue to the Intelligibility of Infinitely Clipped Speech," *Proc. IEE,* **102,** Part C, pp. 88–97, March.

VI Emerson, R. C. (1953), "First Probability Densities for Receivers with Square-Law Detectors," *J. Appl. Phys.,* **24,** pp. 1168–1176, September.

V Fellows, G. E., and D. Middleton (1956), "An Experimental Study of Intensity Spectra after Half-Wave Rectification of Signals in Noise," *Proc. IEE,* **103,** Part C, pp. 243–248.

III French, A. S., and E. G. Butz (1974), "The Use of Walsh Functions in the Wiener Analysis of Nonlinear Functions," *IEEE Trans. Computers,* **C-23,** pp. 225–232, March.

II Fukada, M., and S. Rauch (1966), "On the Closed-Form Output Correlation Function Of Power-Law Devices," *Proc. IEEE,* **54,** pp. 1625–1626, November.

IV Galejs, J. (1959), "Signal-to-Noise Ratios in Smooth Limiters," *IRE Trans. Information Theory,* **IT-5,** pp. 79–85, June.

III George, D. A. (1959), "Continuous Nonlinear Systems," Mass. Inst. Technol. Res. Lab. Electronics Tech. Rept. 355, July.

III Glinski, G. S., and N. U. Ahmed (1964), "Representation of Nonlinear Systems by Orthogonal Functionals," Univ. Ottawa, Dept. Elec. Eng. Tech. Rept. 64-12, July.

II Golding, L. S. (1967), "Bounds on Output Moments of a Class of Nonlinear Devices," *IEEE Trans. Information Theory,* **IT-13,** pp. 142–144, January.

VI Gurland, J. (1953), "Distribution of Quadratic Forms and Ratios of Quadratic Forms," *Ann. Math. Stat.,* **24,** pp. 416–427.

VI ——— (1955), "Distribution of Definite and Indefinite Quadratic Forms," *Ann. Math. Stat.,* **26,** pp. 122–127, March.

III Hause, A. D. (1960), "Nonlinear Least-Square Filtering and Frequency Modulation," Mass. Inst. Technol. Res. Lab. Electronics Tech. Rept. 371, August.

VI Henry, H. E., and P. M. Schultheiss (1962), "The Analysis of Certain Nonlinear Feedback Systems with Random Inputs," *IRE Trans. Information Theory,* **IT-8,** pp. 285–291, July.

IV Hurd, W. J. (1967), "Correlation Function of Quantized Sine Wave Plus Gaussian Noise," *IEEE Trans. Information Theory,* **IT-13,** pp. 65–68, January.

VI Jacobson, M. J. (1963), "Output Probability Distribution of a Multiplier-Averager with Partially Correlated Inputs," *J. Acoust. Soc. Amer.,* **35,** pp. 1932–1938, December.

V Jones, J. J. (1963), "Hard-Limiting of Two Signals in Random Noise," *IEEE Trans. Information Theory,* **IT-9,** pp. 34–42, January.

IV Kaufman, H., and G. E. Roberts (1963), "Correlation Function at the Output of an Error Function Limiter," *J. Electronics and Control.* **15,** pp. 165–170, August.

V ———, and G. E. Roberts (1964), "The Effect of a Power-Law Device on the Correlation Function of a Frequency-Modulated Signal," *J. Electronics and Control,* **16,** pp. 307–324, March.

I Keilson, J., N. D. Mermin, and P. A. Bello (1959), "A Theorem on Cross-Correlation Between Noisy Channels," *IRE Trans. Information Theory,* **IT-5,** pp. 77–79, June.

III Ku, Y. H. (1960), "On Nonlinear Networks with Random Inputs," *IRE Trans. Circuit Theory,*

CT-7, pp. 479–490, December.
III ———, and A. A. Wolf (1966), "Volterra–Wiener Functionals for the Analysis of Nonlinear Systems," *J. Franklin Inst.,* **281,** pp. 9–26, January.
Kuznetsov, P. I., R. L. Stratonovich, and V. I. Tichonov (1965), *Nonlinear Transformations of Stochastic Processes,* Pergamon Press, London.
VI Lampard, D. G. (1956), "The Probability Distribution for the Filtered Output of a Multiplier Whose Inputs Are Correlated, Stationary, Gaussian Time Series," *IRE Trans. Information Theory,* **IT-2,** pp. 4–11, March.
II Langseth, R. E., and R. F. Lambert (1968), "Influence of Bandwidth on Some Nonlinear Transformations of a Gaussian Random Process," *IEEE Trans. Information Theory,* **IT-14,** pp. 88–93, January.
VI Leipnik, R. (1959), "Integral Equations, Biorthonormal Expansions and Noise," *J. Soc. Ind. Appl. Math.,* **7,** pp. 6–30, March.
I Leland, H. R. (1960), "Input-Output Cross-Correlation Functions for Some Memory-Type Nonlinear Systems with Gaussian Inputs," *AIEE Trans.,* Part II, **79,** pp. 219–223, July.
IV McFadden, J. A. (1956), "The Correlation Function of a Sine Wave Plus Noise After Extreme Clipping," *IRE Trans. Information Theory,* **IT-2,** pp. 82–83, June.
IV ——— (1958), "The Fourth Product Moment of Infinitely Clipped Noise," *IRE Trans. Information Theory,* **IT-4,** pp. 159–162, December.
I ——— (1965), "An Alternate Proof of Nuttall's Theorem on Output Cross-Covariances," *IEEE Trans. Information Theory,* **IT-11,** pp. 306–307, April.
II McMahon, E. L. (1964), "An Extension of Price's Theorem," *IEEE Trans. Information Theory,* **IT-10,** p. 168, April.
II Magness, T. A. (1954), "Spectral Response of a Quadratic Device to Non-Gaussian Noise," *J. Appl. Phys.,* **25,** pp. 1357–1365, November.
V Manasse, R., R. Price, and R. M. Lerner (1958), "Loss of Signal Detectability in Band-Pass Limiters," *IRE Trans. Information Theory,* **IT-4,** pp. 34–38, March.
I Masry, E. (1973), "The Recovery of Distorted Band-Limited Stochastic Processes," *IEEE Trans. Information Theory,* **IT-19,** pp. 398–403, May.
IV Max, J. (1960), "Quantizing for Minimum Distortion," *IRE Trans. Information Theory,* **IT-6,** pp. 7–12, March.
VI Mazo, J. E. R., R. F. Pawula, and S. O. Rice (1973), "On a Nonlinear Problem Involving *RC* Noise," *IEEE Trans. Information Theory,* **IT-19,** pp. 404–411, July.
IV Middleton, D. (1946), "The Response of Biased, Saturated Linear and Quadratic Rectifiers to Random Noise," *J. Appl. Phys.,* **17,** pp. 778–801, October.
V ——— (1948), "Rectification of a Sinosoidally Modulated Carrier in the Presence of Noise," *Proc. IRE,* **36,** pp. 1467–1477, December.
V ——— (1949a), "On Theoretical Signal-to-Noise Ratio in FM Receivers—A Comparison with Amplitude Modulation," *J. Appl. Phys.,* **20,** pp. 334–351, April.
V ——— (1949b), "The Spectrum of Frequency-Modulated Waves After Reception in Random Noise — I," *Quart. Appl. Math.,* **7,** pp. 129–173, July.
V ——— (1950), "The Spectrum of Frequency-Modulated Waves After Reception in Random Noise — II," *Quart. Appl. Math.,* **8,** pp. 59–80, April.
——— (1960), *An Introduction to Statistical Communication Theory,* McGraw-Hill, New York. Chapter 5.
IV Mullen, J. A., and D. Middleton (1958), "The Rectification of NonGaussian Noise," *Quart. Appl. Math.,* **15,** pp. 395–419, January.
II Musal, H. M., Jr. (1969), "Logarithmic Compression of Rayleigh and Maxwell Distributions," *Proc. IEEE,* **57,** pp. 1311–1313, July.
IV North, D. O. (1944), "The Modification of Noise by Certain Nonlinear Devices," IRE Winter Meeting, January 28.
I Nuttall, A. H. (1958), "Theory and Application of the Separable Class of Random Processes," Mass. Inst. Technol. Res. Lab. Electronics Tech. Rept. 343, May.

II Papoulis, A. (1965), Comments on "An Extension of Price's Theorem," *IEEE Trans. Information Theory,* **IT-11,** p. 154, January.
II Pawula, R. F. (1967), "A Modified Version of Price's Theorem," *IEEE Trans. Information Theory,* **IT-13,** pp. 285–288, April.
IV —, and A. Y. Tsai (1969), "Theoretical and Experimental Results for the Distribution of a Certain Nonlinear Functional of the Ornstein–Uhlenbeck Process," *IEEE Trans. Information Theory,* **IT-15,** pp. 532–535, September.
IV Pitassi, D. A. (1969), "Second-Order Properties of Products of Clipped Gaussian Processes," *IEEE Trans. Information Theory,* **IT-15,** pp. 535–540, September.
III Pugachev, V. S. (1960), "Statistical Theory of Systems Reducible to Linear," *IRE Trans. Circuit Theory,* **CT-7,** pp. 506–513, December.
V Rice, S. O. (1944), "Mathematical Analysis of Random Noise," *Bell Syst. Tech. J.,* **23,** pp. 282–332 (1944) and **24,** pp. 46–156 (1945). Also in Wax (1954), pp. 133–294.
——— (1948), "Statistical Properties of a Sine Wave Plus Random Noise," *Bell Syst. Tech. J.,* **27,** pp. 109–157, January.
VI Robbins, H. (1948), "The Distribution of a Definite Quadratic Form," *Ann. Math. Stat.,* **19,** pp. 266–270, June.
IV Roe, G. M. (1964), "Quantizing for Minimum Distortion," *IEEE Trans. Information Theory,* **IT-10,** pp. 384–385, October.
IV Rubin, M. (1962), "Comparison of Signal and Noise in Full-Wave and Half-Wave Rectifiers," *IRE Trans. Information Theory,* **IT-8,** pp. 379–380, October.
III Schetzen, M. (1962), "Some Problems in Nonlinear Theory," Mass. Inst. Technol. Res. Lab. Electronics Tech. Rept. 390, July.
IV Schmideg, I. (1969), "The Effects of Limiting on Angle- and Amplitude-Modulated Signals," *Proc. IEEE,* **57,** pp. 1302–1303, July.
V Shaft, P. D. (1965), "Hard limiting of Several Signals and Its Effect on Communication System Performance," *IEEE Trans. Commun. Tech.,* **COM-13,** pp. 504–512, December.
VI Siegert, A. J. F. (1954), "Passage of Stationary Processes Through Linear and Nonlinear Devices," *IRE Trans. Information Theory,* **IT-3,** pp. 4–25, March.
IV Smith, B. (1957), "Instantaneous Companding of Quantized Signals," *Bell Syst. Tech. J.,* **36,** pp. 653–709, May.
Stratonovich, R. L. (1963), *Topics in the Theory of Random Noise,* Gordon & Breach, New York.
Thomas, J. B. (1969), *An Introduction to Statistical Communication Theory,* John Wiley & Sons, New York. Chapter 6.
II Thomson, W. E. (1955), "The Response of a Nonlinear System to Random Noise," *Proc. IEE,* **102,** Part C, pp. 46–48, March.
IV Tufts, D. W., W. Knight, and D. Rorabacher (1969), "Effects of Quantization and Sampling in Digital Correlators and in Power Spectral Estimation," *Proc. IEEE,* **57,** pp. 79–82, January.
Uhlenbeck, G. E., and L. S. Ornstein (1930), "On the Theory of the Brownian Motion," *Phys. Rev.,* **36,** pp. 823–841, September. Also in Wax (1954), pp. 93–111.
III Van Trees, H. L. (1964), "Functional Techniques for the Analysis of the Nonlinear Behavior of Phase-Locked Loops," *Proc. IEEE,* **52,** pp. 894–910, August.
V Van Vleck, J. H., and D. Middleton (1946), "A Theoretical Comparison of the Visual, Aural, and Meter Reception of Pulsed Signals in the Presence of Noise," *J. Appl. Phys.,* **17,** pp. 940–971, November.
Viterbi, A. J. (1963), "Phase-Locked Loop Dynamics in the Presence of Noise by Fokker–Planck Techniques," *Proc. IEEE,* **51,** pp. 1737–1753, December.
III Volterra, V. (1930), *Theory of Functionals,* Blackie & Son Ltd., Glasgow.
Wang, M. C., and G. E. Uhlenbeck (1945), "On the Theory of Brownian Motion II," *Rev. Mod. Phys.,* **17,** pp. 323–342, April–July. Also in Wax (1954), pp. 113–132.
IV Watts, D. G. (1962), "A General Theory of Amplitude Quantization with Applications to Correlation Determination," *Proc. IEE,* **109,** Part C, pp. 209–218, March.

Wax, N. (1954), *Selected Papers on Noise and Stochastic Processes,* Dover Publications, New York.

III Wiener, N. (1958), *Nonlinear Problems in Random Theory,* MIT Press, Cambridge, Mass.

III Wolf, A. A. (1961), "Some Recent Advances in the Analysis and Synthesis of Nonlinear Systems," *AIEE Trans.,* **80,** Part II, pp. 289–300, November.

I Wong, E., and J. B. Thomas (1962), "On Polynomial Expansions of Second-Order Distributions," *J. Soc. Ind. Appl. Math.,* **10,** pp. 507–516, September.

VI Young, G. O. (1958), "Random Function Distributions After a Nonlinear Filter," *IRE WESCON Conv. Rec.,* Part 4, pp. 164–172, August.

III Zadeh, L. A. (1953), "Optimum Nonlinear Filters," *J. Appl. Phys.,* **24,** pp. 396–404, April.

III ——— (1957), "On the Representation of Nonlinear Operators," *IRE WESCON Conv. Rec.,* Part 2, pp. 105–113, August.

Author Citation Index

Subject Index